Fundamentals of Multicomponent High-Entropy Materials

Fundamentals of Multicomponent High-Entropy Materials

Brian Cantor

OXFORD
UNIVERSITY PRESS

OXFORD
UNIVERSITY PRESS

Great Clarendon Street, Oxford, OX2 6DP,
United Kingdom

Oxford University Press is a department of the University of Oxford.
It furthers the University's objective of excellence in research, scholarship,
and education by publishing worldwide. Oxford is a registered trademark of
Oxford University Press in the UK and in certain other countries

Published in the United States of America by Oxford University Press
198 Madison Avenue, New York, NY 10016, United States of America

British Library Cataloguing in Publication Data
Data available

Library of Congress Control Number: 2024940247

ISBN 9780198874270

DOI: 10.1093/9780191986710.001.0001

Printed and bound by
CPI Group (UK) Ltd, Croydon, CR0 4YY

Preface

Everything in our modern world (and indeed in the ancient world) is made up of complex *materials.* We could not have things such as fridges, cars, TVs, aeroplanes, MRI scanners, prosthetic limbs, books, smartphones, skyscrapers, golf clubs, bank cards or pretty much anything else without the materials they are made from. We could not even have much simpler things such as needles and pins, bricks and mortar, or pots and pans. The development of humankind, including our politics, economy, art and culture, is at root a continuing attempt to use new materials in ever-more sophisticated ways to enhance the quality of human life, from early developments such as building mud huts and storing domesticated agricultural crops to modern inventions such as mass transport, keyhole surgery and internet communications. In short, the fields of *materials science* and *materials engineering* are essential and fundamental to our way of living. To maintain and improve human health, wealth and wellbeing we need to understand the physics, chemistry, biology and engineering applications of an enormously wide and rapidly growing range of different materials.

And all materials are *alloys.* In other words, they consist of mixtures of one or more different starting materials or *components* from which they are made. This is just as true of the earliest Stone-Age materials such as polished flints for axe heads and clay pottery for kitchen utensils as it is of the most recent sophisticated modern materials such as compound semiconductors for high-brightness traffic lights and mixed-oxide electrodes for high-energy lithium batteries in electric cars. In this book the words materials and alloys will be used more or less interchangeably.

We have used alloying to manufacture our materials since the emergence of human beings at the beginning of the Stone Age about three million years ago. And throughout the developments of the prehistoric, historic and modern eras, our alloying strategy for manufacturing materials has almost always been to select a main component based on the principal property required, and then make small alloying additions to confer secondary properties on the final material and thus onto the resulting artefact. In technical language, we have always chosen our materials with compositions close to the outer corners and edges of multicomponent phase space. This strategy has led to the development of many successful materials. Typical modern examples include high-temperature nickel superalloys, with nickel as the main component selected for its high melting point, chromium added for corrosion resistance, and titanium or aluminium added for high strength; and soda-lime window glass, with silica as the main component because it is transparent when amorphous, with sodium oxide (soda) added to lower the melting point and improve ease of manufacture, and calcium oxide (lime) added to improve stability and prevent devitrification during use.

But recently, there has been a revolution in our thinking about how to design new materials and how to manufacture them, massively extending our capacity to create useful products. In 2004, two influential papers were published describing a radically new alloying strategy for developing new materials.[1,2] Independently, my own research group in the late 1970s[1] and a research group led by Professor Jin-Wei Yeh in the mid-1990s[2] began to explore the manufacture and properties of materials consisting of several principal components in equal or near-equal proportions, i.e. alloys in the centre of multicomponent phase space, well away from the pure single-component corners, binary edges and ternary faces of any multicomponent phase diagram. The basic concept of *multicomponent alloying* and *multicomponent materials* was first published[3–6] by me and my research colleagues in 2002 and 2003, but the two most important papers launching the new alloying strategy were published by my group[1] and by Professor Wei's group[2] in 2004.

These new materials are called *multicomponent*[1] or *high-entropy*[2] alloys or materials (or, sometimes, *multiple-principal-element* or *compositionally complex* alloys or materials). The new approach overturned millennia-old ideas of how to make materials, and both research groups were forced to make progress despite initial disbelief and, to some extent, disdain from others about the value of such an unusual and innovative approach. In fact, both of the two seminal papers published in 2004 were met initially with little or no scientific interest or attention. Nevertheless, by the end of the decade hundreds of papers were being published on this new methodology, and by the end of the following decade there was an avalanche of many thousands of papers, as scientists worldwide began to explore the enormous potential for finding and developing new materials with exciting and valuable new properties. Multimillion-dollar research programmes have now been initiated and are currently underway in most advanced countries. The first two papers had a total of no more than eight citations in the two years after their publication, yet by 2024 the total number of publications and citations in the field reached almost fifty thousand and over a million respectively, with both figures rising rapidly.[7] A few years ago, a review of the field of multicomponent high-entropy materials[8] said that this new approach had led to the development of 'eight major new categories of materials', despite having still 'only scratched the surface' of the vast potential. And many more new materials and new categories of materials have already been discovered since then.

The field of multicomponent high-entropy alloys has reached a reasonable initial level of maturity, and a number of fundamental issues and questions have been clarified and, to some extent, resolved (although there are still literally trillions of new multicomponent materials yet to be explored, and still many wonderful new phenomena yet to be discovered and/or fully explained). Several detailed research-level reviews and books have already been published,[8–12] attempting to provide at their time of publication a complete picture of our understanding of multicomponent high-entropy materials, but this is becoming impossible given the enormous quantity of research being undertaken. Instead, it seems to be a good time now for a more considered textbook concentrating on the fundamentals of the new field, the main concepts and theories, providing a sober overview, and summarising rather than fully covering the state of play for researchers as well as for students and newcomers entering the field. That is the objective of this book.

Chapter 1 gives a brief history of the major stages in the development of humankind and how they were all driven by the development of (conventional) new alloys and materials, thus setting the scene for the main part of the book on the manufacture, structure and properties of the new multicomponent high-entropy materials. Chapter 2 describes the initial discovery of multicomponent alloying as a new method of manufacturing novel materials and the central importance of a high entropy of mixing, with short biographies of the key scientists involved. Chapter 3 discusses the structure and topology of multicomponent phase space within which unimaginable gazillions of new multicomponent high-entropy materials are located. Chapter 4 describes the basic thermodynamics and underlying atomic and electronic structure of multicomponent high-entropy materials. Chapter 5 covers various theoretical and practical methods of trying to explore the vastness of multicomponent phase space, including the different ways of manufacturing multicomponent high-entropy materials. Chapter 6 describes some of the fundamental and applied properties of multicomponent high-entropy materials. Finally, Chapter 7 summarises the key lessons we have learned about multicomponent high-entropy materials, indicates some aspects that have largely been ignored, and discusses briefly our future need for new alloys and materials so as to maintain humankind's progress, taking more and more people out of poverty and ill-health while avoiding damaging impact on our environment. New materials have always been the key to improvements in our socio-economic conditions or, to put it in plainer English, our health, wealth and wellbeing, and they will continue to be so in the future, with the enormous potential of multicomponent high-entropy materials set to play a critical role.

I have written this book in the last few years, since stepping down in late 2019 from being Vice-Chancellor of the University of Bradford, after a total of the best part of twenty years working as a senior university manager at the Universities of Oxford, York and then Bradford. It was tremendously satisfying and enjoyable to play a role in helping to run such outstanding universities and, in particular, to interact with so many wonderful students and academics in so many diverse fields. But it has been a great pleasure to return to scientific life and rediscover the excitement of scientific research and the emerging fundamental understanding in the rapidly expanding, intellectually challenging and socially important new field of multicomponent high-entropy materials. I have written this book by myself, and all the opinions expressed in it are mine alone. I am, however, particularly grateful for scientific discussions with Professors Roger Doherty, Patrick Grant and Zhongyun Fan, and for academic support from many colleagues in the Brunel Centre for Advanced Solidification Technology (BCAST) at Brunel University and in the Department of Materials at the University of Oxford. I am also grateful for technical support and guidance on publishing matters from Sonke Adlung at Oxford University Press. Throughout my career, I have benefitted from advice and mentoring from three outstanding scientists: Professor Sir Graeme Davies, who was one of my tutors when I was a student at the University of Cambridge; Professor Robert Cahn, who gave me my first scientific job as a Research Fellow at the University of Sussex; and Professor Sir Peter Hirsch, who gave me my first tenured academic job as a Lecturer at the University of Oxford. Graeme and Robert have both sadly passed away, but I am pleased to say that Peter is still well. I have decided to dedicate this book to Peter for his continued help, advice and support over many years as regards my personal development and my career as a scientist, academic and university manager. I am, of course, also eternally indebted to my partner and wife, Gillian Partridge, for putting up with me and all my foibles, and for helping me in innumerable ways. Quite simply, without her I could not cope, and certainly would not have been able to write this book.

References

1. Cantor B, Chang ITH, Knight P, Vincent AJB. (2004). Microstructural development in equiatomic multicomponent alloys. *Materials Science and Engineering A375*: 213–18.
2. Yeh JW, Chen SK, Lin SJ, Gan JY, Chin TS, Shun TT, Tsau CH, Chang SY. (2004). Nanostructured high entropy alloys with multiple principal elements: Novel alloy design concepts and outcomes. *Advanced Engineering Materials 6*: 299–303.
3. Cantor B, Kim KB, Warren PJ. (2002). Novel multicomponent amorphous alloys. *Materials Science Forum 386–388*: 27–32.
4. Kim KB, Warren PJ, Cantor B. (2003). Formation of metallic glasses in novel (Ti33Zr33Hf33)((100-x-y)(Ni50Cu50)(x)Al(y) alloys. *Materials Transactions 44*: 411–13.
5. Kim KB, Warren PJ, Cantor B. (2003). Metallic glass formation in multicomponent (Ti, Zr, Hf, Nb)-(Ni, Cu, Ag)-Al alloys. *Journal of Non-Crystalline Solids 317*: 17–22.
6. Kim KB, Zhang Y, Warren PJ, Cantor B. (2003). Crystallization behaviour in a new multicomponent TiZrHfNiCuAl metallic glass developed by the equiatomic substitution technique. *Philosophical Magazine 83*: 2371–81.
7. Publication and citation count for 'high entropy alloys' and 'multicomponent alloys'. *Web of Science (Clarivate).* https://clarivate.com/products/scientific-and-academic-research/research-discovery-and-workflow-solutions/webofscience-platform/
8. Miracle DB, Senkov ON. (2017). A critical review of high-entropy alloys and related concepts. *Acta Materialia 122*: 448–511.
9. Cantor B. (2021). Multicomponent high-entropy Cantor alloys. *Progress in Materials Science 120*: Article 100754.
10. Gao MC, Yeh JW, Liaw PK, Zhang Y. (eds.). (2016). *High-Entropy Alloys: Fundamentals and Applications.* Basel: Springer.
11. Zhang Y. (2019). *High-Entropy Materials: A Brief Introduction.* Basel: Springer.
12. Murty BS, Yeh JW, Ranganathan S, Bhattacharjee PP. (2019). *High-Entropy Alloys.* Amsterdam: Elsevier.

Contents

1

A History of Alloys

1 Introduction

This chapter gives a brief overview of some of the key developments in human history and explains how they were driven by the development of new materials and alloys. It covers the following: 1) the discovery in the Stone Age of how to use minerals (i.e. oxide alloys), and how this supported the emergence of human beings at the beginning of the last Ice Age and their development of agriculture-based settlements when the Ice Age ended; 2) the discovery of smelting and the extraction of metals, notably copper and iron alloys in the Bronze and Iron Ages, which drove the growth of cities and civilisation, culture, trade and empires; 3) engineering and the manufacture of steel (i.e. low-carbon iron alloys) in the Industrial Revolution, leading to motors and engines, bridges, mills and factories, cars and planes, and most of the accoutrements of our modern technology-dominated economies and associated high quality way of life; and 4) (microalloyed) semiconductors and the silicon chips fuelling the Information Revolution, computers, telecommunications, the internet, and the digital age. To misquote (only slightly) Reed Hastings, the CEO of Netflix: 'Stone Age, Bronze Age, Iron Age: we define entire epics of humanity by the [materials] they use'.[1]

2 The Stone Age and the Agricultural Revolution

Early humans

Human beings, technically the genus *Homo*, emerged in northeast Africa between two and a half and three million years ago, having evolved from our parent genus, the Southern Apes or *Australopithecus*, and spread gradually throughout Asia, Europe and the rest of the world.[1] This took place very roughly at the same time as the onset of the last Ice Age (the geological *Pleistocene* era),[2] when the earth's average temperature was about six degrees centigrade lower than now, and also very roughly at the same time as the start of the Stone Age (the archaeological *Palaeolithic* period),[3] when stone tools first began to be made and used, which is why the first species of *Homo* is called *Homo habilis* (or 'handyman').[4] Our own species, modern humans or *Homo sapiens*, emerged much later, between 250,000 and 300,000 years ago, and we are the only member of the genus *Homo* (and indeed the only descendants of the Southern Apes) to survive to the present day.[5] Other species such as *Homo habilis*, *Homo erectus*, *Homo ergaster*, *Homo rudolfensis*, *Homo denisova* and *Homo neanderthalus* all died out.[5] Why did human beings emerge at the beginning of the Ice Age and the Stone Age? Early humans, like apes, were nomadic hunter-gatherers. But having evolved with larger brains than their predecessors, they were able to learn how to manufacture tools and weapons by shaping stones and flints, which improved their ability to cut down and gather plants as well

Fundamentals of Multicomponent High-Entropy Materials. Brian Cantor, Oxford University Press.
DOI: 10.1093/9780191986710.003.0001

as to hunt down and kill animals, and led to a better diet and a surer supply of food, thus helping them survive the rigours and deprivations of the Ice Age.

The current modern geological era (the *Holocene*) began about 12,000 years ago (10,000 BC), immediately following the end of the Ice Age or *Pleistocene*.[2] Again, this corresponds roughly with the end of the *Palaeolithic* or old Stone Age and the beginning of the *Neolithic* or late Stone Age.[3] And it was a time of massive sociological and technological change: the so-called first Agricultural Revolution, as the ice receded and human beings stopped living as nomadic hunter-gatherers, and began to domesticate animals and crops and live in settled farming communities and villages.[2,6] This led gradually to the development of such diverse but interlinked things as towns and cities, writing and record-keeping, money and industry, trade and commerce, division of labour, and social organisation, with government machinery, kings and queens, armies, judges, priests and the like, i.e. many of the things we now regard, good or ill, as fundamental features of a civilised society.[5,6] It is worth noticing that this later, more organised and enlightened, period of development for human beings is just a minute fraction, less than half a per cent, of the time since humans emerged at the beginning of the Ice Age. In the first 99.5% of the last three million or so years our predecessors lived a ruder, meaner existence in the depths of the Palaeolithic Ice Age.

Human life during the Palaeolithic Ice Age was underpinned and indeed driven by the development of the first human-made weapons and tools, manufactured by cracking, shaping and polishing stones and cutting wood to make rudimentary (by modern standards) axes, knives, hammers and arrows.[3] And human life during the subsequent warmer Neolithic period was underpinned and driven by the first human-made fired ceramics, including bricks and mortar for house building and other construction, as well as clay pots and pans for cooking, storage and other domestic purposes.[3] To understand the development of the associated materials and manufacturing technologies, we need to consider the structure of the earth itself, and the rocks and minerals that provided the first, readily available raw materials or alloys for our stone-age ancestors.

Rocks and minerals

The earth's *inner core* consists[7] of a solid ball of iron with a radius of about 1,200 km, surrounded by an *outer core* of liquid iron–nickel alloy approximately 2,200 km thick, i.e. the total core radius is about 3,400 km. The core is very hot and under extreme gravitational pressure, reaching a temperature of well over 5,000 °C (200 times hotter than room temperature), and a pressure of 100–350 GPa (well over a million times greater than atmospheric pressure). The core is surrounded by a *mantle* of partly liquid (*lower mantle*) and partly solid (*upper mantle*) silicate rock, approximately 2,900 km thick, making up 67% of the earth's mass and 85% of its volume, again at a (not quite so) high temperature and pressure of up to 1,000 °C and 20 GPa.[7] Finally there is a relatively thin *crust*, a mere 70 km thick, made up, broadly speaking, of the cold solid rocks that we are familiar with in the landscapes we see at the surface.[7] Above the earth's crust, the air in our atmosphere is made up mostly of about 78% nitrogen N_2 and 21% oxygen O_2.

Rocks in the earth's crust can be formed in many different ways: *igneous rocks*, such as mountain granite or basalt on the seabed, are formed from solidified volcanic lava;[8]

sedimented rocks, such as sandstone and limestone, are formed from compressed fragments of other eroded rocks and/or deposited biological material such as shells and bones;[9] and *metamorphic rocks*, such as marble or quartz, are formed by transformation from other rocks, caused by a combination of heating and pressure.[10] But in all cases, rocks are made up of *minerals*, which are effectively alloys of different *oxides*. Why is the earth's crust formed of rocks and minerals? Why is it mostly a mixture of oxides? This is simply because oxygen is a very reactive element and is widely available in the gaseous atmosphere, so almost all the other elements react readily with it anywhere near the earth's surface. They all *oxidise* to form one or more oxides. Incidentally, this is also why most materials rust, i.e. why their surfaces react gradually with oxygen when exposed to the atmosphere, forming thin outer layers of oxide or rust, which usually thicken gradually with time.

The abundance of the different elements in the earth's crust is shown[11] in Table 1.1. Oxygen is easily the most abundant element in the earth's crust, and silicon and aluminium are the next two most abundant. Minerals are mostly, therefore, silicates and alumino–silicates, i.e. alloys of silicon oxide or silica SiO_2 with smaller quantities of aluminium oxide or alumina Al_2O_3, together with further, relatively minor additions of other oxides such as calcium oxide or calcia CaO, and magnesium oxide or magnesia MgO. Overall, we can say that the earth's crust is rich in elements such as silicon Si, aluminium Al, iron Fe, calcium Ca and magnesium Mg, which form oxides such as silica SiO_2, alumina Al_2O_3, ferrite FeO, haematite Fe_2O_3, magnetite Fe_3O_4, calcia CaO and magnesia MgO, together with less common elements such as carbon C, copper Cu, zinc Zn, nickel Ni and all the other metals and semi-metals, which form oxides such as carbon monoxide CO, carbon dioxide CO_2, cuprous oxide Cu_2O, cupric oxide CuO, zinc oxide ZnO, nickel oxide NiO, etc. All these oxides, together with small amounts of sulphides and nitrides, are combined chemically in a myriad of different crystalline and amorphous silicates, alumino–silicates, carbonates, magnesites, sulphates and so on, all of which can be regarded as (predominantly) oxide alloys, which are then intermingled on a microscopic scale to form the wide variety of minerals and rocks that make up the earth's crust. Wikipedia lists a total of over 5,000 minerals that are found in our rock formations,[12] and almost 200 different types of macroscopic rock structure.[13] Rocks are effectively complex interwoven physical and chemical aggregates of crystals, particles, fibres and grains of silicate- and alumino–silicate-based oxide alloys.

Table 1.1 Abundance of elements in the earth's crust.[11]

Element	Abundance (wt%)
oxygen	46
silicon	28
aluminium	8.3
iron	5.6
calcium	4.2
sodium	2.5
magnesium	2.4
potassium	2.0

Flints and stone tools

Silica and indeed all silicate materials are based on the tetrahedral silica unit[14] SiO_4 shown in Figure 1.1. Silica itself has a variety of *polymorphic* forms.[15] The equilibrium room-temperature form is hexagonal *quartz*, as shown in Figure 1.2, with corner-sharing tetrahedra spiralling in a threefold pattern along the *c*-axis, with *enantiomorphic* left-handed and right-handed versions.[15] There are two high-temperature forms, *cristobalite* and *tridymite*, each made up from hexagonal sheets of alternating up-and-down, corner-sharing tetrahedra, as shown in Figure 1.3, stacked in either a face-centred cubic (fcc) ABCABC (cristobalite) or hexagonal close-packed (hcp) ABABAB (tridymite) sequence.[15] Cooling to low temperature leads to distorted structures produced by displacive reactions (slight bond bending and stretching) forming, respectively, trigonal (distorted quartz), tetragonal (distorted cristobalite) and orthorhombic (distorted tridymite) structures.[15] And there is a variety of other

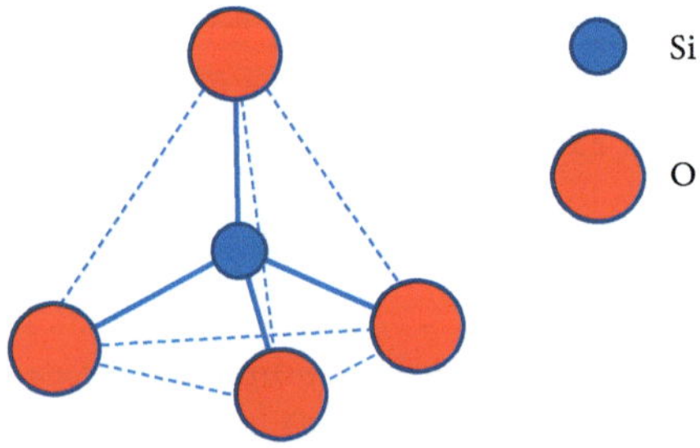

Figure 1.1 Silica tetrahedron SiO_4.

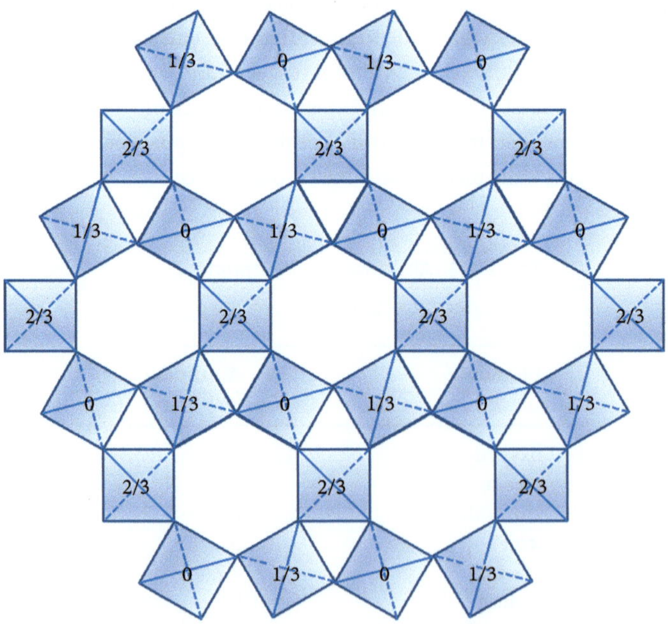

Figure 1.2 Quartz crystal structure, with corner-sharing SiO_4 tetrahedra at vertical heights of 0, 1/3 and 2/3, in a hexagonal array of 3-fold spirals along the *c*-axis.

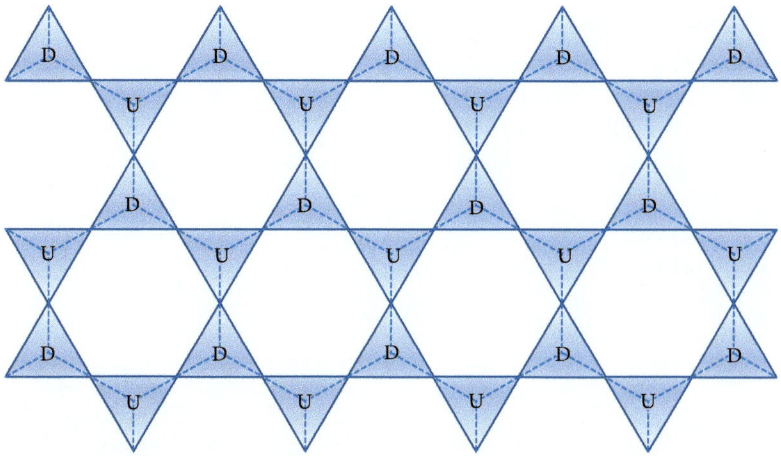

Figure 1.3 Cristobalite and tridymite crystal structures, with alternating up-and-down, corner-sharing SiO_4 tetrahedra in hexagonal sheets, stacked in an fcc ABCABC (cristobalite) or hcp ABABAB (tridymite) sequence along the *c*-axis (U = up; D = down).

non-equilibrium structures, such as monoclinic coesite,[15] tetragonal stishovite[15] and monoclinic moganite.[16] Finally, because of the complexity of the tetrahedral chains and sheets such as shown in Figures 1.2 and 1.3, silica crystallises only sluggishly from the melt, so when it is cooled rapidly it forms instead a glassy *amorphous* or *vitreous* structure.[14]

The first human-made tools and weapons were manufactured by shaping stones and *flints* which were found in the ground or chipped from exposed rocky outcrops and cliffs. *Flints* and *cherts* are examples of *chalcedony* or *cryptocrystalline* silica SiO_2, meaning a hard dense aggregate of microcrystalline, nanocrystalline and amorphous silica, which forms in nodules under high pressure in sedimented and volcanic rocks.[17] Other examples of chalcedony include semi-precious gemstones such as agate, jasper, heliotrope and onyx. When struck in the right way, flintstones fracture with sharp faces and pointed tips, which early humans learned to use as cutting implements, either hand-held directly, or after splicing with twine or bonding with resin to a wooden handle. Small admixtures of iron, magnesium, calcium and other oxides initiate the fractures when making flint tools, and give characteristic grey, cream and brown hues, sometimes with beautiful swirling patterns, particularly when the flints are lapped or polished. Figure 1.4 shows typical microstructures of a series of palaeolithic flints.[18]

The earliest-known stone tools, a crude anvil and some stone flakes, were discovered quite recently,[19] in 2015, at the Lomekwi site near Lake Turkana in northern Kenya, shown in Figure 1.5, dated to 3.3 million years ago, somewhat before the emergence of human beings and probably linked to *Australopithecus*. The earliest-known *Oldowan* human-made stone tools, with deposits of thousands of artefacts, were found further north at Gona in northeastern Ethiopia, in the Afar triangle where the East African Great Rift Valley meets the Red Sea, shown in Figure 1.5, and are dated to about 2.6 million years ago.[20] They are named after the Olduvai Gorge in Tanzania, much further south in the Great Rift Valley, also shown in Figure 1.5, where tools in their distinctive style were first discovered[21] by Louis Leakey in the 1930s. Oldowan tools are manufactured by striking a round stone or flint on the edge of a large square stone platform to create a single conchoidal fracture with sharp edges, which

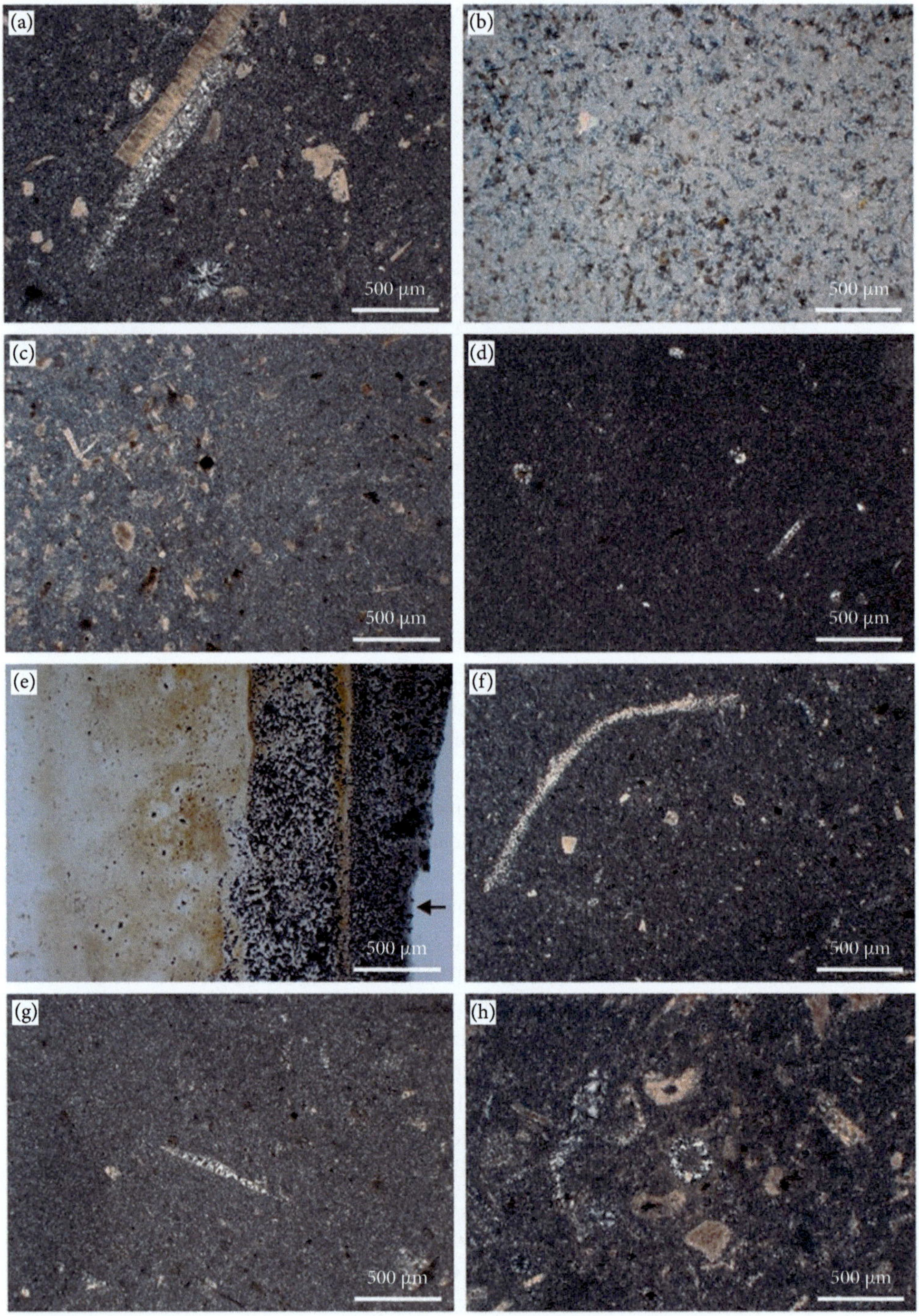

Figure 1.4 Optical micrographs of palaeolithic flints found at a variety of sites in Belgium.
Reproduced from Heaney et al. (1994). Silica: Physical behaviour, geochemistry and materials applications. *Reviews in Mineralogy 29*: 1–82.

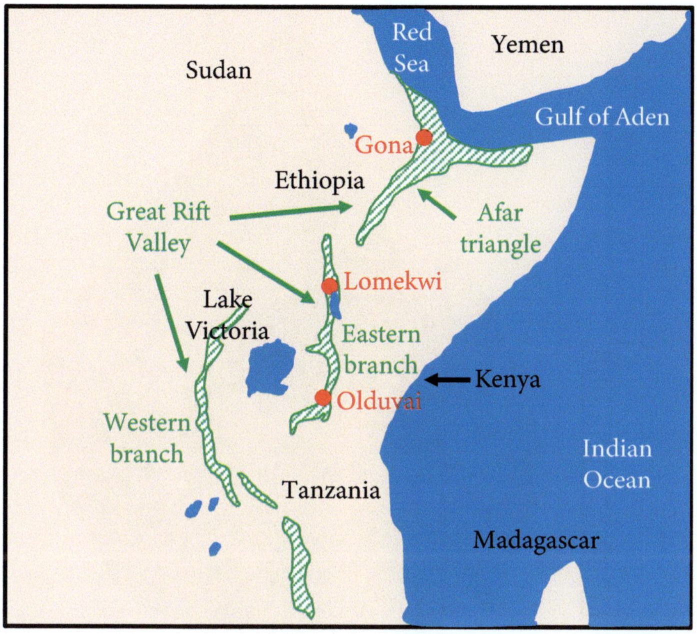

Figure 1.5 The Great Rift Valley

can then be used to make scrapers, awls, chisels, axes, choppers and knives.[20] The Oldowan tool industry continued for at least two million years, with Oldowan tools found throughout Africa, Asia and Europe as humans spread out across the world.[20] The Oldowan tool industry was replaced about 250,000 years ago by the more sophisticated *Acheulean* tool industry, in which fractures on two or more faces are used to make more complex tool shapes,[22] and this was replaced in turn by even more sophisticated multi-stage, multi-flake manufacturing processes such as the *Levallois* technique adopted towards the end of the Palaeolithic period 60,000–70,000 years ago.[23] Typical Oldowan and Acheulean flintstone tools are shown in Figures 1.6 and 1.7 respectively.

Clays, bricks and pottery

The vast family of silicates are all based on silica tetrahedra SiO_4, as shown in Figure 1.1. Individual tetrahedra can be connected via bridging oxygen corners, edges or faces, to produce a bewildering array of different long-chain, sheet and three-dimensional framework structures, all of which can be modified by admixtures of oxides with different covalency, such as Al_2O_3 and MgO, additional cations for electrical balance, and varying levels of hydration with incorporated hydroxyl groups OH^- or hydrogen-bonded adsorbed water molecules.[15]

The equilibrium room-temperature form of alumina[26] is trigonal *corundum* α-Al_2O_3, with an hcp array of oxygen ions O^{2-} and aluminium ions Al^{3+} in two thirds of the octahedral interstices, i.e. it is based on the characteristic octahedral alumina unit AlO_6 shown in Figure 1.8. When corundum is alloyed with trace additions of other oxides, it forms a wide variety of beautiful gemstones, such as *sapphires* and *rubies*.[27] Again there are many non-equilibrium structures, notably the sodium-oxide stabilised, non-stoichiometric

Figure 1.6 Typical Oldowan chopping tool.
Reproduced from Alvarez, Locutus Borg. Permission granted under a creative commons attribution-share alike 2.5 generic license. https://commons.wikimedia.org/wiki/File:Oldowan_tradition_chopper.jpg

β-alumina $xNa_2O - 11Al_2O_3$ ($x = 1.0 - 1.6$), which is used as a solid electrolyte, and cubic *γ-alumina*, which is used as a grinding medium, as well as cubic δ, monoclinic θ, orthorhombic κ, and hexagonal χ forms.[28] Figure 1.9 shows the silica-alumina phase diagram.[29] Only one equilibrium alumino–silicate forms at intermediate compositions, *mullite* $3Al_2O_3$-$2SiO_2$, but this is misleading, since a wide variety of monomer, chain, sheet and framework alumino–silicates can form, intermingling corner-, edge- and face-sharing silica tetrahedra and alumina octahedra, and incorporating other cations and anions, other oxides, and hydroxyl ions and adsorbed water.[15]

When silica SiO_2 is alloyed with a substantial amount of one or more metal oxides MO such that the ratio $MO/SiO_2 \geq 2/1$, the resulting *monomer silicates* or *orthosilicates* have structures based on separate silica tetrahedra SiO_4 linked by ionic bonds with the other cations sitting between the tetrahedra.[15] Minerals in this group include the olivines, garnets, alumino–silicates such as andalusite and sillimanite, and gemstones such as topaz and zircon.[15] *Olivines* are orthorhombic with the general formula M_2SiO_4 or $2MO$-SiO_2, for example Ca_2SiO_4 or $2CaO$-SiO_2, and $(Mg_xFe_y)_2SiO_4$ or $xMgO$-$yFeO$-SiO_2 and are the main component of the earth's solid upper mantle.

With decreasing amounts of metal oxide, i.e. with decreasing $MO/SiO_2 < 2/1$, the silica tetrahedra SiO_4 become dimers, trimers and then chains and double chains, with two and then three bridging oxygen corners, as shown in Figure 1.10, again linked by ionic bonds with the other cations and additional anions such as F^- or OH^- fitting between the tetrahedral chains.[15] These *chain silicates* or *inosilicates* are reminiscent of organic compounds, but with a general backbone -Si-O-Si-O-Si- instead of -C-C-C-. Not surprisingly, they form fibrous,

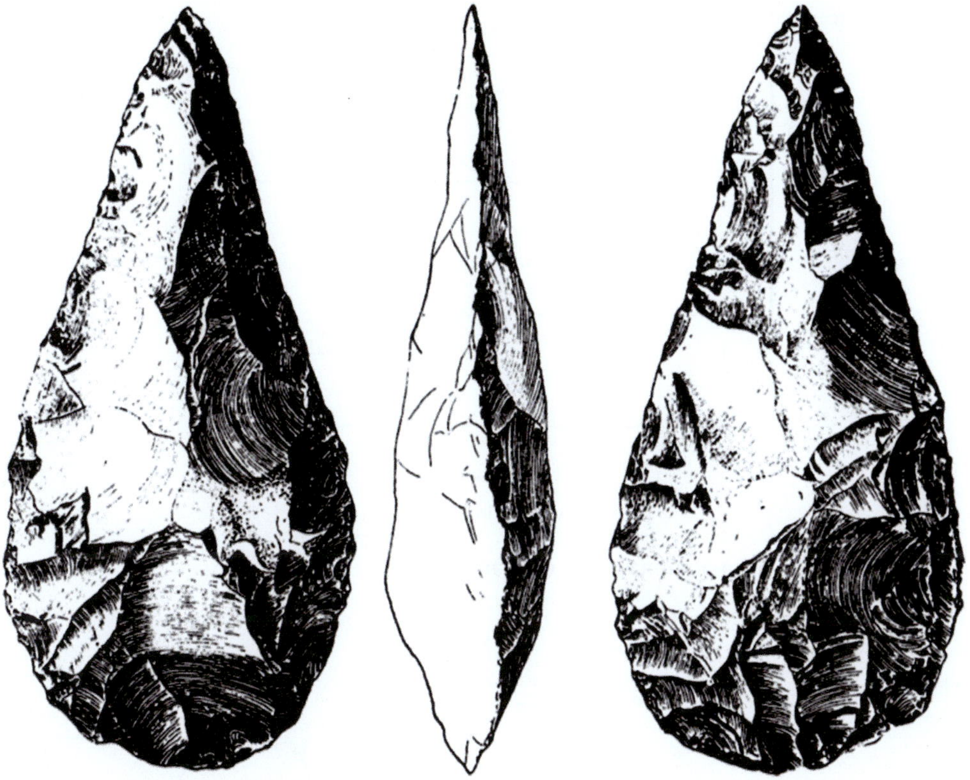

Figure 1.7 Typical Acheulian stone axe from San Isidro, Madrid.

Reproduced from Obermaier. El hombre fosil. Madrid, 1925. Public domain license. https://commons.wikimedia.org/wiki/File:Bifaz_lanceolado-San_Isidro_(Madrid).png

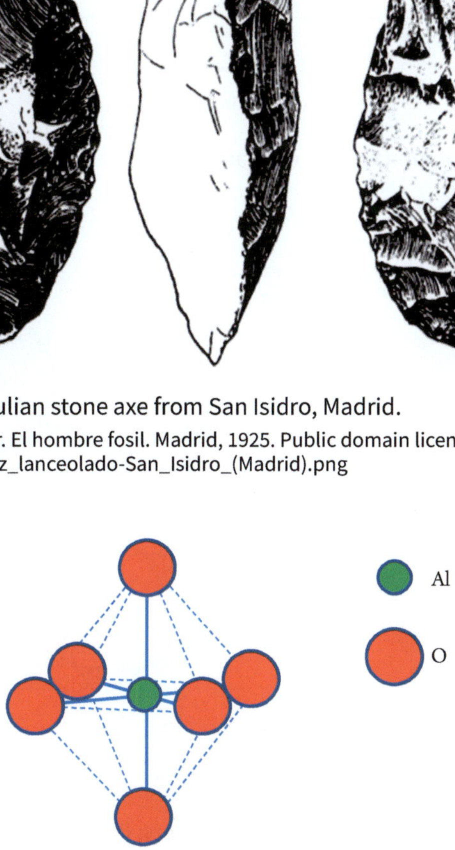

Figure 1.8 Alumina octahedron AlO_6.

prismatic and needle-like crystals. Single-chain silicates include *pyroxenes* such as enstatite $MgSiO_3$ or $MgO\text{-}SiO_2$ with a repeating dimer unit, and *pyroxenoids* such as wollastonite $CaSiO_3$ or $CaO\text{-}SiO_2$ with a repeating trimer unit. Double-chain silicates include *amphiboles* such as hornblende and asbestos, with a generic formula $M_7Si_8O_{22}(OH)_2$, where M can be a wide variety of cations leading to a correspondingly wide variety of amphibole mineral and rock forms.[15]

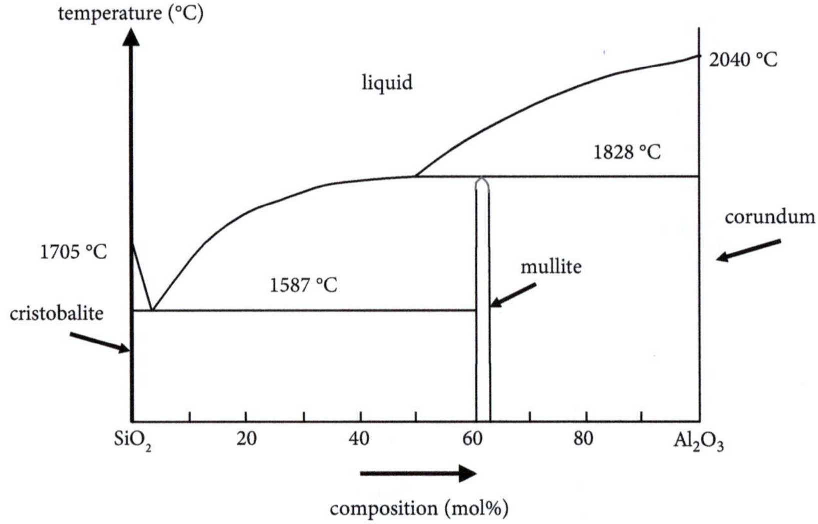

Figure 1.9 Silica-alumina phase diagram.

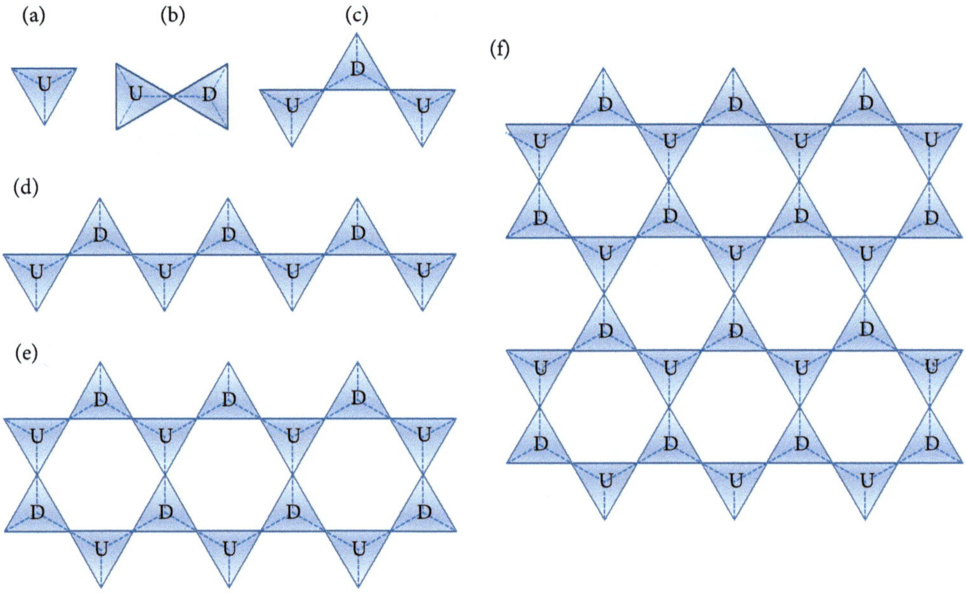

Figure 1.10 Silica tetrahedral structures: (a) monomers; (b) dimers; (c) trimers; (d) single chains; (e) double chains; and (f) sheets (U = up; D = down).

Further decreases in the ratio MO/SiO_2 leads to the formation of two-dimensional *sheet silicates* or *phyllosilicates* and then three-dimensional *framework silicates* or *tectosilicates*[15] with, respectively, three and then four bridging oxygen corners in each silica tetrahedron SiO_4, as shown in Figure 1.10. *Sheet silicates* are made up from alternating hexagonal sheets of silica tetrahedra SiO_4 and cation octahedra MO_6, with van der Waals bonding between the different sheets and a variety of different stacking sequences. The octahedral sites are all filled when the cations are divalent, two-thirds filled when the cations are trivalent, and filled variably with combinations of different cations. Sheet silicates include *clays* with 1/1 tetrahedra/octahedra ratios and *micas* with 2/1 tetrahedra/octahedra ratios.[15] In *framework silicates* all the silica tetrahedra SiO_4 share corners in an open three-dimensional network. Framework silicates include the pure silica structures such as quartz, cristobalite and tridymite, microporous alumino–silicate *zeolites*, and common alumino–silicate minerals such as the *feldspars*, with Al atoms replacing some of the Si atoms at the centres of the tetrahedra, together with interstitial cations to maintain charge neutrality.[15] Feldspars are by far the most abundant minerals in the earth's crust.[15]

Clays are a particular group of sheet alumino–silicate minerals, effectively alloys of aluminium, silicon and hydrogen oxides (alumina Al_2O_3, silica SiO_2 and water H_2O). The most common clay mineral is *kaolinite* or *China clay*, with the chemical composition $Al_2Si_2O_5(OH)_4$ or Al_2O_3-$2SiO_2$-$2H_2O$, consisting of alternating sheets of silica tetrahedra and alumina octahedra, with adsorbed water and hydroxyl groups between the sheets.[15] Clay minerals form microcrystalline aggregates, and when mixed with soil and water the alumina and silica sheets slide over each other, so the material is malleable and formable, and can easily be moulded into useful shapes. On heating and firing,[30] the water is gradually driven off to form meta-kaolinite $Al_2Si_2O_7$ (or Al_2O_3-$2SiO_2$), followed by chemical changes at progressively higher temperatures to form microcrystalline aggregate mixtures of alumina-enriched *spinel* $Al_4Si_3O_{12}$ (or $2Al_2O_3$-$3SiO_2$) and then *mullite* $Al_6Si_2O_{13}$ (or $3Al_2O_3$-$2SiO_2$) embedded in a hard vitreous matrix of silica SiO_2.

$$Al_2Si_2O_5(OH)_4 \rightarrow Al_2Si_2O_7 + 2H_2O$$
$$2Al_2Si_2O_7 \rightarrow Al_4Si_3O_{12} + SiO_2$$
$$3Al_4Si_3O_{12} \rightarrow 2Al_6Si_2O_{13} + 5SiO_2$$

Admixtures of lime CaO help the vitrification of silica rather than forming crystalline quartz; admixtures of soil and stone help control the porosity and therefore friability and strength; and admixtures of other oxides such as iron and magnesium control the characteristic red and terracotta colours of the final pots and bricks.[31,32]

As the Ice Age drew to a close in the late Palaeolithic, the glaciers began to retreat, the days grew warmer, and the early human hunter-gatherers began to settle down and live in small village communities of up to a hundred people. The Natufian culture, for instance, developed initially in the Judaean Hills near Jericho in the Jordan Valley in the Middle East, as shown in Figure 1.11, just before the end of the Palaeolithic, 12,000–9,000 years ago (10,000–7,000 BC).[3] Early villages were typically near the sea, or a river, lake or oasis, so fishing could be added as a source of food to hunting wild animals and collecting wild crops. The use of stone tools and implements became more extensive, including hooks and harpoons for fishing, and sickles and mortars for collecting and grinding wild grains.[3] At first there was no evidence for the cultivation of crops, the domestication of animals (other than dogs, which began to be used for hunting much earlier, 20,000–40,000 years ago), or the development of building materials for permanent dwellings. Soon however, harvestable wheat and other species were being sown, wild sheep and cattle were being domesticated, houses began to be built of mud-dried bricks, and clay storage jars and cooking implements were being manufactured[3]. Larger towns of a thousand or more appeared, signalling the final end of the Palaeolithic and the beginning of the Neolithic period. The first Neolithic city is widely recognised to be Çatalhöyük in Anatolia in southern Mesopotamia, part of the so-called fertile crescent near the Tigris and Euphrates River systems in modern Syria, Turkey and Iraq, as shown in Figure 1.11. The city first appeared about 9,500 years ago (7,500 BC), grew to be a settlement of between 5,000 and 7,000 people, and lasted for about 2,000 years.[3,33] Similar towns and cities grew up throughout the fertile crescent—Jericho on the west bank of the

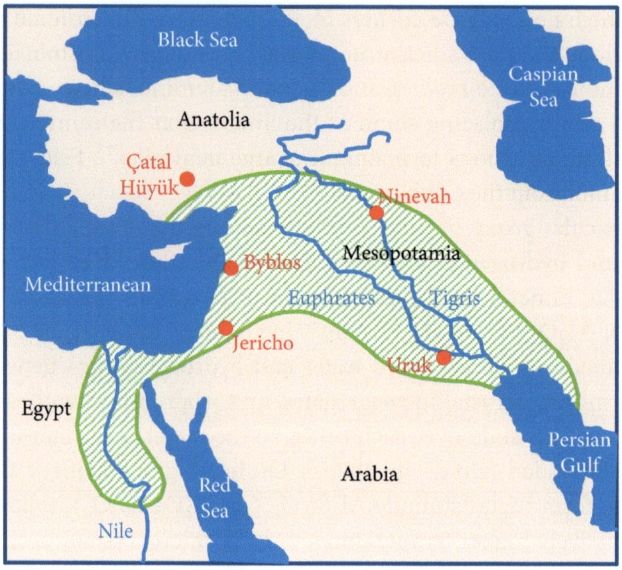

Figure 1.11 The fertile crescent.

Jordan, Sumerian Uruk and Ur, Byblos in Lebanon and Assyrian Ninevah[34]—and then further afield, at places such as Knossos in Crete, Mehrgarh on the Indus, and Jiahu on the Yellow River in China.[35]

The critical technological development supporting the new agriculture and its associated rapid urbanisation was the ability to manufacture fired ceramics to make bricks, without which settlements could not be fully permanent, and clay pots, without which harvested food and cooked meat could not be stored.[3] Neolithic peoples learned how to use chimneys and hearths to increase the temperature of an open fire, kilns to cure moulded bricks or pots, and the right mixture of clays and soils from which to mould them. These features were almost certainly discovered initially by accident, but appeared rapidly around the populated world, feeding further innovations as diverse as the development of clay tablets for writing and extensive links for trading of stone tools, harvested grain and other goods.[3] It used to be thought that these changes spread out geographically in a single wave from their initiation in the Middle East, but it is now believed that the Neolithic Agricultural Revolution and its associated technologies probably developed independently at several places more or less simultaneously, such as in the Yangtze and Yellow River basins in China and in central Mexico in the Americas.[36–38]

3 The Bronze Age and the Iron Age

Smelting

Apart from the special case of gold, metals are found mainly in various oxidised forms as components of different *rocks*, *minerals* and *ores*. The metals have to be *extracted* from their oxide ores, and this is usually achieved using a process known as *smelting*, i.e. heating to a high temperature together with a reducing agent, usually a carbon-containing material such as wood, coke or coal, so as to remove the oxygen in the form of carbon monoxide and/or carbon dioxide.[39]

The process of smelting can be explained by an *Ellingham diagram*, [40] as shown in Figure 1.12, which gives the temperature variation of the free-energy change ΔG per mole of oxygen O_2 on oxidising a range of different elements M to form their oxides M_xO_y:

$$\frac{x}{y}M + O_2 \xrightarrow{\Delta G} \frac{1}{y}M_xO_y$$

Apart from gold, oxide-formation energies are all positive at room temperature,[40] i.e. the free-energy change ΔG is negative, which is why they exist mainly in the form of oxides in minerals and rocks. For metals, the free-energy change ΔG increases (i.e. becomes less negative) with increasing temperature, with almost the same upward slope in all cases,[40] as shown in Figure 1.12. The reason for this can be seen by differentiating the free-energy change ΔG with respect to temperature T:

$$\Delta G = \Delta H - T\Delta S$$

$$\therefore \frac{d\Delta G}{dT} = -\Delta S = -\left(\frac{1}{y}S_{M_xO_y} - \frac{x}{y}S_M - S_{O_2}\right) \approx S_{O_2}$$

The last approximation step arises because metals and their oxides are all condensed phases, i.e. are all solids or liquids with low entropies, $S_{M_xO_y} \approx S_M \approx 0$, and oxygen is a gas with a relatively high entropy S_{O_2}. In other words, the Ellingham diagram slopes for metals are all equal (approximately) to the molar entropy of oxygen gas. The slopes of the different lines on the Ellingham diagram in Figure 1.12 vary slightly from metal to metal, because solid and liquid metal and oxide entropies are not quite zero and are not all exactly the same; and the slopes of the individual lines on the Ellingham diagram change sometimes with increasing temperature, though usually only slightly, when a metal or its oxide melts or (occasionally) boils, as can be seen in a few places in Figure 1.12.

The non-metal carbon C is quite different, because it oxidises to form either carbon monoxide CO or carbon dioxide CO_2, both of which are gases:

$$2C + O_2 \rightarrow 2CO$$

$$C + O_2 \rightarrow CO_2$$

Differentiating again the free-energy change ΔG with respect to temperature T:

$$\text{CO: } \frac{d\Delta G}{dT} = -\Delta S = -(2S_{CO} - 2S_C - S_{O_2}) \approx -S_{O_2}$$

$$\text{CO}_2\text{: } \frac{d\Delta G}{dT} = -\Delta S = -(S_{CO_2} - S_C - S_{O_2}) \approx 0$$

The last approximation steps arise because carbon is a solid with low entropy, $S_C \approx 0$, and the entropies of all gases are relatively high and of similar magnitude, $S_{CO} \approx S_{CO_2} \approx S_{O_2}$. This means that, unlike metals, the free energy change ΔG decreases with increasing temperature for carbon monoxide and is roughly constant with temperature for carbon dioxide, with, respectively, downward-sloping and flat lines on the Ellingham diagram.[40]

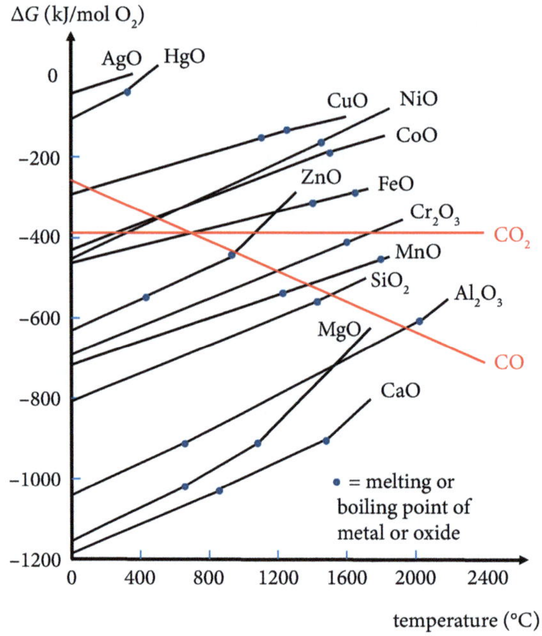

Figure 1.12 Ellingham diagram showing the temperature variation of the free-energy change ΔG per mole of oxygen on oxidising a range of different elements.

The process of *smelting* to extract a metal M from its oxide ore M_xO_y follows a general reaction scheme, with the smelting reaction corresponding to a combination of reduction of the metal oxide M_xO_y and oxidation of carbon to either carbon monoxide or carbon dioxide:[39]

$$\begin{array}{c} M_xO_y \rightarrow xM + \frac{y}{2}O_2 \\ yC + \frac{y}{2}O_2 \rightarrow yCO \\ \hline M_xO_y + yC \rightarrow xM + yCO \end{array}$$

$$\begin{array}{c} M_xO_y \rightarrow xM + \frac{y}{2}O_2 \\ \\ \frac{y}{2}C + \frac{y}{2}O_2 \rightarrow \frac{y}{2}CO \\ \hline M_xO_y + \frac{y}{2}C \rightarrow xM + \frac{y}{2}CO \end{array}$$

Smelting can take place and the metal M can be extracted from its oxide M_xO_y when the free energy change of one or other of these smelting reactions is negative, i.e. at sufficiently high temperatures, when one or other of the decreasing or flat Ellingham diagram lines for carbon monoxide or carbon dioxide respectively have fallen below the rising Ellingham diagram line for the metal M. The higher the oxide-formation energy, i.e. the lower the free-energy change ΔG at room temperature and the lower the Ellingham diagram line, then the higher the temperature needed for successful extraction by smelting, as can be seen on Figure 1.12. This is another way of saying that the more strongly bonded the oxide, the harder it is to extract the metal, and therefore the greater the thermal energy needed to do so.

Bronze

Five or six thousand years after the end of the last Ice Age, i.e. five or six thousand years after the end of the Palaeolithic period and the beginning of the Neolithic, after the first Agricultural Revolution, and after the domestication of plants and animals and the transition from humans as nomadic hunter-gatherers to settled town and city dwellers, i.e. about 6,000 years ago or about 4,000 BC, early humans first learned how to use metallic materials technologically in order to manufacture high quality tools, weapons and implements.[41] As the famous physicist and Nobel Laureate Stephen Chu said, 'The Stone Age did not end because we ran out of stones: we transitioned to better solutions'.[42] *Bronze* was the first useful metallic material discovered, and 4,000 BC marks roughly the transition from the late Stone Age or Neolithic into the *Bronze Age*.[41] Initially, towards the end of the Neolithic and in the *early Bronze Age*, low-tin bronzes were discovered when copper was smelted containing low levels, perhaps 1–2%, of arsenic and/or (sometimes) tin as impurities; later, in the *full Bronze Age*, larger quantities of tin were added to copper intentionally, to improve its strength and hardness; and finally, by the *late Bronze Age*, high-tin bronzes were being used to make high-hardness artefacts such as bells and bearings.[43]

Bronze is an alloy of copper and tin, sometimes with small amounts of other alloying additions, such as lead, aluminium and nickel.[44] Standard bronze alloys typically contain about 10-12 wt% Sn, but low-tin bronzes can contain as little as only one or two wt% Sn, and high-tin bronzes can contain up to 25–30 wt% Sn.[43–45] Figure 1.13 shows the copper-tin phase diagram.[46] The maximum solid solubility of Sn in face-centred cubic (fcc) Cu is almost 16 wt% at about 520°C (793 K).[46] However, the melting point of Cu is depressed strongly by the addition of Sn, with a sizeable solid-liquid freezing range and a partition coefficient well below one, $k = c_s/c_l \approx 0.5 - 0.6$, where c_s and c_l are solid and liquid compositions respectively.[46] When bronze alloys are cast, therefore, they solidify dendritically with significant partitioning of the Sn solute between the solid and liquid phases, producing severe coring of the dendrites.[46,47] This means that single-phase fcc α-Cu (Sn) is only obtained for Sn contents up to about 4 or 5 wt%, depending on the cooling rate.[44,46,47] For higher Sn contents, the interdendritic regions become enriched in Sn and solidify peritectically as β and γ, which transform during cooling to room temperature by a eutectoid decomposition reaction $\gamma \rightarrow \alpha + \delta$, forming a fine-scale interdendritic mixture of α and δ[44,46,47]. As-cast *low-tin bronzes* usually consist, therefore, of relatively soft, cored dendritic fcc α-Cu (Sn); and as-cast *standard tin bronzes* usually consist again of relatively soft dendritic fcc α-Cu (Sn), but with an additional fine-scale interdendritic distribution of hard particles of δ-Cu_4Sn, which has a complex-cubic γ-brass structure.[44,46,47]

Copper ores[43,48] are mostly mixed sulphides, i.e. sulphide alloys, such as *chalcocite* Cu_2S, *chalcopyrite* $CuFeS_2$ or CuS-FeS, and *bornite* Cu_3FeS_3 or Cu_2S-CuS-FeS, as well as some mixed oxides, i.e. oxide alloys, such as *cuprite* Cu_2O, *malachite* $Cu_2CO_3(OH)_2$ or 2CuO-CO_2-H_2O, and *azurite* $Cu_3C_2O_6(OH)_2$ or 3CuO-2CO_2-H_2O. They are usually found at relatively low concentrations in rather lean deposits, so smelted copper frequently contains significant amounts of unwanted *dross*, *gangue* and *slag*, which can be separated by the melting process during smelting and/or by subsequent hammering. Copper deposits are typically associated with iron, nickel and silver ores, with impurities of zinc, arsenic and antimony.[48] The only important tin mineral is *cassiterite* SnO_2, which is only rarely found associated with copper ores, which explains why early bronzes were alloyed with small amounts of

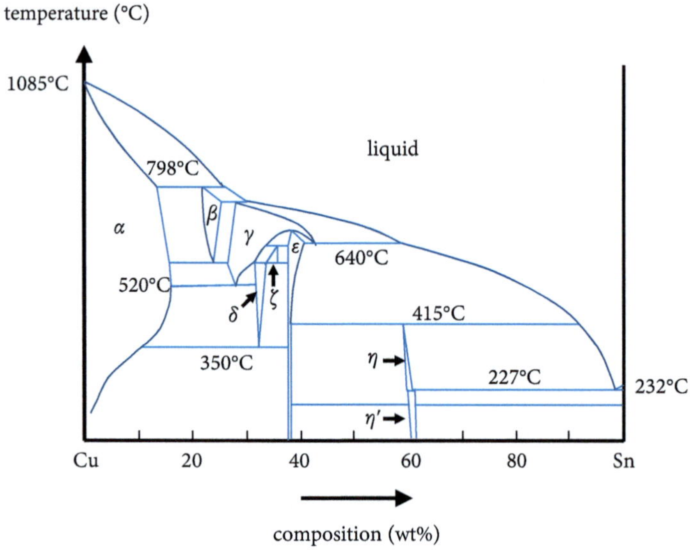

Figure 1.13 Copper-tin phase diagram.

Table 1.2 Typical compositions of early and modern bronze artefacts.[43,44,50]

Provenance	Artefact	Date	Composition (wt%)
Tepe Yahya, Iran	chisel	3,800 BC	Cu-3.7%As
Egypt	axe	3,500 BC	Cu-0.5%As
Tepe Yahya, Iran	dagger	3,000 BC	Cu-3%Sn-1%As
Troy, Turkey	dagger	2,500 BC	Cu-2%Sn-1%As
Bush Barrow, England	dagger	2,000 BC	Cu-10%Sn-0.2%As
El Argar, Spain	axe	1,500 BC	Cu-7%Sn-1%Pb
Anyang, China	bell	770 BC	Cu-13%Sn
modern	sand casting	AD 2,000	Cu-10%Sn-2%Zn
modern	bearing	AD 2,000	Cu-10%Sn-0.3%P

arsenic and other impurities rather than tin, and true tin bronzes were only developed later in the Bronze Age. Table 1.2 shows typical compositions of some early and modern bronzes.[43,44,49]

Pure copper is very soft, but addition of tin confers substantial increased strength, which is important for its use in practical applications, at the same time as still allowing the material to be cast or wrought to shape in an effective way.[44] At low levels of Sn, the material is strengthened by solid solution hardening; and at high levels of Sn, it is strengthened further by the presence of a fine-scale distribution of hard δ-Cu_4Sn particles.[44] Modern tin-bronzes are used to manufacture a diverse variety of products, such as bearings, decorative castings and bells. Small amounts of phosphorus, lead and zinc are often added to improve, respectively, wear resistance, machinability and strength.[44,46,47] The typical dendritic microstructure of as-cast low-tin Cu-5 wt%Sn bronze and higher-tin Cu-10 wt%Sn-0.5 wt%P phosphor-bronze alloys are shown in Figures 1.14 and 1.15 respectively.

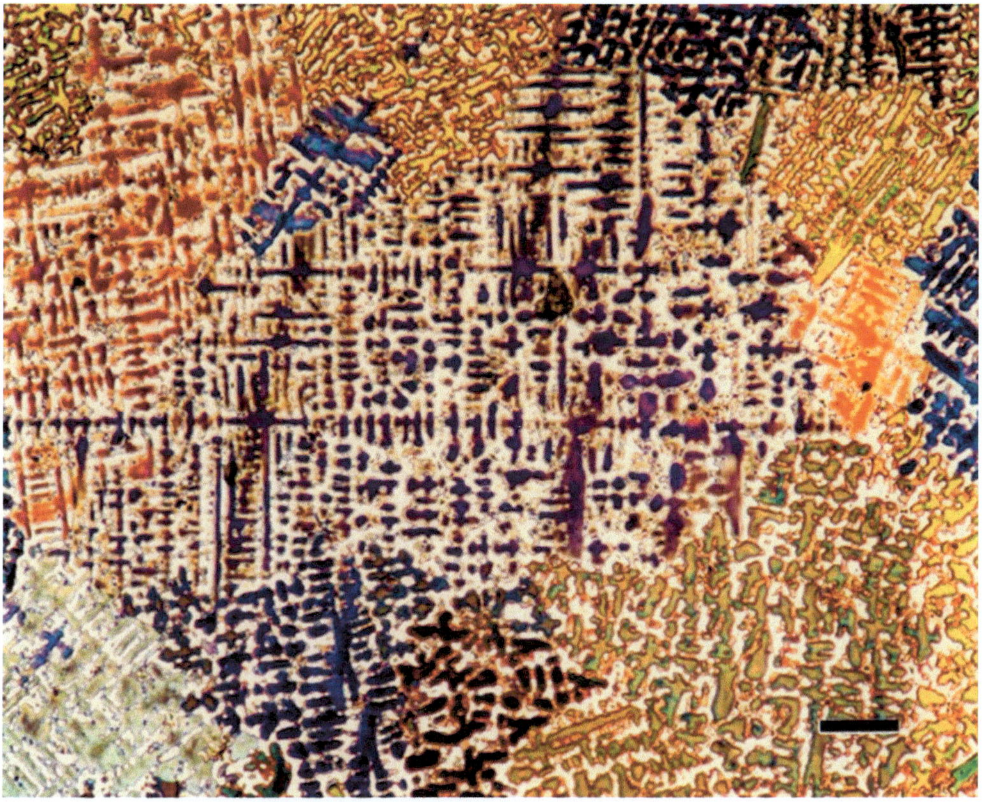

Figure 1.14 Optical micrograph of as-cast Cu-5 wt%Sn bronze alloy showing cored fcc α-Cu(Sn) dendritic structure: 100μm.

Reproduced from Caron RN et al. (2004). Metallography and microstructures of copper and its alloys. In GF VanderVoort ed., *ASM Handbook: Metallography and microstructures*. Materials Park: ASM International, 775–88.

The Bronze Age

Seven *metals of antiquity* were known in prehistoric times: gold, silver, copper, tin, lead, mercury and iron. The other metals remained unknown until much later and were only discovered in the last few hundred years. Gold is not oxidised at room temperature and *native* gold is found, therefore, naturally in its pure form, very soft and not of much use technologically, but rare, expensive and highly prized for its beauty in jewellery and ornaments.[50] Unlike gold, the other six metals are readily oxidised at room temperature, as shown in Figure 1.12, so are normally only found in a variety of different oxide and sulphide minerals and ores.[40] Small amounts of native silver, copper, tin and lead are nevertheless still to be found occasionally in local reducing environments, and native iron can also be found in small quantities in the special form of iron–nickel alloy meteorites.[51] However, none of these metals of antiquity was available in the Stone Age in sufficient quantity to be used to manufacture tools, weapons or implements, so general metallurgical usage was restricted, as with gold, to jewellery and ornaments.[41,43] The ability to extract metals from their ores by smelting must have been discovered accidentally when a wood fire was lit on top of or adjacent to a suitable mineral ore.[52] Tin and lead have very low melting points, 231.9 °C and

Figure 1.15 Optical micrograph of as-cast Cu-10 wt%Sn-0.5 wt%P phosphor-bronze showing fcc α-Cu(Sn) dendritic structure with interdendritic α-Cu(Sn)/δ-Cu_4Sn eutectoid: 100μm.
Reproduced from Caron RN et al. Metallography and microstructures of copper and its alloys. In GF VanderVoort, ed., *ASM Handbook: Metallography and microstructures*. Materials Park: ASM International, 775–88.

327.5 °C (504.9K and 600.5K) respectively, so the smelting process would have happened quite naturally on an open wood fire with a hearth made from rocks and stones containing tin- and lead-bearing minerals. Copper and iron have much higher melting points of 1,085 °C and 1,538 °C (1,358K and 1,811K) respectively, so successful smelting requires much hotter and more effective fires, which were of course being developed gradually throughout the Neolithic period, with the onset and increased use of clay-firing kilns for bricks and pottery.[52] Neolithic men and women must have been amazed at first to see shiny metallic beads of tin and lead, and then later even more beautiful and valuable copper and silver, emerging from the base of a well-stoked fire or well-designed kiln. Tin, lead and silver are soft and, like gold, were not of much use technologically in prehistoric times.[52] Copper was different. The critical discovery that began the transition into the Bronze Age was smelting arsenical and tin-bearing copper bronzes from naturally occurring impure copper ores, which could then be heated and beaten into useful implements such as fish hooks, arrow heads and small tools.[43] This was followed by a second critical discovery, which supercharged the development of a bronze-based economy, namely the ability to re-heat, melt and alloy copper intentionally with controlled additions of tin, in order to manufacture a wider range of weapons and tools from much higher quality bronze materials.[43]

Native-copper artefacts have been found at neolithic cities such as Çatalhöyük in Anatolia, dated as early as 6,000 or 7,000 BC, but tools such as awls, chisels and axes clearly made from smelted copper are not found until much later, between 3,300 and 3,800 BC, in a variety of places in the Middle East, such as Tepe Yahya in southern Iran, the Amuq Valley on the border between Turkey and Syria, and Kfar Monash, north of Tel Aviv in central Israel.[43] As the fourth and then the third millennium BC progressed, the technology for smelting copper spread throughout Mesopotamia, into Egypt and North Africa, along the Black Sea coast to the Danube, across the Aegean Sea to Greece, up north through the Balkans to northern Europe, and out eastwards to Russia, India and the Far East.[43] By about 2,000 BC, early bronze technology had reached as far as Spain in the West and Britain and Ireland in the North, as the migratory Beaker and Wessex cultures spread throughout Europe,[53] as well as India and China in the East, at the famous city of Mohenjodaro, west of the Indus River in Pakistani Sindh, and with the rise of the legendary early Xia and Shang dynasties in Henan province in the Yellow River valley.[54]

At first, metallurgical products were almost all based on relatively pure copper or low-level arsenical and tin bronzes, i.e. alloys of copper with up to no more than 1–2% of arsenic and/or tin, almost certainly derived from smelting copper ores containing arsenic and tin-bearing impurities.[43] This period is sometimes called the *Chalcolithic* or *early Bronze Age*, and marks a transition stage between the late Neolithic and the *full Bronze Age*.[43,52] Tin is superior to arsenic as an alloying addition, partly because it is less toxic, and partly because it can be added in greater quantities, leading to greater improvements in hardness and strength.[41,43] The first true tin bronzes, containing 8–10% Sn, are found in Mesopotamia and the eastern Mediterranean: in the royal graves at Ur not far from the mouth of the Euphrates in Sumer, dated to 2,800 BC; at Troy, just south of the Dardanelles, dated to 2,200 BC; and on the Greek island of Lesbos, just off the Turkish coast, dated to 2,700 BC.[43] Between 2,500–1,500 BC, while arsenical and low-tin bronzes were gradually reaching the far edges of Asia, Europe and Africa, the full Bronze Age was already spreading rapidly throughout the near-East, in Egypt, Iran and Crete, and then on to the rest of the three continents.[43] Since copper and tin ores are rarely found together, the development of mining, alloying and manufacturing as key elements of the new tin–bronze technology acted as a strong driving force for increasing trade between the burgeoning towns and cities which were growing up throughout the bronze-age world. Many of the greatest ancient civilisations of the first two millennia BC were built on the flowering of the full Bronze Age: Minoan Crete and the islands in the Aegean Sea; Hammurabi's kingdom in Babylon; Mycenaean Greece and the city states of Athens and Sparta; Theban Egypt under the Pharaohs Amenhotep III, Tutankhamun and Ramesses I and II; the Maya civilisation on the Yucatan Peninsula and central America; and the Shang dynasty in China. Figures 1.16 and 1.17 show, respectively, a cast and hammered bronze dagger from Cyprus, dated to between 1,650 and 1,200 BC,[55] and the dendritic structure in an as-cast bronze arrowhead from Palestine dated to between 1,500 and 1,300 BC.[56]

Cast and wrought iron

Figure 1.18 shows the iron-carbon phase diagram.[46] Iron has a much higher melting temperature than copper, 1,538 °C (1,811K) compared to 1,085 °C (1,358K). At low and high temperatures, pure iron has a body-centred cubic (bcc) structure called ferrite, δ-ferrite

Figure 1.16 Cast and hammered bronze dagger from Cyprus, dated to between 1,650 and 1,200 bc.

Reproduced from British Museum, object number 1897,0401.1064.

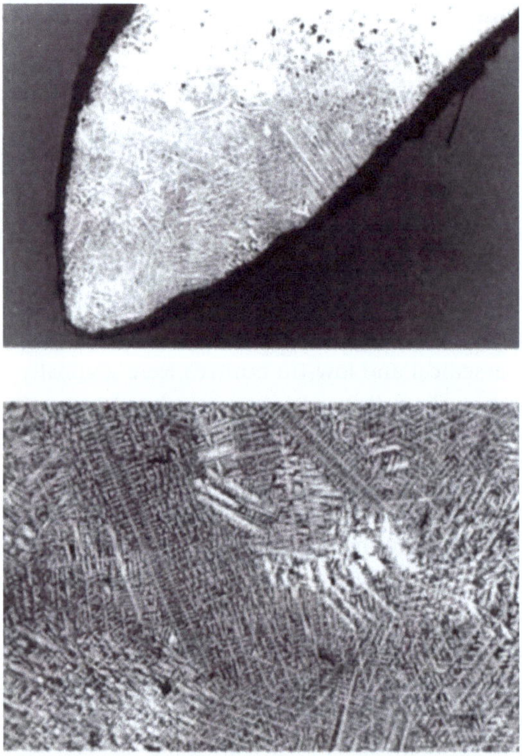

Figure 1.17 Optical micrographs of the dendritic structure in an as-cast Cu-10 wt%Sn-4 wt% As bronze arrowhead from Palestine dated to between 1,500 and 1,300 BC: 200μm.

Reproduced from Scott DA. (1991). *Metallography and microstructure of ancient and historic metals*. London: Getty Conservation Institute and Archetype Books, 88.

between 1,538°C (1,811K) and 1,394 °C (1,667K), and α-ferrite between 912 °C (1,185K) and room temperature; and, at intermediate temperatures, pure iron has an fcc structure called γ-austenite, between 1,394 °C (1,667K) and 912 °C (1,185 K).[46] Iron also has a much higher ability to absorb carbon and form a liquid or solid solution compared to copper, which has

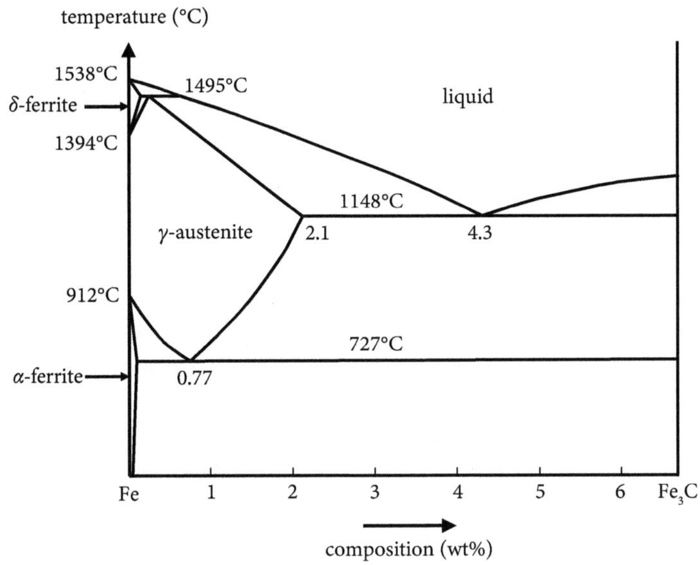

Figure 1.18 Iron-carbon phase diagram.

virtually zero carbon solubility in either the liquid or the solid.[46] When iron is alloyed with carbon, there are three important invariant reactions,[46] as can be seen on Figure 1.18: at low concentrations of carbon, there is a peritectic reaction to form austenite from a mixture of liquid and δ-ferrite, $\delta + L \rightarrow \gamma$ at 0.17 wt% (0.8 at%) C and 1,495 °C (1,768 K), and a eutectoid reaction to form a mixture of α-ferrite and an iron carbide called cementite Fe_3C from austenite, $\gamma \rightarrow \alpha + Fe_3C$, at 0.77 wt% (3.5 at%) C and 727 °C (1,000 K); and at high concentrations of carbon, there is a eutectic reaction to form a mixture of austenite and cementite from the liquid, $L \rightarrow \gamma + Fe_3C$ at 4.3 wt% (17.3 at%) C and 1,148°C (1,421 K). This means that the maximum carbon solubility is 0.09 wt% (0.42 at%) in δ-ferrite, 2.1 wt% (9.1 at%) in γ-austenite, and 0.02 wt% (0.1 at%) in α-ferrite, i.e. much greater in fcc compared to bcc iron,[46] which is why cementite is formed on cooling to room temperature.

Iron is much more abundant in the earth's crust than the other six metals of antiquity, gold, silver, copper, lead, tin and mercury,[11] as shown in Table 1.1. The principal iron-rich ores are haematite Fe_2O_3 and magnetite Fe_3O_4, as well as leaner ores such as goethite FeO(OH), limonite Fe_2O_3-$3H_2O$ and siderite $FeCO_3$ (or FeO-CO_2).[48] Since the eighteenth century, iron has been manufactured by smelting in a *blast furnace*, where the iron oxide ores are heated with coke as a source of the reducing agent carbon C, and with lime CaO as a basic *flux*, to form a calcium-rich *slag* that removes most of the impurities from the ore. Liquid slag floats on the surface of liquid iron at the bottom of the furnace and is tapped off separately as a by-product of the process. The smelted liquid iron dissolves significant amounts of carbon, as well as some manganese, silicon and phosphorus, with a typical composition of Fe-4wt%C-1wt%Mn-1wt%Si-0.5wt%P.[48] It is also tapped off separately, to be cast and solidified via the eutectic reaction to form *pig iron*, which is a mixture of ferrite and cementite Fe_3C or graphite C, and is very hard because of the high concentration of the hard cementite phase.

Modern *cast irons* and *wrought irons* are made by re-melting pig iron and modifying its alloying content, adding small amounts of elements such as chromium and magnesium, to control the cementite and graphite content as well as the microstructure.[44] Cast and wrought irons are hard and abrasion-resistant, and are used widely to manufacture components such as engine blocks, cylinder heads, gear casings and bearings.[44] Modern *steels* are formed when smelted pig iron is subjected to a further process of heating and melting, but this time in an oxidising environment, to drive off most of the dissolved carbon and silicon in the form of gaseous carbon monoxide and/or carbon dioxide and an acid silica-rich slag.[48] This secondary process is carried out in a *steel converter* furnace, and is called *steel conversion* or, more simply, just *steelmaking*. Somewhat paradoxically, therefore, *cast irons* and *wrought irons* are iron-carbon alloys with high carbon contents, typically 1–5 wt%, whereas *steels* are iron-carbon alloys with much lower carbon contents, typically 0–1 wt%,[44] as shown in Figure 1.18. Towards the end of the Bronze Age, furnaces improved, and the smelting of iron ores to manufacture cast and wrought irons was discovered, initiating the *Iron Age*. The discovery of how to manufacture steels was not until much later, so Iron Age peoples learned how to make improved metallurgical tools and implements from smelted, cast and beaten irons with a relatively high carbon concentration, instead of using inferior bronzes. The development of steel technology in the eighteenth and nineteenth centuries is not discussed until the next section of this chapter.

The Iron Age

Native meteoritic iron, usually in the form of an iron-nickel alloy, was used decoratively here and there throughout the Bronze Age, but iron was relatively rare, only used in small quantities, and generally regarded (somewhat surprisingly, given its widespread modern use) as a precious metal similar to gold and silver.[57] The technology of smelting iron is widely agreed to have been developed initially by the Hittites during the second millennium BC in Anatolia in central Turkey.[58–60] The Japanese archaeologist Akanuma[61] discovered the earliest-known (so far) smelted iron fragments, dated to 1,800 BC, at Kaman-Kalehöyük in central Anatolia, about 100 kilometres south-east of modern Ankara; and the earliest known smelted iron dagger, dated to about 1,500 BC, was found 200 kilometres north at Alaca Höyük, close to Boğazkoy, the modern site of Hattusa, the ancient Hittite capital.[58,59] The Hittites continued to manufacture iron tools and weapons, but kept their discoveries secret, presumably to try and maintain advantages in trade and war. The Hittite empire first became dominant in Mesopotamia in about 1,500–1,600 BC, but later in the millennium came increasingly under attack, from the Egyptian Pharaohs to the west, the Assyrians to the south and the Phrygians to the north, finally collapsing in about 1,200 BC, to be replaced by the Assyrian empire. The fall of the Hittites in 1,200 BC led immediately to rapid dissemination of iron technology and the initiation of the Iron Age throughout the rest of the Middle East and subsequently on to the rest of the world. The fall of the Hittite empire was part of a more general, so-called *late Bronze Age collapse* when, in half a century between 1,200 and 1,150 BC, Mycenaean, Aegean, Cretan, Babylonian, Egyptian, Hittite, Levantine, Luwian and Canaan cultures and civilisations all imploded, with virtually all of the principal cities of the eastern Mediterranean violently destroyed and many abandoned never to be occupied again.[62] Many reasons have been put forward for this collapse (as always with major historical events), including volcanoes, drought, migration, warfare and poor-quality, decadent governments, but the underlying cause was undoubtedly the technological transition from a Bronze Age to an Iron Age world economy.[61]

Between about 1,000 BC and 500 BC, knowledge and understanding of how to manufacture iron tools and weapons spread out in different directions: south into Palestine and Egypt, and down into the Sudan, sub-Saharan Africa and Nigeria; west to Greece, the rising Etruscan empire and the Iberian peninsula; north to Austria, the Hallstatt empire and on to Gaul and Britain; and east into northern India and China.[58,60,63] Like the Bronze Age before it, the Iron Age prompted the rise of enormous new empires and cultures: the Hallstatt empire and La Tène culture in middle Europe; the Gauls and the Celts and their colonisation and diaspora throughout western and northern Europe; the kingdoms of Kush and Nubia in Sudan, the rise of Ethiopia and Abyssinia, and the Nok culture in northern Nigeria; the fall of the Zhou and Shang dynasties in China, beaten by Qin (who unified the 'warring states' and from whom China is named) and replaced by the long-lived Han dynasty; and the rise of Alexander the Great followed, perhaps greatest of all, by the Roman republic, the Caesars and the Roman empire.[58,60,63] Figure 1.19 indicates the extent of some of these Iron-Age civilisations and empires.

Early irons were probably not much stronger than the high-quality bronzes that were available in the late Bronze Age, and were more brittle and more difficult to make.[58,60] Their clinching advantage, however, was the widespread availability of iron ore, and, therefore, the relatively low cost of making iron goods, and the ability of each kingdom to access its own independent source material.[60] At first iron was smelted in simple *bowl furnaces*, not much more than clay crucibles on a wood fire, only reaching relatively low temperatures, so that early irons were not fully melted and required repeated reheating and hammering to squeeze out unreacted ore and unwanted dross and gangue, in order to achieve a malleable and, therefore, usable wrought metallic product.[57,58,60] However, the manufacturing technology gradually improved. By about 500 BC, iron was being smelted in taller, *shaft furnaces*;

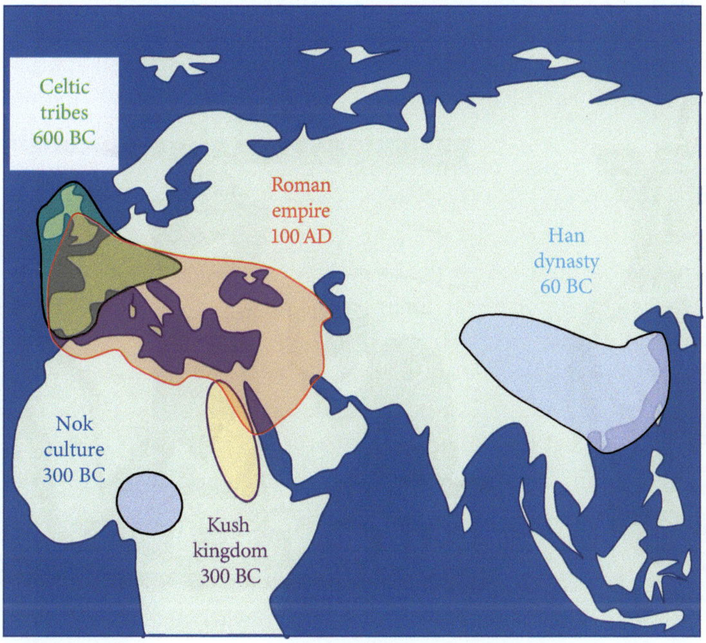

Figure 1.19 Iron age civilisations.

with charcoal as a more efficient fuel and source of reducing carbon (made by heating wood to drive off moisture and organic volatiles); with *tuyères* (holes in the furnace), *bellows* and *chimneys* to promote oxidation and increase furnace temperatures; with additions of lime flux to enhance slag formation and impurity removal; with better control of the tricky balance between oxidation (of the carboniferous fuel to provide heat) and reduction (of the iron ore to iron); with improved *smithing* practices to reheat the smelted iron into the austenite range before working into the desired product form; and with carburising to harden surfaces and edges on a relatively malleable iron tool or weapon core, by heating it in a carboniferous wood or charcoal pack.[57,58] Figure 1.20 shows a drawing of a smith heat-treating an iron bloom in a typical shaft furnace, taken from a Greek vase from about 500 BC.[58,64]

At first smelted wrought iron was only used to make unstressed implements such as pots and pans, and the handles and hilts of bronze swords, axes and hammers to which they were riveted.[58] Crude riveting was used, for instance, as early as 1,400 BC to join a number of small iron parts in order to make the miniature headrest in Tutankhamun's tomb.[58] But soon the superior quality of smelted iron took over from bronze, and iron-age swords and daggers were good enough to cut through bronze armour.[60] A large cache of smelted iron bars and iron weapons dated to about 700 BC was found in the Palace of Sargon II, King of Assyria, Babylon and Sumer, at excavations in Khorsabad near modern Mosul in Syria.[58]

Two key developments in the mid-first millennium BC were the discovery of *quenching and tempering*, and the use of *pattern welding*.[57,58] In the first, *quenching and tempering*, austenitic iron is *quenched* by plunging it into water to prevent cementite formation and retain the carbon in a much harder, distorted-bcc or body-centred-tetragonal (bct) solid solution called *martensite*, followed by re-heating or *tempering*, to allow gradual precipitation of fine-scale cementite, largely maintaining the high hardness of martensite but removing its associated brittleness. In Homer's *Odyssey*, Odysseus describes, rather horrifically, blinding the Cyclops with a hot stake,[57,67] '. . . as when a smith dips a great axe or

Figure 1.20 A shaft furnace from Greece depicted on a vase, showing a smith forging a bloom
Reproduced from Tylecote RF. (1976). *A history of metallurgy*. London: The Metals Society.

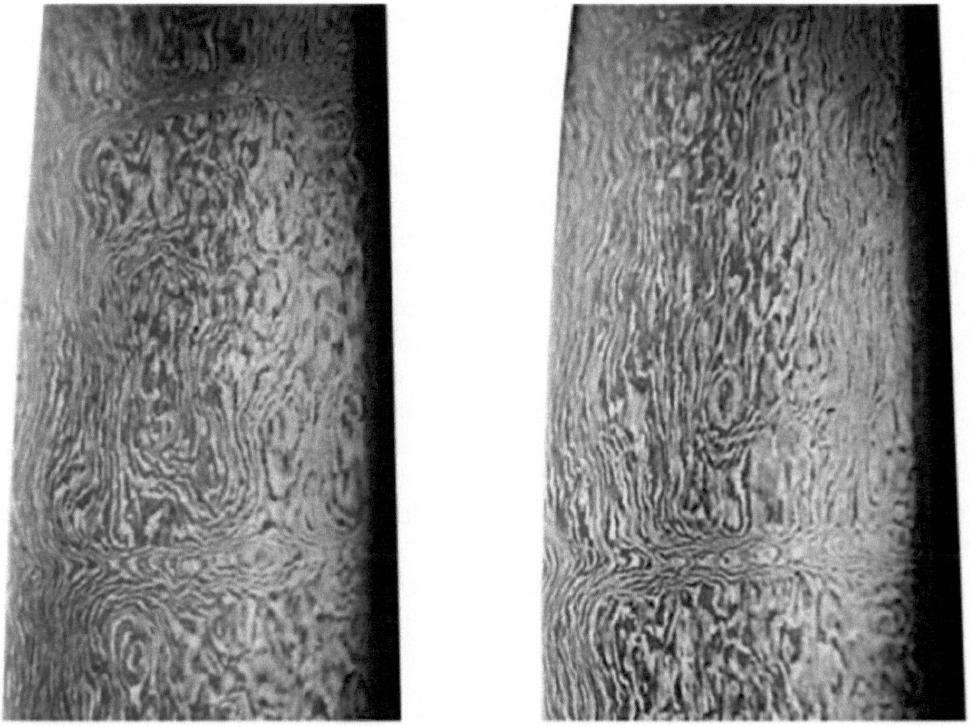

Figure 1.21 Water patterns on the blade of a Persian sword.
Reproduced from Abadi RAA. https://commons.wikimedia.org/wiki/File:Watered_pattern_on_sword_blade1.Iran.JPG.

adze into cold water, amid loud hissing to temper it—for therefrom comes the strength of iron—so hissed his eye about the stake'. In the second, *pattern welding*, austenitic iron is repeatedly folded, beaten and re-welded, to produce nano-scale fibrils and layers of cementite embedded in ferrite, again with outstanding combinations of hardness and ductility.[58,59] This pattern-welding technique led on, in the Middle Ages, to the famous *Damascus steel*, used to make patterned swords and gun barrels, an example of which[66] is shown in Figure 1.21. These and other improvements in metallurgical technology supercharged the growth and success of the Roman empire,[58,60] which began in about 400 BC and lasted about a thousand years, after which it collapsed into the so-called Dark Ages, overrun by barbarians in the middle of the first millennium AD. The Romans were innovative metallurgists, increasing the size of smelting furnaces and the scale of industrial production, and thus expanding the use of wrought iron to make a much wider range of products: ploughshares, hoes and rakes, planes, shears, lamps, furniture, crossbows, catapults, water-wheels and chariots.[60] The Iron Age developed differently in China, with superior furnace designs allowing for the full melting of newly smelted iron and, therefore, the subsequent manufacture of true cast-iron products, perhaps a thousand years before their appearance in the West.[58,59] In the first century AD, the Roman nobleman and philosopher Pliny the Elder said that iron was '. . . the most precious . . . metal for mankind. By its help we cleave the earth . . . build houses, hew stone and so forth';[60] and in the nineteenth century the Scottish writer and chronicler of the Industrial Revolution Samuel Smiles called iron '. . . this extraordinary metal, the soul of every manufacture, and the mainspring perhaps of civilised society'.[68]

4 Steel and the Industrial Revolution

The Scientific Revolution and the Age of Exploration

Many precursor events led up to the *Industrial Revolution* in the eighteenth and nineteenth centuries. Perhaps the two most significant were the *Scientific Revolution* and the *Age of Exploration*, both sparked by the *Renaissance* in Europe in the fifteenth and sixteenth centuries and the *Age of Enlightenment* in the seventeenth century. The *Scientific Revolution* was a deep philosophical change, in which knowledge and understanding of the world began to be derived from experiment and observation rather than from preconceived ideas and beliefs.[69–71] This underpinned an intellectually adventurous spirit in discovering and implementing new ways of doing things, leading to the engineering development of novel industrial machinery and the social development of factories and large cities, both supporting a vastly increased production of goods during the Industrial Revolution. In parallel with the growth of science, the *Age of Exploration* or *Age of Discovery* was also driven by a different kind of restless and physically adventurous spirit, with explorers sailing the high seas, discovering new lands, and building connections between far-flung peoples.[72,73] This provided the context for an enormous increase in trading (albeit in many cases exploitative), with transfers of materials and artefacts all round the world, delivering a strong economic basis for the same increased production of goods during the Industrial Revolution.

Important mathematical and scientific discoveries were made in antiquity, notably in Greece, China and Arabia. But it was not until the fifteenth century that the *scientific method* began to be used explicitly and extensively as a method of generating new knowledge.[70,71] The scientific method involved generating ideas and then (crucially) testing them by experiment and observation of the real world. Bertrand Russell famously remarked that, 'Aristotle maintained that women have fewer teeth than men; although he was twice married, it never occurred to him to verify this statement by examining his wives' mouths'.[74] The *Scientific Revolution* is often said to have started with the publication by Nicolas Copernicus in 1543 of *De revolutionibus orbium coelestium* [On the revolutions of the heavenly spheres], his seminal and revolutionary work on the heliocentric motions of the planets, showing from direct observation that the planets orbit the sun and not the earth.[75] And it is often said to have been completed more than a century later with the publication in 1687 of Isaac Newton's *Philosophiae Naturalis Principia Mathematica* [Mathematical principles of natural philosophy] (or just *Principia*), inventing mathematical calculus and using it to provide a unified explanation of the laws of motion, classical mechanics, gravitational forces, and the motion of the planets and stars, all based firmly on the detailed observations of Johannes Kepler and others.[76]

The *scientific method* was described by Francis Bacon, the father of *empiricism*, as using *inductive reasoning* and *inductive logic* (rather than traditional *deductive reasoning* and *deductive logic*) to argue from observed facts to *hypotheses* and then to *theories* and *physical laws*.[70] As Bacon says in his philosophical treatise *Novum organum: sive indicia vera de interpretatione naturae* [New method: true directions concerning the interpretation of nature] published in 1620:

> There remains simple experience; which, if taken as it comes, is called accident, if sought for, experiment. The true method of experience first lights the candle [hypothesis], and

> then by means of the candle shows the way [arranges and delimits the experiment]; commencing as it does with experience duly ordered and digested, not bungling or erratic, and from it deducing axioms [theories], and from established axioms again new experiments.[77,78]

The resulting *Age of Reason* or *Age of Enlightenment* (or just *the Enlightenment*) in the following two centuries (seventeenth and eighteenth) built an empirically-based scientific understanding of many physical and biological phenomena: light waves, microscopy, telescopes and astronomy, stars and the heavens, elasticity and springs, the atmosphere, pumps and vacuums, clocks and compasses, combustion, basic electricity and magnetism, the classification of plants, and blood and its circulation in the body.[70,71] These advances in experimental investigation and the resulting scientific understanding provided the motivational, methodological and intellectual basis for developing and implementing the new mechanical and engineering equipment that launched the Industrial Revolution.

The Scientific Revolution began with what was seen as the most important problem of the time, namely the attempt to understand the motions of the planets. This was important for the pragmatic objective of helping navigation for sailors at sea. So, curiously, the Scientific Revolution is indissolubly linked with the roughly contemporaneous *Age of Exploration*. In 1498, just before the end of the fifteenth century, the Portuguese explorer Vasco da Gama led four ships, the *São Gabriel*, the *São Rafael*, the *Berrio* and a supply ship, round the Cape of Good Hope and Cape Agulhas on the southernmost tip of Africa, up the coast to Mombasa and Malindi in Kenya, and across the Indian Sea, to reach Calicut and Cochin (modern Kozhikode and Kochi) in Kerala on the west coast of India.[79–81] Half the crew died on the return across the Indian Ocean, which took 132 days running against the fierce winter monsoon winds (compared to twenty-three days on the outward voyage, with the help of the friendlier summer monsoon winds). This epic journey opened up the direct sea route from Europe to Asia and initiated a rapid growth in trade between Europe and the East. A few years earlier, in 1492, the Genoese explorer Christopher Columbus had sailed west on behalf of the Spanish crown, with three ships, the *Pinta*, the *Niña*, and the *Santa Maria*, mistakenly hoping to find an alternative route to the Orient, but instead discovering (from the European perspective) the Americas, landing successively in the Bahamas, Cuba and Hispaniola.[82] In subsequent voyages, Columbus continued to explore the Caribbean and Central and South America, and he was followed by a string of Spanish explorers,[82,83] notably Hernán Cortés who conquered the Aztec Empire in southern Mexico in 1521, and Francisco Pizarro, who similarly conquered the Inca Empire on the western coast of South America eleven years later, in 1532. Figure 1.22 shows the sea routes of some of these early explorers.

Thus began the *Age of Exploration*, and three centuries dominated by colonisation, empire building, trade and exploitation between Europe and dominions in North and South America, Africa, India, China and South-East Asia.[73,82] In the sixteenth century this was mainly undertaken by Portuguese and Spanish freebooters, invading and controlling the Indonesian archipelago and Indian coast in the East and much of Central and South America in the West.[83] But in the seventeenth and eighteenth centuries, Iberian influence waned, and was overtaken by northern Europeans, notably the Dutch and English, who set up a string of fortified trading posts backed by military and naval forces. The Dutch East Indies Company took over parts of Asia from the Portuguese,[82,83] seizing the Moluccas (now the Maluku Islands) in eastern Indonesia in 1605, Malacca, between Singapore and Kuala Lumpur, on

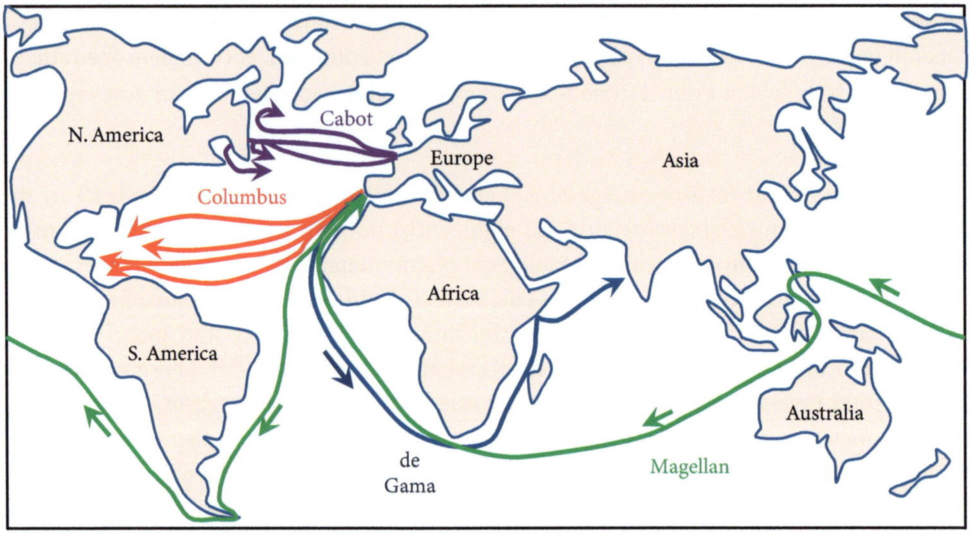

Figure 1.22 Sea routes of some early explorers—Christopher Columbus (1492, 1493, 1502 and 1504), Vasco da Gama (1498), Ferdinand Magellan (1522) and John and Sebastian Cabot (1497, 1498 and 1508).

the west coast of Malaysia in 1641, Ceylon (Sri Lanka) in 1658 and Cochin (Kochi) on the west coast of India in 1662, making Jakarta their capital in Indonesia in 1619. The Dutch also seized[82,83] Pernambuco, Alagoas and Rio Grande do Norte in northern Brazil in the 1630s and 1640s, and founded New York (initially Nieuw Amsterdam) on Manhattan Island in 1625, and Cape Town around Table Mountain and the Cape of Good Hope in South Africa in 1652. Similarly, the English East India Company established trading posts[82,83] in Surat in India in 1612, followed by Madras (Chennai), Bombay (Mumbai), Calcutta (Kolkata), and a total of twenty or more other Indian cities; and British outposts were also set up across North America and the Caribbean, beginning with Jamestown, Virginia in 1607, Plymouth in 1620, Massachusetts Bay in 1630, the Bahamas in 1620, and Jamaica in 1655. This was, of course, just the beginning of European colonial empire building, with Spanish, Portuguese, Dutch, British, French and German colonisation of large parts of Asia, Africa and the Americas continuing throughout the seventeenth, eighteenth and nineteenth centuries, intensifying world trade, mostly through exploiting rich natural resources from around the world to feed industrial production and create wealth in Europe.[73]

The early Industrial Revolution

The *Industrial Revolution* was built on the empirical, scientific and innovative culture engendered by the *Scientific Revolution*, and the free-ranging, worldwide trading opportunities created by the *Age of Exploration*. It took place over a century and a half, from about the middle of the eighteenth century to the end of the nineteenth century.[84–86] It has been described as one of the two largest-ever transformations in human socio-economic affairs, along with the Agricultural Revolution and domestication of plants and animals at the end of the Ice Age. There were three phases:[86]

1. The first phase is sometimes called the *early Industrial Revolution*. New manufacturing methods and processes based on mechanisation and engineering were developed, initially in northern England, radically altering the ways in which goods could be made, leading to an enormous increase in productivity in a number of sectors, notably agriculture, mining and textiles. This period began with the invention of the spinning jenny in 1864 and continued with a wave of new inventions up to the end of the eighteenth century, when the pace of change temporarily slowed down as Europe was engulfed in the *Napoleonic wars* following the *French Revolution* in 1798. The quality, capability and impact of the new industrial machinery was underpinned by two important mid-century inventions: the *coke-fired blast furnace* for large-scale manufacture of *cast iron* and *wrought iron* as basic engineering materials from which to make new machines; and factory workplaces in which to use them for the high-intensity production of goods.[87]
2. In the second phase, during the first half of the nineteenth century after the end of the Napoleonic wars, mechanisation and industrialisation continued apace, gathering momentum and being disseminated across Britain, to other countries in Europe and on throughout the world, vastly increasing world trade. Britain continued to lead, and the end of this period was marked by the *Great Exhibition* of 1851 at the *Crystal Palace* in London, celebrating what proved to be the apogee of British engineering and industrial dominance.[87]
3. The third phase is sometimes called the *late Industrial Revolution*. During the second half of the nineteenth century, a continuing stream of new engineering and technological innovations revolutionised many other sectors, notably transport, construction and energy. The late Industrial Revolution was supercharged by the invention of the *steel converter* by Henry Bessemer in 1856 and the *open-hearth furnace* by William Siemens in 1865, providing reliable and cheap methods of manufacturing high quality steel from which to make industrial machinery, bridges, ships, railways, buildings, and many other products. Over the same period of time, the unification of Germany and the conclusion of the American civil war saw subsequent dynamic economic growth in these two new industrial giants, which rapidly came to rival and then surpass Britain, as its own innovation and power began to wane.[87]

The *early Industrial Revolution* began in the *cotton industry* in Lancashire in the north of England. The British historian Eric Hobsbawm said famously, 'Whoever says Industrial Revolution says cotton'.[85] Raw cotton consists of small round balls of matted irregular fibres that are first spun, i.e. drawn and twisted, to make extended and aligned cotton thread, before being woven into a usable textile material from which to make, for instance, clothes or bedding. Traditionally this was done by hand in India, China, Egypt and Latin America, close to where cotton plants are grown, but as early as in the Middle Ages raw cotton began to be imported into Europe and turned into cloth nearer to its end market, though still by hand. The manufacture of cotton goods in Europe was given a massive boost at the beginning of the Industrial Revolution by a series of inventions which mechanised the process.[85,87] James Hargreaves from Oswaldtwistle near Blackburn in mid-Lancashire patented the *spinning*

Figure 1.23 An early Arkwright water frame with eight bobbins.
Reproduced from Science Museum. Textiles machinery category, B. Fothergill, object number 1860–1865.

jenny in 1764, using a draw-bar driven by a hand-wheel to spin multiple cotton strands or *rovings* onto a series of *spindles* or *bobbins*, initially eight, but later many more; Richard Arkwright from Preston, also in mid-Lancashire, patented the *spinning frame* (or, later, *water frame*) in 1769, using a roller system, initially horse-driven but then with a water mill instead of hand-power, to spin the cotton thread; Samuel Crompton from Bolton, a bit further south in Lancashire, patented the *spinning mule* combining both these developments; and on the other side of the Pennines, in Doncaster in Yorkshire, Edmund Cartwright invented the *power loom* in 1785 to extend mechanisation to the subsequent process of weaving cotton or wool into cloth.[87,88] Figure 1.23 shows an example of an Arkwright spinning frame.[89] Productivity and profitability increased enormously as these inventions were implemented in a wave of new factories and mills, and the textile industry was revolutionised, with 150 Arkwright mills installed in the 1780s, 20,000 spinning machines by the end of the eighteenth century, and at its peak in the mid-nineteenth century 50,000 mules in Lancashire alone.[85]

The first commercially successful *steam engine* was built by Thomas Newcomen[90,91] in 1712, condensing steam to create a vacuum and suck a *piston* into a closed *cylinder*, thus activating a lever to pump water, as shown in Figure 1.24. The engine was first used to drain the mines at the Coneygree Coalworks in Dudley in the English Midlands. The Scottish engineer James Watt improved the efficiency of Newcomen's steam engine with a number of important modifications, beginning in 1759 and continuing through to the end of the eighteenth century:[92,93] firstly by separating the condenser from the cylinder, so the latter did not have to be continually heated and cooled; secondly by introducing a double-acting piston, so power could be derived from both the upstroke and the downstroke; and then, with his business partner Matthew Boulton, inventing a gearing system to convert the reciprocating movement of the piston into a rotary motion, which could be used more generally for grinding, spinning, weaving and milling. Steam engines supported the mining of coke, coal and

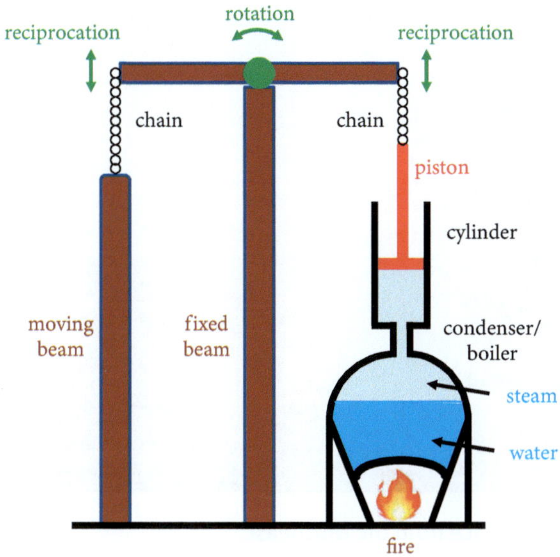

Figure 1.24 Schematic of a Newcomen reciprocating engine.

iron ore as the fundamental industrial materials, as well as allowing textile and other types of manufacturing to be undertaken in factories in many more places, not just near suitable sources of flowing water to drive a water mill, greatly accelerating the progress of the Industrial Revolution. The firm of Boulton and Watt was highly successful, and Watt went on to retire in 1800 as a wealthy man.[92,93] Steam engines continued to evolve, and their use was, of course, enormously enhanced in the nineteenth century, when they were further adapted for transportation by railway locomotives, electricity generation in power stations, and in many other industrial factory settings. According to the historian Anthony Anderson, Watt's improvements to the steam engine[94] 'converted it from a prime mover of marginal efficiency into the mechanical workhorse of the Industrial Revolution'. In 1824, the great French engineer Sadi Carnot said,[90] 'To rob Britain of her steam-engines would be to rob her of her coal and iron, to deprive her of her sources of wealth, to ruin her prosperity, to annihilate that colossal power'.

The Great Exhibition

The Great Exhibition, or to give it its full title 'The Great Exhibition of the Works of Industry of All Nations', was the brain-child of Prince Albert, the husband of Queen Victoria, and took place in 1851 in Knightsbridge, London, on the south side of Hyde Park.[95,97] By the middle of the nineteenth century Britain was at the apex of its industrial strength and economic power. The British Empire covered large parts of the world, and Britain's leadership of the Industrial Revolution was unchallenged and seemed unassailable. The main objective of the Great Exhibition was 'to make clear to the world its [Britain's] role as industrial leader'.[98] It was a remarkable event. It was opened by Queen Victoria on 1 May 1851 and, from May to October, Hyde Park was full of visitors, a total of over six million, including an eclectic range

Figure 1.25 The Crystal Palace from the Northeast.
Reproduced from Nash J. et al. (1854). *Dickinson's comprehensive pictures of the great exhibition of 1851*. London: Dickinson Brothers. https://www.bl.uk/collection-items/dickinsons-comprehensive-pictures-of-the-great-exhibition-of-1851?mobile=on

of luminaries: Charles Darwin, Karl Marx, Charlotte Brontë, Samuel Colt, Michael Faraday, Charles Dickens, Alfred Tennyson, Napoléon III and Lewis Carroll.[95,97] It was housed in and around a purpose-built cast-iron-framed glass-house called the Crystal Palace, almost 2,000 feet long and 500 feet wide (about 563 m × 138 m), three times the size of St Paul's Cathedral, designed by Joseph Paxton and Charles Fox, consisting of over 3,300 cast iron columns, 2,200 cast iron girders, 200 miles (322 km) of wooden sash bars, and almost a million square feet (almost 100,000 m^2) of glazing.[99] An image[95] is shown in Figure 1.25. It was later dismantled and re-erected on Sydenham Hill in Penge, a south London suburb (the area is now known as Crystal Palace), only to be destroyed by fire in 1936.

There were over 100,000 exhibits displayed by more than 15,000 exhibitors.[95,97] Pride of place in the middle of the building was given to an enormous fountain, made of four tons of pink glass, 27 feet (over 8 m) high.[96,97] The biggest exhibit was a massive hydraulic press that had been used to lift and manipulate the thousand-ton (0.9 Gg) wrought-iron box-section tubes used to build Robert Stephenson's famous Britannia Bridge across the Menai Straits at Bangor, connecting the island of Anglesey to Wales and allowing trains to run directly from London to the Dublin ferry at Holyhead.[96] Other exhibits included the first fax machine, the first Colt revolver, a massive steam hammer for forging prop shafts and bearings for steamships, cotton spinning and weaving looms, printing presses, agricultural machines, a cigarette-making machine, the first cast-iron framed piano, a whole gallery of horse-drawn carriages, lots of railway locomotives, early bicycles (or velocipedes), fire engines, the world's first voting machine, an Indian princeling's throne, Cossack armour, a stuffed elephant with a magnificent bejewelled howdah, the Koh-I-Noor (the largest diamond in the world), hundreds of marble columns and statues, millions of opulent products such as carpets, jewellery, Russian and Canadian furs, Swiss gold watches, Sèvres porcelain vases, enamels from Limoges, Axminster carpets, malachite urns, silks from Lyons, and a vast array of modern (then) scientific and industrial equipment—electric telegraphs, microscopes and telescopes, pumps and barometers, theodolites, clocks, musical instruments, early daguerreotypes (photographs), and surgical equipment.[97,98,101] As Queen Victoria put it in her diary,[101] 'every conceivable invention'. Half the exhibits were British, but great displays were also put on by France, Holland, Germany, Switzerland, the United States, Canada, Chile, India and many other nations from all around the world.[97,98] Charlotte Brontë wrote:

> Yesterday I went for the second time to the Crystal Palace . . . It is a wonderful place—vast, strange, new and impossible to describe. Its grandeur does not consist in one thing, but in the unique assemblage of all things. Whatever human industry has created you find there.[102]

The price of admission was at first '£3 for gentlemen and £2 for ladies' (£400 and £270 or \$525 and \$355 respectively in today's currency), but from 24 May 'the masses were let in for a shilling a head' (£6.75 or \$8.88 in today's currency).[101] The exhibition made a profit of £186,000 (roughly £25 million or \$33 million in today's currency),[101] which was handed over to the Royal Commission for the Exhibition of 1851, a charitable trust set up with the mission 'to increase the means of industrial education and extend the influence of science and art upon productive industry'.[103] The money was used to help build the Royal Albert Hall, three great museums—the Victoria and Albert (or V&A), the National History Museum and the National Science Museum—and the Imperial Institute and Royal School of Mines, later combined to form Imperial College of Science and Technology, creating the so-called *Albertopolis* on the original Crystal Palace site in Knightsbridge.[101,103] The Royal Commission continues to this day, awarding fellowships and project funding in line with its original mission.[103]

The late Industrial Revolution

As the nineteenth century progressed, the *late Industrial Revolution* saw innovative mechanisation and industrialisation of production spreading out across the globe and into many other sectors, notably transport, communications and energy. Newcomen and Boulton-Watt steam engines began to be used to drive paddle-steamers towards the end of the eighteenth century, at first mainly on relatively short inland trips, but the mid-nineteenth century saw the rapid introduction of rotary instead of reciprocating steam engines, screw-driven propellors instead of paddles, and hulls made from iron and steel rather than wood.[104] In the 1840s it has been calculated that approximately 90% of the world's shipping was built of wood, but by the 1890s Britain was manufacturing over 80% of all new ships and making them from iron and steel,[104] much more suited to watertight and economical transport of goods over long-haul distances. In 1835, Isambard Kingdom Brunel designed the first purpose-built steamship for crossing the Atlantic and the (then) largest passenger ship in the world, the 236-foot-(72 m)-long, 1,360-tonne-(1.36-Gg)-displacement, wooden-hulled, paddle-wheel-driven steamship,[104–106] the *Great Western*, launched in 1837 with its maiden crossing in 1838; in 1839, he designed an even bigger 322-foot (98-m), 3,400-ton (3,080-tonne or 3.08-Gg) steamship,[104–106] the *Great Britain*, this time using an iron hull and screw-driven propellor, launched in 1843 with its maiden crossing taking just fourteen days in 1845; and in 1854 he designed the *Great Eastern*, at that time by far the largest ship ever built, 692 feet (211 m) long and 17,274 ton (15,670 tonnes or 15.67 Gg) displacement, with a capacity for carrying 4,000 passengers from England to Australia without refuelling.[106,107]

The first steam locomotive was designed by Richard Trevithick in 1804 to run on the plate-track tramway of the Penydarren Ironworks near Merthyr Tydfil in South Wales,[108] followed by Matthew Murray's first rack-and-pinion steam locomotive *Salamanca* in Leeds in 1812, and *Puffing Billy* built by Christopher Blackett and William Hedley for the Wylam Colliery in Northumberland in 1813, the oldest surviving locomotive, currently on display in the Science Museum in Kensington in London.[90,108] The so-called father of the railways, George Stephenson, designed and built *Locomotion* as the first public steam locomotive to be used (though alternating with horse-drawn vehicles) for the Stockton and Darlington Railway in 1825, followed by his famous *Rocket* in 1829, and the first public railway line using only steam locomotives between Liverpool and Manchester in 1830.[91,109] At first these early trains travelled quite slowly but, by the 1840s, railways were being built connecting all the main cities in England and speeds were reaching up to fifty miles per hour (80 km/h).[104] The first underground tube line[104] was the Metropolitan in London, opened in 1863. Cast and wrought iron initially and then steel were essential for making the railway track, the steam engines, the signalling systems, the marshalling and maintenance sheds, and cathedral-like stations with huge arched roofs,[104,110] such as Victoria, Charing Cross and St Pancras in London, opened respectively in 1860, 1864 and 1868.

The famous Iron Bridge, the first bridge to be made out of cast iron, crosses the river Severn at the eponymous Ironbridge Gorge, near the villages of Ironbridge and Coalbrookdale in Shropshire.[111,112] It was designed in 1779, and built and opened in 1781, by the local businessman Abraham Darby III, to allow better access between his ironworks and the nearby coal mines, as well as to both banks of the river for transport of goods to and from other parts of England.[111,112] The locality has been designated as a World Heritage Site, and is sometimes called *the birthplace of the Industrial Revolution* (though not without a certain amount of hyperbole, since there are several other contenders for that title). The bridge itself consists of a beautiful near-semicircular arch, with all its cast-iron girders in compression,[113] as shown in Figure 1.26. It was tremendously successful, aroused

Figure 1.26 The Iron Bridge in Shropshire.
Reproduced from https://commons.wikimedia.org/wiki/File:Iron_Bridge_east_side_in_February_2019.jpg

widespread interest, and set off a worldwide vogue for building bridges from iron and steel instead of the more traditional wood and stone materials. Thomas Telford built a second iron bridge[111] a little further up the Severn in 1795, followed by a much larger one with a span of 236 feet (72 m) across the river Wear in Sunderland,[111] and then by John Rennie's cast-iron bridge over the Thames at Southwark in London,[111] which was completed in 1819. As the nineteenth century wore on, an enormous number of bridges was needed for the growing railway and road networks springing up across the industrialised world, and cast iron was replaced progressively as a building material by wrought iron and then steel, allowing a much wider range of designs because of the greater strength of these materials, and their ability to withstand tensile as well compressive forces. Famous examples include: Isambard Kingdom Brunel's Clifton suspension bridge over the Avon Gorge, not too far from the Severn Estuary, completed in 1863, built using massive wrought-iron suspension chains;[111,114] John A. Roebling's Brooklyn Bridge connecting Manhattan to Brooklyn across the East River, completed in 1883, with a main span of 1,595 feet (486 m), at that time the longest steel suspension bridge in the world;[111] and the first ever steel cantilever bridge, over the Firth of Forth nine miles (14 km) west of Edinburgh in Scotland, designed by John Fowler and Benjamin Baker and completed in 1890.

The Industrial Revolution invaded almost all areas of human endeavour.[115] Agricultural production was enhanced by massive planting, sowing and reaping machines; communications were improved with telegraph cables extended throughout most European and American countries, then across the English Channel, and the Atlantic Ocean; printing presses were installed to mechanise the production of great newspapers, disseminating information widely and rapidly; industrial mining equipment was implemented to speed up the extraction of natural resources; consumer goods began to be produced on a mass basis; and steel frames and then steel-reinforced concrete were used to build multistorey buildings and then soaring skyscrapers, reaching to the heavens.[115] At the time of the Great Exhibition at the Crystal Palace in 1851, it had seemed that British economic and political power and dominance in world affairs was unstoppable, bound to keep going from strength to strength, with its leading role in the Industrial Revolution and the enormous extent of its worldwide empire. But Britain's apparent might was misleading. By the end of the century, the spark of innovation had passed on. France, Germany and the United States were all now vying for leading industrial positions.[86,115]

The motor car industry[116] was initiated by Karl Friedrich Benz, with his Motorwagen, the world's first automobile, built in 1885 and patented in 1886 in Mannheim, Baden-Württemberg in southwest Germany;[117] Nicholas Otto, working with Gottlieb Daimler and William Maybach patented the first four-cycle internal combustion engine in 1876 at their factory in Cologne;[118] and Henry Ford produced his revolutionary Model T, the first mass-produced car, in 1908 in Detroit, Michigan in the United States.[119] Thomas Edison invented the phonograph, the motion-picture camera and the light bulb at his research laboratories in Menlo Park, New Jersey, and launched the world's first electric power station in downtown Manhattan in 1882.[120] The Wright brothers, Orville and Wilbur, inaugurated the aviation industry with their first powered flight in 1903, flying 120 feet (37 m) in twelve seconds, a speed of 6.8 miles per hour (10.9 km/h), in *The Wright Flyer* a few miles outside Kitty Hawk, North Carolina.[121,122] The German chemical industry was becoming dominant, having started by producing dyes for the textile industry based on new organic materials,[123] with the founding of companies such as: Bayer by Friedrich Bayer and Johann Friedrich

Weskott in Barmen (modern-day Wuppertal) in 1863, now one of the biggest pharmaceutical companies in the world, and famous for having developed aspirin;[124] Hoechst (initially Teerfarbenfabrik Meister, Lucius and Co) in Höchst near Frankfurt, also in 1863;[125] and Badische Anilin und Soda Fabrik (BASF) by Friedrich Engelhorn in Mannheim in 1865, currently the biggest chemical company in the world.[126] Lack of good-quality land forced New York and Chicago to build upwards with, for example: the 310-feet-(94-m)-high, eighteen-storey, iron-framed Pulitzer building for the *New York World* newspaper in 1890; Chicago's twenty-one-storey, steel-framed Masonic Temple in 1892; and the 612-foot-(187-m)-high, forty-one-storey Singer Building in Manhattan in 1909.[127,128] And France launched a string of Exposition Universelles (World Fairs) initiated by Napoléon III in Paris in 1855, 1867, 1878, 1889 and 1900, explicitly in order to rival and outdo the Great Exhibition at the Crystal Palace. These included wonders such as: the construction of the 300-metre-high wrought-iron Tour Eiffel (Eiffel Tower) as the entrance to the 1889 Expo, celebrating the centennial of the French Revolution; and the 1900 Expo, celebrating the end of the nineteenth century, occupying 112 hectares of the Rive Droite and Rive Gauche (right and left banks) of the river Seine, connected by a newly constructed bridge Pont Alexandre III (and another 100 hectares in the Bois de Vincennes), with fifty-eight countries represented (more than double the number at the Crystal Palace), and visited overall by almost fifty million people.[129–131] In 1870 Britain still mined 50% more coal to support its industry than France, Germany and the United States put together, but by the end of the century Britain had been overtaken and the relative proportions reversed.[132] Similarly, in 1840 Britain had approximately 2,500 km of railway lines,[133] almost twice as much as the rest of the world; but by 1900, although Britain's railways had multiplied more than tenfold, to over 30,000 km, there was already 200,000 km in the rest of Europe,[133] and yet another 200,000 km in the United States.[134] The growth of German and American industrial strength continued, of course, throughout the twentieth century, joined after the two world wars by Japan, and then in the twenty-first century by China, India and others.

Overall, the Industrial Revolution had enormous consequences for the way people lived. There was a massive increase in productivity and wealth: the population of western Europe roughly trebled, from about eighty million in 1700 to a quarter of a billion in 1900, while its GDP increased tenfold over the same period, from about eighty billion dollars to more than three quarters of a trillion.[135–137] Millions of people were taken out of poverty and starvation. But there was a correspondingly massive social dislocation: by the end of the nineteenth century, cities in America and Europe had grown dramatically, were dominated by multi-storey buildings and skyscrapers, heated and lit by electricity, and connected by telegraph, rail, road and steamship, their populations living in packed city suburbs and employed in industrial factories, instead of being dispersed throughout the countryside working on the land. And enormous regions of Africa, Asia and the Americas had been subjected to colonial exploitation and impoverishment, their natural resources ransacked and expropriated to support the growth of production and wealth in industrialised Europe and (latterly) the United States.

Iron and steel

The Industrial Revolution was driven by the development of two important new processes for making improved materials: first, the invention of the coke-fired *blast furnace* in the early eighteenth century for manufacturing better-quality wrought and then cast irons;[138] and second, the invention in the mid-nineteenth century of the *Bessemer converter* and the *Siemens–Martin open-hearth furnace* for manufacturing high-quality steel.[138] Without reliable, strong, yet ductile and therefore malleable, irons and steels, it would have been impossible to make the beautiful and precise engineering equipment—the machine tools, textile machinery, printing presses, steam engines, dynamos, forges, cranes, transformers, ships, trains and bridges—that powered the ever-growing manufacturing production and associated trade that characterised the Industrial Revolution.

Metallurgical techniques in general (as with almost all technology), and iron and steel-making in particular, evolved relatively little in the period between the late Iron Age and the early Industrial Revolution, i.e. for well over a thousand years, roughly from the Roman era through to the Middle Ages and the Renaissance, and almost right up to the eighteenth century.[139] Throughout these many centuries, iron was smelted in furnaces not much more than a few feet tall, using charcoal, a carboniferous material made by heating wood to drive off water and other volatile (organic) constituents, which acted both as a reducing agent to remove oxygen from the ferrous oxide ores, and also as a fuel to heat the process, speeding it up and softening the iron product so it could subsequently be worked into shape. Temperatures in these relatively rudimentary furnaces were not very high, insufficient to melt fully the smelted iron, and insufficient to remove all the entrained gangue and impurities. Hammering and forging by a blacksmith was, therefore, essential to squeeze out any residual unreduced ore and detritus, and to turn the initial *bloom* of iron into its desired shape. The resulting *wrought iron* was used as a material for making things such as armour, weapons, nails, hand tools, and dies for minting coins.[139] Towards the end of the Middle Ages and on into the Renaissance, furnaces began to grow in height, reaching up to perhaps ten feet tall (over 3 m), and their working temperatures gradually increased above 1,000 °C (1,273K), then 1,100 °C (1,373K) and then 1,200 °C (1,473K), until iron with a few per cent of carbon could be melted fairly well and cast into shape,[139] as shown in the iron–carbon phase diagram in Figure 1.18. By the end of the fifteenth century, *cast iron* was, therefore, also being used fairly widely (it had been in use for much longer in China), though its quality was still not as good as wrought iron.[139]

The last half of the seventeenth century and the first half of the eighteenth century saw a series of innovations which changed this picture dramatically. Furnace height was raised to thirty feet (10 m) or more; water wheels and, later, steam engines were used to replace hand-bellows and create a continuous and intense up-draught of air; iron ore and fuel were loaded continuously at the top to provide a counter-flow to the up-rushing air blast; the waste gases emitted from the top of the furnace, a hot mixture of nitrogen, carbon monoxide and carbon dioxide, were recycled through a heat exchanger to pre-heat the air blast applied through tuyères at the bottom; limestone *flux* was added at the top to the iron ore and coke, to fluidise unreduced oxides in the ore and create a liquid oxide *slag* which could be separated from the reduced iron; a sump or hearth at the bottom of the furnace allowed the molten iron and molten oxide slag to settle, separate and be tapped off semi-continuously; and charcoal as a fuel was replaced, first with cheaper coal, and then with more efficient *coke*, made by pre-heating coal to drive off water and impurities, leaving it sufficiently strong and porous to withstand the intense air blast, at the same time as being rich in carbon for rapid and effective burning and reduction of the ore.[139]

In 1709, Abraham Darby was the first to use coke in the blast furnace at his ironworks in Coalbrookdale, having seen its effectiveness as a fuel for drying malt to make the famous pale ale in the beer industry in Derbyshire.[139,140] It was Darby's son, Abraham Darby II, who later, in 1748, removed embrittlement from cast irons by introducing the use of low-phosphorus ores; and it was Darby's grandson, Abraham Darby III, who even later, in 1781, pushed through the building of the nearby Iron Bridge.[138–142] The development of the *blast furnace* and its associated technology produced much better wrought and cast irons, at first driven largely by the resulting ability to manufacture improved armaments—cannons, rifles and guns—before being adopted more widely for the machinery and equipment of the Industrial Revolution.[139] At first, the Darbys kept their improved iron-making blast furnaces to themselves in order to retain a competitive advantage, but gradually the know-how began to leak out and be copied widely.[138–142] By 1760, there were seventeen coke-fired blast furnaces in operation; by 1775, there were thirty-one; and by 1790, there were eighty-one, producing almost 80% of all British production of iron.[138] Blast furnaces have been used ever since throughout the world as the primary method for smelting iron ore, partly still to make cast iron but mainly now as a precursor to making steel.[143,144] Figure 1.27 shows a schematic diagram of a typical modern blast furnace.

Wrought and cast irons were valuable materials driving the early phases of the Industrial Revolution in the eighteenth century. But the full range of industrial machinery that emerged in the nineteenth century could not have been built without a consistent and high-quality supply of a stronger and more ductile material, namely *steel*. Irons are alloys consisting of iron Fe with several per cent of carbon, as shown in the phase diagram in Figure 1.18. The carbon is usually in the form of a very hard carbide, Fe_3C, called *cementite*. With a few per cent of carbon in the alloy, there is a significant volume fraction of cementite, which means

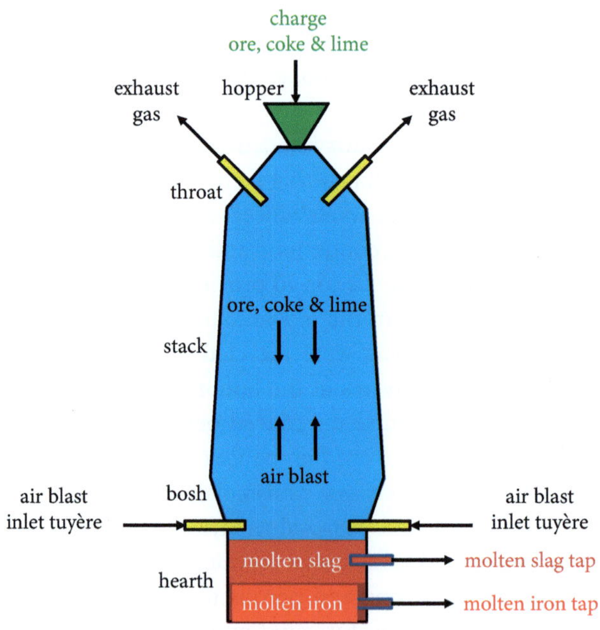

Figure 1.27 Schematic blast furnace.

that wrought and cast irons are hard and strong for engineering applications, but difficult to form into shape, and inclined to fracture during manufacture or in use.[44] However, when the carbon concentration is reduced by a factor of ten to just a few tenths of a per cent, as it is in a steel, the distribution of cementite is much easier to control, so that it is in a very finely-divided form within the material, typically on a microscopic scale or even a nano-scale, making steels strong yet highly formable by either casting or working.[44,145]

Steelmaking and the advantages of a low-carbon concentration were known in the Middle Ages, but without being able to melt the material in a high-temperature furnace the resulting steel was difficult to make in any quantity and was variable in quality. This problem was solved in the middle of the nineteenth century by Henry Bessemer and William Siemens, at roughly the same time but in two quite different ways.[138] Henry Bessemer was born in 1813 in a small village called Charlton, near Hitchen in Hertfordshire in England.[146] His father, Anthony Bessemer, was an inventor, who was born in London but worked mostly in France, where he invented improved optical microscopes, and dies for the Paris Mint, and became a member of the Académie des Sciences.[146] He fled from the horrors of the French Revolution, back to England, and retired to a small estate in Charlton that he bought when he made some money after inventing a novel method of making gold chains.[146] There must have been something strong in the Bessemer family genes, since Henry followed his father as a prolific inventor and businessman, leaving school and setting up his own manufacturing company aged just seventeen, and going on to develop a diverse range of machinery for things such as composing printing type, making graphite for pencils, crushing sugar cane, casting plate glass, embossing postage stamps, copying artwork, and making bronze powder for gold paint.[147] During the Crimean War of 1853–1856, he invented a new kind of artillery shell, which he tried to sell to the French Ministry of Defence, but which turned out to be too powerful for the cast-iron cannons then in use.[148] He immediately spied a business opportunity and set about finding new ways to make better cast iron or cast steel. Throughout 1855, he conducted a wide range of experiments and discovered that blowing air through molten iron promoted a reaction between dissolved carbon in the iron and oxygen in the air, reducing the carbon content to form a good-quality steel, at the same time as heating it dramatically to temperatures well above the melting point, so that it could be poured and solidified easily.[148] In 1856 he patented the process, and read a paper about it at the British Association for the Advancement of Science (BAAS): 'On the Manufacture of Malleable Iron and Steel Without Fuel'.[149] A couple of years later, he invented his famous *Bessemer steel converter*, a tilting furnace into which molten pig iron from the blast furnace was poured, followed by blowing air through it using tuyères in the base of the converter, creating a violent reaction that turned the iron into low-carbon steel in a matter of minutes.[148]

Ironmasters were sceptical about the idea of steelmaking with no fuel, and at first there were indeed teething troubles. Bessemer licensed the process to four ironmasters, who sued him when they failed to get the quality of steel they expected, so he bought back the licences at a substantial premium and set up his own steel company,[138,148] which went on to be one of the largest and most successful in the world. The key to making the process work well lay in two important further developments: first, Robert Forester Mushet's method of adding ferro-manganese, an Fe-Mn alloy, to remove excess absorbed oxygen and sulphur impurities, preventing oxide and sulphide embrittlement of the steel;[138,150] and second, a few years later, Sidney Gilchrist Thomas's discovery of limestone-firebrick furnace linings, so a basic lime CaO flux could be used to remove embrittling phosphorus impurities and

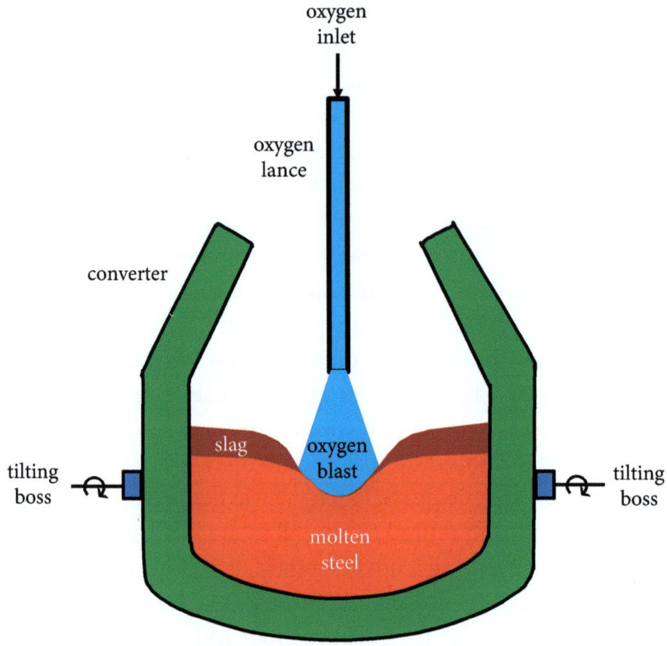

Figure 1.28 Schematic LD steel converter.

allow conversion of the widely-available (and, therefore, much cheaper) basic, phosphorus-containing ores.[138,151] Bessemer was a muscular businessman and, rightly or wrongly, saw off patent claims by Mushet and by William Kelly, an American who claimed to have discovered the process first. By 1870, out of a total of 220,000 tons (200 Gg) of production of British steel, all but 5,000 tons (4.5 Gg), i.e. almost 98%, was made by Bessemer's converter process.[138] In modern times, total world steel production has grown massively, reaching a total in 2019 of approximately 1.9 billion tons (1.7 Pg),[152] of which the majority, approximately 60%, is made by the basic oxygen (BOS) or Linz–Donawitz (LD) process, using a Bessemer converter with an oxygen blast top-blown through a lance, instead of an air blast bottom-blown through tuyères.[39,153] Figure 1.28 shows a schematic diagram of a typical modern LD converter. Bessemer continued to invent new processes and took out a total of more than 130 patents between 1838 and 1883.[154] He was knighted in 1879 and was made a Fellow of the Royal Society (FRS) in the same year. He died in 1898 and is buried in West Norwood cemetery in London, alongside other great inventors and businessmen, including Sir Henry Tate (sugar merchant and philanthropist, after whom the Tate Gallery in London is named), Sir Henry Doulton (inventor and manufacturer of pottery, who, with his father John, founded the Royal Doulton ceramics company), and Baron von Reuter (famous reporter and pioneer of telegraphy, who founded the Reuters news wire agency).[155]

William Siemens was born Carl Wilhelm Siemens in 1823 in Lenthe, a small village just outside Hanover in northern Germany.[156,157] His father was Christian Ferdinand Siemens, a tenant farmer who ran a large estate owned by the King of Hanover, as his forebears had done since 1384, and his mother was Eleonore Deichmann. The Napoleonic Wars created an agrarian crisis in northern Germany (and elsewhere), and William's father had to move to another farm near Mecklenburg-Strelitz, which is where William grew up.[156,157] It was a large family of fourteen children, and William was the seventh child and fourth son. His elder brother, Werner Siemens, was a successful engineer and entrepreneur:[158] the electrical unit of conductance is named after him; and he founded Siemens and Halske, an electric-telegraph company, which later became the multinational Siemens electrical-engineering and telecommunications company, which today is one of the largest engineering companies in the world.[159] William studied at the University of Göttingen, where he worked for a while with the famous physicist Wilhelm Weber, who invented the electric telegraph, and after whom the unit of magnetic flux is named.[156] In 1843, after leaving university and still only nineteen years old, William moved to England to promote a patent for a new electroplating process that his brother Werner had developed.[157,160,161] The trip cost him just £1 and was in many ways wildly successful. He found England to his liking and he stayed there for the rest of his life, changing his name to Charles William Siemens and becoming a naturalised British subject.[160,161] In 1859 he married Anne Gordon, youngest daughter of the Writer to the Signet (a senior Scottish solicitor and notary) and sister to the Professor of Engineering at the University of Glasgow. He was naturalised the same year, and he liked to say that at the same time he took an oath of allegiance to two ladies, Queen Victoria and his wife.[160,161] He was a successful inventor, entrepreneur and businessman, working throughout his career on a variety of innovations with a range of British engineering companies, including Fox, Henderson & Co of Smethwick, John Hicks of Bolton and R.S. Newall's of Dundee.[160,161] He was the first President of the Society of Telegraph Engineers, which later became the Institution of Electrical Engineers (IEE), the forerunner of the Institution of Engineering and Technology (IET).[161] In 1862 he was elected to be a Fellow of the Royal Society (FRS), and in 1871 delivered its prestigious Bakerian Lecture.[161] He was knighted by Queen Victoria in 1883, just a few months before his death. A glass window was installed in his memory in Westminster Abbey.[161]

William Siemens's most successful invention and his greatest achievement was the *open-hearth furnace*, an alternative to Bessemer's converter for turning pig iron from the blast furnace into steel.[162,163] In the early Industrial Revolution in the second half of the eighteenth century, Henry Cort's *puddling process* was used for making wrought iron and later cast steel, with a mixture of pig iron and iron ore melted in a large bath and *puddled* carefully and slowly by hand-stirring.[164,165] The dissolved carbon in the pig iron reacted with oxygen in the oxide ore and was gradually removed into a slag, which floated on the surface of the decarburising molten iron. The Eiffel Tower and Statue of Liberty were both made from puddled wrought iron.[166] But the process was slow and not at first easy to control, relying on the judgement of the puddlers as to when the appropriate level of carbon had been removed throughout the large bath of molten metal. Siemens developed a more efficient *regenerative furnace*, in which the charge was heated from above by the circulation of hot air, with a continual supply of fresh gas heated by the hot exhaust from the furnace.[167,168] In 1865, the French engineer Pierre Émile Martin took out a licence from Siemens to apply this regenerative principle to making iron and steel, and this became known as the *Siemens–Martin process* or *open-hearth process*, similar to the puddling process but without the need for stirring.[167,168] By 1870, steel production had grown dramatically, to over 200,000 tons (1.8 Gg), and the Bessemer process had taken over more or less completely from puddling for its manufacture.[138] And by the end of the century, open-hearth steelmaking was similarly successful if not more so: in just thirty years, steel production had grown by another twenty times, to 4.9 million tons (4.4 Tg), of which 65% was now made by the open-hearth process.[138] Open-hearth steelmaking had three great advantages: first, the regenerative furnace achieved a very high temperature, promoting efficient oxidation and removal of the carbon, and excellent fluidity for casting the resulting molten steel into billets; second, it was financially attractive, easily consuming significant quantities of scrap together with the iron and ore charge, and using low-grade coal as a fuel; and third, because it was slow and, therefore, relatively controllable, with good compositional accuracy making, therefore, high-quality steel.[135] Today, modern steelmaking has, however, largely replaced the open-hearth process with electric-arc furnaces, mainly because of their greater capacity for digesting ever-growing amounts of recycled scrap, but also because of even greater controllability and reproducibility.[39,153]

In the modern era, there continues to be a massive growth in the use of steel, and total world production is approaching two billion tons (1.8 Pg).[152] There is an enormous range of different steels for different applications.[44] Steels have carbon contents ranging between about 0–1 wt%: *low-carbon* or *mild steels* have between 0–0.2 wt%; *medium-carbon steels* between 0.2–0.6 wt%; and *high-carbon steels* between 0.6–1 wt%.[44] Most of these so-called *plain-carbon steels* also contain about 0.5–1.5 wt% manganese, retained from the ferromanganese added during conversion from pig iron to remove adsorbed oxygen and sulphur impurities.[44] As can be seen on the Fe-C phase diagram in Figure 1.18, steels consist of a face-centred-cubic (fcc) solid solution of carbon in iron known as *γ-austenite* at temperatures above 912 °C. When pure iron is cooled slowly to room temperature, γ-austenite transforms into low-temperature, body-centred-cubic (bcc) *α-ferrite*; when a eutectoid composition Fe-0.77 wt%C alloy is cooled to room temperature, it transforms into *pearlite*, which is a lamellar eutectoid mixture of α-ferrite and the hard carbide *cementite* Fe_3C (it is called pearlite because the finely-divided sub-micron lamellae give the phase a pearly look on a polished surface); and at intermediate compositions, it transforms into a mixture of ferrite and pearlite.[145] This means that very low-carbon steels are *ferritic steels*, with a small amount of distributed cementite (at very low levels, the pearlite appears effectively as small denatured particles of cementite embedded in the predominant grains of ferrite); and medium-carbon and high-carbon steels are *ferritic-pearlitic steels*, with a ratio of ferrite/pearlite that increases from 0 to 100% as the composition increases from 0 to 0.77 wt% carbon.[145,169] Figures 1.29 and 1.30 show respectively typical examples of ferritic/pearlitic low-carbon steel and fully pearlitic eutectoid-steel microstructures.[170,171] The hardness and strength of a steel increases and its ductility decreases approximately in proportion to its carbon content, i.e. to the amount of pearlite or cementite in its microstructure, the tensile strength increasing, for instance, from approximately 300–400 MPa at 0.05 wt% carbon to 0.8–1 GPa at the eutectoid composition, depending on the grain size and the extent of deformation and/or annealing and recrystallisation.[145,169,172] When plain carbon steels are *quenched*, i.e. cooled rapidly rather than slowly to room temperature, the carbon has insufficient time to diffuse and precipitate out as cementite, the fcc γ-austenite crystal structure becomes unstable mechanically as well as chemically, transforming via lattice shear rather than chemical diffusion, and the carbon is retained in a supersaturated and distorted body-centred-tetragonal (bct) solid solution, heavily sheared and twinned, known as *martensite*.[169,173] The carbon supersaturation, lattice distortion, shear deformation and fine-scale twinning all contribute to making the resulting *martensitic steels* extremely hard and strong, typically between 1–2 GPa depending on carbon content and speed of quenching, but also brittle and prone to sudden fracture.[169,173] In *quenched and tempered steels*, therefore, martensite is heat treated briefly (*tempered*) to reduce its brittleness, but at the expense of some reduction in strength.

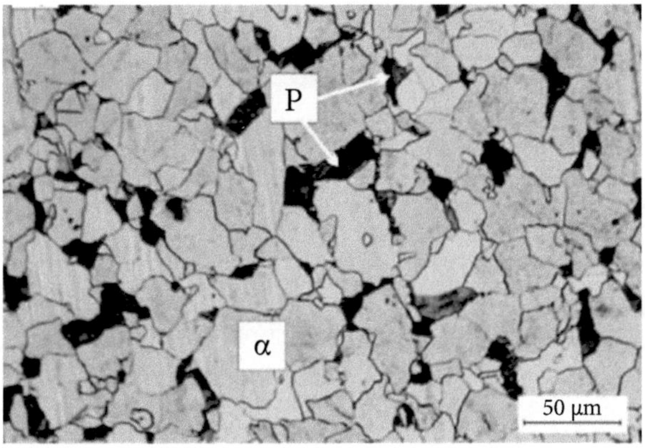

Figure 1.29 Optical micrograph of a ferrite/pearlite microstructure in a low-carbon (0.1 wt% C) microalloyed steel, showing α-ferrite grains (α) with a small amount of distributed pearlite (P).
Reproduced from San Martin et al. (2008). The effect of heating rate on the reaustenitisation of a low carbon niobium microalloyed steel. *Materials Science and Technology 24*: 266–72.

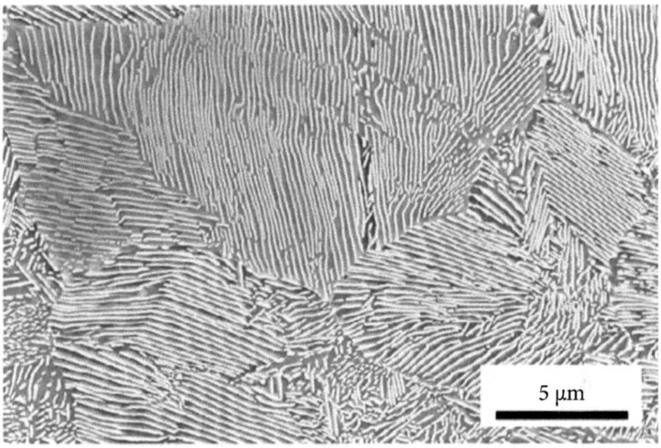

Figure 1.30 Scanning electron micrograph of pearlite in a rail-track high-carbon steel, showing alternating sub-micron thick plates of bcc α-ferrite (white) and cementite Fe_3C (black).
Reproduced from Hamada et al., (2011). Fatigue limit evaluation considering crack initiation for lamellar pearlitic steel. *Procedia Engineering 10*: 1467–72.

In *bainitic steels*, a similar result is achieved by cooling fcc γ-austenite to room temperature at an intermediate rate.[174]

It is often thought that steel is an important material because of its strength and hardness, but the true glory of steel is, in fact, its ductility and associated resistance to fracture. The ubiquity of steel and its wide range of applications and uses derive from the ability to manipulate its malleability and ease of forming into useful shapes, at the same time as adjusting its strength via changes in carbon content, addition of other alloying elements, and *thermomechanical treatment*, using deformation and heat treatment to control the grain size and degree of work hardening, as well as quenching and tempering to control the amount of martensite. Most of the steel used industrially is low carbon or mild steel, with substantial ductility and formability, and only a small amount of cementite to provide sufficient

Figure 1.31 'Golden-era' two-tone cream 1959 Cadillac El Dorado Biarritz convertible.
Reproduced from 'Cadillac 3'. My Auto World. https://myautoworld.com/gm/history/cadillac/cadillac-3/cadillac-3.html, accessed 18 September 2023.

strength and hardness. The mildest low-carbon steel with 0.05–0.1 wt% C is cast into billets, hot forged and hot rolled into thin strip or sheet, and finally cold rolled and then deep-drawn into shapes such as cups, cans and containers, or press-formed into more complex shapes such as car bodies, trains and white goods (fridges, ovens, washing machines, etc.).[145] Figure 1.31 shows a 1959 Cadillac from the 'golden era' of car production, demonstrating the beautiful and complex shapes that can be achieved by press-forming thin mild-steel sheet. Low-carbon steels with higher carbon contents, 0.1–0.2 wt% C, are also cast as billets and then hot forged, rolled or extruded into thicker plates, bars, tubes and shaped sections to be used in ship's hulls, construction beams for buildings and bridges, off-shore rigs and pipelines.[145] Medium-carbon steels are used for heavier duty applications such as rail track, crankshafts, gears, train and truck wheels, and heavy machinery; and high-carbon steels for even more demanding uses such as cutting tools, wire ropes and hawsers, and engineering dies.[145] *Alloy steels* contain small amounts, typically 1–2 wt%, of additional alloying elements such as niobium Nb, molybdenum Mo and chromium Cr, heat treated to increase strength further by forming nano-scale alloy carbide precipitates.[145] *Tool steels* contain even more alloying elements, typically 10–15 wt% tungsten W or molybdenum Mo, to confer extreme high strength and hardness; and *stainless steels* also contain even more alloying elements, in this case up to 15–20 wt% chromium Cr and 8–10 wt% nickel Ni, the former to prevent corrosion and rusting, and the latter to maintain a ductile austenitic structure to room temperature.[145] Overall, there is a total of more than 3,500 different grades of steel, i.e. different detailed alloy specifications, all designed for different specific components and different specific applications.[176]

5 Silicon Chips and the Information Revolution

The Information Revolution

We are currently living through the *Information Revolution*, the third of the three biggest revolutions in human existence, following the *Agricultural Revolution* at the end of the

Ice Age approximately 12,000 years ago, and the *Industrial Revolution* in the eighteenth and nineteenth centuries.[177–179] In the Agricultural Revolution, plants and animals were domesticated, allowing people to stop living marginal lives as nomadic hunter-gatherers and enabling them to move to a more comfortable and secure existence as settled, farming village-dwellers; in the Industrial Revolution, production of goods was mechanised and industrialised, with widespread technological advances in (amongst other things) power generation, manufacturing, transport, printing, communications, sanitation and medicine, dramatically improving peoples' overall health, wealth and wellbeing; and in the Information Revolution, instantaneous access to information is changing fundamentally how people operate and interact, massively enriching their social, cultural and intellectual capabilities and opportunities.[177–179] The Agricultural Revolution was driven by the discovery of how to fire *minerals* and *clays* (oxide alloys) to make both buildings for people to live in and storage containers in which to keep farmed food; the Industrial Revolution was driven by the discovery of how to convert *pig iron* into strong yet malleable *steels* (iron-based alloys) from which to make engineering machinery, industrial plant and consumer goods; and the Information Revolution is driven by the discovery of how to dope (microalloy) tiny regions of *semiconductors* to make *transistors*, and then assemble them into intelligent *silicon chips* that can store and manipulate information on a gigantic scale.

The Information Revolution relies on computing and communication technologies that can store, manipulate and transmit mind-boggling amounts of numerical, textual and pictorial data and information, in many ways equalling and surpassing our human capacity for intelligent thought and calculation.[180] This is achieved by using miniaturised transistors as the basic units, which are built into tiny logic devices, which in turn are connected to make larger calculating machines (or computers), which in turn are networked to provide even greater data storage and thinking capacity.[180] The humble transistor itself relies on using highly controlled *doping* or *microalloying* of minute pieces of extreme-purity silicon and other semiconductor materials with tiny, parts-per-million (ppm) levels of impurities to provide the functionality required.[181,182] We describe the development of semiconductor, transistor and computer technologies one at a time.

Semiconductors

Table 1.3 shows typical values of electrical resistivity ρ and its temperature coefficient $\alpha = \rho^{-1} d\rho/dT$ for some common materials.[183] Metals such as copper Cu, aluminium Al and iron Fe are good *conductors* with very low resistivity, typically $<10^{-7}$ Ωm, so they are highly amenable to electron flow. Ceramics and polymers such as silica SiO_2, alumina Al_2O_3 and polythene $(CH_2)_n$ are *insulators* with very high resistivity, typically $>10^{10}$ Ωm, so they are effectively impermeable to electron flow. Materials such as silicon Si, germanium Ge and gallium arsenide GaAs are an intermediate category called *semiconductors* with intermediate resistivity, typically 10^{-3}–10^{2} Ωm, so electrons can flow but only sluggishly. As shown in Table 1.3, the resistivity of a metal conductor increases slightly with increasing temperature, $\alpha > 0$, because increasing atomic vibrations interfere with the motion of the electrons; insulators do not conduct at any temperature, so $\alpha = 0$; and the resistivity of a semiconductor decreases slightly with increasing temperature, $\alpha < 0$, because the number of available electrons increases as the temperature rises.[183]

Table 1.3 Typical values of room temperature resistivity and temperature coefficient of resistivity for some common materials (after Cantor[184]).

Material	Electrical resistivity $\rho(\Omega m)$	Temperature coefficient α (°C^{-1})
copper	1.7×10^{-8}	0.004
aluminium	2.6×10^{-8}	0.004
iron	10^{-7}	0.006
germanium	$10^{-3} - 0.5$	−0.05
silicon	0.1–60	−0.07
gallium arsenide	$10^{-3} - 100$	
alumina	10^{12}	0
glass	10^{13}	0
polythene	10^{14}	0
silica	10^{17}	0

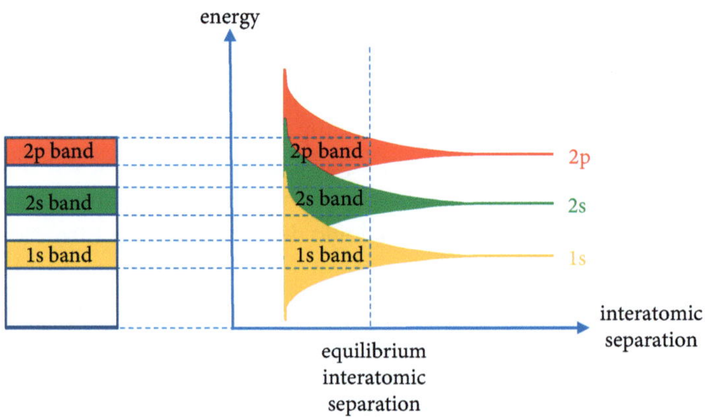

Figure 1.32 Electron energy band structure.

Conductors, insulators and semiconductors show different electrical behaviour because of their different *band structure*. According to the *Pauli exclusion principle*, two electrons cannot share the same quantum numbers or energy level. When a large number of atoms or molecules are brought together to make a material, therefore, the discrete electron energy levels of the individual atoms or molecules spread into energy bands, with a different energy band corresponding to each single-atom electron state and its particular set of quantum numbers (1s, 2s, 2p, 3s, etc.), and with *band gaps* of disallowed energies between the bands,[172–184] as shown schematically in Figure 1.32. The inner core electrons remain bound to their individual atoms or molecules, and form *core bands*; but the outer electrons and any excited states are less well bound and can permeate through the material in the *valence band* and the *conduction band*.

The *Fermi level* E_F is defined as the highest energy an electron can have at absolute zero.[183] For an individual atom, the Fermi level is the energy of the outermost valence electron. For an assembly of atoms or molecules in a material, the Fermi level is the highest energy of all the valence electrons in the valence band.[182–184] In a metal conductor, the Fermi level is in the

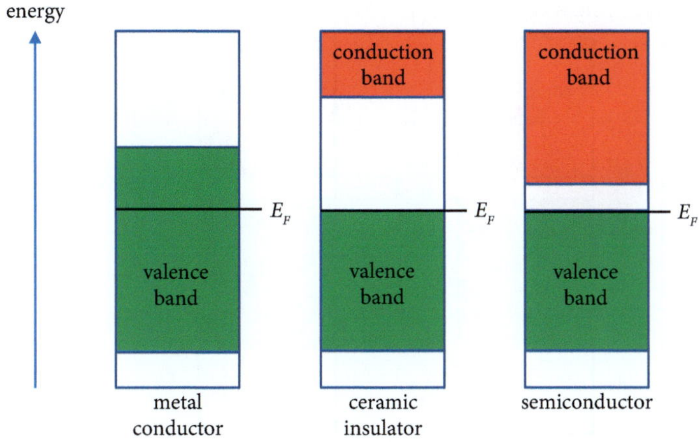

Figure 1.33 Fermi level in a metal conductor, ceramic insulator and semiconductor.

middle of the valence band as shown schematically in Figure 1.33, so that valence electrons can easily gain excess (e.g. thermal or electrical) energy and move readily throughout the material.[182–184] The valence electrons are all *free electrons*, and the resistivity is very low. In a ceramic or polymer insulator, on the other hand, the Fermi level is at the top of the valence band as also shown in Figure 1.33, so that valence electrons cannot gain excess energy and are immobile, because of the band gap of disallowed energies.[182–184] The valence electrons are all bound to their individual atoms or molecules, there are no free electrons, and the resistivity is very high. In a semiconductor, the Fermi energy is again at the top of the valence band, but the band gap of forbidden energies is only small as shown in Figure 1.33, so that electrons at the top of the valence band can sometimes gain enough extra energy to jump across the band gap into the conduction band and become mobile. In other words there are some free electrons, and the resistivity is intermediate.[182–184]

There are two main ways in which conduction can take place in a semiconductor. The first is by thermal excitation of electrons into the conduction band to create what is called an *intrinsic semiconductor*. At a temperature T, the probability p of an electron at the top of the valence band gaining enough thermal energy to jump across the gap into the conduction band is given by the Boltzmann probability:

$$p = \exp\left(\frac{-E_B}{kT}\right)$$

where E_B is the band gap energy and k is Boltzmann's constant. The probability p is only significant when the band gap energy is approximately equal to or less than the thermal energy, $E_B \leq kT$, which only happens at room temperature in a few materials. Every time an electron jumps from the valence band to the conduction band, it creates an *electron-hole pair*, [182–184] as shown schematically in Figure 1.34. A free electron is created near the bottom of the conduction band, which is able to move freely throughout the material; and a hole is left behind near the top of the valence band, which is also able to move freely (and independently) throughout the material, by the mechanism of adjacent electrons jumping into the hole. The electrons and the holes are both *charge carriers*. Each electron carries a negative charge, and responds to an applied voltage by moving from the anode to the cathode; and

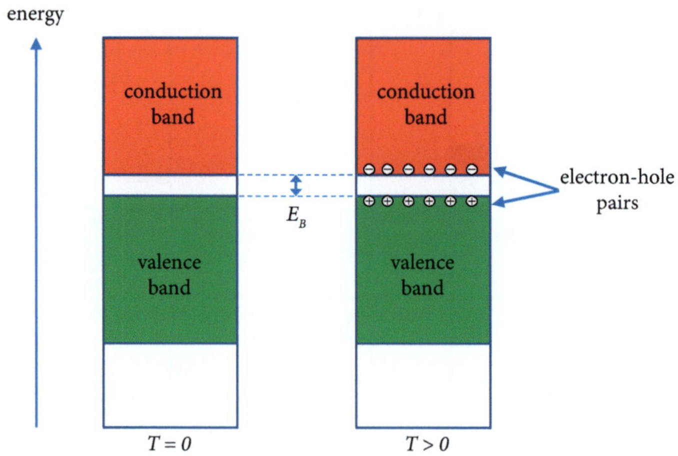

Figure 1.34 Electron-hole pairs in an intrinsic semiconductor.

each hole carries a positive charge, since it is an absence of an electron, with an absence of negative charge, and responds to an applied voltage by moving in the reverse direction from the cathode to the anode.[182–184]

In an intrinsic semiconductor, the number of free electrons equals the number of holes, both determined by a dynamic equilibrium between: 1) continuous creation of electron-hole pairs as electrons randomly acquire sufficient excess thermal energy to jump into the conduction band; and 2) continuous destruction of electron-hole pairs as electrons and holes randomly meet and recombine as they move around the material.[182–184] In one mole of a univalent metal conductor, there are Avogadro's number $N_o \sim 6 \times 10^{23}$ of atoms, and each atom contributes one electron to the valence band, so there are also Avogadro's number $N_o \sim 6 \times 10^{23}$ of free electrons to act as charge carriers. By comparison, in one mole of an intrinsic semiconductor such as pure germanium at room temperature, there are approximately 10^{14} electrons which have enough thermal energy to jump into the conduction band, leaving behind the same number of holes in the valence band, i.e. a total of only about 2×10^{14} charge carriers, many orders of magnitude less than the metal conductor. This is why semiconductors are only poor conductors of electricity.

The second way in which conduction takes place in a semiconductor is by *doping* (or *microalloying*) with a small amount of a component with a different valency, to create what is called an *extrinsic semiconductor*.[182–184] Doping with an element of greater valence injects extra electrons into the conduction band, leading to an excess of free electrons over holes: this is called *n-type doping* (negative-type doping), creating an *n-type semiconductor*.[182–184] On the other hand, doping with an element with a lower valence injects excess holes into the valence band, leading to an excess of holes over free electrons: this is called *p-type doping* (positive-type doping), creating a *p-type semiconductor*.[182–184] Silicon and germanium both have a valency of four, and a diamond-cubic crystal structure, as shown in Figure 1.35. They can both be made n-type by doping with a small amount of 5-valent nitrogen or phosphorus, and they can both be made p-type by doping with a small amount of 3-valent aluminium or gallium. In each case the doping leads to a dilute substitutional solid solution, with the dopant atoms replacing the silicon or germanium atoms on the diamond-cubic lattice. Each

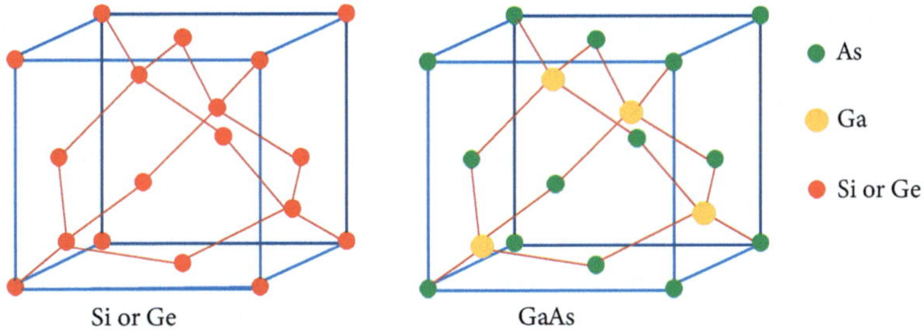

Figure 1.35 Crystal structures of silicon, germanium and gallium arsenide.

n-type dopant atom has one extra valence electron, which has to go into the conduction band, and each p-type dopant atom has one fewer valence electron, leaving a hole in the valence band.[182–184] Even doping a semiconductor with a few parts per million of an n-type or p-type impurity enhances the conductivity dramatically, since each ppm corresponds to an extra $10^{-6} \times 6 \times 10^{23} = 6 \times 10^{17}$ charge carriers, swamping the number of intrinsic electron-hole pairs.

There are intrinsic *compound semiconductors* as well as intrinsic elemental semiconductors.[182–184] For instance equiatomic compounds between 3-valent and 5-valent elements, such as gallium arsenide GaAs or indium antimonide InSb, have an average valency of four, the same as in silicon and germanium, and have the zinc blende structure, similar to diamond-cubic silicon and germanium, as shown in Figure 1.35 They have a similar band structure, with a small energy gap between the valence and conduction bands, and are known as *III-V semiconductors.*[182–184] Equiatomic compounds between 2-valent and 6-valent elements, such as cadmium sulphide CdS and zinc telluride ZnTe also have similar band structures, and are known as *II-VI semiconductors.*[182–184] Compound semiconductors can also be doped to increase their conductivity, in a similar way to elemental semiconductors.[182–184]

Transistors

Early information storage media, calculating machines and computers used large and unreliable *mechanical switches* and/or *electronic valves* to code, store and manipulate the individual bits of data and operational information needed for them to work.[185] Scalability was poor, dramatically limiting the possible sizes and therefore uses of these machines. The invention of semiconductor devices and, in particular, the *transistor* in late 1947 by John Bardeen, Walter Brattain and William Shockley working at Bell Labs in Murray Hill, New Jersey provided a much simpler, smaller and (after much development) more reliable technology, ultimately supercharging the manufacture of general purpose computers and other intelligent machinery through the 1950s and 1960s and on up to the present day (and, in the course of it, completely revolutionising society).[186,187] The first transistors were made individually by hand, and were a few centimetres (~5×10^{-2} m) in size, quite large but still much smaller than their mechanical or valve equivalents.[186,187] Modern transistors are now made in massive numbers by highly sophisticated robotic machinery, and continued miniaturisation has driven down their size to no more than a few nanometres (~5×10^{-9} m), i.e. by a factor of approximately ten million times in just over sixty years.[186,187] Because of the associated increase in the numbers of transistors that can be built into individual devices and their reliability, information-handling capability has been lifted to previously unimaginable levels, with previously undreamt-of machine functionalities, and (despite occasional claims to the contrary) still no end in immediate sight to the relentless process of enhancement. The famous *Moore's Law* predicted in 1965 (correctly) that the number of silicon transistors in a silicon chip would double every couple of years (because of the reduction in transistor size),[188] and this fierce pace of improvement continues to be maintained by innovative developments in materials and manufacturing techniques.[189] To give an idea of the remarkable scale of these developments, the silicon chip company Intel estimated[190] in 2007 that a total of 10^{19} silicon transistors were being manufactured and shipped each year; the total number of MOSFETs (the most common type of silicon transistor) manufactured between 1960 and 2018 has been estimated[191] to be 1.3×10^{22}; and in 2012 the author and historian of technology George Dyson put it more graphically:[192] 'Last time I looked, the digital universe was expanding at the rate of two trillion transistors a second' (a considerable underestimate compared to today's rate).

The first semiconductor device was a simple *p-n junction*, shown schematically in Figure 1.36, discovered by Russell Ohl working at Bell Labs in 1939, just before the outbreak of the Second World War. A p-n junction acts as a diode, a one-way switch that only lets electrical current pass in one direction.[183] With no electrical bias, mobile holes are distributed randomly throughout the p-type region and mobile electrons are distributed randomly throughout the n-type region. When a battery is connected across a p-n junction, however, its behaviour differs sharply depending on the direction of the applied voltage. The junction is *forward biased* when the positive terminal or cathode is connected to the p-type region and the negative terminal or anode is connected to the n-type region, as shown in Figure 1.36. The anode supplies electrons to the n-type region, pushing excess electrons into the conduction band and towards the p-type region; and the cathode removes electrons from the p-type region, pushing excess holes into the valence band and towards the n-type region.[183] The excess electrons and holes meet at the p-n junction interface and recombine, allowing the flow of electricity to be maintained.[183] Forward bias produces, therefore, a steady current which increases with increasing applied voltage.[183] The junction is *reverse biased* when the electrodes are connected the other way round, i.e. when the anode is connected to the p-type region and the cathode is connected to the n-type region, as also shown in Figure 1.36. The anode now supplies electrons to the p-type region, pulling excess holes away from the junction; and the cathode now removes electrons from the n-type region, pulling excess electrons away from the junction.[183] This creates an insulating region, with no charge carriers (neither electrons nor holes) in the vicinity of the p-n junction interface, stifling the continued flow of electricity.[183] Reverse bias prevents, therefore, the flow of electricity, and there is no sustained current.[183] Overall, this produces an asymmetric variation of current versus applied voltage, which only allows electricity to flow in the forward direction.

A *transistor* can be defined as a semiconductor device that can switch or amplify an electrical signal. It usually has at least three contacts (or terminals or electrodes), and a voltage or current applied across one pair of contacts is used either to switch on and off or to amplify the current flowing across another pair of contacts.[186] The first type of transistor developed by Bardeen, Brattain and Shockley was the *point-contact transistor*, but this

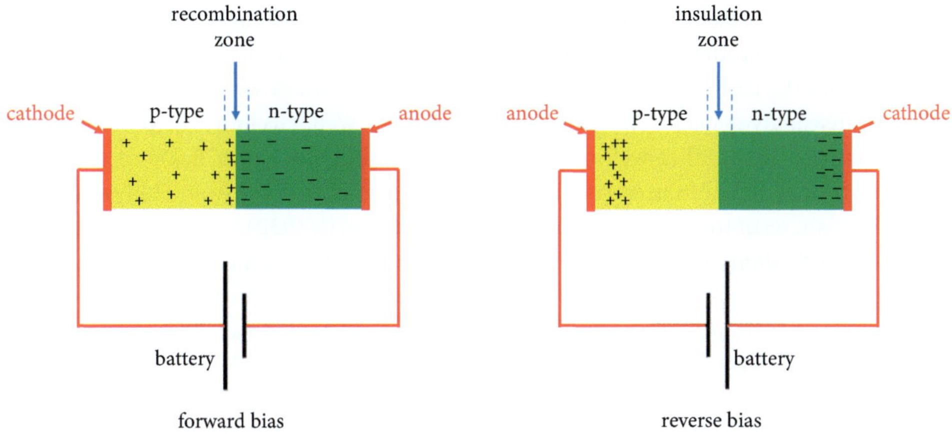

Figure 1.36 Forward and reverse biased p-n junction.

was soon replaced by the *bipolar junction transformer (BJT)*, and then later the *metal-oxide semiconductor field-effect transistor (MOSFET)*, which is overwhelmingly the most common transistor now used in modern silicon chips.[187,193] The different types of transistor all rely on the high sensitivity of electrical conduction in a semiconductor to the number of charge carriers (free electrons and/or holes) that are present in the material. Because of this high sensitivity, the current flowing between two contacts can be controlled by applying a small voltage across two other contacts to inject and/or remove electrons and/or holes. If this control voltage is used to stop or start conduction, the transistor acts as a two-way switch; and if the control voltage is used to increase or decrease conduction, the transistor acts as a tunable amplifier. Together with the simple p-n junction diode, transistor switches and amplifiers supply the key unit functions needed to store and retrieve data, to exercise logic operations such as "and", "or" and "not", and to perform simple arithmetical calculations such as addition and multiplication, all of which are necessary to build a rapid calculating machine or computer.[186,187,193]

William Shockley was born in 1910 in London of American parents.[187] His father, William Hillman Shockley, was an experienced mining engineer and railroad consultant, born and brought up in New Bedford, Massachusetts, the son of a whaling captain with an ancestry that could be traced back to the Pilgrims' *Mayflower*. He was a sharply practical man but also 'a man of culture and refinement, with extensive interests in art, music, languages and literature'.[194] In 1906, aged 51, he left his (then) base in London to visit the remote mining town, Tonopah, Nevada, where he was planning to start a gold mine, and where he met and two years later married Shockley's mother, Cora May Bradford.[187] She was just twenty-seven, bright, able, and feisty, one of the very first female graduates from Stanford University. The gold mine plan failed, and the newly married couple moved back to London in 1908, May falling pregnant and giving birth to William junior shortly thereafter. For his first three years William was brought up in London, and then the family moved back to the United States, buying a house in Palo Alto, California just a few blocks from where May had lived when studying at Stanford.[187]

William was a difficult child, boisterous, morose and introspective, with few friends but with a weird collection of unusual pets, and prone to bouts of uncontrolled rage for no good reason.[187] His parents kept him out of school till he was eight years old, when he was enrolled in Mrs Gurmell's Private School in Palo Alto, transferring two years later to the Palo Alto Military Academy to provide a more disciplined environment. The family moved down the coast to Hollywood near Los Angeles, where William went to Hollywood High School, increasingly concentrating on science and doing well. His father died in 1925, leaving William and his mother well-off, with a goodly inheritance of $75,000 of stocks and bonds (worth about $2.5m in todays' currency).[187] After graduating from Hollywood High School, William went to CalTech to study science, and then in 1932 to do a PhD in physics at MIT, his father's *alma mater*. As a young man, he continued to have an unusual, aggressive, somewhat unstable yet charismatic personality. He drove cross-country from LA to Boston to begin his doctoral studies in a swanky 1929 DeSoto roadster with a fellow physics student Frederick Seitz (later President of Rockefeller University and the US National Academy of Sciences), who reported (somewhat horrified) that William was imbued with Hollywood culture, seeing himself as 'a cross between Douglas Fairbanks, Bulldog Drummond and Ronald Coleman',[195] and that he kept in the glove compartment of the car a loaded pistol that he liked to pull out to take random potshots at passing trees and stray coyote, causing him from time to time to be apprehended and arrested as a 'desperado', sometimes with a night in jail.

In 1936, Shockley completed his PhD supervised by John Slater at MIT on the electronic band structure of sodium chloride.[187] He had married Jean Bailey in California in 1933 and they already had a two-year-old daughter, Alison, so he badly needed a job. Fortunately, that year saw the end of the depression-era hiring freeze at Bell Labs, and Shockley was the first new recruit of their Research Director Mervin Kelly, who had already decided he wanted to pursue the visionary idea of replacing valves made from unreliable and difficult-to-manufacture vacuum tubes with solid-state electronic components. Kelly was quite explicit and 'stressed the importance of this goal so vividly that it made an indelible impression'[196] on the new graduate. At first, Shockley worked unsuccessfully on trying to make useful semiconductor devices from copper oxide, but his research was soon (not surprisingly) interrupted by the onset of the Second World War, and he was diverted to work on radar for bomber planes and anti-submarine techniques. At the end of the war in 1945, Bell Labs was reorganised and Shockley was made head of the new solid-state physics division.[187] The gifted experimentalist Walter Brattain gave up his own military work on radar, joining other Bell scientists and engineers who moved into Shockley's team; and John Bardeen was one of several new hires to bolster the group, recruited from the Naval Ordnance Labs in Washington as an outstanding solid-state theorist.[187]

Earlier that year, Shockley and Kelly had visited Russell Ohl at Bell's Holmdel Labs in New Jersey and discussed his work on p-n junctions. Shockley was very excited and began almost immediately to sketch ideas in his lab notebook about doping silicon and germanium and trying to develop a *field effect*, i.e. using an externally applied field which he reasoned would inject or remove electrons and thus be able to control and amplify the flow of current in the semiconductor.[187] But initial experiments that summer all failed to show the predicted effect. Explaining why was one of the first challenges he set for his new team. Bardeen quickly came up with the correct, albeit at this stage still entirely speculative and untested explanation. The problem was *surface states*, electrical charges that build up on the surface of a semiconductor, effectively masking it electrically from the influence of ambient fields.[187] Sorting out the details of these surface states and finding ways to get through their dampening effect proved to be surprisingly difficult.

Bardeen and Brattain made the breakthrough in late 1947, inventing the *point-contact transistor* which they showed Shockley on 16 December, just over a week before Christmas.[187] Going home that day, Bardeen remarked to his wife Jane, who was peeling carrots in the kitchen for dinner, 'we discovered something important today'.[197] She was shocked 'since John never discussed his work [so] she knew instinctively he must have had a good day'.[197] Shockley immediately arranged a demonstration a week later on 23 December for his boss, Harvey Fletcher, and the Research Director, Ralph Bown (who had taken over a few years earlier, when Kelly was promoted to be Executive Vice-President).[187] Just before they arrived, Brattain made final preparations in the lab. He glued a thin piece of gold foil round the edge of a triangular plastic wedge and delicately used a razor blade to cut a slit through the gold foil along the edge. The edge of the wedge/foil assembly was then carefully pressed into the surface of small block of germanium with a makeshift spring fashioned from a paper clip. Copper wires were soldered to the (now) separate pieces of gold foil on the two sides of the wedge, and to the base of the germanium, and were then attached to batteries and transformers to power the device and to an oscilloscope to measure its output. After Shockley and the top brass had arrived, Brattain powered up the device and they all watched the oscilloscope spot measuring the flow of current through the germanium jumping up and down as Brattain turned on and off the control voltage across the two pieces of gold foil.[188] The amplification effect was obvious:[198] 'boosting the input signal many times ... and there wasn't a single vacuum tube in the entire circuit!'. Voltage and current were both amplified, with an overall power gain of about 2 db (about 60%) at frequencies of up to 1000 Hz.[187] Shockley described it as 'a magnificent Christmas present'.[199] More recently, the President of VLSI Research Inc, Risto Puhakka, has called it: 'probably the greatest invention of the 20th century'.[190] Somewhat officiously, Bown appointed a committee to decide on a name for Bardeen and Brattains' new device and they put a series of six options to Bell scientists for a vote, with *transistor* (a concatenation of the term *trans-resistor*) winning by a landslide.[187] After securing a patent on the invention of the solid-state transistor amplifier, Bell announced their discovery at a (somewhat restrained) press conference on 29 July 1948. The revolutionary significance of the discovery was (perhaps not surprisingly) not well appreciated at the time, and the story was immediately buried by the *New York Times* in a short note on page 46.[187]

Bardeen and Brattain's point-contact transistor is shown schematically in Figure 1.37. The plastic wedge/gold foil arrangement is essentially a way to make two very closely spaced contacts to the semiconductor, which are both pressed into the surface by the spring, pushing through the surface-state electrical charges on the semiconductor surface.[187,200] The germanium block was n-doped, with a thin, chemically infused p-doped *inversion layer* along its top surface. The two gold contacts were referred to as the *emitter* and *collector*, with a third *base* contact at the bottom of the germanium block. A positive voltage applied to the emitter contact injected excess holes into the p-type inversion layer, which flowed along the top surface to be collected at the negatively charged collector, amplifying the main current flowing between the collector and base terminals. The key requirements for the device to work were to have emitter and collector *point contacts* very closely spaced (~ 50 μ m apart) so the injected holes were not lost by recombination in the n-type bulk of the semiconductor, and to have them both pushed firmly into the germanium top surface so the injected holes could break through the blocking surface states.[187] An alternative reverse configuration could also be used, with a p-type semiconductor block (instead of n-type), an n-type inversion layer

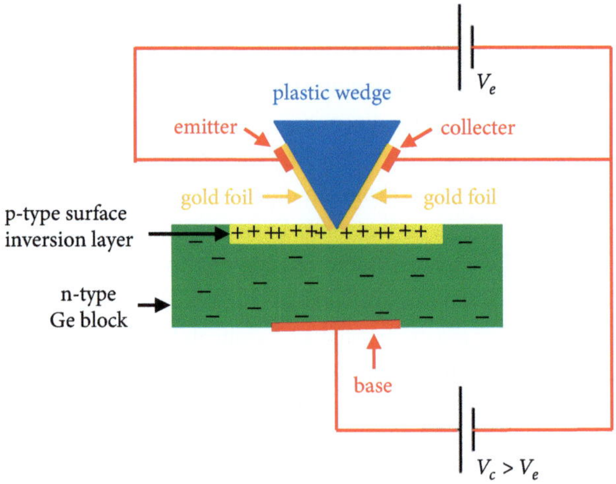

Figure 1.37 Point-contact transistor.

(instead of p-type), and a negative voltage on the emitter contact (instead of positive), so as to amplify the current by injecting excess electrons (instead of excess holes).[187,200]

Bardeen, Brattain and Shockley would some years later (in 1956) be awarded the Nobel Prize for Physics for their invention of the point-contact transistor. But at the time Shockley was not at all happy, feeling that Bardeen and Brattain were getting all the credit.[187] Moreover, Bell decided to drop Shockley and his field-effect transistor from their patent application, partly because his ideas were still only theoretical with no working prototype, but mainly because their patent search threw up a long-lost 1925 patent on a somewhat similar idea by the Austro-Hungarian physicist Julius Edgar Lilienfeld.[201] Shockley was piqued if not livid. He had 'great difficulty sleeping [and] repeatedly tossed and turned in bed, or got up and sat in the kitchen, working out some new ideas'.[202] The day after Christmas 1947, just a couple of days after the first, ground-breaking demonstration of transistor amplification, New York saw 'the greatest snowfall in the city's recorded history' according to the *New York Times*, [203] with a snow blanket of over 25 inches (62.5 cm). Nevertheless, Shockley struggled through the transport gridlock to attend an American Physical Society meeting in Chicago.[187] Wrestling mentally throughout the meeting with the complexity of semiconductor and transistor physics, he became convinced that there was not really any need for an inversion layer, and wondered whether a more straightforward sandwich arrangement of p-n junctions might not work better. He got up early in his hotel room in Chicago on New Year's Day 1948 and wrote up this idea feverishly in thirteen pages of notes.[187] Ruminating later, he 'realised that a new way of making a semiconductor amplifier was staring me in the face, or at least looking me squarely in the eye'. He had come up with the idea of the *bipolar junction transistor* (*BJT*; also called the *junction transistor*, or *sandwich transistor*).[202]

A BJT consists of a pair of adjacent p-n junctions, in an n-p-n configuration[183,193] as shown in Figure 1.38. A thin p region, called the *base* or *gate* separates two n regions, one called the *emitter* or *source* and the other called the *collector* or *drain*. The emitter-base junction is forward biased with a voltage so the anode supplies a flow of electrons to the n-type emitter, pushing excess electrons towards the junction with the p-type base. The base region is, however, very thin, so the excess electrons are not all consumed by recombination near the junction. Instead, they are driven across the junction and into the n-type collector. The base-collector junction is reverse biased so the cathode pulls electrons away from the junction, removing them from the n-type collector. The resulting current in the collector depends

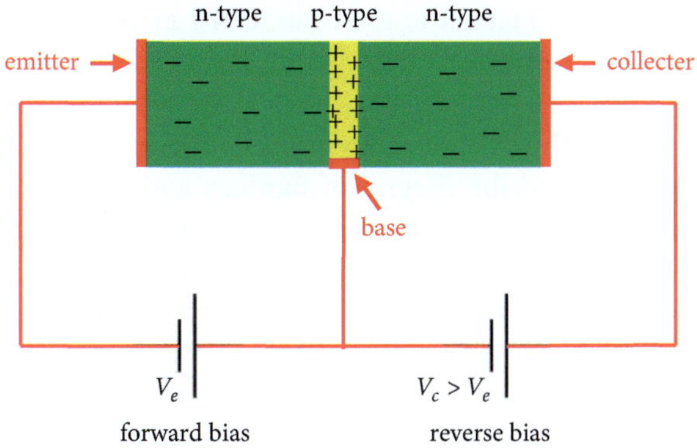

Figure 1.38 Bipolar n-p-n junction transistor (BJT).

exponentially on the emitter voltage, and the transistor again acts as an amplifier but, in this case, the number of injected electrons from a negative voltage at the emitter is modulated by the number captured at the base, so varying the emitter-base potential difference by varying the base voltage reduces or amplifies the current flowing between the emitter and collector. An alternative reverse p-n-p configuration again gives similar results, with injected holes (instead of electrons) providing the current amplification.[183,193] In either case, the linear arrangement of semiconductor elements is much simpler geometrically than a point-contact transistor, and much easier to manufacture, without the need to control either complex closely-separated contacts or a thin surface inversion layer.[187]

Shockley was now keeping his ideas under wraps for as long as possible, essentially paranoid that they would be stolen by other members of the group at Bell or by physicists working on parallel lines at Purdue University and in the US Military.[187] He got another physicist Dick Haynes to witness independently all the key notes in his lab notebook, and recruited his own patent lawyer to take out a second patent on the junction and field-effect transistors. Following Bell's announcement of the point-contact transistor in July 1948, the scientific community, unlike the popular press, had seen the importance of the breakthrough. *Electronics* magazine made it their lead story in September 1948, with a blaring title 'Revolutionary Amplifier: The CRYSTAL TRIODE',[204] and a picture of Bardeen, Brattain and Shockley posing in smart shirts and ties at Brattain's workbench.[204] The picture was obviously selected to convey an image of harmonious teamwork. But this was papering over the cracks. The events of 1948 had 'opened a rift in Bell's semiconductor group that gradually widened into a chasm ... On one side of the fault stood Bardeen and Brattain with their point-contact amplifier; on the other, Shockley and his coterie of assistants working on his sandwich or p-n junction approach'.[205]

In the meantime the Bell scientists were gradually solving a series of key problems:[187] purifying semiconductors to remove the variability of stray impurities, using the fractionating method of *zone melting*; making semiconductor *single crystals* to remove the variability of stray internal crystal boundaries, using the (previously) obscure method of *Czochralski crystal growth*; developing methods to make reliable and regular point contacts on semiconductor surfaces; using inert oxide and metal casings to *pot* the devices and protect them from damage during transport and installation; and modifying circuit designs to make best use of the different electrical characteristics of transistors compared to valves. By mid-1949, Bell had manufactured more than 3,700 of their first *Type A* point-contact transistors and distributed them to other companies, the military, and university researchers.[187] But the point-contact transistor never made it to the commercial market in a big way, apart from limited uses in hearing aids, specialised military equipment and some of Bell's own telephone switching gear.[187]

Shockley was the boss of the solid-state division, getting (rightly) a lot of the credit for developing transistors, much to the chagrin of Bardeen and Brattain, who in effect were frozen out and contributed little more to subsequent transistor research.[187] Bardeen moved to work on the new field of superconductivity (for which he got a second Nobel prize in 1972) and then left Bell to join the University of Illinois at Urbana-Champaign as Professor of Electrical Engineering; Brattain continued to work on surface properties of semiconductors, but was assigned to a different group at Bell.[187] And Shockley was pushing ahead. On 4 July 1951, Bell held another press conference to announce the junction transistor as 'a radically new type of transistor',[206] that was now dramatically outperforming the type A; tiny (just a few mm in size), more reliable, and incredibly sensitive (needing a power input of only a millionth of a watt, compared to a few thousandths of a watt for a point-contact transistor, and a whole watt for most valves[206]).

An avalanche of technical innovations continued throughout the 1950s, initially all at Bell Labs but then increasingly at other companies that were springing up to support and drive the new and growing transistor-based technologies.[187,207] *Zone refining* reduced impurity levels to better than one part in ten billion (< 0.1 ppb); *surface deposition* and *surface diffusion* were used to control doping levels to parts per billion (ppb) and the size of doped regions to (at first) just a few microns; silicon replaced germanium with a larger band gap giving greater sensitivity and control of electron and hole carrier levels, and a higher melting point giving greater resistance to thermal degradation; a thin top-surface layer of silicon oxide was thermally grown or deposited to protect active regions from the ambient atmosphere; a capacitative surface oxide layer with a further conducting metal layer on top finally enabled the manufacture of the *field-effect transistor (FET)*; complex *integrated circuits (ICs)* were made on the surface of a single *silicon chip* using a mixture of photolithography, masking and deposition techniques; and ICs evolved inexorably into *large-scale integration (LSI)*, and then *very-large-scale integration (VLSI)* (the latest iteration reduces the size of individual doped regions to just a few nm, and puts as many as sixty billion[208] MOSFETs on a single very powerful silicon chip).[187,207] Figure 1.39 shows a typical 200-mm-diameter ppb-purity silicon single crystal,[209] and Figure 1.40 shows a typical thin *silicon wafer*, a slice from such a crystal which has been patterned to produce an array of several hundred individual silicon chips each containing thousands of transistors (and costing just $40).[209] Texas Instruments worked with the Regency Division of IDEA to bring out the first *transistor radio*, the Regency TR1 in 1954, shown in Figure 1.41, becoming an icon of the massive growth in rock and roll and teenage pop culture of the late 1950s and 1960s.[187,210] Transistor-based devices were moving into mass production, rapidly taking over in the second half of the twentieth century and early twenty-first century great swathes of human activity, revolutionising (among other things) film and TV, music, phones and communications, management support systems, transport, medical diagnosis, information storage, regulation, and industrial control systems, until by now almost everything is computerised, smart and interconnected, all based on the humble transistor.

The *metal-oxide-semiconductor field-effect transistor (MOSFET)* is an insulated-gate *field-effect transistor (FET)*, discovered by Mohamed (John) Atalla and Dawon Kahng at Bell Labs in 1959, three years after Shockley had left,[187,193] and is shown schematically in Figure 1.42.

Figure 1.39 Typical 200 mm diameter, 1.2 m long ppb-purity silicon single crystal.
'Single Crystal Silicon'. https://www.tf.uni-kiel.de/matwis/amat/iss/kap_5/illustr/i5_3_1.html, accessed 18 September 2023.

A thin silicon oxide layer protects the underlying silicon semiconductor, but also acts as a capacitor, essentially removing any blocking surface states.[187,193] As with the n-p-n BJT, there is a linear geometry of three terminals, *emitter* or *source*, *base* or *gate*, and *collector* or *drain*, and the central gate voltage is used to control the electron flow between the emitter and collector, either to switch the flow on or off, or to amplify it.[193] In this case, however, instead of using the voltage on a p-doped sandwich region to modulate the number of electrons flowing from the emitter to the collector (as in a BJT), electrons are supplied or extracted by an external field applied to a thin aluminium metal contact on top of the protective/capacitative silicon oxide, i.e. by a *field effect* applied via a central *metal-oxide-semiconductor* layer structure (which is why it is called a MOSFET).[193] MOSFETs require even less power than BJTs,

Figure 1.40 Typical thin silicon single crystal wafer patterned with over 300 individual silicon chips, each containing several thousand transistors.
Courtesy of B. Cantor.

and are even easier to manufacture and miniaturise. The key issue is the thickness (or, more accurately, thinness) of the central gate region. It is much easier to deposit a thin metal contact on an oxide layer than to dope a thin p-type sandwich region, so MOSFETs can be made smaller, and can operate up to higher frequencies with higher bandwidths.[182,193] Essentially Atalla and Kahng had finally, after a gap of almost fifteen years, vindicated William Shockley's original 1945 idea of a field-effect transistor. The breakthrough was achieved by using silicon as the base semiconductor, with a silicon oxide layer to control the surface states. Ironically, the Bell team were within a whisker of success at the very outset. Brattain had tried the same geometry with germanium in the early days, but it didn't work, because germanium oxide is not stable and washes off the surface too easily during handling.[187]

In the mid-1950s, Shockley was having a classic mid-life crisis. He felt blocked at Bell, not being promoted, because 'world class physicists had complained bitterly of [his] overbearing competitiveness and ham-handed management',[211] and not allowed royalties, so he couldn't share in the enormous profits that would flow from his inventions.[187] His marriage was on the rocks; his wife Jean contracted cancer; he started an affair with a psychiatrist, Emmy Lanning (to whom he was later re-married); he toyed with taking an academic job; and he bought a fancy Jaguar XK convertible (that didn't help much).[187] In the end he left Bell and persuaded Arnold Beckman, an ex-classmate from Caltech and now a successful science-based businessman, to fund his new company *Shockley Semiconductor Laboratory*, to 'develop and commercialise new and emerging semiconductor technologies',[212] based in Mountain View,

Figure 1.41 Regency TR1 transistor radio.

Reproduced from 'Regency Pocket Radio'. Radio Museum. https://www.radiomuseum.org/r/regency_pocket_radio_tr_1_tr1.html, accessed 18 September 2023.

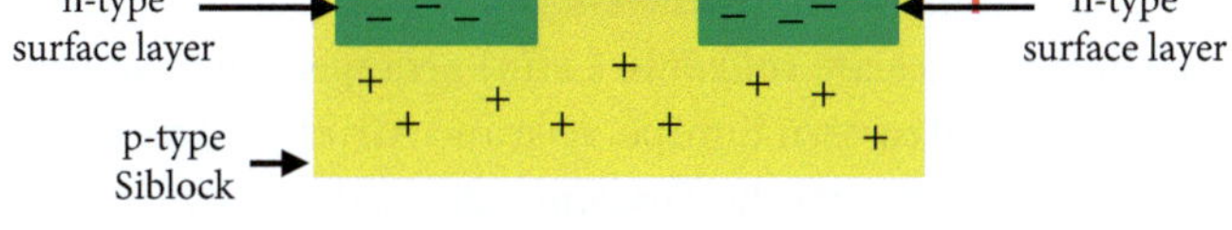

Figure 1.42 Metal-oxide-semiconductor field-effect transistor (MOSFET).

California, close to Stanford University and also close to Shockley's ailing elderly mother, both of which were in Palo Alto. It was the beginning of *Silicon Valley*. The new company was launched in February 1956 at San Francisco's posh Saint Francis Hotel, and Shockley set about recruiting outstanding scientists: 'I plan to build the most creative team in the world for developing and producing transistors and other semiconductor devices.'[213] And in November the news came that he had won the Nobel Prize in Physics. It seemed a wonderful harbinger of all the great successes yet to come. But it was the opposite. That year was the apex of Shockley's success. A year later in September 1957, the '*traitorous eight*', eight of his best scientists, resigned and left the new company, appalled at Shockley's autocratic and paranoid management style, to set up a new company *Fairchild Semiconductors*, just a mile north.[215] Emmy recorded that, when William heard of their intention to do so, she had '... never seen anybody in all my hospital experience look as white as Bill looked—ever. I have never forgotten it. [He] just lay there. Just white'.[215]

Shockley Semiconductor Labs continued to haemorrhage talented scientists and cash, and Beckman's holding company recorded its first-ever loss. In 1961 Beckman lost patience, wrote off his losses and sold the company to Clevite Inc.[187] Shockley's company staggered on, but never made a profit, and he never realised his dream of making millions from his inventions. Clevite fared no better than Beckman, and in 1968 did likewise, wrote off their losses and sold the company to ITT, who soon closed it down.[187] Shockley took a position as Professor of Engineering at Stanford. Sadly, Shockley's later life became mired in controversy as he spoke out increasingly with unpalatable eugenicist and racist views.[187] By the time he died of prostate cancer in 1989, he had fallen out with and become estranged from almost all his previous friends, family and colleagues. The Shockley/Fairchild bust-up became a more-or-less established pattern of development in Silicon Valley, with a total of sixty-five new enterprises being formed over the next twenty years in a kind of rolling process of destructive/creative innovation, all beginning with the founding of Shockley Semiconductor Labs and then Fairchild Semiconductors.

Computers

The Information Revolution that we are currently living through is driven by the continued development of improved *computers*, in turn driven by the continued development of improved *silicon chips*, in turn driven by the continued development of *semiconductors* and *transistors* as discussed in the last two sub-sections (all, of course, based on the discovery of how to microalloy silicon).[180] Modern computers combine several functions in one machine:[180] *calculation, data storage, communication*, and codified instructions or *software*. We discuss each of these briefly before describing how computers were developed and are transforming the world.

Mechanical aids to counting began in ancient times with pebbles (from which the word calculation comes, since pebbles are *calculi* in Latin) arranged on a counting board or abacus, still in use today in parts of eastern Europe, Asia and Africa.[215] The first working example of a mechanical adding machine was the *Pascaline*, devised in Paris in 1642 by the famous French mathematician, philosopher and child prodigy Blaise Pascal when he was still only eighteen years old, to help his father to do the extensive and tedious arithmetical calculations needed in his work as a tax collector.[216,217] Pascalines were beautifully made, operating with a complex system of interlocking and rotating cogged wheels, as shown in Figure 1.43.[218] They were expensive to make and cumbersome to use, so were not widely replicated, but they worked reasonably well, with a carry mechanism, and multiplication achieved by repetition. Pascal continued to improve the design, making some twenty machines in his lifetime, of which a handful have survived to the present day. Precedence is sometimes accorded to the German professor of Hebrew and astronomy Wilhelm Schickard, who made drawings of

Figure 1.43 An early Pascaline six-figure adding machine from the Musée des Arts et Métiers in Paris.

Reproduced from Rama. Musée des Arts et Métiers, Paris. https://commons.wikimedia.org/wiki/File:Pascaline-CnAM_823-1-IMG_1506-black.jpg

a different kind of adding machine almost two decades earlier in Tübingen, but his designs had flaws and he was unable (like others who have tried since) to make a working model from them.[219,220] A series of other arithmometers and comptometers were developed over the following years, but none came into widespread use until the late Industrial Revolution, with the inventions in 1886 of Dorr Eugene Felt's *Comptograph* manufactured by Felt and Tarrant (now Victor Technology),[220] and in 1885 of William Burroughs' adding machine manufactured by the American Arithmometer Company (later the Burroughs Corporation and now part of Unisys).[221] These early commercial adding machines led through the twentieth century to the development of mechanical tills and calculators and ultimately the first computers.

The ability to store, retrieve, use and transport information efficiently relies intrinsically on the development of *digital* techniques for codifying it.[180] The term digital was invented during the Second World War by the Bell mathematician George Stiblitz in 1942 to describe the use of numerical rather than analytical solutions of equations to control the direction of anti-aircraft fire via fast electronic pulses of current.[180] But, like calculating machines, digital methods have an older history. Early long-distance communication using flags, beacons and semaphore relied on relatively short-range line-of-sight between the sender and receiver, and was superseded in the mid-nineteenth century by the invention of the electric telegraph for sending electrical pulses over much longer distances along cables.[180] The archetypal system was developed by Samuel Morse, who first demonstrated it in Morristown, New Jersey in 1838, going on to get US Federal backing for a sum of $30,000 (almost $850,000 in today's currency) to build a thirty-eight-mile (61-km) telegraph line between the Supreme Court in the Capitol in Washington DC and the B&O (Baltimore and Ohio) Mount

Clare Railway Station in Baltimore.[222] The telegraph line was opened on 24 May 1848 with its first successful transmission of the words from the Bible: 'What hath God wrought'.[222] Morse used a forerunner of the modern international Morse code, a system of short and long electric pulses to represent numbers and letters of the alphabet, i.e. to encode information digitally, itself a forerunner of the now universal *binary code* used in computers and microprocessors to represent numbers and letters with characteristic binary sequences of on/off, up/down or one/zero. Each binary digit (each on/off, up/down or one/zero) is called an individual *bit* (contraction of *bi*nary dig*it*) of information, and groups of eight bits are called individual *bytes* (or words) of information.[223] A modern personal computer has typical calculating and storage capacities of 16 and 500 gigabytes respectively (a gigabyte is 1 GB = 10^9 bytes); and total data usage worldwide in 2019 was estimated to be 10–50 zettabytes[224] (a zettabyte is 1 ZB = 10^{21} bytes), expected to grow to 150–200 ZB by 2025.

In 1890, the American inventor Herman Hollerith developed *Hollerith punch cards* with a system of holes in cards to store, retrieve and transmit data.[180] Punch-card methods had been used before, notably by Joseph-Marie Jacquard in France, invented in 1804 to control patterns in the weaving of cloth in *Jacquard looms*, and by Charles Babbage in 1837 using Jacquard's method to control his *Analytical Engine*, widely agreed to be the first-ever real computer, i.e. the first-ever general-purpose, programmable calculating machine.[180] Hollerith went on to found the Tabulating Machine Company to market his inventions,[225] later broadened by acquisition to be the Computing-Tabulating-Recording Company (CTR), and then renamed International Business Machines Corporation or IBM in 1924. Punch cards can store data and text in coded binary form, with the presence or absence of a hole at each place in a card equivalent to a one or a zero or the on and off of a switch, i.e. to one bit of information. But the key new idea was to use strings of binary digits not just for storage and retrieval of numerical and textual information, but also to control the operation of the machine, in Jacquard's case the loom and in Babbage's case the calculations to be performed. This was an early form of *software*.

As already mentioned, the first general-purpose computer is widely regarded as being the Analytical Engine, one of a series of mechanical calculating machines designed by Charles Babbage in the mid-nineteenth century,[180,226] and Babbage has been called 'the father of the computer'.[226] He was born on Boxing Day in 1791 at Crosby Row, Walworth Road about a quarter of a mile (~400 m) south of the Elephant and Castle in Southwark, London. He was one of four children of Benjamin Babbage, a wealthy banker from Totnes in Devon, and Betsy Plumleigh Teape, the local Mayor's daughter.[227] Benjamin became a partner of William Praed of Truro, Cornwall, and together they moved to London to found the bank Praeds, Digby, Box, Babbage and Co. in 1800, with a fine new building opened in 1802 in Fleet Street (Praeds was later taken over by Lloyds).[227] In 1803, Benjamin retired back to Totnes and a few years later bought a fine family house in East Teignmouth, about sixteen miles (25 km) away on the coast, with fine views across the beautiful harbour. Aged eight, Charles went to Mr Halloran's Alphington Academy,[228] 'which for twenty pounds per annum and one guinea entrance gave to sons of gentlemen instruction in English, Latin and Greek [and] necessary qualifications for trade or the sea service',[228] then on to King Edward VI Grammar School in Totness and the Holmwood Academy in Enfield, Middlesex.

In 1810 Babbage went up to Trinity College, Cambridge, enthusiastic to study maths and astronomy, but also finding time to indulge in extra-mural pursuits such as founding a ghost club, playing a lot of chess and whist, and sailing on the rivers in the fens.[227] At that time,

Trinity was the pre-eminent place to study maths, and was firmly in the Newtonian camp (not surprisingly since Newton was the second Lucasian Professor of Maths at Trinity) in an ongoing blazing row over whether calculus was best studied and applied using the methods of Newton or Leibnitz. Babbage was one of a small band of radical British mathematicians who aligned themselves in the early nineteenth century with the French school of Laplace and Lagrange in supporting Leibnitz's methods,[227] a somewhat dangerous thing to do given the intensity of the concurrent Napoleonic Wars of 1800–15. History has, of course, come down pretty decisively on the side of Leibnitz, whose differential methods are now taught more-or-less exclusively worldwide. In 1812 Babbage transferred to Peterhouse College[227] from where he graduated in 1814. In 1815 he lectured to the Royal Institution on astronomy[229] and the following year was elected to the Royal Society. For many years he did not have a permanent job, though he applied for several academic positions, and instead lived on an allowance from his father, who died in 1827 leaving Charles well off with an inheritance of £100,000 (~£8m or $11m in today's money).[230] The following year Charles was appointed to the prestigious Lucasian Chair back at Trinity, from which he resigned a decade later in 1938 to concentrate on politics. He was a remarkable polymath: instrumental in founding[231] the Royal Astronomical Society in 1820 and the British Association for the Advancement of Science (BAAS) in 1831; very active in the Royal Society; trying several times (unsuccessfully) to become a member of parliament; and writing extensively, influentially and controversially[226,227] on topics as diverse as the decline of science, the importance of manufacturing and industry, the value of division of labour, restrictive practices in book publishing, optimum methods of statistical inference, the tabulation of physical constants, the transmutation of species, the validity of miracles (in which he believed), and the sources of divine knowledge.[226,227] As an engineer he invented the cow catcher on the front of railway engines,[232] the dynamometer railway car,[233] the ophthalmoscope,[234] lights for lighthouses, Greenwich time signals, postage rates, and cryptographic methods for setting and breaking codes;[234] he was fascinated by stomach pumps, stamped buttons, locks, volcanic lava, theatre lights, horse-race handicapping, and railway tunnels.[235] He campaigned widely against a somewhat disparate set of activities that he saw as public nuisances, such as organ grinders, drunkenness, hoop-rolling and hereditary peerages.[236,237] He was offered but turned down both a knighthood and a baronetcy. He lived for most of his life at 1 Dorset Street, Marylebone in the centre of London, where he died aged 79 in 1871 from renal inadequacy and cystitis.

Babbage's first mechanical calculating machine was his *Difference Engine No. 1*, which he designed between 1819 and 1822, using a complex interlocking array of mechanical gears, cogs, shafts, spindles, ratchets and dials in order to solve polynomial equations and compute astronomical and mathematical tables.[226,238,239] He received funding from the British Government and began to construct it in 1823 in collaboration with the engineer Joseph Clements.[226,238,239] It was to be made of a total of 25,000 parts and would have been eight feet (~2.5 m) tall and fifteen tons (~14,000 kg) in weight, but it was never completed, with Babbage and Clements falling out over finance in 1831, and the government finally withdrawing support in 1842, having spent a total of approximately £17,000 (~£1m or $1.4m today).[226,238,239] In 1847 he designed an improved *Difference Engine No. 2* but he failed to get any more funding (two of his difference engines were finally built a century and a half later by the Science Museum in Kensington, London and were found to work quite well).[180,240] Between work on these two difference engines, Babbage designed his *Analytical Engine*, the first general-purpose computer, with the important additional feature of

Jacquard punch-card control, making it much more flexible for doing a wide variety of different kinds of calculation.[226,241] It had many features that are found in modern computers. As well as the punch-card system with three card-readers for separate *input streams* of raw data, arithmetical calculations and operational instructions, it also had a *data store* or *memory* for a thousand numbers with forty decimal-point accuracy, an arithmetical unit or *central processor unit (CPU)*, an early kind of *machine language* for the operational instructions, a grid-based physical layout which allowed for multiple loops and conditional branching in the operational logic, and a printing unit to *output* results also on punch cards.[226,241] Lord and Lady Byron's daughter, the talented mathematician Ada Lovelace, was a great friend of Babbage, with whom she communicated extensively, and she is widely considered to have developed the first-ever *software program*, designed to use the Analytical Engine to compute Bernoulli numbers.[242,243] It was the first of several such programs for various mathematical objectives, such as solving polynomials and Gaussian elimination. But, as with Babbage's difference engines, only parts of his analytical engine were ever constructed, and it was never fully completed.

Almost a century later in 1934, Alan Turing graduated aged twenty-one with a first-class honours bachelor's degree in maths from King's College, Cambridge University.[185] Just two years after graduating he published a paper in the *Proceedings of the London Mathematical Society* titled 'On Computable Numbers', which is often regarded as the theoretical foundation of modern computer science[185,244] and has been called 'easily the most influential math paper in history'.[245] It extended Kurt Gödel's work on mathematical axioms and proofs, and invented the concept of a *universal computing machine* (often called a *Turing machine*), which could solve any set of mathematical calculations using an appropriate codified set of logical instructions or *algorithm*.[185,244] The American mathematician Alonzo Church published similar results independently a couple of months earlier[246] using the somewhat less-accessible concept of a *lambda calculus*. The idea that a Turing machine or lambda calculus can be used to compute anything is known as the *Church–Turing thesis* or the *computability thesis*.[247,248] Modern computers use stored programs, i.e. the algorithm or set of computational instructions is stored in the computer memory alongside the input and output data and text rather than fed in separately step at a time. The concept of a *stored-program computer* is usually credited to the Hungarian-born American mathematician John von Neumann, and computers of this type are sometimes called *von Neuman machines*, [186,249], though the idea came out of his collaboration with the engineers J. Presper Eckert and John Mauchly at the University of Pennsylvania, and was first enunciated by the German engineer Konrad Zuse.[187,249,250]

It wasn't until the Second World War that there was sufficient technological drive and funding for the development of the first fully electronic computers, using electrical signals for input and output and to communicate information between different parts of the computer, as well as banks of electronic diode and triode valves to store data and do the computations (earlier hybrid electromechanical computers had already introduced faster electrical signals for communication but retained clunky and slow mechanical switches for data storage and calculations). At Bletchley Park in England,[186,251,252] a series of electronic computers called *Colossus*, designed by the General Post Office (GPO) telephone engineer Tommy Flowers, were made to help with code breaking and military intelligence towards the end of the Second World War between 1943–1945. Ten Colossus machines were in use by the end of the war, successfully intercepting many messages between German High Command and its

Figure 1.44 ENIAC installation at the Moore School of Engineering at the University of Pennsylvania, showing Corporal Irwin Goldstein in the foreground setting switches on one of the input function modules.
Courtesy of Wikipedia. https://commons.wikimedia.org/wiki/File:Classic_shot_of_the_ENIAC.jpg, accessed 18 September 2023.

operating army units throughout Europe, and contributing substantially to shortening the war and preventing a German victory.[186,251,252] They were all fully dismantled after the war and their existence was kept secret until the 1970s. More or less in parallel, between 1943 and 1945, *ENIAC* (the *electronic numerical integrator and computer*) was being designed independently at the University of Pennsylvania in Philadelphia by the engineers J. Presper Eckert and John Mauchly, and was finally commissioned in December 1945 and announced in February 1946 to great public interest and acclaim.[253,254] 'Giant Electronic Brain' blared the press. The installation consisted of a large number of different input, output, storage and calculation modules, as shown in Figure 1.44.[255] By the time it was closed down in 1955, ENIAC had a total of more than 18,000 vacuum tubes, 7,200 crystal diodes, 1,500 electrical relays, 70,000 resistors and five million hand-soldered joints, weighing about thirty tons on a floorplate of 1800 square feet (167 m^2) and consuming 150 kW of electricity.[256] It was rumoured that the lights in Philadelphia went dim whenever ENIAC was switched on. It cost about half a million dollars (~$6m today) and was a thousand times faster than previous hybrid electromechanical machines.[253–256] It was designed initially to calculate artillery firing tables for the US Army, but its first program was a study of the feasibility of a nuclear bomb.[253–256]

For many years it was believed that Colossus and ENIAC were the first programmable electronic computers. It was later discovered, however, that the American physicist John

Vincent Atanasoff with his graduate student Clifford E. Berry had previously designed an electronic computer at Iowa State College (now Iowa State University) between 1937 and 1942 called (not surprisingly) the Atanasoff–Berry Computer (or ABC), but although a small prototype and individual parts of it were built, it was never fully completed.[257] Colossus and ENIAC were followed quite rapidly by the first stored-program electronic computers[186,248] (Colossus and ENIAC had their programs set manually step at a time by trained operators throwing switches), Freddie Williams and Tom Kilburn's *Manchester Baby* in 1948 at Manchester University, followed in 1949 by their *Manchester Mark 1* and Maurice Wilkes's *EDSAC* (*electronic-delay automatic storage computer*) and Eckert and Mauchly's *EDVAC* (*electronic discrete-variable automatic computer*) at the Universities of Cambridge and Pennsylvania respectively.[186]

In 1951, Williams and Kilburn's Manchester Mark 1 evolved into the *Ferranti Mark 1*, and Eckert and Mauchly's ENIAC evolved into *UNIVAC* (*universal automatic computer*),[186,258] the world's first commercial electronic computers, effectively the world's first commercial mainframes. Neither was very successful, Ferranti selling no more than ten machines, and the Eckert–Mauchly Computer Corporation (subsequently taken over by Remington–Rand and then Sperry, now part of Unisys) selling just forty-six at about $1m each to several major US companies and government agencies.[186] However UNIVAC spurred IBM to launch its own large mainframe computer, the IBM-701, which with the business-friendly IBM-702 and lower-cost IBM-650 established IBM as the lead company worldwide for manufacturing electronic computers.[186] The IBM-650 is the first mass-produced electronic computer and a total of over 2000 were made and installed. All these and subsequent early 1950s computers were large valve-based mainframes, expensive and difficult to use, requiring special installation, substantial surrounding physical infrastructure and extensive human support systems.[186,255] Although the IBM-650 and subsequent machines were profitable, the computer business only really began to take off with the introduction in the late 1950s and through the 1960s and 1970s of transistors instead of valves, followed by silicon chips and then microprocessors, using increasingly miniaturised (and therefore higher capacity) data storage and calculation through large scale integration (LSI) and very large scale integration (VLSI).[180] In parallel, the installation of significant numbers of large valve- and then transistor-based mainframes through the 1950s, 1960s and 1970s led to the development of the first commercial high-level computer languages, which rapidly became increasingly sophisticated in order to make software programming easier: *Fortran* for scientific calculations, *Cobol* for business transactions, *Basic* for simplicity, *Pascal* for procedures, *C* for system control, and so on.[259,260]

The first computers using transistors were Manchester University's eponymous *Transistor Computer*[261] at Manchester University in 1953 and Bell Labs' (also eponymous) *TRADIC* (*transistor digital computer*)[262] in 1954. The first fully transistorised computers were the Harwell *Cadet* in 1955 and Burroughs' *Atlas* and the IBM 608[263] in 1957. Once transistors had replaced valves in the mid-1950s and then integrated silicon chips had been introduced in the late 1950s, Moore's Law[188] began to play a key role, with the number of transistors in a computer increasing relentlessly year by year, increasing the power and potential of the technology, and driving a correspondingly relentless series of innovations in computer design, manufacture and application.[186] Figure 1.45 shows Moore's Law graphically,[264] with the number of transistors on a chip increasingly logarithmically with the year of its introduction.[188] In the 1950s and 1960s, computers were mostly sold as large mainframe installations,

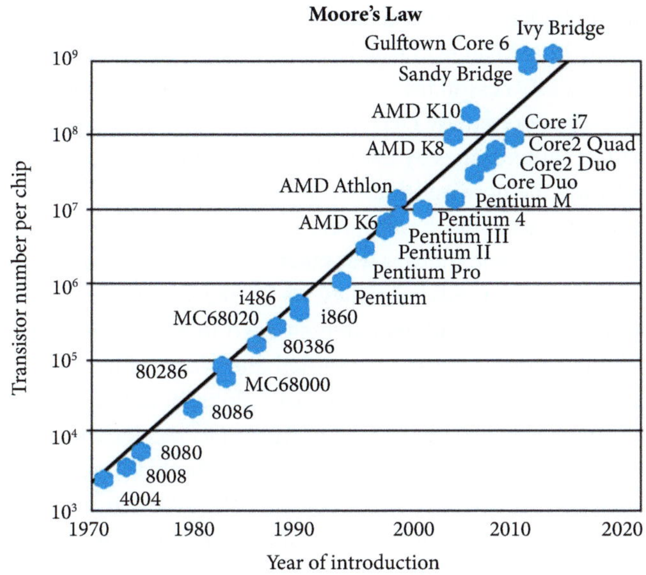

Figure 1.45 Moore's Law, showing the log of the number of transistors on a chip versus the year of its first introduction.

Reproduced from Roser M, Ritchie H. (2020). Moore's Law Transistor Count 1970–2020. https://commons.wikimedia.org/wiki/File:Moore%27s_Law_Transistor_Count_1970-2020.png

with multiple interconnected units and consoles occupying one or more large separate air-controlled rooms, requiring a dedicated series of operators to support machine operation and programming inputs and outputs, usually run as a centralised facility for multiple users with diverse computing needs in a large university, company or government agency. Anyone who remembers visiting a computer centre at that time will remember it as a massive hallowed installation, something like visiting the Kennedy Space Center or the headquarters of the latest Bond villain. But this began to change in 1960, when DEC (the Digital Equipment Corporation) announced the PDP-1, the first minicomputer[186] that 'brought computing out of the computer room and into the hands of the user',[265] followed by the PDP-8 in 1965, a computer just about small enough to sit on a desk or be rack-mounted alongside other pieces of electronic equipment.[186]

In parallel with the development of minicomputers, the technology for networking them together was also getting going. In 1969, Bob Taylor at the US Defense Advanced Research Projects Agency (DARPA) launched the *ARPANET*, based on the ideas of Joseph Licklider, linking four geographically remote mainframe computers at the University of California Los Angeles (UCLA), University of California Santa Barbara (UCSB), University of Utah in Salt Lake City and Stanford Research Institute (SRI) in Palo Alto:[186,266–268] using *modems* to convert digital computer data to waveforms suitable for telecommunication; *packet switching* (developed by Donald Davies at the National Physical Lab or NPL in England) to chop up and coordinate the data being sent and received; specially-designed minicomputers called *interface message processors* (IMPs) or *routers* to control signal messaging protocols and handle the interfaces between the different mainframes; and dedicated telephone lines to carry messages between the different sites. UCLA student programmer Charley Kline typed 'login' at 10.30am on 29 October 1969 to start the network by connecting UCLA and SRI, but the SRI computer immediately crashed.[269] After some adjustments, the two sites were linked later that day with successful transmission of the first two characters 'lo'; the first four planned nodes were all operating by December; there were fifteen computers connected to the network across the United States by 1971; and the same year email was first implemented by Ray Tomlinson using the now well-known '@' terminology.[266–269]

Integrated circuits (*ICs*) and *silicon chips* were invented first in 1958 by Jack Kilby at Texas Instruments in Dallas,[270–272] and the idea was taken up almost immediately by Robert Noyce at Fairchild Semiconductors in Silicon Valley,[270–273] the two inventors and their companies both moving quickly to try and get patents issued in mid-1959. The resulting dispute over ownership of the technology was only settled after years of legal wrangling, when the courts adjudicated with a compromise worthy of Solomon, giving the two scientists and their respective companies equal rights. Kilby went on to receive the 2000 Nobel Prize in Physics but sadly Noyce was not able to share the spoils this time as he had sadly died young, ten years earlier.[270,271] At first, ICs were used almost exclusively in computer guidance and control systems for two major US programmes running throughout the 1960s and early 1970s: the US Air Force's Minuteman intercontinental ballistic missiles (ICBMs), and NASA's Apollo lunar-landing spacecraft.[270,274] But in 1968, Noyce and Gordon Moore (of Moore's Law fame) left Fairchild to found the company Intel to manufacture silicon chips, and in 1970 they introduced the Intel 1103, the first memory chip able to integrate data storage and processing on a single chip, replacing much slower storage systems using separate magnetic loops.[275–277] The next year they brought out the Intel 4004, the first full *microprocessor* or computer on a chip, designed by engineers Federico Faggin, Marcian (Ted) Hoff, Stan Mazor and Masatoshi Shima.[275–277] The associated growth in power and reduction in size led inexorably to a string of at first hobby-style and then business-friendly microcomputers in the early 1970s: the first true personal computer or PC, the MITS Altair,[275,278,279] in 1974; then Steve Jobs and Steve Wozniak's first Apple 1 in 1976; the Commodore, Apple II and TRS-80 all in 1977 (the latter selling over 100,000 units); and in the early 1980s the multi-million-selling ZX-series, as well as the Amstrad and the Amiga.[275,279,280] IBM had no option but to respond, and launched its own PC in 1981, a runaway success followed by clones and look-alikes made by a raft of other companies such as Compaq, Dell and Hewlett Packard.[275,279] And quietly in 1972, Bill Gates and Paul Allen founded Traf-O-Data,[275,279,280] which later became Microsoft, surfing along with all these developments, explicitly marketing software for personal computers and being catapulted to pre-eminence with an exclusive software deal for the IBM PC operating system MS-DOS.[275,279,280] Apple and Microsoft are now two of the biggest companies in the world.[281]

The success of PCs in the 1980s and 1990s, with massive numbers of individual users networked across the world, has initiated the latest string of computer developments in the first two decades of the twenty-first century, with continued growth of even greater machine capacity controlling all kinds of information and infiltrating all kinds of human interaction. The size and convenience of individual computers has been driven down with hand-held portable *smart phones* and *tablets*; the *Internet* and *world-wide-web* have made communication ubiquitous, to even the remotest parts of the world; data storage capacity and application options have soared dramatically, with almost unlimited availability in *the cloud*, i.e. accessed remotely via the network; intelligent microprocessors control every conceivable kind of consumer good, customer service and industrial plant, from fridges and cars to banking and insurance to energy supplies and transport systems; *satellite technologies* such as *GPS* (*global positioning system*) pinpoint and target activities and actions worldwide with accuracies down to better than a millimetre; *artificial intelligence* (*AI*) and *machine learning* programs identify hidden data patterns more effectively than humans, supporting activities as diverse as medical diagnosis and crowd control; the *internet of things* (*IOT*) networks machinery, separating design and manufacture, democratising manufacturing capability to anyone with a smart phone; *search engines* such as Google make information available on every conceivable matter at the drop of a hat; the creative worlds of music, art, drama and film can be *downloaded* by anyone from anywhere at any time; and links between family, friends and work colleagues have been transformed and enriched by *social media* such as Facebook and X, formerly known as Twitter.

The *world-wide-web* was invented by Tim Berners-Lee in 1989 at CERN in Grenoble[282–284] using *web-browser* programs operating with *hypertext transfer protocol* (*HTTP*) to link information stored in *hypertext markup language* (*HTML*) throughout the *Internet*. The first smart phone[282] was IBM's Simon Personal Communicator in 1994, but the technology took off with the launch[285] of the iconic *Apple iPhone* in 2007, shown in Figure 1.46.[286] A modern smart phone is 120 million times faster than the computers that helped Apollo spacecraft to put the first humans on the moon.[287] Over six billion people (more than three quarters of the world's population) now own a smart phone;[288] there are four million data searches every minute;[289] more than 300 billion emails are sent each day;[290] over five hundred hours of video content are uploaded every minute;[291] and there are almost a billion tweets every day.[292] Whole industries have been taken over by computers and been utterly

Figure 1.46 First generation Apple iPhone.
Courtesy of Apple.

transformed, their original technologies effectively consigned to history—cameras, records, videos, film and TV, maps, holidays, shopping, letter post, telephones, typing machines, cash tills and banking—with many more predicted to follow. There is little doubt that we are in the middle of an Information Revolution and an information explosion, showing no sign of abating and with dramatic further developments almost certainly yet to come, all driven at root by continued improvements in beautifully precise microalloying of semiconductors and associated material technologies.[288]

6 Summary

This chapter has shown firstly how all the major events in the evolution of humankind have been both driven by and underpinned by the development of new materials, and secondly how new materials are always alloys, i.e. mixtures of simpler starting materials or components. The evolution of humans from apes at the beginning of the last Ice Age was driven by the discovery of flint tools, made from silica with small admixtures of other oxides that give the characteristic flint appearance and control their fracture behaviour when being shaped into tools and weapons to hunt for food; the settlement of humans into farmsteading villages at the end of the last Ice Age was driven by the discovery of alumino–silicate clays that could be fired to make bricks for dwellings and pots for storage of farmed food from domesticated plants and animals; the growth of cities, trade, business and central governments with the rise of the great ancient Egyptian, Chinese, Latin American and Roman civilisations and empires during the Bronze and Iron Ages was driven by the discovery of how to smelt metals and alloy them to manufacture vastly improved tools, weapons and other artefacts; the building of modern wealthy and healthy societies during the Industrial Revolution, with mechanised mass manufacture of goods, electrical power distribution, medical advances, and massive improvements in travel and communications, was driven by the discovery of improved methods of making strong and ductile iron and steel alloys; and the current Information Revolution transforming human interactions and capabilities via gigantic increases in access to knowledge is being driven by the discovery of wonderful advances in precision microalloying of silicon and other semiconductors on a near-atomic scale. And there are many other examples: nickel superalloys to withstand the fierce heat at the heart of a jet engine, essential for modern air transport; lightweight composite-fibreglass sporting goods that have transformed leisure activities such as golf, tennis, sailing and cycling; lithium oxide alloys used as anodes, cathodes and solid electrolytes in batteries to power handheld smartphones and electrical vehicles; superconducting niobium alloys for high-field precision in X-ray and MRI scanners; doped-silica optical fibres for controlled data transmission and laser applications such as barcode scanners and CD players; steel-reinforced concrete for skyscrapers and other large-scale buildings; steel- and polymer-reinforced rubber for vehicle tyres; soda-lime and borosilicate glasses for windows, bottles and cooking implements; doped silicon and compound semiconductors for solar energy conversion; and so on. The list is endless.

Alloying to create new materials has driven and underpinned all human advances to date (as discussed in this chapter) and will continue to do so in the future (as discussed in Chapter 7). The (relatively) recent discovery of multicomponent high-entropy alloys and materials represents a vast expansion of our potential materials palette that will now be available to drive and underpin all future human advances. The discovery of this enormous range of new materials and their structure, properties and potential uses is the subject of the rest of this book.

References

1. Spoor F, Gunz P, Neubauer S, Stelzer S, Scott N, Kwekason A, Dean MC. (2015). Reconstructed *Homo habilis* type OH 7 suggests deep-rooted species diversity in early *Homo*. *Nature 519*: 83–6.
2. Gradstein FM, Ogg JG, Hilgen FJ. (2012). On the geological time scale. *Newsletters on Stratigraphy 45*: 171–88.
3. Sass SL. (2011). *The Substance of Civilization: Materials and Human History from the Stone Age to the Age of Silicon*. New York: Arcade, 13–37.
4. Spoor et al., Reconstructed, 83–6.
5. Harai YN. (2014). *Sapiens: A Brief History of Humankind*. London: Harvill Secker, 5.
6. Bellwood P. (2003). *First Farmers: The Origin of Agricultural Societies*. Hoboken: Wiley-Blackwell, 44–66.
7. Gupta HK. (2011). *Encyclopedia of Solid Earth Geophysics*. Berlin: Springer Science and Business Media, 137–64.
8. Tarbuck EJ, Lutgens FK, Tasa DG. (2016). *Earth: An Introduction to Physical Geology*, 12th edn. London: Pearson, 134–67.
9. Tarbuck et al., *Earth*, 238–67.
10. Tarbuck et al., *Earth*, 268–99.
11. Lide DR (ed). (2004). *CRC Handbook of Physics and Chemistry*, 85th edn. New York: Taylor and Francis, Section 14: Geophysics, astronomy and acoustics, 4.
12. 'List of minerals', *Wikipedia* (last modified 5 Feb. 2024), https://en.wikipedia.org/wiki/List_of_minerals, accessed 18 September 2023.
13. 'List of rock types', *Wikipedia* (last modified 10 Dec. 2023), https://en.wikipedia.org/wiki/List_of_rock_types, accessed 18 September 2023.
14. Ashby MF, Jones DRH. (1986). *Engineering Materials 2*. Oxford: Pergamon, 156–9.
15. Putnis A. (1992). *Introduction to Mineral Sciences*. Cambridge: Cambridge University Press, 141–84.
16. Ralph J, Ralph I. (2007). Mogánite: A valid IMA mineral species. *Mindat.org*. http://www.mindat.org/min-2739.html
17. Heaney PJ, Prewitt CT, Gibbs GV. (1994). Silica: Physical behaviour, geochemistry and materials applications. *Reviews in Mineralogy 29*: 1–82.
18. Fiers G, Halbrucker E, De Kock T, Laforce B, Vandendriessche H, Messiaen L, Vincze L, Crombe P, Cnudde V. (2019). Preliminary characterization of flint raw material used on prehistoric sites in NW Belgium. *Geoarchaeology 34*: 400–12.
19. Semaw S. (2000). The world's oldest stone artefacts from Gona, Ethiopia: Their implications for understanding stone technology and patterns of human evolution between 2.6–1.5 million years ago. *Journal of Archaeological Science 27*: 1197–1214.

20. Clark JGD. (1969). Advanced Palaeolithic culture. In *World Prehistory: A New Outline*, 2nd edn. Cambridge: Cambridge University Press, 32–69.
21. Leakey M. (1979). *Olduvai Gorge*. London: Book Club Associates, 11–17.
22. Roche H, Harmand S, Prat S, Delagnes A, Brugal J-P, Feibel C, Kibun M, Texier P-J. (2003). Les sites archéologiques plio-pléistocènes de la formation de Nachukui, Ouest-Turkana, Kenya: bilan synthétique 1997–2001. *Comptes Rendus Palevol 2*: 663–73.
23. Marlowe FW. (2005). Hunter-gatherers and human evolution. *Evolutionary Anthropology 14*:54–67.
24. Jose-Manuel Benito Alvarez - Locutus Borg. Permission granted under a creative commons attribution-share alike 2.5 generic license. https://commons.wikimedia.org/wiki/File:Oldowan_tradition_chopper.jpg
25. Hugo Obermaier. (1925). El hombre fosil. Madrid. Public domain license. https://commons.wikimedia.org/wiki/File:Bifaz_lanceolado-San_Isidro_(Madrid).png
26. Dörre E, Hübner H. (1984). *Alumina: Processing, Properties and Applications*. Berlin: Springer-Verlag, 9–11.
27. Ruys AJ. (2018). *Alumina Ceramics: Biomedical and Clinical Applications*. Cambridge: Woodhead, 1–17.
28. Levin I, Brandon D. (1998). Metastable alumina polymorphs: Crystal structures and transition sequences. *Journal of the American Ceramic Society 81*: 1995–2012.
29. Markovska I, Yovkova F, Minov G, Rusev D, Lyubchev L. (2013). Investigation of silane modified ceramic surface of porous mullite ceramics. *International Journal of Environmental and Ecological Engineering 7*: 409–13.
30. Belotto M, Gualtieri A, Artioli G, Clark SM. (1995). Kinetic study of the kaolinite-mullite reaction sequence: Part 1 kaolinite dehydroxylation; and part 2 mullite formation. *Physics and Chemistry of Minerals 22*: 207–22.
31. Home RM. (1952). *Ceramics for the Potter*. Peoria: Chas A Bennett, 16.
32. Taylor L. (2011). *The Ceramics Bible: The Complete Guide to Materials and Techniques*. San Francisco: Chronicle Books, 14–23.
33. Hodder I. (2003). A new phase of excavation at Çatalhöyük. *Anatolian Archaeology 9*: 9–11.
34. Adams RM. (1981). *Heartlands of Cities: Surveys of Ancient Settlement and Land Use on the Central Floodplain of the Euphrates*. Chicago: University of Chicago Press, 52–93.
35. Liu L, Chen X. (2012). *The Archaeology of China: From the Late Paleolithic to the Early Bronze Age*. Cambridge: Cambridge University Press, 75–168.
36. Zohary D, Hopf M, Weiss E. (2012). *Domestication of Plants in the Old World*, 4th edn. Oxford: Oxford University Press, 1–8.
37. Smith BD. (1998). *The Emergence of Agriculture*. New York: Freeman.
38. Smith BD. (2001). Documenting plant domestication: The consilience of biological and archaeological approaches. *Proceedings of the National Academy of Sciences of the United States of America 98*: 1324–6.
39. Gilchrist JD. (1989). *Extraction Metallurgy*, 3rd edn. Oxford: Pergamon, 203–44.
40. Gilchrist,*Extraction*, 129–52.
41. Sass, *The Substance*, 49–67.
42. Chu S. (2013, February 1). Letter from Secretary of State for Energy to Energy Department employees. https://www.azquotes.com/quote/595036?ref=stone-age

43. Tylecote RF. (1976). *A History of Metallurgy.* London: The Metals Society, 5–39.
44. Martin J. (1996). *Materials for Engineering.* London: Institute of Materials, 69–128.
45. Knapp B. (1996). *Copper, Silver and Gold.* Goring: World of Books.
46. Massalski TB, Murray JL, Bennett LH, Baker H. (1986). *Binary Alloy Phase Diagrams.* Materials Park: ASM International, 964–5.
47. Caron RN, Barth RG, Tyler DE. (2004).Metallography and microstructures of copper and its alloys. In GF VanderVoort, ed., *ASM Handbook: Metallography and Microstructures.* Materials Park: ASM International, 775–88.
48. Gilchrist,*Extraction*, 328–68.
49. Tylecote RF. (1986). *The Prehistory of Metallurgy in the British Isles.* London: The Institute of Metals, 11–42.
50. Browne J. (2013). *Seven Elements That Have Changed the World.* London: Weidenfeld and Nicholson, 85–104.
51. Tylecote, *History of Metallurgy*, 1–4.
52. Sass, *The Substance*, 68–81.
53. Cunliffe B. (2010). *Celtic from the West.* Oxford: Oxbow Books, 27–31.
54. Fairbank JK. (1992). *China: A New History.* Cambridge: Harvard University Press, 35.
55. British Museum object number 1897, 0401.1064.
56. Scott DA. (1991). *Metallography and Microstructure of Ancient and Historic Metals.* London: Getty Conservation Institute and Archetype Books, 88.
57. Sass, *The Substance*, 82–97.
58. Tylecote, *History of Metallurgy*, 40–52.
59. Souckova-Siegolova J. (2001). Treatment and usage of iron in the Hittite empire in the 2nd millennium BC. *Mediterranean Archaeology 14*: 189–93.
60. Headrick DR. (2009). *Technology: A World History.* Oxford: Oxford University Press, 35–50.
61. Akanuma H. (2005). The significance of the composition of excavated iron fragments taken from Stratum III at the site of Kaman-Kalehöyük, Turkey. *Anatolian Archaeological Studies, Japanese Institute of Anatolian Archaeology, Tokyo 14*: 147–58.
62. Drews R. (1993). *The End of the Bronze Age: Changes in Warfare and the Catastrophe ca 1200 BC.* Princeton: Princeton University Press, 4.
63. Clark, *World Prehistory*, 160–75.
64. Blümner H. (1886). *Technologie und Terminologie der Gewerbe und Künste bei Griechen und Romern.* Leipzig: Teubner, vol 4, 152.
65. Tylecote, *History of Metallurgy*, 53–63.
66. Abadi RAA. https://commons.wikimedia.org/wiki/File:Watered_pattern_on_sword_blade1.Iran.JPG
67. Homer. (1919). *The Odyssey*, tr. A.T. Murray. London: William Heinemann, Book 9. https://www.perseus.tufts.edu/hopper/text?doc=Perseus%3Atext%3A1999.01.0136%3Abook%3D9%3Acard%3D360
68. Smiles S. (2010). *Men of Invention and Industry.* Bremen: Books on Demand, 87.
69. Cohen IB. (1976). The eighteenth-century origins of the concept of Scientific Revolution. *Journal of the History of Ideas 37*: 257–88.

70. Wootton D. (2016). *The Invention of Science: A New History of the Scientific Revolution.* London: Penguin Books.
71. Whewell W. (1840). *The Philosophy of the Inductive Sciences, Founded upon their History.* London: John W. Parker, 318.
72. Voorhies J. (2000, October). Europe and the Age of Exploration. In *Heilbrunn Timeline of Art History,* New York: Metropolitan Museum of Art, http://www.metmuseum.org/toah/hd/expl/hd_expl.htm
73. Parry JH. (2010). *The Age of Reconnaissance: Discovery, Exploration and Settlement 1450–1650,* 2nd edn. London: Weidenfeld and Nicholson, chs 10 and 12.
74. Russell B. (1968). *The Impact of Science on Society.* New York: Simon and Schuster, 7.
75. Copernicus N. (1976). *De revolutionibus orbium coelestium [On the Revolutions of the Heavenly Spheres]* (1543), tr. AM Duncan. Newton Abbott: David and Charles.
76. Newton I. (1687). *Philosophiae Naturalis Principia Mathematica [Mathematical Principles of Natural Philosophy].* London: Royal Society. Reprinted as *The Principia: The Authoritative Translation and Guide* (2016), tr. IB Cohen, A Whitman, J Budenz. Berkeley: University of California Press.
77. Bacon F. (1620). *Novum organum: Sive indicia vera de interpretatione naturae [New Method: True Directions Concerning the Interpretation of Nature].* Reprinted (2019). Dumfries and Galloway: Anodos Books.
78. Durant W. (1926). *The Story of Philosophy.* New York: Simon and Schuster, 101.
79. Allen RC. (2011). *Global Economic History.* Oxford: Oxford University Press, 14–26.
80. Ames GJ. (2005). *Vasco da Gama: Renaissance Crusader.* New York: Pearson /Longman, 50.
81. Fernandez-Armesto F. (2007). *Pathfinders: A Global History of Exploration.* London: W. W. Norton, 177–9.
82. Jensen D. (1992). *Renaissance Europe: Age of Recovery and Reconciliation.* Lexington: D. C. Heath, 327–64.
83. Allen, *Global Economic History,* 27–39.
84. Diffie BW, Winius GD. (1977). *Foundations of the Portuguese Empire 1415–1850.* Minneapolis: University of Minnesota Press, 177.
85. Ashton TS, Hudson P. (1997). *The Industrial Revolution 1760–1830,* 2nd edn. Oxford: Oxford University Press, 48–75.
86. Derry TK, Williams TI. (1960). *A Short History of Technology: From the Earliest Times to AD 1900.* Oxford: Oxford University Press, 275–310. Reprinted (1993). Mineola: Dover Publications.
87. Derry and Williams, *A Short History,* 557–82.
88. Timmins G. (1996). *Four Centuries of Lancashire Cotton.* Preston: Lancashire County Council, 21–4.
89. Science Museum. Textiles machinery category, B. Fothergill, object number 1860–1865.
90. Derry and Williams, A Short History, 311–42.
91. Rolt LTC. (1963). *Thomas Newcomen: The Prehistory of the Steam Engine.* Newton Abbot: David and Charles.

92. Carnegie A. (2015). *James Watt.* New York: Cosimo (reprinted from 1905 edition).
93. Miller DP. (2019). *The Life and Legend of James Watt: Collaboration, Natural Philosophy and the Improvement of the Steam Engine.* Pittsburgh: University of Pittsburgh Press, 63–115.
94. Anderson A. (1981, December 3). Review: James Watt and the steam engine. *New Scientist*: 685.
95. Shears J. (2017). *The Great Exhibition, 1851: A Sourcebook.* Manchester: Manchester University Press.
96. Clark E. (2012). *The Britannia and Conway Tubular Bridges: General Inquiries on Beams and on the Properties of Materials Used in Construction.* Charleston: Nabu Press.
97. Ffrench Y. (1950). *The Great Exhibition.* London: Harvill Press.
98. Kishlansky M, Geary P, O'Brien P. (2008). *Civilization in the West*, 7th edn. New York: Pearson, *vol C.*
99. Beaver P. (2001). *The Crystal Palace: A Portrait of Victorian Enterprise.* Cheltenham: The History Press.
100. Nash J, Hague L, Roberts D. (1854). *Dickinson's Comprehensive Pictures of the Great Exhibition of 1851.* London: Dickinson Brothers.
101. Picard L. (2009). *The Great Exhibition.* London: British Library.
102. Shorter C. (1908). *The Brontës' Life and Letters. Letter 514, from Charlotte to her Father, the Reverend Patrick Brontë.* New York: Charles Scribner's Sons, *vol 2*, 215. https://babel.hathitrust.org/cgi/pt?id=uc1.31210009905033&view=1up&seq=233&q1=letter%20514
103. https://www.royalcommission1851.org/about-us/
104. Derry and Williams, *A Short History*, 364–402.
105. Corlett E. (2012). *The Iron Ship: The Story of Brunel's SS Great Britain*, 5th edn. Bristol: SS Great Britain Trust.
106. Brindle S. (2006). *Brunel: The Man Who Built the World.* London: Weidenfeld and Nicolson, 119–42.
107. Emerson GS. (1981). *SS Great Eastern.* London: David and Charles.
108. Burton A. (2000). *Richard Trevithick: Giant of Steam.* London: Aurum Press, 87–100.
109. Hamilton Ellis C. (1968). *The Pictorial Encyclopedia of Railways.* London: Hamlyn, 20–30.
110. Jackson A. (1984). *London's Termini*, 2nd edn. Newton Abbot: David and Charles.
111. Derry and Williams, *A Short History*, 429–466.
112. Cossons N, Trinder BS. (2002). *The Iron Bridge: Symbol of the Industrial Revolution*, 2nd edn. Cheltenham: Phillimore.
113. https://commons.wikimedia.org/wiki/File:Iron_Bridge_east_side_in_February_2019.jpg
114. Brindle, *Brunel*, 99–118.
115. Allen, *Global Economic History*, 40–52.
116. Eckermann E. (2001). *World History of the Automobile.* Warrendale PA: Society of Automotive Engineers (SAE) Press.
117. Seidel WA. (2005). *Carl Benz: eine badische Gesichte; Die Vision vom 'Pferdlosen Wagen' verändert die Welt [Carl Benz: A Baden history; The vision of the 'horseless car' changes the world].* Weinheim: Diesbach.

118. Haug G. (2010). *Gottlieb Daimler: Der Traum vom Fahren: historische Roman [Gottlieb Daimler: The dream of travel: A history]*. Stuttgart: Masken Verlag Friedrich Willmann.
119. Curcio V. (2013). *Henry Ford*. New York: Oxford University Press.
120. Stross R. (2007). *The Wizard of Menlo Park: How Thomas Alva Edison Invented the Modern World*. New York: Crown/Random House.
121. Howard F. (1998). *Wilbur and Orville: A Biography of the Wright Brothers*. New York: Dover.
122. Kelly FC (ed). (2002). *Miracle at Kitty Hawk: The Letters of Wilbur and Orville Wright*. New York: Da Capo Press.
123. Derry and Williams, *A Short History*, 531–56.
124. Lewis D, Zitzlsperger U. (2016). *A Historical Dictionary of Contemporary Germany*. Plymouth MA: Rowman and Littlefield, 92.
125. Lewis and Zitzlsperger, *Historical Dictionary*, 340.
126. Lewis and Zitzlsperger, *Historical Dictionary*, 85.
127. Derry and Williams, *A Short History*, 403–28.
128. Landau SB, Condit CW. (1996). *Rise of the New York Skyscraper 1865–1913*. New Haven: Yale University Press, 62–297.
129. Allwood J. (1977). *The Great Exhibitions*. London: Alta Vista.
130. De Tholozany P. *Paris: Capital of the 19th Century*. Doctorate thesis, Brown University. Brown University Library for Digital Scholarship. https://library.brown.edu/cds/paris/worldfairs.html
131. Souto MH. (2012). The 19th century world exhibitions and their photographic memories: Between historicism, exoticism and innovation in architecture. *Quaderns d'Història de l'Enginyeria 13*: 57–80.
132. Derry and Williams, *A Short History*, 306.
133. Fordham University. Modern history sourcebook: Spread of railways in 19th century. https://sourcebooks.fordham.edu/mod/indrev6.asp, accessed 18 September 2023.
134. The Geography of Transport Systems. Rail track mileage and number of class I rail carriers, United States, 1830–2020. https://transportgeography.org/?page_id=1954, accessed 18 September 2023.
135. Maddison A. (2007). Contours of the world economy 1–2030AD. *Essays in Macroeconomic History*. Oxford: Oxford University Press, 376, 379, Tables A1, A4.
136. Bairoch P. (1976). Europe's gross national product 1800–1975. *Journal of European Economic History 5*: 273–340.
137. ' List of regions by past GDP', *Wikipedia* (last modified on 10 Jan. 2024), https://en.wikipedia.org/wiki/List_of_regions_by_past_GDP_(PPP), accessed 18 September 2023.
138. Derry and Williams, *A Short History*, 467–505.
139. Derry and Williams, *A Short History*, 114–56.
140. Raistrick A. (1953). *Dynasty of Iron Founders: The Darbys and Coalbrookdale*. York: Longmans.
141. Mott RA. (1957). The earliest use of coke for ironmaking. *The Gas World 145*(Coking Section Supplement): 7–18.
142. Trinder B. (1974). *The Darbys of Coalbrookdale*. Chichester: Phillimore.
143. Geerdes M, Chaigneau R, Kurunov I, Lingiardi O, Ricketts J. (2015). *Modern Blast Furnace Ironmaking: An Introduction*. Amsterdam: IOS Press, 1–14.

144. Peacey JG, Davenport WG. (2013). *The Blast Furnace: Theory and Practice.* Oxford: Pergamon, 1–15.
145. Bhadeshia HKDH, Honeycombe RWK. (2017). *Steels: Microstructure and Properties*, 4th edn. Oxford: Butterworth-Heinemann, 23–100.
146. Bessemer H. (1989). *Sir Henry Bessemer FRS: An Autobiography; With a Concluding Chapter.* London: Institute of Metals, 1–18. Originally published (1905) London: Offices of Engineering.
147. Bessemer, *Autobiography*, 19–123.
148. Bessemer, *Autobiography*, 130–177.
149. Bessemer, *Autobiography*, 156–161.
150. Anstis R. (1997). *Man of Iron–Man of Steel: The Lives of David and Robert Mushet.* Coleford: Albion House, 140.
151. Chisholm H (ed). (1911). Thomas Sidney Gilchrist. In *Encyclopaedia Britannica*, 11th edn. Cambridge: Cambridge University Press.
152. World Steel Association. 'Global crude steel output increases by 3.4% in 2019. https://worldsteel.org/media-centre/press-releases/2020/global-crude-steel-output-increases-by-3-4-in-2019/, accessed 18 September 2023.
153. McGannon HE (ed). (1971). *The Making, Shaping and Treating of Steel*, 9th edn. Pittsburgh: United States Steel.
154. Bessemer, *Autobiography*, 329–332.
155. Bessemer, *Autobiography*, 375.
156. König W. (2020). *Sir William Siemens 1823–1883: A Biography.* Munich: C. H. Beck, ch 1.
157. Siemens P. (1993). *Sir William Siemens: A Man of Vision.* Ferndown: Siemens/Caric Press.
158. Bähr J. (2016). *Werner von Siemens 1816–1892: eine Biografie (in German).* Munich: C. H. Beck.
159. Feldenkirchen W. (2000). *Siemens: From Workshop to Global Player.* Munich: Piper.
160. König W, *Siemens A Biography*, chs 2–4.
161. 'William Siemens'. *Grace's Guide.* https://www.gracesguide.co.uk/William_Siemens, accessed 18 September 2023.
162. Barraclough K. (1990). *Steelmaking 1850–1900.* London: Institute of Metals, 137–203.
163. Mioche P. (1997). Et l'acier créa l'Europe. *Matériaux Pour l'Histoire de Notre Temps 47*: 29–36.
164. Mott RA. (1977/8). Dry and wet puddling. *Transactions of the Newcomen Society 49*: 153–8.
165. Mott RA. (1983). *Henry Cort: The Great Finer.* London: The Metals Society.
166. Lemoine B. (2019, May 10). What metal the Eiffel Tower is made of? https://www.toureiffel.paris/en/news/130-years/what-eiffel-tower-made, accessed 18 September 2023.
167. Harbison -Walker Refractories Company. (2018). *A Study of the Open Hearth: A Treatise on the Open-Hearth Furnace and the Manufacture of Open-Hearth Steel.* London: Creative Media Partners.
168. Derge G. (1964). *Basic Open-Hearth Steelmaking*, 3rd edn. Seeley W. Mudd series. New York: American Institute of Mining, Metallurgical and Petroleum Engineers.
169. Martin, *Materials*, 100–8.
170. San Martin D, de Cock T, Garcia-Junceda A, Caballero FG, Capdevila C. Garcia de Andres C. (2008). The effect of heating rate on the reaustenitisation of a low carbon niobium microalloyed steel. *Materials Science and Technology 24*: 266–72.

171. Hamada S, Sasaki D, Ueda M, Noguchi H. (2011). Fatigue limit evaluation considering crack initiation for lamellar pearlitic steel. *Procedia Engineering 10*: 1467–72.
172. Gensamer M, Pearsall EB, Pellini WS, Low Jr JR. (2012). The tensile properties of pearlite, bainite and spheroidite. *Metallography, Microstructure and Analysis 1*: 171–89.
173. Bhadeshia and Honeycombe, *Steels*, 135–78.
174. Bhadeshia and Honeycombe, *Steels*, 179–202.
175. 'Cadillac 3'. *My Auto World*. https://myautoworld.com/gm/history/cadillac/cadillac-3/cadillac-3.html, accessed 18 September 2023.
176. Bell T. (2019, January 27). Steel grades and properties. *ThoughtCo*, https://www.thoughtco.com/steel-grades-2340174#:~:text=According%20to%20the%20World%20Steel,%2C%20chemical%2C%20and%20environmental%20properties
177. Wasuda Y. (1985). Three great social revolutions: Agricultural, industrial and informational. *Prometheus 3*: 269–76.
178. Drucker PF. (1999, October). Beyond the information revolution. *The Atlantic*. https://www.theatlantic.com/magazine/archive/1999/10/beyond-the-information-revolution/304658
179. Matthews JT. (2000). The Information Revolution. *Foreign Policy 119*: 63–5.
180. Ceruzzi PE. (2012). *Computing: A Concise History*. Cambridge: MIT Press.
181. Sass, *The Substance*, 265–82.
182. Grovenor CRM. (1989). *Microelectronic Materials*. Bristol: Adam Hilger, 1–48.
183. Cantor B. (2020). *The Equations of Materials*. Oxford: Oxford University Press, 267–99.
184. Solymar L, Walsh D. Syms RA. (2018). *Electrical Properties of Materials*, 10th edn. Oxford: Oxford University Press, 126–61.
185. Ceruzzi, *Computing*, ch 2.
186. Ceruzzi, *Computing*, ch 3.
187. Riordan M, Hoddeson L. (1997). *Crystal Fire*. New York: W.W. Norton.
188. Moore GE. (1965). Cramming more components onto integrated circuits. *Electronics* 38:114–17.
189. Moore GE. (2006). Moore's law at 40. In D Brock, ed., *Understanding Moore's Law: Four Decades of Innovation*. Philadelphia: Chemical Heritage Foundation, 67–84.
190. Gaudin S. (2007, December 12). The transistor: The most important invention of the 20th century? *Computerworld*. https://www.computerworld.com/article/2538123/the-transistor—the-most-important-invention-of-the-20th-century-.html
191. Laws D. (2018, April 2). 13 Sextillion & Counting: The Long & Winding Road to the Most Frequently Manufactured Human Artifact in History. *Computer History Museum* (blog). https://computerhistory.org/blog/13-sextillion-counting-the-long-winding-road-to-the-most-frequently-manufactured-human-artifact-in-history/?key=13-sextillion-counting-the-long-winding-road-to-the-most-frequently-manufactured-human-artifact-in-history
192. Dyson G. (2012, February 17). Q&A: Hacker historian George Dyson sits down with *Wired*'s Kevin Kelly. *Wired*. https://www.wired.com/2012/02/ff-dysonqa/
193. Solymar et al., *Electrical Properties*, 164–224.
194. Riordan and Hoddeson, *Crystal Fire*, 22.
195. Riordan and Hoddeson, *Crystal Fire*, 71.
196. Riordan and Hoddeson, *Crystal Fire*, 82.

197. Riordan and Hoddeson, *Crystal Fire*, 137.
198. Riordan and Hoddeson, *Crystal Fire*, 4.
199. Riordan and Hoddeson, *Crystal Fire*, 140.
200. Hoddeson L. (1981). *The Discovery of the Point-Contact Transistor.* Historical Studies in the Physical Sciences series. Berkeley: University of California Press, *vol 12*, 41–76.
201. Lilienfeld JE. (1930, January 28). Method and apparatus for controlling electric current. US patent 1,745,715 (filed Canada Oct 22 1925, US Oct 8 1926).
202. Riordan and Hoddeson, *Crystal Fire*, 148.
203. Riordan and Hoddeson, *Crystal Fire*, 143.
204. Riordan and Hoddeson, *Crystal Fire*, 166–7.
205. Riordan and Hoddeson, *Crystal Fire*, 155.
206. Riordan and Hoddeson, *Crystal Fire*, 193.
207. Ceruzzi, *Computing*, chs 4–5.
208. GraphCore, https://www.graphcore.ai/products/ipu [accessed on 18/09/2023].
209. 'Single Crystal Silicon'. https://www.tf.uni-kiel.de/matwis/amat/iss/kap_5/illustr/i5_3_1.html, accessed 18 September 2023.
210. 'Regency Pocket Radio'. *Radio Museum.* https://www.radiomuseum.org/r/regency_pocket_radio_tr_1_tr1.html, accessed 18 September 2023.
211. Riordan and Hoddeson, *Crystal Fire*, p 226.
212. Riordan and Hoddeson, *Crystal Fire*, 232.
213. Riordan and Hoddeson, *Crystal Fire*, 237.
214. Riordan and Hoddeson, *Crystal Fire*, 249–251.
215. Boyer CB, Merzbach UC. (1991). *A History of Mathematics*, 2nd edn. New York: Wiley, 252–3.
216. Marguin J. (1994). *Histoire des instruments et machines à calculer, trois siècles de mécanique pensante 1642–1942.* Paris: Hermann, 48.
217. D'Ocagne M. (1893). *Le calcul simplifié.* Paris: Gauthier-Villars et fils, 245.
218. Rama. Musée des Arts et Métiers, Paris. https://commons.wikimedia.org/wiki/File:Pascaline-CnAM_823-1-IMG_1506-black.jpg
219. Gauvin JF. (2008). *Habits of Knowledge: Artisans, Savants and Mechanical Devices in Seventeenth-Century French Natural Philosophy.* Doctorate thesis. Harvard University, 210.
220. Turck JAV. (1921). *Origin of Modern Calculating Machines.* Chicago: The Western Society of Engineers, 75.
221. Turck, *Origin*, 143.
222. Bellis M. (2019, April 2). Samuel Morse and the invention of the telegraph. *Thoughtco.* https://www.thoughtco.com/communication-revolution-telegraph-1991939
223. Anderson JB, Johnneson R. (2006). *Understanding Information Transmission.* Hoboken: Wiley.
224. Chojecki P. (2019, January 31). How big is big data? *Towards Data Science.* https://towardsdatascience.com/how-big-is-big-data-3fb14d5351ba
225. Campbell-Kelly M, Aspray W. *Computer: A History of the Information Machine* (2nd edition). Basic Books, New York, 2016, p 16.
226. Halacy D. (1970). *Charles Babbage: Father of the Computer.* New York: Macmillan.
227. Hyman A. (1982). *Charles Babbage: Pioneer of the Computer.* Princeton: Princeton University Press, 5–30.
228. Hyman, *Babbage*, 13.

229. Essinger J. (2007). *Jacquard's Web*. Oxford: Oxford University Press, 59, 98.
230. Swade D. (2004, September 23). Charles Babbage. *Oxford Dictionary of National Biography*. Oxford University Press. https://www.oxforddnb.com/view/10.1093/ref:odnb/9780198614128.001.0001/odnb-9780198614128-e-962;jsessionid=C441269024F5594D74A043623EB70ACB, updated 21 May 2009.
231. O'Connor JJ, Robertson EF. (2017). Charles Babbage biography. *MacTutor*. https://mathshistory.st-andrews.ac.uk/Biographies/Babbage/, updated October 1998.
232. Lee JAN. *International Biographical Dictionary of Computer Pioneers*. Taylor and Francis, New York, 1995, p 60.
233. Babbage C. *Passages from the Life of a Philosopher*. Longmans, London, 1864, reprinted by Pickering and Chatto, London, 1994, pp 317–318.
234. Franksen OI, Rushton R. General notes, *Journal of the Royal Society of Arts*, 1985, *vol 133*, pp 880–883.
235. Lee JAN. Computer pioneers: Charles Babbage. Institute of Electric and Electronic Engineers, 1995: https://history.computer.org/pioneers/babbage.html
236. Babbage, *Passages*, 342.
237. Babbage, *Passages*, 360
238. Hyman, *Babbage*, ch 4.
239. Swade D. (2001). *The Cogwheel Brain: Charles Babbage and the Quest to Build the First Computer*. London: Abacus, 9–90.
240. Swade, *Cogwheel Brain*, 221–318.
241. Swade, *Cogwheel Brain*, 91–220.
242. Simonite T. (2009, March 24). Short sharp science: Celebrating Ada Lovelace: The world's first programmer. New Scientist. https://www.newscientist.com/blogs/shortsharpscience/2009/03/ada-lovelace-day.html
243. Seymour M. (2018). *In Byron's Wake*. London: Simon and Schuster, ch 12.
244. Turing AM. (1937). On computable numbers, with an application to the entscheidungsproblem. *Proceedings of the London Mathematical Society 42*: 230–65.
245. Wigderson A. (2017). *Mathematics and Computation*. Princeton: Princeton University Press, 15.
246. Church A. (1936). An unsolvable problem of elementary number theory. *American Journal of Mathematics 58*: 345–63.
247. Soare R. (2009). Turing oracle machines, online computing and three displacements in computability theory. *Annals of Pure and Applied Logic 160*: 368–99.
248. Rabin MO. (2012). Turing, Church, Gödel, computability, complexity and randomization: a personal view. https://www.bu.edu/cphs/files/2013/08/Rabin.pdf
249. Von Neumann J. (1945). First draft of a report on the EDVAC. *IEEE Annals of the History of Computing 15*(4): 27–43. http://web.mit.edu/STS.035/www/PDFs/edvac.pdf
250. Zuse K. (2013). *The Computer—My Life*. New York: Springer, 64.
251. Copeland JB. (2006). *Colossus: The Secrets of Bletchley Park*. Oxford: Oxford University Press.
252. Colossus. *The National Museum of Computing*. https://www.tnmoc.org/colossus

253. Haigh T, Priestley M, Rope C. (2016). *ENIAC in Action: Making and Remaking the Modern Computer*. Boston: MIT Press.
254. Eckert Jr JP, Mauchly JW. (1964). Electronic numerical integrator and computer. US Patent Office, patent number 3,120,606, filed 1947.
255. 'Classic shot of the ENIAC'. *Wikipedia*. https://commons.wikimedia.org/wiki/File:Classic_shot_of_the_ENIAC.jpg, accessed 18 September 2023.
256. Farrington G. (1996). ENIAC: Birth of the information age. *Popular Science March*: 74–6.
257. Burks AR. (2003). *Who Invented the Computer? The Legal Battle That Changed Computing History*. Amherst: Prometheus Books.
258. Lavington S. (1998). *A History of Manchester Computers*, 2nd edn. Swindon: The British Computer Society, 20–25.
259. Steinberg J. (2013). *Hello World: The History of Programming*. North Charleston: Createspace Publishing.
260. Wexelblat RL (ed). (1981). *History of Programming Languages*. Cambridge: Academic Press.
261. Anderson DP. (2009). Tom Kilburn: A pioneer of computer design. *IEEE Annals of the History of Computing 31*: 84.
262. Irvine MM. (2001). Early digital computers at Bell Telephone Laboratories. *IEEE Annals of the History of Computing 23*: 22–42.
263. Pugh EW. (1995). *Building IBM: Shaping an Industry and Its Technology*. Boston: MIT Press, 229–30.
264. Roser M, Ritchie H. (2020). Moore's Law Transistor Count 1970–2020. https://commons.wikimedia.org/wiki/File:Moore%27s_Law_Transistor_Count_1970-2020.png
265. Ceruzzi, *Computing*, 71.
266. Abbate J. (2000). *Inventing the Internet*. Cambridge: MIT Press, 36–111.
267. Banks MA. (2008). *On the Way to the Web: The Secret History of the Internet and Its Founders*. Berlin: Springer-Verlag.
268. Hauben M, Hauben R. (1998). Behind the net: The untold story of the ARPANET and computer science. *First Monday 3*: 7.
269. McMillan C. (2019, October 29). Lo and behold: The internet. *University of California Newsroom*. https://www.universityofcalifornia.edu/news/lo-and-behold-internet
270. Ceruzzi, *Computing*, ch 4.
271. Reid TR. (1984). *The Chip: How Two Americans Invented the Microchip and Launched a Revolution*. New York: Simon and Schuster. Revised edn. (2001). New York: Random House
272. Kilby JS. (1964). Miniaturized electronic circuits. US Patent Office, patent number 3,138,743, filed 1959.
273. Noyce RN. (1961). Semiconductor device and lead structure. US Patent Office, patent number 2,981,877, filed 1959.
274. Hall EC. (1996). *Journey to the Moon: The History of the Apollo Guidance Computer*. Reston: American Institute of Aeronautics and Astronautics, 18–19.
275. Ceruzzi, *Computing*, ch 5.
276. Shirriff K. (2016). The surprising story of the first microprocessors. *IEEE Spectrum 53*: 48–54.

277. 1971: Microprocessor integrates CPU function onto a single chip. *Computer History Museum.* https://www.computerhistory.org/siliconengine/microprocessor-integrates-cpu-function-onto-a-single-chip/, updated 30 May 2020.
278. Garland H. (1977). Design innovations in personal computers. *Computer 10*: 24.
279. Swaine M, Freiberger P. (2014). *Fire in the Valley: The Birth and Death of the Personal Computer*, 3rd edn. Raleigh, NC: Pragmatic Bookshelf.
280. Allan RA. (2001). *A History of the Personal Computer: The People and the Technology.* London, ON: Allan Publishing.
281. https://www.gfmag.com/global-data/economic-data/largest-companies
282. Ceruzzi, *Computing*, ch 6.
283. McPherson SS. (2009). *Tim Berners-Lee: Inventor of the World Wide Web.* Milwaukee: Twenty-First Century Books.
284. Gillies J, Cailliau R. (2000). *How the Web Was Born: The Story of the World Wide Web.* Oxford: Oxford University Press.
285. Jobs S. (2007). Keynote address, Macworld, San Francisco, 2007: https://web.archive.org/web/20101222223204/http://www.apple.com/quicktime/qtv/mwsf07/
286. Sharma S. (2021). Birth of a true smartphone: 14 years since the first Apple iPhone. *The Quint.* https://www.thequint.com/tech-and-auto/tech-news/birth-of-a-true-smartphone-eight-years-since-the-first-iphone
287. Puiu T. (2015, October 13). Your smartphone is millions of times more powerful than the Apollo 11 guidance computers. *ZME Science.* https://www.zmescience.com/science/news-science/smartphone-power-compared-to-apollo-432/, updated 11 May 2023.
288. How many smartphones are in the world? *Bank My Cell.* https://www.bankmycell.com/blog/how-many-phones-are-in-the-world, accessed 18 September 2023, updated February 2024.
289. How Many Google Searches Per Day Are There? Useful Search Metrics for Marketers. *Skai.* https://skai.io/monday-morning-metrics-daily-searches-on-google-and-other-google-facts/ accessed 18 September 2023.
290. How Many Emails Are Sent per Day in 2024? *Review 42.* https://review42.com/resources/how-many-emails-are-sent-per-day/, accessed 18 September 2023, updated 13 February 2024.
291. Hours of video uploaded to YouTube every minute as of February 2022. *Statista.* https://www.statista.com/statistics/259477/hours-of-video-uploaded-to-youtube-every-minute/, accessed 18 September 2023.
292. Twitter usage statistics. *Internet Live Stats.* https://www.internetlivestats.com/twitter-statistics/, accessed 18 September 2023.

Further Reading

Allen RC. (2011). *Global Economic History.* Oxford: Oxford University Press.

Ashby MF, Jones DRH. (1986). *Engineering Materials 2.* Oxford: Pergamon.

Bessemer H. (1989). *Sir Henry Bessemer FRS: An Autobiography; With a Concluding Chapter.* London: Institute of Metals. Originally published (1905) London: Offices of Engineering.

Bhadeshia HKDH, Honeycombe RWK. (2017). *Steels: Microstructure and Properties*, 4th edn. Oxford: Butterworth-Heinemann.

Ceruzzi PE. (2012). *Computing: A Concise History*. Cambridge: MIT Press.

Derry TK, Williams TI. (1960). *A Short History of Technology: From the Earliest Times to AD 1900*. Oxford: Oxford University Press. Reprinted (1993). Mineola: Dover Publications.

Gilchrist JD. (1989). *Extraction Metallurgy*, 3rd edn. Oxford: Pergamon.

Grovenor CRM. (1989). *Microelectronic Materials*. Bristol: Adam Hilger.

Harai YN. (2014). *Sapiens: A Brief History of Humankind*. London: Harvill Secker.

Martin J. (1996). *Materials for Engineering*. London: Institute of Materials.

Putnis A. (1992). *Introduction to Mineral Sciences*. Cambridge: Cambridge University Press.

Riordan M, Hoddeson L. (1997). *Crystal Fire: The Invention of the Transistor and the Birth of the Information Age*. New York: W.W. Norton.

Sass SL. (2011). *The Substance of Civilization: Materials and Human History from the Stone Age to the Age of Silicon*. New York: Arcade.

Siemens P. (1993). *Sir William Siemens: A Man of Vision*. Ferndown: Siemens/Caric Press.

Smith BD. (1998). *The Emergence of Agriculture*. New York: Freeman.

Solymar L, Walsh D. Syms RA. (2018). *Electrical Properties of Materials*, 10th edn. Oxford: Oxford University Press.

Tarbuck EJ, Lutgens FK, Tasa DG. (2016). *Earth: An Introduction to Physical Geology*, 12th edn. London: Pearson.

Tylecote RF. (1976). *A History of Metallurgy*. London: The Metals Society.

Wootton D. (2016). *The Invention of Science: A New History of the Scientific Revolution*. London: Penguin Books.

2

Discovery of Multicomponent Materials

1 Introduction

I am often asked how I came up with the idea of *multicomponent materials*. Scientific (and non-scientific) friends and colleagues are (not surprisingly) fascinated by how it took centuries, if not millennia, for such a simple idea to have arisen. The truth is that it was largely serendipity. I was going through a difficult period in my life, having just separated from my first wife and having just been turned down for a permanent lectureship at the university where I was working. But I have always found hard work a great solace in adversity. Looking back I can see that, as a consequence, it was a productive time intellectually.

Nevertheless, the idea of *multicomponent alloys* and *multicomponent materials* (now more often called *high-entropy alloys* and *high-entropy materials*)[i] was not the product of intensive theorising or inspired experimentation. I wish I could say it was. It was more like a random notion that occurred to me and kept nagging away in my head. And the more others said it was a bit of a crazy idea, the more a streak of cussedness made me want to do something about it.

I have always marvelled (as have many others) at the wonderful and somewhat mysterious notion of *entropy* as one of the major driving forces in the world, discovered slowly and painfully by a clutch of great scientists in the nineteenth century. Scientific heroes such as Gibbs, Boltzmann, Kelvin and Maxwell (to name but a few) struggled mightily to understand the behaviour of engines and machinery, the driving forces of the industrial revolution, before fully realising and explaining that entropy (or randomness or disorder) was just as important as energy. And serendipity is just entropy at play in the field of human affairs. So perhaps I don't mind so much that the idea of multicomponent high-entropy materials came from a humdrum random entropic thought and not from a grander scientific vision. After all, there's a neat circularity to it.

This chapter describes the events by which I and my colleagues in England beginning in the late 1970s and Jien-Wei Yeh and his colleagues in Taiwan beginning in the mid-1990s independently discovered and began to explore multicomponent high-entropy materials. It's a personal account, a story, and written therefore (somewhat unusually for a textbook) in the first person. It includes some brief biographical details of those involved.

[i] The definition of terms such as *multicomponent alloys, multicomponent materials, high-entropy alloys and high-entropy materials*, and the differences between them are discussed in detail in Chapter 3.

Fundamentals of Multicomponent High-Entropy Materials. Brian Cantor, Oxford University Press. © Brian Cantor (2024).
DOI: 10.1093/9780191986710.003.0002

2 Discovery of Multicomponent Materials

The first idea

My first wife Maggie and I separated in 1977. More accurately, she sat me down one day after work and told me she had fallen in love with someone she had met who was working as a foreman on a nearby building site, and she was leaving me, taking our two sons with her, to go and live with him in Coventry, about 150 miles away from Brighton where we lived. It was a bolt from the blue. I was distraught. She left the next day. Two years later we were divorced. Maggie and I had married young, both just nineteen, when in a surfeit of youthful love and enthusiasm she had fallen pregnant while I was a first-year science undergraduate and she was a first-year trainee nurse at Cambridge. We were ludicrously poor and we both took on lots of additional part-time jobs to survive. It was tough on student money to feed what rapidly became a family of four, and we just as rapidly built up a pretty impressive portfolio of debts. After completing my bachelor's and doctoral degrees and moving to Sussex on (at last) a regular, though small, salary, our finances were gradually beginning to recover and our debts were beginning to be repaid, but the costs of a divorce wiped me out again. I was back in penury, with still no prospect of a permanent job. The University of Sussex had just refused to approve a full-time lectureship for me, despite several attempts by my boss, the Dean of Engineering, Professor Robert Cahn, to secure one. Worse still, this turned out to be the climax of a slow-burning saga. Robert finally lost patience with the University's continued lack of support for his materials science group and left a couple of years later to take up a Professorship at the Sorbonne in Paris. Without Robert, materials science was washed up at Sussex, and the section closed down fully after a couple more years to let existing students complete their courses.

So as we approached the end of the 1970s, my life seemed to be falling apart personally, professionally and financially. But I threw myself into my research. And things started to look up. In 1980 I took a year's visiting fellowship in India at Banaras Hindu University (BHU), arranged by my great friend Patcha Ramachandra Rao (sadly now deceased), and in 1982 I took another year's industrial fellowship at the General Electric (GE) research labs in Schenectady, New York, arranged by another great friend Fred Luborsky (also sadly now deceased). And finally I was offered and accepted a full-time lectureship at the University of Oxford by the Head of the Department of Metallurgy (now the Department of Materials), Professor Sir Peter Hirsch, which after a brief start in 1981 was deferred for a year while I worked in the US for GE.

For me it was a time of great change. Despite this, or maybe because of it, it was also a time of great intellectual creativity. I had what I believe are three of my best scientific ideas during this period, though none of them made any impact on the scientific community at the time, and all three effectively went off like lead balloons. In 1979 at Sussex I invented the notion of multicomponent alloys; in 1980 in Banaras I invented a tetrahedral interstitial basis for the thermodynamics of glassy metals; and in 1982 at GE I invented a Voronoi polyhedral method for analysing metallurgical microstructures. The idea of multicomponent alloys lay fallow for about thirty years, until it took off in the late 2000s; interstitial thermodynamics of glassy alloys remains (in my view) an undeveloped gem of an idea; and Voronoi polyhedral analysis is now (thanks to others) a fairly standard microstructural technique.

My PhD at Cambridge had been on the solidification and mechanical properties of directionally solidified eutectics. I had enjoyed discovering features of as-solidified eutectic crystallography and microstructure, as well as the deformation and fracture behaviour of what are effectively in-situ aligned composites under controlled tensile and compressive loading conditions. I had demonstrated the complexity of eutectic interface crystallography, discovered the formation of eutectic dendrites, and shown how deformation is dominated by the geometrically necessary dislocations punched out at the eutectic interfaces by differential thermal contraction during cooling. Moving to Sussex I changed field to work on the then quite new field of rapid solidification and metastable materials. I again enjoyed discovering important features of sputter-deposited, melt spun and electro-deposited amorphous alloys, as well as high-strength rapidly solidified aluminium alloys and tool steels. I developed the first theory of glass formation in vapour-deposited materials, made some of the first ever measurements of atomic diffusion in glassy alloys, and showed how the nucleation of martensite transformations was suppressed by the small grain sizes generated during high-speed quenching.

The field of metastable alloys was mushrooming. Unusual microstructures with unusual properties were being manufactured by rapid solidification and other metastable processing methods. We were making previously undiscovered extended solid solutions, novel intermetallic compounds, amorphous alloys, quasicrystals, nanocrystalline materials, nanocomposites, you name it. But in some cases it proved possible to manufacture the same unusual microstructures at relatively low cooling rates, and I began to wonder whether accessing higher alloying compositions wasn't just as important as the direct effect of a high quenching rate. It occurred to me that in most cases conventional casting only allowed the manufacture of rather dilute, two- or three-component alloys, whereas rapid solidification enabled alloy compositions to be pushed higher. I began to realise that all our materials were based on the corners and/or edges of the phase diagrams of mostly binary and ternary alloys, and I began to ask why don't we push compositions even further and mix equiatomic or near-equiatomic multicomponent alloys containing four, five, six or even more components.

I pretty soon learned why. Everyone told me in no uncertain terms. It was a crazy idea, they said. Why would we do such a thing, they said. Some of my scientific colleagues enjoyed discussing the idea but were at best unenthusiastic about its likely value. Research funders were quite clear that it was pure blue-skies thinking and not worth backing. More importantly, potential students and postdocs were uninterested, wanting to work (not unreasonably) on fashionable topics which gave them a good chance of getting a subsequent job. I was stymied. But only for a short while. The idea was growing on me. I wasn't going to give up.

The first experiments

I was talking to everyone in the late 1970s about the idea of multicomponent alloys, albeit with little response or, more accurately, with not much better than a big raspberry. Nothing daunted I decided to plunge in at the deep end. I found time to get one of our technicians to make up our first multicomponent alloy, consisting of twenty components in equal proportions, 5 at% each of the elements Mg, Al, Si, Mn,

Cr, Fe, Co, Ni, Cu, Zn, Ge, Nb, Mo, Ag, Cd, Sn, Sb, W, Pb, and Bi, more usually nowadays written as either MgAlSiCrMnFeCoNiCuZnGeNbMoAgCdSnSbWPbBi or $Mg_5Al_5Si_5Cr_5Mn_5Fe_5Co_5Ni_5Cu_5Zn_5Ge_5Nb_5Mo_5Ag_5Cd_5Sn_5Sb_5W_5Pb_5Bi_5$.[ii] We gave it the more prosaic designation MCA1 (short, fairly obviously, for multicomponent alloy 1). At the time this was (by a country mile) the world record for the most components ever made up into a multicomponent material, and I am pretty sure that it remained so until very recently, i.e. for the best part of forty years (though it never made it into the *Guinness Book of World Records*). I had no idea how to begin to explore what I realised was a vast and so-far completely unexplored alloy space. So this first alloy was made up from more-or-less everything we had in our pure-material cabinet that was not too toxic or difficult to handle, and that we thought would be relatively easy to melt without excessive losses through oxidation or vaporisation.

The alloy was RF (radio-frequency) induction melted in a graphite crucible under argon and allowed to solidify in the crucible after turning off the RF power. The as-solidified alloy proved to be hard and brittle, in fact cracking into several pieces during cooling to room temperature. I had a look at a polished and etched cross-sectional microstructure in the optical microscope, and the result is shown in Figure 2.1. The alloy was a mess, with different phases all jumbled up and with different morphologies. An X-ray diffractometer (XRD) trace showed a similar mess of literally dozens of different peaks, far too many to analyse, as might well be expected from a complex multiphase microstructure such as that shown in Figure 2.1. We also made up two other alloys containing

Figure 2.1 Optical micrograph of the multiphase microstructure of the equiatomic twenty-component alloy MgAlSiMnCrFeCoNiCuZnGeNbMoAgCdSnSbWPbBi.
Courtesy of Brian Cantor.

[ii] Multicomponent alloys are described in this book by listing the elemental components in order of ascending atomic number, i.e. following their (ascending) position in the periodic table. Compositions are written with subscripts for atomic ratio or atomic per cent, or without subscripts for equiatomic compositions. Thus, the original Cantor alloy with an equiatomic mixture of Cr, Mn, Fe, Co and Ni is written as either $Cr_{20}Mn_{20}Fe_{20}Co_{20}Ni_{20}$ or, more simply, CrMnFeCoNi, depending on which notation is more useful.

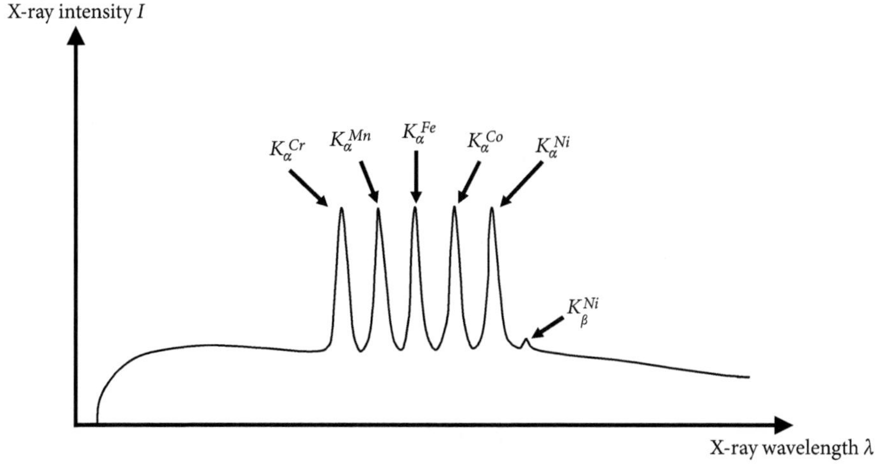

Figure 2.2 Schematic of an EDS spectrum in the SEM from the Cantor alloy CrMnFeCoNi, showing Cr, Mn, Fe, Co, and Ni K_α and Ni K_β lines (Cr, Mn, Fe, and Co K_β lines are hidden below the K_α lines).

sixteen and ten elements in equal proportions, designated MCA2 and MCA3 respectively: MCA2 with the composition MgAlCrMnFeCoNiCuZnNbMoAgCdSnWPb, i.e. the same as MCA1 but without the semi-metals Si, Ge, Sb and Bi; and MCA3 with the composition CrMnFeCoNiCuNbMoAgW, i.e. the same as MCA2 but without the non-transition and low-melting-point elements Mg, Al, Zn, Cd, Sn and Pb. The results were not dissimilar to MCA1, with multiphase brittle as-solidified microstructures. Beyond the optical micrograph in Figure 2.1, little evidence now remains of these first tentative, exploratory experiments.

I wasn't able to take things much further for a while. I was very busy with teaching duties because I had taken on at Sussex the additional job of a temporary lecturer, at the same time as doing my own post-doctoral research and the day-to-day tasks of running Robert Cahn's research group while he was otherwise occupied as Dean. And at the beginning of 1980, as mentioned previously, I went off on sabbatical leave to BHU to discover the wonderful joys, the rich culture, the hurly-burly as well as the peace and meditation of the Indian sub-continent. When I returned from India in time for the October start of the 1980/81 academic year, I found that a young, fresh-faced, devil-may-care undergraduate student named Alain Vincent had signed up to study multicomponent alloys with me for his second-year fifty-hour research project. I don't know whether Alain was foolish or far-sighted, but he certainly wasn't worrying about his future career—he was more interested in being adventurous and having fun. And my crazy multicomponent alloy idea was back up and running.

Cantor alloys

The first thing we discovered, on further investigation of MCA1 by energy-dispersive (EDS) X-ray analysis in a scanning electron microscope (SEM), was that the composition of one of the phases in the alloy consisted of equal proportions of five transition metals from the middle-to-late first transition series in the periodic table, namely Cr, Mn, Fe, Co, and Ni.

In those days, EDS results were displayed as a spectrum on a cathode-ray tube (CRT) scan, which had to be photographed and then analysed by hand, as there was no effective computer control or analysis software available. The X-ray emissions from this phase showed a remarkable and distinctive spectral pattern, looking a bit like a hedgehog or a stegosaurus, as shown schematically in Figure 2.2 (sadly, no XRD chart-recorder traces or EDS photos from these early experiments have survived the intervening forty-year test of time). Alain and I decided to explore this particular phase in more detail. It proved to be, of course, what is now known as the original *Cantor alloy*, an equiatomic five-component single-phase face-centred cubic (fcc) material $Cr_{20}Mn_{20}Fe_{20}Co_{20}Ni_{20}$ or CrMnFeCoNi, which subsequently excited much scientific interest and has now been studied very widely indeed, all round the world.

For his research project, Alain made up three new equiatomic alloys designated V1, V2 and V3 (short, I think, for Vincent 1, 2 and 3). The first alloy, V1, was the same as the equiatomic five-component single phase found in the twenty-component MCA1, i.e. it had the original Cantor alloy composition CrMnFeCoNi so we could study the structure and properties of this material in isolation. The other two alloys, V2 and V3, were modified versions of the Cantor alloy by adding further components to explore its range of solubility: in V2 by the addition of three early transition metals, Nb, Mo and W; and in V3 by the addition of another first-series late transition metal, Cu. Unintentionally, the last of these alloys was found to be contaminated with a small amount of Al. In other words, the compositions of V2 and V3 were CrMnFeCoNiNbMoW and $Al_xCrMnFeCoNiCu$ respectively (where x indicates a small but unknown amount of contamination, probably at the one or two at% level). The three as-cast alloys all showed a segregated dendritic structure. Figure 2.3 shows a relatively low-magnification optical micrograph of the dendrites in the first alloy V1. This is the first-ever record of the structure of the Cantor alloy CrMnFeCoNi. Figure 2.4 shows an even lower magnification optical micrograph of the finer-scale dendrites

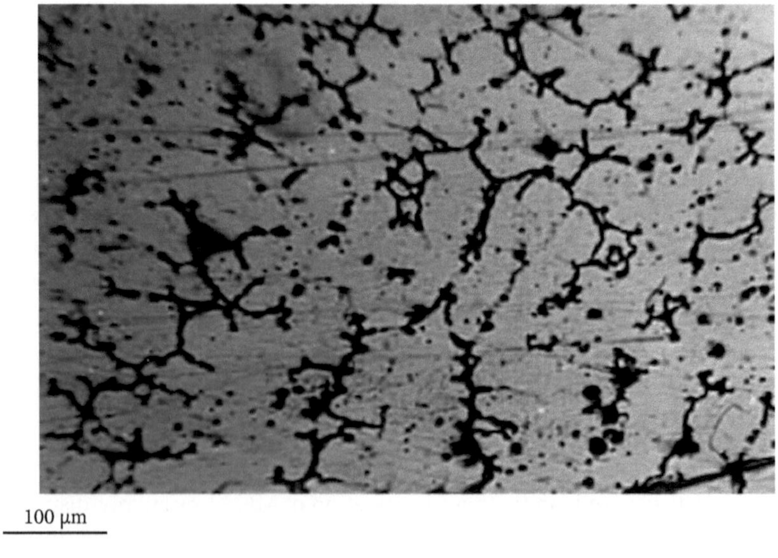

Figure 2.3 First-ever optical micrograph of the as-cast segregated dendritic microstructure of the original Cantor alloy CrMnFeCoNi.
Courtesy of Brian Cantor.

Figure 2.4 Optical micrograph of the as-cast segregated dendritic microstructure of the modified Cantor alloy $Al_xCrMnFeCoNiCu$.
Courtesy of Brian Cantor.

in V3, one of the two other modified Cantor alloys $Al_xCrMnFeCoNiCu$. XRD traces showed quite clearly and (for then) very surprisingly that the Cantor alloy was single-phase fcc with a lattice parameter of 3.595 Å. The eight-component modified Cantor alloy CrMnFeCoNiNbMoW was a mixture of fcc and bcc phases with lattice parameters of 3.637 Å and 2.900 Å, respectively, with the bcc phase (not surprisingly) rich in the early transition elements Nb, Mo and W, and with a small amount of some additional unidentified phases. And the seven-component modified Cantor alloy $Al_xCrMnFeCoNiCu$ was a mixture of two fcc phases with lattice parameters of 3.586 Å and 3.631 Å, respectively. These early multicomponent alloy results are collected together in Table 2.1, and Figure 2.5 shows the (handwritten) header page from Alain Vincent's dissertation.

Confirming the results

By conducting a small number of simple experiments, Alain Vincent and I had quickly uncovered a number of the key features of multicomponent alloys that have all been fully established in recent years, namely:

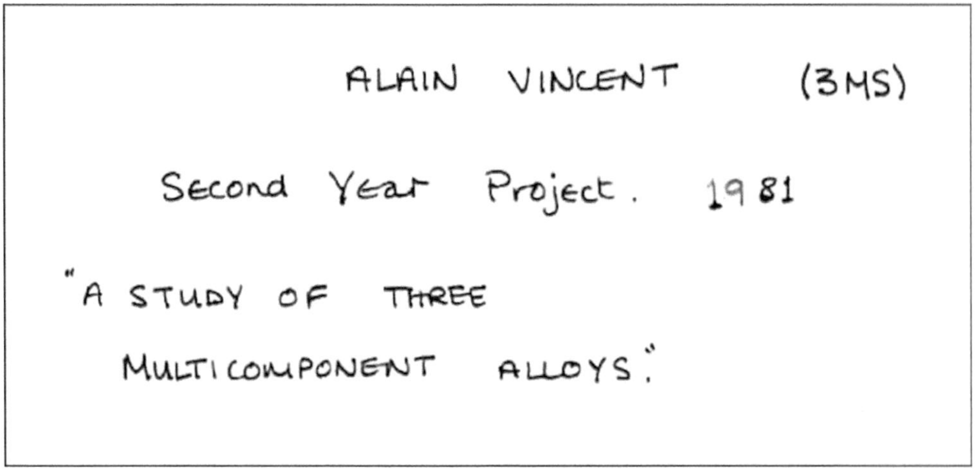
ALAIN VINCENT (3MS)

Second Year Project. 1981

"A STUDY OF THREE

MULTICOMPONENT ALLOYS."

Figure 2.5 Handwritten header page from Alain Vincent's undergraduate dissertation on multicomponent alloys.

Reproduced from Vincent AJB. A Study of Three Multicomponent Alloys, Undergraduate thesis, University of Sussex, 1981.

Table 2.1 Composition, structure and Vickers microhardness of the first series of multicomponent alloys (after Cantor[1] and Vincent[2]).

alloy	composition	structure	hardness
MCA1	MgAlSiCrMnFeCoNiCuZnGeNbMoAgCdSnSbWPbBi	multiphase	hard & brittle
MCA2	MgAlCrMnFeCoNiCuZnNbMoAgCdSnWPb	multiphase	hard & brittle
MCA3	CrMnFeCoNiCuNbMoAgW	multiphase	hard & brittle
V1	CrMnFeCoNi	fcc	180
V2	CrMnFeCoNiNbMoW	fcc + bcc + others	770
V3	Al_xCrMnFeCoNiCu	fcc 1 + fcc 2	220

1. There is a vast number of alloys in multicomponent phase space, and most of these alloys have not yet been studied.
2. As might have been expected, some multicomponent alloys have a multiphase microstructure, but a surprising number consist of no more than one or two phases.
3. There is an fcc single-phase field in multicomponent phase space centred around the middle/late transition metals, and a bcc single-phase field centred around the early transition metals.
4. In particular, there is a five-component single-phase fcc alloy (now known as the original Cantor alloy) consisting of equiatomic CrMnFeCoNi, despite the fact that only one of these five elements is fcc in its pure form at room temperature, and despite the fact that these five elements are a long way from exhibiting complete intersolubility in binary mixtures.
5. The single-phase fcc Cantor alloy and modified single-phase fcc compositions (now known generally as Cantor alloys), must have been stabilised as multicomponent solid solutions by having relatively low enthalpies of mixing and relatively high entropies of mixing.

6. Multicomponent alloys show high hardness and high strength in both single-phase and multiphase microstructures.

I was very excited by the idea of a single-phase fcc solid solution containing as many as perhaps twelve components in equiatomic proportions, in which case every atom would be surrounded on average by twelve different near neighbours. In my mind's eye, this would be a kind of 'averagium' alloy, which would somehow 'smear out' the properties of all its components. It would also have a distorted lattice with a wide variety of strained local atomic environments leading, I surmised, to remarkable and surprising electrical, magnetic, optical and mechanical behaviour.

Although Alain's results were quite clear and exciting, the work was still only preliminary, and the quality of the micrographs, XRD traces and EDS spectra were too poor to be written up and to have any chance of being accepted for publication in a good research journal. I needed to get someone else to repeat the experiments, or I needed to find time to do it myself. For a variety of reasons, this didn't happen for the best part of the next twenty years. The truth is that the pace of my life was hotting up, and my mind (and my body) were becoming otherwise engaged. I arrived back from India towards the end of 1980; later that year I was offered and accepted the industrial fellowship for a year at GE; also later that year I was offered and accepted the lectureship at Oxford; attending a conference in Cambridge, I met and fell in love with my second-wife-to-be, Anne, and after a whirlwind romance we were married in early 1981; Alain's undergraduate project was conducted in late 1980 and early 1981; that summer Anne and I bought and moved into a small house close to the centre of Oxford; and after a hectic term's teaching at Oxford, we went off in early 1982 to Schenectady in beautiful upper New York State for me to work at the GE Central Research Labs. Just before we left for the United States, my father told me he had been diagnosed with lung cancer, which at first seemed to be being treated successfully, only for him to collapse suddenly and die from a secondary brain tumour a couple of years later. And just after we returned from the States, Anne was diagnosed with cervical cancer. Again treatment seemed at first to be successful, but it later turned into bowel cancer and, although Anne fought hard for some years, she gradually subsided and finally succumbed after prolonged illness, dying tragically young ten years later aged just 38.

My main job at GE (studying Voronoi polyhedra to describe material microstructures was a sideline) was supporting two major industrial developments: new electricity distribution transformers using amorphous alloy magnetic cores, which reduced hysteresis losses by more than two-thirds; and magnetic bubble domains in amorphous cobalt alloy thin films as potential high-density data-storage devices for first-generation X-ray CT (computerised-tomography) scanners. Amorphous alloys in distribution transformers were eventually installed in considerable numbers in the USA and Japan; but bubble-domain memory devices were overtaken by several alternative and better high-density data-storage technologies. I found the complex interplay in large-scale industrial R&D between fundamental science, applied engineering, manufacturing scale-up, economic assessment, workflow management and commercial judgement fascinating and challenging, and after returning to Oxford I resolved to concentrate in the future on similar, industry-related research rather than my previous blue-skies approach. In fact this was in accord with Peter Hirsch's vision for his Department of Materials which, after two decades building up its strength and reputation in fundamental materials science, he now wanted to extend to real-world engineering and technology. It was also in accord with the new Thatcher government's vision for university research, as a support facility for wealth-creating private sector companies. The infamous and extremely painful 'Thatcher cuts' reduced university funding sharply and were explicitly intended to persuade university bosses to align their teaching and research with industrial companies. Politically I was opposed to Mrs Thatcher's small-state approach to funding universities (as well as to many of her other policies), but I have to own up to the fact that I did well out of it. Serendipitously (again) it turned out that I was in the right place at the right time. I was effectively offering Oxford's facilities and talent to work with major UK and multinational companies around the world (this was a very novel approach at the time), and my research group flourished dramatically, growing to more than thirty PhD students and postdocs, with massive amounts of semi-industrial-scale processing and analysis equipment, well funded jointly by industry and government on a wide variety of different large-scale projects across the automotive, aerospace, computer, engineering, printing, packaging and other industrial sectors. By the end of the 1980s, I was regularly being offered professorships elsewhere, but I stayed on at Oxford, taking up a Readership, setting up the Oxford Centre for Advanced Materials and Composites (OCAMAC), finally being awarded the (then) second chair, the Cookson Professorship, and becoming Head of Department and then Head of the (newly created) Division of Mathematical and Physical Sciences (effectively head of all maths, physical sciences and engineering at Oxford).

The upshot was that, for both personal and professional reasons, the last thing I was thinking about in any detail was multicomponent alloys, though I did keep an occasional eye out for possible ways to confirm Alain Vincent's results. And in the middle of the 1990s, another feckless young undergraduate student, Peter Knight, signed up to work on this project for his (so-called) part II year. This was a special feature, a kind of quirk, of the Oxford Materials degree course (as well as Oxford Chemistry), which had (unusually at that time in England) a fourth, part II year devoted entirely to a research project that is much more substantial than the usual bachelor's- or even master's-level research project.

Peter and I agreed we would repeat Alain's experiments and try to explore more fully the size and shape of the fcc single-phase field, the range of what are now called the Cantor alloys, in the middle of multicomponent phase space. We decided to stick to equiatomic compositions for simplicity, though we fully realised that the fcc single-phase field must contain many non-equiatomic compositions. Peter made up a series of five-, six, seven- and nine-component alloys as shown in Table 2.2. They all had segregated dendritic as-cast microstructures, as shown in Figure 2.6. Effectively, Peter reconfirmed firstly that there is surprisingly often rather few phases (no more than one or two) in multicomponent alloys; secondly, that there is an enormous fcc single-phase field in multicomponent phase space, centred loosely on the original Cantor alloy composition CrMnFeCoNi, with a wide range of modified single-phase fcc Cantor alloys containing either fewer or more components and with varying equiatomic and non-equiatomic compositions; and thirdly, that multicomponent alloys are strong and hard, in single-phase materials presumably from the high levels of solid solubility and associated lattice distortion and solution hardening, and in multiphase materials from the multiplicity of hardening phases. Peter wrote up his part II thesis,[3] submitted it, and was awarded his degree in 1995. But history sometimes repeats itself. Once

Table 2.2 Composition, structure and Vickers microhardness of modified Cantor alloys (after Knight[3] and Cantor et al.[4]).

alloy	composition	structure	dendrite hardness	Interdendritic hardness
Cantor alloy	CrMnFeCoNi	fcc	300	299
+ Ge	GeCrMnFeCoNi	fcc + bcc	298	1068
+ Ti	TiCrMnFeCoNi	fcc + bcc	998	643
+ V	VCrMnFeCoNi	fcc	1007	287
+ Cu	CrMnFeCoNiCu	fcc	310	314
+ Nb	CrMnFeCoNiNb	fcc	1031	399
+ Ta	CrMnFeCoNiTa	fcc + bcc		
+ TiV	TiVCrMnFeCoNi	fcc + bcc		
+ GeTiV	GeTiVCrMnFeCoNi	fcc + other phases		

again I felt that the quality of the optical and SEM micrographs, XRD traces and EDS analyses were not sufficiently high for me to be confident that a paper would be accepted for publication in a good research journal.

In desperation I turned to one of my very competent senior postdocs, Isaac Chang (now a Professor at Brunel University London), whose experimental skills I knew well and valued highly as he had worked with me for over ten years on a variety of projects. I explained the background and asked him to re-do Alain's and Peter's experiments and for heaven's sake this time do a good job of it. He was very busy, of course, with his own work, but I pointed out that we knew what the results were, we just needed to polish them up and get them into the right shape. All experimentalists like a guaranteed experiment, and even more a guaranteed publishable paper, so after a while Isaac found time to re-do the work, which we wrote up fully for presentation and publication[5] at the Rapidly Quenched Materials Conference RQ11 in Oxford in 2002, which was subsequently re-published[4] by the peer-reviewed research journal *Materials Science and Engineering* in 2004. Its title is 'Microstructural Development in Equiatomic Multicomponent Alloys'. As it happens, I had by then, in 2002 and 2003, already presented the basic idea of multicomponent alloys at several other conferences, and published it in papers on the multicomponent glasses discussed in the next sub-section, in the journals *Materials Science Forum*, *Philosophical Magazine*, *Materials Transactions* and *Journal of Non-Crystalline Solids*.[6–9]

Figures 2.7 and 2.8 show some of Isaac's more professional EDS X-ray maps from the SEM of the as-cast segregated dendritic microstructures in some of Peter's multicomponent alloys from Table 2.2. Figure 2.7 shows segregation of Cr and Fe into the dendrite spines and Mn into the interdendritic regions of the original Cantor alloy CrMnFeCoNi; and Figure 2.8 shows the more complex segregated structure of the equiatomic seven-component modified Cantor alloy with additions of Ti and V, TiVCrMnFeCoNi. Our *Materials Science and Engineering* paper[4] is rightly regarded as one of the two founding papers of the field of multicomponent high-entropy alloys and materials, even though the basic research had been done many years earlier, and the main idea had already been reported several times previously in 2002 and 2003. The paper was cited no more than a couple of times in the two or three years following its publication but has now been cited many thousands of times as the field has taken off.

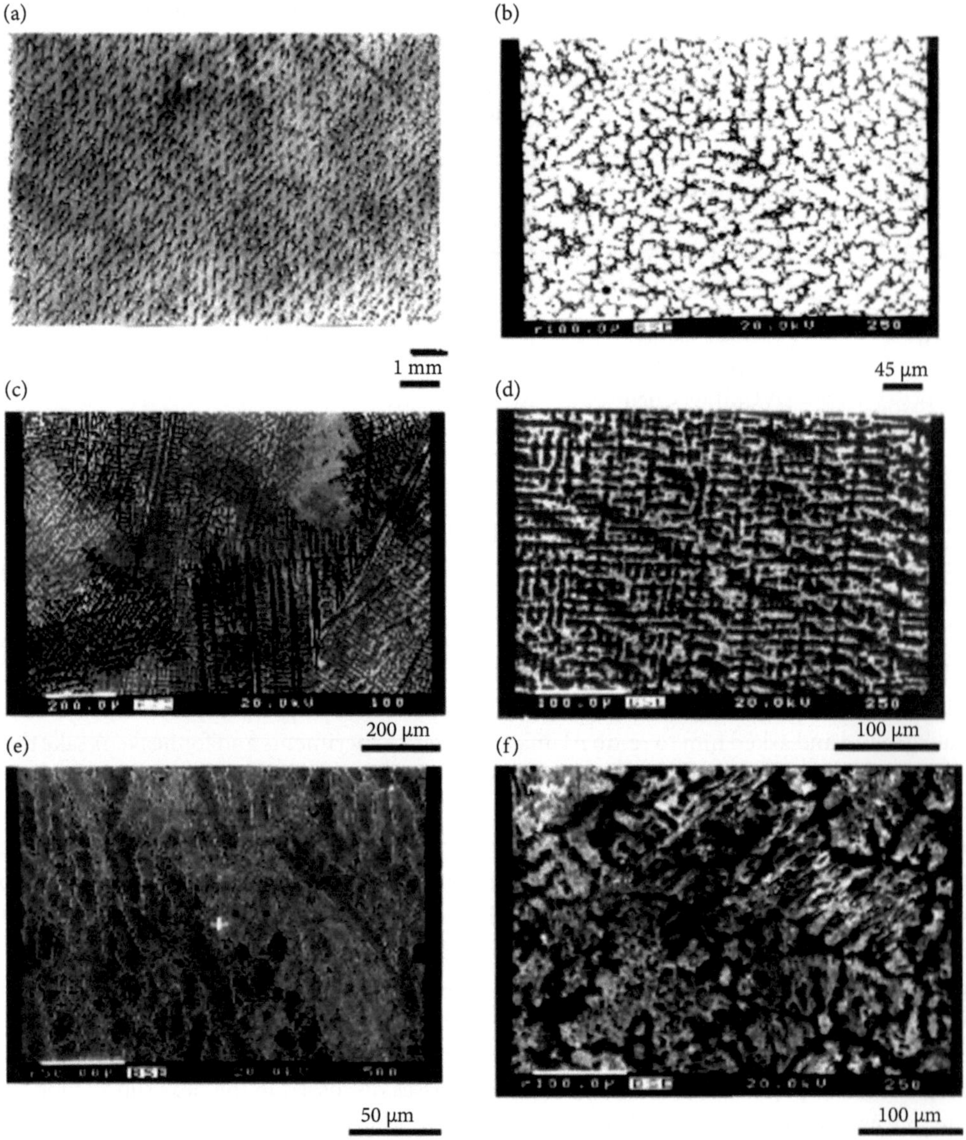

Figure 2.6 Optical micrographs of the as-cast segregated dendritic microstructures of early Cantor alloys: (a) CrMnFeCoNi; (b) CrMnFeCoNiNb; (c) GeCrMnFeCoNi; (d) CrMnFeCoNiCu; (e) TiCrMnFeCoNi; (f) VCrMnFeCoNi.

Reproduced from Knight P. *The Study of Multicomponent Alloy Phases*, Part II thesis, Department of Materials Library, University of Oxford, 1995. Courtesy of Brian Cantor.

Our paper was published in 2004 in *Materials Science and Engineering* and that same year (and unknown to me at the time) Professor Jien-Wei Yeh and his colleagues published their first paper on multicomponent alloys, based on independent research work that they had conducted in the 1990s. Neither of our papers was received with immediate acclaim, and in the next few years there were very few citations to either paper. Given everyone's previous reactions to my multicomponent alloy idea, I was not surprised that our paper received so

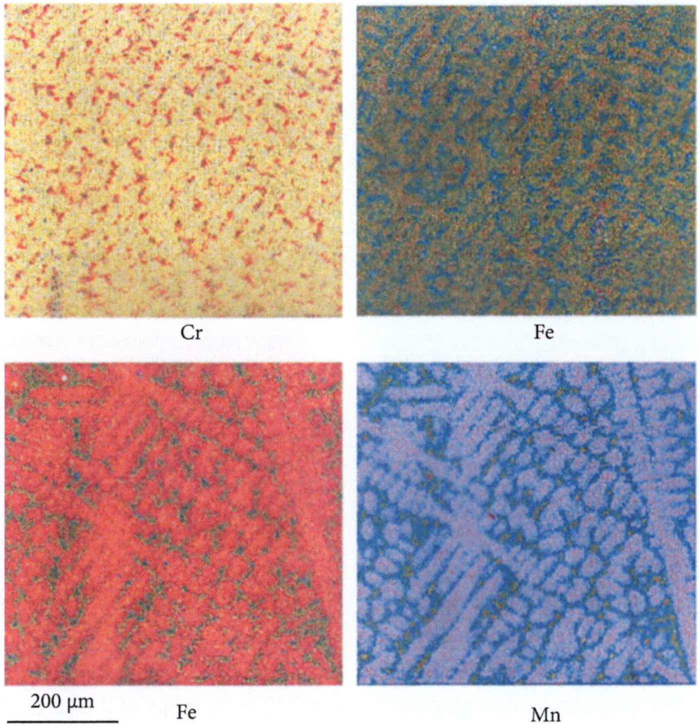

Figure 2.7 Scanning electron micrographs showing EDS X-ray maps of the as-cast segregated dendritic microstructure in the Cantor alloy CrMnFeCoNi.

Reproduced from Cantor B, et al. (2004). Microstructural development in equiatomic multicomponent alloys. *Materials Science and Engineering A375–7*: 213–18.

little notice. I was just thankful that we had finally got the initial work published. Jien-Wei continued to work on multicomponent alloys and his doggedness helped to push them gradually onto the scientific agenda later in the 2000s, and more strongly in the 2010s. His early contribution is described in more detail in the next section. In my case, however, fate intervened and my life again took a rather different turn. In the early 2000s, I was approached by headhunters to consider applying for various positions as a Vice-Chancellor (President) of several British universities. At first I was uninterested, since my research was going so well at Oxford. But in the end, my head was turned. A clever young woman called Fiona MacDonald from one of the headhunter firms said: 'Come on Brian, stop messing around, it's about time you ran your own university'. I couldn't resist, and in 2001 I applied for, was offered and accepted the position of Vice-Chancellor of the University of York, which I took up in the summer of the following year. I knew full well that it would mean giving up my research career in any active way, and so it did, albeit not until after a few more years continuing to supervise my research group at Oxford from afar, as the projects ran down, the postdocs completed their work, and the PhD students got their degrees.

Following Anne's sad death in the mid-1990s, on a visit back to Sussex to do some accelerator experiments, I had re-met Gill Partridge, Robert Cahn's personal assistant in the 1970s when I was his senior postdoc. She had just divorced her husband. We fell in love and a few years later she moved in with me in Oxford (after twenty years living together

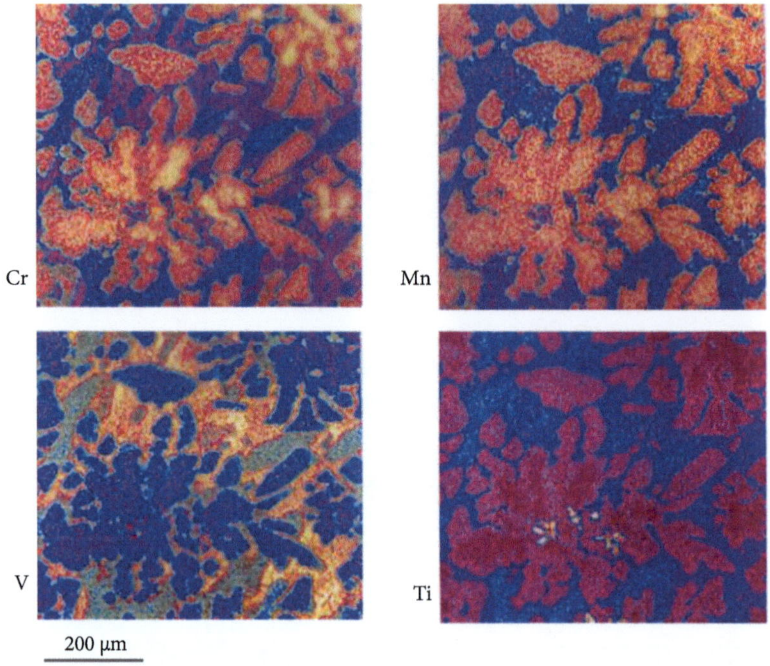

Figure 2.8 Scanning electron micrographs showing EDS X-ray maps of the as-cast segregated dendritic microstructure in the modified Cantor alloy TiVCrMnFeCoNi.
Reproduced from Cantor B, et al. (2004). Microstructural development in equiatomic multicomponent alloys. *Materials Science and Engineering A375–7*: 213–18.

we finally married a few years ago and we continue to live together in wonderful harmony). In 2002, we moved into the Vice-Chancellor's quirky 1960s grace-and-favour house on the campus in York, and I took up my new job. I loved being Vice-Chancellor of a good university. The average time in office for a Vice-Chancellor in Britain is said to be four or five years. I was Vice-Chancellor for eleven years at York, and then went on to do a second stint as Vice-Chancellor of the University of Bradford for a further six years. It is very challenging setting the strategic vision, taking the major decisions and handling the most difficult problems for a complex organisation consisting of ten or twenty thousand students and one or two thousand employees, with a turnover of several hundred million pounds per year, and with important local, national and international roles and partners. But it is also very exciting, dealing with and learning from several hundred world-leading academics, deep-thinking intellectuals in every imaginable field, from philosophy to topology, music to sociology, medieval literature to nursing. I loved it. It is great to realise that you are contributing to improving the life chances of literally millions of people worldwide: through the graduation of the students themselves; the knock-on effects of graduate activities as they build their careers in influential positions; the impact of world-leading research and innovation programmes; and the direct input to local, national and international social, political and business developments. At York we doubled the size of the university, with a near-one billion pound development programme, setting up new departments in management, law, media and arts, and new research institutes in nanotechnology, human rights, renaissance studies, education, infectious diseases and neuroimaging; at Bradford we reinvigorated what was set up as Britain's first technology university, delivering thousands more health professionals, executive managers and real-world engineers to improve the lives of people across the world. I was fully occupied and not really thinking much about multicomponent alloys or materials research. But towards the end of the 2000s, I began to get phone calls and emails out of the blue asking me about the multicomponent high-entropy alloys that Jien-Wei Yeh and I had invented. The field was finally taking off.

Equiatomic substitution

Shortly before being offered the position of Vice-Chancellor of the University of York, I had taken on a young Korean doctoral research student at Oxford called Ki Buem Kim, who agreed, somewhat grudgingly, to study multicomponent amorphous alloys designed using a novel equiatomic-substitution technique. I had been worried about the somewhat *ad hoc* method that Alain Vincent, Peter Knight, Isaac Chang and I had used so far for exploring the vast depths of multicomponent phase space. It was very trial-and-error: select a multicomponent composition and see if it showed up an interesting phase within the material. It was a bit like parachuting into the middle of an unknown continent and just hoping to land near an important feature such as a beautiful lake or a fertile valley. There is nothing wrong with pure blue-skies exploration of unknown scientific territory—it has been instrumental in discovering many important phenomena—but I felt that there should be more organised ways of going about it. After some thought, I came up with the idea of equiatomic substitution. Select a starting material of interest that has two or three different components, each playing a different role, and then replace each component with an equiatomic mixture of several components with broadly similar chemistry. I reasoned that this was the equivalent of finding an estuary at the edge of the unexplored continent and following it into the interior to find the lakes and valleys. Ki Buem was determined to work on bulk metallic glasses, which in the early 2000s had emerged as a hot topic.[iii] Like most other potential research students and postdocs I had talked to over the previous ten or twenty years, he was not really interested in multicomponent alloys, but I persuaded him that we could apply the equiatomic substitution technique to known bulk metallic glasses in order to discover new multicomponent bulk-glass compositions.

Following their discovery by Pol Duwez and co-workers[10] in 1960, the first amorphous alloys were so-called metal-metalloid glasses, mostly based on binary or ternary mixtures of transition metals and metalloids (from main groups III, IV and V in the periodic table), with the metalloid content typically in the range 15–25 at%. Typical examples[7,8] include $Au_{75}Si_{25}$, $Pd_{80}Si_{20}$, $Fe_{80}B_{20}$ and $Ni_{80}P_{20}$. At first, it was thought that a metalloid content in the

[iii] The terms glass, amorphous and non-crystalline will be treated as synonymous and used interchangeably to refer to any solid without a regular periodic crystalline atomic or molecular structure. Some authors use the term non-crystalline to refer generically to any solid without a crystal structure, while reserving the term glass to refer more specifically to a non-crystalline solid manufactured by cooling a liquid fast enough to avoid crystallisation, and the term amorphous to a non-crystalline solid *not* manufactured in that way. We will not adopt that convention.

range 15–25 at% was essential to form a metallic glass, with significant short-range order, the metalloid atoms fitting interstitially between the metal atoms, preventing easy diffusion and crystallisation, and thus stabilising a non-crystalline structure. Quite soon, however, it was shown that metal-metalloid glasses are not the only kind, with the discovery[11,12] of so-called metal–metal glasses, consisting of binary or ternary mixtures of early and late transition metals, such as NiNb and CuZr, with more equal atomic compositions and with limited or no short-range order. At first, all metal–metalloid and metal–metal glasses required very rapid quenching in order to avoid crystallisation and allow the formation of an amorphous structure, using special techniques to spread the liquid into very thin sections so that heat could be removed very rapidly. Later it was discovered that with a judicious choice of ternary alloying additions, bulk metallic glasses[13] could be made using much slower cooling conditions and in much thicker sections, typically up to several mm or even cm. Typical bulk metallic glass compositions[13] included $Pd_{40}Ni_{40}P_{20}$ and $Zr_{50}Cu_{40}Al_{10}$.

Ki Buem and I agreed to concentrate on metal–metal bulk metallic glasses, using the well-known alloy system ZrCu as a starting point. Other binary early–late transition metal alloys such as TiCu, TiNi and NbNi had also been manufactured as metallic glasses in thin sections, and additions of Al up to 10–20 at% had been shown to enhance glass formation, allowing the manufacture of glassy ribbons and rods of somewhat larger section.[13] We agreed, therefore, that for his doctoral work Ki Buem would study a variety of multicomponent alloys composed of mixtures of equiatomic combinations of early transition metals, equiatomic combinations of late transition metals, and aluminium, i.e. alloys with compositions in the range $(TiZrNbHf)_{100\text{-}x\text{-}y}(FeCoNiCuAg)_xAl_y$. Figures 2.9, 2.10 and 2.11 show some of Ki Buem's results. Figure 2.9 shows XRD traces from melt-spun $(Ti_{33}Zr_{33}Hf_{33})_{100\text{-}x\text{-}y}(Ni_{50}Cu_{50})_xAl_y$ alloys, where $x = 10 - 60$ at% and $y = 0 - 30$ at%. Most of the alloys are amorphous, and the extent of the amorphous phase field is shown on the corresponding pseudo-ternary phase diagram in Figure 2.10. The resulting crystalline microstructures after heat treating the same alloys for one hour at 600°C are shown on a similar pseudo-ternary phase diagram in Figure 2.11. Ki Buem completed his doctorate in 2004,[14] and we published a string of papers on equiatomic-substituted multicomponent amorphous alloys and their crystallisation behaviour and properties.[6–9,15,16]

Alain Vincent, Peter Knight and Isaac Chang's work had shown that multicomponent phase space contains enormous fcc and bcc single-phase fields based around multicomponent mixtures of middle/late and early transition metals respectively. Similarly, Ki Buem Kim's work had now shown that multicomponent phase space also contains an enormous region of amorphous alloy formation based on mixtures of early and late transition metals, and that these alloys crystallise on heating to give large multicomponent single-phase fields with intermetallic compound structures $MgZn_2$, $Ni_{10}Zr_7$ and Zr_2Cu. Effectively, equiatomic substitution was a way of manufacturing multicomponent compounds, which had large single-phase fields in multicomponent phase space, similar to the large single-phase fields of close-packed disordered solid solutions such as fcc and bcc.

I summarised our early work on multicomponent materials in two reviews:[17,18] one in the French journal *Annales de Chimie—Science des Matériaux*[17] in 2007 and another somewhat later in the specialist journal *Entropy*[18] in 2014, covering the multicomponent single-phase close-packed solid solutions discovered by Alain, Peter and Isaac; the multicomponent

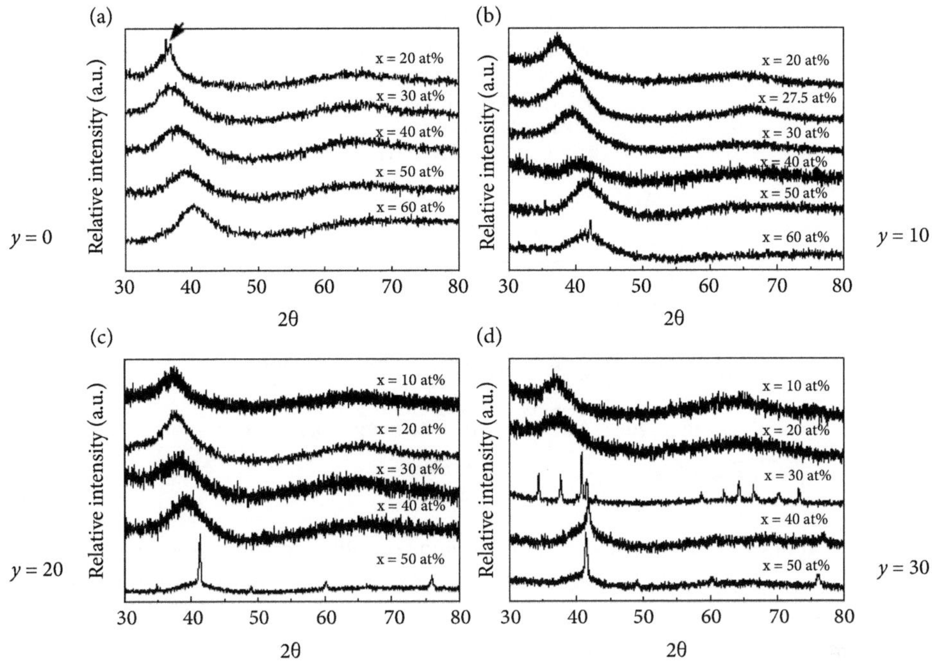

Figure 2.9 XRD traces from melt-spun $(Ti_{33}Zr_{33}Hf_{33})_{90\text{-}x\text{-}y}(Ni_{50}Cu_{50})_xAl_y$ alloys, where x = 10–60 at% and y = 0–30 at%.

Reproduced from Kim KB, et al. (2003). Formation of metallic glasses in novel $(Ti_{33}Zr_{33}Hf_{33})$((100-x-y) $(Ni_{50}Cu_{50})$(x)Al(y) alloys. *Materials Transactions* 44: 411–13; Kim KB. *Formation of Amorphous and Nanocrystalline Phases in Novel Multicomponent Alloys*, DPhil thesis, Oxford, 2004.

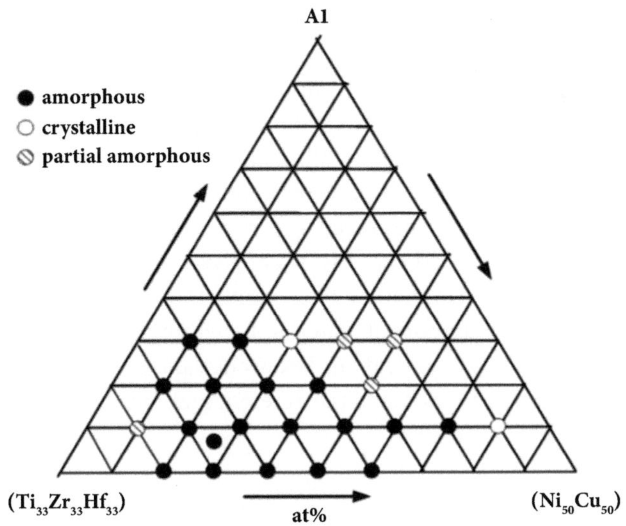

Figure 2.10 Pseudo-ternary phase diagram showing amorphous, partially amorphous, and crystalline regions in melt-spun $(Ti_{33}Zr_{33}Hf_{33})_{90\text{-}x\text{-}y}(Ni_{50}Cu_{50})_xAl_y$ alloys, where x = 10–60 at% and y = 0–30 at%.

Reproduced from Kim KB, et al. (2003). Formation of metallic glasses in novel $(Ti_{33}Zr_{33}Hf_{33})$((100-x-y) $(Ni_{50}Cu_{50})$(x)Al(y) alloys. *Materials Transactions* 44: 411–13; Kim KB. *Formation of Amorphous and Nanocrystalline Phases in Novel Multicomponent Alloys*, DPhil thesis, Oxford, 2004.

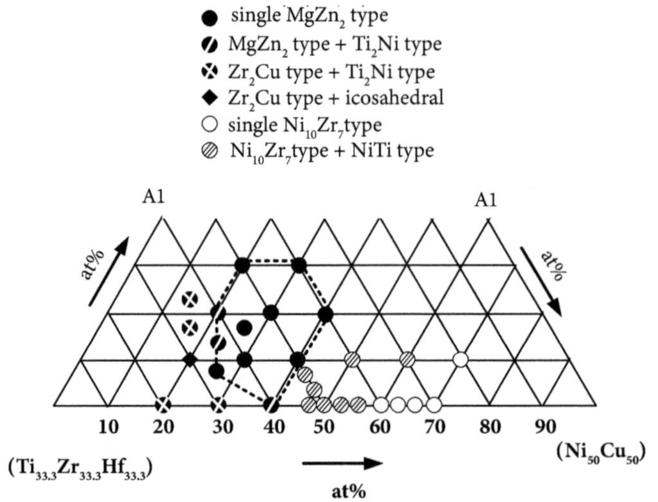

Figure 2.11 Pseudo-ternary phase diagram showing crystalline regions in melt-spun $(Ti_{33}Zr_{33}Hf_{33})_{90-x-y}(Ni_{50}Cu_{50})_xAl_y$ alloys, where x = 10–60 at% and y = 0–30 at% after heat treatment for one hour at 600 °C.

Reproduced from Kim KB, et al. (2003). Formation of metallic glasses in novel $(Ti_{33}Zr_{33}Hf_{33})((100-x-y)$ $(Ni_{50}Cu_{50})(x)Al(y)$ alloys. *Materials Transactions* 44: 411–13; Kim KB. *Formation of Amorphous and Nanocrystalline Phases in Novel Multicomponent Alloys*, DPhil thesis, Oxford, 2004.

metallic glasses and multicomponent single-phase crystallised intermetallic compounds discovered by Ki Buem; and two other issues, namely, some multicomponent dilute alloys being studied by other members of my research group, and my first thoughts on the overall size and shape of multicomponent phase space. I was firmly of the view that, as well as ignoring equiatomic and near-equiatomic multicomponent alloys in the centre of multicomponent phase space, we had also not as a scientific community really explored very effectively multicomponent dilute alloys and materials. Most of our materials use a main component with no more than one or two relatively dilute alloying additions, despite the fact that some of our most successful materials are multicomponent dilute alloys, such as the best nickel superalloys, high-strength steels and engineering ceramics. I also discussed the size of multicomponent phase space and the presence of surprisingly large multicomponent single-phase fields, making the first estimates of the vast total number of potential materials available to us, and the likelihood of multicomponent materials being single-phase or multiphase. Much more work has been done since then, and our knowledge and understanding of the size and shape of multicomponent phase space, and the different methods of exploring its vast depths are covered in detail in later chapters.

3 Discovery of High-Entropy Materials

The first idea

Jien-Wei Yeh has written a detailed description of his discovery of high-entropy materials.[19] Yeh studied the strength, toughness and fracture behaviour of precipitation-hardened aluminium alloys for his master's and doctoral degrees at the National Tsing Hua University (NTHU) in Taiwan. He stayed on as an Associate Professor when he graduated in 1986, was promoted to a Full Professorship in 1994, and has continued to teach and research as a senior academic at NTHU right up to the present day, achieving the title of Distinguished Professor in 2017. He has explained how his early research followed two main tracks: making beautiful coloured nanoglasses by chemically depositing and diffusing silver and copper ions to form nanoscale colloidal particles in silica glass; and making high-strength nanometals by rapid solidification and thermomechanical processing to produce aluminium alloys precipitation-hardened with nanoscale aluminide particles. Yeh made important contributions in both these fields. His scientific advances allowed the development of a number of industrial processes and products: high-hardness, low-flammability yellow protection windows for use in photolithography cells in the liquid-crystal-display (LCD) and integrated-circuit (IC) manufacturing industries; low-cost, high-brightness cat's eye studs for centre and edge delineation of road markings; and twin-disc solidification and reciprocating-extrusion methods for manufacturing high-toughness, corrosion-resistant 7000-series aerospace Al-Zn-Mg alloys.

Yeh believes that this early research work of his provided an essential grounding and launch pad for his subsequent discovery of high-entropy alloys. He explains that at first his broad overall approach was to try and improve material properties by exploring the extremities of alloy microstructures, using high temperatures, high quenching speeds, high diffusion loads, high strains and high deformation rates. He tells the story that, one morning in 1995, the year after he was promoted to his full professorship, he was driving in the Taiwanese countryside from Hsinchu, where NTHU is located, to the capital Taipei, when a thought struck him. Why not try using extreme compositions, rather than extreme temperatures, pressures or strains? Why not move away from manufacturing relatively conventional alloys with unusual processing methods, and instead manufacture unusual alloy compositions with relatively conventional processing methods? This was, of course, a very similar line of thinking to my own thoughts in the late 1970s, of which he was completely unaware since (as described in a previous section) I had not yet been able to publish our early work. Like me, but quite independently, Yeh latched on to the idea of using multiple components in substantial proportions, completely unlike conventional alloys. But he went an important step further, realising immediately that in such alloys there might be a higher-than-normal configurational entropy, which might well lead to interesting and unusual alloy phases, microstructures and properties. What a beautiful idea for a drive through the country. Yeh was theorising on a sunny day, and effectively predicting the results that we had in fact discovered experimentally over a decade earlier. The original Cantor alloy, as described in later chapters, is a near-ideal random fcc solid solution of five equiatomic components, an almost perfect example of high-entropy stabilisation of an extended solid solution as conceived by Jien-Wei Yeh while driving that morning in his car. Yet another wonderful serendipitous (entropic) key event in the story of the discovery of multicomponent high-entropy materials. Jien-Wei Yeh gave his new concept the name *high-entropy alloys*, which has proved to be very effective at capturing scientists' imaginations and has caught on as a very powerful descriptor of the field. As my first Deputy Vice-Chancellor once said to me at York: 'Never underestimate the power of words' (and she was a very distinguished Professor of Medieval English Literature, so she should know).

The first experiments

Yeh experienced similar negative reactions to his first thoughts about multicomponent high-entropy materials as I had received in England. He has described how many of his colleagues were unimpressed by his ideas, how he had to overcome their lack of support, and how some even said he would bring them all into international ridicule. But he had two great advantages: first, he was in a stable and permanent tenured academic professorship position at NTHU; and second, his principal research funder, the Taiwan National Science Council (now the Ministry of Science and Technology) wisely kept faith in him and continued to fund his multicomponent high-entropy materials research programme. In the end, he was aided by a number of colleagues, and collectively they published a total of five papers on this work in 2004, fully nine years after his original idea on the road to Taipei in 1995. This was a long time (though not quite as long as the twenty-five years it took me to publish our early work). Jien-Wei Yeh discovered, as I have, that the gestation period required to bring a breakthrough idea to the point of publication is long and arduous, but getting there is highly gratifying and well worth the effort.

In the lab in 1995, Yeh hypothesised that he would need at least five components in equal or at least substantial proportions to get well away from conventional alloys, and he asked his first-year graduate student Kuo-Hsiung Huang to make up over forty different compositions using this principle, divided into three broad categories: copper-containing, aluminium-containing, and molybdenum-containing. The first paper by Yeh and his colleagues was published[20] in February 2004 in the (then) relatively new research journal *Advanced Engineering Materials* on the microstructure, hardness and wear resistance of two multicomponent alloys, $AlSiTiCrFeCoNiMo_{0.5}$ and $AlSiTiCrFeNiMo_{0.5}$, designated 8E and 7E respectively (i.e. the two alloys were similar, but 7E contained no cobalt). Yeh's co-authors were: Ping-Kang Huang, another of his research students; a researcher called Tao-Tsung Shun (now a professor at Feng Chia University) working at the Industrial Technology Research Institute (ITRI, which is famous for playing an important role in the industrial development of Taiwan, and is located in Zhudong just adjacent to Hsinchu where Jien-Wei worked); and Professor Swe-Kai Chen, a long-time trusted colleague who was Huang's co-supervisor at NTHU. The materials were manufactured by arc melting into small bulk ingots, and then as thin coatings by granulating the ingots followed by atmospheric plasma spraying. XRD traces and SEM micrographs showed that the as-cast alloys consisted of ordered bcc dendrites rich in Cr, Fe, Co and Mo, and a phase-separated interdendritic fcc phase rich in Al and Ni. The more rapidly solidified plasma-sprayed coatings consisted of just a single-phase disordered bcc solid solution. Figure 2.12 shows XRD traces from the single-phase plasma-sprayed coatings, and Figure 2.13 shows SEM micrographs and corresponding EDS point compositions from the as-cast dendritic structures. Yeh and colleagues comment, as we did in our papers, that their multicomponent alloys (or, as they called them in this, their first paper, *multiple-principal-element alloys*[iv]) consist surprisingly often of no more than one or two extended solid-solution phases. Both as-cast alloys had hardnesses of over 900; the plasma sprayed-alloys were (not surprisingly, being single phase) a bit softer, although still very hard for a single bcc phase; and their hardnesses

[iv] The definitions of terms such as *multicomponent materials*, *high-entropy materials* and *multiple-principal-element materials*, and the differences between them, are dealt with in detail in Chapter 3.

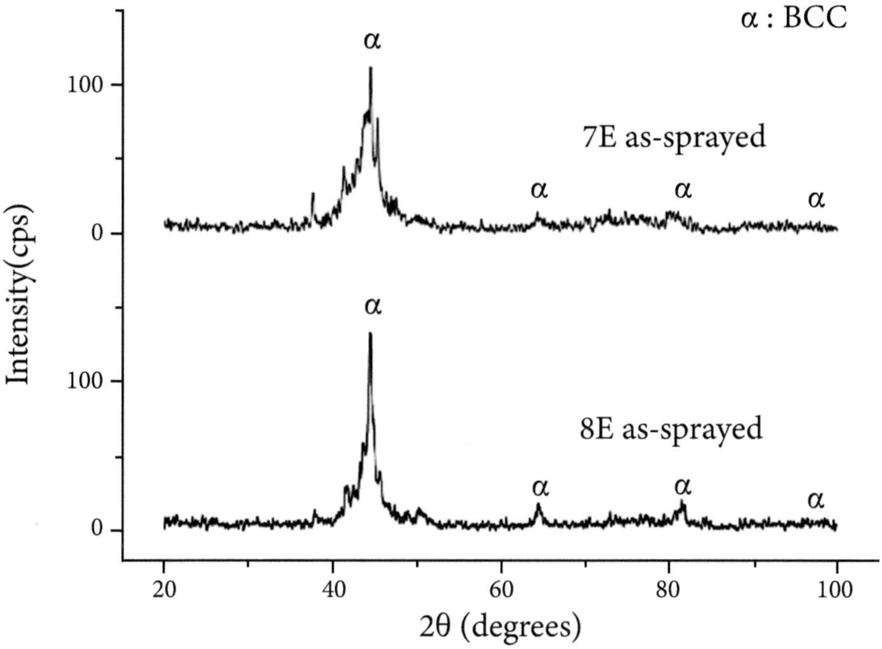

Figure 2.12 XRD traces from plasma-sprayed multicomponent alloys: (a) $AlSiTiCrFeCNiMo_{0.5}$ (7E); (b) $AlSiTiCrFeCoNiMo_{0.5}$ (8E).

Reproduced from Huang PK, et al. (2004). Multi-principal-element alloys with improved oxidation and wear resistance for thermal spray coating. *Advanced Engineering Materials* 6: 74–8.

Table 2.3 Composition, structure and Vickers microhardness of multicomponent plasma-sprayed coatings (after Huang et al.[20]).

alloy	composition	structure	initial hardness	peak-aged hardness
as-cast 8E	$AlSiTiCrFeCoNiMo_{0.5}$	bcc + 2 x fcc	933	
as-cast 7E	$AlSiTiCrFeNiMo_{0.5}$	bcc + 2 x fcc	944	
plasma-sprayed 8E	$AlSiTiCrFeCoNiMo_{0.5}$	bcc	~450	980
plasma-sprayed 7E	$AlSiTiCrFeNiMo_{0.5}$	bcc	~450	1050

increased substantially after heat treatment to induce nanoscale precipitation hardening. The plasma-sprayed coatings showed excellent wear resistance, comparable if not superior to conventional plasma-sprayed, wear-resistant bearing-steel and tool-steel coatings. These results are collected together in Table 2.3.

High-entropy alloys

However, the most important of Jien-Wei Yeh's papers in 2004—his seminal paper, the one most cited, the one in which he invented the name high-entropy alloys—is not his first paper in 2004, but the one he refers to as his '*concept paper*', published a little bit later, in May,

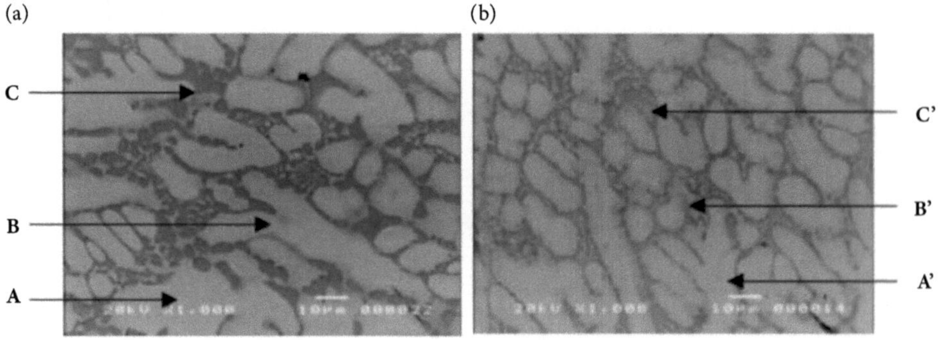

	Al	Si	Ti	Cr	Fe	Co	Ni	Mo
Nominal composition of 8E	13.3	13.3	13.3	13.3	13.3	13.3	13.3	6.7
A	5.1	20.8	17.5	13.2	8.1	13.7	8.2	13.4
B	5.2	19.7	18.2	17.3	8.9	13.2	8.1	9.4
C	39.2	2.2	6.8	6.2	4.2	14.4	26.6	0.4
Nominal composition of 7E	15.4	15.4	15.4	15.4	15.4	N/A	15.4	7.7
A'	6.0	22.0	18.3	14.8	17.1	N/A	9.8	12.1
B'	7.0	20.3	19.8	19.0	17.3	N/A	8.2	8.4
C'	47.6	1.4	3.0	4.2	6.7	N/A	37.0	0.2

In at. %.

Figure 2.13 SEM micrographs and corresponding EDS compositions from as-cast (a) $AlSiTiCrFeCoNiMo_{0.5}$ (8E); (b) $AlSiTiCrFeNiMo_{0.5}$ (7E).

Reproduced from Huang PK, et al. (2004). Multi-principal-element alloys with improved oxidation and wear resistance for thermal spray coating. *Advanced Engineering Materials* 6: 74–8.

again in the research journal *Advanced Engineering Materials*.[21] Its title is: 'Nanostructured High-Entropy Alloys with Multiple Principal Elements: Novel Alloy Design Concepts and Outcomes'. In this paper, Yeh and his colleagues calculate explicitly for the first time the configurational entropy of mixing ΔS_{mix} of a multicomponent solid solution (like for example the Cantor alloy). For any collection of atoms, ΔS_{mix} is given by the Boltzmann equation as a logarithmic function of the probability of the atomic structure as measured by the number of energetically equivalent structures w:

$$\Delta S_{mix} = k \ln w$$

where k is Boltzmann's constant. For an equiatomic ideal random solid solution ΔS_{mix} increases, therefore, logarithmically with the number of components c:

$$\Delta S_{mix} = -R \sum_i x_i \ln x_i = -R \ln (1/c) = R \ln c$$

where x_i is the mole fraction of the ith component, and ΔS_{mix} is now per mole of material so Boltzmann's constant k is replaced by the gas constant R. From this equation, the configurational entropy of mixing has a value of, for instance, $1.1R$ for a material with three

Table 2.4 As-cast and heat-treated Vickers microhardnesses for a variety of different multicomponent materials (after Yeh et al.[21]).

alloy	as-cast hardness	annealed hardness
TiVFeNiCuZr	590	600
AlTiVFeNiZr	800	790
TiVFeNiZrMo	740	760
TiVFeCoNiCuZr	630	620
AlTiVFeCoNiZr	790	800
TiVFeCoNiZrMo	790	790
TiVCrFeCoNiCu	680	680
AlTiVCrFeCoNiZr	780	890
TiVCrFeCoNiZrMo	850	850

components, 1.6R for a material with five components, 1.9R for a material with seven components, and so on. Yeh et al.[21] point out that these configurational entropies are comparable to values of the enthalpy of formation divided by the melting point of 1.4R and 2R, respectively, for even quite strongly bonded intermetallic compounds such as NiAl and TiAl, and they go on to argue that a large number of components will, therefore, often favour a random solid solution over the formation of one or more ordered intermetallic compounds. This is their explanation for the relative paucity of phases and the appearance of large, extended solid-solution, single-phase fields in multicomponent phase space. They therefore introduce the name *high-entropy materials* for their newly discovered multicomponent materials. Somewhat arbitrarily, they define a *high-entropy alloy* or *material* as one containing a least five components in proportions of at least 5 at% (more accurately as between 5–35 at%) for each component.[v]

In his concept paper, Yeh describes a number of specific examples of multicomponent materials, notably a series of alloys with compositions in the range Al_xCrFeCoNiCu where x = 0–3. The alloy structures were again found to consist of only one or two phases, and were single-phase fcc for x = 0–0.5, two-phase fcc + bcc for x = 0.5–0.8, single-phase bcc for x = 0.8–1, and two-phase bcc + ordered bcc for x = 1–3, as shown by the XRD traces in Figure 2.14. Hardness values were high, and increased with increasing aluminium content, as shown in Figure 2.15, caused by a combination of solid solution strengthening in the single-phase solid solutions and precipitation hardening in the two-phase alloys. The high hardnesses found in some of Yeh's other multicomponent high-entropy alloys are given in Table 2.4. As-cast microstructures were again dendritic. Very finely divided, nanoscale second-phase particles were formed in the two-phase alloys by precipitation and/or spinodal decomposition within the dendrites, and the high hardnesses and compressive yield strengths of the alloys were resistant to softening on heat treatment, both of which effects Yeh et al.[21] ascribe to atomic diffusion being restricted by multiple solutes in solid solution. Similarly the multicomponent high-entropy alloys were found to be resistant more generally to degradation by corrosion and/or wear, again because of atomic diffusion being restricted by the presence of multiple solutes in solid solution.

[v] The next chapter gives an alternative, preferable definition.

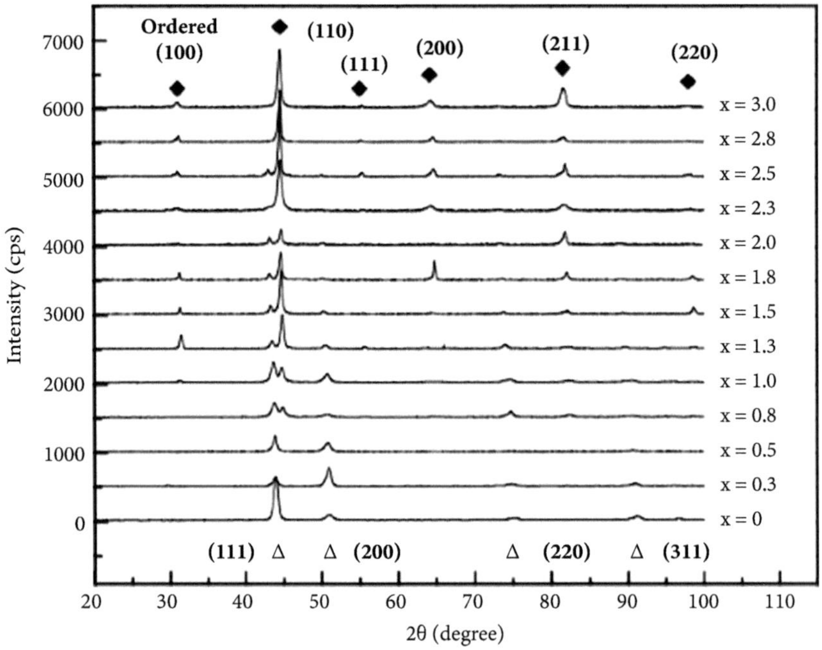

Figure 2.14 XRD traces from as-cast multicomponent alloys with compositions in the range $Al_xCrFeCoNiCu$ where $x = 0 - 3$ (open triangles are fcc reflections and filled diamonds are bcc reflections).

Reproduced from Yeh JW, et al. (2004). Nanostructured high entropy alloys with multiple principal elements: Novel alloy design concepts and outcomes. *Advanced Engineering Materials* 6: 299–303.

Overall, Yeh et al. conclude that:

> The promising properties of nanostructured HE [high-entropy] alloys make them potentially suitable for many applications, such as tools, molds, dies, mechanical parts, and furnace parts that require high strength, thermal stability, and wear and oxidation resistance. They also possess excellent corrosion resistance and can be used as anticorrosive high-strength materials in chemical plants, IC [integrated circuit] foundries, and even marine applications for piping and pump components. In addition, coating technology will further expand the application of HE alloys to functional films, such as hard-facing of golf heads and rollers, diffusion barriers for Cu connections in ultra-large-scale integrated circuits [and] soft magnetic films for ultra-high-frequency communication. Most importantly, the tolerance to impurity elements of HE alloys is quite high, allowing their synthesis from recycled/scrapped metals in a cost effective, and environmentally sound fashion.[21]

What a prescient summary of the potential applications of multicomponent high-entropy materials! At this early stage in his thinking, Jien-Wei Yeh had not only described explicitly the notion of high-entropy stabilisation of extended solid solutions but had also foreseen the wide range of possible practical reasons for studying and using all the exciting new materials that might be discovered by exploring multicomponent phase space. Yeh's concept paper[21]

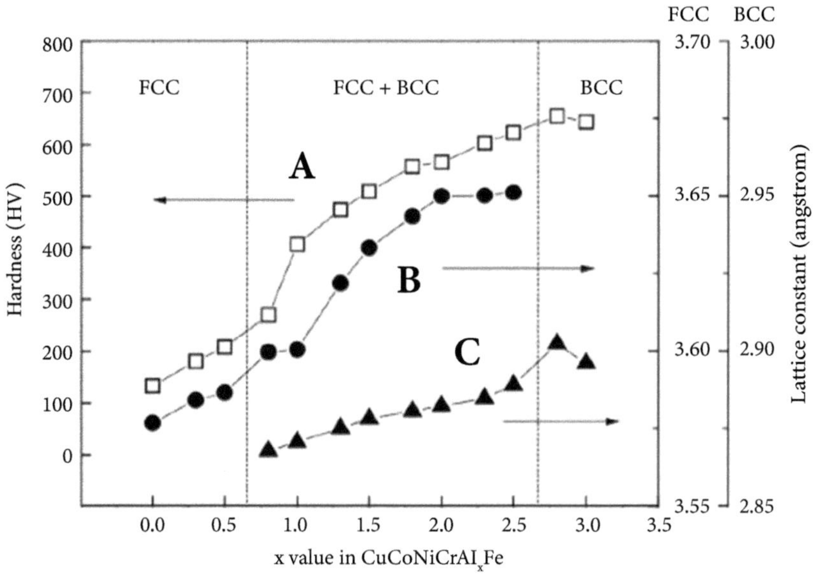

Figure 2.15 Variation of (A) Vickers microhardness and (B) fcc and (C) bcc lattice parameter with aluminium content *x*, for as-cast multicomponent alloys with compositions in the range $Al_xCrFeCoNiCu$ where $x = 0 - 3$.

Reproduced from Yeh JW, et al. (2004). Nanostructured high entropy alloys with multiple principal elements: Novel alloy design concepts and outcomes. *Advanced Engineering Materials* 6: 299–303.

is rightly regarded, alongside my own first fully detailed paper,[4] as one of the two founding papers of the field of multicomponent high-entropy alloys and materials. Like my paper, it was cited only a handful of times in the first few years following its publication but has now been cited many thousands of times.

In 2004, Jien-Wei Yeh and his colleagues published three other papers[22–24] on multicomponent high-entropy alloys, two[22,23] in the journal *Metallurgical and Materials Transactions*, and one[22] in the specialist journal *Surface and Coatings Technology*. One of the papers[21] in *Metallurgical and Materials Transactions* was a short communication, repeating and extending the theme of multicomponent high-entropy extended solid solutions from Yeh's concept paper[21] in *Advanced Engineering Materials*. Four multicomponent alloys, CrFeNiCoCu, $Al_{0.5}CrFeNiCoCu$, $AlCrNiCoCu_{0.5}$ and AlTiVCrFeCoNiCu, were each manufactured in three different ways: in the bulk by arc melting and casting; rapidly solidified by splat quenching; and as thin films by co-sputtering. The as-cast alloys were dendritic, whereas the splat-quenched and co-sputtered alloys showed more finely divided nanoscale structures, and in all cases they consisted of a mixture of one or two simple fcc and bcc phases, as shown in Figure 2.16. Alloy hardnesses were again high for such simple close-packed solid-solution structures, as shown in Table 2.5. The other two papers[23,24] in 2004 concentrated more explicitly on the mechanical properties and wear resistance of multicomponent materials: one[23] using boron additions to as-cast arc-melted $Al_{0.5}CrFeNiCoCu$ to create a multicomponent fcc matrix hardened by multicomponent boride-particle stringers; and the other[24] using reactive sputtering in nitrogen to deposit surface coatings of two multicomponent alloy compositions, $Al_{0.5}CrFeNiCoCu$

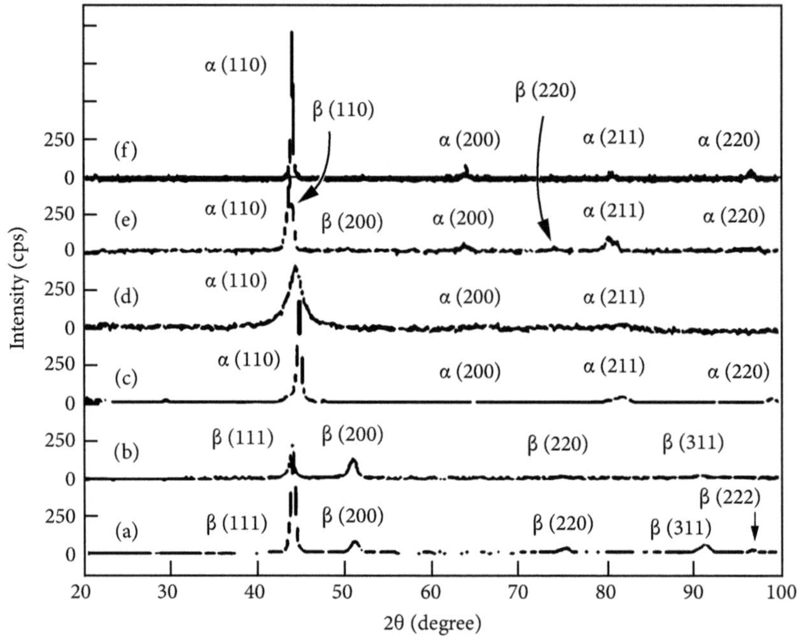

Figure 2.16 XRD traces from multicomponent alloys: (a) as-cast CrFeNiCoCu; (b) as-cast $Al_{0.5}$CrFeCoNiCu; (c) as-cast AlCrNiCo$Cu_{0.5}$; (d) as-sputtered AlCrNiCo$Cu_{0.5}$; (e) as-cast AlTiVCrFeNiCoCu; and (f) as-splat quenched AlTiVCrFeNiCoCu (α= bcc and β = fcc).
Reproduced from Yeh JW, et al. (2004). Formation of simple crystal structures in Cu-Co-Ni-Cr-Al-Fe-Ti-V alloys with multiprincipal metallic elements. *Metallurgical and Materials Transactions* 35A: 2533–6.

Table 2.5 Composition, structure and Vickers microhardness of multicomponent alloys (after Yeh et al.[22]).

alloy	structure	hardness
as-cast CrFeNiCoCu	fcc	133
as-cast $Al_{0.5}$ CrFeNiCoCu	fcc	208
as-cast AlCrNiCo$Cu_{0.5}$	bcc	496
as-sputtered AlCrNiCo$Cu_{0.5}$	bcc	816
as-cast AlTiVCrFeNiCoCu	bcc + fcc	560
as-splat quenched AlTiVCrFeNiCoCu	bcc	666

and AlCrMnFeNiCoCu, hardened by nanoscale multicomponent nitride particles. In all cases these experiments led to high-strength, high-hardness and highly wear-resistant materials.

The four core effects

Professor Yeh and his research students and colleagues continued throughout the 2000s to investigate and publish papers on the structure and properties of a variety of different multicomponent high-entropy materials.[25,26] In 2005, for instance, they published six more papers: two[27,28] in *Metallurgical and Materials Transactions* on more detailed studies of the microstructure and mechanical properties of arc-cast high-entropy $Al_xCrFeCoNiCu$ alloys with $x = 0–3$, similar to those reported previously in 2004; three in *Corrosion Science* and *Applied Physics Letters* on the detailed microstructure and electrochemical properties of another high-entropy alloy, $Al_{0.5}Si_{0.5}CrFeCoNiCu$[29–31], which is mostly glassy as-cast, with a small amount of fine-scale embedded bcc particles; and one[32] on further studies of reaction-sputtered multicomponent nitrided $Al_{0.5}CrFeCoNiCu$, again similar to one of the wear-resistant high-entropy alloys reported previously in 2004.

Just as I had, Yeh summarised his early work on multicomponent high-entropy materials in two reviews:[25,26] one in the French journal *Annales de Chimie—Science des Matériaux*[25] in 2006 and another somewhat later in the *Journal of Metals*[26] in 2013. In particular, he introduced in these reviews his famous *four core properties* of multicomponent high-entropy alloys, namely:

1. high entropy
2. lattice distortion
3. sluggish diffusion
4. cocktail effects.

Yeh argued, as explained in his concept paper in 2004 (and as outlined above), that the configurational entropy of mixing in a multicomponent material can become large enough to counteract the interaction energies between the component atoms, stabilising an extended solid solution rather than the formation of one or more intermetallic compounds. He further argued that a wide range of different atoms with different sizes creates complex local lattice distortions in such a multicomponent extended solid solution, and that this in turn restricts atomic motion by making it difficult for vacancies to move through associated local strain fields. The cocktail effect was proposed initially by Srinivasa Ranganathan,[33] an Emeritus Professor at the Indian Institute of Sciences (IISc) in Bangalore, who pointed out that multimetallic cocktails can be used to make bulk metallic glasses, multi-functional superelastic and superplastic alloys, and nanostructured high-entropy alloys. Multicomponent materials with large numbers of different alloying elements in substantial proportions can exhibit surprising, non-linear and synergistic properties, unlike more conventional materials, where a dilute solution approximation is often reasonable, leading to simple linear or near-linear addition of the separate effects of each alloying element.

Yeh's four core properties have been very influential and have framed much of the subsequent investigation and discussion of the properties of multicomponent high-entropy materials. However, the extent to which multicomponent materials do indeed exhibit high entropy, lattice distortion, sluggish diffusion and cocktail effects has been disputed in recent years.[34,35] In fact, Yeh's four core properties are a bit more complicated than might seem to be the case at first blush, and they need to be defined rather carefully before they can be fully understood. These issues are taken up and considered in more detail in later chapters.

Yeh has described[19] how he gave his first conference presentation, 'High-Entropy Alloys—A New Era of Exploitation', at the XIV International Materials Science Congress (IMRC) in

Cancun in Mexico in the summer of 2005. This was followed by further international presentations of his early work on high-entropy alloys: in India at the International Symposium on Frontiers in Design of Materials at the Indian Institute of Technology (IIT) Madras in November 2005; in China at the 5th Cross Strait Composite Conference in Shanghai in October 2006; in Korea at a symposium on Bulk Metallic Glasses and High Entropy Alloys as part of the International Union of Materials Research Societies—International Conference in Asia (IUMRS-ICA) in Busan in August 2012; and in the USA at the 2012 Materials Science and Technology Conference in Pittsburgh in October 2012. He was working hard as a very effective evangelist for the wonderful new world of multicomponent high-entropy alloys and materials. After exchanging emails for a few years, Professor Yeh and I first met when we both gave plenary talks at the symposium on High Entropy Alloys organised by Professor Peter Liaw from the University of Tennessee (UT), which was part of the Minerals, Metals and Materials Society (TMS) annual spring meeting in San Diego in February 2014.

4 Later Developments

Senkov alloys

In 2010, Oleg Senkov and several colleagues, including Dan Miracle, Senior Scientist in the Materials and Manufacturing Directorate of the US Air Force Research Laboratory (AFRL) at Wright–Patterson Base in Ohio, and Professor Peter Liaw from the University of Tennessee, published an important paper[36] entitled 'Refractory High-Entropy Alloys' in the journal *Intermetallics*. This paper describes the microstructure and mechanical properties of two multicomponent high-entropy materials composed of a mixture of refractory early-transition metals, the four-component alloy NbMoTaW and the five-component alloy VNbMoTaW. The materials were manufactured as vacuum arc-melted buttons, and were both single-phase disordered bcc, as shown in Figure 2.17, with a segregated dendritic structure, as shown in Figure 2.18. By analogy with the five-component single-phase fcc Cantor alloy CrMnFeCoNi and its large surrounding field of single-phase fcc Cantor alloys, the five-component single-phase bcc alloy VNbMoTaW is now often referred to as the *Senkov alloy* and is similarly surrounded by a large field of single-phase bcc Senkov alloys. The dendrites were found to be enriched in W, with the interdendritic regions enriched in V, Nb and Mo. The alloys were extremely hard, with Vickers microhardness values of approximately 4,500 and 5,250 for VNbMoTaW and VNbMoTaW respectively.

High-entropy ceramics

Following the two seminal papers in 2004 by Cantor et al.[4] and Yeh et al.,[21] the first decade of research into multicomponent high-entropy materials concentrated more or less exclusively on metallurgical materials: disordered close-packed solid solutions such as the fcc mixed late-transition metal Cantor alloys[4] and bcc mixed early-transition metal Senkov alloys;[36]

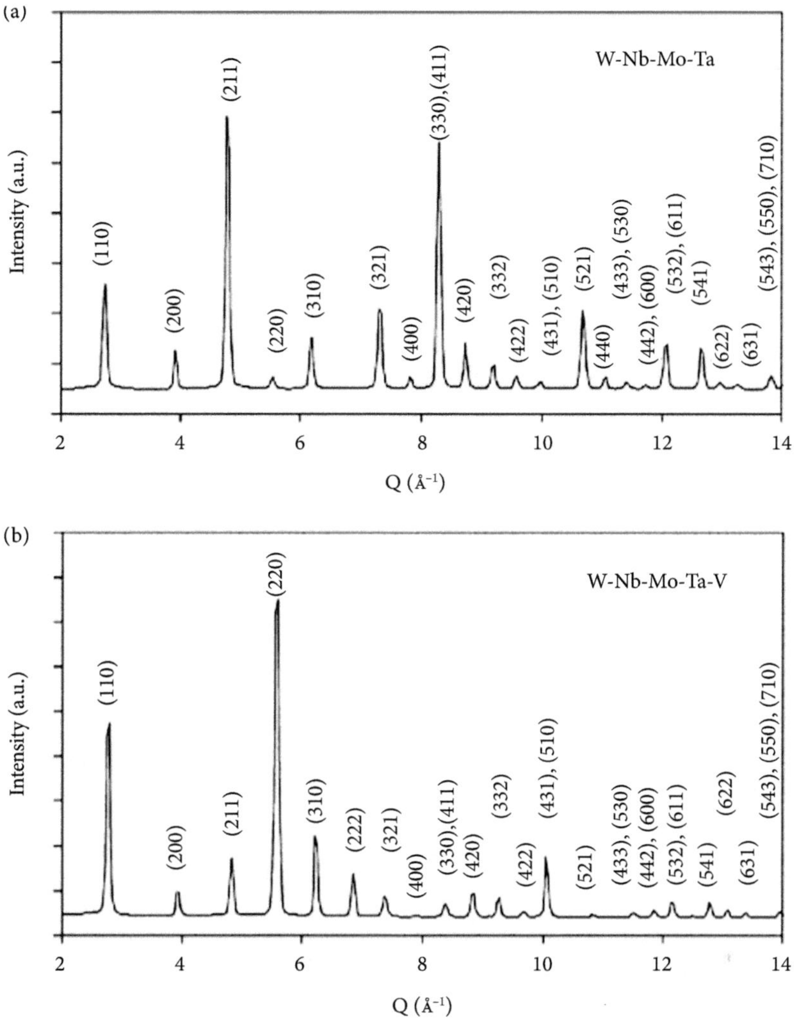

Figure 2.17 XRD traces from refractory multicomponent Senkov alloys: (a) NbMoTaW; and (b) VNbMoTaW.

Reproduced from Senkov ON, et al. (2010). Refractory high-entropy alloys. *Intermetallics* 18: 1758–65.

two-phase fcc + bcc structures in multicomponent alloys containing both early and late transition metals,[4] and when medium levels of aluminium are added to transition-metal mixtures;[21] multicomponent amorphous alloys and bulk metallic glasses;[6–8] and intermetallic compounds such as multicomponent transition metal aluminides and multicomponent Laves phases.[9,15] In 2015, however, Maria and his co-workers published another seminal paper describing how the field of multicomponent high-entropy materials could be expanded to include multicomponent inorganic ceramic compounds, thus opening up another enormous vista of exciting new materials.[37]

Jon-Paul Maria has said[38] that he went to a presentation on multicomponent high-entropy metal alloys, given by a colleague at North Carolina State University (NCSU), and came away entranced and enthused at the prospect of finding similar single-phase structures in

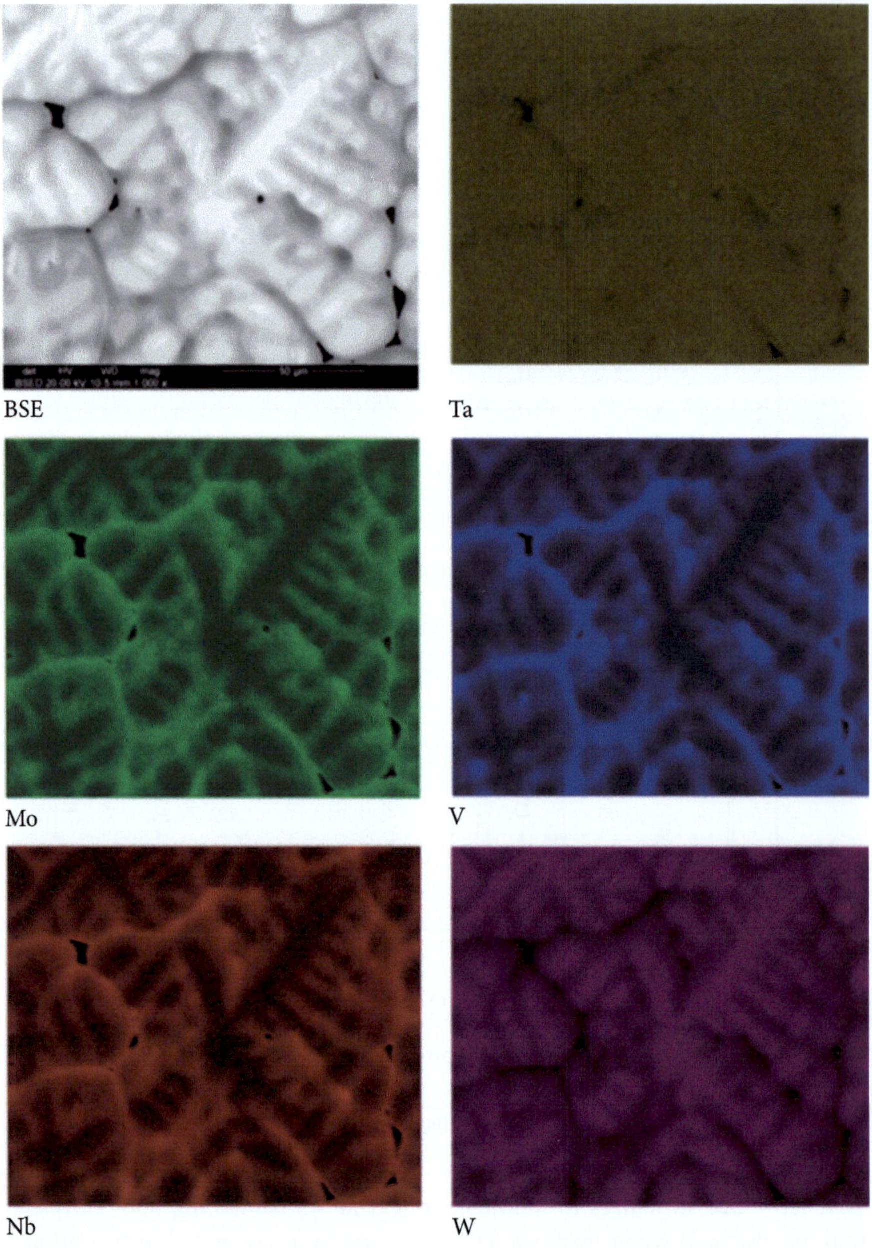

Figure 2.18 SEM micrographs showing a back-scattered image (BSE) and EDS X-ray maps of the as-cast segregated dendritic microstructure in the Senkov alloy VNbMoTaW.
Reproduced from Senkov ON, et al. (2010). Refractory high-entropy alloys. *Intermetallics* 18: 1758–65.

multicomponent inorganic ceramic compounds. On scanning the literature, he was surprised to discover that this possibility had not yet been explored, and he initially asked an undergraduate student, Trent Borman, to investigate the idea, followed up later by a PhD student, Christina Rost. In the end, Maria and colleagues manufactured a large number of different mixtures of metal oxides, explicitly searching for a multicomponent single-phase structure stabilised by a high configurational entropy of mixing in much the same way as had by then been seen many times, as mentioned above, in multicomponent metallurgical materials such as the Cantor and Senkov alloys. Without knowing, they were essentially using the equiatomic substitution technique of Kim et al.[6-9,15,16] applied to metal oxides. In the resulting paper, published in 2015 in *Nature Communications*, Rost et al.[37] showed clearly that mixtures of five different divalent metal mono-oxides—magnesium oxide MgO, cobalt oxide CoO, nickel oxide NiO, cupric oxide CuO, and zinc oxide ZnO—can be sintered to form a multicomponent mono-oxide, with a composition (MgCoNiCuZn)O, with a single-phase rock-salt NaCl crystal structure, and with the different metal cations distributed randomly across the cation sites.

Figure 2.19 shows some of Rost et al.'s[37] X-ray diffraction (XRD) results for the variation of (MgCoNiCuZn)O structure as a function of sintering temperature. The multicomponent mono-oxide had a single-phase rock salt structure when it was manufactured by sintering at temperatures above 850°C (1,123 K), and a multiphase structure when sintered at lower temperatures, suggesting that the single-phase rock-salt structure was indeed stabilised by a high configurational entropy in a similar way to the metallic fcc Cantor and bcc Senkov alloys. This interpretation was reinforced, and effectively confirmed, by three additional experimental observations:

1. The change in structure was reversible when the material was cycled above and below the transition temperature.
2. The minimum transition temperature and, therefore, the maximum range of stability for the single-phase rock-salt structure, corresponded to materials manufactured with an equiatomic mixed oxide composition (equal Mg, Co, Ni, Cu and Zn content), i.e. with the highest configurational entropy of mixing; with higher transition temperatures and, therefore, a lower range of stability, for non-equiatomic compositions (unequal Mg, Co, Ni, Cu and Zn content), i.e. with a lower configurational entropy of mixing.
3. Energy-dispersive X-ray (EDS) composition maps in a transmission electron microscope (TEM) indicated a near-ideal random distribution of the different metal ions on the cation sub-lattice sites of the rock-salt crystal structure.

5 Short Biographies

Alain Vincent

Alain Vincent was born in 1960 in London. His father Louis Vincent was of French/Italian extraction and worked as a restaurant manager, and his mother Huguette Jeanne-Louise Vincent (née Mira) was of Spanish/French extraction and worked as a housewife and a teacher at a junior school. Shortly after Alain was born, the family moved to the seaside resort of Brighton, sixty miles south of London on the south coast of England, where he was brought up and where he still lives and works. He is a died-in-the-wool Brightonian. Brighton was one of the first ever seaside resorts in England, popularised initially in the late-eighteenth century when the Prince Regent (later King George IV) built his holiday residence there,

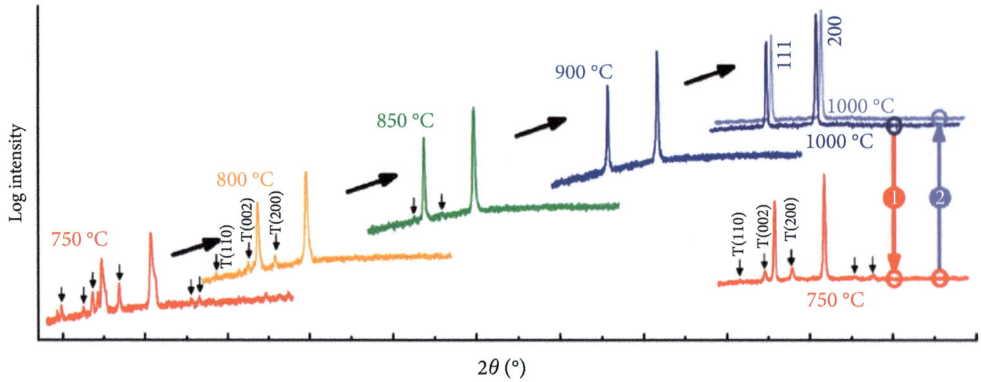

Figure 2.19 XRD traces from multicomponent oxide (MgCoNiCuZn)O sintered at different temperatures from 750 °C to 1,000 °C, showing a single-phase rock-salt structure for T > 850 °C (peaks labelled 111 and 200 at 1,000 °C) and a multiphase structure for T ≤ 850 °C.
Reproduced from Rost CM, et al. (2015). Entropy-stabilized oxides. *Nature Communications* 6, Article 8485.

the architecturally striking (and somewhat kitsch) Royal Pavilion, after which the city was rapidly developed for day-trippers from London when the London–Brighton railway link was built in the mid-nineteenth century. It is an eclectic and attractive mix of fine Georgian squares, decaying Victorian splendour, amusement arcades, fish-and-chip shops and racy seaside tat. It has a reputation on the one hand as being one of the hippest, happiest, free-wheelingest places in England, and on the other hand as being London gangsterland-by-sea, as famously described in the novel *Brighton Rock* by Graham Greene.

Alain went to local primary and secondary schools, majoring in the sciences, and then went on to study materials science and engineering at the (also) local University of Sussex, the first of Britain's post-war new (so-called *plate-glass*) universities, built with typical 1960's brutalist panache a few miles inland in a glorious parkland setting in the beautiful South Downs. Alain's youthful interests were space exploration and the Apollo missions, science fiction, football, film, animation and art; and his more mature (and somewhat more sedate) interests are now home improvements, music, playing the guitar, watching films and gardening. He says that he was attracted to materials science and engineering because 'it dealt with the practicality and application of materials, i.e. what things are made from and why'.[39] After graduating from the University of Sussex, Alain moved to the University of Oxford to study for a doctorate in my research group. He then decided to pursue a business career, training as an accountant, and working as a finance manager and then systems and treasury manager, initially for a publishing company and then for subsidiaries of Boeing and Airbus that provide aeronautical data to the aviation industry. His current employer is NAVBLUE, a subsidiary of Airbus Commercial Services. He recently entered into a formal civil partnership with his long-time partner of more than thirty years, Wanda Wachowski. They met when she was a PA to the partners of the accounting firm where Alain did most of his accountancy training. Alain and Wanda have a daughter and son, Alina and Jean-Luc, both in their twenties.

Alain took on my project on multicomponent alloys at Sussex because 'this option intrigued me as it appeared nothing had been written on the subject before . . . so whatever we found was likely to be new'.[39] He was (and still is) a true explorer and entrepreneur with an adventurous scientific spirit. When I contacted him recently to get biographical information for this book, I found that he had been blissfully unaware for the past forty years of the growth of the multicomponent high-entropy materials field. He commented, 'Who'd have thunk it, that all those years ago [in 1980 and 1981] tinkering around in a lab . . . would be the first known record of work on this subject'.[39] After hearing about the massive worldwide interest in the field, he says he spent a sleepless night surfing the web for information on multicomponent high-entropy alloys. He summed it up in one of his recent emails to me, saying that learning about these developments has brought[39] 'an enormous smile to my face that needs to be seen to be believed. What a thrill . . . it's just a massive hoot to be recognised for being the first to do any work on this topic'.

Isaac Chang

Isaac Chang was born in Hong Kong in 1965. His father Hoo Chang was a chef and restaurant owner and his mother Kwan Chang is a housewife. They emigrated from Hong Kong to London in 1975, when Isaac was just ten years old. Isaac went to junior school in Hong Kong and briefly in London, before attending senior school in Dagenham, a suburb on the Thames Estuary in East London, where he specialised in maths, physics and chemistry. He went on to study for his bachelor's degree in metallurgy and materials at Imperial College in fashionable Knightsbridge just south of Hyde Park in central London, with a final-year project on single-crystal growth of pyroelectric lead tin titanate, followed by a doctorate at the University of Oxford in my research group on rapidly solidified germanium–tin infra-red detector materials. He went into materials science (he says) because it offered 'opportunities to [continue to] combine maths, physics and chemistry, to understand materials and their applications, as well as good employability'.[40]

After completing his doctorate at Oxford, Isaac stayed on as a postdoc in my research group, on a variety of projects on advanced materials manufactured by novel processing methods. When I asked him to repeat Alain Vincent's and Peter Knight's undergraduate project work, he was excited to do so because 'The research topic was very novel . . . [and] I continued to perform detailed studies of equiatomic multicomponent alloys [with] up to 10 principal elements'.[40] In 1995 Isaac left Oxford to become a Lecturer in the School of Metallurgy and Materials at the University of Birmingham, where he was later promoted to Senior Lecturer and then Reader and Head of Education, with responsibility for strategy, development, delivery and accreditation of all metallurgy and materials courses at Birmingham. His research at Birmingham covered many topics in advanced materials and manufacturing: silicide-based intermetallic compounds, structural ceramics, amorphous alloys, metallic nanopowders, aluminium powder metallurgy, high-entropy alloys, laser sintering, direct laser fabrication, freeform fabrication, mechanical alloying and hot isostatic pressing. In 2016, he moved to the Brunel Centre for Advanced Solidification Technology (BCAST) at Brunel University London, to take up a Chair as Professor of Metallurgy and Materials and to become Head of the Liquid Metals Engineering (LiME) Training Centre (LTC). His current research interests continue to be wide and varied: high-strength die casting, multicomponent aluminium alloys, high-entropy alloys, processing of recycled aluminium alloys, machine learning for alloy development, crack healing in defective components, and joining of bimetallic components by electromagnetic pulse technology. He has become a very successful, eclectic and wide-ranging materials-science teacher and researcher. He has been recognised by being appointed as a Fellow of the Institute of Materials, Minerals and Mining (IoM3) and a member of the editorial board for the *Journal of Materials, Chemistry and Physics*.

Isaac has been married for more than twenty-five years to Kwang Leong Choy, who was born in Malaysia, who like Isaac immigrated to England, and who also like Isaac is a Professor of Materials, in her case at University College, London (UCL). She is the Director of the UCL Institute for Materials Discovery and is an expert in advanced ceramics and novel thin and thick film processing of functional materials. They have two children, a daughter and son, Emer and Albert, both in their early twenties and both studying medicine, at Oxford and Imperial College, respectively. Isaac's youthful interests were science fiction, football, table tennis, film, chess and building model planes; his interests now are (again) more sedate: listening to music, watching movies, eating out, cooking, gardening, walking and spending time with his family.

Brian Cantor

I was born in 1948 and brought up in the grimy, immigrant working-class suburbs of inner-city north Manchester, home of the industrial revolution in the eighteenth and nineteenth centuries (one of several cities claiming this honour in the north of England). My father, Oliver Cantor, was a small businessman, also born and brought up in north Manchester, from Ukrainian immigrants who had fled to England from the Russian Tsar's pogroms in the late nineteenth century. My mother, Trudy Cantor, ran a secretarial agency and then a café in the middle of Manchester. She was born and brought up in nearby Liverpool, from Irish immigrants, who had similarly fled to England but in their case from the terrible famines in Ireland in the late nineteenth century. My family was very left-wing, communist, atheist and free-thinking. I was successful at school, passing my final-year exams and graduating two years early from senior school in 1963, aged just fifteen and too young to go to university, so instead I worked as a lab assistant for Exide Batteries and then as a warehouseman for Kellogg's Corn Flakes. I spent all my spare time playing guitar in groups and bands and hiking and climbing in the hills and mountains of northern England and Wales. Two years later, I applied to Christ's College at Cambridge University, somewhat reluctantly because I didn't really want to study for another exam, and was lucky to get in, since the College had just instituted a scheme to broaden its intake by admitting some students directly from their final year school exam results (i.e. without taking the then usually required entrance exam). This is one of several random, entropic events that have impacted decisively on my life and career. As I have said since, 'I wanted to be a rock guitarist or a rock climber, but somehow I became a scientist and an academic'.[35]

Still a teenager at Cambridge, I married my first wife Maggie, and we had our two children, Matthew and Oliver. We were very poor, so while studying I took many other part-time jobs: musician, chef, car mechanic, second-hand car salesman, porter, encyclopaedia salesman, men's outfitter, ice-cream seller, waste-paper sorter—you name it. After completing my bachelor's degree and doctorate at Cambridge, I took a postdoc position at Sussex University, running Robert Cahn's rapid solidification and metastable materials research group while he was Dean of the School of Engineering, and I stayed there throughout the 1970s.

After brief periods working at BHU in Northern India and the GE Research Labs in Schenectady, New York, I moved back to England in 1982 to take up a Lectureship at the University of Oxford, where I remained for over twenty years, being gradually promoted to a Readership, the Cookson Professorship, Head of the Department of Materials and, finally Head of the Division of Mathematical and Physical Sciences (effectively head of science and engineering). In 2002, I moved back to the North of England to become Vice-Chancellor (President) of the University of York and then Vice-Chancellor (President) of the University of Bradford, from where I stepped down a few years ago. During this extended period of almost twenty years in university management, I had the great pleasure of helping thousands of young students graduate, founding new schools, departments and research institutes in many different academic disciplines, promoting local, national and international economic and cultural developments, and supporting research into many of the great issues in the world, both fundamental and applied.

Since leaving Bradford, I have been back working full time as a scientist, concentrating on promoting the field of multicomponent high-entropy materials as an Emeritus Professor at the University of Oxford, and developing a circular economy for the metals-producing and metals-using industries (essential if we are to solve our climate-change problems) as a Senior Advisor and Research Professor at the Brunel Centre for Advanced Solidification Technology (BCAST) at Brunel University London. The founder and Director of BCAST is Zhongyun Fan, who worked as a postdoc with me at Oxford in the 1990s. I have been living for over twenty-five years with my current partner, Gill Partridge (we finally got married a few years ago), whom I first met fifty years ago when I went to Sussex University and we were both married to our first spouses. We live now in a beautiful house tucked away in the rough-hewn but beautiful Pennine hills roughly halfway between Manchester and Leeds. Between us we have three sons (Matthew, Ollie and Simon, all in their forties and fifties), and four grandchildren (Alex, Chris, Ernie and Agatha, all in their teens and twenties: Agatha arrived somewhat unexpectedly after a sustained run of four generations of males only and is therefore widely fawned over). Matthew and Oliver are both computer scientists, IT specialists and managers, Matthew in the insurance and banking industry in Florida (while I have been writing this book, he and his family have just moved back to England after over twenty-five years in Florida) and Oliver in the telecommunications industry in London, and Simon is a web designer and runs a major online and in-person trading business based in Brasov in Romania. I still play the guitar every day, love hiking in the mountains (two of the achievements I am most proud of are climbing Mont Blanc and Mount Fuji, two of the most beautiful mountains in the world), and collecting modern art. I donated over £100k to fund two Cantor lecture series, one at the University of York and one at Bradford, rotating in topic between technology, modern art and business; Gill and I have donated much of our modern art and cartoon collection, over 100 pieces, also worth about £100k to the University of York, where it is all on display; and my two brothers and I set up and funded the Cantor Bridge Tournament at Manchester Bridge Club, in honour of my mum and dad who (like my grandfather) were both great bridge players.

I was fortunate to be the first to discover the wonderful and rich realm of multicomponent and high-entropy materials in the late 1970s and early 1980s[2–9,34,35]. I like the fact that the idea of multicomponent high-entropy materials came to me via a relatively serendipitous, random and entropic thought about how to make novel materials with new and attractive properties. I have also been fortunate throughout my career to be able to make many other contributions to the understanding of materials and manufacturing and the development of academic and other public and private organisations, all to the general enhancement of society and peoples' lives. I am proud to have been elected to be a Fellow of the Royal Academy of Engineering and the Royal Society (i.e. to have been recognised by my scientific and engineering peers), and to have received a CBE (Commander of the British Empire) from the Queen (i.e. to have been recognised by the British public and body politic).

Kuo Hsiung Huang

Kuo Hsiung Huang was born in 1972 in Kaohsiung, a major international port and industrial city in south-west Taiwan. His father, Huang Chao Peng was a small businessman, running a stall selling general goods on a night market in Kaohsiung, and his mother, Chen Ai Zhu, worked on the quality-control line for what later became Fusheng Precision, one of the largest manufacturers of golf clubs in the world (part of the Fusheng Group). Kuo Hsiung attended local elementary, junior-high and then senior schools, graduating from Kaohsiung Senior School in 1990, and going on to do his bachelor's and then master's degrees at NTHU in 1994 and 1996 respectively. He chose materials science and engineering because of its combination of fundamental physics and chemistry and applied engineering, and says that he was fortunate to do his master's research with Professor Jien-Wei Yeh on the novel project of high-entropy alloys: 'I was thankful that Professor Yeh gave me the chance to study high-entropy alloys . . . that's a great part and important experience of my lifetime'.[41] Outside his scientific career, his major interests are ball games, swimming and travelling around visiting different parts of the beautiful island of Taiwan.

After leaving university, Kuo Hsiung worked for four years for his national service in the Taiwan Ministry of Science's National Nano Device Laboratory (NDL: now the Taiwan Semiconductor Research Institute, TSRI), and has continued his career by working progressively for a variety of top level manufacturing companies in Taiwan's world-leading semiconductor industry: Silicon Integrated Systems (SIS), Vanguard International Semiconductor Corporation (VIS) and finally United Microelectronics Corporation (UMC), one of the biggest silicon chipmakers in the world, based in Hsinchu, Taiwan, and a major supplier to companies such as Qualcomm and Sony. He currently works in UMC's Reliability, Technology and Assurance Division, in charge of and responsible for the advanced memory qualification of all their semiconductor products. He met and married his wife Yu Mei Yang in 2001 while he was working at NDL, and they now have three daughters, Shiao-Jia, Shiao-En and Shiao-Qi aged nineteen, sixteen and ten years old respectively. He recognises that being one of the first ever research students to work on multicomponent high-entropy alloys helped him enormously to build his career: 'It let me realise that when you and your professor set a goal [you must] just do your best and try to find the resources you need'.[41]

Swe-Kai Chen

Swe-Kai Chen was born in 1951 in Tuku, a small bucolic town in the centre of Yunlin County, on the northern edge of the Chianan (or Chiayi-Tainan) plain on the western coast of central Taiwan. It is one of the country's least-developed regions, a backwater compared to the dynamic industrial cities further north, with an agricultural economy dominated by farming, forestry and fishing. His parents ran a traditional Chinese medicine store in Tuku, which they started in 1947 and which is still run by his brother. Swe-Kai went to local preliminary and junior high schools in Tuku, before moving to Chieng-Kuo (now Jianguo) School in Taipei, the most famous and most successful senior high school in Taiwan. After leaving school he attended the Fu-Jen Catholic University in New Taipei City to do a bachelor's degree in physics.[vi]

Swe-Kai moved to NTHU in Hsinchu to do his master's degree in materials science, working at the same time as a physics high-school teacher. He graduated in 1975 and did two years' national service in the Taiwan Republic of China (ROC) Army, before joining the China Steel Corporation (CSC) in Kaohsiung as a metallurgical superintendent, where he worked for eight years, including secondments for training at US Steel in Pittsburgh and Chicago in the USA and ISCOR (now part of ArcelorMittal) mainly in Pretoria in South Africa. In 1985, he moved to the magnetic materials laboratory at Taiwan's famous ITRI in Hsinchu, before spending three years doing his doctorate on the structure and properties of iron–manganese–aluminium alloys back in the United States under the supervision of Gerry Byrne at the University of Utah in Salt Lake City. Back in Taiwan after completing his doctorate, he became Vice-President of a magnetic materials company High-End Metals and then joined the faculty of NTHU in Hsinchu in 1991, where he stayed until 2017 when he retired. At NTHU he taught and did research on many materials topics, including palladium-deuterium cold fusion, zinc-air batteries, alumina-magnesia-graphite refractories, high-temperature cermet composites, aluminium nitride ceramics and, of course, multicomponent high-entropy alloys initially in collaboration with his faculty colleague Jien-Wei Yeh.[21–24]

In the field of multicomponent high-entropy materials, Swe-Kai has made many important contributions, developing enthalpy–entropy diagrams to map different types of multicomponent alloy structure, and exploring the thermal, electrical, magnetic, superconducting, electrochemical and hydrogen-storage properties of high-entropy alloys. He tells me[42] that in his laboratory he and his students manufactured a total of over 5,000 different multicomponent high-entropy alloys (this is a remarkably large number, and it is a tribute to their hard work and dedication to have done so much, but it is still, of course no more than a minuscule fraction of the total number of possible multicomponent alloys). Although he has been a senior professor for many years, Swe-Kai still likes to build experimental equipment and do experiments himself, as well as keep a close personal watching eye on all the experimental developments in his lab.

Overall Swe-Kai has taught over fifty master's students and taken out about fifty patents on a wide variety of topics. He likes studying maths, physics and materials, and is also a skilled traditional Chinese medicine (TCM) doctor, having been certified with a Diploma in Chinese Medicine from the US National Certification Commission for Acupuncture and Oriental Medicine (NCCAOM) since 2000. He continues to teach TCM at NTHU. Swe-Kai married his wife, Feng-Lin Suen Chen in 1977, just as he was finishing his army service and

[vi] The large northernmost region of Taiwan that surrounds the capital city Taipei as well as the port city of Keelung ten kilometres further east is the most populous administrative region in Taiwan and, somewhat confusingly, is called New Taipei City: it is not new at all, having been settled since ancient times, and it is not really a city either, but more like a large, densely populated county—indeed it used to be known as Taipei County until it was re-named New Taipei City in 2010.

starting to work for CSC, and they now have two daughters, Jasmine and Nancy, both in their forties, a younger son George, in his thirties, and a young granddaughter Alison, aged just seven years old. He loves gardening, listening to western classical music and reading classical Chinese literature. He is a true polymath.

Jien-Wei Yeh

Jien-Wei Yeh was born in 1954 in the mountain Nan'ao Township in Yilan County in eastern Taiwan, a beautiful region about 60 km south of Yilan City, where the central mountain range runs into the east coast overlooking the East China Sea. Jien-Wei's father was Bi-Duo Yeh, a chief police officer in Fujian province in mainland China until 1949, when he moved to Taiwan where Chiang Kai-Shek was setting up the Republic of China after retreating from Mao Zedong's communist revolution on the mainland. Bi-Duo became Director of Education and Student Affairs at Peng-Lai Elementary school in Nan'ao village, and married Jien-Wei's mother, Wang-Bi-Yun Yeh, who came from Hualien, also on the east coast another 80 km further south. They had seven children: Jien-Wei is their third son, and he also has four younger sisters. Wang-Bi was a housekeeper, ran small businesses and did farm work to support the family finances. Jien-Wei went to school a year early because, he says, 'my parents had the concern that I was naughty and might [otherwise] learn bad habits from other kids'.[19] He studied at Peng-Lai Elementary School and Hualien junior middle and senior schools, graduating in 1971 with high grades and admitted, therefore, initially to study physics at NTHU. After two years of physics, he transferred to become one of the first cohort of students in the (then) new materials science and engineering course at NTHU: 'with an ambition to contribute my engineering expertise to our society....Physics provided me with the basic understanding of the fundamentals ranging from atoms to the cosmos. Materials science provided me with the training to build atoms into materials with various applications'.[43] Doing both subjects gave him: 'the broad background which was a key element in launching my whole career'.[43] After two years of military service, he entered the graduate programme at NTHU, doing a master's degree with his supervisor Kuo-Shung Liu, researching into the strengthening and toughening of aluminium-zinc-magnesium Al–Mg–Zn alloys by thermomechanical treatment. On completing his master's, he was appointed as a Lecturer in the Department of Metallurgy and Materials, and a year later enrolled on the newly established doctoral programme to do a PhD, again supervised by Kuo-Shung Liu, on the toughness and fracture mechanism of precipitation-strengthened aluminium alloys, which he finished in 1986 when he was promoted to Associate Professor.

In the second year of his master's degree, Jien-Wei married his wife, Hsiu-Hui Chang, who was a teacher in the Qiao Ai Elementary School in Taoyuan County, the region on the northern coast of Taiwan halfway between the capital Taipei and NTHU in Hsinchu. He had fallen in love with her some years earlier when they met in 1973 while she was teaching in Jien-Wei's home-town Peng-Lai Elementary School. He says, 'I met her when she and her colleague came to the teachers' office, and I was lonely playing the organ after dinner. After introducing myself, I found that she was the girl I adored'.[43] They now have three grown-up daughters, Yi-Fang Yeh, Yi-Ting Yeh and Shan Yeh, all married and each with two children, so Jien-Wei and Hsiu-Hui have a total of six wonderful grandchildren. Yi-Fang works as an accountant for consumer and multimedia internet companies; Yi-Ting is a co-founder of a company using artificial intelligence (AI) methods to provide online product inspection for manufacturing companies; and Shan Yeh is the manager of a functional glass company. They and their families all live in Hsinchu city not far from Jien-Wei and Hsiu-Hui.

Jien-Wei is not just an excellent scientist and engineer—he also has many cultural interests. His main hobbies are playing football, the harmonica and the flute, singing and travelling. He is a very cultured man. He loves stained glass and has contributed to making beautiful new coating materials that have been used by many stained-glass artists in workshops and factories in Taiwan. He also writes poems and does calligraphy with a brush pen on rice paper. In 1994, eight years after becoming an Associate Professor, he was further promoted to a full professorship, which just a couple of years later gave him the freedom to begin his ground-breaking discoveries and insights into multicomponent high-entropy alloys. He has written that 'I had tried different inventions to overcome the property limits of Al alloys and Mg alloys . . . but all those methods didn't satisfy me. Then I came up with the revolution of high-entropy alloys . . . I directly went to my lab and asked my first-year graduate student [Kuo-Hsiung Huang] to change his original topic to my new idea.'[19] Jien-Wei is rightly proud of his role (alongside my own) in inventing the concept of multicomponent high-entropy materials.[21–24] I like the fact that he (like me) had the idea of multicomponent high-entropy materials as a kind of serendipitous, entropic brainwave while going about his normal business and thinking about how to make new materials. He comments accurately that, 'The invention of high-entropy alloys is in fact a revolution . . . in material history . . . the concept has given the periodic table new life and brought about limitless opportunities in both academic research and industrial applications.'[19] I often use the problems and excitement of exploring a new continent as an analogy for investigating multicomponent phase space.[35] Jien-Wei uses a fishing analogy: 'Conventionally we only fish at or near fishing ports. Now we can fish in the ocean where abundant marine lives can be found, such as large whale, dolphin and shark.'[19] In 2017, in recognition of his eminent and leading role in contributing to establishing the field of multicomponent high-entropy materials and promoting it over more than two decades, he was awarded the honour of becoming a Distinguished Professor at NTHU. Jien-Wei Yeh is known worldwide as a gentleman and a scholar. He has made friends, collaborated with, and supported almost everyone working in the multicomponent high-entropy field, who are all grateful for his help and comradeship. Jien-Wei has explained the significance of multicomponent high-entropy materials by quoting Hillary Clinton's translation of a famous Chinese poem by Lu Yu from the southern Song dynasty 900 years ago: 'After endless mountains and rivers that leave doubt whether there is a path out, suddenly one encounters the shade of a willow, bright flowers and a lovely village.'[19] Jien-Wei comments that, for him, as in the poem: 'In my way of improving nanomaterials and nanoglass I experienced many obstacles and bottlenecks, but I finally found countless material opportunities ahead.'[19]

Srinivasa Ranganathan

Srinivasa Ranganathan was born in 1941 at Cuddalore, an old seaport town almost 200 km south of Chennai on the eastern coast of Tamil Nadu in southern India. His father, Padur Srinivasa Iyengar, was a lawyer and his mother, Sathayabama Ammai was a homemaker. He went to school at the local St Joseph's Higher School in Cuddalore, before going to university to take a chemistry BSc degree in 1960 at Loyola College in Madras followed by a metallurgy BEng degree in 1962 at the Indian Institute of Science (IISc) in Bangalore, a metallurgy PhD in 1965 at the University of Cambridge in England and a two-year postdoc at Berkeley in California. He then moved back to India to become a Lecturer, then Professor, then Chairman of the Department of Metallurgy and finally Chairman of the Division of Mechanical Sciences at Banaras Hindu University (BHU). He retired from BHU in 2006 to become an Indian National Academy of Engineering (INAE) Distinguished Professor at IISc at the same time as being an Honorary Professor at the Jawaharlal Nehru Centre for Advanced Scientific Research and Visiting Professor at the National Institute of Advanced Studies, both also in Bangalore.

Professor Ranganathan (or Rangu as he is known universally by his friends and colleagues) is an extremely distinguished metallurgist and materials scientist, having accumulated many honours throughout his career. He has worked on many fields in physical metallurgy and the structure and mechanical properties of materials, notably contributing fundamental ideas to the coincident site lattice theory of grain boundaries and the structure of quasicrystals and their approximants. He has always been fascinated by the complex interplay between, on the one hand, fundamental geometry, topology, crystallography and group theory and, on the other hand, the structure and properties of materials. He became excited about multicomponent high-entropy materials in 2003 when he saw a preprint of Yeh's first paper on the subject in Advanced Materials Engineering. Like Yeh, he was not at that time aware of my own previous work on the topic. He was sufficiently excited about the ideas expressed to write a short but influential note entitled 'Alloyed Pleasures: Multimetallic Cocktails' in the journal *Current Science*.[33] This note remarked on the use of multiple alloying elements in the manufacture of bulk metallic glasses, gum metals and high-entropy alloys, and speculated on the apparent significance of emergent and surprising properties in a material when a large number of elements are brought together.

Rangu married his wife Sucharita Desiraju in 1968 in Chennai, and they were happy to celebrate their golden wedding anniversary a few years ago. They have two sons, Satyajit and Viswajit, aged in their early fifties and late forties, respectively. They continue to live in Bangalore, which has been the centre of their lives for many years. I had the good fortune to meet Rangu (and later Sucharita) early on in both our careers, briefly in the 1960s in Cambridge when I was an undergraduate and he was a research student in the Department of Metallurgy (as it was then called), and then in 1980 at BHU in Varanasi when I was a Visiting Fellow and he was a young Professor. We became good friends and have met many times since and often shared ideas about materials science and engineering (and life in general). He is a wonderful companion, full of interesting and wide-ranging ideas. I feel privileged to call him a good friend and colleague.

Oleg Senkov

Oleg Senkov was born in 1952 in the small town of Nesterov (now called Zhovkva) in Lviv Oblast (Lviv province) in the far west of Ukraine, roughly halfway between the provincial capital city Lviv and the Polish border (about 30–40 km away from both). His father was Nikolay Senkov, who sadly died just over a year after Oleg was born. Following Nikolay's death, Oleg's mother, Olga Yakimova, took Oleg and his elder sister Lyudmilla away from Nesterov, and the family moved about 1,000 km east so that they could live close to the rest of her family in Mamadysh, another small town on the banks of the Vyatka River in Tatarstan, part of the Russian Federation tucked up just west of the Ural Mountains on the far eastern edge of Europe. Olga Yakimova raised Oleg and Lyudmilla single-handedly, and they both went to the local elementary and high schools. Lyudmilla graduated from high school in 1961, moved to Kazan to study biology at Kazan State University, was married the following year, and worked for over forty years at Kazan Medical Institute until she retired. She and her family still live in Kazan. Oleg was a bright student at school, developing a deep interest in maths and physics, and participating successfully in regional maths and physics Olympiads. In 1969 he too graduated from high school and passed the qualifying exams to enter the Faculty of Physics at M. V. Lomonosov Moscow State University, the most famous and best university in the whole of the (then) Soviet Union. He met and married his first wife Irina and had his first son Aleksandr while he was still a student (like me), and he had to take additional jobs to support his young family (again like me), working in student construction brigades and as a research assistant.

Oleg completed his master's degree in physics in 1975, with a research thesis on the effect of crystal orientation on the surface erosion of ionic crystals, supervised by his professor, A. A. Predvoditelev. He started work as a research engineer in M. M. Myshlyaev's group at the famous Institute of Solid State Physics (ISSP) of the Soviet (now Russian) Academy of Sciences in the town of Chernogolovka, about 60 km northeast of Moscow, studying the microstructure of refractory metals after high-temperature creep. Two years later, while continuing to work at ISSP, he enrolled for a doctorate on the microstructure, deformation and strengthening mechanisms in aluminium alloys, which he completed successfully in 1982. He continued to work for a total of over twenty years at ISSP, becoming progressively a Junior Scientist, Scientist and then Senior Scientist. At the same time as doing research at ISSP, he also taught materials science and physical metallurgy courses at the Moscow State University of Instruments and Engineering, becoming an Associate Professor there in 1993. But geopolitics intervened. As Oleg says, 'In 1992, the Soviet Union collapsed followed by an economic crisis [and] it became very difficult to live and do research'.[44] And in 1994 he was successful in applying for a Canadian National Science and Engineering Council (NSERC) Visiting Professorship to work with John Jonas at McGill University in Montreal, studying the effects of hydrogen as an alloying addition on the crystal structure, elastic properties, dislocation behaviour and plasticity of alpha and beta titanium alloys. In 1996 he moved south and west to the United States to work with Sam Froes at the Institute of Advanced Materials and Processes (IMAP) at the University of Idaho, where he continued his work on thermo-hydrogen processing, as well as studying the microstructure and mechanical properties of novel and nanocrystalline metallic materials. He published a number of important papers during this period in Canada and the United States, on hydrogen in titanium alloys, nanocrystalline aluminium alloys and titanium aluminides manufactured by severe plastic deformation, and multi-step forging to make fine-grain superplastic titanium alloys and intermetallic compounds.

In 1986 while still in Russia, Oleg's first marriage collapsed, and after getting divorced he re-married his second wife Svetlana Senkova in 1989, and his daughter Tatiana was born a year later. He says that, after moving to the west of the United States, his family loved exploring the beautiful countryside in the state of Idaho and the other western states, visiting and hiking in the wonderful Yellowstone, Grand Teton, Olympic and Glacier National Parks. Svetlana was also working as a materials scientist and Tatiana enjoyed being at US schools. In 2000, Oleg was offered and accepted a job working in Dan Miracle's research group in the AFRL at the Wright–Patterson US Air Force base just east of the city of Dayton in Ohio (the base is named, of course, after Wilbur Wright, one of the two Wright brothers who first achieved powered heavier-than-air flight in 1903). Oleg's family moved to Beavercreek, a residential suburb of Dayton, and in 2003 they all became US citizens.

For his first ten years at Wright–Patterson, Oleg worked on amorphous alloys and high-strength aluminium alloys. But in 2009 he and Dan Miracle picked up very early on Yeh's discovery of multicomponent high-entropy alloys, and they were the first[34] to have the idea of investigating the structure and properties of multicomponent high-entropy *refractory alloys*, the first two of which were NbMoTaW and VNbMoTaW, as described earlier in this chapter. They were trying to make refractory alloy bulk metallic glasses but, as Oleg said to me in a recent email, 'I came to an idea of trying a high-entropy alloying approach to produce refractory solid solutions, which could be ductile and strong at temperatures above 1,000 °C [1,273K]'.[44] Oleg has continued to lead worldwide research into multicomponent high-entropy refractory alloys, which as he surmised do indeed show outstanding combinations of strength, low density and fracture resistance at high temperatures. Because of funding cuts at AFRL following the world financial collapse of 2009–10, Oleg's work on amorphous alloys and aluminium alloys was discontinued, but he has instead received special Director's funds to continue working on refractory high-entropy alloys and has also extended his work to a variety of other aerospace materials, including weld materials for nickel-based superalloys and more conventional refractory alloys. Oleg's wife Svetlana also worked as a researcher at AFRL until 2011, when she retired because of the funding cuts, to concentrate on being a housewife and homemaker. His daughter Tatiana graduated with a business degree from Ohio State University in 2012, and now lives and works in Madison, Wisconsin. His first son Aleksandr stayed in Russia, graduated with a forestry degree in 2011, and currently works at the Northern Institute of Forestry in Archangelsk, where he lives with his two young children (Oleg's two grandchildren) Maria and Anton. Oleg still loves to travel around the wonderful countryside of the United States. He clearly also likes good weather, and he tells me that he and the family often spend Christmas vacations on the beach somewhere or other in the Caribbean.

Daniel Miracle

Dan Miracle was born in 1956 in Cleveland, Ohio. His father was Theodore Roosevelt Miracle (named after the famous US President between 1901 and 1909), who came from an Irish immigrant background, and his mother was Annette Eloise Miracle (née Novy), whose family came from Pisek, known as the Athens of Bohemia, about 90 km south of Prague in what used to be the Austro-Hungarian empire and is now part of the Czech Republic. Theodore was born into a poor family in Rockholds, Kentucky in the Appalachian Mountains that moved to Cleveland during the Second World War when he was just six years old. After leaving school he became a carpenter and then a project superintendent for a construction company that made concrete structures for multi-storey buildings. Dan says: 'he had an uncanny ability to get tough-minded people to do what he wanted'[45] (anyone who has worked in the construction industry knows how important yet how difficult that can be). Theodore died sadly quite young in his fifties from lung cancer. Dan's mother was initially 'a stay-at-home mom',[43] but then took a job at a local bank, eventually becoming vice-president. She is now retired and, according to Dan: 'an independent and strong-willed woman'.[43] Dan grew up in Streetsboro, a small town in northeastern Ohio, midway between Cleveland and Akron. He walked to his elementary school (like me) and then went to Streetsboro junior and senior high schools. He played football and wrestled at school, having learned to value physical exercise and competition from his father who was a good athlete. From a young age, he also became a lifetime devotee of fly fishing and cycling, buying both his first fly rod and his first ten-speed bike when he was just twelve years old, from money earned doing a paper round. Aged just fifteen, he went on a 1,200 mile (almost 2,000 km) cycling round-trip with one of his teachers and eleven classmates to visit Disney World in Orlando, Florida.

Dan was always good at maths and science and, after graduating from high school in 1975 with a grade-point average (GPA) of 3.9 (maximum = 4.0), he enrolled for a Bachelor of Science degree at Wright State University in Dayton, also in Ohio, about 200 miles (322 km) south-west of Cleveland. He was the first in his family to do a university degree (again like me). He had to pay his way through college, taking jobs in his father's construction company, for the college on-campus food service, and as a co-op student (lab assistant) in the materials lab at the AFRL at Wright–Patterson US Air Force base just next door to the University. After graduating with his BS degree in 1979, he continued to work full-time in the materials lab and then applied for and won an Air Force scholarship to do graduate studies at Ohio State University in Columbus, about seventy miles (112 km) east of Dayton. After eleven years juggling his PhD research at Ohio State with his full-time duties at AFRL, he completed his PhD in 1990 on the deformation of nickel aluminide bicrystals and the effect of different bicrystal grain-boundary orientations, initially working under the supervision of John Hirth and later with William Clarke and Hamish Fraser. Dan says that he was always going to become some sort of engineer, but he became a materials scientist because 'I have always been drawn by activities that aren't quite mainstream and have a sense of mystery and novelty . . . Materials science was just about right'.[45] And he got an early taste of the breadth of the subject when he was an undergraduate and was working as a lab assistant at the APRL with Harry Lipsitt, who was a great metallurgist and teacher. Over the years, Dan has gradually risen through the ranks at AFRL and is now a Senior Scientist, along the way having won a string of honours and prizes for his research and scientific activities, notably the Presidential Rank Award of Meritorious Professional Service.

Dan met his wife-to-be, Tina Pestona, in his second (sophomore) year at Wright State, while they were both working in the University Rathskeller (the on-campus bar and restaurant). They fell in love almost immediately and were married while he was still an undergraduate (again like me) just thirteen months later. Tina came from a military family. Her father was a Sergeant-Major in the US Army, born in New York from Portuguese parents, and her mother was born and brought up in Tokyo (they had met while he was stationed in Japan after the Second World War). Dan and Tina had their first son, Nathan, in 1981 and their second son, Timothy, four years later, while Dan was a graduate student at Ohio State. While doing his graduate work, he rented a room from his mother-in-law in Columbus, and drove home twice a week to Dayton to see his young wife and child. Nathan and Timothy are both now in their forties, both married, each with two boys, and still living with their families near Dan and Tina in Dayton, and both working as managers, Nathan in the National Air and Space Intelligence Center, and Timothy running a restaurant business. Dan has continued his love of cycling and fly fishing: going on adventurous cycling trips along the west coast of the United States, in the French Alps, along the Blue Ridge trail in the Smoky Mountains in Tennessee and South Carolina, and an off-road tour through the Rocky Mountains in Canada and the United States; and going on fishing trips for trout, steelhead, salmon, permit and tarpon in the Rockies, off the Olympic Peninsula near Seattle, in remote bays on the Yucatan Peninsula in Mexico, and in New Zealand. He also loves woodwork, international travel and photography.

Dan first met Jien-Wei Yeh at the International Symposium on Frontiers in Design of Materials (FDM) at IIT Madras in Chennai in 2005, when at one of the breaks they sat next to each other by a fountain (another serendipitous, random and entropic event). Dan says he asked what Jien-Wei was working on and had no idea what he meant when he heard the reply: 'high-entropy alloys'.[45] He was intrigued. Two and a half years later, in early 2008, there was a call for research proposals from the Air Force Office of Scientific Research (AFOSR) at AFRL looking for 'any new, really far out ideas'.[45] Dan was already working on a proposal on bulk metallic glasses when, with just a week to go before the deadline for submission, he remembered his conversation with Jien-Wei, contacted his colleague Oleg Senkov, and in near-record time they put together a second proposal on high-entropy alloys. Dan says the initial draft was 'inspired by the Cantor alloys',[45] and only at the last moment did he also add further sections on high-entropy refractory alloys and high-entropy light alloys. He told me recently in an email that 'The rationale was simple. Our objective was to develop revolutionary high temperature structural materials with capabilities well beyond current superalloys, and so I felt that using refractory elements was the obvious approach and was surprised that this idea hadn't been tried yet'.[45] It hadn't been tried yet, of course, because only Jien-Wei and I had done any previous work on multicomponent high-entropy materials, so there was still a massive open field to go for. It required adventurous spirits such as Dan and Oleg to see the potential, pick up the baton and run with it (and, by the way, the massive field of multicomponent high-entropy materials continues to remain open for other adventurous spirits). Their proposal was not funded: it was deemed to be 'too ambitious'.[44] They tried again the following year, and failed again, for the same reason. Dan and Oleg were discovering in the United States the same scientific conservatism and disdain for a truly innovative idea that Jien-Wei and I had independently both encountered some years earlier in Taiwan and Britain respectively. Like Jien-Wei and me, however, they were not to be stopped, and in a bloody-minded kind of way (which I really like) just went ahead anyway and did some of the proposed work, finding a small amount of seed-corn funding that they squeezed out of another programme. Their first paper on high-entropy refractory alloys was published in 2010 in the research journal *Intermetallics*.[36] And like my first paper on multicomponent Cantor alloys and Jien-Wei's first paper on high-entropy alloys, both in 2004, their paper is now a classic, receiving hundreds of citations a year.

Peter Liaw

Peter Liaw was born in 1950 in Chiayi City at the northern edge of the Chianan plain in southwestern Taiwan, about 50 km west of Datashan (Mount Daito) in the beautiful Alishan mountain range, with Yushan (Jade Mountain), the tallest mountain in East Asia, a bit further to the east. His father, Duan-Yeh Liaw, was a government officer and his mother, Wang Shi Huang-Liaw, was a housewife. Peter studied at Chiayi High School before moving to NTHU in Hsinchu in the north of Taiwan to take his bachelor's degree in physics. After graduating from NTHU, he did two years of military service, and then won a Royal E Cabell Fellowship from the Department of Materials Science and Engineering at Northwestern University in Evanston, Illinois, moving to the United States in 1974 to do a PhD with Morris Fine on fatigue crack propagation mechanisms in aluminium alloys and steels. He says, 'my grandfather always wanted me to study medicine and become a doctor',[46] but in the end Peter was more attracted to the harder scientific world of physics and then materials science and engineering: 'I thought if I studied materials science and engineering it could be easier to find jobs'.[46]

The end of the 1970s was a hectic time for Peter. He completed his thesis in 1979, returned to Taiwan to marry his fiancée Jin, and then returned to the United States to defend his thesis successfully, finally being awarded the PhD and graduating in 1980. For the next thirteen years, Peter was employed as a senior engineer in the fracture mechanics group at the Westinghouse R&D Research Center in Churchill, just east of downtown Pittsburgh, famous initially for its work on nuclear rocket propulsion systems for the space programme, and then more generally for developing steam turbines and generators, advanced solar energy, lithium battery, magnetohydrodynamic and nuclear fusion technologies, home appliances, consumer electronics and lighting. These programmes provided enormous challenges at that time for developing high-strength, fracture-resistant materials, to which Peter made major contributions, winning a string of internal Westinghouse outstanding performance awards.

In 1993, Peter was appointed to a Professorship in the Department of Materials Science and Engineering at the University of Tennessee (UT) in Knoxville, close to the famous Oak Ridge National Laboratory, where he was worked ever since. He has had a remarkably successful scientific career, having supervised more than 200 postdocs, research students and visiting scholars, having published more than 1,500 papers and reports with over 65,000 citations, and having worked on many different experimental and theoretical aspects of the processing, structure and properties of a wide range of different advanced materials, collaborating widely with scientists at UT, Oak Ridge and throughout the United States and the rest of the world.[46] He holds three different chair appointments as the John Fisher Professor, Ivan Racheff Chair of Excellence and National Alumni Association Distinguished Service Professor, having also won the Moses E and Mayme Brooks Distinguished Professorship in 2003, 2008 and 2011 as well as many other accolades, including the L. R. Hesler Award at UT and the TMS Distinguished Service Award. At UT, he has been Director of the NSF Integrated Graduate Education and Research Training (IGERT), International Materials Institutes (IMI), and Major Research Instrumentation (MRI) Programs. Peter and Jin have two children, both in their forties—Kevin is a doctor, and Karen is an optometrist—and three grandchildren—Skyla aged six, Locke aged five and Lana who is just three years old. Peter has little spare time (fairly obviously), but when he can he loves going for walks with his wife and the rest of the family, particularly his young grandchildren.

Although they both went to NTHU as students, Peter did not meet Jien-Wei Yeh until much later, more than ten years after he had settled into academic life in Tennessee, when he attended Yeh's presentation at the XIV International Materials Science Congress in Cancun, Mexico in 2005. He was immediately excited by the enormous range and potential of Yeh's new multicomponent and high-entropy materials and began to work on them, continuing up to the present day making many important contributions to our understanding in particular of their structure and mechanical properties. He has been very influential in bringing the new field to the attention of scientists and engineers worldwide, but particularly throughout the large and dynamic research community in the United States, collaborating extensively, writing many research papers, actively organising conferences, writing review papers and editing books on the different topics.

Christina Rost

Tina Rost was born in 1987 in the small industrial town of Charleroi on the Monongahela River, about twenty miles south of Pittsburgh in Pennsylvania, USA. Charleroi is famous for the manufacture of glass, following its initial settlement in the nineteenth century by glass-making Walloon immigrants from the Belgian city of Charleroi, hosting one of the largest twentieth-century glass factories in the world owned by the Pittsburgh Plate Glass (PPG) Company, and currently home as it has been for over 100 years to the manufacture of Corning's (later Corelle's) legendary heat-resistant Pyrex borosilicate glass, giving rise to the town's nickname 'Pyrex City'. Tina's family come from German and Rusyn (East Slovakian) forebears, who emigrated to the Pittsburgh area to work in the steel, mining and machining industries traditionally located there. Her father, Allen Rost is a retired metallurgical engineer, having worked for a variety of steel companies in the region. Her mother, Cynthia Tomcanin Rost, is currently a Professor of Nursing at Duquesne University, having been a psychiatric and public health nurse, again working for a variety of hospital and health agencies in the region.

Tina studied at the local Charleroi High School and graduated in 2005. She then took her bachelor's (2009) and master's (2012) degrees in physics at the Indiana University of Pennsylvania, not far away in the small town of Indiana, about fifty miles east of Pittsburgh, and then moved much further south to North Carolina State University (NCSU) to do her PhD supervised by Jon-Paul Maria on multicomponent high-entropy oxides, which she completed in 2016. Given her family background and her exposure to the manufacture of steel and glass, it's hard not to believe that materials science must have been in her blood from an early age. She says, 'I've always enjoyed creative and technical hands-on work. This combined with my life-long passion for learning compelled me toward science ... [I became a materials scientist because of my] interest in the fundamental properties of materials and the governing physics that enables development of technologies'.[47]

Following Jon-Paul Maria's first attempts with Trent Borman to make entropy-stabilised single-phase structures in mixed metal oxides, he asked Tina to do a more detailed study for her Ph.D. She tells me that 'of nearly 100 compositions, one single-phase rock-salt composition stood out (MgCoNiCuZn)O, which I used to investigate the phase behavior of such a system, leading to HEOs [high-entropy oxides]'.[47] After completing her PhD, Tina was a postdoc in the Experiments and Simulations in Thermal Engineering (ExSITE) group in the Department of Mechanical and Aerospace Engineering and a teaching fellow in the Department of Materials Science and Engineering at the University of Virginia in Charlottesville; she left in 2019 to take up a position not far away as an Assistant Professor of Physics at James Madison University in Harrisonburg, Virginia; and she has again recently moved to become an Assistant Professor of Materials Science and Engineering at Virginia Tech in Blacksburg, Virginia. Her main research interests are multicomponent high-entropy ceramics, bulk and thin film synthesis, multi-scale characterisation, and functional materials for extreme environments. She met her husband, Glenn P Barber in 2009, they married in 2013, and they live together in Christiansburg, a town close to Blacksburg. He is a business analyst working for Abbott Laboratories, the multinational medical devices and healthcare company, in their Informatics Division based in Charlottesville. When not working, Tina likes travelling, gaming, and spending time with her husband and their pet rabbits. She says, 'I enjoy both console and PC gaming. I have a particular affinity for all things Nintendo. I also enjoy fantasy role-playing games like Dungeons and Dragons'.[47] She has had two rabbits, a mini-lop called Oreo and a Jersey Wooley called Moxie, who, she says, 'basically runs the house'.[47]

Trent Borman

Trent Borman was born in 1992 in Bismarck, capital city of North Dakota in the rural prairie north of the United States. His mother Karen worked in corporate finance in the grocery industry and his father Todd took a biology degree but became 'a stay-at-home dad' to look after Trent and his sister Brittany while they were growing up,[48] before returning to work as a teacher's aide in special-needs education. They have both now retired. Shortly after Trent was born, the family moved for a year to Des Moines, Iowa, then for a few years to Tacoma, Washington, before finally settling near Minneapolis, Minnesota. Trent graduated in 2011 from Eden Prairie High School in Eden Prairie, a suburb of Minneapolis. He says he 'was fascinated by science from a young age'.[48] His participation in a high school robotics competition 'was a pivotal moment in setting my career trajectory. It re-sparked my wonder in the underpinnings of the world around me, whether it was the physics behind the motors or the materials science behind the alloys we used'.[48]

After leaving high school in 2011, Trent went on to do bachelor's degree in materials engineering with a focus on metallurgy and electronic materials at Iowa State University in Ames, home of the famous US Department of Energy Ames National Laboratory dedicated to 'creating materials and energy solutions'. Just before starting at Iowa State, Trent made a last-minute pivot from electrical engineering to materials science. As he wrote to me (very eloquently):

> I don't just want to know how to use little black boxes (packaged semiconductor integrated circuits) to make a circuit. I want to understand what is happening inside them. Materials science gave me the opportunity to sit at a crossroads and dabble in many fields such as physics, chemistry, mechanics, electronics, and more. Everything in the world must be

> made of something and materials science would give me the background to understand materials down to the atomic level.[48]

During the second summer of his undergraduate course at Iowa State, Trent worked on a solar cell project on secondment in Jon-Paul Maria's research group at Penn State under a National Science Foundation (NSF) research experience for undergraduates (REU) programme. When he returned the following summer, Jon-Paul challenged him to create a five-component single-phase oxide, which he achieved with an (MgTiFeCuZn)O composition in a spinel crystal structure. Jon-Paul then set up a more detailed research programme, as described above, with Tina Rost carrying on the multicomponent high-entropy oxide work for her PhD. After finishing his bachelor's degree at Iowa State, Trent moved in 2014 to Penn State to do a master's degree in materials science and engineering, studying lead zirconate titanate thin films supervised by Susan Trolier-McKinstry (completed in 2016), and then a PhD on physical vapour deposited (PVD) high-entropy carbide thin films supervised again by Jon-Paul Maria. Trent accompanied Jon-Paul when he moved to Penn State, completing his PhD there in 2020. In Trent's own words:

> Ultimately, the opportunity to pursue a PhD that focused on the intersection between equipment design and configuration, processing science, somewhat new materials compositions, and functional properties led me back to Jon-Paul's group. While the group's research on HE oxides frequently leveraged pulsed laser deposition (another PVD technique) I felt that pursuing the HE carbide and nitride work (which we performed by reactive sputtering) would leave me better equipped to pursue a career in a PVD-related field (whether that was in semiconductors, optical coatings, tribological coatings, or some other area). Additionally, it would expand the scope my research experience to new materials.[48]

After finishing his PhD, Trent then took on his current job as a PVD process engineer at Northrop Grumman's Advanced Technology Laboratory (ATL) in Linthicum, Maryland, working in a high mix/low volume semiconductor foundry, where he is enjoying a gradual move from primarily production support to mentoring young engineers and doing R&D into new materials and processes. Trent recently married his wife Sarah, who is an academic advisor for a data science master's programme. They live in Glen Burnie, a small town in Maryland a few miles away from the ATL in Linthicum. When not working as a scientist and engineer, he enjoys cooking, both as a personal skill and as a scientific endeavour, as well as photography, gardening and visiting the many wonderful national parks in the United States.

Jon-Paul Maria

Jon-Paul Maria was born in 1972 in the town of Wilkes-Barre, part of the old industrial Scranton–Wilkes-Barre coal-mining region, a series of towns strung out along the Wyoming Valley in the north-eastern corner of Pennsylvania. His mother and father, Susan and Peter Maria, were both hairdressers and ran a salon called Peter Maria Hair Fashions for more than forty years in Pittston, one of the nearby towns some way further up the valley from Wilkes-Barre. The family lived in Wyoming, another nearby town a bit further still up the valley, where Peter and Susan have been living happily in retirement since 2015. Jon-Paul grew up and went to primary, middle and high school in Wyoming, going on after graduating in 1990 to do his bachelor's (1994), master's (1996) and doctorate (1998) degrees in ceramic science and engineering at Penn State. His first three semesters were at the small Penn State Wilkes-Barre branch campus. He and his brother were the first in his family to go to college (like me) and knew nothing about the process (like me) so they opted for an initial inexpensive local option that as he says: 'turned out to be good choice'.[49] He then moved on to complete his degrees at the main Penn State campus right in the middle of Pennsylvania almost 150 miles to the west, doing his PhD supervised by Susan Trolier-McKinstry on structure–property relationships in ferroelectric relaxor-lead titanate solid-solution epitaxial thin films for piezoelectric applications (single crystals of these materials are now in standard use as sensors and actuators for high-resolution medical ultrasound imaging systems).

As a child, Jon-Paul was fascinated by all types of science. He loved to watch science documentaries and to build things with Lego. He was particularly interested in the weather and decided that he would become a meteorologist when he grew up. As he matured, however, he changed his mind and decided instead to become a materials scientist.[49] He says:

> I am a proud member of the student cohort that was wooed by NASA. I recall some point in high school a presentation by some university person who passed around a white block of stuff with black paint on the surface. It was a space shuttle tile, and I was hooked. The idea that I could make space shuttle parts and become a ceramic engineer was all that I needed to hear.[49]

After completing his PhD at Penn State in 1998, Jon-Paul took up a postdoc position at NCSU working with Angus Kingon on high permittivity oxides that could replace silica in conventional CMOS transistors. staying on to become an assistant professor in 2002, an associate professor in 2006 and finally a full professor in 2011. NCSU was and still is, of course, one of the key component parts of the famous and successful 'research triangle' in the Raleigh–Durham–Chapel Hill area of North Carolina, alongside Duke University and the University of North Carolina as the two other corners of the triangle. In 2018, he returned to Penn State to be closer to his extended family, in particular his ageing parents, as well as to enhance the potential for synergistic collaboration with academic colleagues whose research overlapped more closely with his own. Since 2018 at Penn State, Jon-Paul has continued to work on new materials discovery, synthesis and properties, notably studying ferroelectric thin films, high permittivity materials, novel semiconductor contacts, oxide epitaxy, infra-red plasmonic materials and entropy-stabilised crystals. In 2016, he co-founded Third Floor Materials, a start-up company that endeavours to develop novel infra-red sensor materials and associated technologies.

As mentioned above, Jon-Paul became excited by the idea of multicomponent entropy-stabilised single-phase high-entropy oxides in 2013, when he heard a presentation from a colleague, Carl Koch, on multicomponent entropy-stabilised high-entropy metal alloys. At the time, Jon-Paul was working at NCSU as part of a large team pursuing a government grant to develop new materials to operate in extreme environments, and Koch and his group were proposing that they should investigate high-entropy alloys for that purpose. Jon-Paul had never heard of high-entropy alloys (even as late as 2013, relatively few people had), and he went back to his office wondering if the same concept could apply to what he calls 'his materials',[49] i.e. oxides or, perhaps more generally, inorganic ceramics. He says, 'That night I searched web of science feverishly and could not believe that no-one had yet tried this or at least published it'.[49] The next day he went into his lab and 'rallied his group'[49] to discuss the opportunity and challenge them to manufacture a five-component entropy-stabilised single-phase oxide. In the end, Trent Borman, working in Jon-Paul's lab as an undergraduate researcher seconded from Penn State, accepted the challenge and, as explained in a previous sub-section, managed successfully to make a five-component single phase spinel. Effectively, as Jon-Paul says, 'Trent started the entire experimental program'.[49] I like the important role played by undergraduates in discovering multicomponent high-entropy materials—Alain Vincent and Peter Knight working with me on multicomponent high-entropy metallic alloys, and Trent Borman working with Jon-Paul Maria on multicomponent high-entropy ceramic oxides.

Jon-Paul loves generating new knowledge, publishing meaningful papers and, as he puts it: 'discovering a thing or two'.[49] He sees it as his responsibility as a senior academic to facilitate 'a pipeline of highly trained scientists, engineers, entrepreneurs and researchers'.[49] I think we can all agree with his closing comment in a recent email to me: 'The world needs more enthusiastic and talented technical leaders'.[49] When not doing science and engineering, Jon-Paul enjoys cooking and eating, running, biking and going to the gym, and drawing, painting and 'a little pottery'[49]. As well as being an outstanding and innovative scientist, it sounds to me like he has a wonderful and well-balanced life.

6 Summary

Multicomponent high-entropy materials were discovered independently by myself and my co-workers in the late 1970s and early 1980s, and by Jien-Wei Yeh and his co-workers in the 1990s. Our two seminal papers[4, 21] founding this new field of scientific endeavour were both published after a considerable intervening period at more or less the same time in 2004 (though the basic idea was published briefly somewhat earlier[5–9], in 2002 and 2003). Both papers describe a truly massive increase in the potential palette of new materials that can be manufactured and used for the future development of humankind. We both continued to publish papers on these novel materials through the 2000s. Scientific interest was at first slow to pick up on our early work, but began to accelerate at the beginning of the 2010s, and has mushroomed dramatically over the last few years, with literally thousands of scientists now working on multicomponent high-entropy materials worldwide, multimillion dollar research programmes underway in most developed countries, and thousands of new papers being published every year.

The early work by Jin-Wei Yeh and myself outlined many of the most important features of multicomponent high-entropy materials that have been confirmed by subsequent studies:

1. There is a gigantic number of multicomponent alloys and materials, vastly greater than had been discovered by previous conventional material development methods.
2. Multicomponent alloys and materials frequently consist of no more than one or two phases, often with enhanced solubility of a large number of different components.

3. In other words, unlike binary and ternary systems, multicomponent alloys and materials often absorb high concentrations of many components by dissolving them in existing phases, rather than by forming new phases and compounds.
4. The root cause of this behaviour is a high configurational entropy of mixing in multicomponent alloys and materials relative to the interaction energies between the different components.
5. Nevertheless, there are still many multicomponent alloys and materials that have complex, multiphase structures.
6. Multicomponent single-phase solid solutions and multicomponent multiphase mixtures have complex electronic, atomic, nanoscale and microscopic structures, leading to a variety of new and desirable mechanical, electrical, chemical and other properties.

The structure, properties and potential uses of multicomponent high-entropy alloys and materials are discussed in detail in the rest of this book.

References

1. Cantor B., Personal records.
2. Vincent AJB. *A Study of Three Multicomponent Alloys*, Undergraduate thesis, University of Sussex, 1981.
3. Knight P. *The Study of Multicomponent Alloy Phases*, Part II thesis, Department of Materials Library, University of Oxford, 1995.
4. Cantor B, Chang ITH, Knight P, Vincent AJB. (2004). Microstructural development in equiatomic multicomponent alloys. *Materials Science and Engineering A375–7*: 213–18.
5. Cantor B, Chang ITH, Knight P, Vincent AJB. (2004). Microstructural development in equiatomic multicomponent alloys. In KAQ O'Reilly, P Warren, P Schumacher, B Cantor, eds., *11th International Conference on Rapidly Quenched and Metastable Materials*. Amsterdam: Elsevier, 213–18.
6. Cantor B, Kim KB, Warren PJ. (2002). Novel multicomponent amorphous alloys. *Materials Science Forum 386–88*: 27–32.
7. Kim KB, Warren PJ, Cantor B. (2003). Formation of metallic glasses in novel $(Ti_{33}Zr_{33}Hf_{33})((100\text{-}x\text{-}y)(Ni_{50}Cu_{50})(x)Al(y)$ alloys. *Materials Transactions 44*: 411–13.
8. Kim KB, Warren PJ, Cantor B. (2003). Metallic glass formation in multicomponent (Ti, Zr, Hf, Nb)-(Ni, Cu, Ag)-Al alloys. *Journal of Non-Crystalline Solids 317*: 17–22.
9. Kim KB, Zhang Y, Warren PJ, Cantor B. (2003). Crystallization behaviour in a new multicomponent TiZrHfNiCuAl metallic glass developed by the equiatomic substitution technique. *Philosophical Magazine 83*: 2371–81.
10. Klement Jr W, Willens RH, Dusez P. (1960). Non-crystalline structure in solidified gold–silicon alloys. *Nature 187*: 869–70.
11. Luborsky FE, ed, (1983). *Amorphous Metallic Alloys*. London: Butterworth.

12. Suryanarayana C, Inoue A. (2018). *Bulk Metallic Glasses*, 2nd edn. Boca Raton: CRC Press.
13. Inoue A. (2000). Stabilization of metallic supercooled liquid and bulk metallic glasses. *Acta Materialia 8*: 279–306.
14. Kim KB. *Formation of Amorphous and Nanocrystalline Phases in Novel Multicomponent Alloys*, DPhil thesis, Oxford, 2004.
15. Kim KB, Warren PJ, Cantor B. (2004). Glass-forming ability of novel multicomponent $(Ti_{33}Zr_{33}Hf_{33})$-$(Ni_{50}Cu_{50})$-Al alloys developed by equiatomic substitution. *Materials Science and Engineering A375*:317–21.
16. Kim KB, Warren PJ, Cantor B, Eckert, J. (2005). Crystallization behaviour of novel $(TiZrHf)_{100-x-y}$ $(NiCu)_x$ Al_y with $x = 48$–55. *Journal of Metastable and Nanocrystalline Materials 24–5*: 657–60
17. Cantor B. (2007). Stable and metastable multicomponent alloys. *Annales de chimie science des matériaux 32*: 245–56.
18. Cantor B. (2014). Multicomponent and high entropy alloys. *Entropy 16*: 4749–68.
19. Yeh JW. (2021). My trip from physics to high-entropy materials. In J Brechtl, PK Liaw, eds., *High-Entropy Materials: Theory, Experiments, and Applications.* New York: Springer,37–62.
20. Huang PK, Yeh JW, Shun TT, Chen SK. (2004). Multi-principal-element alloys with improved oxidation and wear resistance for thermal spray coating. *Advanced Engineering Materials 6*: 74–8.
21. Yeh JW, Chen SK, Lin SJ, Gan JY, Chin TS, Shun TT, Tsau CH, Chang SY. (2004). Nanostructured high entropy alloys with multiple principal elements: Novel alloy design concepts and outcomes. *Advanced Engineering Materials 6*: 299–303.
22. Yeh JW, Chen SK, Gan JY, Lin SJ, Chin TS, Shun TT, Tsau CH, Chang SY. (2004). Formation of simple crystal structures in Cu-Co-Ni-Cr-Al-Fe-Ti-V alloys with multiprincipal metallic elements. *Metallurgical and Materials Transactions 35A*: 2533–6.
23. Hsu CY, Yeh JW, Chen SK, Shun TT. (2004). Wear resistance and high-temperature compression strength of fcc $Al_{0.5}CrFeNiCoCu$ alloy with boron additions. *Metallurgical and Materials Transactions 35A*: 1465–9.
24. Chen TK, Shun TT, Yeh JW, Wong MS. (2004). Nanostructured nitride films of multi-element high-entropy alloys by reactive DC sputtering. *Surfaces and Coatings Technology 188–9*:193–200.
25. Yeh JW. (2006). Recent progress in high entropy alloys. *Annales de chimie science des matériaux 31*: 633–48.
26. Yeh JW. (2013). Alloy design strategies and future trends in high-entropy alloys. Journal of Metals *65*: 1759–71.
27. Tong CJ, Chen YL, Chen SK, Yeh JW, Shun TT, Tsau CH, Lin SJ, Chang SY. (2005). Microstructure characterization of $Al_xCoCrCuFeNi$ high-entropy alloy system with multiprincipal elements. *Metallurgical and Materials Transactions 36A*: 881–93.
28. Tong CJ, Chen MR, Chen SK, Yeh JW, Shun TT, Lin SJ, Chang SY. (2005). Mechanical performance of the $Al_xCoCrCuFeNi$ high-entropy alloy system with multiprincipal elements. *Metallurgical and Materials Transactions 36A*: 1263–71.

29. Chen YY, Duval T, Hung UD, Yeh JW, Shih HC. (2005). Microstructure and electrochemical properties of high entropy alloys—a comparison with type 304 stainless steel. *Corrosion Science 47*: 2257–79.
30. Chen YY, Hong HC, Shih HC, Yeh JW, Duval T. (2005). Electrochemical kinetics of the high entropy alloys in aqueous environments—A comparison with type 304 stainless steel. *Corrosion Science 47*: 2679–99.
31. Chen YY, Hong UT, Yrh JW, Shih HC. (2005). Mechanical properties of a bulk $Cu_{0.5}NiAlCoCrFeSi$ glassy alloy in 288 degrees C high-purity water. *Applied Physics Letters 87*: Article 261918.
32. Chen TK, Wong MS, Shun TT, Yeh JW. (2005). Nanostructured nitride films of multi-element high-entropy alloys by reactive DC sputtering. *Surface and Coatings Technology 200*: 1361–5.
33. Ranganathan S. (2003). Alloyed pleasures: Multimetallic cocktails. *Current Science 85*: 1404–6.
34. Cantor B. (2021). Multicomponent high-entropy Cantor alloys. *Progress in Materials Science 120*: Article 100754.
35. Cantor B. (2021). A personal perspective on the discovery and significance of multicomponent high-entropy alloys. In J Brechtl, PK Liaw, eds., *High-Entropy Materials: Theory, Experiments, and Applications.* New York: Springer,1–36.
36. Senkov ON, Wilks GB, Miracle DB, Chuang CP, Liaw PK. (2010). Refractory high-entropy alloys. *Intermetallics 18*: 1758–65.
37. Rost CM, Sachet E, Borman T, Moballegh A, Dickey EC, Hou D, Jones JL, Curtarolo S, Maria JP. (2015). Entropy-stabilized oxides. *Nature Communications 6*: Article 8485.
38. Maria JP., Private communication.
39. Vincent AJB., Private communication.
40. Chang ITH., Private communication.
41. Huang KH., Private communication.
42. Chen SK., Private communication.
43. Yeh JW., Private communication.
44. Senkov ON., Private communication.
45. Miracle DB., Private communication.
46. Liaw PK., Private communication.
47. Rost, CM., Private communication.
48. Borman T., Private communication.
49. Maria J-P., Private communication.

Further Reading

Cantor B, Chang ITH, Knight P, Vincent AJB. (2004). Microstructural development in equiatomic multicomponent alloys. *Materials Science and Engineering A375–7*: 213–18.

Cantor B. (2021). Multicomponent high-entropy Cantor alloys. *Progress in Materials Science 120*: Article 100754.

Cantor B. (2021). A personal perspective on the discovery and significance of multicomponent high-entropy alloys. In J Brechtl, PK Liaw, eds., *High-Entropy Materials: Theory, Experiments, and Applications.* New York: Springer,1–36.

Ranganathan S. (2003). Alloyed pleasures: Multimetallic cocktails. *Current Science 85*: 1404–6.

Senkov ON, Wilks GB, Miracle DB, Chuang CP, Liaw PK. (2010). Refractory high-entropy alloys. *Intermetallics 18*: 1758–65.

Yeh JW, Chen SK, Lin SJ, Gan JY, Chin TS, Shun TT, Tsau CH, Chang SY. (2004). Nanostructured high entropy alloys with multiple principal elements: Novel alloy design concepts and outcomes. *Advanced Engineering Materials 6*: 299–303.

Yeh JW. (2021). My trip from physics to high-entropy materials. In J Brechtl, PK Liaw, eds., *High-Entropy Materials: Theory, Experiments, and Applications.* New York: Springer, 37–62.

3

Multicomponent Phase Space

1 Introduction

In 2004 two papers were published[1,2] that expanded massively our knowledge and understanding of the potential for making different materials. Almost all materials are *alloys*, i.e. mixtures of different starting materials or *components*. And almost all materials that had been made previously were *dilute alloys*, i.e. materials manufactured usually with one main component (occasionally two) and usually with no more than one or two other components added in relatively low (i.e. dilute) proportions. Cantor et al.[1] and Yeh et al.[2] both described work conducted some years earlier in which the authors had taken a very novel approach and, independently of each other, had begun to make *multicomponent materials*, i.e. alloys consisting of many different components in substantial (i.e. *concentrated* or *non-dilute*) proportions. As described and discussed in Chapter 2, this early work by Cantor et al.[1] and Yeh et al.[2] demonstrated clearly that there is a vast range of such multicomponent materials that have never previously been made (or even thought about), and that they have a wide range of unusual and surprising atomic structures, microstructures, associated properties and potential applications.[i]

The set of all materials that can be manufactured with different compositions from a given set of components is called an *alloy system*.[3–5] Thus, the set of all binary alloys or materials that can be manufactured from two components A and B is called the A–B *binary alloy system*; the set of all ternary alloys and materials that can be manufactured from three components A, B and C is called the A–B–C *ternary alloy system*; and the set of all multicomponent alloys or materials that can be manufactured from multiple components A, B, C, D, . . . , etc., is called the A–B–C–D– . . . *multicomponent alloy system*. The structure and properties of any particular material depend upon its composition, external conditions such as the ambient temperature, pressure, electric field, magnetic field, etc. and its previous history, i.e. the route by which it was manufactured. It is often convenient to represent the structure and properties of all the materials in a given alloy system (i.e. all the materials which can be manufactured from a given set of components) on a *phase diagram*, which is a geometrical representation of the *phase space* of that alloy system.[3–5] The phase space covers the range of different materials in the alloy system, i.e. those that can be made with different compositions from that particular set of components; and the associated phase diagram shows the conditions (the compositions, temperatures, pressures, etc.) over which materials in that alloy system can have particular equilibrium or non-equilibrium structures and/or properties. The phase space of any particular A–B binary alloy system is represented on the corresponding A–B *binary phase diagram*; the phase space of any particular A–B–C ternary alloy system is represented on the corresponding

[i] The basic idea of *multicomponent alloying* to manufacture *multicomponent materials* had already been published briefly a year or two earlier in 2002 and 2003, as described in detail in Chapter 2.

Fundamentals of Multicomponent High-Entropy Materials. Brian Cantor, Oxford University Press. © Brian Cantor (2024).
DOI: 10.1093/9780191986710.003.0003

A–B–C *ternary phase diagram*; and the phase space of any A–B–C–D– . . . multicomponent alloy system is represented on the corresponding A–B–C–D– . . . *multicomponent phase diagram.*

The whole of *multicomponent phase space* is the alloy system (or the set of materials) that can be manufactured from the set of all possible starting materials or components. By convention, starting materials or components can be elements (such as, for instance, carbon C, iron Fe or chlorine Cl) or compounds (such as, for instance, water H_2O, rock salt NaCl or titanium carbide TiC). Which components we use to describe an alloy or alloy system is partly a matter of practical convenience: sometimes compounds are more convenient starting materials than elements for the practical process of manufacturing particular materials of interest. But compounds can themselves be made by chemical combination from their constituent elements, so from a fundamental point of view we can always select the elements as our components with no loss of generality. In other words, multicomponent phase space covers all the possible materials that can conceivably be manufactured and is the alloy system obtained by using all the elements as the starting materials or components.

The objective of this chapter is to describe and discuss the size, shape and *geography* of multicomponent phase space. This will be important when we discuss in a later chapter different strategies for exploring multicomponent phase space in order to try and find particular materials with interesting and/or valuable structures and properties. Chapter 2 explained how the initial work on multicomponent high-entropy materials by Cantor et al.,[1] Yeh et al.[2] and others demonstrated that multicomponent phase space contains large regions consisting of materials with single-phase structures—single-phase multicomponent close-packed solid solutions, single-phase multicomponent amorphous alloys, and single-phase multicomponent compounds—but that it also contains other large regions consisting of materials with more complex, multiphase structures. This chapter considers the size, shape and occurrence of these different regions in more detail. First, however, we need to set out some precise definitions of the different kinds of materials that can be found in multicomponent phase space in order to create a clear taxonomy with which to describe the size and structure of multicomponent phase space accurately.

2 Definitions and Terminology

Alloy types

In their first papers, Cantor et al.[1] referred to their newly discovered concentrated alloys containing many components as *multicomponent alloys*. Yeh et al.[6] first referred to their newly discovered concentrated alloys containing many components as *multiple-principal-element alloys*, and later that year[2] moved to using yet another name and referred to them as *high-entropy alloys*. Other names have also been used in more recent years, including *complex concentrated alloys*, *compositionally complex alloys*, *medium-entropy alloys* and so on.[7–9] Overall, the name high-entropy alloys or high-entropy materials has been widely adopted, and is the most common way to describe the field.[7–9] Nevertheless, the name high-entropy alloys has been criticised on the grounds that many multicomponent materials do not have a particularly high entropy[7–9] (a not unreasonable objection). And the name multicomponent alloys has also been criticised on the grounds that it includes some conventional materials, for example high-temperature nickel superalloys or high-strength alumino-silicates[7–9] (a less-reasonable objection since conventional multicomponent materials are clearly important and there seems to be no good reason why they should be excluded). This variety of different names is, however, fairly confusing in its own right, and is further confounded by the different names sometimes being used as alternatives for the *same* set of materials, and sometimes being used to refer to *different* types of material.

The proliferation of names and the resulting confusion has arisen from three main causes. First, we need to separate carefully several different underlying concepts: namely, the *multiplicity* of the number of components, the *concentration* of several or all of the components, and the resulting changes of *entropy* (these different concepts have previously been somewhat mixed up). Second, we need to separate carefully the description of different types of *material* as determined just by the composition, and the description of different types of *material structure* (i.e. atomic structure or microstructure) as determined again by the composition, but also by external conditions such as the temperature and pressure, as well as the manufacturing route (these different descriptions have also previously been somewhat mixed up). Third, we need to use names that mean what they say in a plain and simple way (this has not previously always been the case). As a fourth guiding principle, we need, nevertheless, to respect as far as possible current usage as it has grown up with the field. We use these four guiding principles to set out some definitions, and then we look at the relationships between the different alloy types, as well as their relative positions in multicomponent phase space.

This book is entitled *Multicomponent High-Entropy Materials*, and we, therefore (not surprisingly) concentrate on these two names and their corresponding alloy types, namely, *multicomponent alloys or materials* and *high-entropy alloys or materials.* And we also concentrate on a third name and corresponding alloy type, namely, *multiple-principal-element alloys or materials.* These three were, after all, the first names to be used for the field[1,2,6] and are still the ones most commonly used.[7–9] They also separate out very neatly the three underlying concepts mentioned of, respectively, multiplicity, entropy and concentration. Detailed definitions are developed later in this chapter, but it is helpful first to sketch out in rather broad terms that we use the name multicomponent alloys to refer (rather obviously) to the set of *all* multicomponent alloys; we use the name multiple-principal-element alloys to refer (again rather obviously) to the set of all those multicomponent alloys that have several components in substantial proportions; and we use the name high-entropy alloys to refer (yet again rather obviously) to the set of all those multicomponent alloys that have a high configurational entropy of mixing. This all seems a bit tautological and obvious but, as mentioned, definitions are best when they say what they mean in a plain and simple way (and unavoidably, therefore, in a somewhat tautological and obvious way). As defined this way, high-entropy alloys are a subset of multiple-principal-element alloys, which in turn are a subset of multicomponent alloys (since not all multicomponent alloys have multiple principal components, and not all multiple-principal-element alloys have a high configurational entropy of mixing, or indeed a high entropy of any other kind). More accurate and detailed definitions of these three material types and some others, and the relationships between them, are given in the following sub-sections.

Multicomponent alloys

A *multicomponent alloy or material* is any alloy or material manufactured from more than two components (or elements). The set of multicomponent alloys contains, therefore, all alloys and materials except for single components (or single elements) and binary alloys. The name multicomponent alloys emphasises the first of the three underlying concepts, i.e. the *multiplicity* of the number of components. This is the underlying concept first discovered by Cantor et al.,[1] which opened the eyes of the scientific community to the massively enhanced potential for new alloys to be found by exploring multicomponent phase space.[7–9] The threshold for an alloy to be regarded as multicomponent is taken for several reasons to be the transition from two to three components, i.e. from a binary to a ternary alloy: partly for the structural reason that non-linear additive effects of different solutes require a minimum of two solutes in a solvent, i.e. an alloy with at least three components; partly for the thermodynamic reason that a configurational entropy of mixing higher than 1 R, where R is the gas constant, requires a minimum of three components; partly for the practical reason that our knowledge of most binary alloy systems is relatively high, but our knowledge of many ternary alloy systems is relatively sketchy; and partly for the simple etymological reason that the prefix *multi-* usually implies anything more than one or two.

The set of elements, the set of binary alloys, and the set of multicomponent alloys form what is called technically a *space-filling, disjoint partition*[ii] of the set of all possible materials.[10–12] In other words, every possible material is either a single element, a binary alloy or a multicomponent alloy. And the partition of the set of all possible materials into the three disjoint subsets of single elements, binary alloys and multicomponent alloys is clearly based on the different multiplicity of components in each subset: it depends, therefore, on the alloy or material type as determined by its composition alone. The vast majority of materials are multicomponent alloys (as already discussed in Chapter 2, and also discussed more fully later in this chapter); and there are more binary alloys than single elements, since a total of c^2 binary alloy systems can be made from a total of c components.

Multiple-principal-element alloys

A *multiple-principal-element alloy or material* is any alloy or material manufactured from more than two components in substantial proportions, where substantial means concentrated rather than dilute, i.e. at a concentration level above a few at%, so that atoms of that component interact with each other rather than being widely separated, and a dilute-solution approximation is unlikely to be reasonable in determining the effect of that component on the alloy structure and properties. The name multiple-principal-element alloys emphasises the second of the three underlying concepts, i.e. the *concentration* of some or all of its multiple components. This is the underlying concept that leads to non-linear and, therefore, surprising behaviour in the structure and properties of many multicomponent alloys and materials.[7–9]

[ii] A family of *subsets* $P = \{P_i\}$ of a *superset* S is called a *partition* when it fulfils three conditions:[10–12] (1) P does not contain the empty set Φ, i.e. $\Phi \notin P$; (2) the *union* of the subsets in P is equal to the superset S, i.e. $\cup P_i = S$, and the partition P is said to *cover* or *space-fill* S; and (3) the *intersection* of any two of the subsets in P is empty, i.e. $P_i \cap P_j = \Phi$, and the subsets of P are said to be *pairwise disjoint*.

A *multicomponent dilute alloy or material* is any alloy or material manufactured from more than two components, but with no more than two components in substantial proportions, where substantial again means concentrated rather than dilute, i.e. at a concentration level above a few at%, so that atoms of that component interact with each other rather than being widely separated, and a dilute-solution approximation is unlikely to be reasonable in determining the effect of that component on the alloy structure and properties. Multicomponent dilute alloys are, therefore, all those multicomponent alloys that are not multiple-principle-element alloys. The name multicomponent dilute alloys again emphasises (this time in a negative sense) the second of the three underlying concepts, i.e. the *dilution* (or the lack of concentration) of its multiple components. Multicomponent dilute alloys, unlike multiple-principal-element alloys are not expected, therefore, to show strong non-linear or surprising behaviour in their structure and properties, which should instead be relatively predictable using a combination of the dilute-solution approximation and linear superposition of the effects of different solutes (though this does not necessarily mean that they are not interesting or potentially valuable).

The set of multiple-principal-element alloys and the set of multicomponent dilute alloys form a space-filling, disjoint partition of the set of multicomponent alloys. In other words, every possible multicomponent alloy is either a multiple-principal-element alloy or a multicomponent dilute alloy. And the partition of the set of multicomponent alloys into the two disjoint subsets of multiple-principal-element alloys and multicomponent dilute alloys is clearly based on the different concentration of the components in each subset: it depends again on the alloy or material type as determined by its composition alone. The majority of multicomponent alloys are almost certainly multiple-principal-element alloys rather than multicomponent dilute alloys, since the former occupy the massive central bulk of multicomponent phase space, whereas the latter are restricted to relatively restricted regions near its single-element corners and binary edges (though the field of multicomponent dilute alloys still contains an extremely large number of different materials, because of the enormous magnitude of multicomponent phase space).

High-entropy alloys

A *high-entropy alloy or material* is any multicomponent alloy or material in which the configurational entropy of mixing plays a major role in stabilising the alloy structure. The configurational entropy in these alloys and materials is high relative to other factors that affect the alloy structure (such as the interaction energies between the different atoms and the non-configurational entropy) and acts to stabilise a simple alloy structure of no more than one or two solid-solution phases with broad and variable stoichiometry, as opposed to multiple compound phases with fixed stoichiometric compositions. The name high-entropy alloys as defined in this way emphasises the third of the three underlying concepts, i.e. the role of *high entropy* in stabilising multicomponent single-phase structures. This is, of course, the underlying concept responsible for the surprisingly common occurrence of a single-phase structure in multicomponent alloys, first seen by Cantor et al.[1] in their discovery of the Cantor alloy CrMnFeCoNi, and first enunciated as an important principle by Yeh et al.[2] This is also the underlying concept that has been so influential in exciting scientists worldwide and promoting the field of multicomponent high-entropy alloys.[7–9]

Yeh et al.[2] originally used a more functional definition of high-entropy alloys or materials, as consisting of all alloys with five or more components in substantial proportions (with the word 'substantial' defined by Yeh et al.[2] as ranging between 5–35 at%). This definition has been widely quoted and used, but it suffers from two serious problems: first, the limits on the number of components and their composition are clearly somewhat arbitrary; and second, it includes (paradoxically and, therefore, misleadingly) many multiphase alloys consisting of mixtures of fixed-composition stoichiometric compounds, which simply do not have high entropy, and where the configurational entropy is not so significant in determining the alloy structure. The conceptual definition we are adopting here, as given above in terms of the role of a high configurational entropy of mixing, is, therefore, preferable. Compared to the original often-used definition, it excludes multicomponent alloys consisting of multiphase fixed-composition stoichiometric compounds, and includes alloys with only three or four components, where the configurational entropy of mixing is still high and plays a dominant role in determining the alloy structure. Both of these changes are desirable, since they reflect more accurately and respect more accurately the underlying scientific concept of the role of high entropy in stabilising single-phase structures and determining, therefore, the properties of many multicomponent alloys and materials.

A *multicomponent multiphase alloy or material* (or just a *multiphase alloy or material*, since a single component or binary alloy is extremely unlikely to be multiphase) is any multicomponent alloy or material in which the configurational entropy of mixing does *not* play a major role in stabilising the alloy structure. The configurational entropy in these alloys is low relative to other factors that affect the alloy structure (such as the interaction energies between the different atoms and the non-configurational entropy), which therefore act to stabilise a structure of multiple compound phases with relatively fixed stoichiometric compositions, as opposed to a simple alloy structure of no more than one or two solid-solution phases with broad and variable stoichiometry. Multiphase alloys are, therefore, all those multiple-principal-element alloys that are not high-entropy alloys. The name multiphase alloys again emphasises (this time in a negative sense) the third of the three underlying concepts, i.e. the role of a relatively *low entropy* in stabilising multiphase fixed-stoichiometry compound structures.

The set of high-entropy alloys and the set of multiphase alloys form a space-filling, disjoint partition of the set of multiple-principal-element alloys. In other words, every possible multiple-principal-element alloy is either a high-entropy alloy or a multiphase alloy. And the partition of the set of multiple-principal-element alloys into the two disjoint subsets of high-entropy alloys and multiphase alloys is clearly based on a different configurational entropy of the alloy structure: it depends this time on the type of alloy structure (single phase or multiphase) as determined not just by the alloy composition, but also by the external conditions of temperature and pressure, and the manufacturing route (in principle, any particular material with a given composition can be either a high-entropy alloy or a multiphase alloy, depending on the ambient conditions and/or the manufacturing route).[iii] The relative proportions of high-entropy alloys and multiphase alloys are not really known, since there has been only limited exploration of the vast depths of multicomponent phase space, and what exploration has taken place has been subject to quite a strong selection bias in favour of high-entropy alloys.

[iii] We are using the word alloy or material to refer to a *particular composition* with a *particular microstructure* (in this case, single phase or multiphase).

High-entropy alloy structures

Multicomponent solid solutions

A *multicomponent or high-entropy solid solution* is any alloy or material manufactured from more than two components that has a single-phase random solid-solution crystal structure, usually close packed such as face-centred cubic (fcc), body-centred cubic (bcc) or hexagonal close packed (hcp). A multicomponent solid solution is, therefore, a high-entropy alloy in which the high configurational entropy of mixing stabilises a single-phase close-packed fcc, bcc or hcp crystal structure, with typical metallic bonding and a metallic free-electron band structure.

Multicomponent compounds

A *multicomponent or high-entropy compound* is any alloy or material manufactured from more than two components that has a single-phase ordered compound crystal structure, with two or more sublattices, and a random distribution of chemically similar components on at least one sublattice (but sometimes on several or all sublattices). A multicomponent compound is, therefore, a high-entropy material in which the high configurational entropy of mixing on at least one sublattice stabilises a single-phase compound structure. Multicomponent compounds can have different kinds of chemical bonding, leading to different kinds of compound structure: *multicomponent intermetallic compounds* with typical metallic bonding and a free-electron band structure, such as the ordered bcc transition-metal mono-aluminides MAl (with a CsCl crystal structure), where M represents a combination of different transition metals; *multicomponent ionic compounds* with ionic bonding and localised electronic transfer between different ions, such as the ordered simple-cubic alkali halides MX (with a rock-salt NaCl crystal structure), where M is a combination of positively charged alkali metal cations and X is a combination of negatively charged halide anions; and *multicomponent covalent compounds* with covalent bonding and localised electron sharing between different atoms, such as the ordered bcc transition-metal mono-carbides MC (again with a rock-salt NaCl crystal structure), where M is a combination of transition metals.

Multicomponent glasses

A *multicomponent or high-entropy glass or amorphous alloy* is any alloy or material manufactured from more than two components that has a single-phase amorphous structure. A multicomponent glass is, therefore, a high-entropy material in which the high configurational entropy of mixing stabilises a single-phase amorphous structure. Multicomponent glasses can again have different kinds of chemical bonding, leading to different kinds of glassy material: *multicomponent glassy metals* with typical metallic bonding and a free-electron band structure; *multicomponent glassy ceramics* with random network structures and covalent bonding via localised electron sharing between the different atoms; and *multicomponent glassy polymers* with van der Waals bonding between tangled long-chain covalently bonded polymer molecules.

The set of multicomponent solid solutions, the set of multicomponent compounds, and the set of multicomponent glasses form a space-filling, disjoint partition of the set of high-entropy alloys. In other words, every possible high-entropy alloy is either a multicomponent solid solution, a multicomponent compound or a multicomponent glass. And the partition of the set of high-entropy alloys into three disjoint subsets of multicomponent solid solutions, compounds and glasses is again based clearly on a different type of alloy structure (random solid solution, compound or glass) as determined not only by the alloy composition but also by the external conditions of temperature and pressure, and the manufacturing route (in principle again, any particular material with a fixed composition can be either a multicomponent solid solution, compound or glass, depending on the ambient conditions and/or the manufacturing route).[iv] The relative proportions of multicomponent solid solutions, compounds and glasses are again not really known, since there has only been limited exploration of multicomponent phase space.

Somewhat confusingly, some authors reserve the name high-entropy alloys exclusively for what we have defined here as multicomponent solid solutions, and some other authors reserve the name high-entropy alloys for multicomponent solid solutions and multicomponent glasses, in both cases preferring not to use the name alloys for compounds. This is illogical. And it is not in the spirit of this book, which defines alloys as mixtures of different components of all kinds. It is preferable to define high-entropy alloys as we do here, as all alloys and materials with a high configurational entropy, which are then subdivided into multicomponent solid solutions, multicomponent glasses and multicomponent compounds.

Other alloy types

Medium-entropy alloys

A medium-entropy alloy or material is any ternary or quaternary alloy in which the configurational entropy of mixing is relatively high, and stabilises a single-phase solid solution structure, in just the same way as for a high-entropy alloy or material with a larger number of components. The name medium-entropy alloy was introduced[7–9] when it was realised that there are indeed some ternary and quaternary alloys of this type, which have a fairly high entropy but lie outside the original definition of high-entropy alloys, since they have fewer than five components. Intuitively, however, it seems more reasonable to include together all alloys or materials in which the configurational entropy of mixing is at least moderately high, and significant in stabilising a single-phase alloy structure (solid solution, amorphous or compound). This is achieved with the definition of high-entropy alloys given above. With this definition of high-entropy alloys, the need for the term medium-entropy alloys has essentially disappeared.

Complex concentrated alloys and compositionally complex alloys

These terms were introduced to overcome the problem that the original definition of high-entropy alloys includes multicomponent alloys with low configurational entropy of mixing, while the name (taking the words at their face value) excludes them.[7–9] However, the two terms are rather clunky and do not seem to add anything to the straightforward and

[iv] We are again using the word alloy or material to refer to a *particular composition* with a *particular microstructure* (in this case, solid solution, compound or glass).

well-established use of the names multicomponent alloys and multiple-principal-element alloys. With the definitions above of multicomponent alloys, multiple-principal-element alloys and high-entropy alloys, the need for the terms complex concentrated alloys and compositionally complex alloys has again essentially disappeared.

The overall Venn diagram

The different alloy types represented by single elements, binary alloys, multicomponent alloys, multiple-principal-element alloys, multicomponent dilute alloys, high-entropy alloys, multiphase alloys, multicomponent solid solutions, multicomponent compounds and multicomponent glasses are all subsets of the set of all materials. The relationships between these different alloys and alloy types are shown schematically on an overall Venn diagram[v] in Figure 3.1. The Venn diagram summarises the relationships between the different alloys and different alloy types as follows: the set of elements, the set of binary alloys, and the set of multicomponent alloys partition the set of all materials, i.e. every possible material is either a single element, a binary alloy or a multicomponent alloy; the set of multiple-principal-element alloys and the set of multicomponent dilute alloys partition the set of multicomponent alloys, i.e. every possible multicomponent alloy is either a multiple-principal-element alloy or a multicomponent dilute alloy; the set of high-entropy alloys and the set of multiphase alloys partition the set of multiple-principal-element alloys, i.e. every

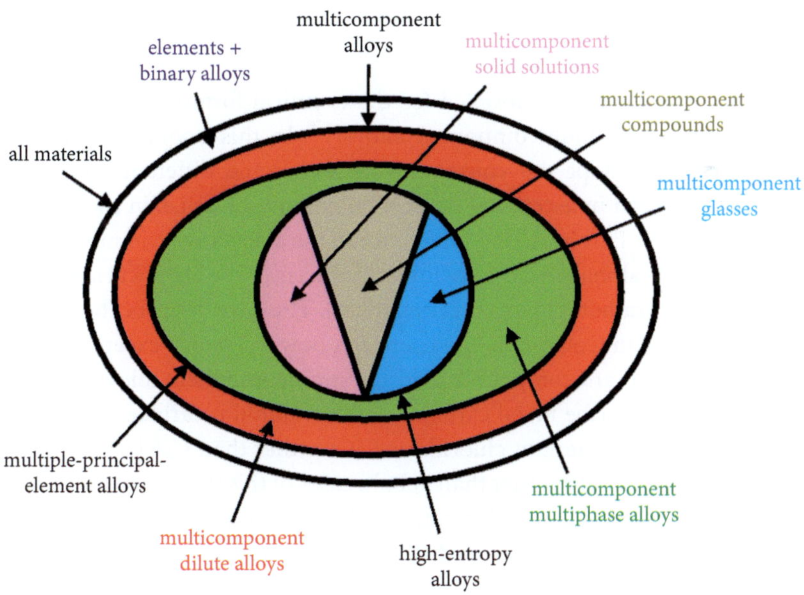

Figure 3.1 Overall Venn diagram showing schematically the relationships between different alloy types as subsets of the universal set of all materials.

[v] Note that the shapes of the different regions in the overall Venn diagram in Figure 3.1 are irrelevant, and the Venn diagram, according to the usual convention, represents schematically only the *relationships* between the different sets as shown by the relative positions of the different regions.

multiple-principal-element alloy is either a high-entropy alloy or a multiphase alloy; and the set of multicomponent solid solutions, the set of multicomponent compounds and the set of multicomponent glasses partition the set of high-entropy alloys; in other words, every high-entropy alloy is either a multicomponent solid solution, a multicomponent compound or a multicomponent glass. The Venn diagram also summarises the relative sizes of the different alloys and different alloy types as follows: the vast majority of materials are multicomponent alloys; the vast majority of multicomponent alloys are multiple-principal-element alloys; the numbers of high-entropy alloys and multiphase alloys are, as far as we know, more or less similar; and the numbers of multicomponent solid solutions, compounds and glasses are also, as far as we know, more or less similar.

Acronyms

It is quite common to abbreviate some of the names for different alloy types and use acronyms instead, namely, *MCAs*, *MPEAs*, *HEAs*, *MEAs* and *CCAs* for multicomponent alloys, multiple-principal-element alloys, high-entropy alloys, medium-entropy alloys and complex concentrated alloys or compositionally complex alloys, respectively. We do not use these acronyms any further in this book, preferring to avoid the rather indigestible and unreadable alphabetti-spaghetti text that then results.

Alloy compositions

Different conventions have been adopted for the chemical formulae used to describe the chemical compositions of multicomponent alloys.[7–9] In this book, we adopt two main conventions. First, we adopt the convention of listing the different components of a multicomponent material in ascending order of atomic number, i.e. in ascending position in the periodic table.[7–9] This is used in preference to the alternative common convention of listing the components alphabetically, because it emphasises the chemical similarities of the different components (rather than obfuscating them). Thus, we describe the Cantor alloy as CrMnFeCoNi rather than CoCrFeMnNi. There are, however, some exceptions to this rule, mostly for materials that are compounds and not multicomponent alloys. For instance, it would be perverse to describe metal carbides such us titanium carbide as CTi rather than TiC, alkali halides such as caesium chloride as ClCs rather than CsCl, or metal–metalloid glasses such iron–boron as BFe rather than FeB. In cases of this type we use the existing well-established chemical nomenclature. Second, we adopt the convention of using atomic ratio or atomic per cent as subscripts to designate the chemical composition of a multicomponent alloy, leaving out the subscripts in cases of equiatomic composition.[7–9] Thus, we describe (for instance) the Cantor alloy as either CrMnFeCoNi or $Cr_{20}Mn_{20}Fe_{20}Co_{20}Ni_{20}$, and an equiatomic iron-cobalt-nickel mono-aluminide as (FeCoNi)Al or $Al_{50}Fe_{16.7}Co_{16.7}Ni_{16.7}$ (in both cases usually using the former for simplicity). Notice that with this convention the composition of (for instance) the two aluminium-modified Cantor alloys $Al_{0.5}$CrMnFeCoNi and $Al_{0.5}$(CrMnFeCoNi) are different: the first is $Al_9Cr_{18.2}Mn_{18.2}Fe_{18.2}Co_{18.2}Ni_{18.2}$ and the second is $Al_{33.33}Cr_{13.33}Mn_{13.33}Fe_{13.33}Co_{13.33}Ni_{13.33}$.

3 The Size of Multicomponent Phase Space

Number of materials

The development of the field of multicomponent high-entropy materials began slowly in the late 1970s, with the first two seminal papers[1,2] not published until 2004[vi] after independent and painstaking initial work by Cantor et al.[1] and Yeh et al.[2] After the first papers were published and after a continuing slow start throughout the 2000s, there followed more than a decade of increasingly intensive research into multicomponent high-entropy materials that has led to a number of major new scientific and technological insights, as well as the discovery of many new and exciting materials,[7–9] all of which are discussed in later sections and chapters of this book. However, the very first major insight, discovered in the late 1970s and published in 2004 by Cantor et al.,[1] is probably the most significant and paradigm-shifting of all, namely, the discovery of the enormous size, scope and scale of multicomponent phase space, the literally stupendous number of associated materials that are potentially available for manufacture and use, and the correspondingly enormous number of exciting new material structures and properties available to be used in technological applications.

The size and scale of multicomponent phase space had not been realised before the work by Cantor et al.[1] and Yeh et al.[2] because, as described in Chapter 1, materials had previously been made using what seemed to be a natural and obvious strategy, namely, keeping down the number of alloying components and adding each component in relatively small proportions. This was essentially a received view about how to develop new alloys—an unspoken assumption in favour of parsimony in using alloying additions, underlain perhaps by a subconscious idea that increasing the number of components in a material and increasing their proportions should be avoided because it might make it difficult to manufacture the material satisfactorily. We now know that this approach to alloy development was somewhat blinkered. It had always been done that way. As a consequence, however, we had seemed to be running out of material options, leading some commentators to worry about a future paucity of materials, about the possibility of a failure to keep up the regular supply of new and better materials required by the ever-escalating demands of current (and expected future) improvements in technology.[vii]

We show later in this section that on the contrary it is in fact almost impossible to comprehend the vastness of multicomponent phase space and the mind-boggling number of multicomponent materials that are yet to be even thought of, let alone made or studied. The situation is in fact quite the reverse of previous received wisdom, and we are very far indeed from being in any danger of running out of new materials. It is not overstating the case to say that the discovery of multicomponent phase space has endowed us with an almost inexhaustible supply of new materials that we can continue to mine (metaphorically) in order

[vi] The basic idea of *multicomponent alloying* to manufacture *multicomponent materials* had already been published briefly a year or two earlier in 2002 and 2003, as described in detail in Chapter 2.

[vii] This refers to society's ever-increasing need for new *kinds* of materials with ever *better* properties—stronger, lighter, more ductile, more conducting, more insulating, more magnetic, better at solar energy conversion, more corrosion-resistant, more wear-resistant, etc.—to drive and support technological advances. This need is fulfilled by discovering new material formulations with new and improved functionalities. The possibility of exhausting some or any of the world's *raw* materials and, therefore, the supply of components from which we can manufacture materials is another matter, though this too is likely to be alleviated by discovering better and more efficient material formulations, including those more resilient to recycling, reshaping, reusing and life extension.

to support and drive improvements in peoples' health, wealth and wellbeing well into the far future.

As discussed in some detail in Chapter 2, the vastness of multicomponent phase space was first discovered in Cantor et al.'s early studies in the late 1970s and early 1980s,[1] was subsequently and independently supported by Yeh et al.'s early studies in the 1990s,[2] and emerged into the open literature in 2002 and then in detail in 2004.[1,2] The size of multicomponent phase space is enormous, and it is natural to ask the question just how big is it? And the corresponding number of multicomponent materials is also enormous, and it is again natural to ask just how many are there? Cantor[9,13] has shown how to calculate the total number of materials that could possibly be made. This number is, as mentioned, an important socio-political measure of the depth of new material resource and functionality potentially available to humankind.

The number of materials in any alloy system depends firstly on the number of components c and secondly on the material specification x, i.e. the difference in composition that separates two adjacent but different materials in the alloy system. For a material specification x in at% or mole%, there are $n = 100/x$ different compositions available for each of the components in the material. The material specification x is usually used for practical purposes,[14–16] to define the limits within which the composition of a given material can vary and it can still be treated as the same material with an acceptable and well-defined structure and properties. The material specification reflects essentially the precision within which we are able to manufacture a given material, as well as the precision we wish to achieve for consistency and, therefore, useful application of its properties.[14–16] It is important for anyone making a material upstream to know how tight the composition needs to be controlled during manufacture, and it is also important for anyone selling or purchasing a material downstream to know if they have supplied or received the correct as-specified material with its appropriate and guaranteed design properties.

Calculating the total number of materials in a given alloy system is a matter of combinatorial maths.[17,18] If one out of $n = 100/x$ atoms is changed from one kind to another, the composition changes by x at%, the material goes out of specification, and it is effectively changed into a different material. We need, therefore, to calculate how many different ways we can pick $n = 100/x$ atoms from c different kinds of atom (i.e. c different components), with the order of picking irrelevant so that the number of different ways of picking is a *combination* not a *permutation*, and with multiple picking of atoms of each type allowed so that the number of different ways of picking is *with repetition* and not *without repetition*. For c components and $n = 100/x$ composition points for each component, the total number of materials that can be made is given by the *law of combinations with repetition*:[17,18]

$$N = \binom{c+n-1}{n} = \frac{(c+n-1)!}{(c-1)!n!}$$
$$= \exp\{(c+n-1)\ln(c+n-1) - (c-1)\ln(c-1) - n\ln n\}$$

where the first expression on the right-hand side uses the bracket notation for a binomial coefficient, and the last expression comes from using Stirling's approximation for large numbers.

We can generate the whole of multicomponent phase space, containing all the possible materials that could conceivably be made, by using all the known elements in the periodic

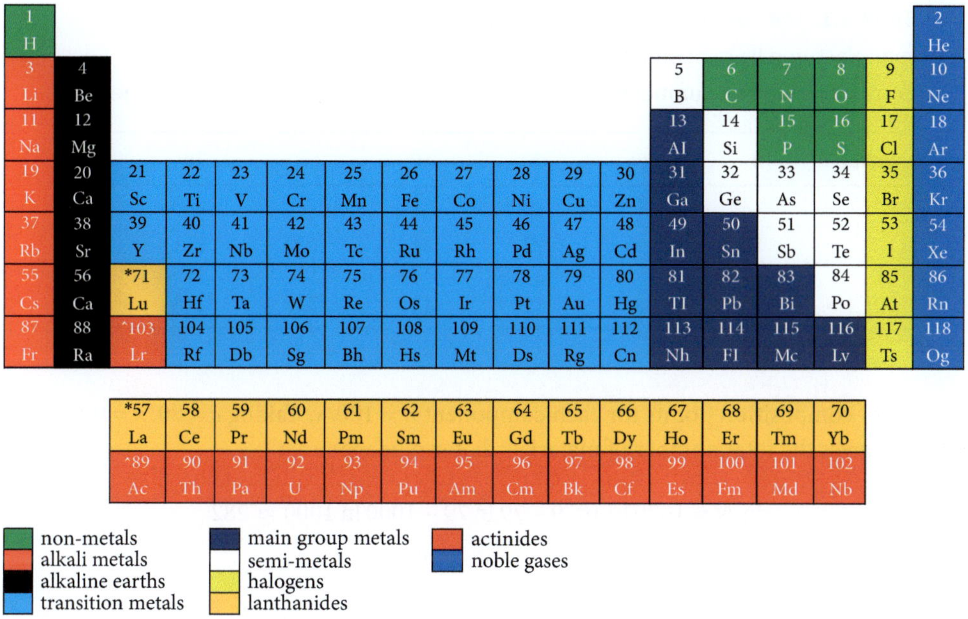

Figure 3.2 The periodic table.

table as our set of starting materials or components. Figure 3.2 shows the periodic table, including all the current known elements[19] up to a total number of $c = 118$. This covers the complete set of elements in the first seven rows, ranging from hydrogen with atomic number $z = 1$ to oganesson with $z = 118$ (the last few only discovered quite recently).[19] All the elements in row seven from francium Fr ($z = 87$) to oganesson ($z = 118$), including the actinides and the trans-actinides (or heavy elements) are radioactive, often with short lifetimes, and are, therefore, not available for manufacturing useful materials. Similarly, all the noble gases in column 18 from helium He to oganesson Og ($z = 2$, 10, 18, 36, 54, 86 and 118) are too chemically inert to be of much value in manufacturing useful materials. Excluding these elements (there are thirty-eight of them), we are left with a total of $c = 80$ possible components from which we can manufacture all possible materials. Cantor[9,13] has suggested that we might consider being more conservative, and leave out periods 1 and 2, and columns 15–17, which would reduce the number of components to $c = 64$, but he has also pointed out that this might seem rather unreasonable, since it would exclude important anionic compound-formers such as fluorine F and chlorine Cl, interstitial elements and covalent compound-formers such as carbon C, nitrogen N and boron B, light metals such as lithium Li and beryllium Be, and semi-metals such as selenium Se and tellurium Te.

Most engineering materials are specified to $x = 0.1\%$, though lower-grade materials are often specified to no better than $x = 1\%$. However, many high-performance materials require a greater degree of specification, down to $x = 0.01\%$, or 0.001%,[14–16] and in some special cases materials are designed and specified to the nearest part per million (ppm) or $x = 0.0001\%$ (for instance, to control very tightly the levels of interstitial impurities in high-strength steels[20] or dopants in semiconductors[21]).

Table 3.1 shows the total number of materials that can be made using different assumptions for the number of components c, the material specification x, and the corresponding

Table 3.1 The number of different materials N, in the form $\log_{10}(N)$, that can be manufactured from c components, with a material specification of x at%, corresponding to a number of composition points $n = 100/x$ for each component

x (at%)	n	$c =$	40	50	60	70	80
1	10^2		36	41	46	50	53
0.1	10^3		72	86	100	111	123
0.01	10^4		111	134	157	179	200
0.001	10^5		150	183	216	248	279

number of composition points for each component $n = 100/x$. Taking $c = 80$, $x = 0.1\%$, and $n = 1000$ gives the total number of materials that can be made as:

$$\ln N = 1079 \ln 1079 - 79 \ln 79 - 1000 \ln 1000 \approx 282$$

$$\therefore N = 10^{123}$$

The total number of materials that can be made is on the order of $N = 10^{123}$, which as already mentioned is an unimaginably large number. For comparison, the number of atoms in the galaxy is estimated[22] to be approximately 10^{68}, the number of atoms in the universe is estimated[23–25] to be approximately 10^{80} and the size of the universe is estimated[22] to be approximately 9.3 billion light years or 10^{36} nm. Even if we use Cantor's suggestion[9,13] of a more limited (though clearly over-conservative) number of components, $c = 64$, we get $N = 10^{104}$, and if in addition we also use a more limited (though clearly even more over-conservative) specification of $x = 1\%$, we get $N = 10^{47}$. Cantor[9,13] took an ultra-conservative approach, using $c = 60$, $x = 0.1\%$, and $n = 1000$, giving the total number of possible materials as 10^{100}, which is one googol (a convenient round number to remember), and with $c = 60$ and $x = 1\%$, we get $N = 10^{46}$. Some of the multicomponent alloy compositions would be difficult to manufacture and unlikely to be of much use, so the different estimates should be taken only, therefore, as broad approximations. Nevertheless, it is quite clear that, whatever assumptions are made, there is an unimaginably large number of materials populating multicomponent phase space.

As can be seen from the calculations given above and the results in Table 3.1, the number of materials that can be made is truly enormous. These calculations and results emphasise quite dramatically the ground-breaking significance of the first discovery of multicomponent alloys and the first appreciation of their significance by Cantor et al.,[1] subsequently supported strongly and independently by Yeh et al.[2]

Number of known materials

We can use the same methods of combinatorial maths[17,18] to calculate approximately how many materials have ever actually been made and, therefore, how many materials we actually know anything about at all. The number of known materials N_k is less than the number of all possible ternary materials, $N_k < N_t$ because: on the one hand there are still many ternary alloy systems for which little or nothing is known, at least in the central composition field in the middle of many ternary phase diagrams; and on the other hand despite the extent of recent research into multicomponent high-entropy materials, we still know very little about the vast majority of possible quaternary and/or higher order alloy systems. With c components, the number of different ternary systems is $c(c-1)(c-2)$, and the number of materials in each ternary system is given (again) by the law of combinations with repetition,[17,18] this time for just three components. Taking (again) the number of components as $c = 80$, and the material specification as $x = 0.1\%$, corresponding to $n = 100/x = 1000$ composition points for each component, the number of known materials N_k is given by:

$$N_k < N_t = c(c-1)(c-2)\binom{3+n-1}{n}$$

$$= c(c-1)(c-2)\binom{n+2}{n} = \frac{c(c-1)(c-2)(n+2)!}{2!n!}$$

$$= \frac{80 \times 79 \times 78 \times 1002 \times 1001}{2} = 2.4722 \times 10^{11} \approx 2.5 \times 10^{11}$$

In other words the number of known materials is in the order of $N_k \approx 2.5 \times 10^{11}$ and the fraction f_k of all materials that are known is infinitesimally small:

$$f_k = N_k/N < 10^{-111}$$

The number of alloy systems

We can again use similar methods of combinatorial maths[17,18] to calculate the number of different alloy systems and the number of different materials in each alloy system. For instance, many studies of multicomponent alloys have focused on 5-component alloy systems, partly because of Cantor et al.'s early and surprising discovery[1] of an extended single-phase solid solution in the equiatomic 5-component Cantor alloy CrMnFeCoNi, and partly because of Yeh et al.'s original definition[2] of high-entropy alloys as consisting of a minimum of five components. It is natural to ask, therefore, how many 5-component alloy systems are there, how many materials are there in each 5-component alloy system, and what is the total number of possible 5-component materials? It is also natural to ask similar questions about 6-component alloy systems, 7-component alloy systems and so on, i.e. about c-component systems generally.

To calculate how many different c-component alloy systems there are, we need to work out how many different ways we can pick c components from a total number of C possible components, with the order of picking irrelevant so that the number of different ways of picking is a *combination* not a *permutation*, and with multiple picking of components not (in this case) allowed so that the number of different ways of picking is *without repetition* and not *with repetition*. The number of c-component alloy systems s_c is given by the *law of combinations without repetition*:[17,18]

$$s_c = \binom{C}{c} = \frac{C!}{c!\,(C-c)!}$$
$$= \exp\{C \ln C - c \ln c - (C-c) \ln (C-c)\}$$

where the first expression on the right-hand side again uses the bracket notation for a binomial coefficient, and the last expression again comes from using Stirling's approximation for large numbers. To calculate how many different materials n_c there are in each c-component alloy system, we can use our previous expression obtained from the law of combinations with repetition for an alloy system with c components, a material specification x and a number of possible composition points for each component $n = 100/x$:

$$n_c = \binom{c+n-1}{n} = \frac{(c+n-1)!}{(c-1)!n!}$$
$$= \exp\{(c+n-1) \ln (c+n-1) - (c-1) \ln (c-1) - n \ln n\}$$

where the first expression on the right-hand side again uses the bracket notation for a binomial coefficient, and the last expression again comes from using Stirling's approximation for large numbers. Finally, the total number of different c-component materials N_c is given by the number of c-component alloy systems s_c multiplied by the number of materials n_c in each c-component alloy system:

$$N_c = s_c n_c$$

Table 3.2 shows the resulting number of c-component alloy systems s_c, the number of materials in each c-component alloy system n_c, and the total number of c-component materials $N_c = s_c n_c$, for different numbers of components c, using a total number of possible components $C = 80$, a material specification $x = 0.1\%$, and a number of composition points

Table 3.2 The number of c-component alloy systems s_c, number of materials in each c-component alloy system n_c, and total number of c-component materials $N_c = s_c n_c$, for a variety of different numbers of components c, using a total number of components $C = 80$, a material specification $x = 0.1\%$, and a number of composition points for each component $n = 1000$.

number of components c	number of alloy systems s_c	number of materials in each system n_c	total number of c-component materials $N_c = s_c n_c$
4	7.9×10^{6}	7.5×10^{8}	5.9×10^{15}
5	1.3×10^{8}	2.1×10^{11}	2.7×10^{19}
6	1.8×10^{9}	4.8×10^{13}	8.6×10^{22}
7	2.0×10^{10}	8.8×10^{15}	1.8×10^{26}
10	1.2×10^{13}	2.1×10^{22}	2.5×10^{35}
20	3.4×10^{19}	1.1×10^{41}	3.7×10^{50}

Table 3.3 The number of c-component alloy systems s_c, number of materials in each c-component alloy system n_c, and total number of c-component materials $N_c = s_c n_c$, for a variety of different numbers of components c, using a total number of components $C = 60$, a material specification $x = 1\%$, and a number of composition points for each component $n = 100$.

number of components c	number of alloy systems s_c	number of materials in each system n_c	total number of c-component materials $N_c = s_c n_c$
4	2.4×10^6	7.8×10^5	1.9×10^{12}
5	3.0×10^7	2.3×10^7	6.9×10^{14}
6	3.0×10^8	5.4×10^8	1.6×10^{17}
7	2.4×10^9	1.0×10^{10}	2.4×10^{19}
10	5.5×10^{11}	3.1×10^{13}	1.7×10^{25}
20	3.8×10^{16}	4.9×10^{22}	1.9×10^{39}

for each component $n = 1000$. There are, for example, eight million 4-component alloy systems, each containing three-quarters of a billion different materials; over a hundred million 5-component alloy systems, each containing twenty billion different materials; almost two billion 6-component alloy systems, each containing almost fifty trillion different materials; and, of course, even more insanely large numbers of different alloy systems and materials for ten, twenty or more components. Table 3.3 shows equivalent results for the ultra-conservative assumption of a number of possible components $C = 60$, a material specification $x = 1\%$, and a number of composition points for each component $n = 100/x = 100$. The corresponding numbers of different alloy systems and materials are (not surprisingly) a bit smaller but are still quite gigantic. The results in Tables 3.2 and 3.3 serve yet again to emphasise strongly the vastness of multicomponent phase space and the enormous number of multicomponent alloys that are yet to be made and investigated.

4 Multicomponent Phase Diagrams

The Gibbs phase rule

The Gibbs phase rule describes the fundamental relationship between the number of components c and the number of phases p that co-exist at equilibrium within any given material under specified ambient conditions.[3–5] A *component* is defined as a starting material from which the material is manufactured, from a fundamental point of view effectively one of the elements from the periodic table. And a *phase* is defined as a region of space within a material that is homogeneous in its structure and properties. The Gibbs phase rule says that at equilibrium:

$$p + f = c + 2$$

where f is called the number of *thermodynamic degrees of freedom* in the material, i.e. the number of external defining conditions—the material composition, the pressure and the temperature—that can be varied independently.[viii]

We can understand this as follows.[5] In any phase at equilibrium, the temperature, pressure and composition must all be constant and uniform. Otherwise, there would be a flow of heat, volume (expansion or contraction) or matter within the phase. If the temperature varied from place to place, the phase would not be at equilibrium, and heat would flow from the hotter to the colder place until equilibrium was established at an intermediate constant temperature. Similarly, if the pressure varied from place to place, the phase would again not be at equilibrium, and the higher pressure would expand against the lower pressure until equilibrium was established at an intermediate constant pressure. And, finally, if the composition varied from place to place, the phase would again not be at equilibrium, and matter would flow from the higher concentration to the lower concentration until equilibrium was established at an intermediate constant composition.

The number of variables needed to specify the condition or state of each phase is, therefore, $c - 1$ independent composition variables (the composition of all the components must add up to 100%, so it is only necessary to specify $c - 1$ composition variables) plus the temperature and pressure, i.e. $(c - 1) + 2 = c + 1$. The total number of variables $n_{variables}$ is, therefore, the number of phases p multiplied by the number of variables needed to specify each phase $c + 1$:

$$n_{variables} = p\,(c + 1)$$

But at equilibrium the temperature, pressure and chemical potential of each component must not only be constant within each phase, but also be the same in all phases. Otherwise there would again be a flow of heat, volume (expansion or contraction) or matter, this time from one phase to another, but in the same way as described above within each phase. The total number of equations $n_{equations}$ connecting all the variables is, therefore, the number of quantities which must be equal in all phases $c + 2$ multiplied by the number of equations needed for equality across all phases $p - 1$:

$$n_{equations} = (p - 1)\,(c + 2)$$

The number of degrees of freedom (i.e. the number of independent variables) f is then the number of variables minus the number of connecting equations:

$$f = n_{variables} - n_{equations}$$

$$= p\,(c + 1) - (p - 1)\,(c + 2) = (pc + p) - (pc + 2p - c - 2) = c - p + 2$$

$$\therefore p + f = c + 2$$

viii The Gibbs phase rule above describes the relationship between the number of components c and the number of phases p that co-exist at equilibrium as $p + f = c + 2$ when, as well as the material composition, there are just two ambient conditions that can be varied independently, namely, the temperature and pressure. If we are interested in the effect of additional independent ambient conditions such as, for instance, electric field or magnetic field, then the Gibbs phase rule is modified to become $p + c = f + 3$ or $p + c = f + 4$, adding one to the right-hand side of the equation for each additional independent ambient condition.

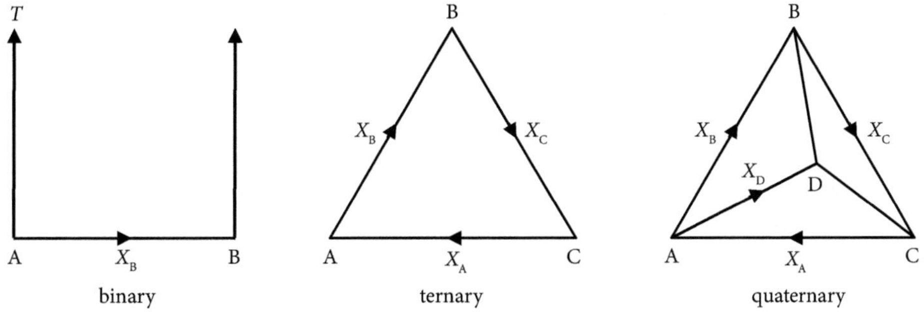

Figure 3.3 Schematic Cartesian, triangular and tetrahedral phase spaces representing binary, ternary and quaternary phase diagrams respectively.

which is the Gibbs phase rule. At constant (e.g. atmospheric) pressure, the phase rule becomes $p + f = c + 1$; and at constant (e.g. atmospheric) pressure and constant (e.g. room) temperature, it becomes $p + f = c$.

Phase diagrams

As explained at the beginning of this chapter, *phase diagrams* are used to show graphically[3–5] the range of structures and properties for all the materials in a given alloy system under different conditions of material composition, temperature and pressure.[ix]

Binary phase diagrams are usually[3–5] plotted under conditions of constant pressure, with two variables, namely, temperature T and composition x_B. There is only one composition variable: the sum of the concentrations of the two components must add up to 100% ($x_A + x_B = 1$) where x_A and x_B are atom fractions of the two components A and B. A Cartesian phase space plotting temperature versus composition on the two axes is used for any particular binary phase diagram to represent graphically the range of different binary alloy structures, as shown schematically in Figure 3.3. Each of the points separated by x at% along the horizontal composition axis represents a different material. According to the Gibbs phase rule at constant pressure $p + f = c + 1 = 3$, so single-phase materials have two degrees of freedom, appearing in single-phase fields on the phase diagram within which the temperature and composition can be varied independently; two-phase materials have only one degree of freedom, appearing in two-phase fields on the phase diagram, which are regions consisting of bundles of constant-temperature tie lines along which only the composition can be varied; and three-phase materials have no degrees of freedom, appearing only at invariant points on the phase diagram, which are fixed in space and allow for no variation in either temperature or composition.[3–5]

Ternary phase diagrams are usually[3,4] plotted under conditions of constant temperature and pressure, with two composition variables x_B and x_C. There are only two composition variables because the sum of the concentrations of the three components must add up to

[ix] The discussion of phase diagrams that follows ignores possible variations in material structure and properties with additional (independent) ambient conditions such as, for instance, electric or magnetic field, each of which would require an additional dimension for the relevant phase diagram.

100% ($x_A + x_B + x_C = 1$) where x_A, x_B and x_C are atom fractions of the three components A, B and C). A triangular phase space plotting the compositions of the three components on the three axes is used for any particular ternary phase diagram to represent graphically the range of different ternary alloy structures at any given temperature and pressure, as shown schematically in Figure 3.3. Each of the points separated by x at% along any of the three composition axes represents a different material. According to the Gibbs phase rule at constant temperature and pressure $p + f = c = 3$, so single-phase materials have two degrees of freedom, appearing in single-phase fields on the phase diagram within which the concentration of two components can be varied independently; two-phase materials have only one degree of freedom, appearing in two-phase fields on the phase diagram, which are regions consisting of bundles of tie lines along which the concentration of only one component can be varied independently; and three-phase regions have no degrees of freedom, appearing on the phase diagram at invariant points that are fixed in space and allow for no variation in the concentration of either component.[3–5]

Quaternary phase diagrams can in principle (but with difficulty) be plotted under conditions of constant temperature and pressure, with three composition variables x_B, x_C and x_D. There are only three composition variables because the sum of the concentrations of the four components must add up to 100% ($x_A+x_B+x_C+x_D = 1$) where x_A, x_B, x_C and x_D are atom fractions of the four components A, B, C and D). A three-dimensional tetrahedral phase space plotting the compositions of the four components on the four axes is used for any particular quaternary phase diagram to represent graphically the range of different quaternary alloy structures at any given temperature and pressure, as shown schematically in Figure 3.3. Each of the points separated by x at% along the four composition axes represents a different material. According to the Gibbs phase rule at constant temperature and pressure $p + f = c = 4$, so single-phase regions have three degrees of freedom, appearing in three-dimensional single-phase fields in the phase diagram (we say "in" instead of "on" the phase diagram because of its three (or more)-dimensional nature for $c > 2$) within which the concentration of three components can be varied independently; two-phase regions have two degrees of freedom, appearing in two-phase fields in the phase diagram, which are three-dimensional regions consisting of bundles of tie triangles within which the concentration of only two components can be varied independently; three-phase regions have only one degree of freedom, appearing in three-phase fields in the phase diagram, which are two-dimensional regions consisting of bundles of tie lines along which the concentration of only one component can be varied independently; and four-phase regions have no degrees of freedom, appearing at invariant points in the phase diagram, which are fixed in space and allow for no variation in the concentration of any of the components. Quaternary phase diagrams are not often plotted in this way, because of the difficulty of representing a three-dimensional tetrahedral phase space on the two-dimensional page.

By extension from the descriptions of binary, ternary and quaternary phase diagrams above, *multicomponent phase diagrams* with any (greater) number of components c can in principle[9,13] be plotted under conditions of constant temperature and pressure with $c - 1$ compositional variables, $x_B, x_C, \ldots,$ etc. There are only $c - 1$ composition variables because the sum of the concentrations of the c components must add up to 100% ($\sum x_i = 1$ where x_i is the atom fraction of the ith component). A c-cornered hyper-polyhedron phase space in $(c - 1)$ dimensions (technically known as a *c-simplex*) plotting the compositions of the c components on the c axes is needed for any particular multicomponent phase diagram to

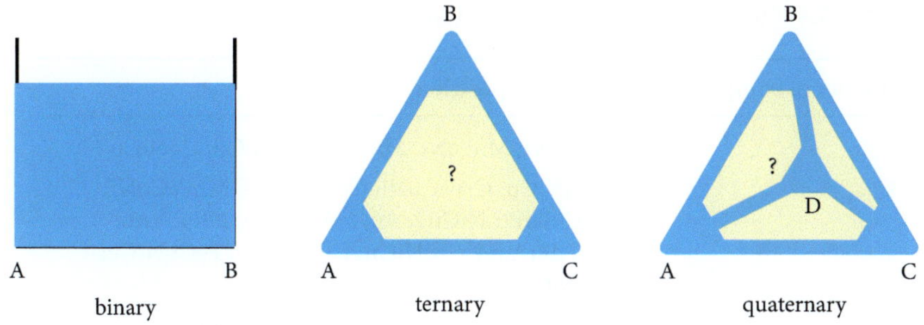

Figure 3.4 Schematic binary, ternary and quaternary phase diagrams showing well-known materials (in blue) near the single element corners and binary edges, and poorly known or unknown materials (in yellow) in the central regions.

represent graphically the range of different multicomponent alloy structures at any given temperature and pressure. Each of the points separated by x at% along the c composition axes represents a different material. According to the Gibbs phase rule at constant temperature and pressure $p + f = c$, so single-phase regions have $c - 1$ degrees of freedom, appearing in $(c - 1)$-dimensional single-phase fields in the phase diagram within which the concentration of $(c - 1)$ components can be varied independently; two-phase regions have $(c - 2)$ degrees of freedom, appearing in $(c - 2)$-dimensional regions in the phase diagram within which the concentration of only $(c - 2)$ components can be varied independently; and so on, down to c-phase regions with no degrees of freedom, appearing at invariant points in the phase diagram, which are fixed in space and allow for no variation in the concentration of any of the components. Such multicomponent phase diagrams can only be plotted in c-dimensional space, i.e. can only be plotted in principle and not in practice, but it is not too difficult to conceptualise multicomponent phase diagrams as extensions of our more familiar binary, ternary and quaternary phase diagrams.

The whole of multicomponent phase space can be represented[9,13] on a multicomponent phase diagram where the number of components c is taken to be the full set of all possible components, i.e. the full set of all usable elements C. In other words, the full range of all possible materials is represented in multicomponent phase space by the set of all the different composition points in a multicomponent phase diagram consisting of $c = C$ components, i.e. all points that are separated by at least x% in at least one of the C dimensions within a C-cornered hyper-polyhedron or C-simplex. As already pointed out, it is not possible in practice to draw this hyper-polyhedron, but it is not too difficult to conceptualise it as an extension of our familiar binary, ternary and quaternary phase diagrams, in which all possible compositions and different materials are represented as composition points along a line or within a triangle or a tetrahedron respectively.

Figure 3.4 shows typical binary, ternary and quaternary phase diagrams, demonstrating schematically our good knowledge about dilute materials near the single element corners and along the binary edges (shaded in blue), but our poor knowledge of concentrated materials in the central ternary and quaternary regions (shaded in yellow).[9,13] Similarly, our knowledge of multicomponent materials is restricted to dilute materials near the single element corners and along the binary material edges of any higher-order multicomponent phase diagram.[9,13]

Table 3.4 Equiatomic single-phase fcc Cantor alloys.

number of components	alloy composition
3	CrCoNi; MnFeNi; MnCoNi; FeNiCo; CoNiCu
4	AlTiNiCu; CrFeCoNi; CrMnCoNi; MnFeCoNi; NiCuPdPt; NiCuPdAu; NiCuPtAu; NiPdPtAu; CuPdPtAu; CrCoNiCu; FeCoNiCu; FeCoNiCu; CoNiCuZn
5	AlCoNiCuZn; TiCrFeCoNi; TiFeCoNiCu; VFeCoNiCu; CrFeCoNiCu; CrFeCoNiCu; CrFeNiCuMo; MnFeCoNiCu; MnFeNiCuPt; NiCuPdPtAu
6	VCrMnFeCoNi; CrMnFeCoNiCu; CrMnFeCoNiNb

5 Multicomponent Phases

Cantor alloys

As first discovered by Cantor et al.,[1] there is an enormous region in multicomponent phase space consisting of materials with a single-phase fcc structure, made up mostly (but not exclusively) of middle-to-late transition-metal components, loosely based around the original Cantor alloy composition CrMnFeCoNi, and including alloys with a wide range of different components and a wide range of different compositions.[8,9] Materials in this region of multicomponent phase space are now often referred to as *modified Cantor alloys* or, more simply, just *Cantor alloys*.[9] The size of this region of multicomponent phase space is very large, and there are literally millions of different Cantor alloys, forming a truly *extended* multicomponent fcc solid-solution single-phase field[8,9] (with the emphasis on the word *extended*).

The wide range of single-phase fcc Cantor alloys in multicomponent phase space is stabilised at least partly by a high configurational entropy of mixing, as first suggested by Yeh et al.[2] (this stabilisation and the entropy of mixing are discussed in more detail, along with other thermodynamic issues, in Chapter 4). The discovery of the original Cantor alloy and its surrounding extended region of single-phase fcc solid-solution alloys undoubtedly played a critical role in enthusing and exciting scientists worldwide, leading to the rapid build-up of the multicomponent high-entropy materials field throughout the 2010s and 2020s. Many single-phase fcc Cantor alloys have now been made by different processing methods and, at least to some extent, had their structure and properties studied (probably a few hundred different alloys have been made, with as many as perhaps several thousand independent reports).[8,9] They are almost certainly the most heavily studied type of multicomponent material so far. But it is quite clear that there are still many, many more Cantor alloys yet to be manufactured and investigated.

Some typical examples of equiatomic and non-equiatomic Cantor alloys (by no means all) are collected together in Tables 3.4 and 3.5, respectively. It is not, of course, possible easily (or, indeed, at all) to draw or visualise the complex multidimensional shape of the fcc

Table 3.5 Non-equiatomic single-phase fcc Cantor alloys.

number of components	alloy composition	compositionx (at %) range x (at %)
4	$Al_xFeNiCu$	0–14
	$CrFeCoNi_x$	20–40
	$CrCoNiFe_x$	0–17
	Mn_x FeCoNi	10
5	$Al_xCrFeCoNi$	0–14
	$Al_xSi_xFeCoNi$	0–8
	$CrMn_xFeCoNi$	0–14
	$CrMnFeCoNi_x$	0–100
	$CrFeCoNiCu_x$	0–11
	$CrMnFeNiCu_x$	12–24
	$CrFeCoNiMo_x$	0–20
6	$Al_xCrMnFeCoNi$	0–8
	$Ti_xCrMnFeCoNi$	0–2
	$V_xCrMnFeCoNi$	0–16
	$CrMnFeCoNiCu_x$	0–16
	$CrMnFeCoNiNb_x$	0–16

single-phase field within the C-dimensional hyper-polyhedral phase diagram that represents multicomponent phase space. The best that can be done is perhaps to indicate graphically the approximate extent of the single-phase fcc composition range for some of the different components, as shown in Figure 3.5. The composition range for each component varies, of course, depending on the temperature and on what other components and how much of each are present in the particular alloy, and Figure 3.5 gives only an approximate maximum width of the composition range that has been reported in each case (i.e. each component will dissolve in the single-phase fcc structure across approximately this or a smaller composition range, depending upon the temperature and the overall alloy composition—the quoted composition ranges are almost certainly underestimates of the real maximum composition ranges, since not all temperatures and alloy compositions (by a long way) have yet been investigated).

As well as the original Cantor alloy CrMnFeCoNi, Tables 3.4 and 3.5 and Figure 3.5 show that the single-phase fcc region of Cantor alloys includes a wide range of different materials,[7–9] for instance: equiatomic 5- and 6-component alloys such as TiCrFeCoNi,[26] MnFeNiCuPt,[27] NiCuPdPtAu[28] and VCrMnFeCoNi;[1] equiatomic 3- and 4-component alloys such as CrCoNi,[29] CoNiCu,[30] CrFeCoNi,[29] CoNiCuZn[30] and NiPdPtAu;[31] and non-equiatomic 5- and 6-component alloys such as (CrFeCoNi)-11at%Cu,[32] (CrFeCoNi)-10at%Mo,[33] (CrMnFeCoNi)-8at%Al[34] and (CrMnFeCoNi)-2at%Ti.[35] More details of the range of single-phase fcc Cantor alloys are given in the review paper by Cantor.[9]

Two of the most heavily studied multicomponent materials are the original Cantor alloy CrMnFeCoNi and the Al-modified Cantor alloy $Al_xCrMnFeCoNi$, mainly because these were the two multicomponent single-phase fcc alloys first discovered by Cantor et al.[1] and Yeh et al.,[2] respectively. Bracq et al.[36] varied independently each of the five components of the original Cantor alloy CrMnFeCoNi, and found single-phase fcc alloys over the composition

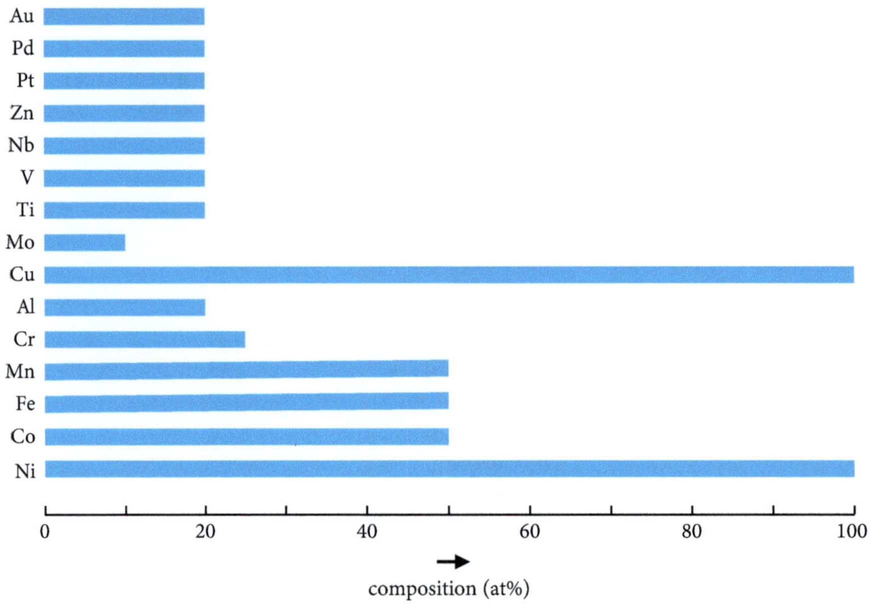

Figure 3.5 Approximate maximum composition ranges for different components in single-phase fcc Cantor alloys.

ranges 0–25 at % Cr, 0–50 at% Mn, 0–50 at % Fe, 0–50 at % Co and 0–100 at% Ni, in reasonable agreement with Calphad[4,37] thermodynamic calculations of single-phase fcc stability at 1,100 °C (1,373K), but not at room temperature (where Calphad predicts that the alloys should all become multiphase). A summary of these results is shown on a set of pseudo-binary phase diagrams in Figure 3.6. This is one example of many studies that confirm, as shown in Tables 3.4 and 3.5 and Figure 3.5, the large size and complex shape of the fcc Cantor alloy single-phase field in multicomponent phase space; and that also show moderate agreement between experimental observation and thermodynamic calculation (though there is an element of circularity in this agreement, since thermodynamic calculations are themselves based initially on experimentally determined phase fields and their associated properties, as discussed in more detail in Chapter 4).

As discovered in the very first studies by Cantor et al.,[1] fcc Cantor alloys exhibit dendritic microstructures when manufactured by melting and casting. The scale of the dendritic segregation patterns is similar to conventional binary and ternary alloys, ranging typically from 5–50 μm depending on the alloy composition and the cooling rate during casting.[1,9,38] Laurent-Brocq et al.[38] used high-resolution electron-probe microanalysis (EPMA) to make detailed composition measurements on arc melted/suction cast ingots of the original Cantor alloy CrMnFeCoNi, and found enrichment of chromium, iron and cobalt in the as-solidified dendritic spines, and enrichment of manganese and nickel in the interdendritic regions, as can be seen in Figure 3.7. More careful analysis of the EPMA results, together with differential scanning calorimetry (DSC) on re-heating,[38] demonstrated that solidification of the Cantor alloy follows the near-pseudo-binary phase diagram shown in Figure 3.8. At the Cantor alloy composition CrMnFeCoNi (corresponding to a pseudo-binary composition of $(CrFeCo)_{60}(MnNi)_{40}$ or c_o = 40 at% on Figure 3.8), the pseudo-binary liquidus and solidus temperatures are T_l ~ 1340 °C (1613K) and T_s ~ 1290 °C (1563K), i.e. there is a solid-liquid

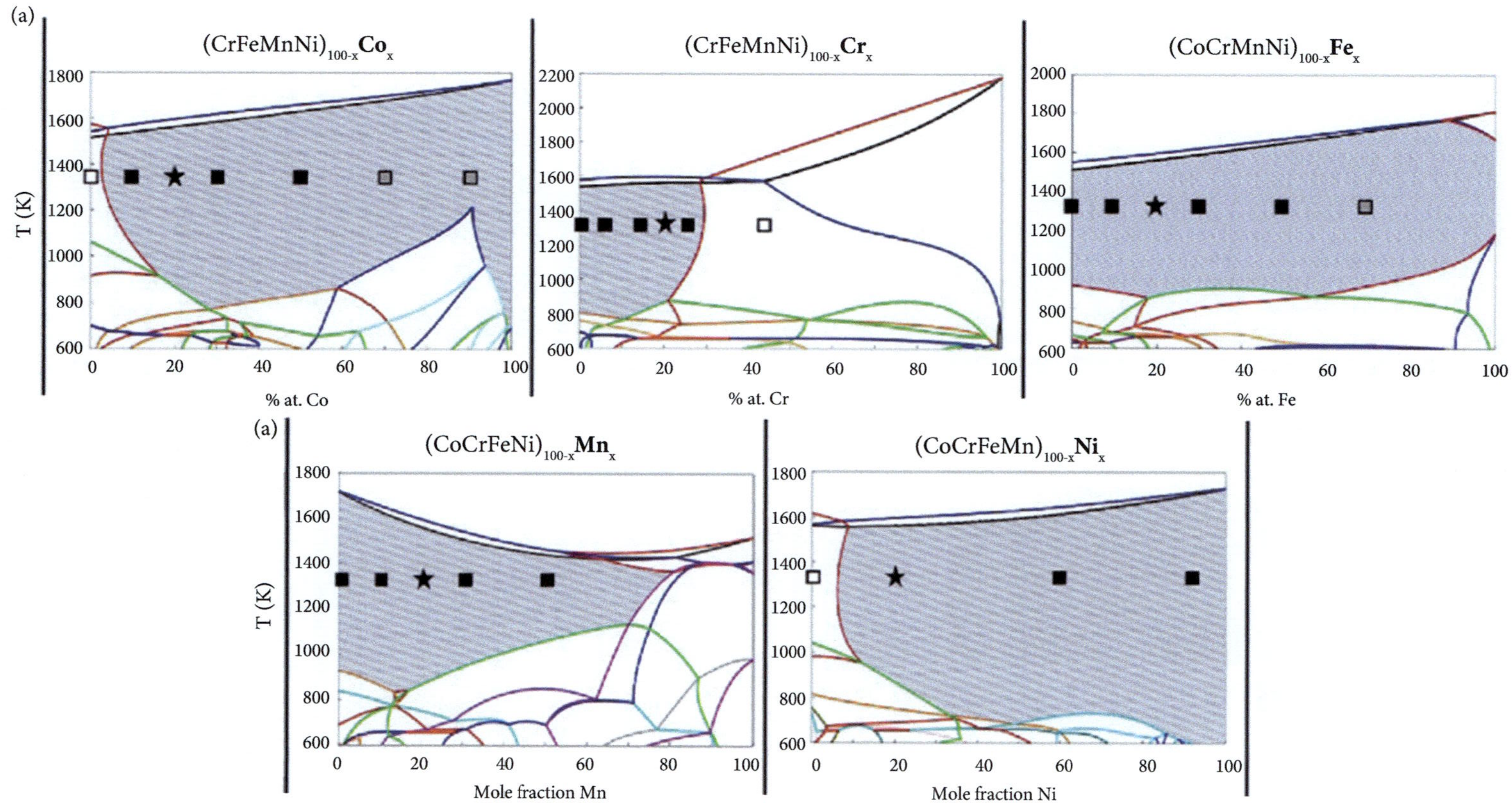

Figure 3.6 Pseudo-binary phase diagrams showing experimental results for the effect of varying independently the individual components of the Cantor alloy CrMnFeCoNi, compared with Calphad thermodynamic calculations.

*Solid symbols = single-phase fcc at room and elevated temperatures; open symbols = two-phase at room and elevated temperatures; grey symbols = single-phase fcc at elevated temperature + two-phase at room temperature; star = original Cantor alloy composition; grey-shaded region = Calphad predicted fcc single-phase field shaded grey).

Reproduced from Bracq et al. (2015). Insights into the phase diagram of the CrMnFeCoNi high entropy alloy. *Acta Materialia 88*: 355–65.

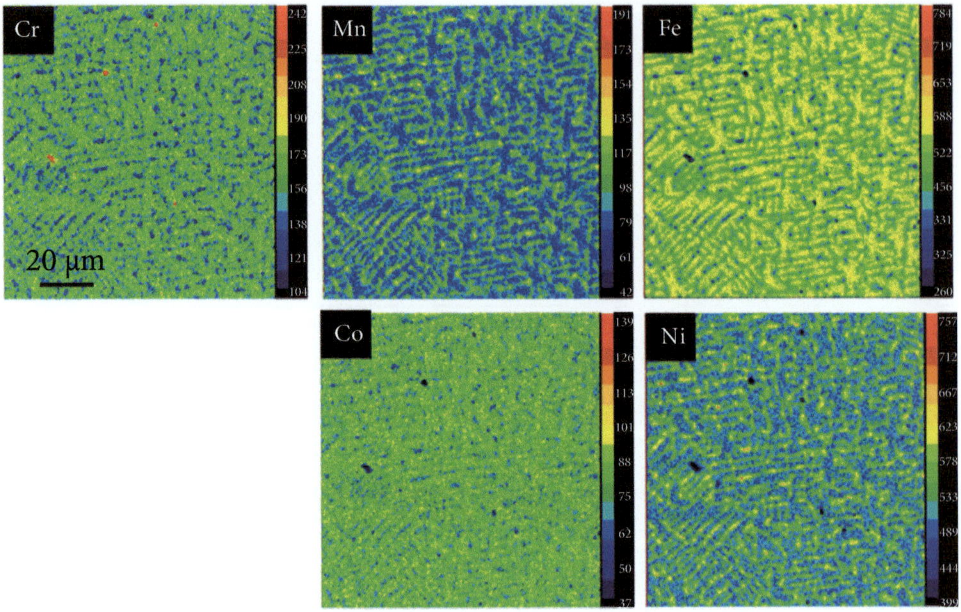

Figure 3.7 EPMA composition maps of the as-cast single-phase fcc Cantor alloy CrMnFeCoNi, showing segregation of Cr, Fe and Co into the dendrite spines and Mn and Ni into the interdendritic regions.

*Yellow = high solute concentration; blue = low solute concentration.
Reproduced from Bracq et al. (2015). Insights into the phase diagram of the CrMnFeCoNi high entropy alloy. *Acta Materialia 88*: 355–65.

thermal gap of $T_l - T_s = 50$ °C (50K); and from the Scheil equation,[5] the interdendritic microsegregation corresponds to a partition coefficient of $k \approx 0.75$, with pseudo-binary initial and final solidification compositions in the middle of the dendrite spines and the interdendritic spaces of $c_o = 30$ at% and $c_o/k = 53$ at%, respectively.[38]

In general, Cantor alloys exhibit high strength and extensive ductility (as discussed in more detail in later chapters), and thermomechanical processing using heavy deformation followed by soaking at high temperature leads to homogenisation of the alloy composition and recrystallisation of the alloy microstructure, resulting in beautiful equiaxed grain structures with occasional annealing twins, as shown in Figure 3.9 in the original Cantor alloy CrMnFeCoNi,[39] with grain sizes ranging from 1–500 μm depending on the detailed deformation and heat-treatment conditions. Similar overall microstructural features are found in many Cantor alloys with a wide range of different compositions.[5,8] Atom-probe (AP) composition maps of the homogenised and recrystallised Cantor alloy are shown in Figure 3.10, with no detectable variations in composition at the atomic scale.[38] Detailed statistical analysis of AP results of this type shows that all five elements exhibit a near-perfect random distribution at the atomic level across the fcc lattice, with no evidence of any significant clustering or short-range ordering of the different components, i.e. the AP results show that the Cantor alloy is thermodynamically a near-perfect ideal multicomponent solid solution.[38]

The Al-modified Cantor alloys $Al_xCrMnFeCoNi$ and $Al_xCrFeCoNi$ have been studied in detail by He et al.,[40] Hsu et al.[41] and Cieslak et al.[42] At high temperatures, the alloys are single-phase fcc for compositions ≤ 8 at% Al, two-phase fcc + bcc for compositions

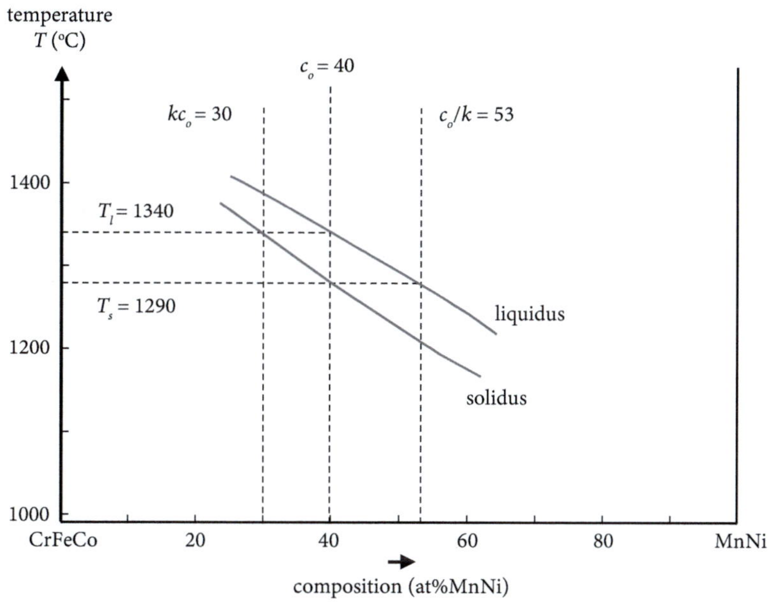

Figure 3.8 Pseudo-binary phase diagram for solidification of the Cantor alloy CrMnFeCoNi.

Reproduced from Bracq et al. (2015). Insights into the phase diagram of the CrMnFeCoNi high entropy alloy. *Acta Materialia 88*: 355–65.

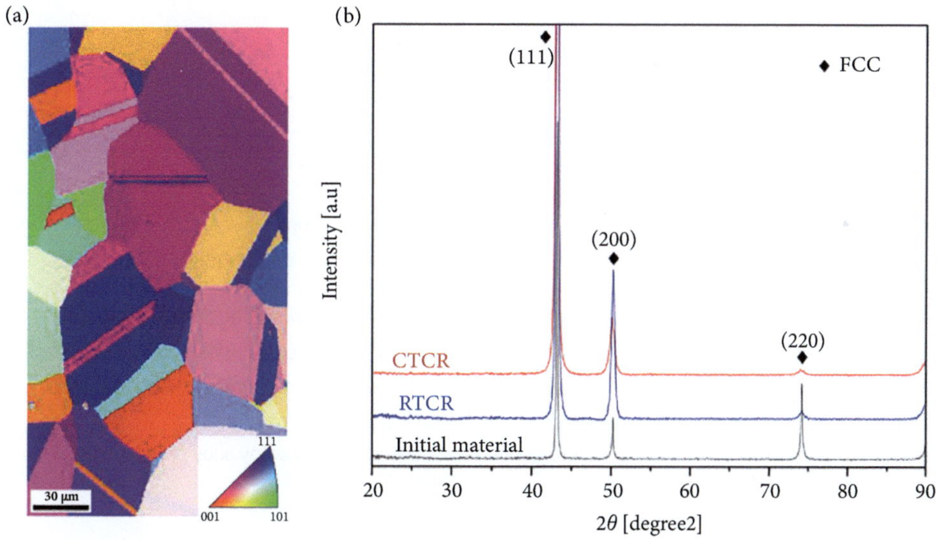

Figure 3.9 Microstructure of homogenised and recrystallised single-phase fcc Cantor alloy CrMnFeCoNi: (a) electron back-scattered diffraction (EBSD) image of equiaxed grains containing occasional annealing twins; and (b) single-phase fcc X-ray diffraction (XRD) patterns as-homogenised and after room temperature (RTCR) and cryogenic temperature (CTCR) rolling.

Reproduced from Won et al. (2018). Ultrafine-grained CoCrFeMnNi high-entropy alloy produced by cryogenic multi-pass caliber rolling, *Journal of Alloys and Compounds* 74: 290–5.

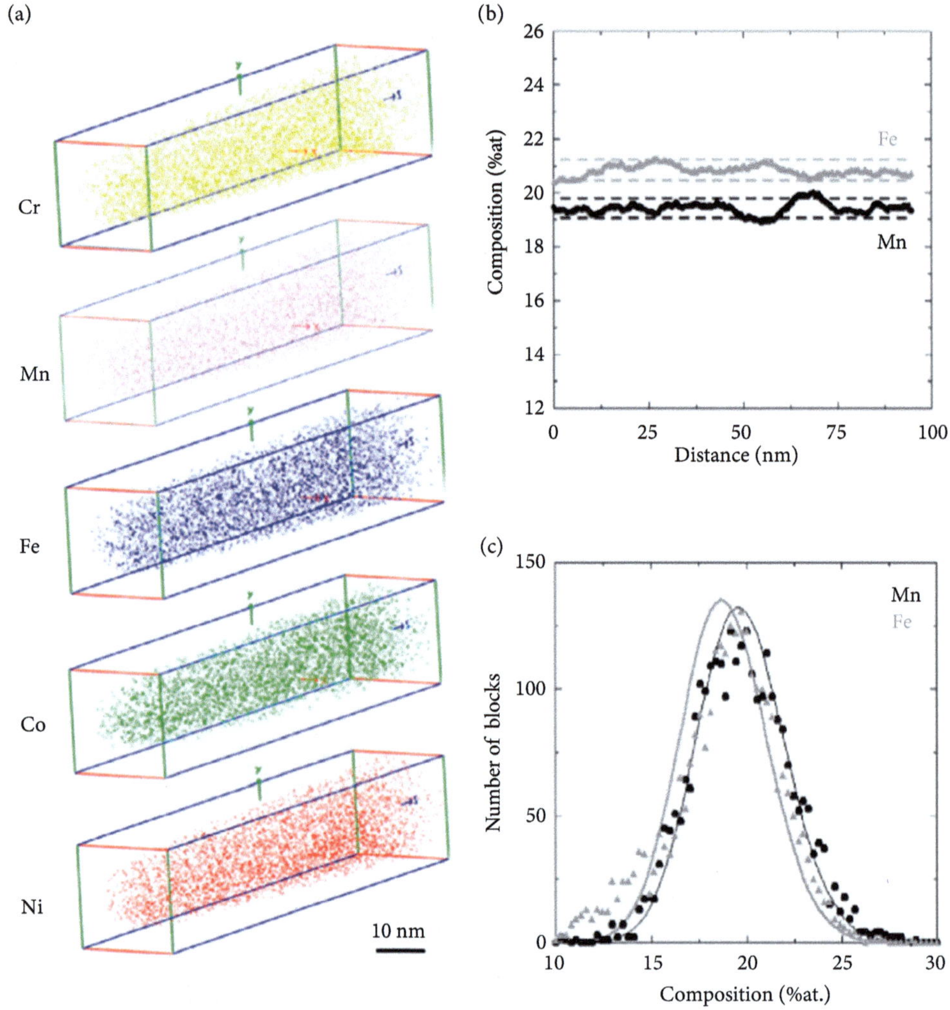

Figure 3.10 Atom probe (AP) measurements from the homogenised and recrystallised single-phase fcc Cantor alloy CrMnFeCoNi, showing a random atomic-scale distribution of all five components: (a) composition maps (each point is a single atom); (b) composition profiles for Mn and Fe; and (c) composition distributions for Mn and Fe.

Reproduced from Won et al. (2018). Ultrafine-grained CoCrFeMnNi high-entropy alloy produced by cryogenic multi-pass caliber rolling, *Journal of Alloys and Compounds* 74: 290–5.

between 8–16 at% Al, and single-phase bcc for compositions ≥ 16 at % Al, as shown on the X-ray diffractometer (XRD) traces for $Al_xCrMnFeCoNi$ alloys with x = 0–20 at% in Figure 3.11. On cooling to room temperature the disordered bcc phase shows partial ordering to form a CsCl-type ordered bcc structure, similar to and isomorphous with the well-known binary intermetallic compounds FeAl, CoAl, and NiAl,[43,44] i.e. there is a transition from a multicomponent bcc solid solution $Al_xCrMnFeCoNi$ to a multicomponent CsCl-type mono-aluminide compound (CrMnFeCoNi)Al. This is one example of many studies that show how extended single-phase fields in multicomponent phase space (in this case, the single-phase fcc Cantor alloys and the single-phase bcc multicomponent

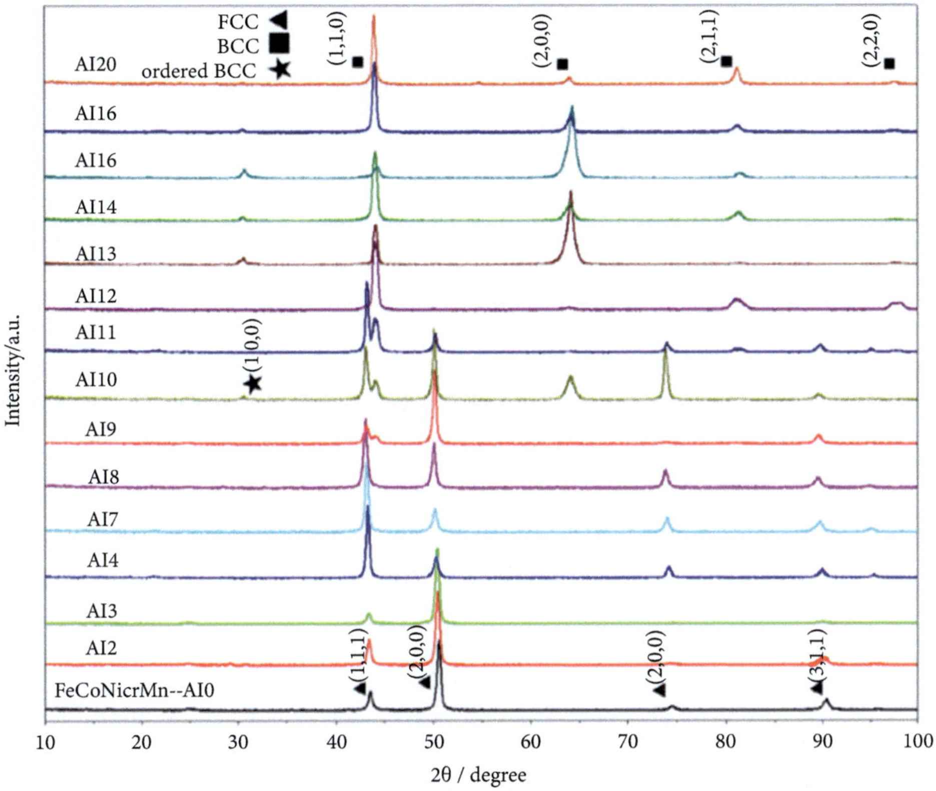

Figure 3.11 XRD traces from $Al_xCrMnFeCoNi$ alloys, showing fcc peaks only for ≤8 at% Al, fcc + bcc peaks for 8–16 at% Al and bcc peaks only for ≥16 at% Al.

Reproduced from He et al. (2014). Effects of Al addition on structural evolution and tensile properties of the FeCoNiCrMn high-entropy alloy system. *Acta Materialia 62*: 105–13.

mono-aluminides) are often separated by similarly extended two-phase regions (in this case fcc + bcc), though they can also be separated by more complex multiphase regions; and also show how cooling single-phase multicomponent solid solutions to room temperature can lead to the formation of ordered multicomponent intermetallic compounds (as indeed happens in a similar way in many simpler, well-known binary and ternary alloy systems).

In summary, the Cantor alloy region in multicomponent phase space has the following characteristics:[8,9]

1. It is largely based around components from the middle and late part of the first transition series, i.e. chromium, manganese, iron, cobalt, nickel and copper, as represented by the original Cantor alloy CrMnFeCoNi.
2. It extends towards the earlier part of the first transition series, with the addition of components such as titanium and vanadium; towards other transition series, with the addition of components such as niobium, molybdenum, palladium, platinum and gold; and to some extent towards non-transition-metal main-group elements, notably with the addition of aluminium and (to a much lesser extent) silicon.

3. It extends towards ternary and quaternary sub-systems such as CrCoNi and CrFeCoNi, breaking out at the edge of multicomponent phase space in the form of the pure single-element nickel, and (at high temperatures) iron and cobalt, and well-known binary fcc solid-solution alloys such as Fe–Ni and Co–Ni.
4. Its extent is higher at high temperatures, because of the enhanced significance at high temperature of high configurational entropy (or, indeed, any kind of entropy).
5. Its extent is greater when alloys are manufactured by metastable processing methods such as rapid solidification or vapour deposition that do not allow sufficient time for the formation of more complex, ordered, stoichiometric compounds and/or multiphase structures.
6. Most of the middle-to-late transition-metal components mentioned above show very extended solubility levels in the single-phase fcc structure, exhibiting composition ranges up to 10–100% wide, depending on the temperature and overall alloy composition.
7. Other components, such as the non-metals carbon, boron and nitrogen also dissolve in the single-phase fcc structure, but only in much smaller quantities (similar to the small amounts of these components that typically dissolve in many conventional binary and ternary alloys).
8. Multicomponent materials near and just beyond the edge of the single-phase fcc composition range, i.e. in the vicinity of the solvus for a variety of different components, can form two-phase fcc matrix + second-phase precipitate structures, similar to those formed in conventional binary and ternary alloys.
9. Multicomponent materials further beyond the single-phase fcc composition range, i.e. further beyond the solvus for a variety of different components, can form coarser, dispersed two-phase fcc + second phase or multiphase structures, again similar to those formed in conventional binary and ternary alloys.

Senkov alloys

As first discovered by Senkov et al.,[45] there is another enormous region in multicomponent phase space consisting of materials with a single-phase bcc structure, this time made up mostly (but not exclusively) of early transition-metal components, and this time loosely based around the original Senkov alloy composition VNbMoTaW, including alloys with a wide range of different components and a wide range of different compositions.[8,46] Materials in this region of multicomponent phase space are sometimes called *refractory high-entropy alloys, RHEAs, modified Senkov alloys* or, more simply, just *Senkov alloys*.[9,46] The size of this region of multicomponent phase space is again very large, with again literally millions of different Senkov alloys, forming another truly *extended* multicomponent (this time) bcc solid-solution single-phase field.[8,46]

The wide range of single-phase bcc Senkov alloys in multicomponent phase space is again stabilised at least partly by a high configurational entropy of mixing, as first suggested by Yeh et al.[2] (as mentioned before, this stabilisation and the entropy of mixing is discussed in more detail, along with other thermodynamic issues, in Chapter 4). The discovery of the Senkov alloys in 2010, following the earlier discovery of the Cantor alloys, re-emphasised strongly the initial, somewhat surprising, general feature of the appearance of very large single-phase

Table 3.6 Equiatomic single-phase bcc Senkov alloys.

number of components	alloy composition
3	TiVCr; ZrNbHf
4	AlTiVNb; TiVZrMo; TiVNbTa; TiVMoTa; TiZrNbHf; TiZrNbTa; VCrFeMo; VNbTaW; NbMoTaW
5	AlTiVNbMo; AlTiVNbTa; TiVZrNbHf; TiVZrNbTa; TiVZrMoTa; TiVNbMoW; TiZrNbMoHf; TiZrMoHfTa; TiZrNbHfTa; TiNbMoTaW; VNbMoTaW
6	TiVZrNbHfTa; TiVZrNbMoTa; TiVNbMoTaW; TiZrNbMoHfTa;
7	TiVCrZrNbMoTa; VCrNbMoTaWRe
8	TiVCrZrNbMoTaW

fields in multicomponent phase space.[9] Many single-phase bcc Senkov alloys have now been made by different processing methods and, at least to some extent, had their structure and properties studied (probably one or two hundred different alloys, with one or two thousand independent reports).[8,46] But it is again quite clear that there are still many, many more Senkov alloys yet to be manufactured and investigated.

Some typical examples of equiatomic and non-equiatomic Senkov alloys (again, by no means all) are collected together in Tables 3.6 and 3.7 respectively. Once again, it is not possible to show easily the complex multidimensional shape of the bcc single-phase field within the *C*-dimensional hyper-polyhedral phase diagram that represents multicomponent phase space, and Figure 3.12 again gives an approximate maximum width of the composition range that has been reported for a variety of different components (i.e. each component will dissolve in the single-phase bcc structure across approximately this or a lower composition range, depending upon the temperature and the overall alloy composition—the quoted composition ranges are again almost certainly underestimates of the real maximum composition ranges, since not all temperatures and alloy compositions have yet been investigated).

As well as the original Senkov alloy VNbMoTaW, Tables 3.6 and 3.7 and Figure 3.12 show that the single-phase bcc region of Senkov alloys includes a wide range of different materials,[8,46] for instance: equiatomic 5-, 6-, 7- and 8-component alloys such as AlTiVNbMo,[47] TiVZrNbHf,[48] TiVZrNbHfTa,[49] TiVNbMoTaW,[50] VCrNbMoTaWRe[51] and TiVCrZrNbMoTaW;[52] equiatomic 3- and 4-component alloys such as TiVCr,[53] ZrNbHf,[54] TiVZrMo,[52] VCrFeMo[53] and NbMoTaW;[45] and non-equiatomic 4-, 5- and 6-component alloys such as (TiZrHf)–15 at% Ta,[56] (TiZrNbTa)–5 at% Mo,[57] (TiZrNbHfTa)–13 at% Al[58] and (VNbMoTaW)–9 at% Cr.[58] More details of the range of single-phase bcc Senkov alloys are given in the review papers by Senkov et al.[46] and Miracle et al.[60]

As with the single-phase fcc Cantor alloys discussed in the last sub-section, single-phase bcc Senkov alloys exhibit dendritic microstructures when manufactured by melting and casting, with the scale of the dendritic segregation patterns again not dissimilar to those found in conventional binary and ternary alloys, ranging from 5–50 μm depending on the alloy composition and the cooling rate during casting. Senkov et al.[45] used energy-dispersive X-ray spectra (EDS) in a scanning electron microscope (SEM) as well as EPMA to make detailed

Table 3.7 Non-equiatomic single-phase bcc Senkov alloys.

number of components	alloy composition	composition range x (at%)
4	$TiZrHfTa_x$	15
5	$Al_xTiVNbMo$	0–27
	$Al_xTiVNbTa$	0–20
	$Ti_xZrNbMoHf$	11–27
	$TiZr_xNbMoHf$	11–27
	$Ti_xZrNbMoHf_x$	11–27
	$Ti_xZrNbMoTaW$	0–20
	$TiZrNbMo_xTa$	0–5
	$TiVZrNbMo_x$	7–25
6	$Al_xTiZrNbHfTa$	0–13
	$TiZrNbMo_xHfTa$	0–17
	$VCr_xNbMoTaW$	9–23

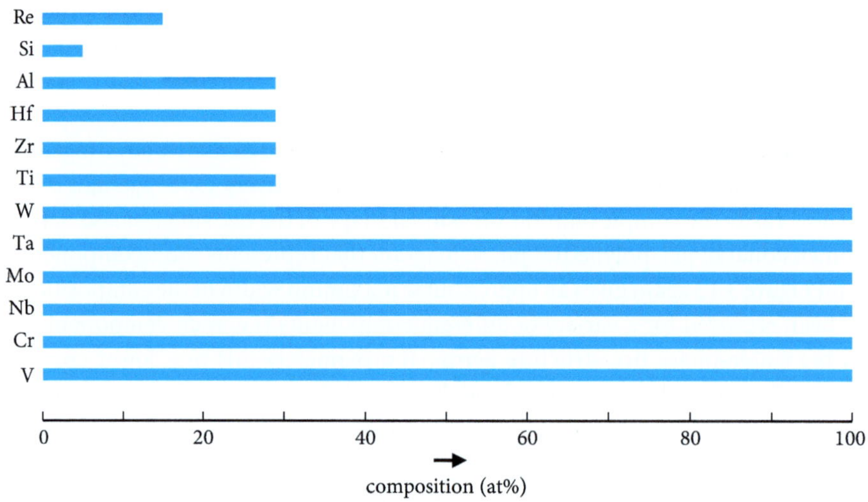

Figure 3.12 Approximate maximum composition ranges for different components in single-phase bcc Senkov alloys.

composition measurements on arc-melted buttons of the original Senkov alloy VNbMoTaW, as shown in Figure 3.13 and Table 3.8, and found enrichment of tantalum (slightly) and tungsten in the as-solidified dendritic spines, and enrichment of vanadium, niobium and molybdenum (slightly) in the interdendritic regions. Calculated partition coefficients for the different components ranged from k = 0.61 for vanadium to k = 1.82 for tungsten, with only very limited segregation of molybdenum (k = 0.97) and tantalum (k = 1.19). Similar dendritic microstructures are found in many as-cast Senkov alloys with a wide range of different compositions.[60,61]

In general, Senkov alloys exhibit extremely high strengths but little ductility at room temperature (discussed in more detail in later chapters), and it is not usually possible, therefore,

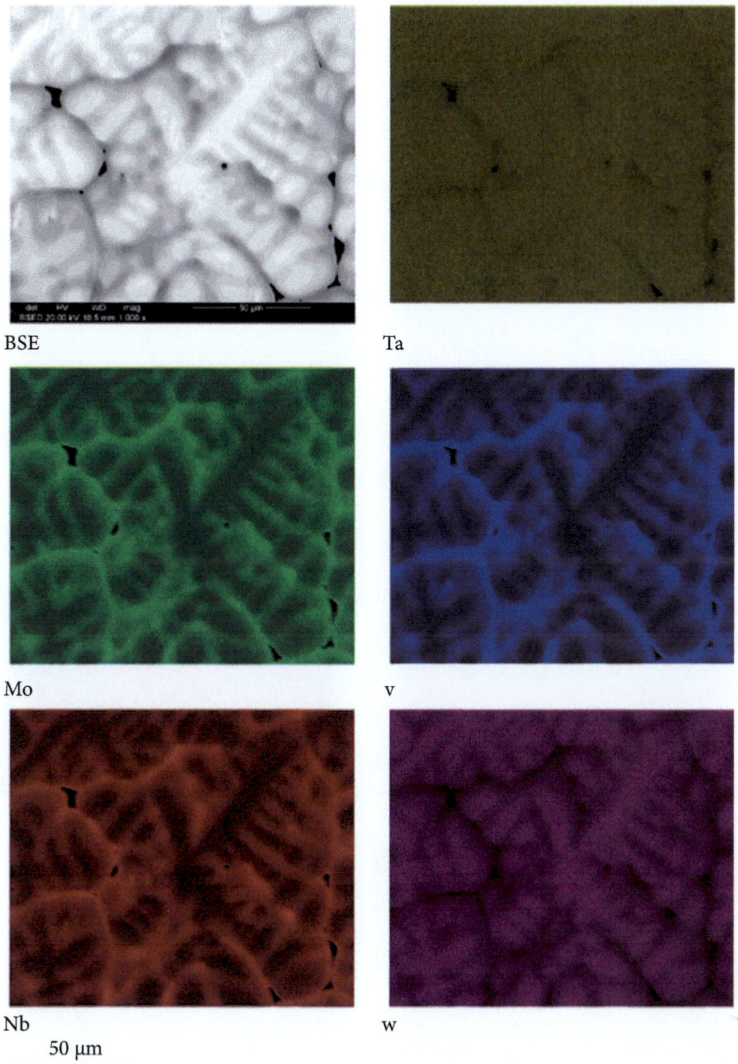

Figure 3.13 EDS composition maps of the as-cast single-phase bcc Senkov alloy VNbMoTaW, showing segregation of W into the dendrite spines and V, Nb and Mo into the interdendritic regions.

Reproduced from Senkov et al. (2010). Refractory high-entropy alloys. *Intermetallics* 18: 1758–65.

to use thermomechanical processing to homogenise the alloy composition and/or recrystallise the alloy microstructure.[61,62] The equiatomic 5-component Senkov alloy TiZrNbHfTa is an exception, however, and exhibits significant ductility, with tensile elongations of 10–20% depending on the cooling rate during casting and the corresponding grain size, and can be cold rolled to between 85–90% reduction in thickness.[63,64] As with the single-phase fcc Cantor alloys, therefore, single-phase bcc TiZrHfNbTa can be thermomechanically processed by cold rolling followed by soaking at high temperature, leading to homogenisation and recrystallisation, resulting again in beautiful equiaxed grain structures but (in this case) without any annealing twins,[64] as shown in Figure 3.14.

Table 3.8 Electron-probe microanalysis (EPMA) composition measurements and corresponding calculated partition coefficients for each of the components in the as-cast Senkov alloy VNbMoTaW (after Senkov et al[45]).

	V	Nb	Mo	Ta	W
concentration in dendrite spines c_d (at%)	16.0	19.2	19.0	19.1	26.7
concentration in interdendritic regions c_i (at%)	26.3	23.5	19.5	16.0	14.7
partition coefficient $k = c_d/c_i$	0.61	0.82	0.97	1.19	1.82

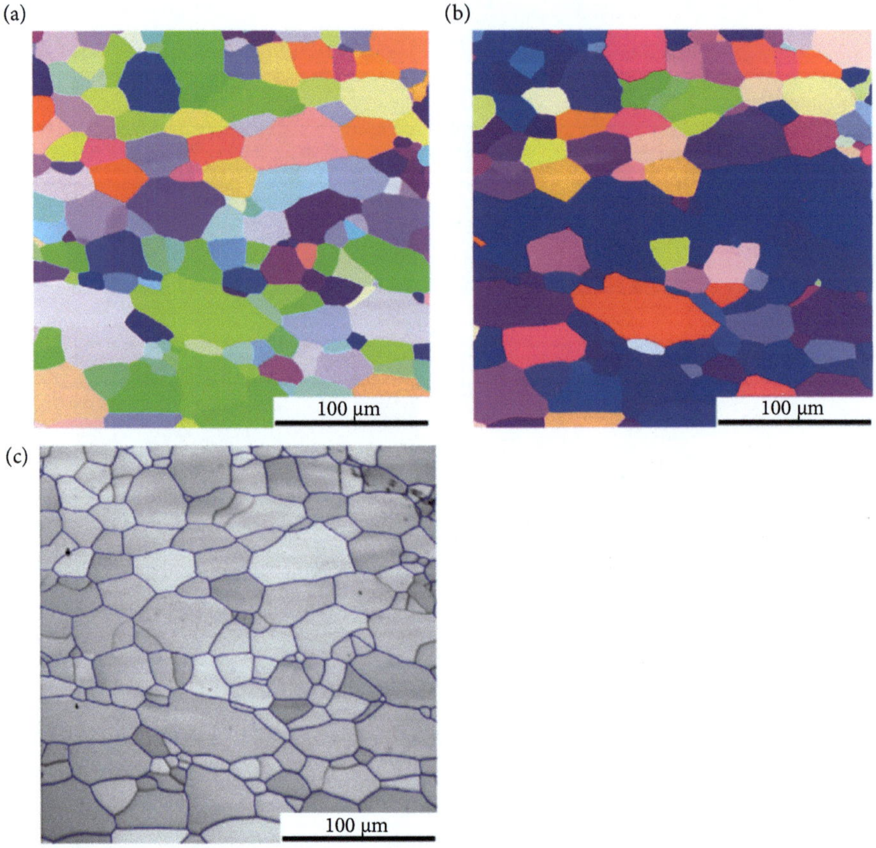

Figure 3.14 Microstructure of homogenised and recrystallised single-phase bcc Senkov alloy TiZrNbHfTa: (a) electron back-scattered diffraction (EBSD) inverse pole figure map; (b) grain misorientation map (> 10° = blue); and (c) back-scattered SEM image.

Reproduced from Senkov and Semiatin, (2015). Microstructure and properties of a refractory high-entropy alloy after cold working. *Journal of Alloys and Compounds* 649: 1110–23.

In summary, the Senkov-alloy region in multicomponent phase space has the following characteristics (in many ways similar to the Cantor alloy region):[8,46,61]

1. It is largely based around components from the early part of the first three transition series, i.e. titanium, vanadium, zirconium, niobium, molybdenum, hafnium, tantalum and tungsten, as represented by the original Senkov alloy VNbMoTaW.
2. It extends towards the middle and late parts of the three transition series, with the addition of components such as chromium, manganese, iron and rhenium; and to some extent towards non-transition-metal main-group elements, notably with the addition of aluminium.
3. It extends towards ternary and quaternary sub-systems such as TiVCr, ZrNbHf, VNbTaW and NbMoTaW, breaking out at the edge of multicomponent phase space in the form of the pure single elements vanadium, chromium, niobium and molybdenum, and well-known binary bcc solid-solution alloys such as Cr-V, V-Nb, Nb-Mo and (at high temperatures) Cr-Mo and Cr-W.
4. Its extent is again higher at high temperatures, because of the enhanced significance of high configurational entropy (or, indeed, any kind of entropy).
5. Its extent is again greater when alloys are manufactured by metastable processing methods such as rapid solidification or vapour deposition that do not allow sufficient time for the formation of more complex, ordered, stoichiometric compounds and/or multiphase structures.
6. Most of the early-to-middle transition-metal components mentioned above show very extended solubility levels in the single-phase bcc structure, exhibiting composition ranges often up to 100% wide, depending on the temperature and overall alloy composition.
7. Other components, such as the non-metals carbon, boron and nitrogen again dissolve in the single-phase bcc structure, but only in much smaller quantities (similar to the small amounts of these components that typically dissolve in many conventional binary and ternary alloys).
8. Multicomponent materials near and just beyond the edge of the single-phase bcc composition range, i.e. in the vicinity of the solvus for a variety of different components, can form two-phase bcc matrix + second-phase precipitate structures, similar to those formed in conventional binary and ternary alloys.
9. Multicomponent materials further beyond the single-phase bcc composition range, i.e. further beyond the solvus for a variety of different components, can form coarser, dispersed two-phase bcc + second phase or multiphase structures, similar to those formed in conventional binary and ternary alloys.

Multicomponent hcp alloys

Although there are very large disordered fcc and bcc solid-solution single-phase regions in multicomponent phase space (the Cantor and Senkov alloys respectively), there have been rather few reports of multicomponent disordered hcp solid solutions. Takeuchi et al.[65] were the first to find a single hcp phase in multicomponent YGdTbDyLu and GdTbDyTmLu alloys, i.e. mixtures of hcp yttrium and a number of late-lanthanide

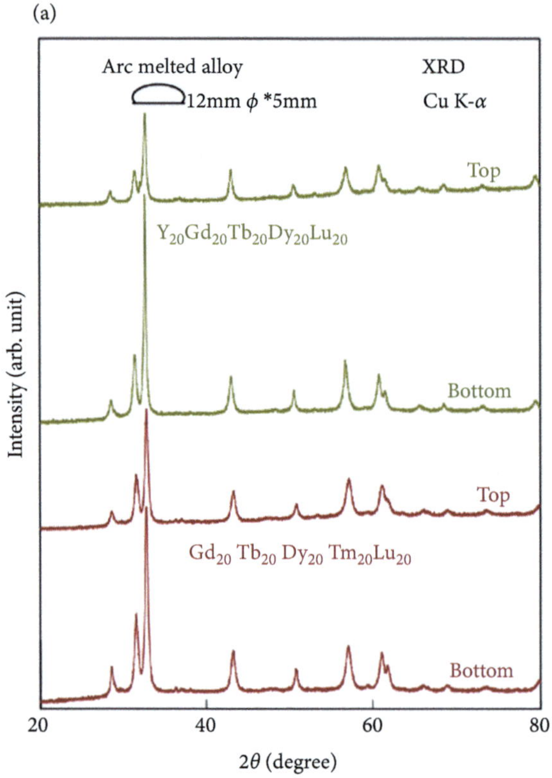

Figure 3.15 XRD traces showing hcp single-phase structure in multicomponent YGdTbDyLu and GdTbDyTmLu alloys.

Reproduced from Takeuchi et al. (2014). High-entropy alloys with a hexagonal close-packed structure designed by equi-atomic alloy strategy and binary phase diagrams. *Journal of Metals 66*: 1984–92.

rare-earth hcp elements, manufactured by arc melting followed by quenching, as shown in the XRD traces in Figure 3.15. Similar single-phase hcp structures have been seen in other multicomponent alloys based mostly on late-lanthanide rare-earth hcp elements[66–68] such as YGdTbDyHo, YLaGdTbHo and YLaGdTbHoEr; and also in some multicomponent second- and third-period late-transition-metal-based alloys[69,70] such as RuRhReOsIr and $Mo_{20}Ru_{20}Rh_{22.5}W_{11.5}Ir_{26}$; and a dual-hcp structure has been found[71] in a quaternary early-transition metal alloy ScTiZrHf. Youssef et al.[72] showed that a single fcc phase in a mechanically alloyed 5-component $Li_{20}Mg_{10}Al_{20}Sc_{20}Ti_{30}$ alloy transformed to a single hcp phase after annealing for one hour at 500 °C; and there have been several reports[73,74] of the original Cantor alloy CrMnFeCoNi transforming from fcc to hcp at high pressures, as shown in the XRD traces in Figure 3.16. The hcp phase has also been seen occasionally as a secondary phase in a number of other multiphase materials. It seems, therefore, that there is again a large single-phase hcp region in multicomponent phase space, based around the late-lanthanide rare-earth hcp elements and extending to some of the main transition metals, but the full extent and complex shape of this region is nowhere near as well investigated as its fcc and bcc equivalents. More details of multicomponent single-phase multiphase hcp materials are given in the review by Li et al.[75]

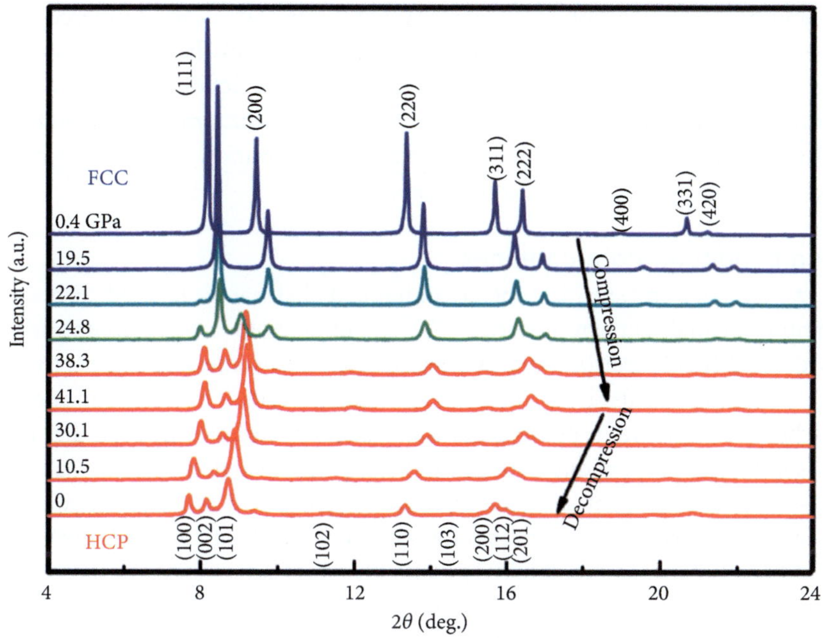

Figure 3.16 XRD traces from the Cantor alloy CrMnFeCoNi, showing a transition from single-phase fcc to single-phase hcp at high-pressures.

Reproduced from Zhang et al. (2017). Polymorphism in a high-entropy alloy. *Nature Communications 8*: Article 15687.

Multicomponent intermetallic compounds

As described in the last few sub-sections, there are certainly two (and probably three) very large disordered solid-solution single-phase regions in multicomponent phase space: an fcc single-phase field of Cantor alloys based on combinations of mostly late transition metals; a bcc single-phase field of Senkov alloys based on combinations of mostly early transition metals; and an hcp single-phase field based on combinations of mostly late-lanthanide rare-earth metals. In broad terms, we can understand this by noticing that, throughout these phase fields, the multiple alloying elements are quite similar *chemically*, so there are no very strong chemical interactions between them, preventing the formation of strongly bonded stoichiometric compounds and thus enabling a relatively high configurational entropy of mixing to stabilise a random or near-random multicomponent single-phase solid solution. It is not surprising that this stabilises a disordered fcc solid solution for mixtures of late transition metals, many of which are fcc in their elemental state (nickel, copper, silver, palladium, platinum and gold, as well as iron and cobalt at high temperatures); it is also not surprising that it stabilises a disordered bcc solid solution for mixtures of early transition metals, many of which are bcc in their elemental state (vanadium, chromium, niobium, molybdenum, tantalum and tungsten); and again it is not surprising that it stabilises a disordered hcp solid solution for mixtures of late-lanthanide rare-earths, many of which are hcp in their elemental state (gadolinium, terbium, dysprosium, holmium, erbium and thulium). When the chemical range of the alloying components is somewhat broadened, however, the resulting stronger chemical interactions disrupt the disordered fcc, bcc and hcp solid-solution structures, leading to the formation of ordered *multicomponent intermetallic compounds*;[7,8] and further broadening of the chemical range of the alloying components produces even stronger chemical interactions and the formation of even more strongly bonded covalent or ionic *multicomponent ceramic compounds*.[7,8] This sub-section discusses multicomponent intermetallic compounds, and the next sub-section discusses multicomponent ceramic compounds.

Multicomponent mono-aluminides

The most popular non-transition metal component that has been used to manufacture multicomponent high-entropy materials[7–9] is aluminium. As a main-group metallic element with no d-band electrons, aluminium is, of course, somewhat different chemically from the transition metals. Nevertheless, aluminium is a true free-electron metal with a close-packed fcc structure in its elemental state, and we have already seen that it can dissolve in substantial proportions both in multicomponent single-phase fcc Cantor alloys[9] (up to 5–10%, i.e. somewhat less than equiatomic levels) and in multicomponent single-phase bcc Senkov alloys[46] (often up to fully equiatomic levels of 20–25%). With increasing aluminium levels beyond the solubility limit, the fcc Cantor alloys begin to exhibit a bcc phase, first as a secondary phase in the fcc matrix, then as the majority phase, and finally as a single-phase material.[8,9] The bcc phase in Al-modified Cantor alloys is sometimes in the form of a multicomponent disordered random solid-solution; but, because of the chemical difference between aluminium and the transition metals, the bcc phase has a marked tendency to form a multicomponent mono-aluminide intermetallic compound, i.e. a B2 CsCl-type ordered bcc structure MAl, where M is a multicomponent mixture of different transition metals, with two distinct aluminium and transition metal sub-lattices,[8,9] as shown in Figure 3.17, isomorphous with the well-known binary B2 CsCl-type ordered bcc mono-aluminides FeAl, CoAl and NiAl found in conventional binary alloys.[44] At high aluminium concentration levels, the bcc Senkov alloys also show a tendency, though somewhat less marked, to form a similar B2 CsCl-type ordered bcc mono-aluminide intermetallic compound at particular compositions and temperatures,[46,60] notably around the composition AlX_2Y, where X is a multicomponent mixture of early transition metals such as titanium, zirconium and hafnium, and Y is another multicomponent mixture of middle transition metals such as chromium, iron, molybdenum and tungsten, corresponding to the ternary B2 CsCl-type ordered bcc mono-aluminides Ti_2CrAl and Ti_2MoAl[60,76] (the site occupancy of the three components on the two CsCl-structure sub-lattice sites is not well established at these compositions, and seems to be somewhat variable).

Table 3.9 lists some (but by no means all) of the multicomponent alloys that have been reported to exhibit a B2 CsCl-type ordered bcc mono-aluminide structure, either in single-phase or two-phase form. Quite a large number of particular alloys have been reported in different studies to exhibit both disordered bcc and ordered B2 structures. In some cases, this is because ordered X-ray reflections can be weak and difficult to detect on XRD traces; but in other cases, alloys are disordered at high temperatures where relatively high configurational entropies of mixing are dominant, but ordered at low temperatures where entropy effects weaken and the chemical interaction energies of ordering dominate. The disordered state is often retained by quenching to room temperature, and the ordered state appears on either slow cooling to room temperature or re-annealing at lower

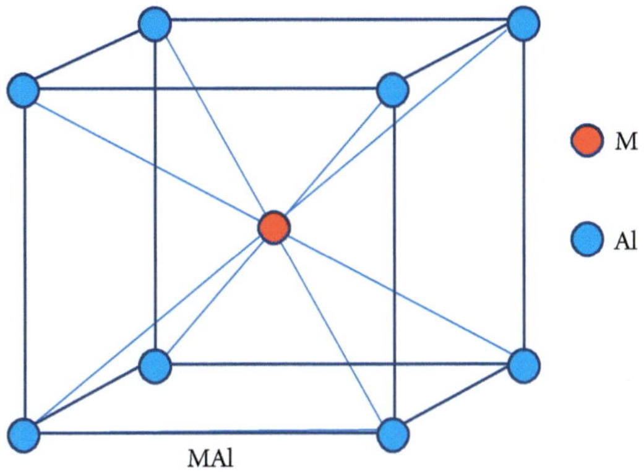

Figure 3.17 B2 CsCl-type ordered bcc multicomponent mono-aluminide MAl with body-centred and cube-corner sub-lattice sites for the aluminium atoms Al and a random distribution of multicomponent transition metal atoms M respectively.

Table 3.9 Multicomponent alloys containing a multicomponent B2 CsCl-type mono-aluminide intermetallic compound.

phases	alloy composition
B2	Al(MnFeCoNi); Al(MnFeCoNiCu); AlCrFeCoNi; $Al_2CrFeCoNi$; $Al_2CrFeCoNiCu$; AlTiVCr; AlCoNi; AlTiVNb; AlTiNbMo; AlTiNbHf; AlTiHfTa
B2 + bcc	AlFeCoNi; AlCrFeCoNi; AlMnFeCoNi; AlCrMnFeCo; AlCrFeCoNiCu; AlTiCrFeCoNi $AlTiZrNbMo_{0.5}Ta_{0.5}$; $Al_{0.3}Ti_{1.4}Zr_{1.3}NbTa$; $Al_{0.25}TiZrNbTa$
B2 + fcc	$Al_{0.5}CrFeCoNi$; $Al_3CrFeCoNi$;

temperatures. Table 3.9 shows that the single-phase B2 multicomponent mono-aluminide region includes a wide range of materials, such as Al(MnFeCoNiCu),[34] AlCrFeCoNi,[77] AlTiVNb[78] and AlTiNbHf;[79] and the two-phase B2 multicomponent mono-aluminide + disordered bcc region also includes a wide range of materials, such as AlFeCoNi,[80] AlCrFeCoNiCu[81] and $Al_{0.25}TiZrNbTa$.[82] Overall, there is a another very large single-phase region in multicomponent phase space, with a B2 CsCl-type ordered bcc multicomponent mono-aluminide structure, partly separate from and partly overlapping with the disordered bcc single-phase Senkov alloy and Al-modified disordered fcc Cantor alloy regions. More details of the range of multicomponent mono-aluminides are given in the review papers by Senkov et al.[46] and Miracle et al.[60] As with the fcc Cantor alloys and the bcc Senkov alloys, it is quite clear that, though many B2 multicomponent mono-aluminides have been found, there are still many, many more yet to be made and yet to be investigated.

Muralikrishna et al.[83,84] used the equiatomic substitution method invented by Kim et al.[85] to manufacture a series of multicomponent transition-metal mono-aluminides with

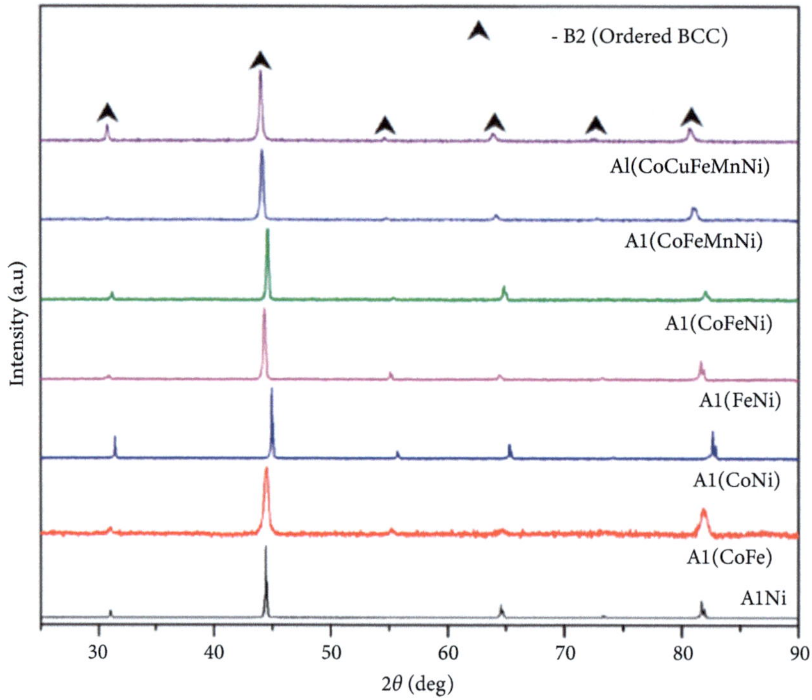

Figure 3.18 XRD traces showing the B2 CsCl-type ordered bcc crystal structure in a series of multicomponent transition-metal mono-aluminides MAl, where M is a multicomponent mixture of different transition metals.

Reproduced from Muralikrishna et al. (2020). Tracer diffusion in ordered pseudo-binary multicomponent aluminides. *Scripta Materialia 178*: 227–31.

the following compositions: NiAl, (FeCo)Al, (NiCo)Al, (FeCoNi)Al, (MnFeCoNi)Al, and (MnFeCoNiCu)Al. In all cases, the as-cast and homogenised materials were single-phase B2 CsCl-type ordered bcc,[83,84] as shown in the XRD traces in Figure 3.18. The as-cast alloys were all dendritic, as shown for the 6-component alloy (MnFeCoNiCu)Al in Figure 3.19, and could be homogenised without deformation by extended soaking at high temperature, as shown again for the 6-component alloy (MnFeCoNiCu)Al in Figure 3.20. Detailed high-resolution imaging (HREM) and EDS analysis showed that in all cases the aluminium atoms occupied one set of sub-lattice sites in the CsCl structure, and the transition metal atoms occupied the other set of sub-lattice sites and were distributed near-randomly throughout the material.[83,84] Figure 3.21 shows a typical example of a high-angle annular dark-field scanning transmission electron micrograph (HAADF-STEM) from the 6-component (MnFeCoNiCu)Al alloy, and the corresponding crystal structure with its two distinct (dark and light) interpenetrating aluminium and multicomponent transition-metal sub-lattices.[84]

The single-phase region of B2 multicomponent transition-metal mono-aluminides has a stoichiometry MAl where M is a multicomponent mixture of different transition metals, so fairly obviously it corresponds to relatively high aluminium contents centred around c_{Al} = 50 at%. Unlike the single-phase multicomponent mono-aluminides, multiphase materials are more often found at lower aluminium concentrations (Table 3.9). One of the most studied multicomponent alloy systems in this region of multicomponent phase space

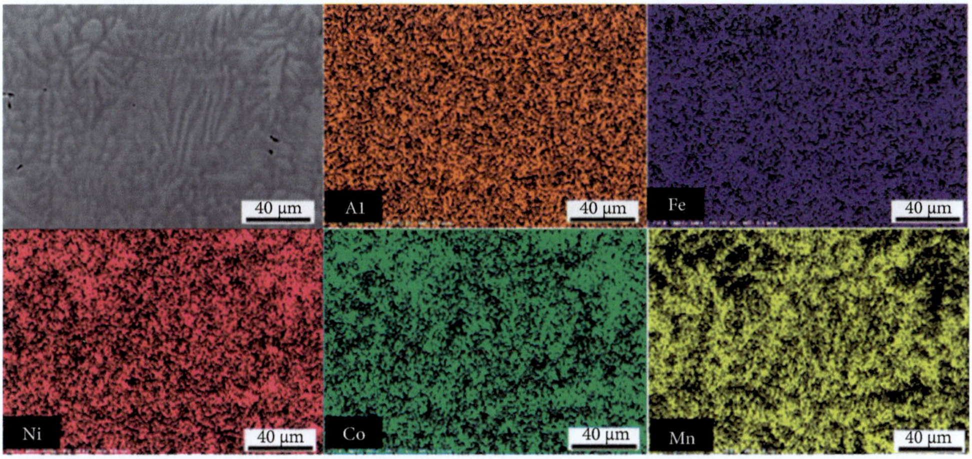

Figure 3.19 EDS composition maps of as-cast 6-component single-phase (MnFeCoNiCu)Al mono-aluminide, showing segregation of Fe, Co and Ni into the dendrite spines and Mn into the interdendritic regions.

Reproduced from Muralikrishna et al. (2020). Novel multicomponent B2-ordered aluminides: Compositional design, synthesis, characterization and thermal stability. *Metals 10*: Article 1411.

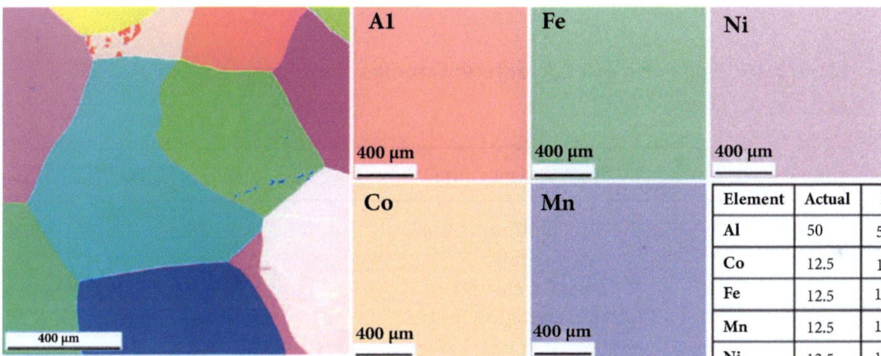

Element	Actual	Measured
Al	50	50.50 ± 0.3
Co	12.5	12.12 ± 0.1
Fe	12.5	12.35 ± 0.2
Mn	12.5	12.85 ± 0.2
Ni	12.5	12.19 ± 0.1

Figure 3.20 Electron back-scattered diffraction (EBSD) and energy-dispersive X-ray (EDS) composition maps of homogenised and equiaxed single-phase 6-component single-phase (MnFeCoNiCu)Al mono-aluminide.

Reproduced from Muralikrishna et al. (2020). Tracer diffusion in ordered pseudo-binary multicomponent aluminides. *Scripta Materialia 178*: 227–31.

is AlCrFeCoNi, and a range of different phases has been reported, depending on the detailed alloy composition, manufacturing method and subsequent thermomechanical treatment.[8,46,60] Some of these results are listed in Table 3.10. There is clearly a fine balance of free energies in this region of multicomponent phase space between the disordered bcc, disordered fcc and ordered B2 phases, and a wide range of different microstructures can be developed, again depending on the detailed composition, manufacturing process and subsequent thermomechanical treatment.[8,46,60]

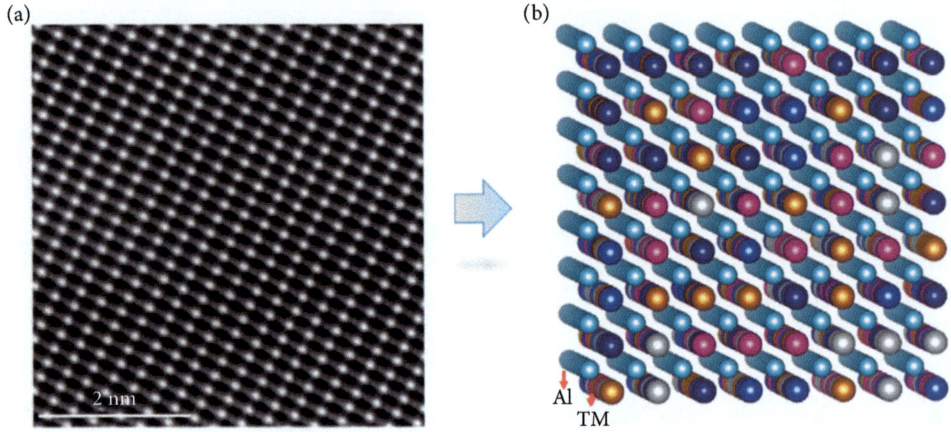

Figure 3.21 B2 CsCl-type ordered bcc crystal structure in a 6-component single-phase (MnFeCoNiCu)Al mono-aluminide alloy: (a) Fourier-filtered HAADF-STEM image of aluminium (dark) and transition metal (bright) atom columns; and (b) corresponding crystal structure with aluminium atoms (light blue) on one set of sub-lattice sites, and transition metal atoms (other colours) distributed randomly across the other set of sub-lattice sites. Reproduced from Muralikrishna et al. (2020). Novel multicomponent B2-ordered aluminides: Compositional design, synthesis, characterization and thermal stability. *Metals 10*: Article 1411.

Table 3.10 Phases reported in different studies of multicomponent AlCrFeCoNi alloys.

alloy	processing route	phases
AlCrFeCoNi	arc melting	bcc + B2
	arc melting	B2
	suction casting	B2
	induction melting	$B2 + \sigma$
	arc melting	$B2 + L1_2$
$Al_xCrFeCoNi$		
$x = 0.3$	induction melting	fcc + B2
$x = 0.3$	sputtering	fcc + B2
$x = 0.44$	arc melting	fcc + B2
$x = 0.5–1.2$	arc melting	fcc + bcc + B2
$x = 0.75–1.75$	arc melting	B2 + bcc
$x = 1.16$	arc melting	bcc + B2
$x = 1–3$	arc melting	B2
$x = 1.5–3$	arc melting	B2
$Al_{1.7}CrFeCoNi_{0.3}$	laser deposition	bcc + B2
$Al_{1.7}CrFeCoNi_{0.3}$	laser deposition	fcc + B2
$Al_2(CrFeCoNi)_{14}$	arc melting	fcc + bcc + B2
$AlCrFeCoNi_{2.1}$	induction melting	fcc + B2
$Al_2(Cr_2FeCoNi)_{14}$	suction casting + annealing	$bcc + \sigma$

Other multicomponent intermetallic compounds

As well as the B2 ordered bcc mono-aluminides, a number of other multicomponent intermetallic compounds have been found in alloys from nearby regions of multicomponent phase space,[8] mostly in multiphase mixtures with other phases.[8] A selection of these intermetallic compounds and some of the corresponding alloys in which they have been found is shown in Table 3.11. The multicomponent intermetallic compounds include (among others) $L1_2$ Ni_3Al-type ordered fcc, topologically complex phases (TCPs) such as complex tetragonal sigma σ and complex rhombohedral mu μ, C14 $MgZn_2$-type hexagonal and C15 $MgCu_2$-type cubic Laves phases, and quasicrystalline phases. Not surprisingly, $L1_2$ ordered fcc is found in some multicomponent mixtures of late transition metals, close to the single-phase fcc Cantor alloys with relatively small additions of aluminium, i.e. with compositions similar to the $L1_2$ intermetallic compounds found in conventional binary and ternary transition metal-aluminium alloys, but it is also found in many other aluminium-free alloys. Also not surprisingly, sigma, mu and Laves phases are found in some multicomponent mixtures of early and late transition metals, again with compositions similar to these phases when found in conventional binary and ternary transition metal-based alloys. More details of the wide range of intermetallic compounds found in multicomponent multiphase materials, as well as the associated alloy compositions, manufacturing conditions and thermomechanical treatments needed to form them, are given in the book by Murty et al.[8]

Single-phase $L1_2$ Ni_3Al-type ordered fcc materials have been found in 5-component MnFeNiCuPt by Takeuchi et al.[86] and in 4-component VFeCoNi by Wang et al.[87] In both cases, the alloys were manufactured by induction melting and formed single-phase fcc as-cast dendritic structures, as shown for MnFeNiCuPt in Figure 3.22, very similar to the single-phase as-cast dendritic structures found in the fcc Cantor alloys, bcc Senkov alloys and B2 mono-aluminide alloys as shown in Figures 3.8, 3.14 and 3.20, respectively. In both cases, the alloys could be homogenised and recrystallised by deformation and annealing at high temperature, forming single-phase fcc equiaxed structures after

Table 3.11 Intermetallic compounds found in multicomponent materials.

phases	alloy composition
fcc + $L1_2$	AlCoNiCuZn; $Al_{0.5}CrFeNiCoCu$; $Al_8Cr_{17}Fe_{17}Co_{17}Ni_{33}Cu_8$; $Al_4Ti_2(CrFeCoNi)_{94}$; TiVCrMnFe; $Ti_{0.2}CrFeCoNi$; $TiMo_{0.1}W_{0.1}$
fcc + σ	VCrFeCoNi; VMnFeCoNi; $V_{0.5}CrFeNiW_{0.5}$ VCrMnFeCoNi; CrFeCoNiMo; $(CrMnFeNi)_{95}Co_5$; $(CrMnFeCo)_{87.5}Ni_{12.5}$
bcc + σ	$Al_2(Cr_2FeCoNi)_{14}$; $AlCrFeCoNiMo_{0.5}$; TiCrFeMoW; VCrMnFeMo
fcc + Laves	$Ti_{0.7}CrFeCoNi$; CrFeCoNiNb; CrFeCoNiZr; CrFeCoNiTa; $(CrFeCoNiCu)_{88}Nb_{12}$; CrFeCoNiCuNb; CrMnFeCoNiNb
bcc + Laves	AlTiVCrNb; $AlCrFeCoNiNb_{0.75}$; $AlTi_{0.5}CrFeCoNi$; TiVZrNbHf; TiCrZrNbHf; TiCrZrNbHfTa

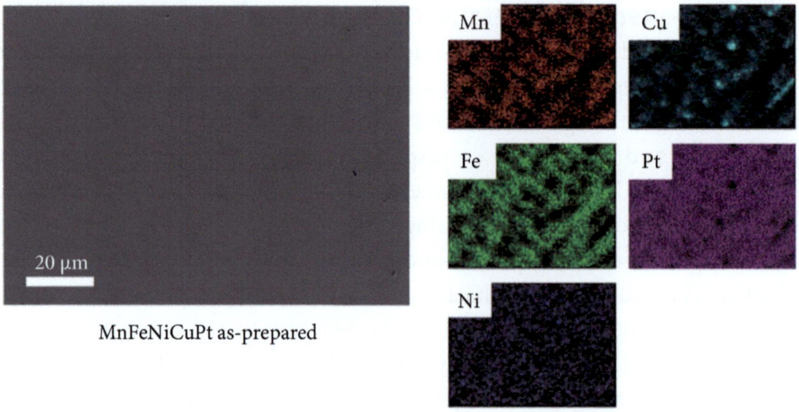

Figure 3.22 Back-scattered scanning electron micrograph (SEM) and energy-dispersive X-ray (EDS) composition maps of as-cast single-phase fcc MnFeNiCuPt, showing segregation of Fe and Ni into the dendrite spines and Mn and Cu into the interdendritic regions.

Reproduced from Takeuchi et al. (2017). MnFeNiCuPt and MnFeNiCuCo high-entropy alloys designed based on $L1_0$ structure in Pettifor map for binary compounds. *Intermetallics 82*: 107–15.

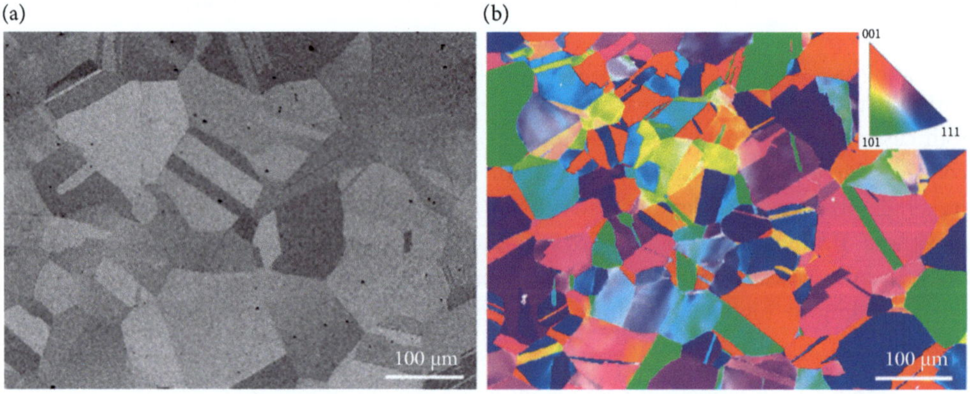

Figure 3.23 Microstructure of homogenised and recrystallised single-phase fcc VFeCoNi showing equiaxed grains with occasional annealing twins: (a) back-scattered SEM image; and (b) electron back-scattered diffraction (EBSD) inverse pole figure map.

Reproduced from Wang et al. (2019). Fcc-$L1_2$ ordering transformation in equimolar FeCoNiV multi-principal element alloy. *Materials and Design 168*: Article 107648.

quenching to room temperature, as shown for VFeCoNi in Figure 3.23, again very similar to the single-phase homogenised and equiaxed structures found in the fcc Cantor alloys, bcc Senkov alloys and B2 mono-aluminide alloys as shown in Figures 3.10, 3.15 and 3.21 respectively. With slow cooling to room temperature or with heat treatment at a lower temperature, the alloys underwent ordering reactions to form single-phase equiaxed $L1_2$ Ni_3Al-type ordered fcc structures, as shown by the selected area electron diffraction patterns (SADP) and high-resolution transmission electron micrographs (HRTEM) from VFeCoNi in Figure 3.24, corresponding to the ordered $L1_2$ structure shown in Figure 3.25.

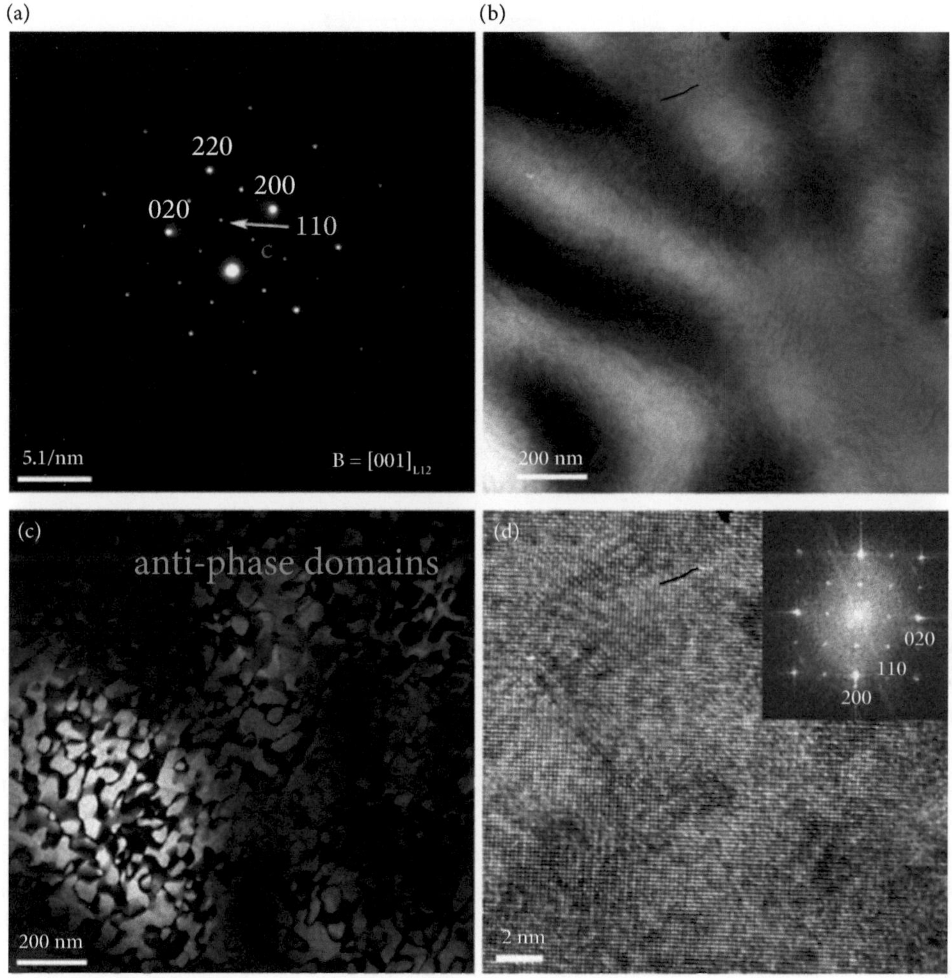

Figure 3.24 Microstructure of single-phase $L1_2$ ordered fcc VFeCoNi: (a) selected area diffraction pattern (SADP) showing ordered $L1_2$ diffraction spot (arrowed); and (b) bright-field (BF), (c) dark-field (DF) and (d) high-resolution (HRTEM) transmission electron micrographs showing ordered domains and anti-phase domain boundaries.
Reproduced from Wang et al. (2019). Fcc-$L1_2$ ordering transformation in equimolar FeCoNiV multi-principal element alloy. *Materials and Design 168*: Article 107648.

Takeuchi et al.[86] were not able to determine the stoichiometry of the $L1_2$ phase in the 5-component MnFeNiCuPt alloy, but it seems likely to be the multicomponent tri-transition-metal platinide $(MnFeCoNi)_3Pt$, isomorphous with the $L1_2$ phases Mn_3Pt, Fe_3Pt, Co_3Pt, and Ni_3Pt found in binary Mn–Pt, Fe–Pt, Co–Pt, and Ni–Pt alloy systems respectively. Unlike Takeuchi et al.,[86] Wang et al.[87] were, however, able to use a combination of *ab initio* density functional theory calculations, thermodynamic calculations and magnetic and elastic constant measurements to demonstrate that the $L1_2$ phase in quaternary VFeCoNi has a stoichiometry of $(FeCoNi)_3V$, isomorphous with the $L1_2$ phases Fe_3V, Co_3V and Ni_3V found in binary Fe–V, Co–V and Ni–V alloys systems, respectively, with the vanadium atoms distributed on the cube-corner $L1_2$ sub-lattice sites and the other transition metal atoms

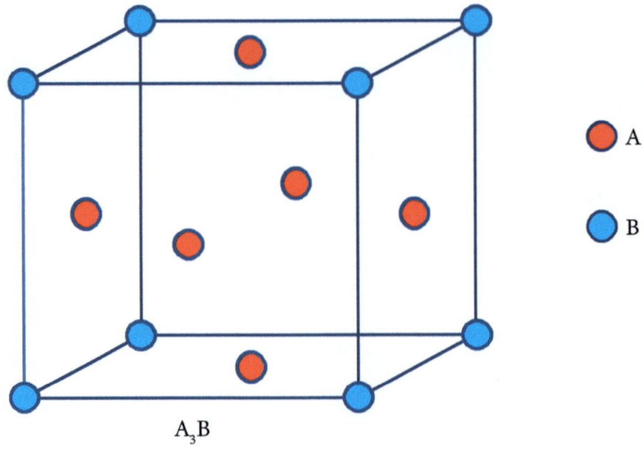

Figure 3.25 L1$_2$ Ni_3Al-type ordered fcc multicomponent intermetallic compound A_3B with face-centred and cube-corner sub-lattice sites for random distributions of multicomponent A and B atoms respectively.

distributed randomly across the face-centred L1$_2$ sub-lattice sites. These results together with earlier studies of binary Fe_3V, Co_3V and Ni_3V intermetallic compounds by Liu and co-workers[88,89] indicates that there is a large L1$_2$ Ni_3Al-type ordered fcc single-phase region in multicomponent phase space, centred around the quaternary composition $(FeCoNi)_3V$, as shown in the pseudo-quaternary phase diagram in Figure 3.26. Whether this region is contiguous with Takeuchi et al.'s L1$_2$ $(MnFeCoNi)_3Pt$ alloy,[86] the binary L1$_2$ Mn_3Pt, Fe_3Pt, Co_3Pt, and Ni_3Pt intermetallic compounds,[44,88,89] and/or the well-known L1$_2$ γ'-$(NiTi)_3Al$ strengthening precipitate phases found in nickel-based superalloys[90] is unclear, though it seems quite likely that it is. Overall, therefore, there is another very large single-phase region in multicomponent phase space, with an L1$_2$ Ni_3Al-type ordered fcc structure. It clearly has a very complex shape within the *C*-dimensional hyper-polyhedral phase diagram that represents multicomponent phase space, but its full extent remains to be determined.

Karpets et al.[91] investigated a series of multicomponent $VCrMnFeCoNi_x$ alloys with nickel compositions ranging from $x = 0 - 3$ (corresponding to $x = 0 - 37.5$ at%). Without nickel, the base equiatomic 5-component VCrMnFeCo alloy had a single-phase sigma σ-CrFe-type complex tetragonal structure. When nickel was added to the base alloy to manufacture 6-component $VCrMnFeCoNi_x$ alloys, the resulting materials were two-phase fcc + σ, with the fcc content gradually increasing from 0 to 100% and the σ content gradually reducing from 100 to 0% with increasing nickel content, as shown in Figure 3.27. The alloys were 100% sigma for $x \leq 0.5$ (9.1 at% Ni), two-phase fcc + σ for $0.5 < x < 2$, and single-phase fcc for $x \geq 2$ (28.6 at% Ni). The sigma phase has also been seen as a secondary or precipitate phase in a number of other multicomponent alloys,[8] as can be seen in Table 3.11, but there has not yet been any serious attempt to map the extent of the single-phase sigma region in multicomponent phase space.

As discussed in Chapter 2, the early studies of Kim et al.[85,92] used the equiatomic substitution method to manufacture a wide range of multicomponent glasses with the generic composition $X_{100-x-y}Y_xAl_y$, where X is a multicomponent mixture of early transition metals such as titanium, zirconium, niobium and hafnium, Y is a multicomponent mixture of late transition metals such as iron, cobalt, nickel, copper and silver, and where $x = 10 - 60$ at%

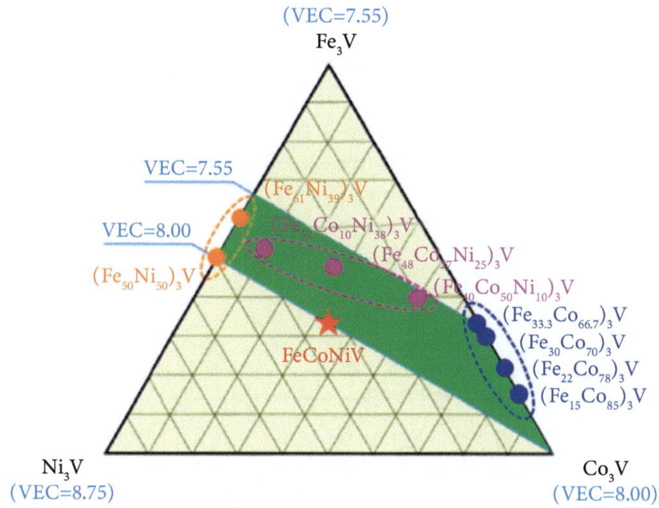

Figure 3.26 Pseudo-quaternary phase diagram showing L12 ordered fcc single-phase region (green) in multicomponent phase space (VEC = valence electron concentration).

Reproduced from Wang et al. (2019). Fcc-$L1_2$ ordering transformation in equimolar FeCoNiV multi-principal element alloy. *Materials and Design 168*: Article 107648.

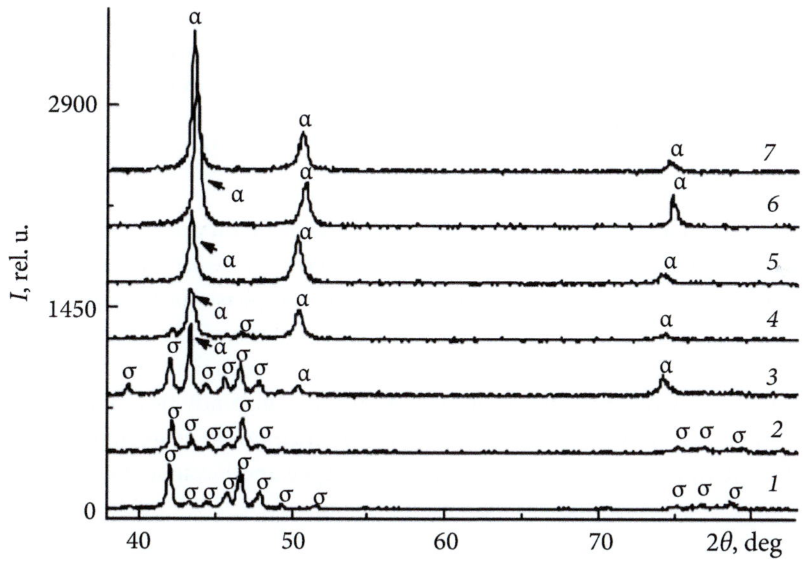

Figure 3.27 XRD traces from $VCrMnFeCoNi_x$ alloys showing a transition from single-phase σ to two-phase fcc + σ and then single-phase fcc as the nickel content increases from x = 0 to 3 (alloys 1–7 correspond to x = 0, 0.5, 1, 1.5, 2, 2.5 and 3 respectively).

Reproduced from Karpets et al. (2015). Effect of nickel on the structure and phase composition of the $VCrMnFeCoNi_x$ high-entropy alloy. *Journal of Superhard Materials 37*: 182–8.

and $y = 0 - 30$ at%. Many of these multicomponent glasses crystallised to form a single C14 $MgZn_2$-type hexagonal Laves phase or in some cases a two-phase structure containing the C14 phase,[85,92] as shown schematically on the pseudo-ternary phase diagram in Figure 3.28.

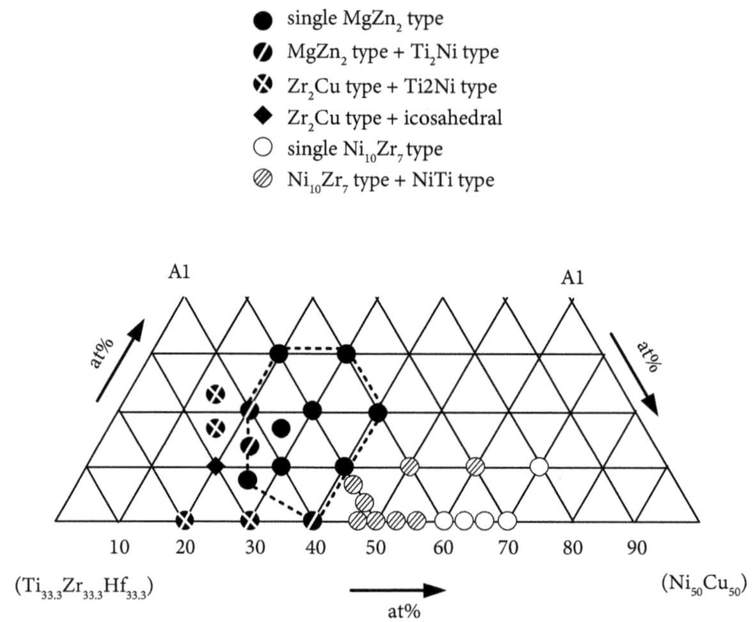

Figure 3.28 Pseudo-ternary phase diagram showing crystalline regions including large single C14 $MgZn_2$-type Laves-phase region in melt-spun glassy $(TiZrHf)_{90-x-y}(NiCu)_xAl_y$ alloys, where $x = 10$–60 at% and $y = 0$–30 at% after crystallisation by heat treatment for one hour at 600 °C.
Reproduced from Kim et al. (2003). Formation of metallic glasses in novel (Ti33Zr33Hf33)((100-x-y)(Ni50Cu50)(x)Al(y) alloys. *Materials Transactions 44*: 411–13.

The C14 and other Laves phases have also been seen as secondary or precipitate phases in a number of other multicomponent alloys,[8] as can be seen in Table 3.11. It is clear that there is another very large single-phase region in multicomponent phase space with a C14 $MgZn_2$-type hexagonal Laves phase structure, but the full extent of this single-phase region remains to be mapped out, as is the detailed stoichiometry and the occupancy of the multiple components across the two Laves-phase sub-lattice sites. It is also not clear and remains to be determined whether or not there are similar large multicomponent single-phase regions of the other two main Laves phases, cubic C15 $MgCu_2$-type and hexagonal C36 $MgNi_2$-type.

Multicomponent ceramics

Following the discovery of multicomponent high-entropy alloys by Cantor and his colleagues in the late 1970s/early 1980s and by Yeh and his colleagues in the 1990s, and their seminal papers[1,2] on these new materials in 2004,[x] further research grew only slowly through the 2000s and then accelerated increasingly rapidly in the 2010s, but it concentrated almost exclusively on the multicomponent metallurgical alloys discussed in the last few sub-sections

[x] The basic idea of *multicomponent alloying* to manufacture *multicomponent materials* had already been published briefly a year or two earlier in 2002 and 2003, as described in detail in Chapter 2.

of this chapter: the fcc Cantor alloys at first, then the bcc Senkov alloys, then the intermetallic compounds and then associated nearby (in multicomponent phase space) multiphase metallic materials. Quite early in the 2000s, Kim et al.[85,92] had developed the equiatomic substitution technique for exploring multicomponent phase space that is particularly suitable for investigating multicomponent and high-entropy chemical compounds. Nevertheless, if research was at first slow to build up into multicomponent high-entropy metallurgical alloys, it was much slower still to get going into multicomponent high-entropy ceramic compounds, beginning very slowly in the mid/late 2010s and not increasing significantly until the end of the 2010s and the early 2020s.

In fact, the first full study of high-entropy ceramics was not until 2015 by Rost et al.,[93] who used Kim et al.'s equiatomic substitution technique[85,92] to investigate a multicomponent oxide (MgCoNiCuZn)O, made by sintering an equimolecular mixture of five component mono-oxides (magnesium oxide MgO, cobalt oxide CoO, nickel oxide NiO, cupric oxide CuO and zinc oxide ZnO), selected because of their similar stoichiometry and cationic valency, and because three of them (MgO, CoO and NiO) independently exhibit the rock-salt NaCl-type crystal structure shown in Figure 3.29, though with only limited solubility for the other two (CuO and ZnO). Unlike metallurgical materials that can usually be made relatively conveniently by melting and casting, most ceramic materials have to be manufactured by high-temperature sintering of pellets or preforms made from their constituent component powders. Multicomponent sintered (MgCoNiCuZn)O was found to have a single-phase rock-salt NaCl-type ordered cubic structure when the sintering temperature was above 850 °C, and to have a multiphase structure when the sintering temperature was below 850 °C, as shown by the XRD traces[93] from quenched samples in Figure 3.30. The formation of the single-phase rock-salt structure was reversible and, when a sintered pellet was repeatedly re-annealed at temperatures above and below 850 °C, the structure alternated reversibly between single-phase rock salt and multiphase,[93] as also shown in Figure 3.30. A combination of extended X-ray fine structure (EXAFS) measurements and atomic-scale EDS composition maps demonstrated that the five cations (magnesium Mg^{2+}, cobalt Co^{2+}, nickel Ni^{2+}, copper Cu^{2+} and zinc Zn^{2+}) were distributed near-randomly on the cationic

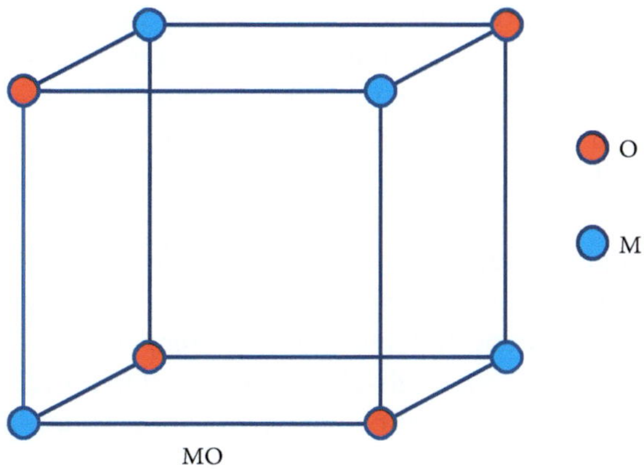

Figure 3.29 Rock-salt NaCl-type ordered cubic multicomponent mono-oxide MO with alternate cube-corner sub-lattice sites for the oxygen atoms O and a random distribution of multicomponent transition metal atoms M respectively.

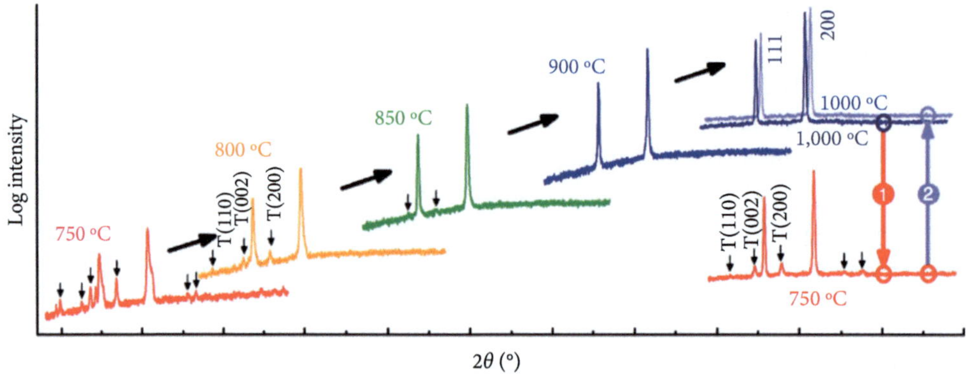

Figure 3.30 XRD traces from multicomponent oxide (MgCoNiCuZn)O sintered at different temperatures from 750 °C to 1,000 °C, showing a single-phase rock salt structure for T > 850 °C (peaks labelled 111 and 200 at 1,000 °C) and a multiphase structure for T ≤ 850 °C.
Reproduced from Rost et al. (2015). Entropy-stabilized oxides. *Nature Communications* 6: Article 8485.

sub-lattice,[93] as shown for example by the high-angle annular dark field (HAADF) image and atomic-scale EDS maps in Figure 3.31. Varying the composition of each of the individual oxide components one at a time showed that the minimum sintering temperature required to form a single-phase rock-salt structure was always at 850 °C, corresponding to the equimolecular composition where the configurational entropy of mixing different cations on the cationic sub-lattice is at its highest,[93] as shown in Figure 3.32. These results all indicated that the single-phase rock-salt structure was stabilised by the high configurational entropy associated with a near-random distribution of multiple cations on the cation sub-lattice (as mentioned, this stabilisation and the entropy of mixing is discussed in more detail, along with other thermodynamic issues, in Chapter 4).

Figure 3.32 indicates that there is a large rock-salt NaCl-type mixed mono-oxide single-phase field in multicomponent phase space, centred loosely around (MgCoNiCuZn)O, and extending out to the binary mono-oxide solid solution alloys MgO–CoO, MgO–NiO and CoO–NiO, as well as to the individual mono-oxides MgO, CoO and NiO. The full extent of this single-phase field has again not yet been mapped out in any detail. It seems unlikely (but not completely impossible) that the mono-oxide rock-salt single-phase field is contiguous with two other large rock-salt single-phase fields, namely the multicomponent alkali halides (LiNaK)(ClBr) and the multicomponent mono-boro-carbo-nitrides (TiZrNbHfTaW)(BCN).

Since Rost et al.'s study of multicomponent mixed divalent oxides,[93] there have been, at first gradually and then more rapidly, a growing number of investigations into many other multicomponent single-phase and multiphase ceramic compounds,[94–96] and Table 3.12 lists some typical (but by no means all) single-phase examples. Table 3.12 shows how the range of multicomponent high-entropy ceramic compounds has the following characteristics:

1. The range of multicomponent single-phase ceramics includes strong covalent compounds such as the mono-carbides and diborides as well as strong ionic compounds such as the mono-oxides and spinels.

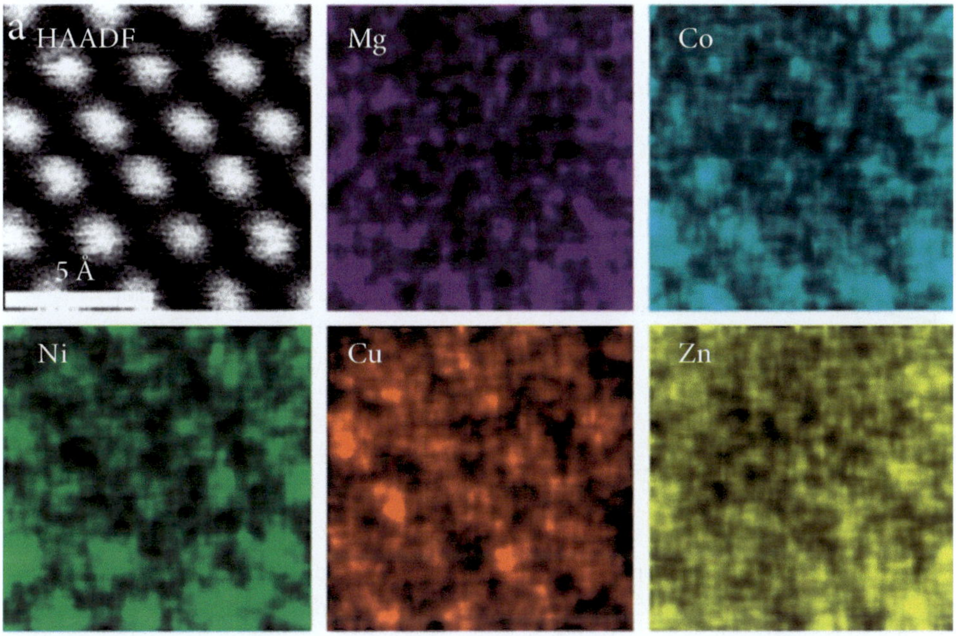

Figure 3.31 HAADF-STEM atomic image of the single-phase multicomponent oxide (MgCoNiCuZn)O, and corresponding atomic-scale energy-dispersive X-ray (EDS) composition maps for each of the five cations, showing even intensities from a near-random distribution of cations on the cation sub-lattice.

Reproduced from Rost et al. (2015). Entropy-stabilized oxides. *Nature Communications* 6: Article 8485.

2. It also includes compounds with one multicomponent cationic sub-lattice such as the dioxides, two multicomponent cationic sub-lattices such as the perovskites, and multicomponent cationic and anionic sub-lattices such as the oxy-fluorides and boro-carbo-nitrides.
3. It also includes equimolar and non-equimolar compounds, with modified compositions obtained by changing cation and anion concentrations at either constant or varying valency,[94,96] though in the latter case composition variations are more limited, and require corresponding charge-compensation mechanisms such as balanced oxidation and reduction of different components and/or creation of cationic or anionic vacancies and/or interstitials.
4. The different multicomponent ceramic single-phase fields have a wider composition range at higher temperatures, and when the materials are manufactured by rapid cooling from high temperature.[93]
5. Multiphase ceramics are formed, as expected, in a similar way to multiphase metallurgical materials, at compositions and temperatures beyond the single-phase fields.[94–96]

More details of the multicomponent single-phase and multiphase covalent and ionic ceramic compounds that have been found, including their compositions and the manufacturing conditions needed to make them, are given in the reviews by Zhang and Reece[94] and Xiang et al.[96]

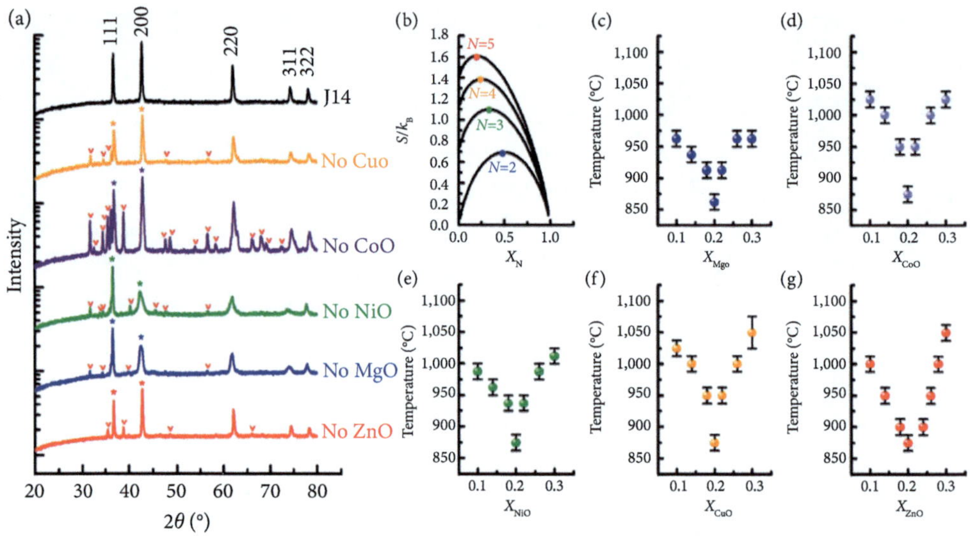

Figure 3.32 Effect of varying composition in multicomponent oxide (MgCoNiCuZn)O: (a) XRD traces showing multiphase structures obtained by removing individual binary oxides one at a time; (b) configurational entropy of mixing versus composition for binary, ternary, quaternary and quinary materials, showing maximum at equiatomic compositions; (c)–(g) composition-dependence of transition temperature above which sintered multicomponent oxide is single-phase, showing lowest transition temperature for equiatomic material.
Reproduced from Rost et al. (2015). Entropy-stabilized oxides. *Nature Communications* 6: Article 8485.

Table 3.12 Multicomponent ceramic compounds.

phase	ceramic composition
rock salt	$(Li_{6.5}Ti_{0.5}MnNb)(O_{1.7}F_{0.3})$ (MgCoNiCuZn)O (TiVZrNb)C; (TiZrNbHfTa)C
fluorite	$(ZrYHfCe)O_2$; $(ZrYHfCeGd)O_2$ $(TiZrSnCeHf)O_2$ $(Mo_{0.5}CePrNdSmGd)O_2$
pyrochlore	$(SmEuTbDyLu)_2Zr_2O_7$ $(LaCeNdSmEu)_2Zr_2O_7$
perovskite	$(TiMnZrSnHf)SrO_3$; $(TiGeZrSnSn)SrO_3$ $(TiZrHfNbSn)(SrBa)O_3$ $(YLaNdSmGd)(CrMnFeCoNi)O_3$
spinel	$(AlMnFeCoNi)_3O_4$; $(AlCrMnFeNi)_3O_4$; $(CrMnFeCoNi)_3O_4$; $(CrMnFeNiZn)_3O_4$ $(CrMnFeCoNi)Fe_2O_4$
mono-borides	(VCrNbMoTa)B; (VCrNbMoW)B (CrMnFeCoMo)B; (CrNiMoTaW)B
boro-carbo-nitrides	(TiZrNbHfTa)(CN); (TiZrNbTaW)(BCN) (TiZrNbHfTa)(BCN); (ZrNbHfTaW)(BCN)
hexaborides	$(YNdSmEuYb)B_6$; $(YCeSmErYb)B_6$
disilicides	$(TiZrNbMoW)Si_2$; $(TiNbMoTaW)Si_2$

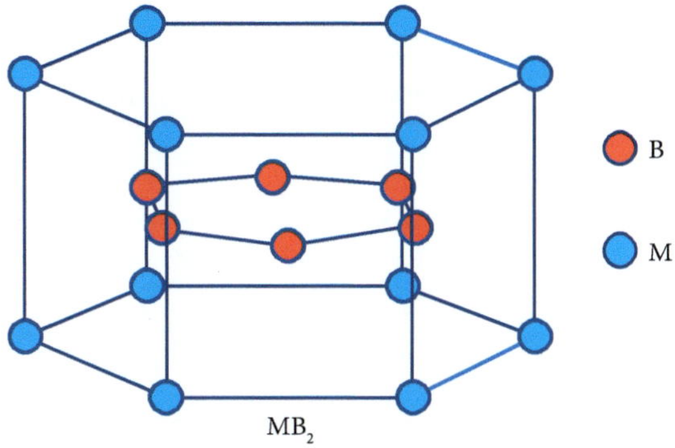

Figure 3.33 C32 hexagonal AlB_2-type multicomponent diboride MB_2 with centre-plane and hexagonal-corner sub-lattice sites for boron B atoms and a near-random distribution of multicomponent metal M atoms respectively.

Not long after Rost et al.'s first detailed and ground-breaking study[93] of a single-phase high-entropy *ionic* ceramic compound, namely, the multicomponent rock-salt NaCl-type mixed mono-oxide (MgCoNiCuZn)O, it was followed by a similar detailed investigation by Gild et al.[97] into a single-phase high-entropy *covalent* ceramic compound, namely, the multicomponent hexagonal C32 AlB_2-type mixed diboride. Gild et al.[97] again used Kim et al.'s equiatomic substitution technique[85,92] to make seven equimolecular 5-component mixed diborides, $(TiCrZrHfTa)B_2$, $(TiZrNbHfTa)B_2$, $(TiZrMoHfTa)B_2$, $(TiZrNbMoTa)B_2$, $(TiZrNbMoHf)B_2$, $(TiZrMoHfW)B_2$ and $(TiNbMoHfTa)B_2$, by ball milling mixtures of the individual boride powders followed by compaction and sintering at 2000 °C. The starting materials, titanium diboride TiB_2, zirconium diboride ZrB_2, niobium diboride NbB_2, molybdenum diboride MoB_2, hafnium diboride HfB_2, tantalum diboride TaB_2, and tungsten diboride WB_2, were again selected because of their similar stoichiometry and valency, and because they each independently exhibit the C32 AlB_2-type crystal structure shown in Figure 3.33. With the exception of the tungsten-containing material, the sintered materials all formed a single-phase multicomponent diboride with the C32 AlB_2-type crystal structure,[96] as shown by the XRD traces in Figure 3.34 (though in some cases there was a small admixture of an oxide impurity). Annular bright-field (ABF) and HAADF atomic-scale imaging in a high-resolution scanning transmission electron microscope (HRSTEM), as shown in Figure 3.35, and EDS composition maps at low magnification in an SEM and at atomic-scale resolution in the HRSTEM all indicated that the six single-phase multicomponent diborides were compositionally homogeneous with a near-random distribution of metal atoms across the metal sub-lattice of the C32 AlB_2 structure.[96] The results shown in Figure 3.34 indicate that there is a large C32 AlB_2-type mixed diboride single-phase field in multicomponent phase space, centred loosely around $(TiZrNbMoHfW)B_2$, and extending out to binary diboride solid solution alloys such as TiB_2–ZrB_2, ZrB_2–NbB_2, etc., as well as to the individual diborides TiB_2, ZrB_2, NbB_2, MoB_2, HfB_2, TaB_2 and WB_2. Once again, the full extent of this single-phase field remains to be mapped out in detail.

More recently, a wide variety of other multicomponent ionic and covalent ceramic compounds have been explored,[94–96] as shown in Table 3.12. Large single-phase field

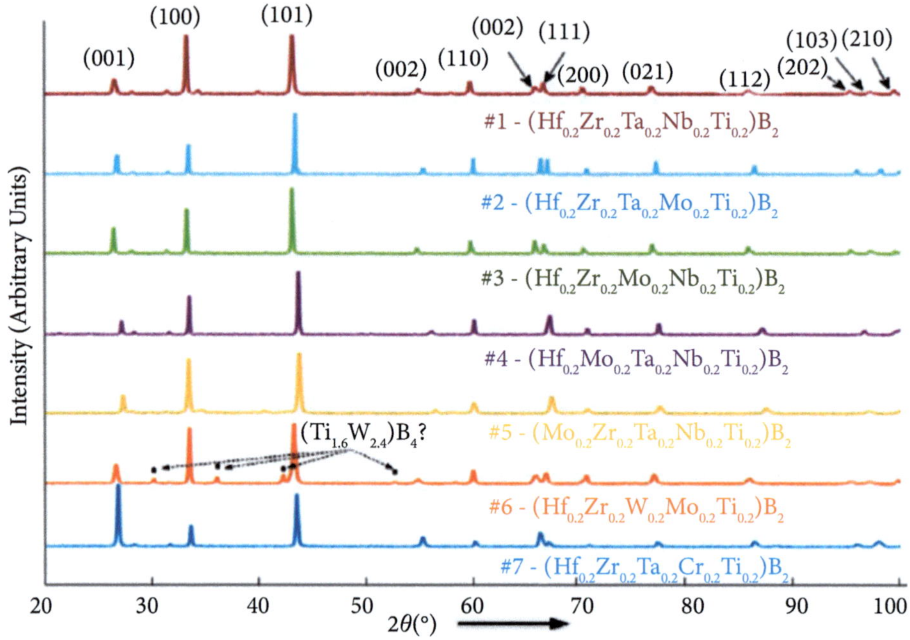

Figure 3.34 XRD traces showing a single-phase C32 hexagonal AlB_2-type crystal structure in six multicomponent mixed transition-metal diborides (indexed peaks), with an additional tetraboride phase in the W-containing diboride.

Reproduced from Gild et al. (2016). High-entropy diborides: A new class of high-entropy materials and a new type of ultrahigh temperature ceramics, *Nature Scientific Reports* 6: Article 37946.

in multicomponent phase space have been identified for a number of different well-known chemical stoichiometries and crystal structures including, for instance, spinels, perovskites, fluorites, mono-borides, mono-carbides and tetra-borides, in each of which the composition can be varied widely within and, to a lesser extent, beyond a fixed balance of valencies between the different components (Figure 3.36).[94–96] As two other, more recent examples of such multicomponent ceramic compounds: Qin et al.[98] made a series of three superhard single-phase B33 CrB-type orthorhombic mono-borides, (VCrNbMoTa)B, (VCrNbMoW)B and (VCrNbTaW)B, as shown in the XRD traces and EDS composition maps in Figures 3.37 and 3.38, respectively; and Zhu et al.[99] made a series of tunable magnetic multicomponent fcc oxide spinels, namely 7-component $(AlCrMnFeCoNi)_3O_4$ and 6-component $(AlCrMnFeCo)_3O_4$, $(AlCrMnFeNi)_3O_4$, $(AlCrFeCoNi)_3O_4$, $(AlMnFeCoNi)_3O_4$, and $(AlMnFeCoZn)_3O_4$, as shown in the XRD traces in Figure 3.38.

Overall, we can conclude that there are many very large high-entropy single-phase ceramic compound regions in multicomponent phase space, separated by similarly large regions of multiphase materials. As with multicomponent metallurgical materials, these large multicomponent single-phase ceramic fields have complex shapes within the C-dimensional hyper-polyhedral phase diagram that represents multicomponent phase space, but the extent of the single-phase fields are greater when the materials have equimolecular or near-equimolecular compositions, when they are manufactured by sintering at high rather than

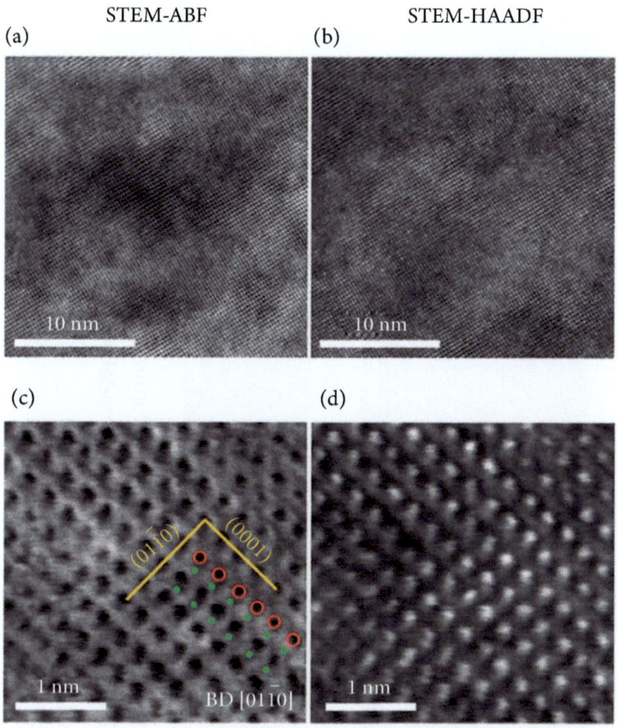

Figure 3.35 Low and high magnification (a) and (c) ABF-STEM and (b) and (d) HAADF-STEM atomic images of the single-phase multicomponent diboride $(TiZrMoHfTa)B_2$, with even lattice spacings and image contrast from a near-random distribution of metal atoms on the metal sub-lattice.

Reproduced from Gild et al. (2016). High-entropy diborides: A new class of high-entropy materials and a new type of ultrahigh temperature ceramics, *Nature Scientific Reports 6*: Article 37946.

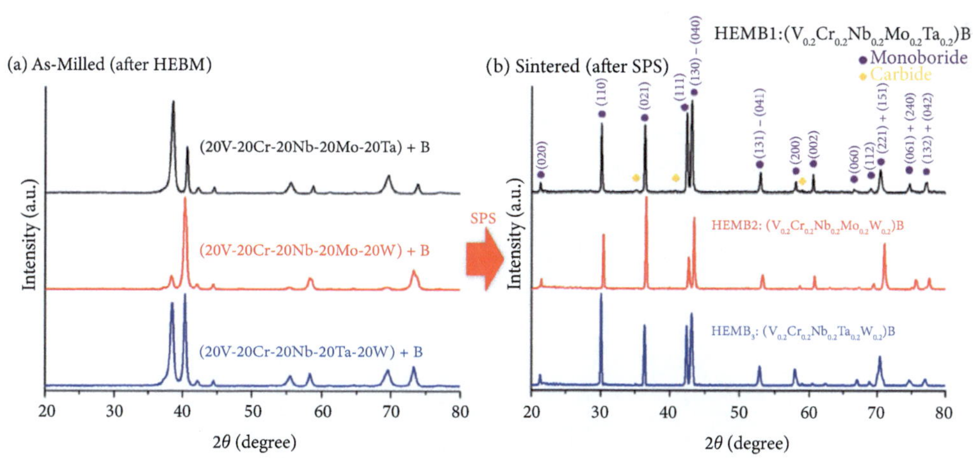

Figure 3.36 XRD traces from (a) as-milled mixed transition metal and boron powders and (b) single-phase multicomponent (VCrNbMoTa)B, (VCrNbMoW)B and (VCrNbTaW)B mono-borides after sintering with, in each case, a single-phase B33 CrB-type orthorhombic crystal structure (indexed peaks).

Reproduced from Qin et al. (2020). High-entropy monoborides: Towards superhard materials. *Scripta Materialia 189*: 101–5.

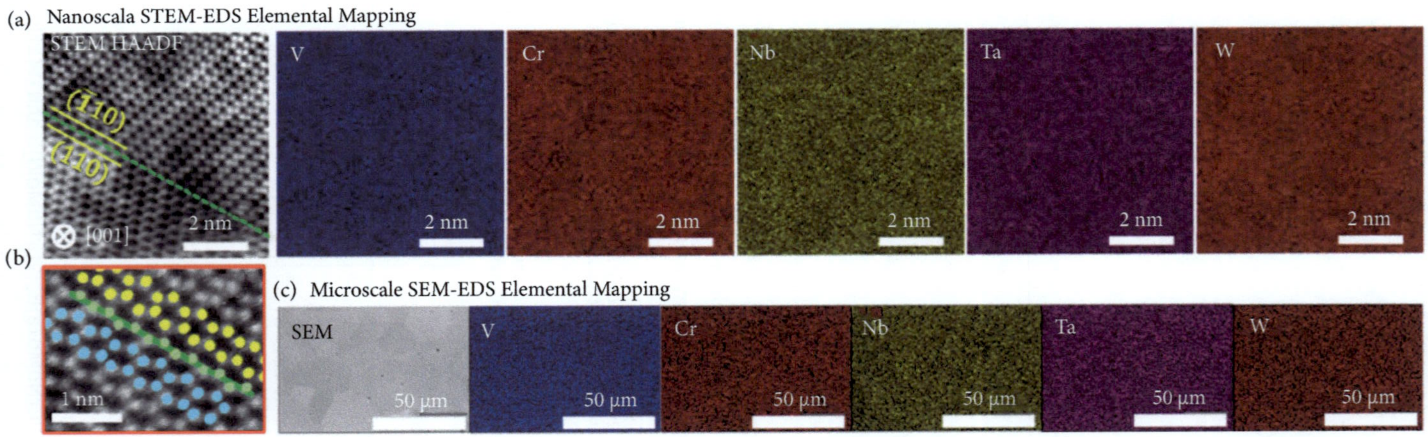

Figure 3.37 (a) and (b) low- and high-magnification HAADF-STEM atomic images of the single-phase multicomponent mono-boride (VCrNbTaW)B mono-boride, and (c) corresponding atomic-scale EDS composition maps for each of the five metal atoms, showing even intensities from a near-random distribution of metal atoms on the metal sub-lattice.

Reproduced from Qin et al. (2020). High-entropy monoborides: Towards superhard materials. *Scripta Materialia 189*: 101–5.

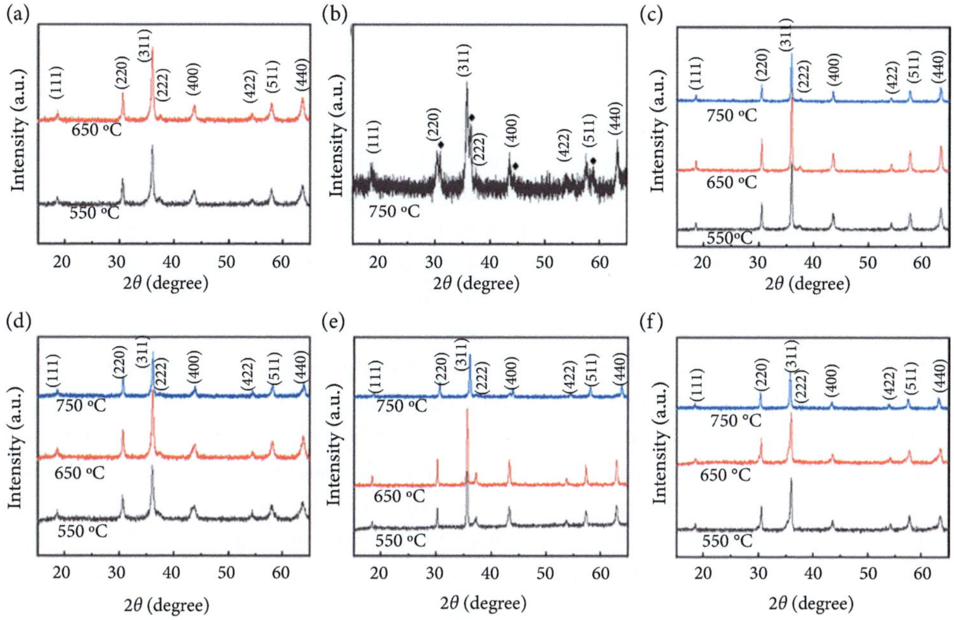

Figure 3.38 XRD traces from (a) $(AlCrMnFeCoNi)_3O_4$; (b) $(AlMnFeCoZn)_3O_4$; (c) $(AlCrMnFeNi)_3O_4$; (d) $(AlCrMnCoNi)_3O_4$; and (e) $(AlCrFeCoNi)_3O_4$, $(AlCrMnFeCo)_3O_4$, made by sintering at different temperatures between 550 °C and 750 °C, with, in each case, a single-phase AB_2O_4 fcc spinel crystal structure (indexed peaks).

Reproduced from Zhu et al. (2021). Structure and magnetic properties of a class of spinel high-entropy oxides. *Journal of Magnetism and Magnetic Materials 535*: Article 168063.

intermediate temperatures, and when they are quenched rapidly to room temperature at the end of the sintering process.

Multiphase materials

We have seen throughout this section that there are many very large complex-shaped single-phase fields in multicomponent phase space, such as the disordered fcc Cantor and bcc Senkov alloys, disordered hcp alloys, multicomponent mono-aluminides and trialuminides, Laves phases, TCPs such as σ and μ, multicomponent ceramic mono-oxides, spinels and perovskites, mono-boro-carbo-nitrides and diborides. It seems clear that there is not a wealth of new phases and compound structures that are developed and appear as we explore deeper into multicomponent phase space (in fact, there are hardly any) and, instead, the addition of more and more components is absorbed by well-known binary and ternary phase fields expanding to form extended solid-solution regions, with multiple components randomly or near-randomly distributed across either all the lattice sites (in multicomponent disordered solid solutions) or one or more of the sub-lattice sites (in multicomponent intermetallic and ceramic compounds).

The formation of such large single-phase fields was, of course, one of the first (and initially surprising) discoveries about multicomponent materials,[1,2] and is explained at least partly by the above-mentioned paucity of new multicomponent compound structures, and partly by the stabilising effect of a high configurational entropy of mixing when many different components are distributed on a single set of lattice or sub-lattice sites (which is why the field became known as high-entropy alloys or high-entropy materials).[7–9] The formation of large multicomponent single-phase fields, the paucity of new multicomponent compounds and the high configurational entropy of mixing in multicomponent solid solutions are all discussed more fully in Chapter 4. There is no doubt that the early discovery of such large multicomponent solid-solution single-phase fields[1,2] excited scientists' curiosity worldwide and stimulated the rapid growth of the field over the last decade or so. Much of the extensive scientific work on multicomponent high-entropy materials since the early 2000s has concentrated naturally, therefore, on discovering and beginning to chart the number and extent of multicomponent single-phase fields, initially investigating equiatomic and equimolecular compositions, and only expanding slowly into other regions of multicomponent phase space.[7–9] But we have seen earlier in this section that, between the very large *single-phase* fields, there are equally large regions of multicomponent phase space, also containing many millions of materials with *multiphase* structures that have so far been studied rather less intensively. Multicomponent *single-phase* materials have already been found to have many interesting and potentially valuable new mechanical, electrical, magnetic, chemical and other properties,[7–9] and the same is almost certainly true for multicomponent *multiphase* materials, but they have only recently begun to receive much scientific attention.

Multiphase materials can form in a variety of ways,[7–9] including:

1. Solidification of secondary and tertiary phases in segregated regions during manufacture by solidification or vapour deposition.
2. Formation of multiple eutectic and/or peritectic phases during manufacture by solidification or vapour deposition.
3. Formation of multiphase composite structures during manufacture by mechanical alloying and/or chemical reaction processing.
4. Precipitation of secondary and tertiary phases from initial extended solid solutions on cooling from high temperature during either initial manufacture or subsequent heat treatment.
5. Ordering reactions during cooling from high temperature during either initial manufacture or subsequent heat treatment.
6. Eutectoid and/or peritectoid reactions during cooling from high temperature during either initial manufacture or subsequent heat treatment.

The Gibbs phase rule is $p + f = c + 2$, or $p + f = c$ at a fixed pressure and temperature (such as atmospheric pressure and room temperature).[3–5] The maximum number of phases in any material at room temperature and atmospheric pressure is given at *equilibrium*, therefore, as $p_{max} = c$ (corresponding to $f = 0$), but can be somewhat higher, $p_{max} > c$, under *non-equilibrium* manufacturing conditions. Many of the multicomponent materials that have been investigated contain five and sometimes six, seven or even more components, but studies of multiphase materials have so far been almost exclusively on two-phase or (occasionally) three-phase materials, and there have been almost no studies of materials containing four, five, six, seven or more different phases. This is a rich area for future investigation. There are many examples of multicomponent multiphase microstructures formed by the different methods listed above, and more details are given in the reviews by Miracle and Senkov[7] and Cantor,[9] as well as the book by Murty et al.[8] As mentioned, however, most of these examples are from materials close to the single-phase fields that have been the major focus of scientific investigation so far, containing, therefore, usually no more than one or (occasionally) two additional phases. The next two sub-sections describe two of the most common types of multicomponent multiphase (i.e. mostly so far two-phase) materials, formed during manufacture by eutectic solidification and solid-state precipitation respectively.

Multicomponent eutectics

Between the large number of large single-phase regions in multicomponent phase space, there must be a complex network of *multicomponent eutectic, peritectic, eutectoid and peritectoid* valleys, channels and saddle-points, corresponding to the many multiphase eutectic, peritectic, eutectoid and peritectoid reactions[xi] that can take place.[3–5] This network will include multiphase binary, ternary, quaternary etc. reactions, i.e. with reaction schemes that can be described in general as follows:[3–5]

$$\text{eutectic: L} \rightarrow \sum_{i=1}^{n} \alpha_i$$

$$\text{peritectic: L} + \sum_{i=1}^{k} \alpha_i \rightarrow \sum_{i=k+1}^{n} \alpha_i$$

$$\text{eutectoid: } \alpha_0 \rightarrow \sum_{i=1}^{n} \alpha_i$$

$$\text{peritectoid: } \alpha_0 + \sum_{i=1}^{k} \alpha_i \rightarrow \sum_{i=k+1}^{n} \alpha_i$$

where L is the liquid phase, α_i are different solid phases, and n + 1 is the total number of phases involved in the reaction (n is the final number of phases, since the initial liquid L or solid α_0 phase is fully consumed by each of these reactions).

From the Gibbs phase rule at fixed temperature and pressure (such as room temperature and atmospheric pressure),[3–5] the maximum number of phases at equilibrium in a *c*-component material is $p_{max} = c$, so at equilibrium the combination of multiphase eutectic, peritectic, eutectoid and peritectoid reactions in any particular material and any given alloy system can produce any number of phases up to a maximum of $n = p_{max} = c$. When $n = p_{max} = c$, the corresponding *c*-phase eutectic, peritectic, eutectoid or peritectoid reaction is invariant in that alloy system, i.e. takes place only at a fixed temperature and composition in the multicomponent alloy phase space; and when $n < p_{max} = c$, the corresponding ($< c$)-phase eutectic, peritectic, eutectoid or peritectoid reaction is not invariant, but still takes place in only very limited regions of the multicomponent alloy phase space.[3–5] The range of reaction schemes (i.e. the range of combinations of multiple invariant and additional lower-order non-invariant eutectic, peritectic, eutectoid and peritectoid reactions) that are possible during manufacture of different multicomponent multiphase materials is enormous and the

[xi] And other less common reactions such as monotectic, metatetic and syntectic.[3]

vast majority of possibilities in different alloy systems and materials have not yet been investigated at all. We concentrate in this section on simple-eutectic multicomponent alloy systems, i.e. multicomponent alloy systems that contain only eutectic reactions, and no more than one invariant eutectic reaction. This seems very restrictive, but not that much is known even about these multicomponent alloy systems, let alone all the others with multiple invariant eutectics, multiple other invariant reactions (such as peritectic, eutectoid and peritectoid), and multiple lower-order non-invariant reactions.

In a binary alloy system, eutectic reactions produce two solid phases during solidification:[3,5,100,101]

$$L \rightarrow \alpha + \beta$$

and we call this a *binary eutectic reaction*. From the Gibbs phase rule, a binary eutectic reaction in a binary alloy system is invariant at constant (atmospheric) pressure, i.e. it takes place only at a given temperature T_E and composition c_E. This means that almost all materials in the alloy system (all alloys that do not have the invariant eutectic composition, i.e. that have compositions $c \neq c_E$) are either hypoeutectic ($c < c_E$) or hypereutectic ($c > c_E$) and solidify initially by forming single-phase primary α or β, before completing solidification with the invariant two-phase binary eutectic reaction:[5,100,101]

$$L \rightarrow \alpha \text{ or } L \rightarrow \beta$$

$$\text{then } L \rightarrow \alpha + \beta$$

This produces hypoeutectic and hypereutectic microstructures that are a mixture of primary α or β dendrites interspersed with interdendritic regions of fine-scale (sub-micron-scale) two-phase eutectic $\alpha + \beta$, usually with a lamellar or rod-like morphology, though other morphologies are possible. For most alloy compositions the primary dendritic and interdendritic regions are of roughly comparable size and scale (typically several tens of microns); however, for compositions far from the invariant eutectic composition and with, therefore, only very limited terminal eutectic reaction, the microstructures look essentially like large α or β dendritic grains with small amounts of distributed secondary β or α particles respectively; and for compositions close to or equal to the invariant eutectic composition, the microstructures look like (solely) fine-scale two-phase eutectic mixtures of $\alpha + \beta$.

In a ternary alloy system, eutectic reactions can produce two or three solid phases:[3,5,100,101]

$$L \rightarrow \alpha + \beta$$

$$L \rightarrow \alpha + \beta + \gamma$$

i.e. we can have *binary eutectic reactions* and *ternary eutectic reactions*, producing two or three solid phases respectively. From the Gibbs phase rule, a ternary eutectic reaction in a ternary alloy system is again invariant, i.e. it takes place at a given temperature and composition; and binary eutectic reactions in a ternary alloy system have one degree of freedom so they take place along particular lines of temperature and composition in the ternary alloy phase space. This means that almost all materials in the alloy system (all alloys that do not have an invariant ternary eutectic composition or a composition on the binary eutectic lines) solidify initially by forming single-phase primary α, β or γ, before undergoing a two-phase binary eutectic reaction, and then completing solidification with the invariant ternary eutectic reaction:[3,5,100,101]

$$\begin{aligned}&L \rightarrow \alpha \text{ or } L \rightarrow \beta \text{ or } L \rightarrow \gamma\\ &\text{then } L \rightarrow \alpha+\beta \text{ or } L \rightarrow \alpha+\gamma \text{ or } L \rightarrow \beta+\gamma\\ &\text{then } L \rightarrow \alpha+\beta+\gamma\end{aligned}$$

Again, this produces hypoeutectic and hypereutectic microstructures that are a mixture of primary α, β or γ dendrites interspersed with interdendritic regions of fine-scale (sub-micron-scale) two-phase $\alpha+\beta$, $\alpha+\gamma$ or $\beta+\gamma$ and three-phase $\alpha+\beta+\gamma$ eutectics. For most alloy compositions the primary dendritic and interdendritic regions are again of roughly comparable size and scale (typically several tens of microns); however, for compositions far from the binary or ternary eutectic compositions and with, therefore, only very limited terminal eutectic reaction, the microstructures look essentially like large α, β or γ dendritic grains with small amounts of distributed secondary and tertiary α, β and/or γ particles; and for compositions close to or equal to the binary or ternary eutectic compositions, the microstructures look like (solely) fine-scale two- and three-phase eutectic mixtures of $\alpha + \beta + \gamma$.

In general, therefore, in a c-component alloy system, eutectic reactions can produce any number of solid phases up to a maximum equal to the number of components c:

$$\begin{aligned}&L \rightarrow \alpha+\beta\\ &L \rightarrow \alpha+\beta+\gamma\\ &\quad\vdots\\ &L \rightarrow \alpha+\beta+\gamma+\ldots\,\alpha_c\end{aligned}$$

i.e. we can have *binary eutectic reactions, ternary eutectic reactions, quaternary eutectic reactions* and so on, up to *c-phase eutectic reactions*, producing two, three, four and so on up to c solid phases respectively. From the Gibbs phase rule: a c-phase eutectic reaction in a c-component alloy system is again invariant, i.e. it takes place at a given temperature and composition; $(c-1)$-phase eutectic reactions in a c-component alloy system have one degree of freedom so they take place along particular lines of temperature and composition in the multicomponent-alloy phase space; $(c-2)$-phase eutectic reactions in a c-component alloy system have two degrees of freedom so they take place in particular (very limited) regions in the multicomponent-alloy phase space; and so on. This means that almost all materials in the alloy system (all alloys that do not have the invariant c-phase eutectic composition, or a composition that lies on the $(c-1)$-phase eutectic lines, or in the limited $(c-2)$-phase regions, $(c-3)$-phase regions, etc.) solidify initially by forming single-phase primary α, β, . . . or α_c, before undergoing a two-phase binary eutectic reaction, followed by a three-phase ternary eutectic reaction, and so on, finally completing solidification with the invariant c-phase eutectic reaction:

$$\begin{aligned}&L \rightarrow \alpha \text{ or } L \rightarrow \beta \text{ or } L \rightarrow \gamma \text{ or } \cdots\ L \rightarrow \alpha_c\\ &\text{then } L \rightarrow \alpha+\beta \text{ or } L \rightarrow \alpha+\gamma \text{ or } \cdots\\ &\text{then } L \rightarrow \alpha+\beta+\gamma \text{ or } L \rightarrow \alpha+\beta+\delta \text{ or } \cdots\\ &\quad\vdots\\ &\text{then } L \rightarrow \alpha+\beta+\gamma+\ldots\,\alpha_c\end{aligned}$$

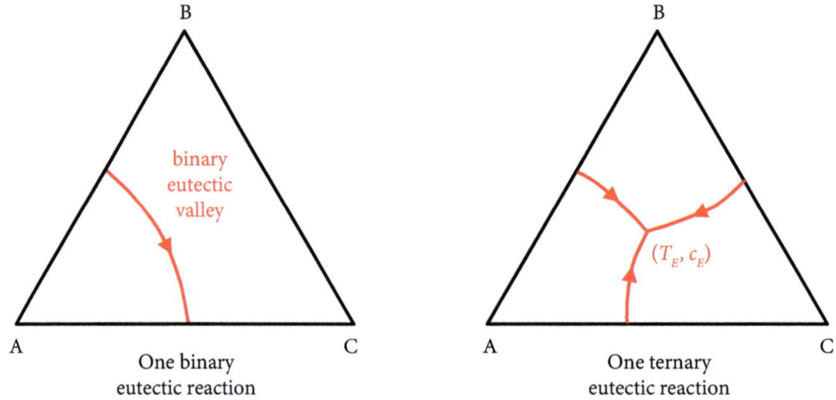

Figure 3.39 Eutectic reactions in a ternary alloy system: (a) one non-invariant binary eutectic reaction $L \rightarrow \alpha + \beta$ along a binary eutectic valley extending from pure A-B alloys to pure A-C alloys; (b) three non-invariant binary eutectic reactions, $L \rightarrow \alpha + \beta$, $L \rightarrow \alpha + \gamma$ and $L \rightarrow \beta + \gamma$, along binary eutectic valleys that meet at the invariant ternary eutectic reaction $L \rightarrow \alpha + \beta + \gamma$ at (T_E, c_E).

Again, this produces hypoeutectic and hypereutectic microstructures that are a mixture of primary α, β, . . . or α_c dendrites interspersed with interdendritic regions of fine-scale (sub-micron-scale) two-, three-, four-, up to *c*-phase eutectics. For most alloy compositions the primary dendritic and interdendritic regions are again of roughly comparable size and scale (typically several tens of microns); for compositions far from the eutectic compositions, however, the microstructures look essentially like large primary dendritic grains with small amounts of distributed secondary, tertiary . . . etc., α, β, γ... α_c particles; and for compositions close to or equal to the eutectic compositions, the microstructures look like (solely) fine-scale multiphase eutectic mixtures of $\alpha + \beta + \gamma + \ldots \alpha_c$.

The range of options is shown in Figure 3.39 for the very simple case of a ternary alloy system, with either one non-invariant binary eutectic reaction ($L \rightarrow \alpha + \beta$), or three non-invariant binary eutectic reactions ($L \rightarrow \alpha + \beta$, $L \rightarrow \alpha + \gamma$ and $L \rightarrow \beta + \gamma$) and a three-phase invariant ternary eutectic reaction ($L \rightarrow \alpha + \beta + \gamma$).[3–5] There are clearly many more options when we extend our interest to higher-order 4-, 5-, 6- and (in general) *c*-component alloy systems.

Many of the multiphase materials that have been studied so far were formed by a eutectic reaction during solidification, and almost all of these have relatively simple two-phase binary eutectic structures,[7–9,102–105] i.e. the alloy compositions either lie on a pseudo-binary eutectic section within the multicomponent alloy phase space, or are sufficiently far from an invariant multiphase eutectic (or other invariant) reaction such that solidification is largely completed by the end of the first two (primary and binary eutectic) stages of solidification. In either case, most of the multicomponent eutectic alloys studied so far show typical two-phase hypoeutectic, eutectic or hypereutectic microstructures, very similar to those found in conventional binary alloys, with a mixture of a single-phase primary regions and interdendritic coupled two-phase eutectic regions. The main difference is that the two eutectic phases in multicomponent materials are multicomponent solid solutions or multicomponent intermetallic or ceramic compounds rather than simple one- or two-component phases.

Table 3.13 Multicomponent two-phase eutectic alloys.

alloy	phases
$CrFeCoNiNb_{0.5}$	fcc + C14 $MgZn_2$ Laves phase
$V_{0.5}FeCoNi_2Nb_{0.75}$	fcc + C14 $MgZn_2$ Laves phase
$CrFeCoNiZr_{0.45}$	fcc + C15 $MgCu_2$ Laves phase
$CrFeCoNiTa_{0.4}$	fcc + C15 $MgCu_2$ Laves phase
$CrFeCoNi_2Zr_{0.6}$	fcc + $Ni_7(ZrHf)_2$
$CrFeCoNi_2Hf_{0.55}$	fcc + $Ni_7(ZrHf)_2$
CrMnFeCoNiPd	fcc + Mn_7Pd_9
$VFeCoNiMo_{0.6}$	fcc + $CoMo_2Ni$
$CrFeCoNi_{2.2}$	fcc + B2
$AlCrFeCoNi_{2.1}$	$L1_2$ + B2
$AlCrFeCo_2Ni_2$	$L1_2$ + B2
$AlCrFe_2CoNi_2$	$L1_2$ + B2
AlCrFeNi	bcc + B2
$Al_{32.5}V_{20}Cr_{10}Ni_{32.5}Mo_5$	bcc + B2
$Al_{32.5}Cr_{30}Ni_{32.5}Mo_5$	bcc + B2
$Al_{32.5}V_{30}Ni_{32.5}Mo_5$	bcc + B2
ScTiZrNb	bcc + hcp
$NbMoTaWRe_{0.5}(TiC)$	bcc + MC

Table 3.13 lists some (by no means all) of the multicomponent two-phase eutectic structures that have been studied. Table 3.13 shows that there is a wide variety of different pseudo-binary multicomponent eutectic alloys, including, for instance: hcp + bcc[106] in 4-component ScTiZrNb; fcc + C14 Laves phase[107] in 5-component $CrFeCoNiNb_{0.5}$; $L1_2$ + B2[108] in 5-component $AlCrFeCoNi_{2.1}$; bcc + B2[109] in 5-component $Al_{32.5}V_{20}Cr_{10}Ni_{32.5}Mo_5$; and bcc + MC where M is a mixture of transition metals[110] in 7-component $NbMoTaWRe_{0.5}(TiC)$. Figure 3.40 shows scanning electron micrograph (SEM) images of typical microstructures in pseudo-binary multicomponent $CrFeCoNiNb_x$ alloys:[107] primary fcc dendrites and interdendritic fcc + C14 Laves-phase eutectic in a hypoeutectic alloy with $x = 0.45$; fully developed coupled fine-scale lamellar fcc + C14 Laves-phase eutectic with $x = 0.5$; and primary C14 Laves-phase dendrites and interdendritic fcc + C14 Laves-phase eutectic in two hypereutectic alloys with $x = 0.55$ and 0.65. More details of the different multicomponent pseudo-binary eutectics that have been investigated and their structures and properties are given in the reviews by Chanda et al.,[102] Lu et al.,[103] Tiwari et al.[104] and Chang et al.[105]

The different pseudo-binary multicomponent eutectic alloys in Table 3.13 exhibit different types of coupled lamellar, rod-like, Chinese-script and other structures,[102] but in general the size and spacing of the two eutectic phases are usually very fine scale (in the tens and hundreds of nanometres range).[102–105] Coupled eutectic solidification takes place at a speed and scale determined by lateral diffusion of alloying elements in the liquid, just ahead of the growing two-phase solid/liquid interface,[5] as shown in Figure 3.41. In a binary alloy, this involves lateral cross-diffusion of just two components, but in multicomponent materials

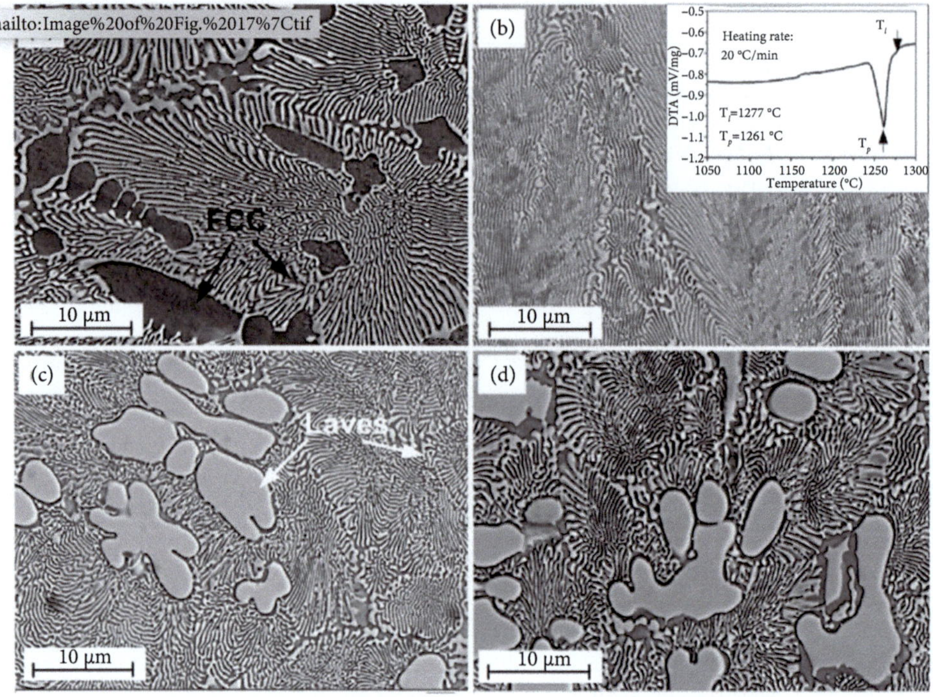

Figure 3.40 SEM images of pseudo-binary fcc + C14 Laves-phase eutectic microstructures in multicomponent $CrFeCoNiNb_x$ alloys: (a) hypoeutectic ($x = 0.45$); (b) eutectic ($x = 0.5$); (c) hypereutectic ($x = 0.55$); (d) hypereutectic ($x = 0.65$); insert = differential thermal analysis (DTA) trace showing eutectic melting point.

Reproduced from Chanda et al. (2020). A review on nano-/ultrafine advanced eutectic alloys. *Journal of Alloys and Compounds* 827: Article 154226.

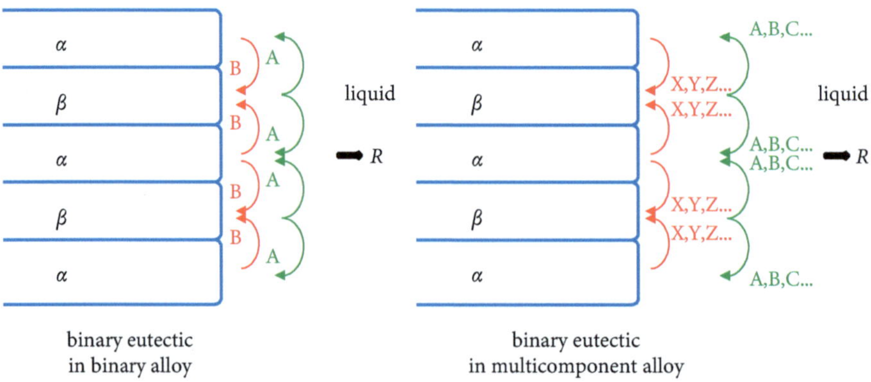

Figure 3.41 Coupled binary eutectic solidification: (a) in a binary alloy, showing lateral diffusion of components A and B in opposite directions ahead of the solid/liquid interface; and (b) in a multicomponent alloy, showing lateral diffusion of multiple components A, B, C . . . and X, Y, Z . . . in opposite directions ahead of the solid/liquid interface.

it involves lateral cross-diffusion of many components, as shown in Figure 3.41. This slows down the rate of solidification, which is determined by the speed of the slowest diffusing species and, therefore, reduces the size and spacing of the lamellae or rods of the different phases under any given set of solidifying conditions (i.e. for any given rate of cooling and, therefore, rate of extraction of latent heat).

Most of the multicomponent eutectics that have been studied are pseudo-binary eutectics with simple two-phase hypoeutectic, eutectic and hypereutectic microstructures, similar to those found in binary alloy systems, as shown for example in Figure 3.40. There have, however, also been a few reports[102–105] of multicomponent invariant eutectic reactions producing ternary and quaternary multiphase microstructures, and most of these are listed in Table 3.14. In general, it does not seem easy (perhaps understandably) in ternary, quaternary, quinary and higher-order multiphase eutectic reactions for the three, four or more eutectic phases to grow side-by-side in a straightforward coupled way similar to that shown for binary and quasi-binary eutectics in Figure 3.41, and the microstructures that have been observed in these higher-order eutectic reactions are generally still very fine scale, but usually much more complex.[102–105] The bright-field transmission electron micrograph (TEM) and associated selected area diffraction patterns (SADPs) in Figure 3.42 show an example from the detailed study by Cai et al.[109] of the coupled four-phase microstructures formed during solidification by the invariant quaternary eutectic reaction L → fcc Al + θ-Al_2Cu + Q-AlCuSiMg + diamond-cubic (dc)-Si in quaternary Al-Mg-Si-Cu alloys,[111] with a complex mixture of plate, rod and particle morphologies for the four eutectic phases.

Multicomponent precipitation-hardened alloys

Thermomechanical treatments, i.e. appropriate combinations of deformation and annealing, can be used to create multicomponent supersaturated solid solutions followed by *multicomponent precipitation reactions*, to form multiphase matrix-precipitate microstructures

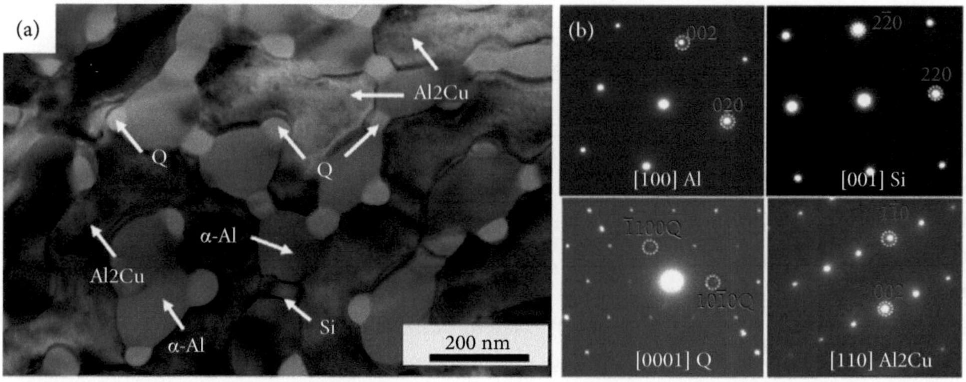

Figure 3.42 Complex four-phase coupled microstructure formed by the quaternary eutectic reaction L → fcc Al + θ-Al_2Cu + Q-AlCuSiMg + dc-Si in quaternary Al-Mg-Si-Cu alloys: (a) bright-field TEM image; and (b) SADPs of the four phases.

Reproduced from Cai et al. (2022). Microstructure and mechanical properties of ultrafine quaternary Al-Cu-Si-Mg eutectic alloy. *Materials Science and Engineering 12*(1): 7.

Table 3.14 Multicomponent multiphase eutectic alloys.

	alloy system	phases
ternary	Mg-Si-Ca	dc-Si + Mg_2Si + $MgSi_3Ca_2$
	Mg-Si-Cu	dc-Si + Mg_2Si + $Mg_6Si_7Cu_{16}$
	Al-Mg-Ni	fcc Al + Mg_2Al_3 + Al_3Ni
	Al-Si-Cu	fcc Al +θ-Al_2Cu + dc-Si
	Al-Cr-Ni	fcc Al +θ-Al_2Cr + Al_3Ni
	Al-Fe-Ni	fcc Al + Al_3Ni + Al_9FeNi
	Ti-Fe-Zr	hcp Ti + TiFe + $FeZr_2$
	Ti-Ni-Nb	hcp Ti + TiNi + bcc Nb
	Ti-B-Si	hcp Ti + Ti_6Si_2B + Ti_5Si_3
	Cu-Zn-P	fcc Cu +β-CuZn + Cu_3P
	Cd-Sn-Pb	hcp Cd + bct Sn + fcc Pb
	Ni-Al-Zr	B2-NiAl +γ-Ni_3Al + Ni_7Zr_2
quaternary	Al-Mg-Si-Cu	fcc Al +θ-Al_2Cu + Q-AlCuSiMg + dc-Si
	Al-Mg-Si-Mn	fcc Al + Mg_2Si +α -AlFeMnSi + dc-Si
	Al-Mg-Si-Ni	fcc Al + Mg_2Si + Al_3Ni + dc-Si
	Al-Si-Cu-Fe	fcc Al +θ-Al_2Cu +β-$Al_9Fe_2Si_2$ + dc-Si
	Zn-Cd-Sn-Pb	hcp Zn + hcp Cd + bct Sn + fcc Pb
quinary	Al-Mg-Si-Fe-Ni	fcc Al + Mg_2Si + Al_3Ni + π-AlFeNiMgSi + dc-Si

in multicomponent materials with compositions in the vicinity of the solvus boundaries surrounding the large single-phase fields in multicomponent phase space. The general reaction scheme is:

$$\alpha_{ss} \rightarrow \alpha_{eq} + \sum_{i=2}^{n} \alpha_i$$

where α_i are different solid phases, α_{ss} and α_{eq} are the initial supersaturated and final equilibrium matrix phase respectively, and n is the total (and final) number of phases involved in the reaction. From the Gibbs phase rule at fixed temperature and pressure (such as room temperature and atmospheric pressure),[3–5] the maximum number of phases at equilibrium in a c-component material is $p_{max} = c$, so at equilibrium the precipitation reactions in any particular material and any given alloy system can, in principle, produce any number of phases up to a maximum of $n = p_{max} = c$. The maximum number of phases can be somewhat greater, $n > p_{max} = c$, when precipitation takes place under non-equilibrium conditions; and precipitation reactions often take place via one or more intermediate metastable precipitate phases, which can form more rapidly than the equilibrium phase(s), i.e. are less-favoured thermodynamically but more favoured kinetically:

$$\alpha_{ss} \rightarrow \alpha''_{eq} + \sum_{i=2}^{n} \alpha''_i \rightarrow \alpha'_{eq} + \sum_{i=2}^{n} \alpha'_i \cdots \rightarrow \alpha_{eq} + \sum_{i=2}^{n} \alpha_i$$

where α'_{eq}, α''_{eq}, α'_i and α''_i are metastable solid phases that are precursors to their corresponding equilibrium phases α_{eq} and α_i. In practice, however, precipitation behaviour is not quite so complex: even in multicomponent materials most precipitation reactions lead to a single preferred lowest-energy stable precipitate (i.e. $n = 2$); and there has been little study of metastable intermediate precipitates in multicomponent materials. In other words, most work so far has concentrated on simple precipitation reactions, producing two-phase matrix-precipitate microstructures:

$$\alpha_{ss} \rightarrow \alpha_{eq} + \beta$$

Multicomponent single-phase fcc Cantor alloys and bcc Senkov alloys have excellent mechanical properties, with attractive combinations of strength and ductility. There has been quite a lot of scientific interest in further improving these properties, and one of the popular (and very effective) routes that has been used has been to modify alloy compositions, taking them near to the boundary of the single-phase region, to allow for heat treatment and precipitation of strengthening phases. Not surprisingly, as with precipitation in conventional binary and ternary alloys, a wide range of different *matrix-precipitate microstructures* can form in multicomponent alloys, even when there is only one precipitate phase, depending in detail on the particular alloy composition, grain size, deformation, heat treatment and initial manufacturing process. This enables a wide range of processing schedules that can be used to develop a correspondingly wide range of microstructures that can in turn be used to tune the final multicomponent alloy properties (as is done routinely in conventional binary and ternary alloys).[100,101] The mechanical and other properties of precipitation-hardened multicomponent alloys are discussed in detail in later chapters. Here we note only the enormous variety of possible multiphase materials and microstructures that can form by precipitation reactions throughout multicomponent phase space, for compositions lying between the different large single-phase fields, in regions adjacent to their solvus limits.

Most alloys that have been investigated so far undergo no more than one or two precipitation reactions, with resulting two- or three-phase matrix-precipitate structures. As with eutectic solidification, there has been almost no investigation of precipitation reactions involving four, five, six or more phases. Precipitation mechanisms can still vary widely, however, and here we will just show one or two examples. Figure 3.43 shows an example[110] of a two-phase matrix-precipitate microstructure, with three variants of nanoscale laths of B2 ordered multicomponent MAl mono-aluminide precipitated in a multicomponent fcc matrix, in a non-equiatomic aluminium-modified fcc Cantor alloy $Al_{0.3}CrFeCoNi$ after homogenisation and solution-treatment for one hour at 1,150 °C (1,423 K), 50% reduction in thickness by cold rolling, and then precipitation annealing for fifty hours at 800 °C (1,073 K). Another example[111] of a different kind of two-phase (so-called) *basket-weave* matrix-precipitate microstructure is shown in Figure 3.44, with precipitation of even finer nanoscale fibres of B2 multicomponent MAl mono-aluminide in a multi-component bcc matrix, in a non-equiatomic Al-modified bcc Senkov alloy $Al_{0.5}TiZrNbTa_{0.5}$, manufactured from elemental powders by sintering and homogenising for ten hours at 1,400 °C (1,673 K) and then precipitation-annealing for two hours at 800 °C (1,073 K). A final example[112] of a three-phase matrix-precipitate microstructure is shown in Figure 3.45, with precipitation of a combination of nanoscale blocky M_7C_3 and $M_{23}C_6$ carbides in a series of carbon-modified fcc Cantor

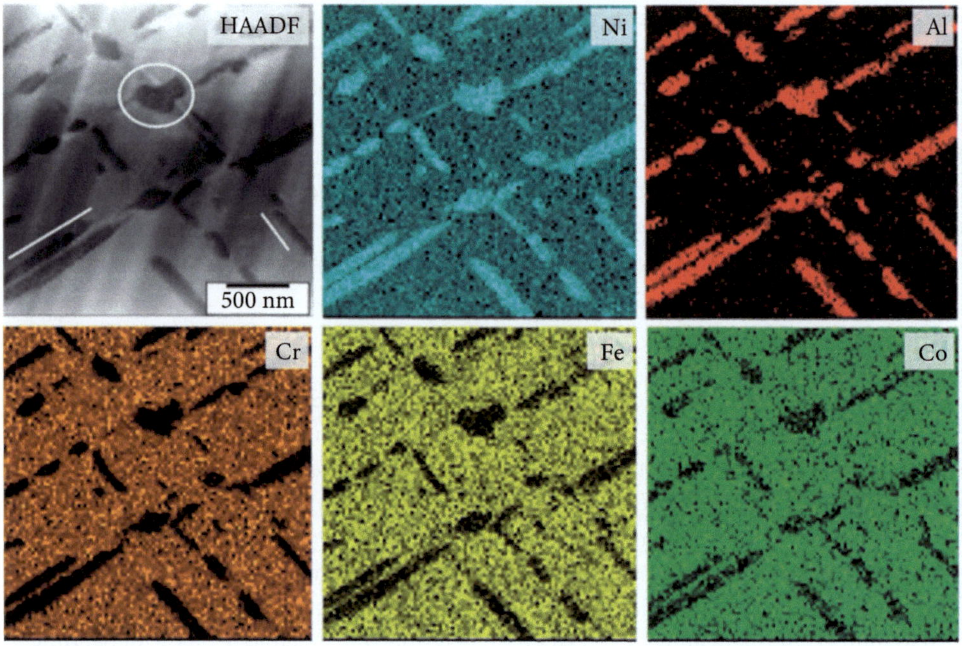

Figure 3.43 Microstructure of three lath variants (labelled 1, 2 and 3) of B2 NiAl-type ordered bcc mono-aluminide precipitates in an fcc matrix in a multicomponent $Al_{0.3}CrFeCoNi$ alloy: HAADF-STEM image (upper left) and corresponding EDS composition maps showing Ni and Al segregation into the B2 precipitates and Al, Cr and Fe segregation into the matrix.
Reproduced from Choudhuri et al. (2018). Crystallographically degenerate B2 precipitation in a plastically deformed fcc-based complex concentrated alloy. *Materials Research Letters* 6: 171–7.

alloys C_xCrMnFeCoNi where $x = 0$–0.25, after casting and then precipitation-annealing for fourteen hours at 800 °C (1,073 K).

Multicomponent dilute alloys

As discussed earlier in this chapter, the critical insight that launched the discovery and exploration of the field of multicomponent high-entropy materials was the realisation that we can find literally billions of exciting new materials by expanding our alloying options to cover all of multicomponent phase space, rather than being restricted to the limited alloying options previously available when material compositions are kept close to individual single components and binary alloys (i.e. when we stick to compositions near the corners and edges of multicomponent phase space).[7–9] As further explained earlier in this chapter, there are two main dimensions to this massive expansion of possible materials options: on the one hand the increased *multiplicity* of alloying components, and on the other hand their increased *concentration*.[7–9] When we explore deep into multicomponent phase space, we are throwing off two different excessively conservative shackles on our alloying strategy in order to find interesting and valuable new materials: not staying restricted to only *a few* alloying components; and not staying restricted to only relatively *dilute* amounts of those alloying additions.

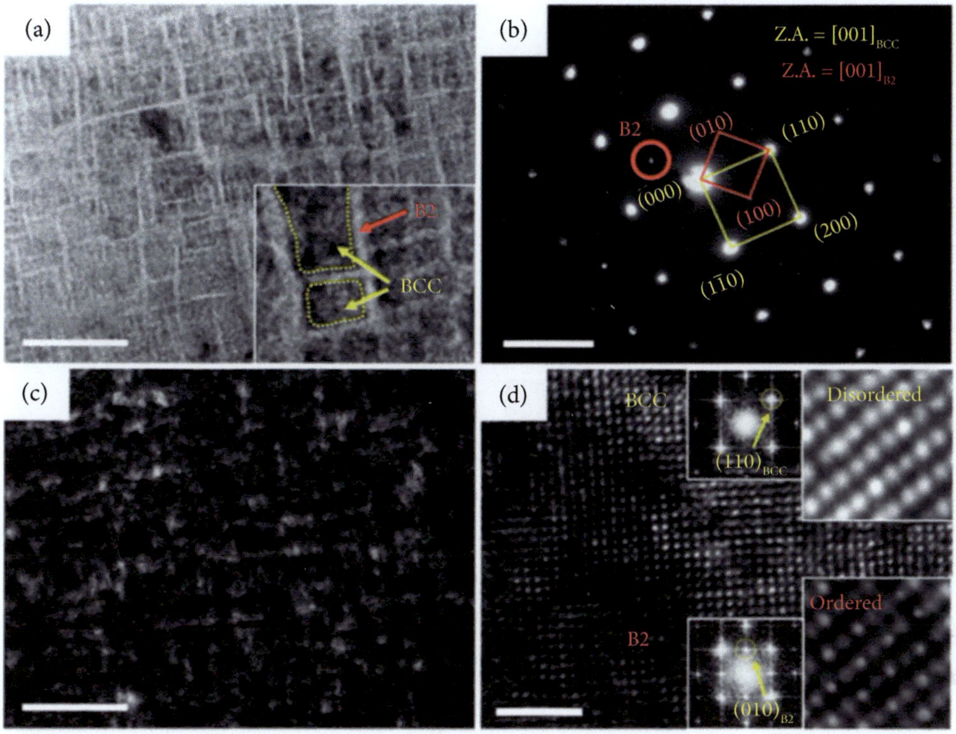

Figure 3.44 Microstructure of basket-weave B2 NiAl-type ordered bcc precipitates in a bcc matrix in a multicomponent Al-modified Senkov alloy $Al_{0.5}TiZrNbTa_{0.5}$: (a) bright-field TEM image; (b) corresponding SADP; and (c) and (d): dark-field TEM and HREM atomic images of B2 precipitates.

Reproduced from Wen et al. (2021). Effects of Al on precipitation behaviour of TiNbTaZr refractory high-entropy alloys. *Metals* 11: Article 1030514.

In most of this section we have discussed the different types of single-phase and multi-phase multicomponent materials that have begun to be discovered when we remove both these restrictions and start making materials with many components in relatively concentrated proportions. For some unknown reason, however, our collective enthusiasm for investigating more concentrated materials has surpassed our enthusiasm for adding more components. Thus, as already remarked, although there have now been many studies of equiatomic and near-equiatomic (i.e. concentrated) 4- and 5-component materials, there have still been rather few studies of concentrated 6-component materials, and virtually none of concentrated 7- or 8-component (or more) materials.[7–9] But not only have we still remained cautious in increasing the number of components too far, we have also shown surprisingly little collective interest in exploring *multicomponent dilute alloys*, i.e. 4-, 5-, 6-, or 7-component (and more) materials, where there is one major component and many dilute alloying additions.[7–9]

Of course, some of our most well-known and important engineering materials are examples of *multicomponent dilute alloys*, based on a single major component with in some cases increasing numbers of (small amounts of) alloying components added consciously as we have sought to squeeze out improved material properties and extra performance in service,

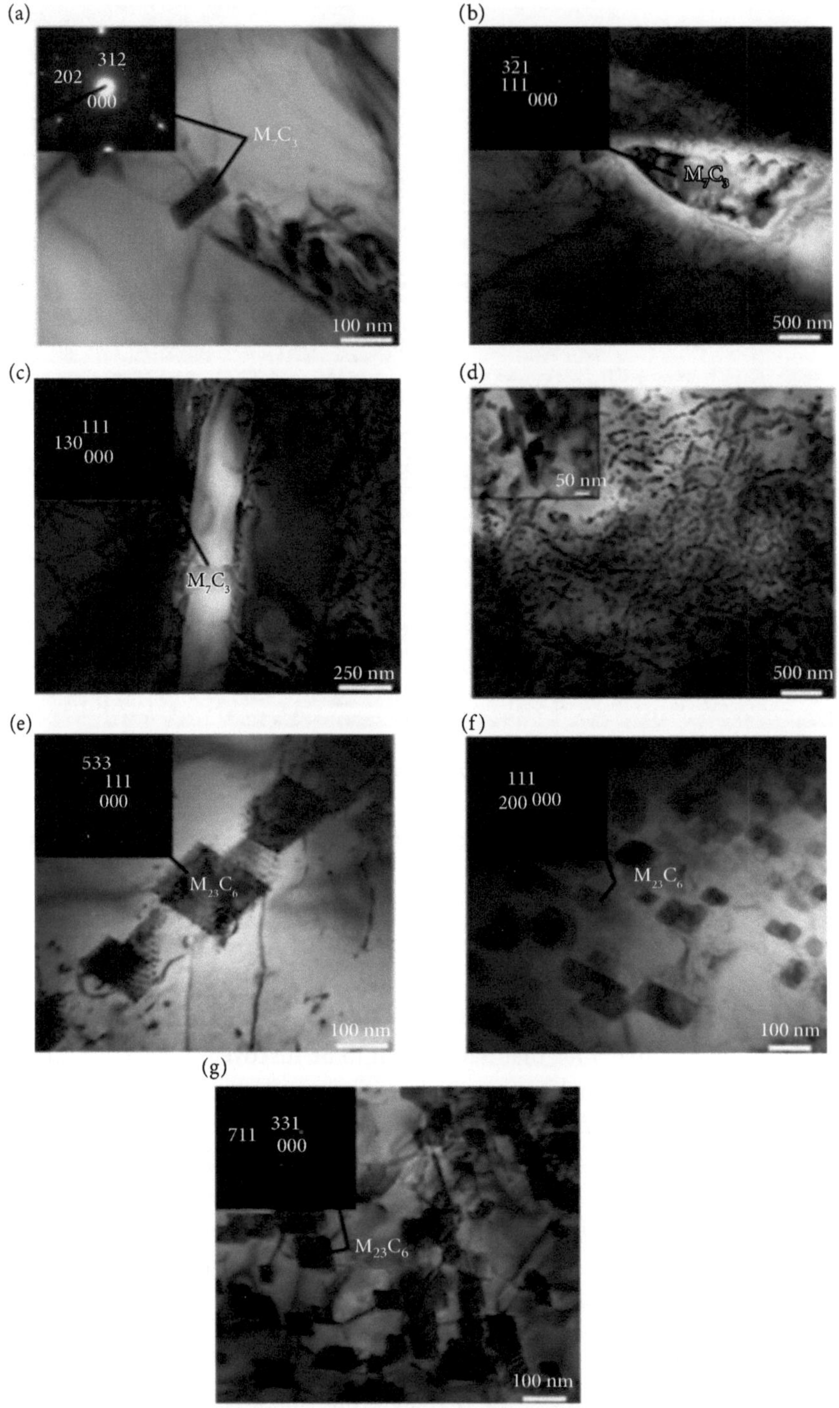

Figure 3.45 Microstructure of blocky M_7C_3 and $M_{23}C_6$ carbide precipitates in an fcc matrix in multicomponent carbon-modified Cantor alloys C_xCrMnFeCoNi: bright-field TEM images and SADP patterns from (a)–(c) as-cast and (d)–(j) heat treated alloys; with $x = 0.1$ in (a), (d) and (e); $x = 0.175$ in (b) and (f); and $x = 0.25$ in (c) and (j).

Reproduced from Choudhuri et al. (2018). Crystallographically degenerate B2 precipitation in a plastically deformed fcc-based complex concentrated alloy. *Materials Research Letters* 6: 171–7.

alongside increasingly careful control of other unwanted but unavoidable low-level impurity components as again we have sought to prevent deleterious effects on material properties and in-service behaviour. High-performance steels, engineering ceramics, modern electronic materials and nickel-based superalloys all provide examples of *multicomponent dilute alloys* that can sometimes have material specifications covering five, six, seven and more relatively dilute alloying additions.[20,21,90] But these are just a few very specialised cases that have arisen as particularly important engineering materials have been gradually improved, i.e. we have only investigated a few very special dilute regions of multicomponent phase space in an *ad hoc* and somewhat serendipitous way. Extension of material compositions to larger numbers of (even dilute) components has been done parsimoniously if not grudgingly, and there has been no conscious attempt to explore materials containing large numbers of dilute alloying additions in a rigorous and scientific way. This is another major opportunity for future investigation.

Nickel-based superalloys are probably the best example of a class of well-known multicomponent dilute alloys that have been developed over many years to meet a high-performance engineering need, in this case the ability to withstand the extremely demanding combination of high temperature, high stress, and aggressive oxidation and corrosion required at the heart of a modern aeroengine. Table 3.15 shows the compositions of some typical modern (so-called second-, third- and fourth-generation) nickel-based superalloys, used for the single-crystal turbine blades in aeroengines, all based on nickel Ni as the principal high-temperature material, and each containing up to about ten other components added in relatively dilute amounts.[90] These alloys are carefully solidified as single crystals to reduce the creep at high temperatures associated with the presence of grain boundaries, and all have a basic (so-called) γ/γ' microstructure, with an fcc Ni-based solid-solution γ-matrix strengthened by dense arrays of nanoscale coherent cuboidal $L1_2$ Ni_3Al-type ordered-fcc γ'-precipitates,[115] as shown in Figure 3.46. The compositions of nickel-based single-crystal superalloys for turbine blades have been developed and refined over many years to meet in

Table 3.15 Compositions (wt%) of some second-, third- and fourth-generation nickel-based single-crystal turbine-blade superalloys.[90]

alloy	Cr	Co	Mo	Re	Ru	W	Al	Ti	Ta	Nb	Hf	other
second generation												
CMSX-4	6.5	9	0.6	3		6	5.6	1	6.5		0.1	
PW1484	5	10	2	3		6	5.6		8.7		0.1	
René N5	7	8	2	3		5	6.2		7		0.2	
third generation												
CMSX-10	2	3	0.4	6		5	5.7	0.2	8	0.1	0.03	
René N6	4.2	12.5	1.4	5.4		6	5.7		7.2		0.15	
TMS-75	3	12	2	5		6	6		6		0.1	
fourth generation												
MC-NG	4	<0.2	1	4	4	5	6	0.5	5		0.1	
PW1497	2	16.5	2	5.95	3	6	5.5		8.25		0.15	0.03C
TMS-138	2.9	5.8	2.9	5.1	1.9	6.1	5.8		5.6		0.05	0.04B

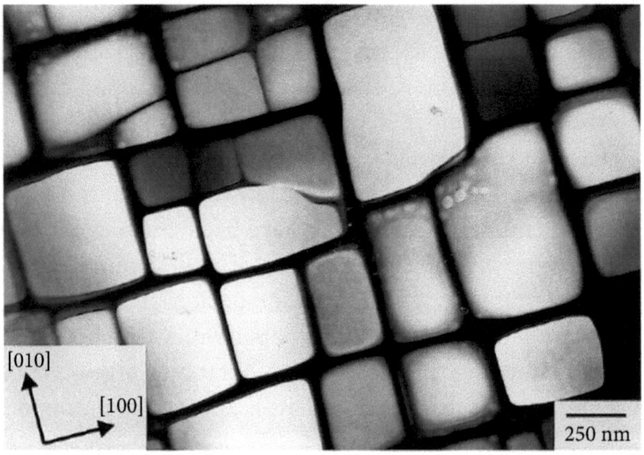

Figure 3.46 TEM image of cast and heat treated CMSX-10 nickel-based single-crystal turbine-blade superalloy showing densely packed arrays of cuboidal γ'-precipitates (light) embedded in an fcc γ-matrix (dark).

Reproduced from Schulze and Feller-Kniepmeier (2000). Transmission electron microscopy of phase composition and lattice misfit on the Re-containing nickel-base superalloy CMSX-10. *Materials Science and Engineering* A281: 204–12.

a balanced way a variety of different and complex design requirements: increasing components such as aluminium, titanium and tantalum to provide a high fraction (typically about 70%) of the strengthening γ'-precipitates, but without adding too much so as to ensure that they are embedded in a continuous γ-matrix; addition of creep-strengthening elements dissolved in the γ matrix such as rhenium, tungsten, molybdenum and ruthenium, but again without adding too much so as to prevent precipitation of deleterious TCP σ and μ phases; overall not increasing these secondary components too much, so as to allow for initial homogenisation and solution treatment in the γ solid-solution region at high temperature; maintaining a balanced secondary-component composition to adjust the γ and γ' lattice parameters, keep a small γ/γ' lattice mismatch, and thus achieve a high degree of matrix/precipitate coherency and, therefore, good microstructural stability; adding enough chromium to provide basic oxidation and corrosion resistance with a protective chromium oxide Cr_2O_3 surface layer, again balanced against oxidation of all the other secondary components; and maintaining a stable surface microstructure suitable for deposition of the complex layered thermal-barrier coatings (TBCs) that are used to provide further protection for single-crystal turbine blades against extreme high temperature and corrosion conditions.[90]

Despite our detailed knowledge and understanding of a few special highly engineered cases of multicomponent dilute alloys, such as nickel-based superalloys, there are still some very simple, general questions yet to be answered about multicomponent dilute alloys. We know that solubility is enhanced in multicomponent concentrated alloys compared with that found in conventional binary and ternary alloys, leading to the formation of large single-phase regions in multicomponent phase space. Is this also true for multicomponent dilute alloys? In most binary alloys, solubility is fairly restricted, but does the solubility of each alloying element increase in the presence of multiple other solutes in dilute as well as concentrated alloys? More accurately, if c_{ib} is the maximum solubility of component i

in a binary alloy, and c_{im} is the corresponding maximum solubility of component i in a multicomponent dilute alloy, do we expect $c_{im} > c_{ib}$, $c_{im} \approx c_{ib}$, or $c_{im} < c_{ib}$? The solvus between the binary edges of the multicomponent phase diagram can either bulge outwards, remain flat or bulge inwards. The evidence from concentrated multicomponent transition metal-based alloys such as the Cantor and Senkov alloys[9,45,46] suggests that in many cases $c_{im} > c_{ib}$. But Hou and Fan[116] have shown that there is no general rule in aluminium alloys, and that all three relations $c_{im} > c_{ib}$, $c_{im} \approx c_{ib}$, and $c_{im} < c_{ib}$ are possible for different alloying components.

Overall, we can conclude that there is still much experimental study and theoretical analysis to be done to explore the full range of possible material microstructures and properties in almost all (all but a few) multicomponent dilute alloys.

Multicomponent glasses

There are several different kinds of *glass* (also called *amorphous material* or *non-crystalline material*).[xii] *Ceramic glasses* or *inorganic glasses* are usually based on alloy compositions in which a mixture of different oxides or other inorganic compounds form complex network molecular structures that become tangled and intertwined in the liquid phase, hampering atomic and molecular diffusion within the liquid and therefore preventing easy crystallisation.[117] These materials are sometimes called *network glasses*. *Polymer glasses* are similarly based on long-chain single polymer or co-polymer (i.e. alloyed) molecules that again are tangled and intertwined in the liquid phase, again hampering molecular diffusion within the liquid and again preventing easy crystallisation.[118] For a long time, it was thought that metallic materials could not form glasses, since they do not form such network or long-chain atomic or molecular structures. In the 1960s, however, Duwez and co-workers[119] showed that this was not the case when they quenched a binary gold–silicon Au–Si alloy and obtained a material with an amorphous, non-periodic atomic structure. Since then many metallic glasses have been made by a variety of different manufacturing methods.[120,121] The first metallic glasses were so-called *metal–metalloid glasses*, mixtures of metals and semi-metals at compositions that produce significant short-range ordering, again hampering atomic diffusion and preventing easy crystallisation[120,121] in a similar, though somewhat more limited, way to network oxide glasses and amorphous polymers. Later, however, so-called *metal–metal glasses* were also discovered with little or no short-range order and with more regular or near-ideal atomic structures.[120,121]

In binary and ternary alloys, glassy materials are usually found in the vicinity of eutectic compositions,[120,121] i.e. at compositions between two (or sometimes three) different crystalline-solid single-phase fields, usually roughly midway between the solubility limit of the terminal solid solution formed by dissolving secondary components in the principal component and the stoichiometric composition of any intermetallic or ceramic compounds

[xii] The terms *glass, amorphous material* and *non-crystalline material* are sometimes used interchangeably, so they all refer to any solid in which the atoms or molecules are not arranged in a regular periodic crystalline structure. Sometimes, however, only the term *non-crystalline material* is used generically in this way to refer to any solid without a crystalline structure; the term *glass* is then reserved exclusively to refer to those non-crystalline materials that are manufactured by cooling from the liquid phase; and the term *amorphous material* is reserved exclusively to refer to those non-crystalline materials that are *not* manufactured by cooling from the liquid phase. We do not adopt this latter convention, and we use all three terms synonymously.

that form by reaction between the principal and secondary components. Because the eutectic liquid has a lower melting point than the surrounding single-phase crystalline solids, it can be cooled to relatively low temperatures where diffusion becomes sluggish, the kinetics of solidification are slow, and crystallisation can be avoided. This effect is amplified when the eutectic valley is deep so that the liquid phase (and, therefore, the metastable non-crystalline glass structure) is more strongly favoured kinetically over competing stable crystalline solid phases.[120,121]

In multicomponent alloys, glasses form similarly in near-eutectic compositions roughly midway between the various different large solid-solution and intermetallic- or ceramic-compound single-phase fields.[122] Overall, this means that multicomponent glasses are favoured in materials containing (similar to binary and ternary glasses) network-forming, long-chain molecule and metalloid components, and for compositions in the vicinity of a deep eutectic valley, i.e. where the chemical characteristics of the components are sufficiently similar to allow complete solubility in the liquid, but sufficiently different to prevent full solubility in any single crystalline solid-solution or compound solid phase, and thus allow access for the liquid to low temperatures and correspondingly sluggish atomic or molecular diffusion and, therefore, crystallisation.[121]

Ma et al.[122] and Kim et al.[85,92] discovered the first multicomponent high-entropy glasses in, respectively, melt-spun TiCuZrHfM alloys with M = Fe, Co or Ni, and melt-spun

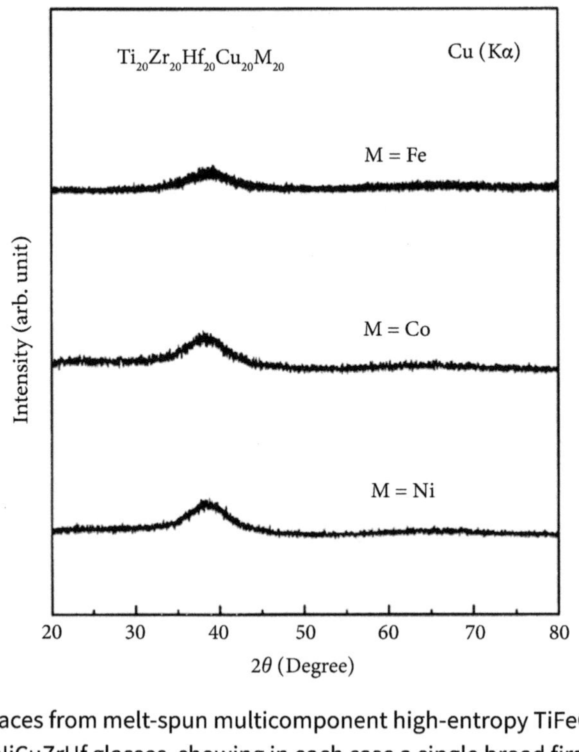

Figure 3.47 XRD traces from melt-spun multicomponent high-entropy TiFeCuZrHf, TiCoCuZrHf and TiNiCuZrHf glasses, showing in each case a single broad first-near-neighbour diffraction peak.

Reproduced from Ma et al. (2002). Bulk glass formation of Ti-Zr-Hf-Cu-M (M = Fe, Co, Ni) alloys. *Materials Transactions* 43: 277–80.

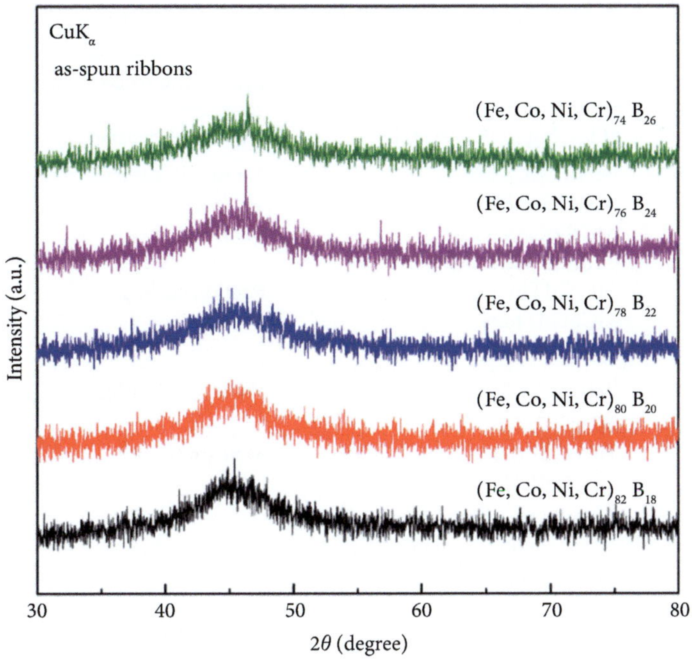

Figure 3.48 XRD traces from melt-spun multicomponent high-entropy $(CrFeCoNi)_{100-x}B_x$ glasses with $x = 18$–26, showing in each case a single broad first-near-neighbour diffraction peak.

Reproduced from Ding et al. (2017). High-entropy effect on structure and properties of (FeCoNiCr)-B amorphous alloys. *Journal of Alloys and Compounds* 696: 345–52.

$(TiZrHf)_{90-x-y}(NiCu)_xAl_y$ alloys with $x = 10$–60 at% and $y = 0$–30 at%. Figure 3.47 shows typical XRD traces from melt-spun TiFeCuZrHf, TiCoCuZrHf and TiNiCuZrHf alloys,[122] showing in each case the single broad diffraction peak from atomic first near-neighbours that is characteristic of a glassy material.[120,121] Figure 3.48 shows similar results[123] from melt-spun $(CrFeCoNi)_{100-x}B_x$ alloys with $x = 18$–26. Table 3.16 shows some (by no means all) of the multicomponent high-entropy glasses that have been manufactured by different methods. Table 3.16 shows that there is a wide variety of different materials and manufacturing methods, including melt-spun[124] $AlCrFeCoNiZr_{0.6}$ and $CrFeCoNiCuZr_{0.6}$, suction-cast[125] AlFeGdTbDy, AlCoGdTbDy and AlNiGdTbDy, copper-mould-cast[126] $BeNi_xCu_{1-x}TiZrHf$ with $x = 0.125$–0.875, and water-quenched PNiCuPdPt.[127] More details of the different multicomponent high-entropy glasses that have been investigated and their structures and properties are given in the book by Suryanarayana and Inoue.[121]

6 Summary

The size of multicomponent phase space is mind-bogglingly large. The total number of different multicomponent alloys and materials that can be made is (very roughly) in the range 10^{40}–10^{120} depending on how sharply the number of components is restricted and how precisely the composition of each material is specified. Only a miniscule fraction of the total

Table 3.16 Multicomponent high-entropy glasses.

alloy	processing route
$(Li_{0.55}Mg_{0.45})CaZnSrYb$	Cu mould casting
$BeNi_xCu_{1-x}TiZrHf$ x=0.125–0.875	Cu mould casting
$AlCrFeCoNiZr_{0.6}$ $CrFeCoNiCuZr_{0.6}$	Melt spinning
AlFeGdTbDy AlCoGdTbDy AlNiGdTbDy	Suction casting
PNiCuPdPt	Water quenching
TiFeCuZrHf TiCoCuZrHf TiNiCuZrHf	Melt spinning
$(TiZrHf)_{90-x-y}(NiCu)_xAl_y$ $(TiZrNbHf)_{90-x-y}(NiCuAg)_xAl_y$ x=10-60; y=0-30	Melt spinning
$(CrFeCoNi)_{100-x}B_x$ x=18–26	Melt spinning

number of multicomponent materials have ever been investigated or even made, despite the intensive scientific effort of exploration of multicomponent materials in recent years. To put it another way, there are many wonderful new materials with exciting structures and properties yet to be discovered.

Multicomponent phase space is characterised by enormous single-phase fields of multicomponent solid solutions and multicomponent compounds, separated by equally enormous multicomponent multiphase fields. Large fcc and bcc solid-solution single-phase fields (the Cantor and Senkov alloys respectively) have been investigated quite a lot, but their full extents remain to be defined, and there are still literally millions and millions of Cantor and Senkov alloys yet to be made, let alone studied. And there are equally large disordered-hcp and diamond-cubic solid-solution single-phase fields and ordered aluminide, oxide, boride, carbide and other intermetallic- and ceramic-compound single-phase fields that have hardly been mapped out at all. Only a mere handful of multicomponent multiphase materials have been studied, concentrating so far almost exclusively on some simple multicomponent two-phase materials with eutectic and matrix/precipitate microstructures.

In summary, although at the broadest level the geography and topology of multicomponent phase space is becoming clear, its detailed structure is still very hazy, and the vast majority of multicomponent materials are yet to be studied. Despite more than a decade of increasingly intensive scientific work, there has still been very little investigation of multicomponent materials with more than five components, multicomponent multiphase materials, multicomponent dilute alloys or multicomponent compounds. To use a non-technical term, there are still gazillions of exciting materials to discover for our future benefit.

References

1. Cantor B, Chang ITH, Knight P, Vincent AJB. (2004). Microstructural development in equiatomic multicomponent alloys. *Materials Science and Engineering* A375–7: 213–18.
2. Yeh JW, Chen SK, Lin SJ, Gan JY, Chin TS, Shun TT, Tsau CH, Chang SY. (2004). Nanostructured high entropy alloys with multiple principal elements: Novel alloy design concepts and outcomes. *Advanced Engineering Materials* 6: 299–303.
3. Hillert M. (2008). Phase equilibria and potential phase diagrams. In *Phase Equilibria, Phase Diagrams and Phase Transformations*, 2nd edn. Cambridge: Cambridge University Press, 155–84.
4. West DRF, Saunders N. (2017). *Ternary Phase Diagrams in Materials Science*, 3rd edn. New York: CRC Press
5. Cantor B. (2020). The Gibbs phase rule. In *The Equations of Materials.* Oxford: Oxford University Press, 45–65.
6. Huang PK, Yeh JW, Shun TT, Chen SK. (2004). Multi-principal-element alloys with improved oxidation and wear resistance for thermal spray coating. *Advanced Engineering Materials* 6: 74–78.
7. Miracle DB, Senkov ON. (2017). A critical review of high-entropy alloys and related concepts. *Acta Materialia 122*: 448–511.
8. Murty BS, Yeh JW, Ranganathan S, Bhattacharjee PP. (2019). *High-Entropy Alloys.* Amsterdam: Elsevier.
9. Cantor B. (2021). Multicomponent high-entropy Cantor alloys. *Progress in Materials Science 120*, art 100754.
10. Halmos PT. (2019). *Naïve Set Theory.* Los Angeles:Bow Wow Press.
11. Lucas JF. (1990). Elementary logic. In *Introduction to Abstract Mathematics*, 2nd edn. Lanham:Rowman and Littlefield, 1–42.
12. Jech T. (2002). *Set Theory: The Third Millenium Edition.* Berlin: Springer, ch 1.
13. Cantor B. (2021). A personal perspective on the discovery and significance of multicomponent high-entropy alloys. In J Brechtl, PK Liaw, eds., *High-Entropy Materials: Theory, Experiments, and Applications.* New York: Springer,1–36.
14. ASM International (2006). *Worldwide Guide to Equivalent Irons and Steels*, 5th edn. Metals Park: ASM International.
15. ASM International (2001). *Worldwide Guide to Equivalent Non-Ferrous Metals and Alloys*, 4th edn. Metals Park: ASM International.
16. ASTM (2020). *Annual Book of ASTM Standards.* West Conshohocken: ASTM International.
17. Koh KM, Chen CC. (1992). Permutations and combinations. In *Principles and Techniques in Combinatorics.* Singapore: World Scientific, 1–50.
18. Brualdi RA. (2010). *Introductory Combinatorics*, 5th edn. Hoboken: Pearson Prentice Hall.
19. Scerri ER. (2019). *The Periodic Table: Its Story and Its Significance.* Oxford: OUP.
20. Bhadeshia HKDH, Honeycombe RWK. (2017). *Steels: Microstructure and Properties*, 4th edn. Amsterdam: Elsevier Butterworth-Heinemann.
21. Grovenor CRM. (1998). *Microelectronic Materials.*New York: Taylor and Francis.

22. Oxford Educational Blog (2015, November 24). Number of atoms in the universe. Oxford Education Blog, https://educationblog.oup.com/secondary/maths/numbers-of-atoms-in-the-universe, accessed 18 September 2023.
23. Barrow JD. (2002). Eddington's unfinished symphony. In *The Constants of Nature.* New York: Vintage Books, 77–97.
24. ThoughtCo. (2019, August 8). How Many Atoms Exist in the Universe? https://www.thoughtco.com/number-of-atoms-in-the-universe-603795, accessed 18 September 2023.
25. Oxford Educational Blog, https://educationblog.oup.com/secondary/maths/numbers-of-atoms-in-the-universe, accessed 18 September 2023.
26. Zhang KB, Fu ZY, Zhang JY, Wang WM, Wang H, Wang YC. (2009). Microstructure and mechanical properties of CoCrFeNiTiAl high-entropy alloys. *Materials Science and Engineering A508*: 214–19.
27. Takeuchi A, Wada T, Zhang Y. (2017). MnFeNiCuPt and MnFeNiCuCo high-entropy alloys designed based on $L1_0$ structure in Pettifor map for binary compounds. *Intermetallics 82*: 107–15.
28. Freudenberger J, Rafaja D, Geissler D, Giebeler L, Ullrich C, Kauffman A, Heilmaier M, Nielsch K. (2017). Face centred cubic multi-component equiatomic solid solutions in the Au-Cu-Ni-Pd-Pt system. *Metals 7*: 135.
29. Laplanche G, Gadauad P, Bärsch C, Demtröder K, Reinhart C, Schreuer J, George EP. (2018). Elastic moduli and thermal expansion coefficients of medium-entropy subsystems of the CrMnFeCoNi high-entropy alloy. *Journal of Alloys and Compounds 746*: 244–55.
30. Varalakshmi S, Kamaraj M, Murty BS. (2010). Formation and stability of equiatomic and nonequiatomic nanocrystalline CuNiCoZnAlTi high-entropy alloys by mechanical alloying. *Metallurgical and Materials Transactions A41*: 2703–9.
31. Freudenberger J, Rafaja D, Geissler D, Giebeler L, Ullrich C, Kauffman A, Heilmaier M, Nielsch K. (2017). Face centred cubic multi-component equiatomic solid solutions in the Au-Cu-Ni-Pd-Pt system. *Metals 7*: 135.
32. Dahlborg U, Cornide J, Dahlborg MC, Hansen TC, Leong Z, Dominguez LA, Chamberland S, Cunliffe A, Goodall R, Todd I. (2015). Crystalline structures of some high entropy alloys obtained by neutron and X-ray diffraction. *Acta Physica Polonica A128*: 552–6.
33. Ming K, Bi X, Wang J. (2017). Precipitation strengthening of ductile CrFeCoNiMo alloys. *Scripta Materialia 137*: 88–93.
34. He JY, Liu WH, Wang H, Wu Y, Liu XJ, Nieh TJ, Lu ZP. (2014). Effects of Al addition on structural evolution and tensile properties of the FeCoNiCrMn high-entropy alloy system. *Acta Materialia 62*: 105–13.
35. Shahmir H, Nili-Ahmadabadi M, Shafiee A, Langdon T. (2018). Effect of a minor titanium addition on the superplastic properties of a CoCrFeNiMn high-entropy alloy processed by high pressure torsion, *Materials Science and Engineering A718*: 468–76.
36. Bracq G, Laurent-Brocq M, Varvenne C, Perrière L, Curtin WA, Joubert J-M, Guillot I. (2019). Combining experiments and modelling to explore the solid solution strengthening of high and medium entropy alloys. *Acta Materialia 177*: 266–79.
37. Lukas HL, Fries SG, Sundman B. (2007). *Computational Thermodynamics: The Calphad Method.* Cambridge: Cambridge University Press.

38. Laurent-Brocq M, Akhatova A, Perrière L, Chebini S, Sauvage X, Leroy E, Champion Y. (2015). Insights into the phase diagram of the CrMnFeCoNi high entropy alloy. *Acta Materialia 88*: 355–65.
39. Won JW, Lee S, Park SH, Kang M, Lim KR, Park CH, Na YS. (2018). Ultrafine-grained CoCrFeMnNi high-entropy alloy produced by cryogenic multi-pass caliber rolling. *Journal of Alloys and Compounds 74*: 290–5.
40. He JY, Liu WH, Wang H, Wu Y, Liu XJ, Nieh TJ, Lu ZP. (2014). Effects of Al addition on structural evolution and tensile properties of the FeCoNiCrMn high-entropy alloy system. *Acta Materialia 62*: 105–13.
41. Hsu YC, Li CL, Hsueh CH. (2020). Effects of Al addition on microstructure and mechanical properties of $CoCrFeMnNiAl_x$ high entropy alloy films. *Entropy 22*: Article 2.
42. Cieslak J, Tobola J, Berent K, Marciszko M. (2018). Phase composition of $Al_xFeNiCrCo$ high entropy alloys by sintering and arc-melting methods. *Journal of Alloys and Compounds 740*: 264–72.
43. Villars P, Calvert LD. (1991). *Pearson's Handbook of Crystallographic Data for Intermetallic Phases.* Russell Township: ASM International.
44. Westbrook JH, Fleischer RL, eds. (2000). *Intermetallic Compounds: Vol 1: Crystal Structures of Intermetallic Compounds.* New York: Wiley-Blackwell.
45. Senkov ON, Wilks GB, Miracle DB, Chuang CP, Liaw PK. (2010). Refractory high-entropy alloys. *Intermetallics18*: 1758–65.
46. Senkov ON, Miracle DB, Chaput KJ. (2018). Development and exploration of refractory high entropy alloys—A review. *Journal of Materials Research 33*: 3092–3218.
47. Chen SY, Yang X, Dahmen KA, Liaw PK, Zhang Y. (2016). Microstructures and crackling noise of $Al_xNbTiMoV$ high entropy alloys. *Entropy 16*: 870–84.
48. Sobol OV, Andreev AA, Gorban VF, Krapivka NA, Stolbovoi VA, Serdyuk IV. (2012). Reproducibility of the single phase structural state of the multi-element high-entropy alloy TiVZrNbHf system and related superhard nitrides formed by the vacuum arc method. *Technical Physics Letters 38*: 616–19.
49. Gao MC, Zhang B, Yang S, Guo SM. (2016). Senary refractory high-entropy alloy HfNbTa-TiVZr. *Metallurgical and Materials Transactions A47*: 3333–45.
50. Han ZD, Chen N, Zhao SF, Fan LW, Yang GN, Shao Y, Yao KF. (2017). Effect of Ti additions on mechanical properties of NbMoTaW and VNbMoTaW refractory high-entropy alloys. *Intermetallics 84*: 153–7.
51. Zhang B, Mu Y, Gao MC, Meng WJ, Guo SM. (2017). On single phase status and segregation of an as-solidified septenary refractory high entropy alloy. *MRS Communications 7*: 78–83.
52. Mu Y, Liu H, Liu Y, Zhang X, Jiang Y, Dong T. (2017). An ab-initio and experimental studies of the structure, mechanical parameters and state density on the refractory high entropy alloy systems. *Journal of Alloys and Compounds 714*: 668–80.
53. Tsai DC, Shieu FS, Chang SY, Yao HC, Deng MJ. (2010). Structures and characterizations of TiVCr and TiVCrZrY films deposited by magnetron sputtering under different bias powers. *Journal of the Electrochemical Society 157*: K52–8.

54. Guo W, Dmowski W, Noh JY, Rack P, Liaw PK, Egami T. (2013). Local atomic structure of a high-entropy alloy: An X-ray and neutron scattering study. *Metallurgical and Materials Transactions A44*: 1994–97.
55. Guo J, Huang X, Huang W. (2017). Microstructure and room temperature mechanical properties of $FeCrMoVTi_x$ high-entropy alloys. *Journal of Materials Engineering and Performance 26*: 3071–8.
56. Zhang Z, Zhang H, Tang Y, Zhu L, Ye Y, Li S, Bai S. (2017). Microstructure, mechanical properties and energetic characteristics of a novel high-entropy alloy $HfZrTiTa_{0.53}$. *Materials Design 133*: 435–43.
57. Wang S, Xu J. (2018). (TiZrNbTa)-Mo high-entropy alloys: Dependence of microstructure and mechanical properties on Mo concentration and modelling of solid solution strengthening. *Intermetallics 95*: 59–72.
58. Lin CM, Juan CC, Chang CH, Tsai CW, Yeh JW. (2015). Effect of Al addition on mechanical properties and microstructure of refractory $Al_xHfNbTaTiZr$ alloys. *Journal of Alloys and Compounds 624*: 100–7.
59. Zhang B, Gao MC, Zhang Y, Guo SM. (2015). Senary refractory high-entropy alloy $Cr_xMoNbTaVW$. *Calphad 51*: 193–201.
60. Miracle DB, Tsai MH, Senkov ON, Soni V, Banerjee R. (2020). Refractory high entropy superalloys. *Scripta Materialia 187*: 445–52.
61. Couzinié J-P, Dirras G. (2019). Body-centred cubic high-entropy alloys: From processing to underlying deformation mechanisms. *Materials Characterization 147*: 533–44.
62. George EP, Raabe D, Ritchoe RO. (2019). High-entropy alloys. *Nature Reviews Materials 4*: 515–34.
63. Senkov ON, Semiaton SL. (2015). Microstructure and properties of a refractory high-entropy alloy after cold working.*Journal of Alloys and Compounds 649*: 1110–23.
64. Dirras G, Lilenstein L, Djemia P, Laurent-Brocq M, Tingaud D, Couzinié J-P, Perrière L, Chauveau T, Guillot I. (2016). Elastic and plastic properties of as-cast equimolar TiHfZrTaNb high-entropy alloy. *Materials Science and Engineering A654*: 30–8.
65. Takeuchi A, Amiya K, Wada T, Yubuta K, Zhang W. (2014). High-entropy alloys with a hexagonal close-packed structure designed by equi-atomic alloy strategy and binary phase diagrams. *Journal of Metals 66*: 1984–92.
66. Luznik J, Kozelj P, Vrtnik S, Jelen A, Jaglicic Z, Meden A, Feuerbacher M, Dolinsek J. (2015). Complex magnetism of HoDyYGdTb hexagonal high-entropy alloy. *Physical Review B 92*: Article 224201.
67. Zhao YJ, Qiao JW, Ma SG, Gao MC, Yang HJ, Chen MW, Zhang Y. (2016). A hexagonal close-packed high-entropy alloy: The effect of entropy. *Materials and Design 96*: 10–15.
68. Qiao JW, Bao ML, Zhao YJ, Yanh HJ, Wu YC, Zhang Y, Hawk JA, Gao MC. (2018). Rare-earth high entropy alloys with hexagonal close-packed structure. *Journal of Applied Physics 124*: Article 195101.
69. Yusenko KV, Riva S, Carvalho PA, Yusenko MV, Arnaboldi S, Sukhikh AS, Hanfland M, Gromilov SA. (2017). First hexagonal close packed high-entropy alloy with outstanding stability under extreme conditions and electroactivity for methanol oxidation. *Scripta Materialia 138*: 22–7.

70. Takeuchi A, Wada T, Kato H. (2019). High-entropy alloys with hexagonal close-packed structure in IrMoRhRuW alloys designed by sandwich strategy for the valence electron concentration of constituent elements in the periodic chart. *Materials Transactions 60*: 1666–73.
71. Rogal L, Czerwinski F, Jochym PT, Litynska-Dobrzynska L. (2016). Microstructure and mechanical properties of the novel HfScTiZr equiatomic alloy with hexagonal solid solutions. *Materials and Design 92*: 8–17.
72. Youssef KM, Zaddach AJ, Niu C, Irving DL, Koch CC. (2014). A novel low-density, high-hardness. High-entropy alloy with close-packed single-phase nanocrystalline structures. *Materials Research Letters 3*: 95–9.
73. Zhang F, Wu Y, Lou H, Zeng Z, Prakapenka VB, Greenberg E, Ren Y, Yan J, Okasinski JS, Liu X, Zeng Q, Lu Z. (2017). Polymorphism in a high-entropy alloy. *Nature Communications 8*, Article 15687.
74. Tracy CL, Park S, Rittmen DR, Zinkle SJ, Bei H, Lang M, Ewing RC, Mao WL. (2017). High pressure synthesis of a hexagonal close-packed phase of the high-entropy alloy CrMnFeCoNi. *Nature Communications 8*, Article 15634.
75. Li RX, Qiao JW, Liaw PK, Zhang Y. (2020). Preternatural hexagonal high-entropy alloys: A review. *Acta Metallurgica Sinica (English Letters) 33*: 1033–45.
76. Naka S, Khan T. (1997). Designing novel multiconstituent intermetallics: Contribution of modern alloying theory in developing engineered materials. *Journal of Phase Equilibria 18*: 635–49.
77. ZhangY, Ma SG, Qiao JW. (2012). Morphology transition from dendrites to equiaxed grains for AlCoCrFeNi high entropy alloys by copper mold casting and Bridgman solidification. *Metallurgical and Materials Transactions A43*: 2625–30.
78. Yurchenko NY, Stepanov ND, Zherebtsov SV, Tikhnovsky MA, Salischev GA. (2017). Structure and mechanical properties of B2 ordered refractory AlNbTiVZrx (x=0-1.5) high-entropy alloys. *Materials Science and Engineering A704*: 82–90.
79. Coury FG, Butler T, Chaput K, Saville A, Copley J, Foltz J, Mason P, Clarke K, Kaufman M, Clarke A. (2018). Phase equilibria, mechanical properties and design of quaternary refractory high entropy alloys. *Materials Design 155*: 244–56.
80. Borkar T, Chaudhury V, Gwalani B, Choudhuri D, Mikler CV, Soni V, Alam T, Ramanujan RV, Banerjee R. (2017). A combinatorial approach for assessing the magnetic properties of high entropy alloys: The role of Cr in $AlCo_xCr_{1-x}FeNi$. *Advanced Engineering Materials 19*: Article 1700048.
81. Xiao DH, Zhou PF, Wu WQ, Diao HY, Gao MC, Song M, Liaw PK. (2017). Microstructure, mechanical and corrosion behaviors of AlCoCuFeNi–(CrTi) high entropy alloys. *Materials and Design 116*: 438–47.
82. Senkov ON, Jensen JK, Pilchak AL, Miracle DB, Fraser HL. (2018). Compositional variation effects on the microstructure and properties of a refractory high-entropy superalloy AlMo0.5NbTa0.5TiZr. *Materials and Design 139*: 498–511.
83. Muralikrishna GM, Vaidya M, Murty BS, Divinski SV. (2020). Tracer diffusion in ordered pseudo-binary multicomponent aluminides. *Scripta Materialia 178*: 227–31.

84. Muralikrishna GM, Esther ACM, Guruvidyathri K, Watermeyer P, Liebscher CH, Kulkarni KN, Wilde G, Divinski SV, Murty BS. (2020). Novel multicomponent B2-ordered aluminides: Compositional design, synthesis, characterization and thermal stability. *Metals 10*: Article 1411.
85. Kim KB, Warren PJ, Cantor B. (2003). Formation of metallic glasses in novel (Ti33Zr33Hf33)((100-x-y)(Ni50Cu50)(x)Al(y) alloys. *Materials Transactions 44*: 411–13.
86. Takeuchi A, Wada T, Zhang Y. (2017). MnFeNiCuPt and MnFeNiCuCo high-entropy alloys designed based on $L1_0$ structure in Pettifor map for binary compounds. *Intermetallics 82*: 107–15.
87. Wang S, Chen S, Jia Y, Hu Z, Huang H, Yang Z, Dong A, Zhu G, Wang D, Shu D, Tian F, Dai Y, Sun B. (2019). Fcc-$L_1$2ordering transformation in equimolar FeCoNiV multi-principal element alloy. *Materials and Design 168*: Article 107648.
88. Liu CT, Inouye H. (1979). Control of ordered structure and ductility of $(Fe,Co,Ni)_3V$ alloys. *Metallurgical Transactions A10*: 1515–25.
89. Liu CT. (1984). Physical metallurgy and mechanical properties of ductile ordered alloys $(Fe,Co,Ni)_3V$. *International Metallurgical Reviews 29*: 168–94.
90. Reed RC. (2006). *The Superalloys: Fundamentals and Applications.* Cambridge: Cambridge University Press.
91. Karpets MV, Myslyvchenko OM, Makarenko OS, Gorban VF, Krapivka MO, Degula AI. (2015). Effect of nickel on the structure and phase composition of the $VCrMnFeCoNi_x$ high-entropy alloy. *Journal of Superhard Materials 37*: 182–8.
92. Kim KB, Warren, Cantor B. (2005). Crystallization behaviour of novel $(TiZrHf)_{100\text{-}x\text{-}y}$ $(NiCu)_x$ Al_y with $x = 48–55$. *Journal of Metastable and Nanocrystalline Materials 24–25*: 657–60.
93. Rost CM, Sachet E, Borman T, Moballegh A, Dickey EC, Hou D, Jones JL, Curtarolo S, Maria JP. (2015). Entropy-stabilized oxides. *Nature Communications 6*: Article 8485.
94. Zhang RZ, Reece MJ. (2019). Review of high entropy ceramics: Design, synthesis, structure and properties. *Journal of Materials Chemistry A 7*: 22148–62.
95. Sarkar A, Breitung B, Hahn H. (2020). High entropy oxides: The role of entropy, enthalpy and synergy. *Scripta Materialia 187*: 43–8.
96. Xiang H, Xing Y, Dai FZ, Wang H, Su L, Miao L, Zhang G, Wang Y, Qi X, Yao L, Wang H, Zhao B, Li J, Zhou Y. (2021). High-entropy ceramics: Present status, challenges, and a look forward. *Journal of Advanced Ceramics10*: 385–441.
97. Gild J, Zhang Y, Harrington T, Jiang S, Hu T, Quinn MC, Mellor WM, Zhou N, Vecchio K, Luo J. (2016). High-entropy diborides: A new class of high-entropy materials and a new type of ultrahigh temperature ceramics. *Nature Scientific Reports 6*: Article 37946.
98. Qin M, Yan Q, Wang H, Hu C, Vacchio KS, Luo J. (2020). High-entropy monoborides: Towards superhard materials. *Scripta Materialia 189*: 101–5.
99. Zhu H, Xie H, Zhao Y, Dai S, Li M, Wang X. (2021). Structure and magnetic properties of a class of spinel high-entropy oxides. *Journal of Magnetism and Magnetic Materials 535*: Article 168063.
100. Chadwick GA. (1972). *Metallography of Phase Transformations.* London: Butterworths, chs 4 and 6.

101. Porter DA, Easterling KE, Sherif MY. (2009). *Phase Transformations in Metals and Alloys.* Boca Raton: CRC, chs 4 and 5.
102. Chanda B, Potnis G, Jana PP, Das J. (2020). A review on nano-/ultrafine advanced eutectic alloys. *Journal of Alloys and Compounds 827*: Article 154226.
103. Lu Y, Dong Y, Jiang H, Wang Z, Cao Z, Guo S, Wang T, Li T, Liaw PK. (2020). Promising properties and future trend of eutectic high entropy alloys. *Scripta Materialia 187*: 202–9.
104. Tiwari CS, Pandey P, Sarkar S, Das R, Samal S, Biswas K, Chattopadhyay K. (2021). Five decades of research on the development of eutectic as engineering materials. *Progress in Materials Science 122*: Article 100793.
105. Chang I and Cai Q. (2021). From simple binary to complex multicomponent eutectic alloys. *Progress in Materials Science 122*: Article 100779.
106. Rogal L, Morgiel J, Swiatek Z, Czerwinski F. (2016). Microstructure and mechanical properties of the new NbScTiZr eutectic high entropy alloy. *Materials Science and Engineering A651*: 590–7.
107. Chanda B, Das J. (2018). Composition dependence on the evolution of nanoeutectic in $CoCrFeNiNb_x$ ($0.45<x<0,65$) high entropy alloys. *Advanced Engineering Materials 20*: Article 1700908.
108. Lu Y, Gao X, Jiang L, Chen Z, Wang T, Jie J, Kang H, Zhang Y, Guo S, Ruan H, Zhao Y, Cao Z, Li T. (2017). Directly cast bulk eutectic and near-eutectic high entropy alloys with balanced strength and ductility in a wide temperature range. *Acta Materialia 124*: 143–50.
109. Ye X, Xiong J, Wu X, Liu C, Xu D, Zhang W, Fang D, Li B. (2021). A new infinite solid solution strategy to design eutectic high entropy alloys with B2 and bcc structure. *Scripta Materialia 199*: Article 113886.
110. Wei Q, Luo G, Zhang J, Jiang S, Chen P, Shen Q, Zhang L. (2020). Designing high entropy alloy-ceramic eutectic composites of $MoNbRe_{0.5}TaW(TiC)_x$ with high compressive strength. *Journal of Alloys and Compounds 818*: Article 152846.
111. Cai Q, Cantor B, Tong VS, Wang F, Mendis CL, Chang ITH, Fan ZY. (2021). Microstructure and mechanical properties of ultrafine quaternary Al-Cu-Si-Mg eutectic alloy. *Materials Science and Engineering 12*. https://doi.org/10.3390/met12010007
112. Choudhuri D, Shukla S, Green WB, Gwalani B, Ageh V, Banerjee R, Mishra RS. (2018). Crystallographically degenerate B2 precipitation in a plastically deformed fcc-based complex concentrated alloy. *Materials Research Letters 6*: 171–3.
113. Wen J, Chu X, Cao Y, Li N. (2021). Effects of Al on precipitation behaviour of TiNbTaZr refractory high-entropy alloys. *Metals 11*, Article 1030514.
114. Stepanov ND, Yurchenko NY, Tikhonovsky MA, Salishchev GA. (2018). Effect of carbon content and annealing on structure and hardness of the CoCrFeNiMn-based high entropy alloys. *Journal of Alloys and Compounds 687*: 59–71.
115. Schulze C, Feller-Kniepmeier M. (2000). Transmission electron microscopy of phase composition and lattice misfit on the Re-containing nickel-base superalloy CMSX-10. *Materials Science and Engineering A281*: 204–12.
116. Hou N and Fan ZY., Unpublished research.
117. Shelby JE. (2020). *Introduction to Glass Science and Technology*, 3rd edn. London: Royal Society of Chemistry.

118. Young RJ, Lovell PA. (2011). *Introduction to Polymers*, 3rd edn. Boca Raton: CRC Press, 383–97.
119. Klement W Jr, Willens RH, Duwez P. (1960). Non-crystalline structure in solidified gold-silicon alloys. *Nature 187*: 869–70.
120. Luborsky FE. (1983). *Amorphous Metallic Alloys.* London: Butterworths.
121. Suryanarayana C, Inoue A. (2018). *Bulk Metallic Glasses.* Boca Raton: CRC Press.
122. Ma L, Wang L, Zhang T, Inoue A. (2002). Bulk glass formation of Ti-Zr-Hf-Cu-M (M = Fe, Co, Ni) alloys. *Materials Transactions 43*: 277–80.
123. Ding J, Inoue A, Han Y, Kong FL, Zhu SL, Wang Z, Shalaan E, Al-Marzouki F. (2017). High-entropy effect on structure and properties of (FeCoNiCr)-B amorphous alloys. *Journal of Alloys and Compounds 696*: 345–52.
124. Guo S, Hu Q, Ng C, Liu CT. (2013). More than entropy in high-entropy alloys: Forming solid solution or amorphous phase. *Intermetallics 41*: 96–103.
125. Huo JT, Huo L, Men H, Wang XM, Inoue A, Wang JQ, Chang CT, Li RW. (2015). The magnetocaloric effect of GdTbDyAlM (M = Fe, Co, Ni) high-entropy bulk metallic glasses. *Intermetallics 58*: 31–5.
126. Zhao SF, Shao Y, Liu X, Chen N, Ding HY, Yao KF. (2015). Pseudo-quinary TiZrHfBe(CuNi)high-entropy bulk metallic glasses with large glass-forming ability. *Materials and Design 87*: 625–31.
127. Takeuchi A, Chen N, Wada T, Yokoyama Y, Kato H, Inoue A, Yeh JW. (2011). PdPtCuNiP high-entropy alloy as a bulk metallic glass in the centimeter. *Intermetallics 19*: 1546–54.

Further Reading

Cantor B. (2021). Multicomponent high-entropy Cantor alloys. *Progress in Materials Science 120*, Article 100754.

Cantor B. (2020). *The Equations of Materials.* Oxford: Oxford University Press.

Hillert M. (2008). *Phase Equilibria, Phase Diagrams and Phase Transformations*, 2nd edn. Cambridge: Cambridge University Press.

Miracle DB, Senkov ON. (2017). A critical review of high-entropy alloys and related concepts. *Acta Materialia 122*: 448–511.

Murty BS, Yeh JW, Ranganathan S, Bhattacharjee PP. (2019). *High-Entropy Alloys.* Amsterdam: Elsevier.

Senkov ON, Miracle DB, Chaput KJ. (2018). Development and exploration of refractory high entropy alloys—A review. *Journal of Materials Research 33*: 3092–3218.

Xiang H, Xing Y, Dai FZ, Wang H, Su L, Miao L, Zhang G, Wang Y, Qi X, Yao L, Wang H, Zhao B, Li J, Zhou Y. (2021). High-entropy ceramics: Present status, challenges, and a look forward. *Journal of Advanced Ceramics 10*: 385–441.

4
Structure of Multicomponent Materials

1 Introduction

This chapter discusses several aspects of the structure of multicomponent high-entropy materials: the fundamental thermodynamics of single-phase and multiphase multicomponent materials; the atomic and electronic structure of multicomponent single-phase solid solutions; and defects such as vacancies, dislocations and grain boundaries in multicomponent single-phase solid solutions, as well as their effect on fundamental properties such as atomic diffusion plastic flow and recrystallisation.

2 Thermodynamics of Multicomponent Materials

Free energy of multicomponent materials

In a multicomponent single-phase solid solution, the *Gibbs free energy* of the material G at any temperature T is given by:[1–5]

$$G = \sum_i x_i \mu_i^{ss} = \sum_i x_i \mu_i + \Delta G_{mix} = \sum_i x_i \mu_i + \Delta H_{mix} - T\Delta S_{mix}$$

where x_i is the molar fraction of the ith component, μ_i and μ_i^{ss} are, respectively, the *chemical potential* (or *molar free energy*) of the ith component before and after mixing to form the solution, and ΔG_{mix}, ΔH_{mix} and ΔS_{mix} are, respectively, the *free energy of mixing, heat (or enthalpy) of mixing* and *entropy of mixing*. The first term on the right-hand side is the free energy of the unmixed components, and the last two terms are the free energy change caused by mixing them to form the single-phase solid solution.

If the solid solution is *ideal*, there is no *interaction energy* between the different atoms, which act in the solid solution, therefore, effectively like a packing of non-interacting billiard balls. The heat of mixing is then zero, i.e. $\Delta H_{mix} = 0$, and the entropy of mixing is given by the configurational entropy of a random distribution of the different atoms within the solid solution:

$$\Delta S_{mix} = -R \sum_i x_i \ln x_i$$

where R is the gas constant.[i] The overall ideal free energy of the material is then:[1–5]

[i] Boltzmann's equation gives the entropy of a particular atomic arrangement as $k \ln w$ where w is its probability as given by the number of equivalent atomic arrangements and k is Boltzmann's constant. For one mole of an

Fundamentals of Multicomponent High-Entropy Materials. Brian Cantor, Oxford University Press. © Brian Cantor (2024).
DOI: 10.1093/9780191986710.003.0004

$$G = \sum_i x_i \mu_i^{ss} = \sum_i x_i \mu_i + RT \sum_i x_i \ln x_i$$

and the chemical potential of each component i in the ideal solid solution μ_i^{ss} differs from its pure chemical potential μ_i because of its individual contribution $RT \ln x_i$ to the ideal (random) configurational entropy of mixing:

$$\mu_i^{ss} = \mu_i + RT \ln x_i$$

If the solid solution is not ideal but instead is *near-ideal* or *regular*, there are small pair-wise, three-way, four-way, etc., interaction energies ω_{ij}, ψ_{ijk}, χ_{ijkl}, etc., between the different atoms, so the heat of mixing is not zero, i.e. $\Delta H_{mix} = f(\omega_{ij}, \psi_{ijk}, \chi_{ijkl} \ldots) \neq 0$,[ii] but the atomic interactions are insufficiently strong to affect the entropy of mixing significantly, which is still, therefore, given by the configurational entropy of a random distribution of the different atoms within the solid solution. The overall regular free energy of the material is then:[1–5]

$$G = \sum_i x_i \mu_i^{ss} = \sum_i x_i \mu_i + \Delta H_{mix}(\omega_{ij}, \psi_{ijk}, \chi_{ijkl} \ldots) + RT \sum_i x_i \ln x_i$$

If the solid solution deviates further from ideality and is *non-regular*, there are larger interaction energies ω_{ij}, ψ_{ijk}, χ_{ijkl}, etc., between the different atoms, so the entropy of mixing is no longer given only by the configurational entropy of a random distribution of the different atoms within the solid solution. The overall non-regular free energy of the material is then:[1–5]

$$G = \sum_i x_i \mu_i^{ss} = \sum_i x_i \mu_i + \Delta H_{mix}(\omega_{ij}, \psi_{ijk}, \chi_{ijkl} \ldots) + RT \sum_i x_i \ln x_i - T\Delta S_{mix}^{xs}(\omega_{ij}, \psi_{ijk}, \chi_{ijkl} \ldots)$$

where $\Delta S_{mix}^{xs} = f'(\omega_{ij}, \psi_{ijk}, \chi_{ijkl} \ldots) \neq 0$ is the excess entropy of mixing over and above the random configurational entropy. Notice that there are several contributions to the excess entropy ΔS_{mix}^{xs}:

1. Excess *configurational* entropy of mixing arising from a non-random distribution of the different atoms, caused by their interactions;
2. Excess *vibrational* entropy of mixing arising from changes in the vibrational spectra of the different atoms caused by their interactions; and
3. Excess *magnetic* and *electronic* entropy of mixing arising from changes in the magnetic and electronic interactions between the different atoms.

ideal (i.e. random) multicomponent solid solution, $\ln w = -N_o \sum_i x_i \ln x_i$ where N_o is Avogadro's number, so the configurational entropy of mixing is[1–5] $\Delta S_{mix} = -R \sum_i x_i \ln x_i$, where R is the gas constant.

ii The heat of mixing is a function of *all* the interaction energies, i.e. the functional dependence of $\Delta H_{mix} = f(\omega_{ij}, \psi_{ijk}, \chi_{ijkl} \ldots)$ is a shorthand notation for its dependence on $\omega_{AB}, \omega_{BC}, \omega_{AC} \ldots, \psi_{ABC}, \psi_{ABD}, \psi_{ACD} \ldots, \chi_{ABCD}, \chi_{ABCE}, \chi_{ACDE} \ldots$, etc., where A, B, C, D, E, etc., are the different individual components. The same notation is also used in the next paragraph for the functional dependence of the excess entropy of mixing ΔS_{mix}^{xs}.

4. Excess *configurational* entropy arising from the presence of defects such as vacancies and dislocations.

Notice also that the excess entropy of mixing ΔS_{mix}^{xs} can be positive or negative, since the entropy of mixing in the non-regular solution can be lower or higher than the ideal random configurational entropy.

The free energy of non-ideal (i.e. regular and non-regular) solid solutions is often described in an alternative way, using the formalism[1,2] of *activities* a_i and *activity coefficients* γ_i, where the activity of each component i is related to its molar fraction by $a_i = \gamma_i x_i$. Each component has a chemical potential in the solid solution μ_i^{ss} that now differs from its pure chemical potential μ_i according to its activity a_i instead of its molar fraction x_i:

$$\mu_i^{ss} = \mu_i + RT \ln a_i$$

The molar Gibbs free energy of the material is then:[1–4]

$$G = \sum_i x_i \mu_i^{ss} = \sum_i x_i (\mu_i + RT \ln a_i) = \sum_i x_i (\mu_i + RT \ln \gamma_i x_i) = \sum_i x_i \mu_i + RT \sum_i x_i \ln x_i + RT \sum_i x_i \ln \gamma_i$$

so any deviation from ideality is given by the sum of the activity coefficients in the last term on the right-hand side, which are collectively equivalent, therefore, to the heat of mixing and the excess entropy of mixing:

$$\Delta H_{mix}(\omega_{ij}, \psi_{ijk}, \chi_{ijkl} \ldots) - T\Delta S_{mix}^{xs}(\omega_{ij}, \psi_{ijk}, \chi_{ijkl} \ldots) = RT \sum_i x_i \ln \gamma_i$$

Non-ideal mixing effects are thus distributed among the different components via their individual activity coefficients. This activity formalism treats solute effects independently, one at a time, and is convenient for relatively dilute solutions with relatively few components, when solute atoms are widely separated, their interactions can be regarded as small deviations from ideality, and the independent effects of the different components can be assessed separately and then summed linearly. The formalism of heats and entropies of mixing treats solute effects collectively and is more suitable for concentrated solutions and large numbers of components, when solute interactions are extensive and multiple, so it is not reasonable to use a dilute solution approximation nor to sum the effects of the different solutes linearly.

There have been many theories, analyses and approximations used to assess and calculate values of the activity coefficients γ_i and/or the heat of mixing ΔH_{mix} and the excess entropy of mixing ΔS_{mix}^{xs} for different types of interatomic interactions ω_{ij} and ψ_{ijk} in binary and ternary solid solutions, allowing for factors such as long-range order, short-range order, spinodal demixing, ferromagnetic coupling and differences in molar volume.[1,2,6,7] In general, however, these thermodynamic calculations become increasingly unwieldy, difficult to use and, therefore, unhelpful as the number of components increases in a multicomponent system. Thus the total number of different interaction energies in a single-phase solid solution with

c components is:

$$\frac{c(c-1)}{2} + \frac{c(c-1)(c-2)}{3 \times 2} + \frac{c(c-1)(c-2)(c-3)}{4 \times 3 \times 2} + \ldots$$

which is 11, 26 and 57 for 4-, 5- and 6-component materials respectively, and all of these interaction energies need to be estimated together with their independent effects on the internal energy (or enthalpy), non-randomness of mixing, and atomic vibrational spectra of the material. Moreover, there have been very few direct measurements of the fundamental thermodynamic properties (free energy, enthalpy or entropy) of multicomponent solid solutions. Suffice it to say in summary, therefore, that in the vast majority of cases we do not know (either theoretically or experimentally) in any detail the extent of the multiple interaction energies in multicomponent materials, the corresponding activity coefficients or heats and entropies of mixing, or their overall free energies.

Entropy of multicomponent materials

Chapter 3 showed that multicomponent phase space contains a series of very large single-phase solid solutions: single-phase multicomponent fcc and bcc regions of Cantor alloys and Senkov alloys; single-phase regions of multicomponent intermetallic compounds such as the CsCl-type B2 mono-aluminides MAl, Ni_3Al-type L12 tri-metal aluminides M_3Al, $MgZn_2$-type C14 and $MgCu_2$-type C15 Laves phases, and topologically close-packed (TCP) mu and sigma phases μ and σ; and single-phase regions of multicomponent ceramic compounds such as rock-salt NaCl-type mono-oxides, mixed spinels and perovskites, mono-, tri- and hexa-borides, mixed boro-carbo-nitrides; and so on. In each of these large regions of multicomponent phase space, the configurational entropy of mixing seems to dominate in a multicomponent material over the interaction energies between the different atoms, the associated heat of mixing and any resulting overall tendency towards phase separation and the formation of fixed-composition stoichiometric compounds. This sub-section explores the configurational entropy of mixing in such multicomponent solid solutions and considers when and where it dominates over the interaction energies and heat of mixing in the material to promote the formation of a single-phase solid solution.

Equiatomic solid solutions

In an equiatomic multicomponent single-phase solid solution with c components (for instance, the original fcc Cantor alloy CrMnFeCoNi and bcc Senkov alloy VNbMoTaW), the molar fractions of all the components are the same, $x_i = 1/c$ for all i, and the ideal entropy of mixing, ideal free energy of mixing and overall ideal free energy are then:[4,5]

$$\Delta S_{mix} = -R\sum_i x_i \ln x_i = -R \ln (1/c)$$

$$\Delta G_{mix} = -T\Delta S_{mix} = RT \ln (1/c)$$

$$G = \sum_i x_i\mu_i^{ss} = \sum_i x_i\mu_i + RT \ln (1/c)$$

Table 4.1 shows the ideal entropy of mixing ΔS_{mix} and the corresponding ideal free energy of mixing ΔG_{mix} at 1,000 K (1,273 °C) in such an ideal equiatomic multicomponent single-phase solid solution for different numbers of components ranging from c = 2–50, and the

Table 4.1 Ideal entropy and Gibbs free energy of mixing $\Delta S_{mix} = -R \ln(1/c)$ and $\Delta G_{mix} = RT \ln(1/c)$ versus number of components c in an equiatomic multicomponent solid solution (taking R = 8.3145 J/K/mol).

number of components c	ln (1/c)	ideal entropy of mixing ΔS_{mix} (J/K/mol)	ideal free energy of mixing ΔG_{mix} at 1,000K (kJ/mol)
2	−0.69	5.76	−5.76
3	−1.10	9.13	−9.13
4	−1.39	11.53	−11.53
5	−1.61	13.38	−13.38
6	−1.79	14.90	−14.90
7	−1.95	16.18	−16.18
8	−2.08	17.29	−17.29
9	−2.20	18.29	−18.29
10	−2.30	19.12	−19.12
20	−3.00	24.91	−24.91
30	−3.40	28.28	−28.28
40	−3.69	30.67	−30.67
50	−3.91	32.53	−32.53

same results are shown graphically in Figure 4.1. The ideal entropy of mixing in J/K/mol and the ideal free energy of mixing at 1,000 K (1,273 °C) in kJ/mol are numerically equal and opposite, the first positive and the second negative. Differentiating:

$$\frac{d\Delta G_{mix}}{dc} = -T\frac{d\Delta S_{mix}}{dc} = -\frac{RT}{c} \rightarrow 0 \text{ as } \quad c \rightarrow \infty$$

so the ideal entropy of mixing increases monotonically and the ideal free energy of mixing decreases monotonically with increasing number of components c, but the rate at which the entropy of mixing increases and the rate at which the free energy of mixing decreases both slow down gradually but noticeably as the number of components increases beyond $c = 10$. Thus, the ideal entropy of mixing more than doubles from 5.8 to 13.4 J/K/mol as the number of components increases from $c = 2$ to 5; it fails to double again and instead goes from 13.4 to 24.9 J/K/mol as the number of components increases further to $c = 20$; and it then increases by less than 50% and goes from 24.9 to 32.5 kJ/K/mol as the number of components increases further to $c = 50$. The ideal free energy of mixing at 1,000 K (1,273 °C) decreases in a parallel way.

Table 4.1 and Figure 4.1 explain why Yeh et al. coined the term[8] *high-entropy alloys* for multicomponent materials such as the original Cantor alloy and, as mentioned in previous chapters, this has become the accepted name for the field.[9–11] And, of course, *high entropy* was the first of Yeh's *four core effects* in these materials. In fact the term high-entropy alloy applies very well to the original Cantor and Senkov alloys, which are equiatomic 5-component near-random solid-solution materials,[12] close to ideal, with a free energy of mixing dominated, therefore, by the ideal configurational entropy of mixing.

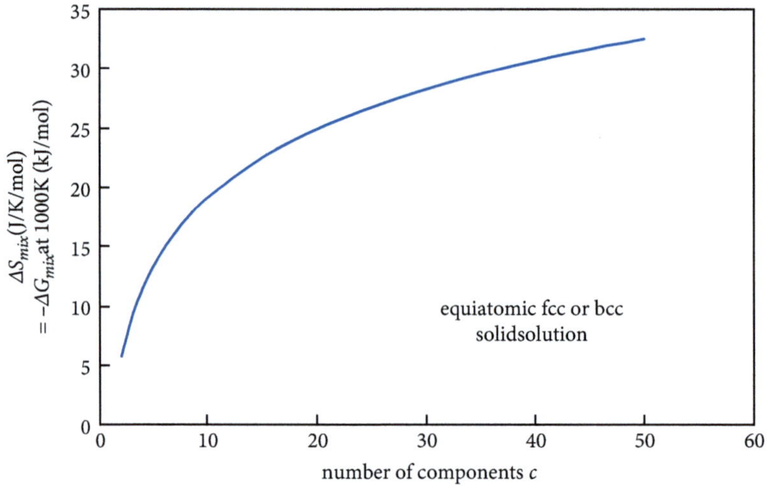

Figure 4.1 Ideal entropy of mixing ΔS_{mix} and ideal free energy of mixing ΔG_{mix} at 1,000 K for an equiatomic multicomponent fcc or bcc solid solution (such as the original Cantor alloy CrMnFeCoNi and original Senkov alloy VNbTaMoW) versus the number of components *c*.

Equiatomic compounds

In an equiatomic multicomponent intermetallic or ceramic mono-compound MX (such as the Cs-Cl-type B2 multicomponent mono-aluminides MAl or the rock salt NaCl-type multicomponent mono-oxides MO), one sub-lattice is fixed with a single anionic species X (Al or O) and the other sublattice consists of a random arrangement of multicomponent cations M. The molar fraction of the fixed anion X is ½, and for an equiatomic mixture of *c* different cations M their molar fractions are all the same $x_i = 1/2c$. The ideal entropy of mixing, ideal free energy of mixing and overall ideal free energy of the multicomponent mono-compound are then:[4,5]

$$\Delta S_{mix} = -R\sum_i x_i \ln x_i = -\tfrac{1}{2}R\ln\,(1/2c)$$

$$\Delta G_{mix} = -T\Delta S_{mix} = \tfrac{1}{2}RT\ln\,(1/2c)$$

$$G = \sum_i x_i\mu_i^{ss} = \sum_i x_i\mu_i + \tfrac{1}{2}RT\ln\,(1/2c)$$

More generally, for any equiatomic multicomponent intermetallic or ceramic compound with stoichiometry M_aX_b, one sub-lattice is again fixed with a single anionic species X and the other sublattice consists again of a random arrangement of multicomponent cations M. The molar fraction of the fixed anion X is $b/\,(a+b)$, and for an equiatomic mixture of *c* different cations M their molar fractions are all the same $x_i = a/\,(a+b)\,c$. The ideal entropy of mixing, ideal free energy of mixing and overall ideal free energy of the multicomponent compound are then:[4,5]

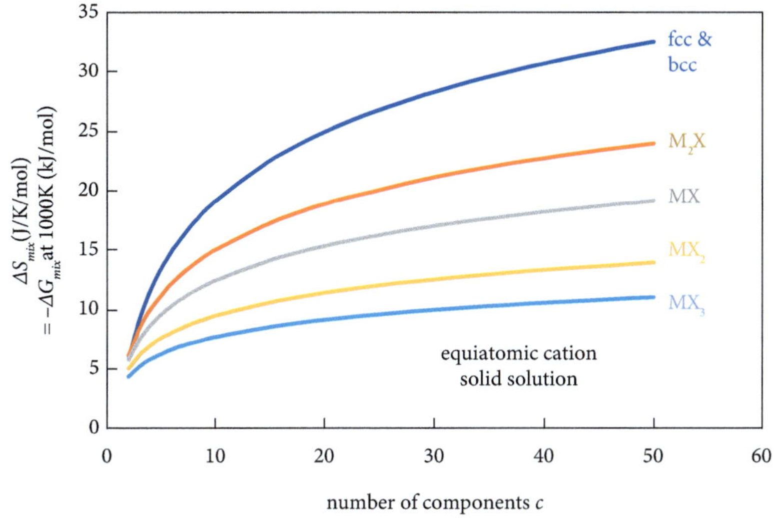

Figure 4.2 Ideal entropy of mixing ΔS_{mix} and ideal free energy of mixing ΔG_{mix} at 1,000 K for an equiatomic multicomponent fcc or bcc solid solution (such as the original Cantor alloy CrMnFeCoNi and original Senkov alloy VNbTaMoW) and for equiatomic multicomponent intermetallic or ceramic compounds with stoichiometries M_2X, MX, MX_2 and MX_3 versus the number of components *c*.

$$\Delta S_{mix} = -R\sum_i x_i \ln x_i = -\frac{aR}{(a+b)} \ln\left(\frac{a}{(a+b)\,c}\right)$$

$$\Delta G_{mix} = -T\Delta S_{mix} = \frac{aRT}{(a+b)} \ln\left(\frac{a}{(a+b)\,c}\right)$$

$$G = \sum_i x_i \mu_i^{ss} = \sum_i x_i \mu_i + \frac{aRT}{(a+b)} \ln\left(\frac{a}{(a+b)\,c}\right)$$

Figure 4.2 shows the ideal entropy of mixing ΔS_{mix} and the corresponding ideal free energy of mixing ΔG_{mix} at 1,000 K (1,273 °C) in equiatomic intermetallic or ceramic compounds with stoichiometries M_2X, MX, MX_2, and MX_3 for different numbers of cation components ranging from $c = 2$ to 50. For any given number of components *c*, the ideal entropy of mixing decreases and the ideal free energy of mixing increases similarly with increasing stoichiometric ratio b/a. This picture is, however, somewhat misleading. The reduction in ideal entropy of mixing and the corresponding increase in ideal free energy of mixing both arise essentially because the entropies and free energies are per mole of compound, and there are fewer moles of cation components in a mole of the compound the higher the stoichiometric ratio b/a. In fact, the ideal entropies and ideal free energies of mixing are identical for all stoichiometries when re-calculated per mole of cations M. Interaction energies and heats of mixing per mole of compound reduce in a similar way to the increase shown in Figure 4.2 for the ideal entropy of mixing, so the balance of compound-forming interaction energies and solution-forming configurational entropies of mixing remains essentially the same, independent of stoichiometry.

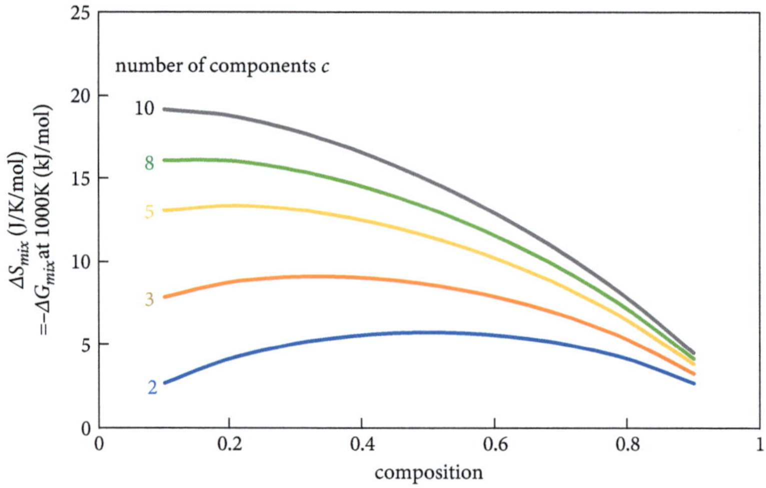

Figure 4.3 Ideal entropy of mixing ΔS_{mix} and ideal free energy of mixing ΔG_{mix} at 1,000K for a non-equiatomic multicomponent solid solution $(M_i)_a(M_jM_kM_l\ldots)_{1-a}$ versus molar fraction a of one of the components for different numbers of components c.

Non-equiatomic compositions

To investigate composition effects, consider varying the molar fraction of just one of the components in a non-equiatomic multicomponent single-phase solid solution with c components (such as modified fcc Cantor and modified bcc Senkov alloys). The overall material composition is $(M_i)_a(M_jM_kM_l\ldots)_{1-a}$, the molar fractions of the different components are $x_i = a$ and $x_j = x_k = x_l \ldots = (1-a)/(c-1)$, and the ideal entropy of mixing, ideal free energy of mixing and overall ideal free energy are then:[4,5]

$$\Delta S_{mix} = -R\sum_i x_i \ln x_i = -R\left(a \ln a + (1-a)\ln\left(\frac{(1-a)}{(c-1)}\right)\right)$$

$$\Delta G_{mix} = -T\Delta S_{mix} = RT\left(a \ln a + (1-a)\ln\left(\frac{(1-a)}{(c-1)}\right)\right)$$

$$G = \sum_i x_i\mu_i^{ss} = \sum_i x_i\mu_i + RT\left(a \ln a + (1-a)\ln\left(\frac{(1-a)}{(c-1)}\right)\right)$$

Figure 4.3 shows the ideal entropy of mixing ΔS_{mix} and the corresponding ideal free energy of mixing ΔG_{mix} at 1,000 K (1,273 °C) in a non-equiatomic multicomponent single-phase solid solution $(M_i)_a(M_jM_kM_l\ldots)1\text{-}a$ (such as modified fcc Cantor and modified bcc Senkov alloys) as a function of varying mole fraction of one of the components $x_i = a$ for different numbers of components ranging from $c = 2$ to 10. The maximum ideal entropy of mixing and the corresponding minimum ideal free energy of mixing both occur at mole fractions of $x_i = a = 1/c = \frac{1}{2}, \frac{1}{3}, \frac{1}{5}, \frac{1}{8}$ and $\frac{1}{10}$ for $c = 2$, 3, 5, 8, and 10 respectively because these correspond to equiatomic compositions. However, the ideal entropy of mixing stays relatively high for quite a wide range of compositions either side of its maximum equiatomic value (and the corresponding ideal free energy of mixing stays relatively low for quite a wide range of compositions either side of its minimum equiatomic value).

In summary, therefore, for a multicomponent high-entropy material with interaction energies between the different components that are not too large, the configurational entropy of mixing is most likely to dominate and lead to the formation of a solid solution at the equiatomic composition, and this effect is likely to persist over a reasonable composition range either side of the equiatomic composition. This explains the wide composition ranges seen in the Cantor and Senkov alloy regions, as shown in Figures 3.6 and 3.13.

Effect of temperature

To investigate temperature effects, consider the original Cantor alloy CrMnFeCoNi, which has been shown by, for example, atom-probe measurements such as shown in Figure 3.10 to have a near-random arrangement of its five components on a single-phase fcc lattice. In other words the original Cantor alloy is near-ideal or regular and, as discussed in the previous sub-section, this means that there may be very small pairwise, three-way, four-way, etc., interaction energies ω_{ij}, ψ_{ijk}, χ_{ijkl}, etc., between the different atoms, so the heat of mixing is not zero, i.e. $\Delta H_{mix} = f(\omega_{ij}, \psi_{ijk}, \chi_{ijkl} \ldots) \neq 0$, but the atomic interactions are insufficiently strong to affect the entropy of mixing significantly, which is still, therefore, given by the configurational entropy of a random distribution of the different atoms within the solid solution. Tsai et al.[13] have calculated binary interaction energies ω_{ij}^b from the known binary heats of mixing ΔH_{mix}^b for all the *i-j* binary sub-systems of the original Cantor alloy CrMnFeCoNi (i.e. the ten binary systems Cr-Mn, Cr-Fe, Cr-Co, Cr-Ni, Mn-Fe, Mn-Co, Mn-Ni, Fe-Co, Fe-Ni, and Co-Ni), using the *quasichemical approximation* for a regular solution:[iii]

$$\Delta H_{mix}^b = -N_o z x_i x_j \omega_{ij}^b$$

The results are shown in Table 4.2. Tsai et al.[13] then use two crude approximations: firstly assuming that the pairwise interaction energies in the multicomponent solid solution are approximately equal to the interaction energies in the corresponding *i-j* binary sub-systems; and secondly ignoring multiple ternary and higher-order interaction energies, i.e.:

$$\omega_{ij} = \omega_{ij}^b$$

$$\psi_{ijk}, \chi_{ijkl} \ldots = 0$$

so the heat of mixing is given approximately by:

$$\Delta H_{mix} = -N_o z \sum_{i,j} x_i x_j \omega_{ij} \approx -\left(N_o z / c^2\right) \sum_{i,j} \omega_{ij}^b$$

[iii] In the quasichemical approximation for a regular binary solid solution, the atomic arrangement is random so the entropy of mixing is ideal, and the heat of mixing is given by the difference in the numbers of *ii*, *jj* and *ij* bonds n_{ii}, n_{jj} and n_{ij} multiplied by their respective bond energies ϵ_{ii}, ϵ_{jj} and ϵ_{ij} before and after mixing mole fractions x_i and x_j to form the solid solution: $\Delta H_{mix}^b = -n_{ij}\left(2\epsilon_{ij} - \epsilon_{ii} - \epsilon_{jj}\right) = -N_o z x_i x_j \omega_{ij}^b$ where N_o is Avogadro's number, z is the coordination number (= 12 for fcc Cantor alloys) and $\omega_{ij}^b = 2\epsilon_{ij} - \epsilon_{ii} - \epsilon_{jj}$. When $\omega_{ij}^b > 0$, $2\epsilon_{ij} > \epsilon_{ii} + \epsilon_{jj}$, *ij* bonds are preferred, and there is a negative heat of mixing, which favours the formation of a solution; when $\omega_{ij}^b < 0$, $2\epsilon_{ij} < \epsilon_{ii} + \epsilon_{jj}$, *ii* and *jj* bonds are preferred, and there is a positive heat of mixing, which favours the separation of *i* and *j*.

Table 4.2 Binary interaction energies ω^b_{ij} and heats of mixing ΔH^b_{mix} for all the binary sub-systems[13] of the original Cantor alloy CrMnFeCoNi, using the quasichemical approximation for a regular solution.[1–5]

alloy	interaction energy ω^b_{ij} (meV)	regular heat of mixing ΔH^b_{mix} (kJ/mol)
Cr-Mn	11.3	−3.8
Cr-Fe	−17.5	5.9
Cr-Co	−23.5	6.8
Cr-Ni	−22.1	6.4
Mn-Fe	16.2	−4.7
Mn-Co	−17.6	2.2
Mn-Ni	48.4	−14.0
Fe-Co	4.5	−1.3
Fe-Ni	15.9	−4.6
Co-Ni	0	0

The entropy of mixing, free energy of mixing and overall free energy of the material are then:[4,5]

$$\Delta S_{mix} = -R\sum_i x_i \ln x_i = -R\ln\,(1/c)$$

$$\Delta G_{mix} = \Delta H_{mix} - T\Delta S_{mix} \approx -\left(N_o z/c^2\right)\sum_{i,j}\omega^b_{ij} + RT\ln\,(1/c)$$

$$G = \sum_i x_i\mu_i^{ss} \approx \sum_i \mu_i/c - \left(N_o z/c^2\right)\sum_{i,j}\omega^b_{ij} + RT\,\ln\,(1/c)$$

Using the values for ω^b_{ij} in Table 4.2, $\sum_{ij}\omega^b_{ij} = 25.6$ meV, and taking $N_o = 6.022 \times 10^{23}$, $z = 12$, $c = 5$, and 1.602×10^{-22} kJ/eV, the heat of mixing for the alloy is then calculated as $\Delta H_{mix} \approx -N_o z\sum_{ij}\omega^b_{ij}/25 = -7.4 \times 10^{21}$eV/mol = −1.2 kJ/mol. This is fairly close to zero (as expected for a near-ideal solution) and is negative (i.e. supports compound formation, but only rather weakly). Also from Table 4.2, the strongest binary interaction is between Mn and Ni, with an interaction energy of $\omega^b_{MnNi} = 48.4$ meV, indicating therefore an individual contribution to the heat of mixing of approximately $-N_o z\omega^b_{MnNi}/25 = -1.4 \times 10^{22}$ eV/mole = −2.24 kJ/mole, i.e. almost twice as large as the overall heat of mixing, and again negative (i.e. again supports compound formation). For comparison, however, the regular configurational entropy of mixing for the alloy is given in Table 4.1 as $\Delta S_{mix} = -R\ln\,(1/c) = 13.4$ J/mol/K, which makes a contribution to the free energy of mixing of −13.4 kJ/mol and −4.0 kJ/mol at 1,000 K and 298 K (room temperature) respectively, both much greater than the heat of mixing, i.e. much greater than the contribution from the interaction energies between the atoms, either individually or collectively.

It is not surprising, therefore, that the original Cantor alloy forms a single-phase fcc solid solution so readily and that it remains stable down to relatively low temperatures (i.e. down to room temperature) where it becomes kinetically stable (and the same for fcc Cantor alloys generally and other multicomponent single-phase solid solutions and compounds). At sufficiently low temperatures, entropy terms approach zero and, according to the third law of thermodynamics, only mixtures of pure single crystals are stable at equilibrium at absolute zero. In other words, any solid solution ultimately becomes thermodynamically unstable as the temperature is decreased: the only question is whether or not this happens at such a low temperature that atomic diffusion has become very sluggish and a solid solution is stabilised kinetically. As mentioned, the calculations above are built on a strong simplifying approximation, taking the pairwise interactions between the different atoms in the multicomponent alloy as equal to the binary interactions in the corresponding binary alloy, ignoring higher-order three-, four- and five-way interactions, and ignoring any excess entropy of mixing, almost certainly, therefore, underestimating the overall atomic interaction energy and the magnitude of the resulting heat of mixing. But they still give results that are essentially similar to the more detailed (though heuristic and more opaque) Calphad calculations of Bracq et al.,[14] as discussed in Chapter 3 and shown in Figure 4.4, also predicting a single-phase fcc solid solution at high temperature, with several phases forming at low temperature. The calculations based on Tsai et al.'s approximation[13] and Bracq et al.'s Calphad results[14] explain in broad terms why fcc Cantor alloys form so readily as a single-phase fcc solid solution over an enormous region in multicomponent phase space; why a single-phase fcc structure is fully stable at high temperatures and is retained readily on cooling to low temperatures; but why precipitation of other phases is seen after very prolonged annealing at intermediate temperatures.[15,16]

The simple calculations described above based on Tsai et al.'s[13] approximation for pairwise and higher-order atomic interaction energies suggest that a single-phase fcc solid-solution is the thermodynamically stable structure for the original Cantor alloy CrMnFeCoNi at temperatures down to room temperature or even below, and is then kinetically stable at even lower temperatures; Bracq et al.'s[14] more complex Calphad calculations suggest that a single-phase fcc solid solution is the thermodynamically stable structure for temperatures down to only about 900 K (623 °C), as can be seen in Figure 4.4, suggesting an intermediate temperature region where precipitation might take place (albeit probably somewhat sluggishly) before the structure again becomes kinetically stabilised at lower temperatures. In fact, the original Cantor alloy CrMnFeCoNi has been subjected to a number of detailed deformation and annealing studies to investigate its stability as a function of thermomechanical treatment,[15–20] and it is now well established that:

1. A single phase fcc structure is fully (thermodynamically) stable, independent of mild or extremely severe deformation[17–20] and/or heat treatment[15,16] for extended periods of time (up to several years), at all temperatures above approximately 800 °C (1,073 K), as shown in Figure 4.5(a).
2. A single phase fcc structure also remains stable for prolonged periods of time at lower temperatures below 800 °C (1,073 K)[15,16].
3. However, heat treatment for very extended periods of 500–1,000 days (i.e. several years) at intermediate temperatures in the range ~600–700 °C (~873–973 K) leads to grain-boundary precipitation of small Cr-rich, Ni-depleted sigma-phase particles,[17] as shown in Figure 4.5(b).
4. Heat treatment for very extended periods of 500–1,000 days (i.e. several years) at slightly lower temperatures in the range ~450–550 °C (~723–823 K) leads to grain-boundary precipitation of a mixture of small bcc Cr, $L1_0$-NiMn and fcc CoFe particles,[16] as shown in Figure 4.5(c).

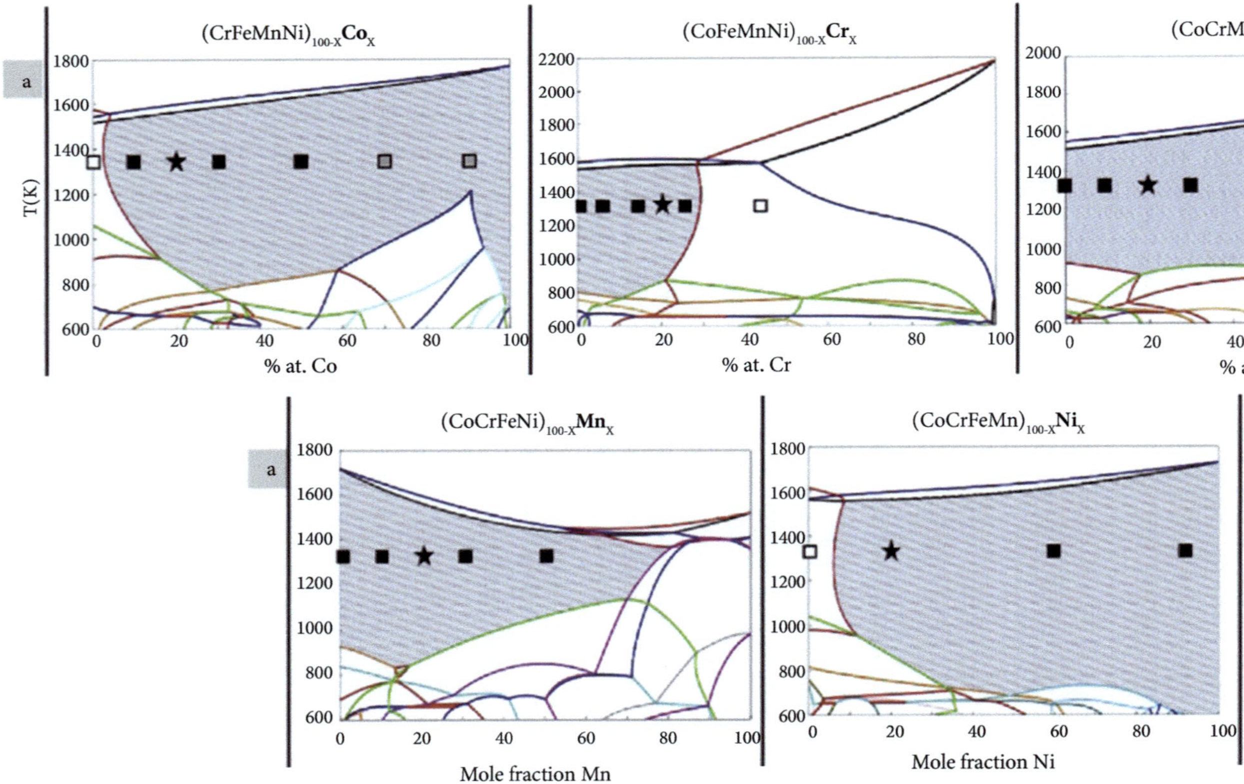

Figure 4.4 Pseudo-binary phase diagrams showing experimental results for the effect of varying independently the individual components of the Cantor alloy CrMnFeCoNi compared with Calphad thermodynamic calculations.

*Solid symbols = single-phase fcc at room and elevated temperatures; open symbols = two-phase at room and elevated temperatures; grey symbols = single-phase fcc at elevated temperature + two-phase at room temperature; star = original Cantor alloy composition; grey shading = Calphad predicted fcc single-phase field.
Reproduced from Bracq et al. (2019). Combining experiments and modelling to explore the solid solution strengthening of high- and medium-entropy alloys. *Acta Materialia* 177: 266–79.

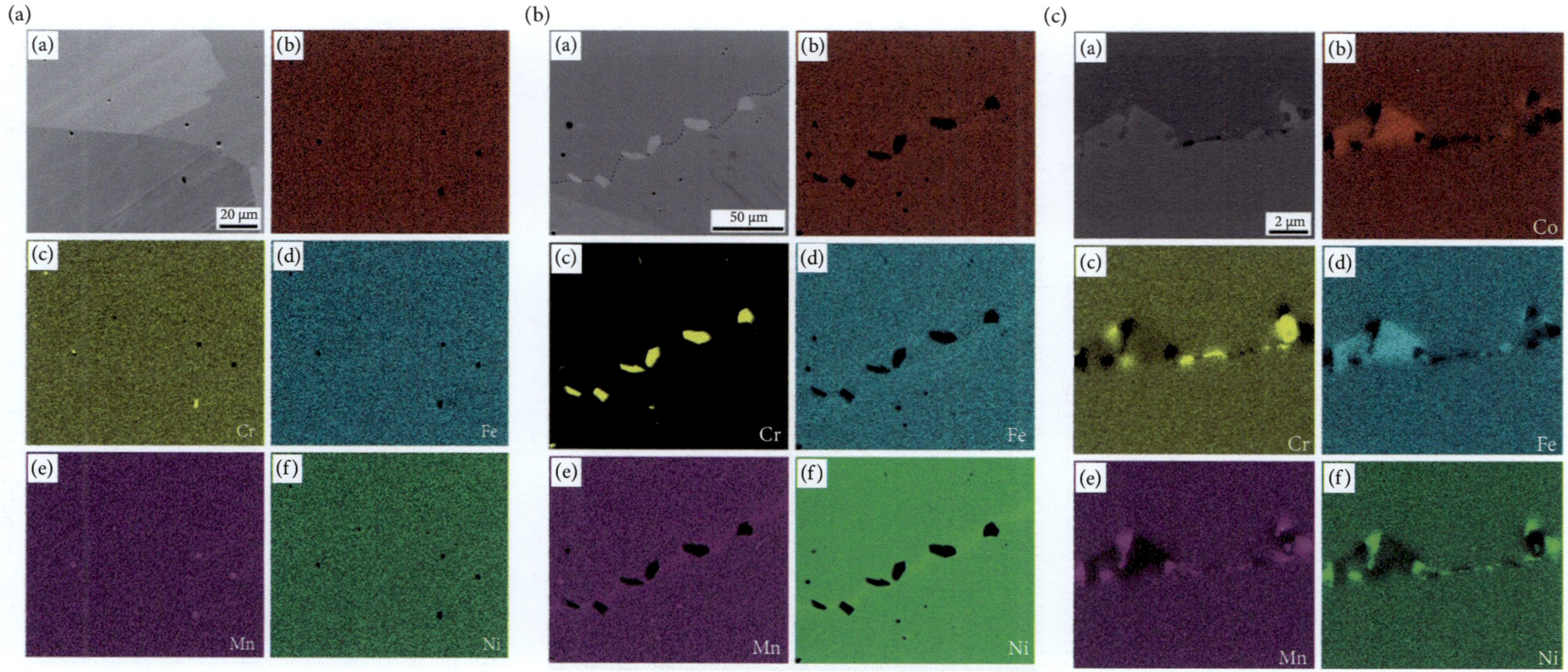

Figure 4.5 Back-scattered scanning electron (BSE) images and Co, Cr, Fe, Mn and Ni energy-dispersive X-ray (EDX) maps, showing grain boundary precipitation in the original Cantor alloy CrMnFeCoNi after extended annealing at intermediate temperatures: (a) no precipitates after 500 days at 900 °C (the small dark particles are extraneous oxide and sulphide inclusions); (b) Cr-rich, Ni-depleted σ-particles after 500 days at 700 °C; and (c) bcc Cr, $L1_0$-NiMn and fcc FeCo particles after 500 days at 700 °C.

Reproduced from Otto et al. (2016). Decomposition of the single-phase high-entropy alloy CrMnFeCoNi after prolonged anneals at intermediate temperatures. *Acta Materialia* 112: 40–52.

5. A single-phase fcc structure is again fully (now kinetically) stable, independent of mild or extremely severe deformation and/or heat treatment for extended periods of time (up to several years), at all temperatures below approximately 400 °C (673 K).[16]
6. Severe deformation in the intermediate temperature region in the form of tensile creep, hot torsion or hot compression accelerates the precipitation reactions via enhanced diffusion along the resulting extensive strain-induced dislocation, small-angle sub-boundary and nanoscale high-angle grain boundary networks.[17–20]
7. Similar precipitation reactions also take place in some non-equiatomic Cantor alloys.[21,22]

High-entropy materials

As shown in Chapter 3, multicomponent phase space contains a series of very large single-phase solid solutions: single-phase multicomponent fcc and bcc regions of Cantor alloys and Senkov alloys and single-phase regions of various different multicomponent intermetallic and ceramic compounds. As explained earlier in this sub-section, a single-phase structure is formed in each of these large regions of multicomponent phase space because the configurational entropy of mixing dominates over the interaction energies between the different atoms, the associated heat of mixing, and any resulting overall tendency, therefore, towards phase separation and the formation of fixed-composition stoichiometric compounds. Essentially, a high multicomponent configurational entropy of mixing acts together with a paucity of new multicomponent compounds to increase intersolubility between the different components and stabilise the formation of single-phase solid solutions, which explains why they are found to be so prevalent and extended in multicomponent phase space.[4,9–11] The term high-entropy alloys applies well to all these single-phase materials. Furthermore, many other multicomponent materials are multiphase mixtures of such extended solid solutions,[4,9–11] e.g. fcc + bcc, fcc + hcp, fcc + σ, bcc + B2, etc., and also have substantial configurational entropy of mixing from the sum of the individual configurational entropies of mixing of the individual phases. Once again, the term high-entropy alloys applies well to all these materials.

However, not all multicomponent materials have a free energy that is dominated by a high configurational entropy of mixing, since:

1. Many multicomponent solid-solution compositions are destabilised with substantial excess vibrational entropies of mixing[9] ΔS^{xs}_{mix} and/or heats of mixing[23,24] ΔH_{mix}, larger than and, therefore, dominating over the configurational entropy of mixing.
2. Many multicomponent materials are multiphase with phases of limited stoichiometry range,[9–11,25,26] i.e. are not extended solid solutions, and have only limited configurational entropies of mixing.

The term *high-entropy alloys* is less appropriate for these materials.

In summary, therefore, there is an enormous number of multicomponent materials. Many have single-phase structures that are strongly influenced by and stabilised by the configurational entropy of mixing, and the materials are, therefore, appropriately designated as high-entropy alloys or materials. Many consist of several non-stoichiometric phases strongly influenced by and stabilised by their configurational entropies of mixing, and the materials are again, therefore, appropriately designated as high-entropy alloys or materials. However, many consist of one or more stoichiometric fixed-composition phases that are not strongly influenced by or stabilised by the configurational entropy of mixing, and the materials are, therefore, less appropriately designated as high-entropy alloys or materials and are better described as multicomponent multiphase alloys or materials (as discussed in more detail in Chapter 3).

3 Atomic Structure of Multicomponent Materials

Local atomic clusters

Figure 4.6 shows a schematic random distribution of ten equiatomic components A–J on a two-dimensional square lattice, similar to a figure in the book by Murty et al.,[10] demonstrating in very simplified form two important features of multicomponent single-phase solid-solution materials such as the fcc Cantor alloys and bcc Senkov alloys (and, in a slightly modified form, all other multicomponent single-phase materials including intermetallic and ceramic compounds):

1. There is a very large number of different *local atomic environments* or *atomic clusters* with, for example, different A atoms surrounded by a wide range of different distributions of the other A, B, C, D, E, F, G, H, I and J atoms, and in the same way different distributions around all the other nine types of atom.
2. There are significant *local atomic distortions* and *lattice strains*, caused by the different sizes and electronic structures of the different atoms and their varied local atomic environments.

We deal with the first of these issues in the present sub-section and the second of these issues in the next sub-section.

It is rather obvious that multicomponent single-phase materials, for instance the equiatomic 5-component fcc Cantor alloy CrMnFeCoNi, the equiatomic 5-component bcc Senkov alloy VNbMoTaW or the equiatomic 10-component schematic in Figure 4.6, are different from corresponding single-component single-phase materials such as pure fcc Cu or pure bcc Fe (or, indeed, pure diamond cubic Si or pure crystoballite silica SiO_2) for the very simple reason that there are five or ten (in the particular cases of the Cantor and Senkov alloys and Figure 4.6) different atoms with five or ten (in these particular cases) different properties distributed in a random or near-random way across the different lattice points, rather than the same atom with identical properties at every lattice point. But the variation in local structure and properties in multicomponent single-phase materials is much more extensive and complex than implied just by noticing that there are five or ten (or however many) rather than one different kind of atom and, therefore, five or ten (or however many) rather than one different kind of lattice point. This is because the local material structure and properties at any lattice point depend not just on which particular atom is located at that particular lattice point, but also on its interactions with its neighbouring atoms at adjacent and nearby lattice points. In other words, the variation in properties from lattice point to lattice point depends on the range of local atomic environments, i.e. on the range of different local atomic clusters of each atom at that lattice point and its surrounding near-neighbours.

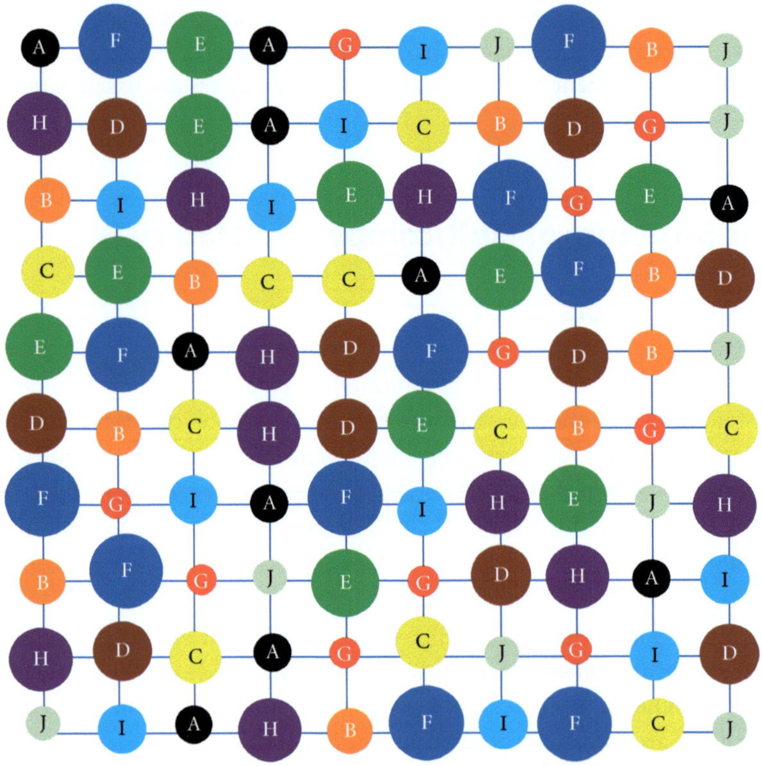

Figure 4.6 Schematic representation of ten components distributed randomly on a square lattice.
Similar to a figure from Murty et al. (2019). High-entropy alloys. Amsterdam: Elsevier.

Consider a general lattice with a single atom at each lattice point, and with n_1, n_2 and n_3 first, second and third near neighbours respectively.[iv] The *cluster size* of each atom together with its first near neighbours is $n_1 + 1$ atoms, increasing to $n_2 + n_1 + 1$ atoms including second as well as first near neighbours, and to $n_3 + n_2 + n_1 + 1$ atoms including third as well as first and second near neighbours. The number of different clusters N_1 of $n_1 + 1$ atoms, N_2 of $n_2 + n_1 + 1$ atoms, and N_3 of $n_3 + n_2 + n_1 + 1$ atoms in a multicomponent equiatomic single-phase material with a random (ideal or regular) arrangement of c components is given[27] by the law of permutations with repetition:[v]

$$N_1 = c^{n_1+1}$$

$$N_2 = c^{n_2+n_1+1}$$

$$N_3 = c^{n_3+n_2+n_1+1}$$

[iv] Note that $n_1 = z$, the coordination number.

[v] We want to calculate how many ways we can pick $n_1 + 1$ (or, including second or second and third near neighbours, $n_2 + n_1 + 1$ or $n_3 + n_2 + n_1 + 1$) atoms from a total of c components. Let the atomic sites in the cluster be labelled 1, 2, 3, etc., up to $n_1 + 1$ (or $n_2 + n_1 + 1$ or $n_3 + n_2 + n_1 + 1$). Different *sets* of the c components distributed across the different atomic sites are clearly different, but different *arrangements* of any given set of c components across the different atomic sites are also different. In other words we need an *ordered permutation* rather than a *non-ordered combination*. And we need a *permutation with repetition* since each of the different component atoms can be picked more than once.[27]

These are *orientation-dependent* cluster numbers (i.e. equivalent clusters with a different orientation are regarded as different), as is appropriate when we are dealing with properties that depend on orientation, such as dislocation slip, diffusion or magnetisation, when the material is responding to an oriented vector of applied stress, concentration gradient or magnetic field respectively. If cluster orientation is not significant, however, the number of different clusters is smaller because of the number of *self-similarity operations s* associated with the crystal symmetry:

$$N_1 = (c^{n_1+1})/s$$

$$N_2 = (c^{n_2+n_1+1})/s$$

$$N_3 = (c^{n_3+n_2+n_1+1})/s$$

with $s = 24$ for instance for a cubic crystal.[28,29] These are *orientation-independent* cluster numbers (i.e. equivalent clusters with a different orientation are regarded as the same), as is appropriate when we are dealing with properties that are independent of orientation, such as density, bulk modulus or chemical potential.

Table 4.3 shows the numbers of oriented and non-oriented local atomic clusters N_1, N_2 and N_3 out to first, second and third near neighbours respectively for multicomponent equiatomic single-phase fcc Cantor alloys with c components, taking for fcc $n_1 = 12$, $n_2 = 6$ and $n_3 = 24$, so the cluster sizes are 13, 19 and 43 atoms respectively. Table 4.4 shows equivalent numbers for multicomponent equiatomic single-phase bcc Senkov alloys, taking for bcc $n_1 = 8$, $n_2 = 6$ and $n_3 = 12$, so the cluster sizes are 9, 15 and 27 atoms respectively. The same results are shown graphically in Figures 4.7 and 4.8 for fcc and bcc respectively. It is obvious from Tables 4.3 and 4.4 and Figures 4.7 and 4.8 that the numbers of different local atomic clusters in multicomponent single-phase materials are very large indeed. Most (though not all) of the important properties of a material are determined by interactions between first and second near-neighbour atoms, and many (though not all) are orientation-dependent. The number, therefore, of oriented, first and second near-neighbour, local atomic clusters for the original single-phase fcc Cantor alloy CrMnFeCoNi with five components is, for instance, just under twenty trillion (1.9×10^{13}), and the equivalent number of local atomic clusters for the original single-phase bcc Senkov alloy also with five components is just over thirty billion (3.1×10^{10}). With non-oriented clusters and first near neighbours only, the numbers of local atomic clusters are, of course, somewhat smaller but still very large, almost fifty million (4.8×10^7) for the original fcc Cantor alloy and eighty thousand (8.0×10^4) for the original bcc Senkov alloy. With six components, the number of oriented, first and second near-neighbour, local atomic clusters increases to just over six hundred trillion (6.1×10^{14}) for fcc Cantor alloys and almost half a trillion (4.7×10^{11}) for bcc Senkov alloys; and with eight components to well over a quadrillion (1.4×10^{17}) for fcc Cantor alloys and over thirty trillion (3.5×10^{13}) for bcc Senkov alloys. The number of local atomic clusters is, of course, very much larger again, and indeed becomes astronomical, if we include third near neighbours ($>10^{30}$ or $>10^{18}$ for the 5-component Cantor and Senkov alloys respectively) and/or increase the number of components to (say) ten or more ($>10^{19}$ or $>10^{15}$ for Cantor and Senkov alloys respectively), as shown in Tables 4.3.and 4.4 and Figures 4.7 and 4.8.

Table 4.3 Number of different oriented and non-oriented local atomic clusters N_1, N_2 and N_3 out to first, second and third near neighbours respectively for a multicomponent single-phase fcc material with *c* components.

number of components *c*	oriented clusters 1st	oriented clusters 2nd	oriented clusters 3rd	non-oriented clusters 1st	non-oriented clusters 2nd	non-oriented clusters 3rd
3	1.6×10^{6}	1.2×10^{9}	3.3×10^{20}	6.4×10^{4}	4.8×10^{7}	1.3×10^{19}
4	6.7×10^{7}	2.7×10^{11}	7.7×10^{25}	2.7×10^{6}	1.1×10^{10}	3.1×10^{24}
5	1.2×10^{9}	1.9×10^{13}	1.1×10^{30}	4.8×10^{7}	7.6×10^{11}	4.4×10^{28}
6	1.3×10^{10}	6.1×10^{14}	2.9×10^{33}	5.2×10^{8}	2.4×10^{13}	1.2×10^{32}
8	5.5×10^{11}	1.4×10^{17}	6.8×10^{38}	2.2×10^{10}	6.0×10^{15}	2.7×10^{37}
10	10^{13}	10^{19}	10^{43}	4.0×10^{11}	4.0×10^{17}	4.0×10^{41}
20	8.2×10^{16}	5.2×10^{24}	8.8×10^{55}	3.3×10^{15}	2.1×10^{23}	3.5×10^{54}
50	1.2×10^{22}	1.9×10^{32}	1.1×10^{73}	4.8×10^{20}	7.6×10^{30}	4.4×10^{71}

Table 4.4 Number of different oriented and non-oriented local atomic clusters N_1, N_2 and N_3 out to first, second and third near neighbours respectively for a multicomponent single-phase bcc material with *c* components.

number of components *c*	oriented clusters 1st	oriented clusters 2nd	oriented clusters 3rd	non-oriented clusters 1st	non-oriented clusters 2nd	non-oriented clusters 3rd
3	2.0×10^{4}	1.4×10^{7}	7.6×10^{12}	8.0×10^{2}	5.6×10^{5}	3.0×10^{11}
4	2.6×10^{5}	1.1×10^{9}	1.8×10^{16}	1.0×10^{4}	4.4×10^{7}	7.2×10^{14}
5	2.0×10^{6}	3.1×10^{10}	7.5×10^{18}	8.0×10^{4}	1.2×10^{9}	3.0×10^{17}
6	1.0×10^{7}	4.7×10^{11}	1.0×10^{21}	4.0×10^{5}	1.9×10^{10}	4.0×10^{19}
8	1.3×10^{8}	3.5×10^{13}	2.4×10^{24}	5.2×10^{6}	1.4×10^{12}	9.6×10^{22}
10	10^{9}	10^{15}	10^{27}	4.2×10^{7}	4.2×10^{13}	4.2×10^{25}
20	5.1×10^{11}	3.3×10^{19}	1.3×10^{35}	2.0×10^{10}	1.3×10^{18}	5.2×10^{33}
50	2.0×10^{15}	3.1×10^{25}	7.5×10^{45}	8.0×10^{13}	1.2×10^{24}	3.0×10^{44}

The number of local atomic clusters is in all these cases, by any stretch of the imagination, very large indeed, and this has an important effect on the spatial consistency of the properties of multicomponent single-phase solid solutions such as the fcc Cantor and bcc Senkov alloys. Taking the cube root of the number of local atomic clusters gives the size of a piece of the material sufficiently large to include all possible local atomic configurations, i.e. a piece of the material big enough to average reasonably over all the different local atomic environments and, therefore, big enough to represent fully the material and its properties. The linear dimension of such a piece of material is $\sqrt[3]{1.9 \times 10^{13}} = 2.7 \times 10^{4}$ clusters or ~8 μm for oriented, first and second near-neighbour clusters in the original single-phase fcc Cantor alloy, and $\sqrt[3]{3.1 \times 10^{10}} = 3.1 \times 10^{3}$ clusters or ~1 μm in the original single-phase bcc Senkov alloy, in both cases taking the atomic separation very roughly as 0.3 nm. In other words, the properties of the original fcc Cantor and bcc Senkov alloys will vary from grain to grain if their polycrystalline grain size is below ~8 μm or ~1 μm respectively. To put it another way, the grain size needs to be above ~8 μm or ~1 μm respectively to have a material

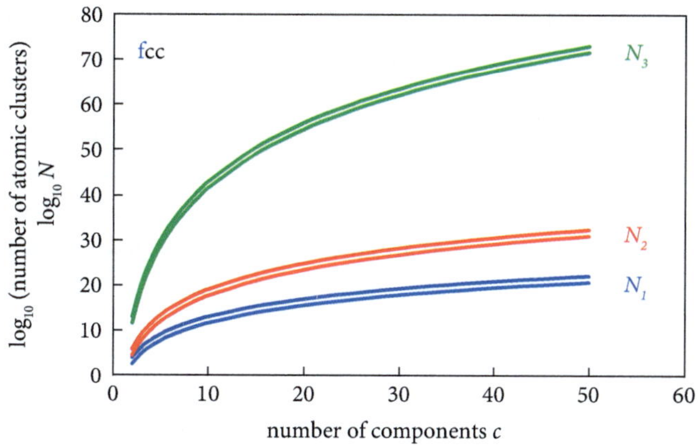

Figure 4.7 Number of different oriented (upper lines) and non-oriented (lower lines) local atomic clusters out to first (N_1), second (N_2) and third (N_3) near neighbours for a multicomponent single-phase fcc material with c components.

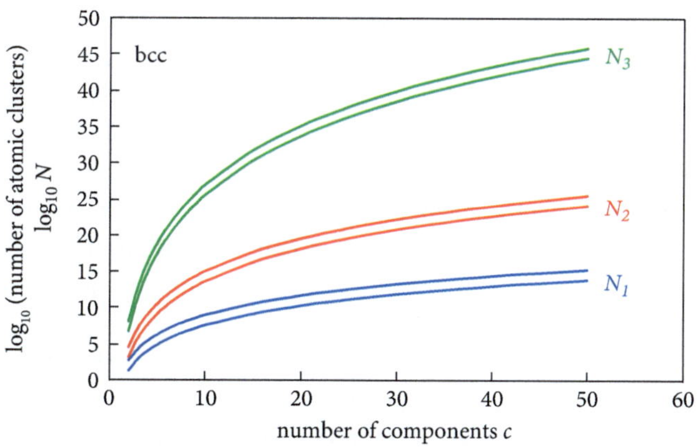

Figure 4.8 Number of different oriented (upper lines) and non-oriented (lower lines) local atomic clusters out to first (N_1), second (N_2) and third (N_3) near neighbours for a multicomponent single-phase bcc material with c components.

with consistent properties. For equiatomic 10-component single-phase fcc Cantor or bcc Senkov alloys, the polycrystalline grain size needs to be above $\sqrt[3]{10^{19}} = 2.2 \times 10^6$ clusters ≈ 0.7 mm or $\sqrt[3]{10^{15}} = 10^5$ clusters ≈ 30 μm respectively. And, if third near neighbours are important, the grain size needs to be even bigger again, above $\sqrt[3]{1.1 \times 10^{30}} = 10^{10}$ clusters $\approx$ 3 m or $\sqrt[3]{7.5 \times 10^{18}} = 2 \times 10^6$ clusters ≈ 0.6 mm for the original equiatomic 5-component single-phase fcc Cantor or bcc Senkov alloy respectively.

This is clearly a very different situation from that found in a conventional material consisting of either a single component or a single main component with one or more dilute alloying additions. The extremely large number of different local atomic environments and atomic clusters in multicomponent single-phase fcc and bcc solid solutions (such as the Cantor and Senkov alloys—and indeed in other crystal structures such as multicomponent intermetallic and ceramic compounds) plays, therefore, an important role in material properties that depend strongly on local atomic interactions, such as vacancy migration and diffusion, or dislocation slip and plastic flow, as discussed in more detail later in this chapter. The extremely large number of different local atomic environments and atomic clusters in these materials also makes it extremely challenging (if not in many cases near impossible), as discussed in some detail in Chapter 5, to determine their structure and properties with any degree of confidence by fundamental techniques such as *ab initio* molecular dynamics (MD) or quantum-mechanical modelling that are limited to no more than a few thousand atoms at best.

Lattice distortions

Causes of lattice distortion

At absolute zero, all the atoms in a single-component single-phase material are located exactly on their respective crystal lattice points.[vi] The lattice is then *undistorted.* With an undistorted lattice at absolute zero, the variation of diffracted intensity with diffraction angle $I(\theta)$ for X-rays, electrons or neutrons consists in each case of a series of perfectly sharp lines at specific Bragg angles θ_{hkl} corresponding to specific diffracting crystal planes $\{hkl\}$ that are characteristic of the particular crystal structure of the material, as shown schematically for an fcc crystal in Figure 4.9. And the *radial distribution function* or *pair-distribution function* (*RDF* or *PDF*) $\rho(r)$, derived by Fourier transformation of the intensity function $I(\theta)$, consists similarly of a set of perfectly sharp lines at specific distances r_n where $n = 1, 2, 3$, etc., corresponding to the first, second, third, etc., near-neighbour shells that are again characteristic of the particular crystal structure of the material, as shown schematically again for an fcc crystal in Figure 4.10.

A crystal lattice becomes *distorted* when one or more atoms is displaced from its lattice point. At realistic temperatures above absolute zero, the atoms in a single-component single-phase material have thermal energy and vibrate about their crystal lattice positions. As a result the lattice expands, the characteristic diffraction angles and near-neighbour distances increase, and the diffraction lines and RDF peaks broaden, approximately symmetrically, as the atoms spend time at a variety of different positions around their lattice points, as also shown in Figures 4.9 and 4.10. The atoms are still located *on average* exactly on their respective crystal lattice points, but their positions deviate stochastically with time as they oscillate elastically about their lattice points. In other words, because of thermal vibrations at realistic temperatures, a snapshot of the atomic structure of the material at any particular time shows a *stochastically distorted* lattice, with most of the atoms displaced from their lattice points, but the time-averaged structure remains virtually identical to the undistorted lattice.[vii] At the same time, thermally induced crystal defects such as vacancies and dislocations are also formed at realistic temperatures above absolute zero, creating a small number of local regions where the lattice is *permanently distorted*, with atoms in the near vicinity

[vi] Or, for a molecular crystal or a lattice with a basis, exactly on points that are all well defined in geometric relation to the lattice points.

[vii] It is not completely identical because of thermal expansion, i.e. the potential field within which the atoms vibrate is not perfectly symmetrical so there is a gradual slight increase in lattice parameter with increasing atomic vibration at increasing temperature.

of the defects deviating permanently from their respective lattice points. Permanent lattice distortions caused by atoms in the vicinity of crystal defects contribute to a further, usually small and, in this case, usually asymmetric broadening of the diffraction lines and the corresponding RDF peaks and near-neighbour shells shown in Figures 4.9 and 4.10.

Addition of solute atoms to a single-component single-phase material provides additional sources of lattice distortion because the size and electronic structure of the solute atoms are different to those of the solvent atoms. For dilute solutions, the solute atoms are widely separated and are effectively independent and non-interacting. There are then two basic effects:

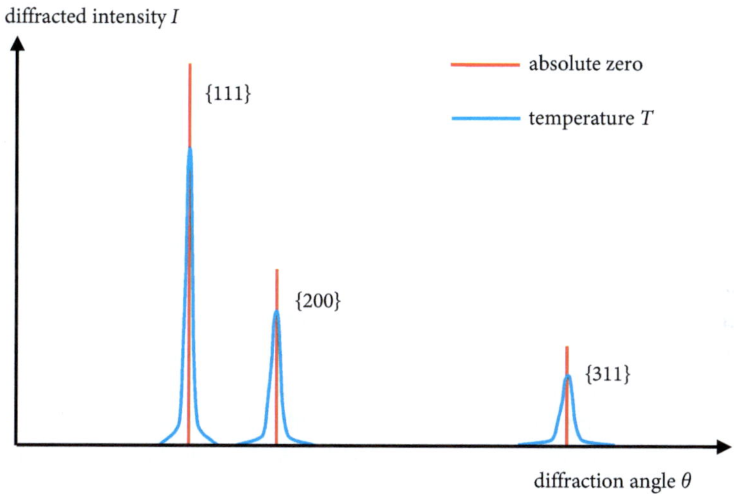

Figure 4.9 Schematic variation of diffracted intensity I versus diffraction angle θ, showing the {111}, {200} and {311} diffraction lines in an fcc crystal at absolute zero (red) and at a realistic temperature T (blue).

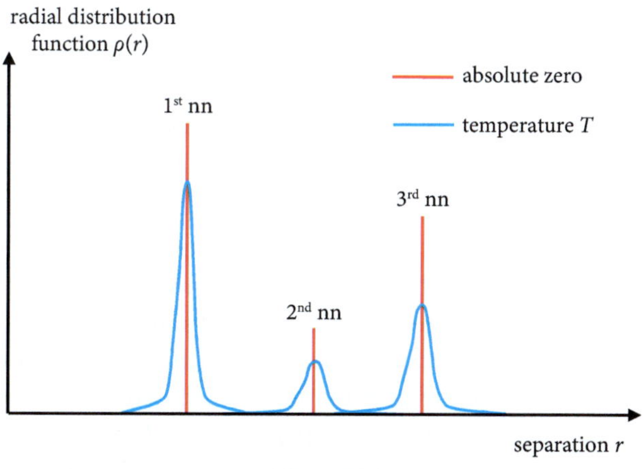

Figure 4.10 Schematic radial distribution function $\rho(r)$ for the probability variation of finding two atoms at a separation r, showing the first, second and third near-neighbour shells in an fcc crystal at absolute zero (red) and at a realistic temperature T (blue).

first, the lattice as a whole expands or contracts; and second, each solute atom provides an additional local region of permanent lattice distortion. The first effect, the lattice expansion or contraction, is often linear, and dilute binary solutions usually obey *Vegard's law*, with the lattice constant a (or constants a, b and c) proportional to the solute concentration x:

$$a \propto x = kx$$

where k is the Vegard's law constant.[30] And in multicomponent dilute solutions with two or more different solutes, the lattice expansion or contraction is usually linearly additive:[30]

$$a = \sum_i k_i x_i$$

where k_i and x_i are respectively the Vegard's law constant and the concentration of the ith solute.

Lattice distortions are more complex in concentrated multicomponent single-phase materials, because the different solute atoms interact with each other, often producing non-linear effects. The overall expansion or contraction of the crystal lattice and the variation of lattice parameter with the concentration of the different components do not obey Vegard's law so well. Alonso and Alonso[31] have compiled data for a total of seventy-nine different cases of lattice parameter measurements in concentrated multicomponent single-phase cubic alloys, and a selection of their results are shown in Table 4.5. Overall, Alonso and Alonso's data[31] show that:

(1) Measured lattice parameters deviate from Vegard's law by up to 4–5%.
(2) Deviations from Vegard's law can be positive or negative.
(3) The average deviation from Vegard's law is ±0.64% across all seventy-nine sets of measurements.
(4) The original Cantor alloy CrMnFeCoNi and the original Senkov alloy VNbMoTaW deviate relatively little from Vegard's law.
(5) Deviations from Vegard's law are somewhat smaller in multicomponent alloys restricted to transition metals only, and somewhat larger when they contain a mixture of transition and main-group elements.

As well as not obeying Vegard's law, local lattice distortions around individual solute atoms are also more complex in concentrated multicomponent single-phase materials for another reason. This is because every atom acts effectively as a centre of local lattice distortion, caused by the wide variety of local atomic configurations as discussed in the last sub-section. This is why *lattice distortion* was the second of Yeh's *four core effects* in multicomponent high-entropy materials. Every atom i is effectively displaced from its lattice point, by an atom displacement vector $\boldsymbol{d_i} = (d_{ix}, d_{iy}, d_{iz})$, where d_{ix}, d_{iy} and d_{iz} are its displacements in the three directions x, y and z; and the sum total of all the atomic displacements causes each atom to be a centre of local stress and strain, as given by local atom stress and strain tensors $\boldsymbol{\sigma_i}$ and $\boldsymbol{\epsilon_i}$:

Table 4.5 Measured a and calculated a' lattice parameters (using Vegard's law) for selected concentrated multicomponent single-phase cubic materials (after Alonso and Alonso[31]).

alloy	measured lattice parameter a (nm)	calculated lattice parameter from Vegard's law a' (nm)	deviation $100 \times (a - a')\ /\ a(\%)$
CrMnFeCoNi	0.3597	0.3594	0.083
CrFeCoNiPd	0.36803	0.36455	0.946
CrCoNiCuZn	0.28831	0.29012	−0.628
AlCrCoNiZn	0.287	0.296	−3.136
$Al_{42.2}Cr_{8.8}Fe_{14}Co_{17.2}Ni_{17.8}$	0.2882	0.3005	−4.268
VNbMoTaW	0.31832	0.31849	−0.053
TiVZrNbHf	0.33663	0.33613	0.149
TiVNbMoTa	0.31945	0.32153	−0.651
$Ti_2V_{0.5}ZrMo_{0.2}Hf$	0.34584	0.33805	2.252

$$\boldsymbol{\sigma_i} = \begin{bmatrix} \sigma_{ixx} & \sigma_{ixy} & \sigma_{ixz} \\ \sigma_{iyx} & \sigma_{iyy} & \sigma_{iyz} \\ \sigma_{izx} & \sigma_{izy} & \sigma_{izz} \end{bmatrix}$$

$$\boldsymbol{\epsilon_i} = \begin{bmatrix} \epsilon_{ixx} & \epsilon_{ixy} & \epsilon_{ixz} \\ \epsilon_{iyx} & \epsilon_{iyy} & \epsilon_{iyz} \\ \epsilon_{izx} & \epsilon_{izy} & \epsilon_{izz} \end{bmatrix}$$

where the individual stress and strain tensor components σ_{iab} and ϵ_{iab}, are individual local stresses and strains on the face of the ith atom normal to axis $a = x, y$ or z in the direction of axis $b = x, y$ or z. There are different stress and strain tensors for each atom i, so the material supports a complex and very extensive internal spatial distribution of hydrostatic and shear stresses and strains, each fluctuating throughout the material from zero at some atom positions undisplaced from their lattice points up to maximum positive and down to minimum negative values at some other atom positions displaced substantially from their lattice points.

Lattice distortions in Cantor alloys

Table 4.6 shows 12-coordinated metallic Goldschmidt atomic radii r for some of the components that have been used to manufacture multicomponent single-phase fcc Cantor alloys.[32] The maximum difference in atomic size Δr among the five different components, the corresponding *maximum atomic misfit* δ_{max} and the root-mean-square (RMS) *average atomic misfit* δ_{av} in the original Cantor alloy CrMnFeCoNi are:

$$\Delta r = r_{\mathrm{Cr}} - r_{\mathrm{Ni}} = 128 - 125\mathrm{pm} = 3\ \mathrm{pm}$$

$$\delta_{max} = (r_{\mathrm{Cr}} - r_{\mathrm{Ni}})\,/r_{\mathrm{Ni}} = (128 - 125)\,/125 = 2.4\%$$

$$\delta_{av} = \sqrt{\sum_i x_i \left(\frac{r_i - r_{av}}{r_{av}}\right)^2} = \sqrt{\sum_i 0.2\left(\frac{r_i - 126.2}{126.2}\right)^2} = 0.9\%$$

Table 4.6 Goldschmidt atomic radii *r* for components used to manufacture fcc Cantor alloys[32].

component	Goldschmidt atomic radius *r* (pm)
Cr	128
Mn	127
Fe	126
Co	125
Ni	125
Al	143
Cu	128

where summation is over all the components $i = 1$ to n, x_i is the mole fraction of the ith component ($x_i = 0.2$ for all five components in the Cantor alloy), n is the number of components ($n = 5$ for the Cantor alloy) and $r_{av} = \sum_i x_i r_i$. The values of Δr, δ_{max} and δ_{av} remain unchanged when Cu is included in a modified CrMnFeCoNiCu$_x$ Cantor alloy; but they are somewhat larger when Al instead of Cu is included in a modified Al$_x$CrMnFeCoNi Cantor alloy:

$$\Delta r = r_{\mathrm{Al}} - r_{\mathrm{Ni}} = 143 - 125 \text{ pm} = 18 \text{ pm}$$

$$\delta_{max} = (r_{\mathrm{Al}} - r_{\mathrm{Ni}})/r_{\mathrm{Ni}} = (143 - 125)/125 = 14.4\%$$

$$\delta_{av} = 1.4\%$$

for the maximum Al solubility with $x_{\mathrm{Al}} = 8\%$ and $x_{i \neq \mathrm{Al}} = 18.4\%$.

The relationship between the stability of a single-phase solid solution in a multicomponent material and the maximum atomic size difference among its components Δr or the corresponding atomic misfits δ_{max} and δ_{av} is not simple and straightforward, as explained in the discussion of multicomponent Hume-Rothery rules in Chapter 5. Nevertheless, the relatively low values of Δr, δ_{max} and δ_{av} in the original Cantor alloy CrMnFeCoNi explain in general terms why it is a near-ideal random solid solution;[12] unchanged values of Δr, δ_{max} and δ_{av} in the modified Cantor alloy CrMnFeCoNiCu$_x$ explain in general terms why the maximum solubility of Cu reaches the equiatomic level of 16.67% for a 6-component alloy (though only when the material is manufactured by melt spinning),[33] as shown in Tables 3.4 and 3.5 and Figure 3.6; and somewhat higher values of Δr, δ_{max} and δ_{av} in the modified Cantor alloy Al$_x$CrMnFeCoNi explain in general terms why the maximum solubility of Al is only 8 at%,[34] approximately half the equiatomic level of 16.67 at% for a 6-component alloy, as also shown in Tables 3.4 and 3.5 and Figure 3.6.

We might expect, therefore, to find lattice distortions on a scale of somewhat under approximately 1 pm or 1% on average, rising to a maximum of 3 pm or 3% in the original Cantor alloy, and quite a bit higher in some of the modified Cantor alloys such as those containing Al. Clearly these calculations are rather simplistic, since they ignore effects such as short-range order, non-ideality, relaxation, and electronic distortions (short-range order and electronic distortions are discussed in later sub-sections). More detailed analyses of the expected lattice distortion have also taken account of measured lattice parameters of the fcc

Cantor alloys (instead of r_{av}), have used a variety of alternative definitions of atomic misfit and lattice distortion δ_{max} and δ_{av}, and have also used *ab initio* calculations of the single-phase fcc Cantor alloy structures[35–38] (the latter are more satisfying theoretically but, as mentioned previously, are not able to cover the enormous number of different local environments). Nevertheless they all lead to similar though slightly higher predicted local lattice distortions of $\Delta r \approx 3.3–6.6$ pm.[35–38]

Lattice distortions have been measured in a variety of multicomponent solid solutions (but mainly in Cantor alloys) by a variety of different techniques, including synchrotron X-ray diffraction (XRD),[35] neutron diffraction (ND)[39] and extended X-ray absorption fine-structure analysis (EXAFS).[36,40] Measurements are difficult because of the need to separate static local atomic displacements caused by lattice distortions, dynamic displacements caused by thermal vibrations, and bulk lattice strain. Essentially, sophisticated fitting software has to be used with very high-quality diffraction data to separate out these different effects within the overall broadening of diffraction peaks such as those shown schematically in Figures 4.9 and 4.10, in order to calculate the local lattice distortions caused specifically by the variation in atom sizes, electronic structure and local cluster configurations.

Okamoto et al.[35] used a large cylindrical image-plate camera installed on the BLO2B1 multiaxial single-crystal diffractometer at the Spring-8 synchrotron facility at Sayo, Japan to obtain precision X-ray diffraction (XRD) traces at both 25 K and 300 K (–248 °C and 27 °C) from a (111)-oriented, 20-μm-long, 18-μm-diameter high-purity single-crystal specimen of the equiatomic CrMnFeCoNi Cantor alloy, made by focused ion-beam (FIB) machining after melting the alloy and drop-casting it into a cold cylindrical Cu mould, followed by cold rolling to 60% and recrystallisation for three hours at 1,100 °C (1,373 K) to produce a grain size of 70 μm. The root mean square (RMS) average atomic displacement (or distortion) in the single crystal alloy specimen was determined by refining and profiling the resulting X-ray peaks, using full-matrix, least-squares, SHELXL-97 fitting software, giving values of 4.8 and 7.7 pm within an estimated error of ±0.5 pm at 25 K and 300 K respectively, indicating an average static atomic displacement (at 25 K) of $\Delta r \approx$ 4.8 pm and an average room-temperature dynamic (i.e. thermal) displacement of 7.7–4.8 = 2.9 pm.

Owen et al.[39] used the POLARIS powder diffractometer installed on the ISIS Neutron and Muon Source at the Rutherford Appleton Laboratory (RAL) at Harwell, England to obtain high-quality total neutron diffraction (ND) traces from powder specimens of the equiatomic CrMnFeCoNi Cantor alloy, made by gas atomisation followed by heat treatment for two hours at 1,200 °C (1,473 K) to homogenise the powder particles and sinter them into a handleable specimen. Comparison results were obtained in a similar way using similar gas-atomised powder specimens from pure fcc Ni and fcc Ni-20, 25 and 33 at% Cu, and Ni-37.5 at% Cu-25 at% Cr alloys. In each case, the neutron scattering data were refined and then Fourier-transformed using GSAS, GudrunN and StoG software, to produce high-precision partial distribution functions (PDFs) as shown in Figure 4.11, with local atomic displacements (distortions) measured by Gaussian-fitted full peak-widths at half maximum height (FWHM) as shown in Figure 4.12. The largest distortions are seen in the Cantor alloy, though not by much. The differences in FWHM values between the Cantor alloy and pure Ni were attributed to static atomic distortions in the Cantor alloy of $\Delta r \approx 2$ pm, with the remaining shared component of peak width representing dynamic (thermal) distortions of

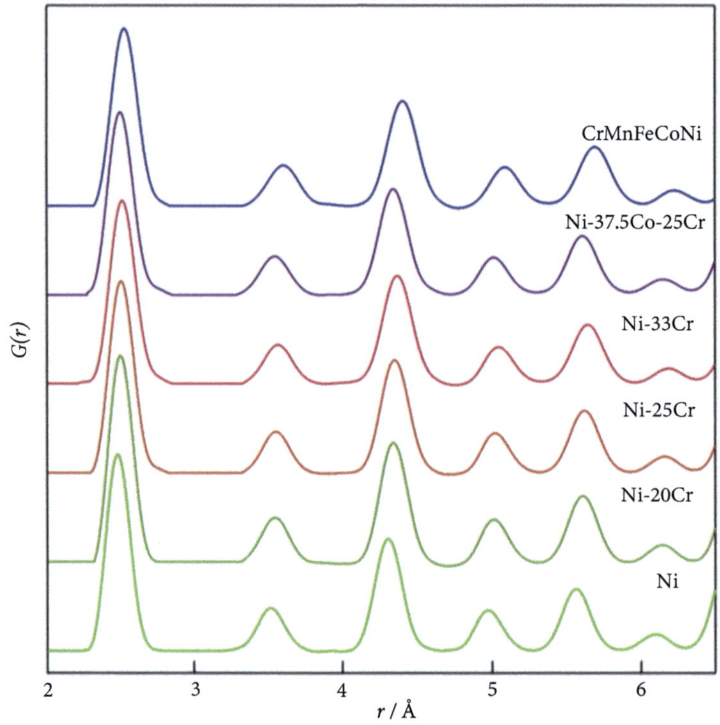

Figure 4.11 Normalised pair-distribution functions (PDFs) $G(r)$ obtained by neutron diffraction for the first six coordination shells in the fcc Cantor alloy CrMnFeCoNi, pure fcc Ni, and fcc Ni-20, 25 and 33 at%Cr, and Ni-37.5 at%Co-25 at%Cr alloys.

Reproduced from Owen et al. (2017). An assessment of the lattice strain in the CrMnFeCoNi high-entropy alloy. *Acta Materialia 122*: 11–18.

~18 pm, again claimed to be within an error of approximately ±0.5 pm, obtained from the first near-neighbour peaks as shown in Figure 4.12.

Oh et al.[36] used the 7D beamline of the 3 GeV PLS-II accelerator at Pohang, Korea, with a 1.5 × 4 mm cross-section X-ray beam and a Si (111) monochromator in transmission mode, to obtain individual elemental extended X-ray absorption fine spectra (EXAFS) spectra for each of the components Cr, Mn, Fe, Co and Ni in an equiatomic CrMnFeCoNi Cantor alloy made by vacuum induction-melting from >99.8% purity starting elements, followed by casting under vacuum, 50% rolling at 900 °C (1,173 K), homogenisation for three hours at 1,200 °C (1,473 K) under Ar, quenching into water, and finally mechanical grinding to make a 15-μm-thick ribbon sample. The resulting datasets from each element were aligned and processed, smoothing the pre-edge onset, removing instrumental background and absorption from other edges, and normalising to fit single atomic-absorption profiles, using Athena and Artemis programs from the IFEFFIT 1.2.11d software suite, for first scattering events from photoelectron waves with wave numbers in the range 30–105 nm^{-1} and interatomic distances of 0.1–0.3 nm. Mean individual elemental distortions relative to the average interatomic bond length were found to be approximately +0.1%, +0.5%, −0.1%, −0.4% and −0.1% for Cr, Mn, Fe, Co and Ni respectively, where positive and negative values represent increased and decreased bond lengths respectively, corresponding to local atomic

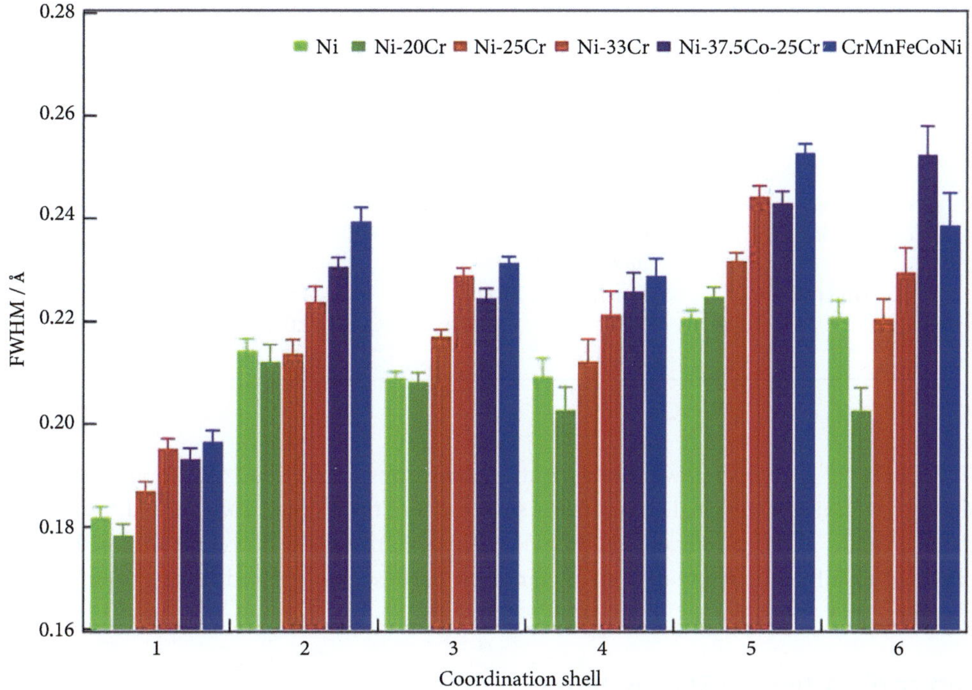

Figure 4.12 Gaussian-fitted full peak widths at half maximum height (FWHM) for the first six coordination shells, obtained from normalised pair-distribution functions (PDFs) in the fcc Cantor alloy, pure fcc Ni, and fcc Ni-20, 25 and 33 at%Cr, and Ni-37.5 at%Co-25 at%Cr alloys.
Reproduced from Owen et al. (2017). An assessment of the lattice strain in the CrMnFeCoNi high-entropy alloy. *Acta Materialia 122*: 11–18.

expansion and compression strains respectively, with individual fluctuations quoted to be an order of magnitude larger, up to a maximum of 2–3%. In other words, local distortions were greatest for Cr and Mn atoms, intermediate for Fe and Ni atoms, and smallest for Co atoms, in rough agreement with the Goldschmidt atomic radii in Table 4.6, with an overall maximum distortion of 3% of r_{av} (= 126.2 pm), i.e. $\Delta r \approx 3.8$ pm. Using very similar EXAFS methods, Tan et al.[40] measured, somewhat surprisingly, rather different mean individual elemental distortions of −0.1%, −0.2%, +0.3%, −0.9%, and +1.1% for Cr, Mn, Fe, Co and Ni respectively, correlating poorly with the component Goldschmidt atomic radii (unlike Oh et al.'s results[36]), and with no indication of the maximum size of the measured distortions.

The different experimental and calculated results for the maximum magnitude of local atomic distortions Δr in the equiatomic fcc Cantor alloy CrMnFeCoNi are collected together in Table 4.7. In summary, we can conclude that there are small but significant local lattice distortions in the equiatomic fcc Cantor alloy CrMnFeCoNi that fluctuate randomly over very large distances (because of the enormous number of different local atomic cluster configurations, as discussed in the previous sub-section), with an average value of just under 1 pm (corresponding to just under ~1% strain), reaching a maximum value of ~2–6 pm (corresponding to ~2–6% strain). Similar measurements and calculations indicate, as expected, somewhat smaller lattice distortions in simpler Cantor alloy compositions such as

Table 4.7 Experimental and calculated values for the maximum local atomic distortion Δr in the equiatomic fcc Cantor alloy CrMnFeCoNi, obtained by a variety of different methods.

	method	maximum local atomic distortion Δr (pm)
calculations	Goldschmidt radii[4]	3.0
	ab initio[35–38]	3.3–6.6
experiments	synchrotron XRD[35]	4.8
	ND[39]	2.0
	EXAFS[36]	3.8

CrFeCoNi,[37,40–42] but somewhat larger lattice distortions in modified Cantor alloy compositions containing additional components such as VFeCoNi[43] and FeNiCoCrPd.[42] Overall, the fcc Cantor alloys clearly exhibit significant local lattice distortions and strains that are expected to influence many of their material properties, as discussed further later in this chapter.

Lattice distortions in Senkov alloys

Table 4.8 shows 12-coordinated metallic Goldschmidt atomic radii r for some of the components that have been used to manufacture multicomponent single-phase bcc Senkov alloys.[32] The maximum difference in atomic size Δr among the five different components, the corresponding *maximum atomic misfit* δ_{max} and the root mean square (RMS) *average atomic misfit* δ_{av} in the original Senkov alloy VNbMoTaW are:

$$\Delta r = r_{\mathrm{Nb}} - r_{\mathrm{V}} = 145 - 135 \text{ pm} = 10 \text{ pm}$$

$$\delta_{max} = \left(r_{\mathrm{Nb}} - r_{\mathrm{V}}\right)/r_{\mathrm{V}} = (143 - 135)/135 = 7.4\%$$

$$\delta_{av} = \sqrt{\sum_i x_i \left(\frac{r_i - r_{av}}{r_{av}}\right)^2} = \sqrt{\sum_i 0.2\left(\frac{r_i - 141}{141}\right)^2} = 3.5\%$$

The values of Δr, δ_{max} and δ_{av} remain unchanged when either Al or Ti are included in modified Al_xVNbMoTaW or Ti_xVNbMoTaW Senkov alloys respectively; they are somewhat larger when either Zr or Hf are included in modified VNbZr_xMoTaW or VNbMoHf_xTaW Senkov alloys respectively:

$$\Delta r = r_{\mathrm{Zr}} - r_{\mathrm{V}} = 155 - 135 \text{ pm} = 20 \text{ pm}$$

$$\delta_{max} = \left(r_{\mathrm{Zr}} - r_{\mathrm{V}}\right)/r_{\mathrm{V}} = (155 - 135)/135 = 14.8\%$$

$$\delta_{av} = 4.8\%$$

for an equiatomic 6-component alloy with $x_i = 16.67\%$.

The relationship between the stability of a single-phase solid solution in a multicomponent material and the maximum atomic size difference among its components Δr or the corresponding atomic misfits δ_{max} and δ_{av} is not simple and straightforward, as mentioned above

Table 4.8 Goldschmidt atomic radii r for components used to manufacture bcc Senkov alloys[32].

component	Goldschmidt atomic radius r (pm)
V	135
Nb	145
Mo	145
Ta	145
W	135
Al	143
Ti	140
Zr	155
Hf	155

in the discussion of lattice distortions in Cantor alloys, and as explained in more detail in the discussion of multicomponent Hume-Rothery rules in Chapter 5. Nevertheless, the somewhat higher values of Δr, δ_{max} and δ_{av} in the original Senkov alloy VNbMoTaW (compared with the original Cantor alloy CrMnFeCoNi) explain in general terms why Senkov alloys are not usually found to be ideal random solid solutions and, instead, often exhibit some short-range order;[44] unchanged values of Δr, δ_{max} and δ_{av} in the modified Senkov alloys Al_xVNbMoTaW and Ti_xVNbMoTaW explain in general terms why the maximum solubility of Al and Ti both reach the equiatomic level of 16.67% for a 6-component alloy,[10,45,46] as shown in Table 3.6 and Figure 3.13; and, somewhat surprisingly, still higher values of Δr, δ_{max} and δ_{av} in the modified Senkov alloys $VNbZr_x$MoTaW and $VNbMoHf_x$TaW do not seem to prevent the maximum solubility of Zr and Hf from both also reaching the equiatomic level of 16.67 at% for a 6-component alloy[10,45,45], as also shown in Table 3.6 and Figure 3.13.

We might expect, therefore, to find lattice distortions on a scale of approximately 4 pm or 3% on average rising to a maximum of 7–8 pm or 10% in the original Senkov alloy, and still higher in some of the modified Senkov alloys such as those containing Zr or Hf. Clearly these calculations are again rather simplistic, since they ignore effects such as short-range order, non-ideality, relaxation, and electronic distortions (short-range order and electronic distortions are discussed in later sub-sections). More detailed analyses of the expected lattice distortion have again used more fundamental and *ab initio* calculations of the single-phase bcc Senkov alloy structures, leading to similar predicted local lattice distortions of $\Delta r \approx 5$–15 pm.[37]

Zou et al.[47] used a conventional Oxford Instruments Xcalibur single-crystal X-ray diffractometer with a fine-focus Mo Kα radiation source and CCD detector to obtain precision X-ray diffraction (XRD) traces from a 40-μm-sized single crystal of the quaternary equiatomic bcc Senkov alloy NbMoTaW, extracted by focused ion-beam (FIB) milling from a sample made by arc-melting high-purity elemental powders under Ar with a Ti getter to remove oxygen, quadruple re-melting to ensure homogeneity, casting on a copper chill mould, and further homogenising for seven days at 1,800 °C (2,073 K) under Ar in a sealed Ta ampoule. The root mean square (RMS) average atomic displacement (or distortion) in the single crystal alloy specimen was determined by refining and profiling the resulting X-ray peaks, using full-matrix, least-squares, SHELXL-97 fitting software, giving a static distortion of $\Delta r \approx 7.4$

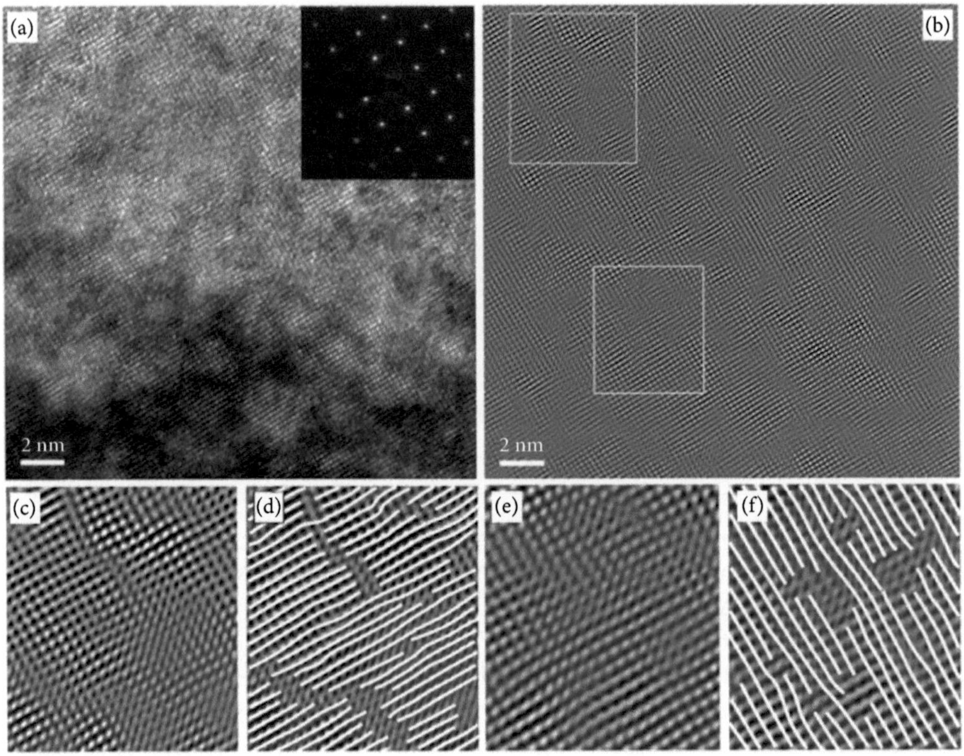

Figure 4.13 HRTEM images of the equiatomic NbMoTaW Senkov alloy: (a) bright-field (BF) image with [100] zone axis; (b) corresponding inverse fast-Fourier transform; (c) and (e) enlarged images of the boxes in (b); and (d) and (f) traces corresponding to (c) and (e) to show the lattice distortions.

Reproduced from Zou et al. (2014). Size-dependent plasticity in an $Nb_{25}Mo_{25}TA_{25}W_{25}$ refractory high-entropy alloy. *Acta Materialia 65*: 85–97.

pm, with an associated dynamic (thermal) distortion of 6.1 pm, within an error of ±0.5 pm. Lattice distortions on a scale of tens of pm were also observed directly in the same alloy by high-resolution transmission electron microscopy (HRTEM), with [100] zone axis images containing distorted {110} sets of lattice planes, as shown in Figure 4.13.

Guo et al.[48] used high-energy synchrotron X-rays with a wavelength of 12.488 pm from the beamline ID6-C at the Advanced Photon Source at the US Argonne National Lab in Illinois to obtain transmission X-ray diffraction (XRD) traces from a sputtered 1.5-μm-thick sample of the equiatomic modified ternary Senkov alloy ZrNbHf on a 1.5-mm-thick silica substrate, and the high-intensity powder diffractometer (HIPD) at the US Los Alamos National Lab in New Mexico to obtain time-of-flight (TOF) neutron diffraction (ND) traces, from a fine-grained 0.5-mm-cube sample of the same alloy made by arc-melting and casting in Ar with a Ti getter, followed by electrical-discharge slicing and then sealing in a V container. The resulting X-ray and neutron diffraction results were processed with a pdfgetN software package to obtain radial distribution functions (RDFs), which showed an overlap of the first and second near-neighbour peaks, corresponding to local lattice distortions of approximately 9.5 pm. Similar combined X-ray and neutron diffraction results

Table 4.9 Experimental and calculated values for the maximum local atomic distortion Δr in equiatomic bcc Senkov alloys, obtained by a variety of different methods.

	method	alloy	maximum local atomic distortion Δr (pm)
calculations	Goldschmidt radii[4]	VNbMoTaW AlVNbMoTaW TiVNbMoTaW	10.0
	Goldschmidt radii[4]	VNbZrMoTaW VNbMoHfTaW	20.0
	ab initio[37]	TiVNb AlTiVNb TiZrNbTaHf	5–15
experiments	XRD[47]	NbMoTaW	4.8
	synchrotron XRD & ND[48]	ZrNbHf	9.5
	synchrotron XRD & ND[49]	ZrNbTaHf	13.7

on an equiatomic quaternary modified Senkov alloy ZrNbTaHf showed static and dynamic (thermal) lattice distortions of 13.7 and 2.4 pm respectively.[49]

The different experimental and calculated results for the maximum magnitude of local atomic distortions Δr are collected together in Table 4.9. In summary, we can conclude that there are significant local lattice distortions in the bcc Senkov alloys, considerably larger than those found in the fcc Cantor alloys, that again fluctuate randomly over very large distances (because of the enormous number of different local atomic cluster configurations, as discussed in the previous sub-section), with average values of approximately 4 pm (corresponding to ~3% strain), reaching maximum values of ~5–20 pm (corresponding to >5% strain). The bcc Senkov alloys, like the fcc Cantor alloys clearly exhibit significant local lattice distortions and strains that are expected to influence many of their material properties, as discussed further later in this chapter.

Lattice distortions in multicomponent mono-oxides

Table 4.10 shows ionic radii[50] for some of the metallic cations that have been used to manufacture multicomponent single-phase NaCl-type rock-salt mono-oxide ceramics MO. The maximum difference in ionic size Δr among the five different metallic cation components, the corresponding *maximum atomic misfit* δ_{max} and the root mean square (RMS) *average atomic misfit* δ_{av} in the multicomponent single-phase NaCl-type mono-oxide (MgCoFeNiCu)O are:

$$\Delta r = r_{\mathrm{Zn}} - r_{\mathrm{Co}} = 88 - 79\mathrm{pm} = 9\ \mathrm{pm}$$

$$\delta_{max} = \frac{r_{\mathrm{Zn}} - r_{\mathrm{Co}}}{r_{\mathrm{Co}}} = \frac{88-79}{79} = 11.4\%$$

$$\delta_{av} = \sqrt{\sum_i x_i \left(\frac{r_i - r_{av}}{r_{av}}\right)^2} = \sqrt{\sum_i 0.2\left(\frac{r_i - 84.6}{84.6}\right)^2} = 3.9\%$$

Table 4.10 Bivalent ionic radii r for components used to manufacture multicomponent single-phase rock-salt mono-oxides.[50]

component	bivalent ionic radius r (pm)
Mg	86
Co	79*
Ni	83*
Cu	87
Zn	88

(*low-spin non-magnetic values)

though these values are reduced to $\Delta r = 6$ pm, $\delta_{max} = 7.5\%$ and $\delta_{av} = 0.6\%$ respectively for aligned ferromagnetic electronic spins on the Co and Ni cations. Yet again, the relatively low values of Δr and δ in the single-phase multicomponent mono-oxide ceramic (MgCoFeNiCu)O explain in general terms why, like the original Cantor alloy, it is a near-ideal random solid solution.[51] Yet again, multicomponent interatomic and ceramic compounds such as (MgCoFeNiCu)O clearly exhibit significant local lattice distortions and strains that are expected to influence many of their material properties, as discussed further later in this chapter.

More details of experimental and modelling studies of local lattice distortions and associated atomic strains are given in the reviews by Song et al.[37] and Owen and Jones.[52]

Short-range order

Causes of short-range order

In an ideal or regular solid solution, the atoms are distributed at random on the lattice, and the probability of finding an atom of one of the components at any particular lattice point is proportional to its atom or molar fraction, independent of the surrounding environment, i.e. independent of its neighbouring atoms. In a non-regular solid solution, however, this is not the case. The probability of finding an atom of one of the components at any particular lattice point is not now proportional to its atom or molar fraction and is not independent of its neighbouring atoms.

To be more specific, the probability p_{ij}^{n} of two atoms of components i and j being nth near neighbours in a solid solution is given by:

$$\text{ideal or regular: } p_{ij}^{n} = x_i x_j$$

$$\text{non-regular: } p_{ij}^{n} \neq x_i x_j$$

where x_i and x_j are the molar fractions of the ith and jth components respectively.[viii] In other words, non-regular solid solutions exhibit short-range order, with an increased probability for either association or disassociation of any particular pair of component atom types

[viii] Notice that this applies to pairs of like atoms when $i = j$ as well as unlike atoms when $i \neq j$.

i and j in any particular near-neighbour shell n. The increased and decreased probabilities are, of course, driven by the non-zero pairwise, three-way, four-way, etc., interaction energies between the different atoms ω_{ij}, ψ_{ijk}, χ_{ijkl}, etc. that contribute to the heat of mixing in a non-regular solution, and the corresponding deviation from a random distribution of atoms. Instead of using the probabilities p_{ij}^n themselves, short-range order effects are usually quantified[53] by the *Warren-Cowley short-range order parameters* (or *short-range order parameters* or just *SRO parameters*) α_{ij}^n:

$$\alpha_{ij}^n = 1 - \frac{p_{ij}^n}{x_i x_j}$$

In an ideal or regular solution, the short-range-order parameters α_{ij}^n are all zero; and in a non-regular solution they are either positive or negative for, respectively, the disassociation or association of atoms of components i and j as nth near neighbours.

Short-range order is fairly straightforward in a binary A-B alloy. To maintain the overall composition for alloys with a tendency towards *demixing*, negative SRO parameters α_{AA}^1 and α_{BB}^1 for first near-neighbour like-atoms AA and BB are balanced by a positive SRO parameter α_{AB}^1 for first near-neighbour unlike-atoms AB; these are in turn balanced by somewhat smaller positive SRO parameters α_{AA}^2 and α_{BB}^2 for second near-neighbour like-atoms AA and BB, and a somewhat smaller negative SRO parameter α_{AB}^2 for second near-neighbour unlike-atoms AB; and these are again in turn balanced by even smaller negative SRO parameters α_{AA}^3 and α_{BB}^3 for third near-neighbour like-atoms AA and BB, and an even smaller positive SRO parameter α_{AB}^3 for unlike-atoms AB; and so on. And vice versa, to maintain the overall composition for alloys with a tendency towards *mixing*, a negative SRO parameter α_{AB}^1 for first near-neighbour unlike-atoms AB is balanced by positive SRO parameters α_{AA}^1 and α_{BB}^1 for first near-neighbour like-atoms AA and BB; these are in turn balanced by a somewhat smaller positive SRO parameter α_{AB}^2 for second near-neighbour unlike-atoms AB and somewhat smaller negative SRO parameters α_{AA}^2 and α_{BB}^2 for second near-neighbour like-atoms AA and BB; and these are again in turn balanced by an even smaller negative SRO parameter α_{AB}^3 for unlike-atoms AB, and even smaller positive SRO parameters α_{AA}^3 and α_{BB}^3 for third near-neighbour like-atoms AA and BB; and so on. In each case, there are waves of alternating positive and negative SRO parameters radiating from one near-neighbour shell to the next, at their strongest in the first near-neighbour shell, and decaying gradually and dying out as the number of atoms increases and the interaction energies weaken in higher-order near-neighbour shells.

In a binary alloy, a tendency towards demixing with negative like-atom short-range order, $\alpha_{AA}^1 < 0$ and $\alpha_{BB}^1 < 0$, is called *short-range clustering*, and is typical of an alloy at temperatures not too far above a critical point below which the demixing interaction energies become dominant, leading to *phase separation* or *precipitation* into two phases, one A-rich and the other B-rich. And in a binary alloy, vice versa, a tendency towards mixing with negative unlike-atom short-range order, $\alpha_{AB}^1 < 0$, is called *short-range ordering*, and is typical of an alloy at temperatures not too far above a critical point below which its mixing interaction energies become dominant, leading to an *order-disorder transformation* and the formation of a fully ordered stoichiometric intermetallic or ceramic compound A_xB_y, with one or more different sub-lattices for the A and B atoms.

Short-range order is more complex (as is everything else, of course) in multicomponent solid solutions because of the large number of different SRO parameters corresponding to a large number of different interaction energies for each near-neighbour shell. In a binary alloy, there are only pairwise interactions to consider and, therefore, only pairwise short-range order effects and pairwise deviations from a random distribution; and there are only three pairwise interactions to be considered, namely AA, BB and AB in an A-B binary alloy. As just explained, interaction energies that promote association of like-atoms AA and BB in a particular near-neighbour shell inevitably promote the opposite and balancing disassociation of unlike-atoms AB in that shell; similarly, interaction energies that promote disassociation of like-atoms AA and BB in a particular near-neighbour shell inevitably promote the opposite and balancing association of unlike-atoms AB in that shell. In a multicomponent alloy, on the other hand, there are many interaction energies and many short-range-order parameters, with a complex interplay of the different interaction energies and a correspondingly complex interweaving of different associations and disassociations of the different component atoms in each near-neighbour shell. As with so many other properties of multicomponent solid solutions, this multiplicity of defining parameters and corresponding effects causes considerable difficulties with both measuring and modelling of short-range order in particular materials of interest.

There have been various attempts[54–56] to develop more generic short-range-order parameters in order to capture these multiple interactions and corresponding multiple short-range order effects, but in general this has proved difficult to do in any reasonably simple and useful way. So-called *general multicomponent short-range order parameters* (or *GM-SRO parameters*) are typically of the form:[55,56]

$$\beta^{n}_{ijk\ldots|pqr\ldots} = 1 - \sum_{i,p} \frac{p^{n}_{ijk\ldots|pqr\ldots}}{x_i x_p}$$

where $\beta^{n}_{ijk\ldots|pqr\ldots}$ and $p^{n}_{ijk\ldots|pqr\ldots}$ are the GM-SRO parameter and probability respectively for finding one or other atoms of type i, j, k, etc., and one or other of atoms of type p, q, r, etc., as nth near neighbours. Such GM-SRO parameters are defined so as to capture the potential for pairwise co-association of different groups of atoms rather than individual atoms in particular near-neighbour relationships. There are, however (as can be seen by inspection), very many GM-SRO parameters for any given multicomponent material, and they are also (as can again be seen by inspection) quite complex to use. Even then, they still only deal with pairwise interactions, and fail to capture the potential for any three-way, four-way, etc. interactions and corresponding triple, quadruple, etc. associations of similar or different atoms and their related short-range order effects.

It is important to realise that all multicomponent materials (in fact, all materials) have the potential to exhibit short-range order depending on the temperature and the manufacturing conditions. The structure of a material is determined by its molar Gibbs free energy G at any temperature T as given in the last section by[1–5].

$$\begin{aligned} G = \sum_i x_i \mu_i^{ss} = \sum_i x_i \mu_i + \Delta G_{mix} &= \sum_i x_i \mu_i + \Delta H_{mix} - T\Delta S_{mix} \\ &= \sum_i x_i \mu_i + \Delta H_{mix}\left(\omega_{ij}, \psi_{ijk}, \chi_{ijkl} \ldots\right) + RT \sum_i x_i \ln x_i \\ &\quad - T\Delta S^{xs}_{mix}\left(\omega_{ij}, \psi_{ijk}, \chi_{ijkl} \ldots\right) \end{aligned}$$

where x_i is the molar fraction of the i'th component, μ_i and μ_i^{ss} are, respectively, the *chemical potential* (or *molar free energy*) of the i'th component before and after mixing to form the solution, ΔG_{mix}, ΔH_{mix} and ΔS_{mix} are, respectively, the *free energy of mixing*, *heat (or enthalpy) of mixing* and *entropy of mixing*, and ω_{ij}, ψ_{ijk}, χ_{ijkl} etc. are pairwise, three-way, four-way etc. *interaction energies* between the different atoms. The first term in the last expression on the right hand side is the free energy of the unmixed components, the second term is the heat of mixing as a function of the interaction energies $\Delta H_{mix}\left(\omega_{ij}, \psi_{ijk}, \chi_{ijkl}\ldots\right)$, the third term is the ideal or regular (random) configurational entropy of mixing $RT\sum_i x_i \ln x_i$, and the last term is the excess (additional non-ideal, non-regular, non-random) entropy of mixing as a function of the interaction energies $\Delta S^{xs}_{mix}\left(\omega_{ij}, \psi_{ijk}, \chi_{ijkl}\ldots\right)$. The entropy can be ignored at very low temperatures: the Gibbs free energy is then dominated by the heat of mixing and the interaction energies between the atoms, and the equilibrium structure is either a fully ordered stoichiometric compound or a fully demixed combination of pure components and/or stoichiometric compounds. On the other hand, the interaction energies can be ignored at very high temperatures, the Gibbs free energy is then dominated by the configurational entropy of mixing, and the equlibrium structure is a fully disordered regular solution, solid or liquid depending on whether the melting point is respectively high or low. The transition between these two regimes takes place at some intermediate critical temperature for ordering or demixing, with a temperature range just above the critical temperature where the interaction energies cannot be ignored completely and the solution exhibits short-range order. With very strong interaction energies, the critical temperature is above the melting point and the equilibrium solid structure consists at all temperatures of a mixture of pure components and stoichiometric compounds; with lower interaction energies, the critical temperature is below the melting point and the equilibrium solid structure consists of a mixture of pure components and stoichiometric compounds at low temperatures and a solid solution at higher temperatures, with short-range order just above the critical temperature, but fully disordered near the melting point; and with sufficiently weak interaction energies, the critical temperature is well below the temperature at which diffusion can take place, and a (non-equilibrium) disordered or short-range ordered solid solution can often then be retained at low temperatures, depending on the cooling rate during manufacture.

The long-standing traditional method for measuring short-range order in conventional materials has been by deconvoluting X-ray and neutron diffraction (XRD and ND) traces to remove other peak-broadening effects in a similar way to that described in the last sub-section for measuring local lattice distortions. However, as with trying to measure local lattice distortions, the deconvolution and fitting procedures required to extract short-range order effects and determine SRO parameters become rapidly much more arduous and uncertain as the number of components increases. An alternative and, therefore, popular method of measuring short-range order in multicomponent materials has increasingly been to use, instead, direct composition measurements from streams of individual atoms obtained by *atom-probe tomography* (AP), in which the atoms are field-emitted progressively and one at a time by electrical pulses from a sharp needle-shaped sample of the material to be investigated, with their nature individually determined by TOF mass spectroscopy and their position within the sample simultaneously imaged with a position-sensitive (PoS) crystal detector. Alongside the difficulties of making effective direct *measurements* of short-range order in multicomponent materials, *theoretical* approaches using *ab initio* and other fundamental calculations have their own difficulties, hampered by the need to limit calculations to no more than a few thousand or well below the number needed to capture the full range of local atomic clusters, as discussed in detail in an earlier sub-section.

Short-range order in Cantor alloys

Laurent-Brocq et al.[12] obtained atom-probe (AP) composition measurements using a Cameca Atom Probe with an electrical pulse rate of 30 kHz and a pulse rate (average number of atoms emitted/pulse) of 20%, from regions typically 10 × 100 nm in size, i.e. containing between one and two million atoms in each measured region, from samples held at 80 K of the original single-phase fcc Cantor alloy CrMnFeCoNi made by suction-casting and then homogenisation by annealing for six hours at 1,100 °C (1,373 K). Figure 4.14 shows an example of the resulting distributions of each of the five component atoms in part of one

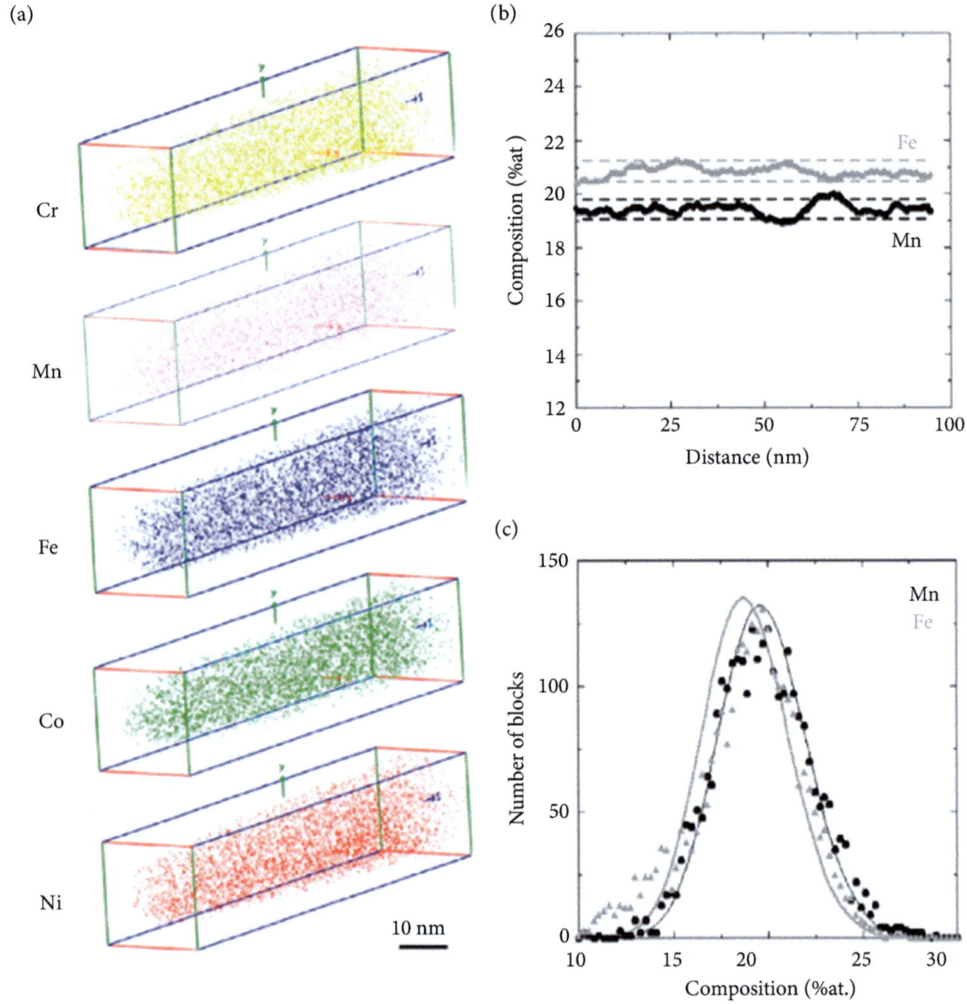

Figure 4.14 AP composition measurements from the original Cantor alloy CrMnFeCoNi: (a) individual distributions of Cr, Mn, Fe, Co and Ni atoms; (b) variation of numbers of Mn and Fe atoms with distance, showing fluctuations within expected random error bars; and (c) distribution of Mn and Fe compositions from nm blocks fitted to random binomial functions.

Reproduced from Laurent-Brocq et al. (2015). Insights into the phase diagram of the CrMnFeCoNi high-entropy alloy. *Acta Materialia 88*: 355–65.

measured region. Histograms of the molar fractions measured in individual nm-scale blocks of the material were found to fit accurately to a random binomial function for all five of the component atom types, as shown in Figure 4.14 for Mn and Fe, indicating that there was no short-range order in the material, with a completely homogeneous distribution of all five component atoms. Similar complete homogeneity and an absence of any short-range order have been measured again by Muniandy et al.[57] using similar AP techniques in well-annealed samples of the equiatomic 5-component single-phase fcc Cantor alloy, re-confirming that it has essentially a near-ideal random solid-solution structure. A similar homogeneity, lack of short-range order, and random (near-ideal) atomic distribution was also found using similar AP measurements[58] in non-equiatomic single-phase fcc modified Cantor alloys with reduced Co and Cr contents: $Cr_2Mn_{26}Fe_{40}Co_6Ni_{26}$ and $Cr_2Mn_{27}Fe_{41}Co_6Ni_{24}$.

Interestingly, however, a combination of extended X-ray absorption fine spectra (EXAFS) results[59] and analysis of streaking on electron diffraction patterns[60] indicates that the simpler equiatomic 3-component single-phase fcc modified Cantor alloy CrCoNi is not so homogeneous, and contains small ~1-nm-sized domains of short-range ordering, as shown in Figure 4.15, with enhanced Cr-Co and Cr-Ni and depleted Cr-Cr pairs in the first near-neighbour shell, i.e. with negative α^1_{CrCo} and α^1_{CrNi}, and positive α^1_{CrCr}.

This behaviour and the difference between the 5- and 3-component alloys is thought to be caused by differences in magnetic interactions between the component Cr, Mn,

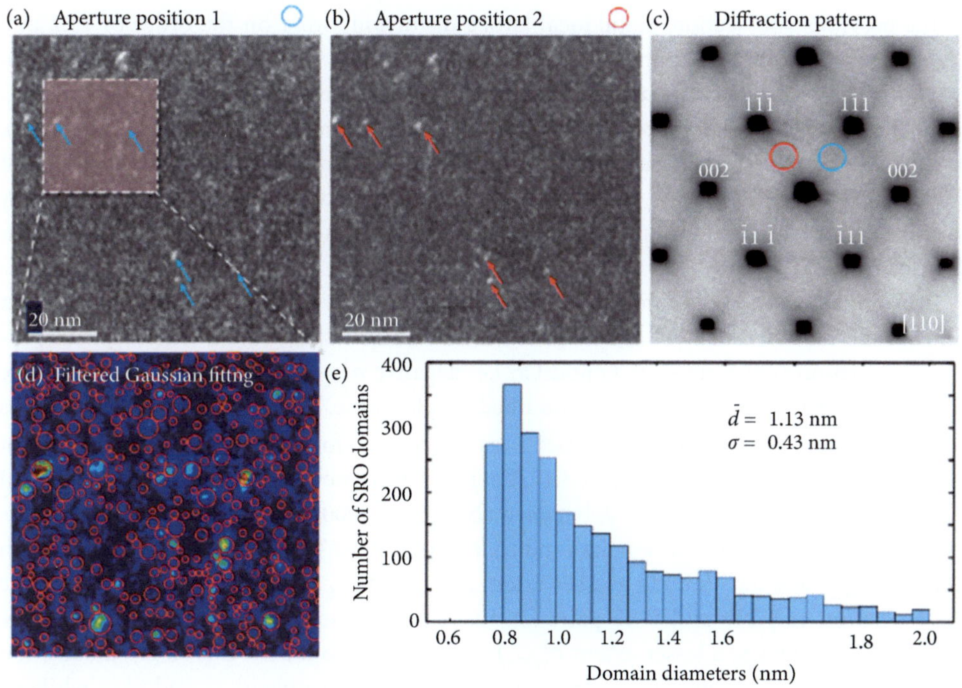

Figure 4.15 HRTEM of modified Cantor alloy CrCoNi after homogenisation: (a) and (b) filtered dark-field images using red and blue selected area diffraction apertures respectively; (c) diffraction pattern; (d) filtered Gaussian fitting showing nm-sized domains; (e) domain size distribution.

Reproduced from Zhang et al. (2020). Short-range order and its impact on the CoCrNi medium-entropy alloy. *Nature 581*: 283–7.

Fe, Co and Ni atoms.[60] Some modified Cu-containing single-phase fcc Cantor alloys,[61–63] such as $CrFeCoNiCu_{0.5}$, $Al_{0.3}CrFeCoNiCu_{0.5}$ and $Al_{0.3}CrFeNi_2Cu$, exhibit short-range clustering of Cu atoms, i.e. there is a negative α^1_{CuCu} and positive α^1_{XCu} where X = Cr, Fe, Co and Ni, that gradually builds up with annealing time to form small spherical nm-sized Cu-rich clusters. *Ab initio* modelling broadly confirms these experimental results for the presence and/or absence of short-range order in different fcc Cantor alloy compositions.[57–63]

Short-range order in Senkov alloys

As discussed in the previous sub-section, single-phase bcc Senkov alloys exhibit somewhat larger lattice distortions than single-phase fcc Cantor alloys; and, similarly, they are found to exhibit somewhat stronger short-range order effects. Maiti and Steurer[49] investigated short-range order in the equiatomic quaternary single-phase bcc modified Senkov alloy ZrNbTaHf made by arc-melting, remelting five times to ensure homogeneity, chill-casting and then annealing for various times up to eight days at 1,800 °C (2,073 K), using a combination of: single-crystal synchrotron X-ray diffraction (XRD) with 68.36-pm wavelength X-rays and a PILATUS 2M detector on the Swiss–Norwegian beamline (SNBL) at the European Synchrotron Research Facility (ESRF) in Geneva, Switzerland; powder ND with 115.45-pm-wavelength thermal neutrons on the HRPT diffractometer at the Paul Scherrer Institute in Villigen, Switzerland; atom-probe (AP) tomography on a LEAP 4000X-HR with laser pulsing at 80–100 pJ/pulse at 160 kHz; and high-resolution transmission electron microscopy (HRTEM) on an FEI Tecnai F30 operating at 300 keV. As mentioned in the previous sub-section, static local lattice distortions were on the order of 13.7 pm, and this was shown by AP and HRTEM to be associated with short-range clustering of Zr and Hf atoms, i.e. negative α^1_{ZrZr}, α^1_{HfHf} and α^1_{ZrHf} and corresponding positive α^1_{ZrNb}, α^1_{ZrTa}, α^1_{NbHf} and α^1_{TaHf} (in other words negative $\beta^1_{ZrHf|ZrHf}$ and corresponding positive $\beta^1_{ZrHf|NbTa}$), which was not present in the as-cast alloy, but built up progressively with annealing time at 1,800 °C (2,073 K), beginning with small nm-sized regions of short-range order and finally forming a three-dimensional phase-separated network on a scale of ~10 nm, as shown in Figure 4.16.

In a similar way, AP atomic composition measurements on magnetron-sputtered thin films of the modified Senkov alloy $C_xTiCrNbTaW$ exhibit short-range ordering with positive association (clustering) of Ti and Cr, i.e. negative α^1_{TiTi}, α^1_{CrCr} and α^1_{TiCr} (or negative $\beta^1_{TiCr|TiCr}$).[64] *Ab initio* modelling of modified single-phase bcc Senkov alloys also shows significant short-range ordering effects,[44,65] as can be seen for the equiatomic 5-component Senkov alloy VNbMoTaW[44] in Figure 4.17. There is little or no short-range order above about 1,000 °C (1,273 K) with all $\alpha^1_{ij} \approx 0$; between about 500–1000 °C (773–1273 K), short-range association of W-Nb and W-V pairs (and to a lesser extent Ta-Nb and Mo-Nb pairs) with negative α^1_{WNb} and α^1_{WV} (and to a lesser extent α^1_{TaNb} and α^1_{MoNb}) is balanced by short-range disassociation of V-Nb and V-Ta pairs (and to a lesser extent W-Mo pairs) with positive α^1_{VNb} and α^1_{VTa} (and to a lesser extent α^1_{WMo}); and below about 500 °C (773 K), association of W-Nb and W-V becomes even stronger with strongly negative α^1_{WNb} and α^1_{WV}, while all other pairs become strongly disassociated with all other SRO parameters α^1_{ij} strongly positive, probably caused by the formation of a phase-separated structure at these low temperatures. Similar behaviour is seen in *ab initio* modelling of other Senkov alloys, such as VMoTaW[44] and MoNbTaW.[65] On the other hand, AP atomic composition measurements in the as-cast equiatomic quaternary single-phase bcc modified Senkov alloy AlTiVCr exhibited random

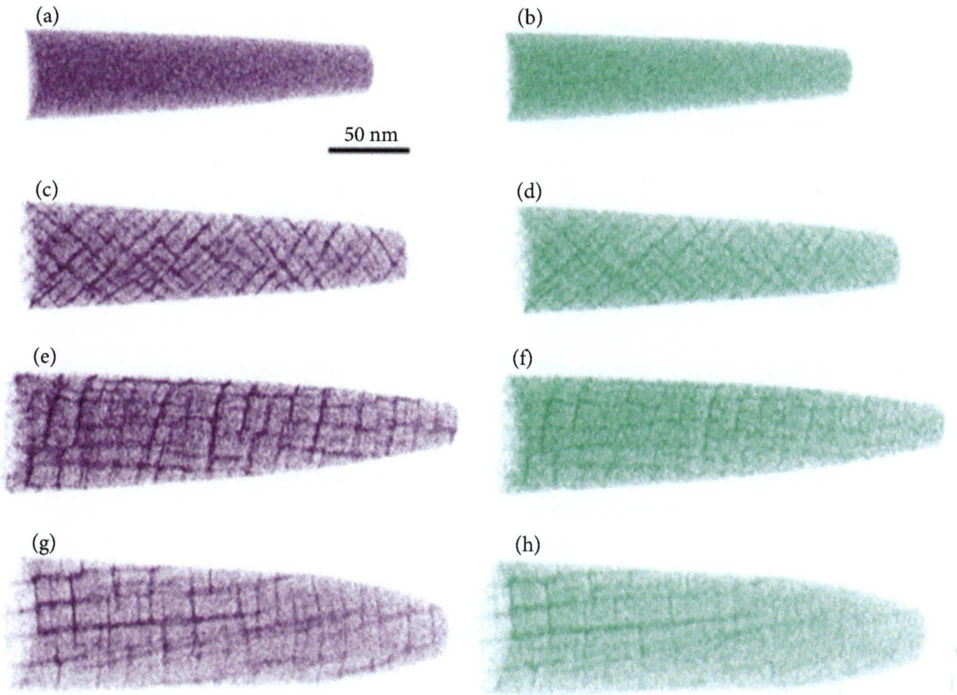

Figure 4.16 AP composition profiles for Zr (purple) and Hf (green) from modified Senkov alloy ZrNbTaW: (a) and (b) as-cast; and after annealing at 1,800 °C for (c) and (d) six hours; (e) and (f) one day; and (g) and (h) four days.

Reproduced from Maiti and Steurer. (2016). Structural disorder and its effect on mechanical properties in single-phase TaNbHfZr high-entropy alloy. *Acta Materialia 106*: 87–97.

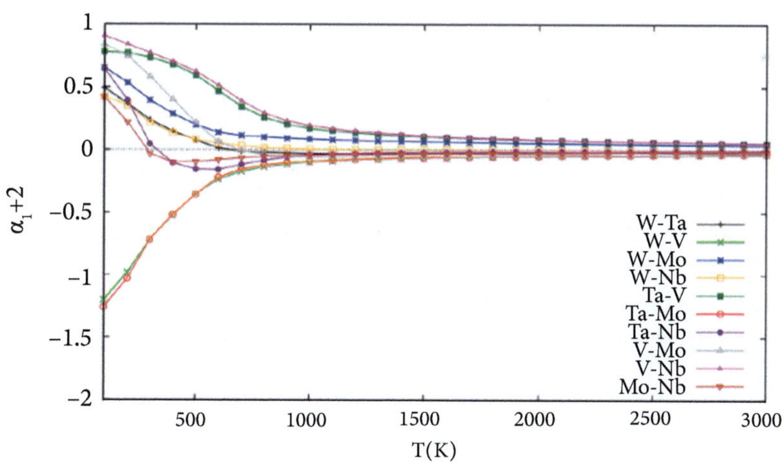

Figure 4.17 First near-neighbour pairwise short-range order parameters α_{ij}^{1} versus temperature, calculated by *ab initio* modelling for equiatomic Senkov alloy VNbMoTaW.

Reproduced from Fernandez-Caballero et al. (2017). Short-range order in high-entropy alloys: Theoretical formulation and application to MoNbTaVW system. *Journal of Phase Equilibria and Diffusion 38*: 391–403.

atomic distributions of all the components, with no evidence of any short-range order, and all the α and β values close to zero.[66] Further details of atom-probe (AP) studies and *ab initio* modelling of short-range order in multicomponent materials can be found in the reviews by Hu et al.[67] and Ferrari et al.[68] respectively.

3 Electronic Structure of Multicomponent Materials

Electronic structure of atoms

The structure of an atom is described[69–72] by its *many-body time-dependent wave function* $\Psi(\{\boldsymbol{r}_i\}, t)$, which gives the probability[ix] of finding the set of fundamental particles (its component protons, neutrons and electrons) $\{i\} = i, j, k \ldots$, etc. at positions $\{\boldsymbol{r}_i\} = \boldsymbol{r}_i, \boldsymbol{r}_j, \boldsymbol{r}_k \ldots$, etc. respectively at any given time t. The wave function Ψ is obtained by solving the *time-dependent Schrödinger equation:*[69–72]

$$i\hbar\frac{\partial\Psi}{\partial t} = \mathcal{H}\Psi$$

and the ground and excited state energies of the material E can then be found by integrating the wave function:

$$E = \langle\Psi|\mathcal{H}|\Psi\rangle = \int\Psi^*\mathcal{H}\Psi dV$$

where $\hbar = h/2\pi$, h is Planck's constant, Ψ^* is the complex conjugate of Ψ, $\mathcal{H}$ is the *Hamiltonian operator*[x] (or just the *Hamiltonian*), the integration is over the volume V occupied by the atom, and the expression in angular brackets is the compact Dirac notation for the full integral on the right hand side.[xi] The *Hamiltonian* $\mathcal{H}$ is given by:

$$\mathcal{H} = \mathcal{T} + \mathcal{V} = -\sum_i \frac{\hbar^2}{2m_i}\nabla_i^2 + \frac{1}{2}\sum_{i\neq j}\frac{e^2}{4\pi\epsilon_o}\frac{z_i z_j}{(\boldsymbol{r}_i - \boldsymbol{r}_j)}$$

where $\mathcal{T}$ and $\mathcal{V}$ are kinetic and potential energy operators given respectively on the right hand side by the sum of the individual kinetic energies of all the constituent fundamental particles i and the sum of the individual electrostatic attractions or repulsions of all pairs of constituent fundamental particles i and j, and where m_i and $\boldsymbol{r}_i$ are the mass and position of the ith particle, e is the charge on the electron, $z_i = -1, +1$ and 0 for electrons, protons and neutrons respectively, and $\nabla_i^2 = \partial/\partial x_i^2 + \partial/\partial y_i^2 + \partial/\partial z_i^2$ is the Laplacian operator for the ith particle.[69–72]

[ix] More accurately, the probability of finding the set of particles $\{i\}$ at positions $\{r_i\}$ at time t is given by the square of the wave function $\Psi\Psi^*$ where Ψ^* is the complex conjugate of Ψ. The particles are all somewhere in the atom at time t, so $\int\Psi\Psi^* dV = 1$ when the integral is over the volume V occupied by the atom.

[x] An operator (denoted here by curly capitals) is a function of a function, its domain and range are both sets of functions, and it maps any given function in its domain to another function in its range. In *quantum mechanics*, an operator $\mathcal{P}$ represents a property p in the sense that it maps any given wave function Ψ defined over some region of space and time $(\boldsymbol{r}, t)$ to give the value of the property as a function p over the same region of space and time $(\boldsymbol{r}, t)$. The Hamiltonian operator $\mathcal{H}$ is a particular case of an operator that maps the wave function Ψ to give the energy E over $(\boldsymbol{r}, t)$.

[xi] We will not use the Dirac notation any further in what follows. It is given here simply in the interest of completeness.

Solving these equations is fiendishly difficult even for just a single, simple atom because of the large number (except for a hydrogen atom) of constituent electrons, protons and neutrons, and also because of the complex structure of the equations themselves. We can simplify the problem by: 1) assuming the atom is in a stationary state, i.e. by removing any time dependence;[xii] 2) concentrating on the electrons alone, i.e. by treating the protons and neutrons in the nucleus as a fixed, central ionic charge; and 3) treating the electrons as independent, i.e. moving separately under the influence of the central ionic charge and the mean field of the other electrons.[73,74] The electronic structure of the atom is then described by $\Psi(\boldsymbol{r_i}) = \prod_i \psi_i(\boldsymbol{r})$, the product of z *single-body wave functions* $\psi_i(\boldsymbol{r})$ (one for each of the z electrons, where z is the atomic number), obtained from the *time-independent Schrödinger equation*:

$$E\psi = \mathcal{H}\psi$$

with the Hamiltonian now given by:

$$\mathcal{H} = \mathcal{T} + \mathcal{V} = -\frac{\hbar^2}{2m}\nabla^2 + V(\boldsymbol{r})$$

where the first and second terms on the right-hand side are now respectively the kinetic energy and the potential energy $V(\boldsymbol{r})$ of each electron individually in the combined electrostatic field of the nucleus and the mean field of the other electrons.[73,74] Solving these equations subject to the *Pauli exclusion principle* (no two electrons can be in the same state) gives a set of wave functions $\{\psi_i\}$ and a set of corresponding energies $\{E_i\}$ for the electrons. This is still not easy to do, requiring a *self-consistent mean field* $V(\boldsymbol{r})$, to be found iteratively by estimating $V(\boldsymbol{r})$ calculating $\{\psi_i\}$ and $\{E_i\}$, re-estimating $V(\boldsymbol{r})$, re-calculating $\{\psi_i\}$ and $\{E_i\}$, and so on,[73,74]

The stationary state electronic wave functions $\psi_i(\boldsymbol{r})$ and their corresponding energies E_i are quantised automatically[xiii] by electron confinement within the atom by the field $V(\boldsymbol{r})$ since each of the states $\psi_i(\boldsymbol{r})$ must have nodes at the edge of the confinement region, and only states with integral wave numbers within the atom between these nodes are allowed.[69,70] Electrons can move in three dimensions within an atom, so their wave functions $\psi_i(\boldsymbol{r})$ can be separated into three independent components:

$$\psi_i(\boldsymbol{r}) = \psi_{ir}(r)\,\psi_{i\theta}(\theta)\,\psi_{i\phi}(\phi)$$

using spherical coordinates $\boldsymbol{r} = (r, \theta, \phi)$ because of the spherical symmetry of an atom. The electrons have, therefore, three quantum numbers corresponding to their motion and confinement in each of the three dimensions, with a fourth quantum number corresponding to the electron spin.[69,70] The principal quantum number $n = 1, 2, 3$, etc. defines the electron shell; the *azimuthal quantum number* $l = 0, 1 \ldots (n-1)$ (corresponding to s, p, d, f…) defines the electron orbital within each shell; the *magnetic quantum number* $m = -l, -(l-1) \ldots 0 \ldots (l-1), l$ defines the orbital orientation; and the *spin quantum number* $s = \pm \frac{1}{2}$ defines the electron spin. Because of the Pauli exclusion principle, the z electrons in an atom fill up the stationary states and their associated wave functions $\psi_i(\boldsymbol{r})$ with increasing quantum numbers, beginning with two electrons in the lowest energy state 1s ($n = 1$; $l = 0$; $s = \pm \frac{1}{2}$),

[xii] Essentially this corresponds to setting the temperature to absolute zero, $T = 0$.
[xiii] This is the so-called quantum particle in a box effect.

then two electrons in the next lowest energy state 2s ($n = 2$; $l = 0$; $s = \pm 1/2$), then six electrons in the next lowest energy state 2p ($n = 2$; $l = 1$; $m = -1, 0$ and 1; $s = \pm 1/2$), and so on.[69,70]

Band structure

A pure solid material is formed when a large number of atoms of the same type is brought together (say one mole or Avogadro's number $N_o = 6.022 \times 10^{23}$). Individual electron states in each atom overlap and form bands of states with a range of energies,[69–72] often with gaps in energy between the different bands caused by strong Bragg reflection of electrons at diffracting planes within the solid.[xiv] According to *Bloch's theorem*, the wave functions $\psi_i(\boldsymbol{r})$ in a crystalline material can be written as a set of *Bloch waves*, which are plane waves $\phi_i(\boldsymbol{r}) = e^{i\boldsymbol{kr}}$ modulated by a periodic function $u_i(\boldsymbol{r}) = u_i(\boldsymbol{r} + \boldsymbol{R})$ with the periodicity of the crystal lattice:[69–72]

$$\psi_i(\boldsymbol{r}) = \phi_i(\boldsymbol{r})\, u_i(\boldsymbol{r}) = e^{i\boldsymbol{kr}} u_i(\boldsymbol{r})$$

where $\boldsymbol{k}$ is the wave vector and $\boldsymbol{R}$ is any symmetry operation in the crystal lattice.[xv] The *density of states* (*DOS*) curve $n(E)$ plots the overall variation in the number of states n with their energy E. Again because of the Pauli exclusion principle, the zN_o electrons in the material fill up the states in the energy bands from the lowest energy to the highest, which is called the *Fermi energy* or *Fermi level* E_F. The material is a conductor when the Fermi level is somewhere in the middle of an energy band or there are no gaps in energy between the different energy bands, since there are then nearby states just above the Fermi level for electrons to gain extra energy easily and move through the material. The material is, however, an insulator or a semiconductor when the Fermi level is at the top of an energy band and there is a gap to the next band, since there are then no nearby states just above the Fermi level for electrons to gain extra energy easily and move through the material.[69–72]

When the temperature is raised above zero, or when a dilute binary solid solution is formed with a small alloying addition, the resulting atomic vibrations, thermal vacancies, interstitials, and substitutional solute atoms affect the local electronic band structure of the pure solid material, injecting or removing electronic charge, distorting the band geometry and energy levels, and influencing the *electrical conductivity* σ and *thermal conductivity* κ by acting as scattering centres for the propagation of electrons and phonons. In dilute alloys, these effects can be regarded as local perturbations, with a local density of states (LDOS) deviating slightly in some places from the overall DOS of the material.[69,70] Applying the kinetic theory of gases to the nearly free conduction electrons in a dilute metallic alloy gives the electrical and thermal conductivity σ and κ as:[69,70]

$$\sigma = ne^2\tau/m$$

$$\kappa = n\pi^2k^2T\tau/3m$$

[xiv] The energy of an electron E depends on its frequency ν according to de Broglie's equation $E = h\nu$, and increases, therefore, with increasing magnitude of the wave vector $\boldsymbol{k}$, but there are forbidden electron energies (or energy gaps) when $\boldsymbol{k}$ fulfils Bragg's law and electrons are diffracted by the solid structure.

[xv] i in the exponential is the imaginary number: $i = \sqrt{-1}$. Elsewhere i is a label for different electrons: $\{i\} = i, j, k$, etc.

respectively, where n is the number of conduction electrons per unit volume, e and m are the charge and mass of an electron, $\tau = v/\lambda$ is the *relaxation time* or average time between electron scattering events, v and λ are the average electron velocity and electron *mean free path* respectively, k is Boltzmann's constant and T is the temperature. This leads to the well-known empirical Wiedemann–Franz law for metals above their Debye temperature, giving a constant ratio of thermal to electrical conductivity $\kappa/\sigma = L$ where L is the Lorentz number:[69,70]

$$\frac{\kappa}{\sigma} = \frac{\pi^2 k^2 T}{3e^2} = L = 2.45 \times 10^{-8}\ \mathrm{W\Omega/K^2}$$

which arises essentially because in metal alloys the electrical and thermal conductivities both depend on the average time τ between electron scattering events as the electrons move through the material. The Wiedemann-Franz law does not apply to ceramic alloys, which are usually insulators without free electrons and which have, therefore, very low electrical conductivity so the thermal conductivity is dominated by the motion of phonons rather than electrons.

A perturbation methodology is, of course, not possible for multicomponent high-entropy solid-solution materials in which local atomic structures (clusters of individual atoms and their immediate near neighbours) vary widely at all atomic sites over very large distances. We have already seen in Chapter 3 that there are literally billions and trillions of different local atomic structures in multicomponent solid solutions with four, five or more components, extending over many microns if not millimetres in distance. Not surprisingly, the electronic band structures of multicomponent high-entropy materials exhibit correspondingly widespread distortions, with wide variations in local charge, geometry and electron energy, as well as extensive scattering of electrons and phonons.[75–89] From Tables 4.3 and 4.4 and Figures 4.7 and 4.8, the number of different oriented first and second near-neighbour clusters and, therefore, the number of different distorted local electronic band structures is just under twenty trillion and just over thirty billion respectively for 5-component single-phase fcc Cantor and bcc Senkov alloys, and very many more for alloys with six or more components or if third near neighbours are also taken into account. This leads, of course, to rather ill-defined Bloch waves, an ill-defined band structure, a widespread variation from site to site in the LDOS, a corresponding spreading of the overall (averaged) DOS for the material, and correspondingly reduced electrical and thermal conductivity.

It is very difficult, unfortunately, to *measure* directly the electronic band structure of a multicomponent high-entropy material, and there have been, therefore, very few attempts to do so.[9–11] There have, however, been a number of attempts to *calculate* the electronic band structure of multicomponent high-entropy materials using *ab initio* atomic modelling techniques,[75–89] but, as with calculations of their atomic structure discussed previously, these attempts are bedevilled by the difficulty of representing reasonably the very widespread and therefore large number of variations in local structure. Computer techniques can deal very accurately with some tens of atoms and, with increasing difficulty and uncertain approximations, some hundreds and thousands of atoms, but they come nowhere near managing the trillions of different local atomic (and, therefore, electronic) structures that are needed to describe multicomponent high-entropy materials. We discuss this limitation in considerably more detail in Chapter 5 when we look at the different methods that have been used to try to model multicomponent high-entropy materials. For the moment, we describe in the rest of this section some of the calculations of electronic band structure that have been made,[75–88]

Table 4.11 Some multicomponent high-entropy materials whose electronic band structures and density of states curves have been calculated by *ab initio* atomic modelling techniques.[75–88]

material	structure
CrMnFeCoNi	fcc
CrFeCoNiPd	fcc
CrFeCoNi	fcc
RuRhPdAgOsIrPtAu	fcc
$AlCrMnFeNi_x$	bcc
$Al_{0.5}TiNiCuZrPd$	bcc
TiVCrZrNbMoTa	bcc
TiVCrZrNbMoTaW	bcc
CrNbMoTaW	bcc
VNbMoTaW	bcc
AlTiVCrMnFeCoNiZrNbMoHf	Laves

since they are all that is available, but with a strong reservation about their limited scope and, therefore, rather uncertain validity. Table 4.11 lists some examples of the multicomponent high-entropy materials whose electronic band structures have been calculated using *ab initio* atomic modelling techniques.[75–88]

Electronic structure of fcc Cantor alloys

A number of studies[82–85] have calculated the density of states (DOS) and investigated the resulting electrical and thermal conductivities for a variety of multicomponent single-phase fcc Cantor alloys, including the original equiatomic 5-component Cantor alloy CrMnFeCoNi, the modified equiatomic Cantor alloy CrFeCoNiPd (i.e. with Pd replacing Mn) and a range of binary, ternary and quaternary equiatomic and non-equiatomic sub-system alloys, such as FeNi, CoNi, NiPd, CrCoNi, MnFeNi, FeCoNi, CrMnCoNi, CrFeCoNi, and MnFeCoNi. Measurements from this wide range of different single-phase fcc alloys fall somewhat surprisingly into two distinct and clearly defined groups, exhibiting different electrical and thermal conduction behaviour corresponding to two different types of electronic band structure. Figure 4.18 shows the *electrical resistivity* (inverse electrical conductivity) $\rho = 1/\sigma$ for some of the single-phase fcc Cantor alloys as a function of temperature between 4 K and 300 K (−269 °C and 27 °C), as measured by Jin et al.[84] using a Quantum Design Physical Property measurement system on single crystals of the alloys manufactured by arc melting the pure elements, re-melting five times to ensure homogeneity, casting into a 12-mm-diameter water-cooled copper mould, and finally re-melting and solidifying in a xenon arc-lamp powered optical floating zone single-crystal furnace.

Alloys in Figure 4.18(a) containing only late transition-metal components such as Fe, Ni and Co exhibit low values of electrical resistivity (and correspondingly high values of electrical conductivity); whereas alloys in Figure 4.18(b) containing early transition-metal components such as Cr and Mn as well as late transition-metal components such as Fe, Co and Ni exhibit much higher values of electrical resistivity (and correspondingly low values

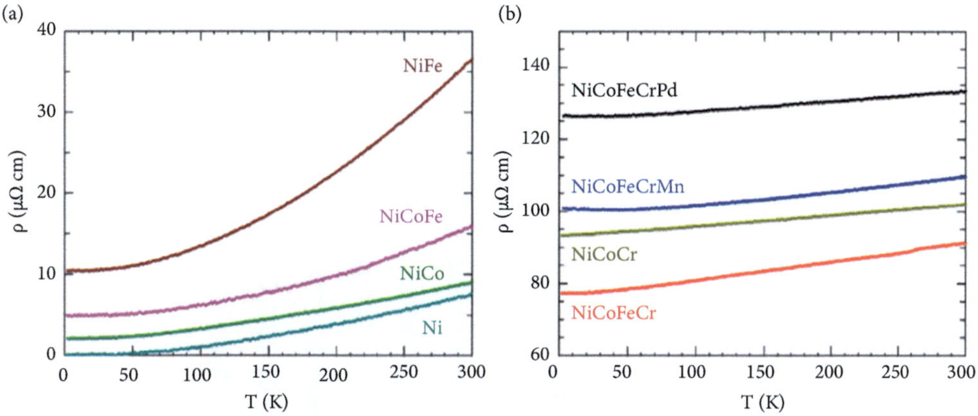

Figure 4.18 Electrical resistivity ρ versus temperature T for a variety of equiatomic single-phase fcc Cantor alloys.

Reproduced from Jin et al. (2016). Tailoring the physical properties of Ni-based single-phase equiatomic alloys by modifying the chemical complexity. *Scientific Reports* 6: Article 20159.

of electrical conductivity).[85] The difference is most marked near absolute zero when there is little or no contribution to electron scattering from thermal vibrations, with a difference in *residual electrical resistivity* ρ_o between the two groups of alloys of approximately two orders of magnitude, as can be seen in the measured and calculated values[84] for a variety of single-phase fcc Cantor alloys in Figure 4.19. Similar results have also been obtained by Bag et al.[82] and Mu et al.[84] Typical values of residual electrical resistivity for the two groups of alloys[84] are $\rho_o = 1.7$ and 92.7 $\mu\Omega$cm respectively for the two ternary (i.e. nominally similar) single-phase fcc alloys FeCoNi and CrCoNi, or even more starkly $\rho_o = 1.3$ and 124.8 $\mu\Omega$cm respectively for the somewhat more different alloys CoNi and CrFeCoNiPd.

The difference in measured and calculated electrical resistivity between the two groups of single-phase fcc Cantor alloys, i.e. those containing only late transition-metal components such as Fe, Co and Ni, and those containing early transition-metal components such as Cr and Mn as well as late transition-metal components such as Fe, Co and Ni, is mirrored in a similar clear difference in electronic band structure, as shown by *ab initio* atomic modelling calculations.[83–86] Figure 4.20 shows typical examples from Jin et al.[85] of magnetically resolved Bloch wave band spectra $E(\boldsymbol{g})$ (where crystal orientation is represented by reciprocal lattice vector $\boldsymbol{g}$) and density of states (DOS) curves $n(E)$ for the two ternary alloys FeCoNi and CrCoNi, obtained by density functional theory (DFT) calculations using the Korringa–Kohn–Rostoker (KKR) method, implemented by the Hutsepot code, with a mean-field coherent-potential approximation (CPA), local spin-density approximation (LDA) for the exchange-correlation term and atomic-sphere approximation (ASA) for the crystal potential.

The top two Figures 4.20(a) and (b) are for the late transition-metal ternary alloy FeCoNi and the bottom two Figures 4.20(c) and (d) are for the early and late transition-metal ternary alloy CrCoNi; the left-hand two Figures 4.20(a) and (c) are for minority spin-down electrons only and the right-hand two Figures 4.20(b) and (d) are for majority spin-up electrons only; all four Figures 4.20(a)–(d) show calculated Bloch wave spectra $E(\boldsymbol{g})$ (on a black background), and corresponding density of states (DOS) curves $n(E)$ (in blue); and the Fermi level is indicated in each case with green horizontal lines as the (arbitrary) zero point for

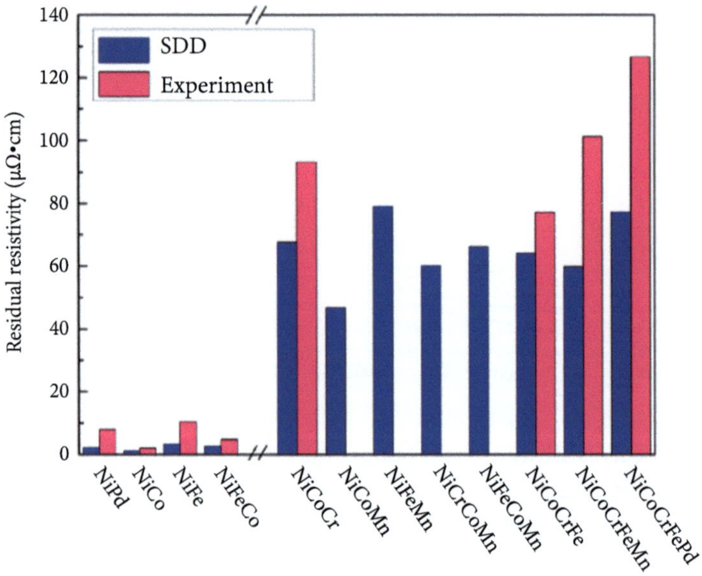

Figure 4.19 Measured (red) and calculated (blue) values of residual electrical resistivity ρ_o for a variety of equiatomic single-phase fcc Cantor alloys.

Reproduced from Mu et al. (2019). Uncovering electron scattering mechanisms in NiFeCoCrMn derived concentrated solid solution and high entropy alloys. *NPJ Computational Materials* 5: 1–8.

energy. The electron band structures of both alloys are, not surprisingly, metallic in character, with overlapping s, p and d bands, and the Fermi level is near the top of the 3d band. As expected, clear blurring of the Bloch wave spectra and associated spreading of the density of states above and below the Fermi energy can be seen in Figures 4.20(c) and (d) respectively for both minority spin-down and majority spin-up electrons in the mixed early and late transition-metal ternary alloy CrCoNi, leading to strong scattering of all electrons, a very small mean free path close to the atomic size, $\lambda \sim a$, and high electrical resistivity ρ (low electrical conductivity σ) as shown in Figures 4.18 and 4.19. However, although blurring of the Bloch wave spectra and spreading of the density of states above and below the Fermi energy is again seen for the minority spin-down electrons in the late transition-metal ternary alloy FeCoNi in Figure 4.20(a), no spreading in the density of states and instead only sharp Bloch waves, sharp energy bands and a sharp Fermi level are seen for the majority spin-up electrons in Figure 4.20(b). Jin et al.[85] interpret this behaviour as caused by local alignment of individual atomic spins in ferromagnetic Fe, Ni and Co atoms, suppressing majority spin-up electron scattering and thus leading to a much longer mean free path and low electrical resistivity ρ (high electrical conductivity σ) as shown in Figures 4.18 and 4.19. Jin et al.[85] describe this behaviour as majority spin-up electrons *short-circuiting* the expected scattering from multicomponent compositional disorder and local atomic distortions, with addition of early transition-metal antiferromagnetic Cr and Mn atoms disrupting majority spin-up alignment, preventing short-circuiting, and reinstating strong local electron scattering.

Similar band structures have been found in a series of other single-phase fcc materials,[86] as shown in Figure 4.21, again obtained by density functional theory (DFT) calculations using the KKR method, implemented by the Hutsepot code, with a mean-field coherent-potential

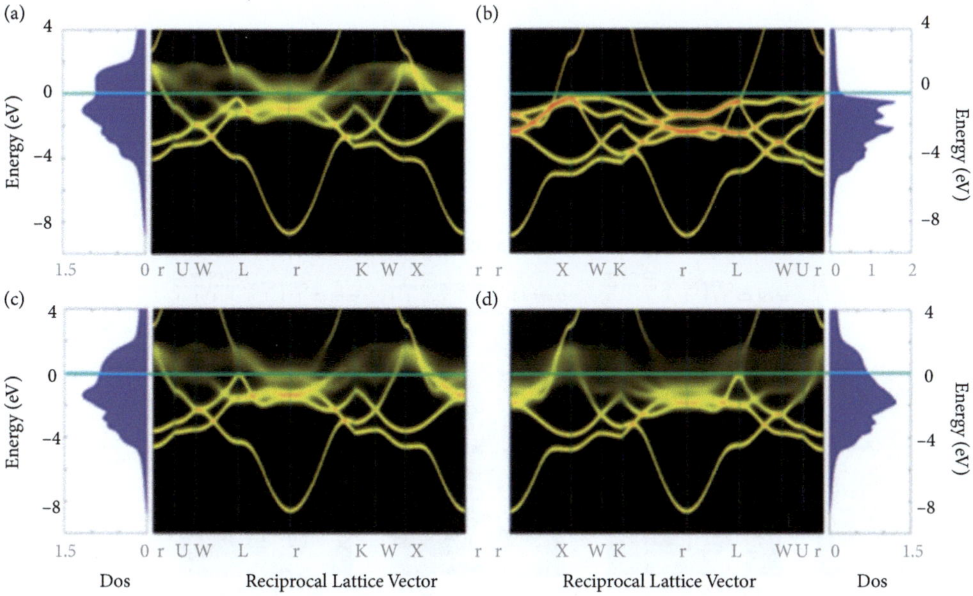

Figure 4.20 Calculated Bloch wave spectra $E(g)$ (black background) and DOS curves $n(E)$ (blue) for minority spin-down (a,c) and majority spin-up (b,d) electrons in ternary single-phase fcc Cantor alloys FeCoNi (a,b) and CrCoNi (b,d) with the Fermi level (green horizontal line) at zero energy.

Reproduced from Jin et al. (2016). Tailoring the physical properties of Ni-based single-phase equiatomic alloys by modifying the chemical complexity. *Scientific Reports 6*: Article 20159.

approximation (CPA), local spin-density approximation (LDA) for the exchange correlation and atomic-sphere approximation (ASA) for the crystal potential. Not surprisingly, the electron band structures are again all metallic in character, with overlapping s, p and d bands, and the Fermi level is again near the top of the 3d band. In pure Ni, there are sharp Bloch waves, energy bands and Fermi levels for both majority spin-up and minority spin-down electrons; in binary single-phase fcc CoNi and FeNi, there are sharp Bloch waves, sharp energy bands and a sharp Fermi level for majority spin-up electrons, but blurred Bloch waves and spread energy bands and Fermi levels for minority spin-down electrons; and in quaternary single-phase fcc CrFeCoNi there are blurred Bloch waves and spread energy bands and Fermi level for both majority spin-up and minority spin-down electrons. As shown in Figures 4.18 and 4.19, these different electronic band structures again correspond[84,85] to high electrical conductivity and long mean free paths when there are only late transition-metal components, in pure Ni, CoNi and FeNi, as opposed to low electrical conductivity and short mean free paths when there are both early and late transition-metal components, as in CrFeCoNi. Typical mean free paths for the two groups of alloys have been calculated[84] to be 169, 17 and 48 nm respectively for late transition-metal alloys CoNi, FeNi and FeCoNi, and 4.1 and 4.0 nm respectively for the mixed early and late transition-metal alloys $Cr_{0.2}Ni_{0.8}$ and CrCoNi.

Figures 4.22 and 4.23 show experimental and calculated values[83] of electrical resistivity (inverse electrical conductivity) $\rho = 1/\sigma$ and thermal conductivity κ versus temperature for the same two equiatomic ternary single-phase fcc Cantor alloys FeCoNi and CrCoNi whose

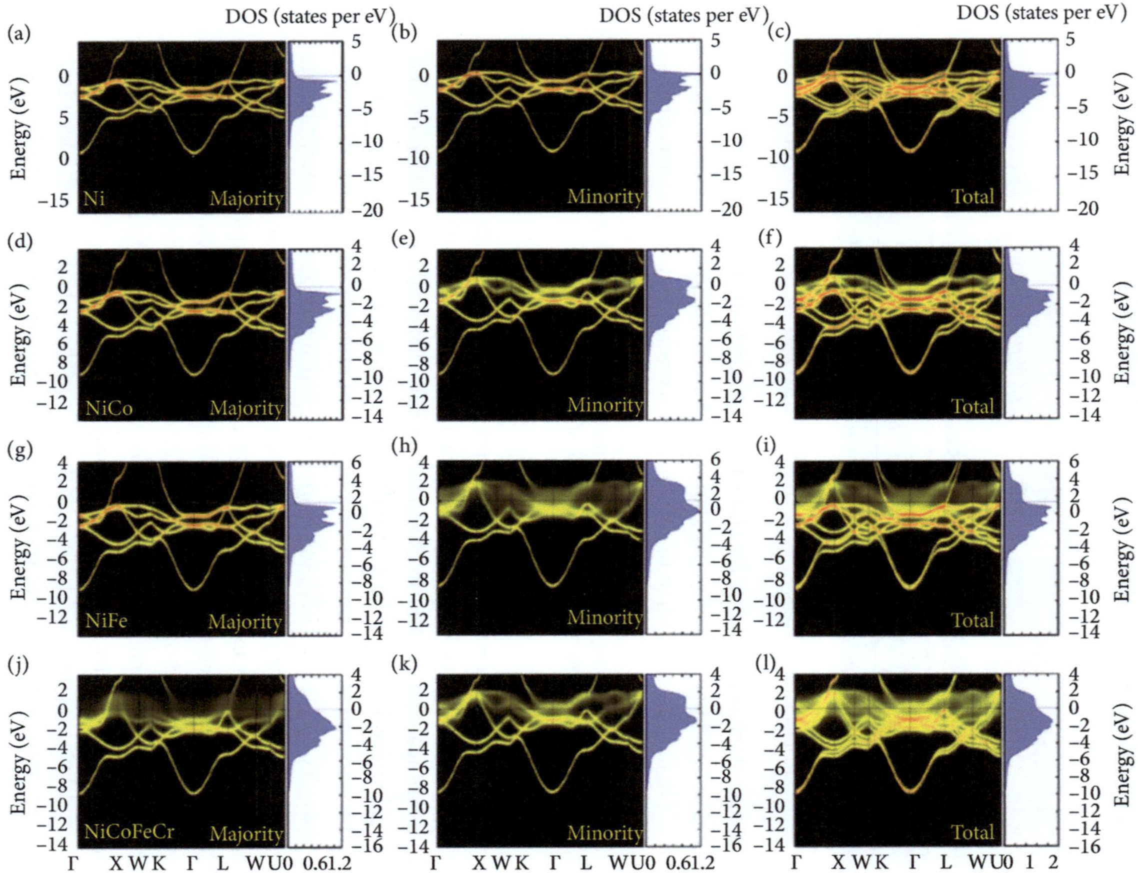

Figure 4.21 Calculated Bloch wave spectra $E(g)$ (black background) and DOS curves $n(E)$ (blue) for majority spin-up (a,d.g,j), minority spin-down (b,e,h,k) and all (c,e,i,l) electrons in single-phase materials Ni (a,b,c), CoNi (d,e,f), FeNi (g,h,i) and CrFeCoNi (j,k,l), with the Fermi level at zero energy.

Reproduced from Zhang et al. (2015). Influence of chemical disorder on energy dissipation and defect evolution in concentrated solid solution alloys. *Nature Communications 6*: Article 8736.

Bloch wave spectra $E(\mathbf{g})$ and density of states curves $n(E)$ are shown in Figure 4.20. Calculated and measured electrical resistivities are not in particularly good agreement in either alloy, although in both cases they show a similar general increasing trend with increasing temperature. Thermal conductivity calculated from measurements of electrical resistivity using the Wiedemann–Franz law gives quite good agreement with direct thermal conductivity measurements for non-magnetic CrCoNi, but gives much poorer agreement for magnetic FeCoNi, with a more complex range of different scattering processes in the ferromagnetic material.[83]

The calculated electronic band structures, density of states curves, electrical resistivities and electrical and thermal conductivities discussed above and shown in Figures 4.19–4.23 were obtained by *ab initio* atomic modelling,[83–86] typically using: (a) a few hundred atoms, nowhere near the number required to capture the range of different local atomic and electronic environments in multicomponent materials, as already discussed earlier in

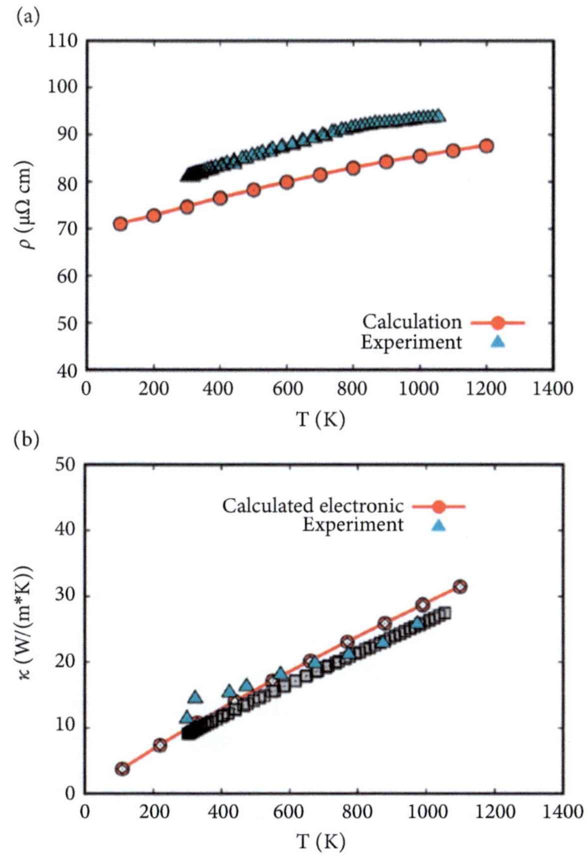

Figure 4.22 (a) Electrical resistivity ρ and (b) thermal conductivity κ for equiatomic ternary single-phase fcc CrCoNi: calculated (red circles), measured (light blue triangles) and Wiedemann–Franz law (grey squares).

Reproduced from Samolyuk et al. (2018). Temperature-dependent electronic transport in concentrated solid solutions of the 3d-transition metals Ni, Fe, Co and Cr from first principles. *Physical Review B 98*: Article 165141.

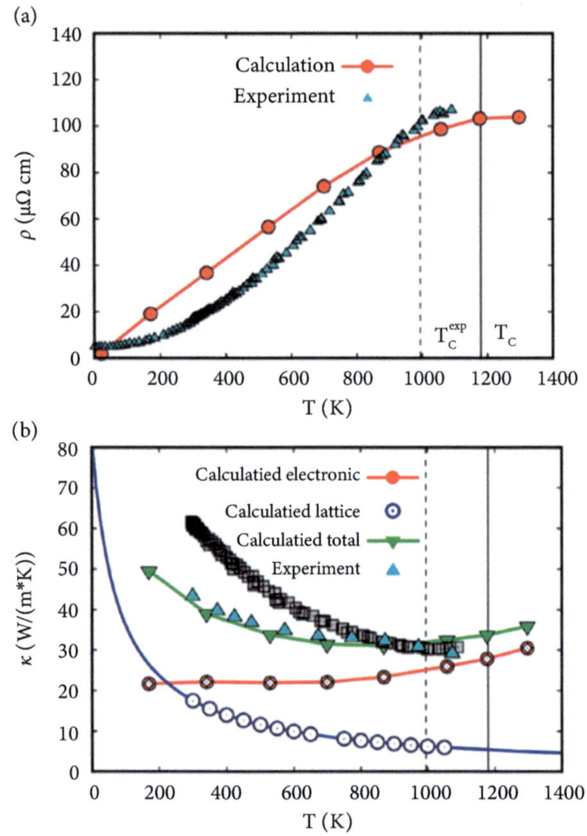

Figure 4.23 (a) Electrical resistivity ρ and (b) thermal conductivityκ for equiatomic ternary single-phase fcc FeCoNi: (a) calculated (red circles) and measured (light blue triangles); (b) calculated electronic (red circles), calculated phonon (blue circles) calculated total (green triangles), measured (light blue triangles) and Wiedemann-Franz law (grey squares).
Reproduced from Samolyuk et al. (2018). Temperature-dependent electronic transport in concentrated solid solutions of the 3d-transition metals Ni, Fe, Co and Cr from first principles. *Physical Review B 98*: Article 165141.

this chapter; and (b) a mean field CPA for the potential field surrounding individual atoms, effectively smearing out and, therefore, losing the variation in different local atomic and electronic environments even for the few hundred atoms considered, as discussed in more detail in Chapter 5. The details of the calculated electronic band structures and density of states curves are not, therefore, likely to be too accurate; and, unsurprisingly, the calculated and measured electrical and thermal conductivities are not in particularly good agreement, as already mentioned and as can be seen clearly in Figures 4.19, 4.22 and 4.23. Nevertheless, the general trends are captured fairly well, showing clear metallic conduction behaviour in multicomponent single-phase fcc Cantor alloys, with broad electron energy bands and density of states curves, and in most cases strong scattering from local chemical and structural disorder leading to low electrical and thermal conductivity and short electron mean free paths, but with short circuiting in some cases, when electron spins on ferromagnetic atoms are well aligned, leading to higher electrical and thermal conductivity and much longer electron mean free paths.

Electronic structure of bcc Senkov alloys

There has also been a number of studies[75–78] using similar methods to calculate the density of states and to investigate the resulting electrical and thermal conductivities for a variety of multicomponent single-phase bcc alloys, including the original equiatomic 4- and 5-component Senkov alloys NbMoTaW and VNbMoTaW, similar equiatomic 5-component modified Senkov alloys such as NbMoTaWX where X = Cr, Zr, Hf and Re, TiZrVMoX where X = Cr, Nb, Ta and W, and other non-equiatomic alloys such as $AlCrMnFeNi_x$. Figure 4.24 shows a series of examples of partial and total DOS curves $n(E)$ for each of the s, p and d bands from each of the components in equiatomic single-phase bcc NbMoTaW and NbMoTaWX Senkov alloys where X = Cr, V and Zr, calculated[77] by *ab initio* atomic modelling with the CASTEP code, using a special quasi-random supercell (SQS) method and a generalised gradient approximation (GGA) and Perdew–Burke–Ernzerhof (PBE) function for the exchange-correlation term. Figure 4.25 shows another series of examples of Bloch wave spectra $E(\boldsymbol{g})$ (left-hand side) and corresponding density of states (DOS) curves $n(E)$ (right-hand side) for non-equiatomic single-phase bcc $AlCrMnFeNi_x$ alloys, where $x = 0$, 0.25 and 0.5, again calculated[76] by *ab initio* atomic modelling with the CASTEP code, using a virtual crystal approximation (VCA) method and again a GGA and PBE function for the exchange-correlation term. The electron band structures $E(\boldsymbol{g})$ and density of states curves $n(E)$ calculated for single-phase bcc Senkov alloys shown in Figures 4.24 and 4.25 are similar to those calculated for single-phase fcc Cantor alloys shown in Figures 4.20 and 4.21, again metallic in character, with overlapping s, p and d bands, but with the Fermi level closer to the middle of the 3d band. Chemical and structural disorder associated with the widespread variation in local atomic and electronic structure again leads to spreading of electron states above and below the Fermi level, strong electron scattering, high electrical resistivity and low electrical and thermal conductivity. Typical values of electrical resistivity are ~90 and ~150 $\mu\Omega$cm for equiatomic TiZrNbHf and AlCrFeCoNi respectively at room temperature.[89,90]

4 Defects in Multicomponent Materials

Vacancies and diffusion

Vacancy migration

A *vacancy* is a vacant lattice site, i.e. a lattice site where there is no atom, where the expected atom is missing or absent.[3,91,92] Vacancies are one of a number of different kinds of *point defect* in the crystal structure of a material. According to the third law of thermodynamics, all crystals are perfect at equilibrium at absolute zero and there are, therefore, no vacancies (or any kind of defect) at absolute zero. At any realistic temperature, however, the excess energy of creating a vacancy can be balanced by the entropy associated with mixing vacancies throughout the crystal, so there is an equilibrium number of vacancies in any crystal that increases with increasing temperature. In a single component pure material, there is a characteristic and fixed excess energy required to create a vacancy, the *vacancy formation energy* ΔE_v, which is independent of where the vacancy is in the crystal, since all the lattice sites

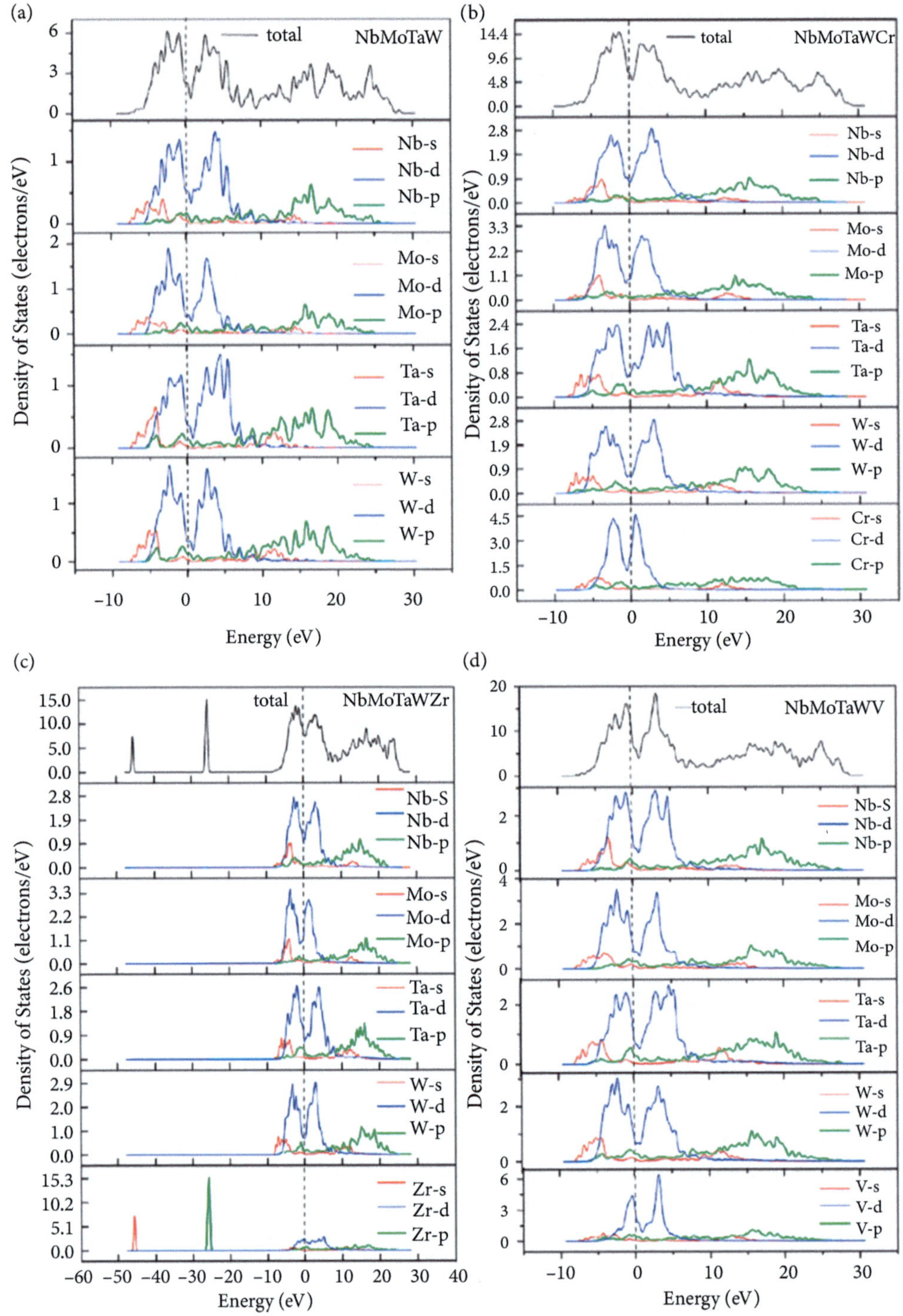

Figure 4.24 Calculated total and partial DOS curves for each of the s, p and d bands from each of the components in multicomponent single-phase bcc Senkov alloys, with the Fermi level at zero energy: (a) NbMoTaW; (b) CrNbMoTaW; (c) ZrNbMoTaW; and (d) VNbMoTaW.

Reproduced from Tong et al. (2020). Influence of alloying elements on mechanical and electronic properties of NbMoTaWX (X = Cr, Zr, V, Hf and Re) refractory high-entropy alloys. *Intermetallics 126*: Article 106928.

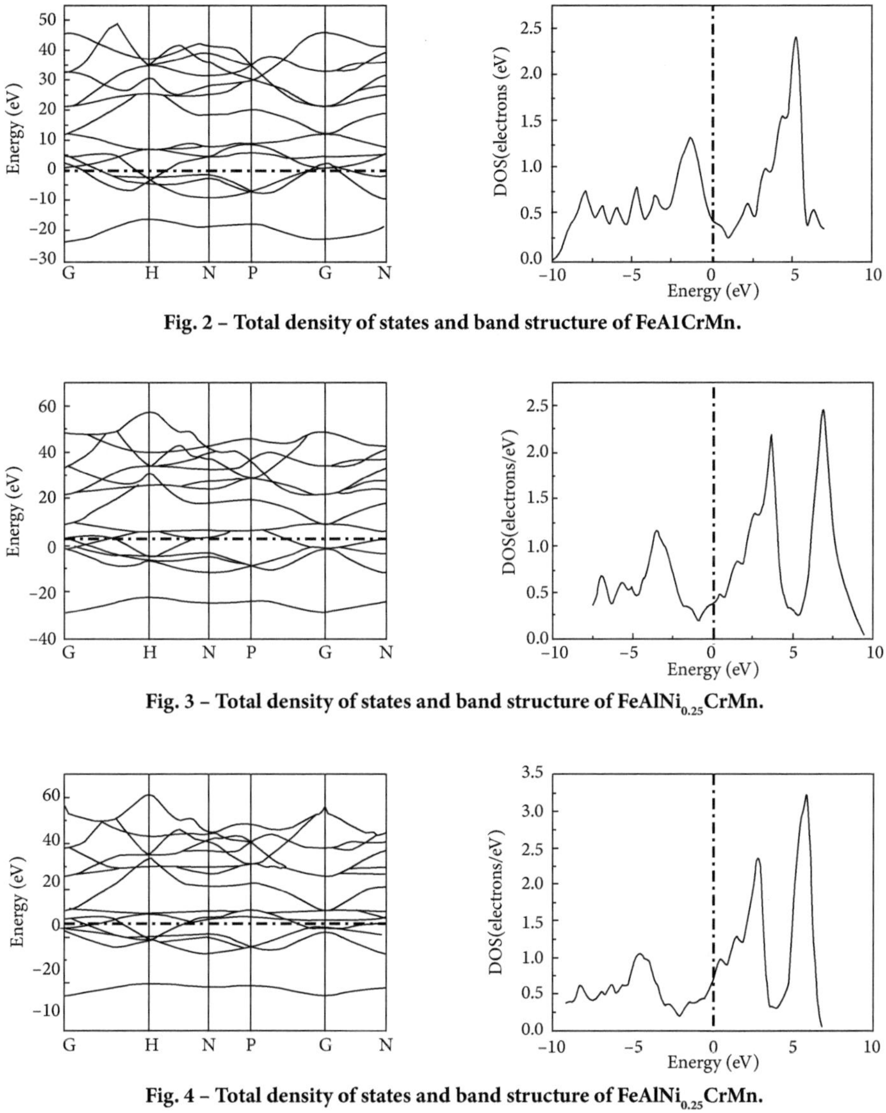

Figure 4.25 Calculated Bloch wave spectra $E(g)$ (left-hand side) and DOS curves $n(E)$ (right-hand side) for quaternary single-phase bcc Senkov alloys, with the Fermi level at zero energy: (1) AlCrMnFe; (2) $AlCrMnFeNi_{0.25}$ and (3) $AlCrMnFeNi_{0.5}$.

Reproduced from Chang and Zhang (2022). First-principles calculations to investigate elastic and thermodynamic properties of $FeAlNi_xCrMn$ quintenary alloys. *Journal of Materials Research and Technology 18*: 1322–32.

are identical. The probability of finding a vacancy at any particular lattice point p_v and the equilibrium number of vacancies in the crystal N_v are given at any temperature T, therefore, by Boltzmann's equation:[3,91]

$$p_v = N_v/N_o = \exp\ (-\Delta E_v/kT)$$

where N_o is Avogadro's number and k is Boltzmann's constant.[xvi]

The most common form of atomic motion or *diffusion* in solid crystalline materials is by a mechanism called *vacancy migration* or *vacancy hopping*, in which an atom adjacent to a vacancy jumps into the vacant lattice site.[3,91,92] As near-neighbour atoms continue to jump progressively into adjacent vacancies, the vacancies appear to migrate or hop in the reverse direction throughout the crystal, hence the name *vacancy migration* or *vacancy hopping*. Each time an atom jumps into a vacancy, it has to overcome an energy barrier in order to squeeze between surrounding nearby atoms. In a single component pure material, there is a characteristic and fixed energy barrier for such atomic jumps and, therefore, for the process of vacancy hopping, the *vacancy migration energy* ΔE_m, which is again independent of where the vacancy is in the crystal, since all the lattice sites are identical. The probability of an atom adjacent to a vacancy jumping into it by acquiring enough energy to overcome the barrier and squeezing through the surrounding atoms, i.e. the probability of vacancy migration[xvii] p_m, is given again by Boltzmann's equation:[3,91,92]

$$p_m = \exp\left(-\Delta E_m/kT\right)$$

Atomic motion in a single-phase pure material is, then, fairly straightforward. From time to time, a solute atom adjacent to a vacancy acquires enough energy to squeeze past its neighbours and jumps into it. Since the probability of a vacancy being present and the probability of an adjacent solute atom acquiring enough energy to jump into it are both independent of their position in the lattice (because all lattice sites are identical), the resulting process of vacancy hopping and reverse solute diffusion is completely random and stochastic, both spatially and temporally. In other words, solute diffusion takes place via what is called a *random walk*. This means that it obeys *Fick's laws*: the flux of solute atoms J (the rate dn/dt at which a number of atoms dn migrate in time dt through unit cross-sectional area A) is proportional to the solute concentration gradient $\partial c/\partial x$ (or $\nabla \mathbf{c}$ in three dimensions); the change of solute concentration with time $\partial c/\partial t$ is proportional to its spatial curvature $\partial^2 c/\partial x^2$ (or $\nabla^2 c$ in three dimensions); and the constant of proportionality in both cases is called the *diffusion coefficient* or *diffusivity* D, which is a material parameter that effectively measures the speed of diffusion:[3,91,92]

$$J = (dn/dt)/A = -D\,(\partial c/\partial x)$$

$$\partial c/\partial t = D\left(\partial^2 c/\partial x^2\right)$$

The diffusion coefficient D depends on the frequency of vacancy hopping, i.e. the frequency with which individual solute atoms jump into adjacent vacancies, so it is proportional to the probability of a vacancy being present p_v and the probability of an adjacent solute atom jumping into it p_m:

[xvi] More accurately the probability of finding a vacancy at any lattice point $p_v = N_v/N_o = \exp\left(-\Delta G_v/kT\right) = \exp\left(-(\Delta H_v - T\Delta S_v)/kT\right) = \exp\left(-(\Delta E_v + p\Delta V_v - T\Delta S_v)/kT\right)$ where ΔG_v, ΔH_v, ΔS_v and ΔV_v are the excess Gibbs free energy, enthalpy, entropy and volume of forming a vacancy respectively, and p is the pressure. But the pressure term $p\Delta V_v$ can be neglected as insignificant at atmospheric pressure compared to the formation energy ΔE_v, and $\Delta S_v/k \approx 1$ so the pre-exponential factor $\exp\left(\Delta S_v/k\right)$ is a small numerical factor and can also be neglected (i.e. taken as $\exp\left(\Delta S_v/k\right) \approx 1$) to a first approximation.

[xvii] Similar considerations apply to the probability of vacancy migration p_m as to the probability of vacancy formation p_v, as described in footnote xvi.

Table 4.12 Typical vacancy formation and migration energies ΔE_v and ΔE_m respectively and activation energies for diffusion $Q = \Delta E_v + \Delta E_m$ for some pure fcc and bcc metals.[91,92]

metal	vacancy formation energy ΔE_v (eV)	vacancy migration energy ΔE_m (eV)	activation energy for diffusion Q (eV)
fcc			
Al	0.68	0.68	1.36
Fe	1.4	1.26	2.66
Ni	1.78	1.32	3.1
Cu	1.29	0.78	2.07
Au	0.89	0.84	1.73
bcc			
V	2.1	1.3	3.4
Nb	2.65	1.0	3.65
Mo	3.0	1.5	4.5
Ta	2.8	1.45	4.25
W	4.0	1.5	5.5

$$D \propto p_v p_m = D_o \exp\left(-Q/kT\right)$$

where D_o and $Q = \Delta E_v + \Delta E_m$ are, respectively, the *frequency factor* and the *activation energy for diffusion*.[3,91,92]

Vacancy formation and migration energies can be measured by a combination of dilatometry and positron annihilation as functions of time and temperature, and from Arrhenius plots of diffusion rates. Table 4.12 gives some typical examples[91,92] of measured vacancy formation and migration energies in pure metals ΔE_v and ΔE_m respectively, and corresponding activation energies for diffusion $Q = \Delta E_v + \Delta E_m$. Typical equilibrium numbers of vacancies are then $p_v = N_v/N_o \approx 10^{-3} - 10^{-5}$, i.e. roughly one in ten thousand lattice sites are vacant for most pure metals at their melting point, and $p_v = N_v/N_o < 10^{-16}$ at room temperature, although such low values are never achieved since high-temperature vacancies are inevitably quenched in and retained during cooling to room temperature. Typical values of diffusion coefficient are $D \approx 10^{-12} - 10^{-13}\,\mathrm{m^2/s}$, again for most pure metals at their melting point.

Vacancies in multicomponent materials

Atomic motion in a concentrated multicomponent solid solution is (not surprisingly) much more complex, essentially because of the wide variety of different local atomic structures and associated lattice distortions surrounding different vacancies, leading to both a wide range of different vacancy formation energies ΔE_v and a wide range of different vacancy migration energies ΔE_m, depending upon where the vacancy is and what exactly are its precise surrounding atoms. Just as there is an enormous number of different local atomic clusters surrounding each individual type of atom in a multicomponent solid solution (as discussed in an earlier section of this chapter), there is a similarly enormous number of different local atomic clusters surrounding the vacancies in a multicomponent solid solution or, to put it in another way, there is an enormous number of different vacancy structures. Consider a general lattice with a single atom at each lattice point, and with

n_1, n_2 and n_3 first, second and third near neighbours respectively.[xviii] The *cluster size* of atoms surrounding any vacancy is then n_1 if we only consider its first near neighbours, $n_2 + n_1$ if we also include its second near neighbours, $n_3 + n_2 + n_1$ if we also include its third near neighbours, and so on. The number of different orientation-dependent clusters N_1 of n_1 atoms, N_2 of $n_2 + n_1$ atoms, and N_3 of $n_3 + n_2 + n_1$ atoms surrounding a vacancy in a multicomponent equiatomic single-phase material with a random (ideal or regular) arrangement of c components is given[27] (as mentioned in a previous section when discussing clusters surrounding individual atoms) by the law of permutations with repetition:[xix]

$$N_1 = c^{n_1}$$

$$N_2 = c^{n_2+n_1}$$

$$N_3 = c^{n_3+n_2+n_1}$$

These are *orientation-dependent* cluster numbers and vacancy structures (i.e. equivalent vacancy structures of a different orientation are regarded as different), as is appropriate when we are dealing with properties or processes that depend on orientation. This is sometimes the case for atomic diffusion, when it takes place driven by a fixed concentration gradient in a fixed direction, such as during oxidation or carburisation at the free surface of the material. If cluster orientation is not significant, however, the number of different clusters is smaller (as discussed previously when discussing clusters surrounding individual atoms) because of the number of *self-similarity operations s* associated with the crystal symmetry:

$$N_1 = (c^{n_1})/s$$

$$N_2 = (c^{n_2+n_1})/s$$

$$N_3 = (c^{n_3+n_2+n_1})/s$$

with $s = 24$ for instance for a cubic crystal.[28,29] These are *orientation-independent* cluster numbers and vacancy structures (i.e. equivalent vacancy structures of a different orientation are regarded as the same), as is appropriate when we are dealing with properties or processes that are independent of orientation. This is also sometimes the case with atomic diffusion, when it takes place in all directions simultaneously, driven by a general overall reduction in the bulk free energy of the material, such as during the process of precipitation or grain growth, and is also the case when we are dealing with the properties of vacancies themselves, such as their volume or electronic structure.

Table 4.13 shows the numbers of oriented and non-oriented local atomic clusters surrounding vacancies N_1, N_2 and N_3, i.e. the numbers of oriented and non-oriented vacancy structures out to first, second and third near neighbours respectively, for multicomponent equiatomic single-phase fcc Cantor alloys with $c = 3$–50 components, taking for fcc $n_1 = 12$, $n_2 = 6$ and $n_3 = 24$, so the cluster sizes are 12, 18 and 42 atoms respectively. Table 4.14

[xviii] Note that $n_1 = z$, the coordination number.

[xix] We want to calculate how many ways we can pick n_1 (or, including second or second and third near neighbours, $n_2 + n_1$ or $n_3 + n_2 + n_1$) atoms from a total of c components. Let the atomic sites in the cluster be labelled 1, 2, 3, etc., up to n_1 (or $n_2 + n_1$ or $n_3 + n_2 + n_1$). Different *sets* of the c components distributed across the different atomic sites are clearly different, but different *arrangements* of any given set of c components across the different atomic sites are also different. In other words we need an *ordered permutation* rather than a *non-ordered combination*. And we need a *permutation with repetition* since each of the different component atoms can be picked more than once.[27]

Table 4.13 Number of different oriented and non-oriented local atomic clusters N_1, N_2 and N_3 around a vacancy out to first, second and third near neighbours respectively for a multicomponent single-phase fcc material with c components.

number of components c	oriented clusters			non-oriented clusters		
	1st	2nd	3rd	1st	2nd	3rd
3	5.3×10^5	3.9×10^8	1.1×10^{20}	2.2×10^4	1.6×10^7	4.6×10^{18}
4	1.7×10^7	6.9×10^{10}	1.9×10^{25}	7.1×10^5	2.9×10^9	7.9×10^{22}
5	2.4×10^8	3.8×10^{12}	2.2×10^{29}	10^7	1.5×10^{11}	9.4×10^{27}
6	2.2×10^9	10^{14}	4.8×10^{32}	9.1×10^7	4.2×10^{12}	2.0×10^{31}
8	6.9×10^{10}	1.8×10^{16}	8.5×10^{37}	2.9×10^9	7.5×10^{14}	3.5×10^{36}
10	10^{12}	10^{18}	10^{42}	4.2×10^{10}	4.2×10^{16}	4.2×10^{40}
20	4.1×10^{15}	2.6×10^{23}	4.4×10^{54}	1.7×10^{14}	1.1×10^{22}	1.8×10^{53}
50	2.4×10^{20}	3.8×10^{30}	2.2×10^{71}	10^{19}	1.6×10^{29}	9.5×10^{69}

Table 4.14 Number of different oriented and non-oriented local atomic clusters N_1, N_2 and N_3 around a vacancy out to first, second and third near neighbours respectively for a multicomponent single-phase bcc material with c components.

number of components c	oriented clusters			non-oriented clusters		
	1st	2nd	3rd	1st	2nd	3rd
3	6.6×10^3	4.8×10^6	2.5×10^{12}	2.7×10^2	2.0×10^5	1.1×10^{11}
4	6.6×10^4	2.7×10^8	4.5×10^{15}	2.7×10^3	1.1×10^7	1.9×10^{14}
5	3.9×10^5	6.1×10^9	1.5×10^{18}	1.6×10^4	2.5×10^8	6.2×10^{16}
6	1.7×10^6	7.8×10^{10}	1.7×10^{20}	7.0×10^4	3.3×10^9	7.1×10^{18}
8	1.7×10^7	4.4×10^{12}	3.0×10^{23}	7.0×10^5	1.8×10^{11}	1.3×10^{22}
10	10^8	10^{14}	10^{26}	4.2×10^6	4.2×10^{12}	4.2×10^{24}
20	2.6×10^{10}	1.6×10^{18}	6.7×10^{33}	1.1×10^9	6.8×10^{16}	2.8×10^{32}
50	3.9×10^{13}	6.1×10^{23}	1.5×10^{44}	1.6×10^{12}	2.5×10^{22}	6.2×10^{42}

shows equivalent numbers for multicomponent equiatomic single-phase bcc Senkov alloys, taking for bcc $n_1 = 8$, $n_2 = 6$ and $n_3 = 12$, so the cluster sizes are 8, 14 and 26 atoms respectively. The same results are shown graphically in Figures 4.26 and 4.27 for fcc and bcc respectively. It is obvious from Tables 4.13 and 4.14 and Figures 4.26 and 4.27 that the numbers of different vacancy structures in multicomponent single-phase materials are very large indeed. This is, of course, essentially the same result as discussed previously for the numbers of different local atomic clusters surrounding individual atoms in a multicomponent solid solution, as shown in Tables 4.3 and 4.4 and Figures 4.7 and 4.8. The only difference is that the numbers of vacancy structures are slightly smaller, since the cluster sizes are reduced by one, with a fixed vacancy at the centre of each cluster rather than a variety of c different components.

The number of oriented, first and second near-neighbour, vacancy structures in the original single-phase fcc Cantor alloy CrMnFeCoNi with five components is, for instance, just under four trillion (3.8×10^{12}), and the equivalent number of vacancy structures in the original single-phase bcc Senkov alloy also with five components is just under ten billion

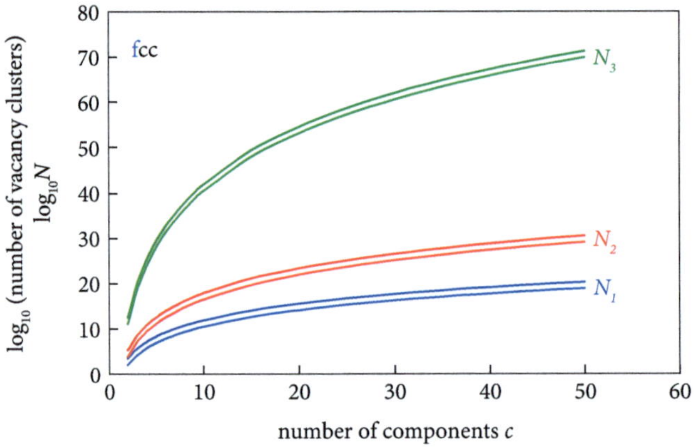

Figure 4.26 Number of different oriented (upper lines) and non-oriented (lower lines) local atomic clusters around a vacancy out to first (N_1), second (N_2) and third (N_3) near neighbours for a multicomponent single-phase fcc material with *c* components.

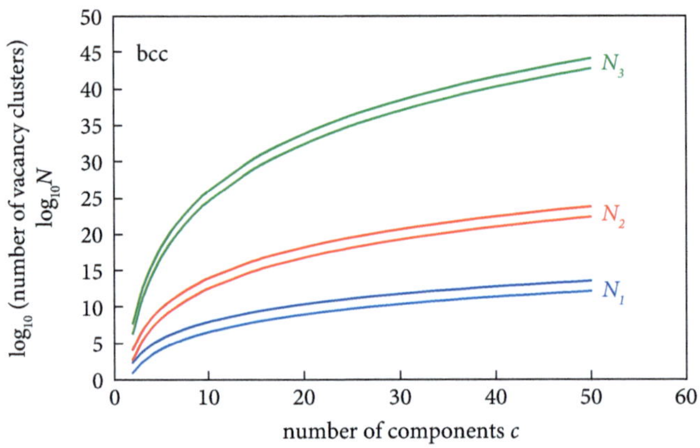

Figure 4.27 Number of different oriented (upper lines) and non-oriented (lower lines) local atomic clusters around a vacancy out to first (N_1), second (N_2) and third (N_3) near neighbours for a multicomponent single-phase bcc material with *c* components.

(6.1×10^9). With non-oriented clusters and first near neighbours only, the numbers of vacancy clusters are, of course, somewhat smaller but still very large: ten million (10^7) for the original fcc Cantor alloy and sixteen thousand (1.6×10^4) for the original bcc Senkov alloy. With six components, the number of oriented, first and second near-neighbour, vacancy clusters increases to a hundred trillion (10^{14}) for fcc Cantor alloys and almost a hundred billion (7.8×10^{10}) for bcc Senkov alloys; and with eight components to well over a quadrillion (1.8×10^{16}) for fcc Cantor alloys and well over a trillion (4.4×10^{12}) for bcc Senkov alloys. The number of vacancy clusters is, of course, very much larger again, and indeed becomes astronomical, if we include third near neighbours ($>10^{29}$ or $>10^{18}$ for the 5-component Cantor and Senkov alloys respectively) and/or increase the number of components to (say) ten or more ($>10^{18}$ or $>10^{14}$ for Cantor or Senkov alloys respectively), as shown in Tables 4.13 and 4.14 and Figures 4.26 and 4.27.

Vacancy migration in multicomponent materials

In a multicomponent solid solution material, it is important to realise that the excess energy required to create a vacancy or the *vacancy formation energy* ΔE_v is different for each of the enormous number of different vacancy structures because of the differences in local lattice distortions. It is also important to realise that the energy barrier to an adjacent atom jumping into a vacancy or the *vacancy migration energy* ΔE_m is also different for each of these different vacancy structures, again because of the differences in local lattice distortions. In other words, there is a wide spread of values for both the vacancy formation and vacancy migration energies ΔE_v and ΔE_m. There is little or no lattice distortion in a single-component solid-solution pure material, and the variation of energy as an individual vacancy hops from one lattice site to another consists of a series of *identical* jumps, as shown schematically in Figure 4.28, from identical wells at each of the lattice sites all with the same well energy E_w, over identical saddle points midway between each of the lattice sites all with the same saddle-point energy E_s. The well energy E_w is constant at all points in the hopping process because the vacancy structure, and therefore its formation energy ΔE_v, is independent of where the vacancy is in the lattice; similarly, the saddle point energy E_s is constant at all points in the hopping process, because the vacancy structure, and therefore its migration energy $\Delta E_m = E_s - E_w$, is independent of where the vacancy is in the lattice. In a concentrated multicomponent solid solution, however, the situation is very different, because of the large number of vacancy structures and associated local lattice distortions at different lattice points, the corresponding range of vacancy formation and migration energies ΔE_v and ΔE_m, and the corresponding spread of well and saddle-point energies E_w and E_s. In a concentrated multicomponent solid solution material, the variation in energy of an individual vacancy as it hops from one lattice site to another consists instead, therefore, of a series of *variable* jumps, as also shown schematically in Figure 4.28, from variable wells at the different lattice sites all with different well energies E_w, over variable saddle points midway between each of the lattice sites all with different saddle-point energies E_s.

The overall rate of atomic diffusion and, therefore, the diffusion coefficient in concentrated multicomponent solid-solution materials is affected in a variety of ways[93] by the variation in vacancy formation and migration energies ΔE_v and ΔE_m, and the corresponding spread of well and saddle point energies E_w and E_s. High-energy barriers in some places

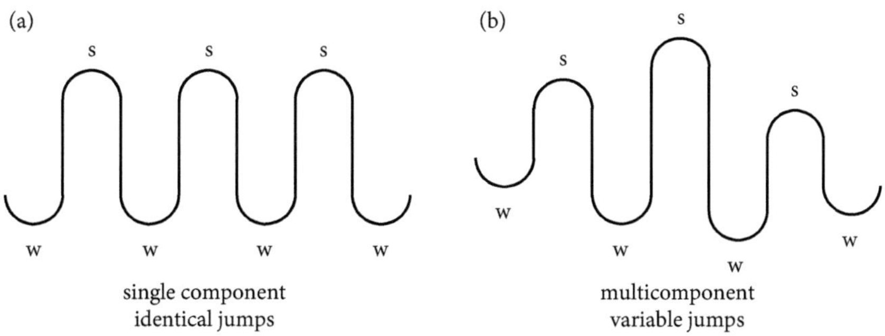

Figure 4.28 Schematic variation of energy as a vacancy hops from lattice site to lattice site: (a) in a single component material with all jumps identical; and (b) in a multicomponent material with variable jumps.
*s = saddle-point energy; w = energy well.

make it particularly difficult for a vacancy to hop from one site to the next; low energy barriers in other places make it particularly easy for a vacancy to hop from one site to the next; and a vacancy can be trapped in some places, hopping backwards and forwards between adjacent or near-adjacent lattice points, or going round and round in circles. The overall impact of these different effects is quite complex, and strictly speaking diffusion is no longer truly random spatially, since the lattice points are no longer all identical. Thomas and Patala[93] have shown that, with some simplifying assumptions, the different effects can be separated. Simple Boltzmann expressions for the probability of a vacancy being present p_v and the probability of an adjacent atom jumping into it p_m can then be replaced by expectation values $\langle p_v \rangle$ and $\langle p_m \rangle$ obtained by (rather complex) integrations[xx] over the range of different formation and migration energies ΔE_v and ΔE_m, and their corresponding well and saddle-point energies E_w and E_s, allowing for correlations between different jumps with different energies, so the diffusion coefficient D is now given by:

$$D \propto \langle p_v \rangle \langle p_m \rangle \langle F_{ij} \rangle = \langle \exp\ (-\Delta E_v/kT) \rangle \times \langle \exp\ (-\Delta E_m/kT) \rangle \times \langle F_{ij} \rangle$$
$$= \langle \exp\ (-\Delta E_v/kT) \rangle \times \langle \exp\,(-E_s/kT) \rangle \times \langle \exp\,(E_m/kT) \rangle \times \langle F_{ij} \rangle$$

where $\langle F_{ij} \rangle$ is another expectation value obtained by integration over the series of different correlation factors between different ith and jth jumps.

Figure 4.29 shows the spread of well and saddle-point energies E_w and E_s obtained by Thomas and Patala[93] for vacancies hopping between adjacent lattice sites at 1,000 °C (1,273 K) in the original 5-component single-phase fcc Cantor alloy CrMnFeCoNi, using a nudged elastic band (NEB) method with a spring constant of 10 eV/nm, relaxing the structure at intermediate points between adjacent sites by molecular-dynamics (MD) calculations with a modified embedded-atom method (MEAM) potential within the LAMMPS software. As shown in Figure 4.29, Thomas and Patala[93] obtained mean well and saddle-point energies of $\overline{E_w} = 0$ (taken as an arbitrary zero-energy reference point) and $\overline{E_s} = 0.81$ eV, i.e. the average barrier height and, therefore, vacancy migration energy was $\overline{\Delta E_m} = \overline{E_s} - \overline{E_w} = 0.81$ eV, with standard deviations of $\sigma_w = 0.08$ eV corresponding to a relatively narrow distribution of well energies, and $\sigma_s = 0.31$ eV corresponding to a relatively broad distribution of saddle-point energies. The diffusion coefficient D was found to be fairly sensitive to the spread of energy values during the hopping process, varying between a half and five times a reference diffusion coefficient D^* in an equivalent material with the same barrier energy of 0.81 eV but with constant well and saddle-point energies, as shown in Figure 4.30.

The exact results obtained by these calculations should be treated with considerable caution (as with other modelling results), since they average over only about 3,000 vacancy hops, well below the total number of almost four trillion different vacancy structures (including first and second near neighbours). In other words, they fail to capture anywhere near the full range of different vacancy structures, local lattice distortions, and corresponding formation and migration energies and well and saddle-point energies, which strictly speaking should all be sampled in order to obtain accurate results. This explains why a range of different results have been obtained by similar *ab initio* and MD calculations by different investigators, for instance $\overline{\Delta E_m} = \overline{E_s} - \overline{E_w} = 0.60 \pm 0.05$ eV by Kottke et al.[94] and 0.97 ± 0.28 eV

[xx] Notice that these integrations are not simply over the range of wells and saddle points that *exist* in the material, but instead have to be over the somewhat different range of wells and saddle points actually *sampled* by a vacancy as it hops throughout the material.

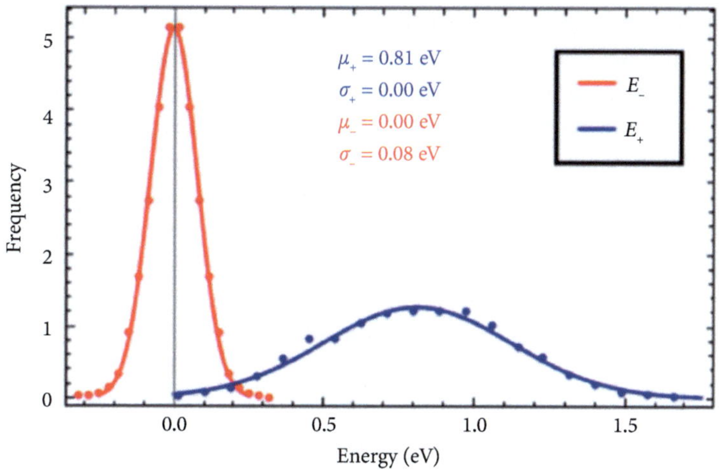

Figure 4.29 NEB calculations of the distribution of well and saddle-point energies $E_S = E_+$ (blue) and $E_W = E_-$ (red) for vacancy hopping in the original five-component fcc Cantor alloy CrMnFeCoNi, showing mean values $\overline{E_W}$ (= 0 as a reference zero energy point) and $\overline{E_S}$, and standard deviations $\sigma_S = \sigma_+$ and $\sigma_W = \sigma_-$ respectively.

Reproduced from Thomas and Patala. (2020). Vacancy diffusion in multi-principal element alloys: The role of chemical disorder in the ordered lattice. *Acta Materialia* 196: 144–53.

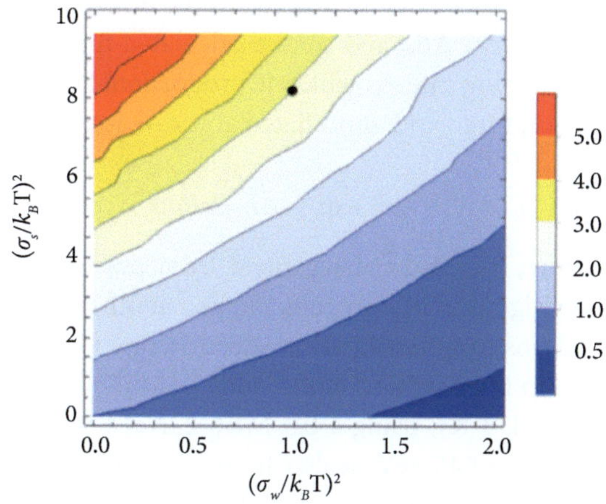

Figure 4.30 NEB calculations of the normalised diffusion coefficient (colour scale) in the original 5-component fcc Cantor alloy versus standard deviations σ_W and σ_S of distributions in well and saddle-point energies.

Reproduced from Thomas and Patala. (2020). Vacancy diffusion in multi-principal element alloys: The role of chemical disorder in the ordered lattice. *Acta Materialia 196*: 144–53.

obtained by Choi et al.[95] and Li et al.,[96] compared with 0.81 ± 0.32 eV obtained by Thomas and Patala.[93] Nevertheless, the results in Figure 4.30 do show clearly that diffusion is slower when there is a wide spread of well energies (high σ_W) because vacancies can become trapped in the lowest energy wells; but it is faster when there is a wide spread of saddle-point energies (high σ_S) because vacancies can take migration paths that simply avoid the highest energy barriers.

Positron annihilation spectroscopy or *PAS* is a high-sensitivity method for determining vacancy concentrations and, therefore, measuring vacancy formation energies.[97,98] A material of interest is exposed to high-energy positrons from a positron-emitting radioactive isotope, and they are then trapped at different positions in the crystal structure of the material. *Positron lifetimes*, i.e. the trapping times between positron capture and annihilation, depend on the local atomic structure where they are trapped, and are longest for crystal sites with relatively open local structures, typically at defects such as vacancies, dislocations, dislocation loops, etc. The resulting signal I after a counting time t consists of the sum of a series of exponential decays, as positrons in different sites are annihilated with different intensities and lifetimes, depending on the site efficiencies for capturing and annihilating the positrons:[97,98]

$$I = \sum_i I_i \exp\,(t/\tau_i)$$

where I_i and τ_i are intensities and time constants for capture and annihilation at different sites i in the crystal. In a well-annealed material, most of the positrons are trapped within the interstices of the crystal structure, providing a background signal with a characteristic trapping time constant τ_{xl}, but at high temperatures a significant number are also trapped in vacancies with a longer characteristic trapping time τ_v. The concentration of vacancies c_v at any given temperature T is equal to the probability p_v of a vacancy being present at a given lattice site which, as discussed before, is given by the Boltzmann equation as a function of the vacancy formation energy ΔE_v, and can be calculated from an Arrhenius plot of the corresponding vacancy trapping rate κ_v, which is proportional to the intensity I_v and time constant τ_v for positrons captured and annihilated at vacancies:[97,98]

$$c_v = p_v = N_v/N_o = \exp\,(-\Delta E_v/kT) \propto \kappa_v \propto I_v/\tau_v$$

where I_v, τ_v and, therefore, κ_v can be determined from positron annihilation results by stripping out the vacancy signal from positrons trapped at other sites.

Sugita et al.[99,100] used positron annihilation spectroscopy (PAS) to investigate vacancy formation and migration in the original Cantor alloy CrMnFeCoNi at temperatures in the range 20–1,200 °C (293–1,473 K), using a 0.3 MBq 22NaCl positron source emitting β^+ positrons with energies up to 545 keV, sealed in an electrical-discharge wire-cut tube of the alloy, which had been manufactured initially by arc melting followed by homogenisation for twenty-four hours at 1,100 °C (1,373 K). Positron lifetimes were determined from the time delay between initial detection of 1,274 keV γ-rays emitted concurrently with β^+ positron creation in the ^{22}NaCl source and subsequent detection of 511 keV γ-rays emitted concurrently with β^+ positron annihilation by electrons in the alloy. As shown in Figure 4.31, positron annihilation was dominated by background trapping in crystal interstices below a transition temperature of $T = 1{,}181$ K, but became dominated by vacancy trapping at higher temperatures. Typical positron annihilation lifetimes were $\tau_{xl} = 108$ ps at room temperature, and $\tau_v = 149$ ps at 1,100–1,150 °C (1,373–1,423 K). The mean vacancy formation energy was calculated to be $\overline{\Delta E_v} = 1.69 \pm 0.13$ eV from the transition temperature $T_t = 1{,}181$ K using Schulte et al.'s empirical relationship:[99,101]

$$\overline{\Delta E_v} = (-0.06 \pm 0.07) + (14.8 \pm 0.9)\, T_t \times 10^{-4}\text{eV}$$

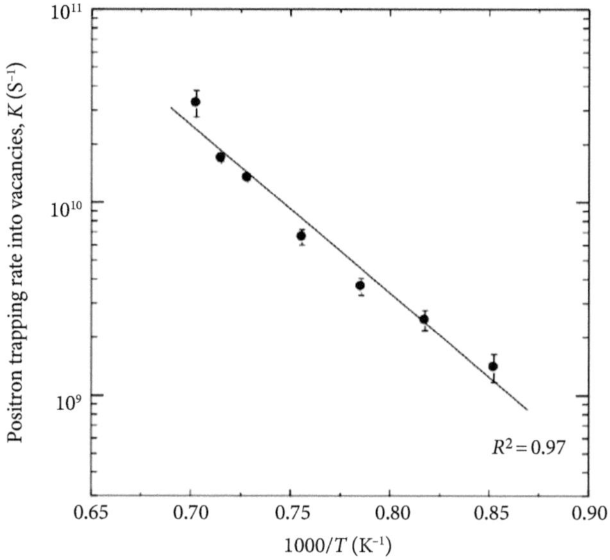

Figure 4.31 Mean positron lifetime $\bar{\tau}$ versus temperature T in the original Cantor alloy CrMnFeCoNi.

Reproduced from Sugita et al. (2020). Vacancy formation enthalpy in CoCrFeMnNi high-entropy alloy. *Scripta Materialia 176*: 32–5.

in good agreement with $\overline{\Delta E_v} = 1.72 \pm 0.18$ eV as calculated from an Arrhenius plot of ln positron trapping rate $\ln(\kappa_v)$ versus inverse temperature $1/T$, as shown in Figure 4.32. Together with Thomas and Patala's molecular-dynamics calculations[93] of the mean vacancy migration energy of $\overline{\Delta E_m} = \overline{E_s} - \overline{E_w} = 0.81$ eV, this corresponds to an overall mean activation energy for diffusion $\overline{Q}$ of:

$$\overline{Q} = \overline{\Delta E_v} + \overline{\Delta E_m} = 1.72 + 0.81 = 2.53 \text{ eV} = 244 \text{ kJ/mole}$$

somewhat lower than the measured tracer diffusion activation energies of $Q = 270$–313 kJ/mole for the different alloy components by Vaidya et al.,[102] with the underestimate probably caused by the limited sampling capability of Thomas and Patala's MD modelling of vacancy hops,[93] as discussed previously. Notice that these values of mean vacancy formation and migration energies and activation energy for diffusion are of broadly similar magnitude to those for self-diffusion in pure fcc iron Fe and nickel Ni, as shown in Table 4.12, i.e. ΔE_v = 1.72, 1.4 and 1.78 eV, ΔE_m = 0.81, 1.26 and 1.32 eV and Q = 2.53, 2.66 and 3.1 eV for the Cantor alloy, pure Fe and pure Ni respectively.

Jodi et al.[103] have used positron annihilation spectroscopy (PAS) to study a range of Cantor alloys with compositions $(CrMnCoNi)_{1-x}Fe_x$ where $x = 0.2$–0.8 (equivalent to $CrMnFe_xCoNi$ where $x = 1$–16, i.e. varying the composition by increasing the Fe content from equiatomic 20 at% up to 80 at%). The time constant for positrons trapped and then annihilated in vacancies τ_v remained constant independent of alloy composition, as shown in Figure 4.33(a), but their intensity (number annihilated per second) increased dramatically at high Fe contents, as shown in Figures 4.33(b) and (c), indicating a corresponding dramatic increase in the proportion and, therefore, concentration of vacancies at Fe sites in the crystal

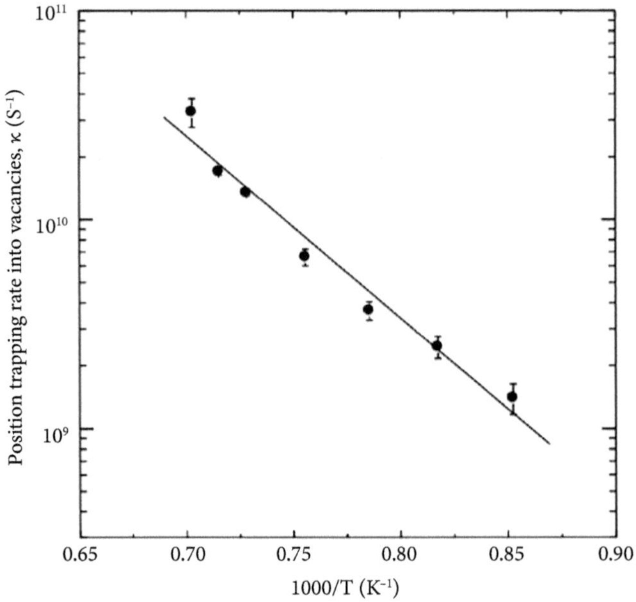

Figure 4.32 Arrhenius plot of ln positron trapping rate κ_v versus inverse temperature $1/T$ in the original Cantor alloy CrMnFeCoNi.

Reproduced from Sugita et al. (2020). Vacancy formation enthalpy in CoCrFeMnNi high-entropy alloy. *Scripta Materialia 176*: 32–5.

structure. This behaviour agrees with modelling results, indicating lower formation energies in the original Cantor alloy CrMnFeCoNi for vacancies at Fe and Mn sites compared to Co and Ni sites.[93,103]

Interdiffusion in multicomponent materials

In a simple binary A-B alloy, there are two *intrinsic diffusion coefficients* D_A and D_B corresponding to the intrinsic rate at which A and B atoms respectively jump into adjacent vacancies.[3,91,92] The formation and migration energies ΔE_v and ΔE_m and the corresponding well and saddle-point energies E_w and E_s are different for A and B atoms adjacent to and squeezing into a vacancy, so the intrinsic diffusion coefficients are also different, $D_A \neq D_B$. This means that, when diffusion takes place to equalise the concentrations of A and B in a *diffusion couple* consisting of two compositions of the A-B alloy, one A-rich and the other B-rich,[xxi] $D_A \neq D_B$ so the rate at which A atoms migrate from the A-rich alloy into the B-rich alloy is different from the rate at which B atoms migrate in the opposite direction from the B-rich alloy into the A-rich alloy.[3,91,92] And, of course, the corresponding reverse flow of vacancies from the B-rich alloy into the A-rich alloy is similarly different from the

[xxi] Strictly speaking, the description of interdiffusion in the main text applies accurately only in an ideal binary solid solution. The intrinsic diffusion coefficients D_A and D_B describe the rates at which A and B atoms respectively jump into vacancies in an A–B solid-solution at (dynamic) equilibrium, i.e. with a constant composition at all points throughout the material. In a diffusion couple, the composition is not constant and the rates at which A and B atoms jump into vacancies are given instead by the respective *chemical diffusion coefficients* $D_A^c = \alpha D_A$ and $D_B^c = \beta D_B$, where α and β are *thermodynamic factors* to allow for the variation in chemical activity of A and B atoms respectively with composition. In an ideal solid solution $\alpha = \beta = 1$.

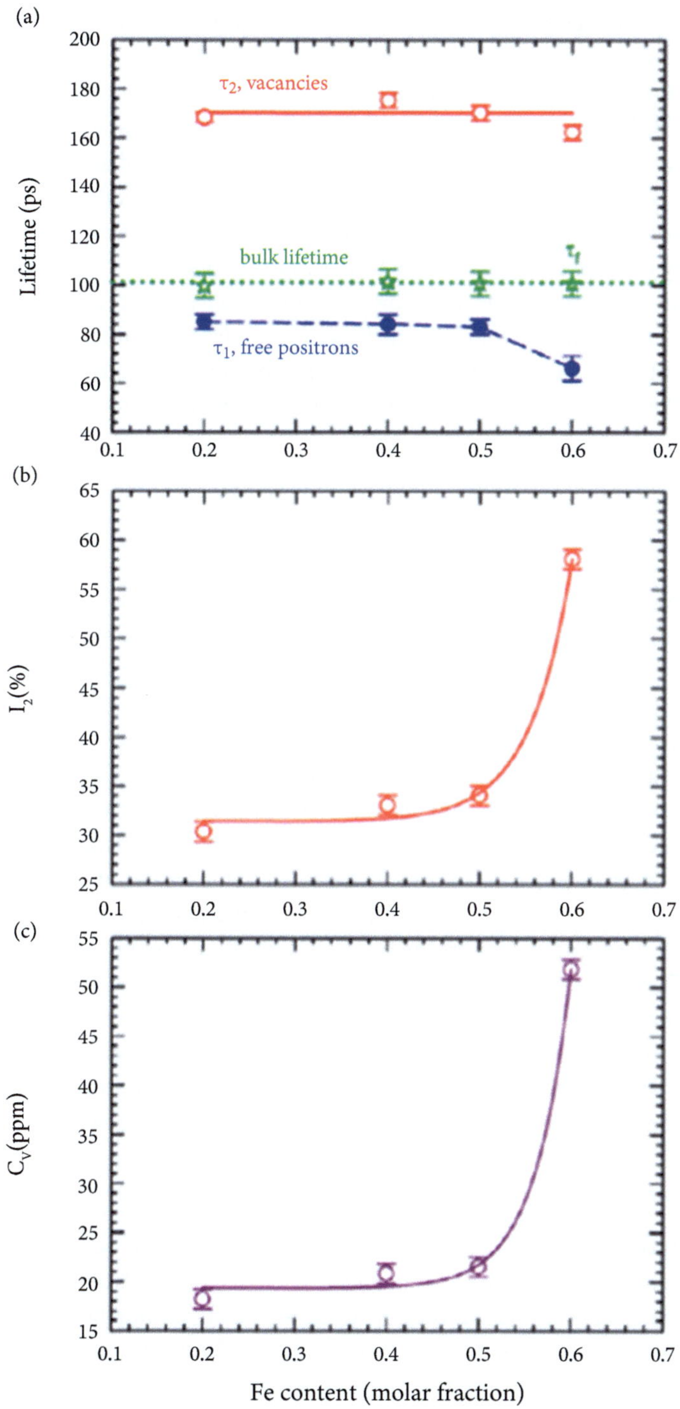

Figure 4.33 PAS measurements of positron annihilation lifetimes (a) and intensities (b) and vacancy concentrations (c) in Cantor alloys with varying Fe contents ($\tau_1 = \tau_{xl}$, $\tau_2 = \tau_v$, $\tau_f = \bar{\tau}$ and $I_2 = I_v$).

Reproduced from Jodi et al. (2021). Study of vacancies in $Fe_x(CoCrMnNi)_{100-x}$ medium- and high-entropy alloys by positron annihilation spectroscopy. *Scripta Materialia 194*: Article 113654.

reverse flow of vacancies from the A-rich alloy into the B-rich alloy. Diffusion of the two components in opposite directions trying to even up the composition throughout the diffusion couple is unbalanced, so the resulting composition profile or *diffusion profile* after some period of diffusion is asymmetric about the *Matano interface* (where the composition is halfway between the two initial alloy compositions). And an unbalanced diffusion of the two components in opposite directions produces an overall mass flow in the direction of the faster diffusing component, matched by a reverse vacancy flow in the direction of the slower diffusing component. This is called the *Kirkendall effect*.[3,91,92] The *Matano interface* moves in the direction of the vacancy flow with a velocity v, and the rate at which the two alloys intermix is determined by an *interdiffusion coefficient* $\tilde{D}_{AB}$, intermediate between the two intrinsic diffusion coefficients D_A and D_B. To be more accurate, the *Matano interface velocity* v and the *interdiffusion coefficient* $\tilde{D}_{AB}$ are given by *Darken's equations*:[3,91,92]

$$v = (D_A - D_B)\frac{dx_A}{dx} = (D_B - D_A)\frac{dx_B}{dx}$$

$$\tilde{D}_{AB} = x_A D_B + x_B D_A$$

Diffusion couples can be used quite effectively, therefore, to produce *diffusion profiles* from which the interdiffusion coefficient $\tilde{D}_{AB}$ can be obtained, with Darken's equations then being used to extract the underlying intrinsic diffusivities D_A and D_B.[3,91,92] This is not always entirely straightforward, however, since the intrinsic diffusion coefficients D_A and D_B are not necessarily constant with alloy composition, complicating the shape of the diffusion profile, the measurement of the interdiffusion coefficient $\tilde{D}_{AB}$, and the extraction of the two intrinsic diffusion coefficients D_A and D_B. A better and more direct method[3,91,92] of measuring the intrinsic diffusion coefficients D_A and D_B, though more difficult experimentally, is to use concentration profiles for interdiffusion between two isotopes of each component (one at a time) to determine *tracer diffusion coefficients*, D_A^* and D_B^*, which are effectively equal to the intrinsic diffusion coefficients at the fixed measurement composition and temperature, $D_A^* = D_A$ and $D_B^* = D_B$.

As always, the situation is much more complicated in multicomponent materials with more than two components. In general, the extension of Fick's second law to a multicomponent system with c components leads to large number of interdiffusion coefficients $\tilde{D}_{ij}$ between components i and j, each of which is composition dependent:[4]

$$\tilde{D}_{ij} = f(x_k : k = 1 \text{ to } c)$$

For any 5-component material such as the original fcc Cantor or bcc Senkov alloys, there are, for example, fifteen interdiffusion coefficients,[xxii] each of which depends on five component compositions, all of which can vary in any general composition field (though, of course, only four compositions are independent since $\sum_i x_i = 1$).[4] And the number of interdiffusion coefficients grows rapidly with increasing number of components c: reaching twenty-one in a 6-component material, for instance, with each a function of five independent component compositions; and twenty-eight in a 7-component material, with each a function of six independent component compositions. This large number of variables makes it fiendishly

[xxii] Strictly ten interdiffusion coefficients $\tilde{D}_{ij} = \tilde{D}_{ji}$ with $i \neq j$, and five self-diffusion coefficients $D_i = \tilde{D}_{ii}$ with $i = j$.

difficult, if not virtually impossible, to deconvolute general diffusion composition profiles and obtain useful diffusion coefficients. It becomes even more critical to concentrate on using tracer diffusion experiments to measure intrinsic diffusion coefficients, but of course these measurements are only relevant, as mentioned above, at a particular multicomponent alloy composition. Any real diffusion field of interest represents a complex interaction between the large number of individual intrinsic diffusion coefficients and their variation with composition throughout the composition field.

Sluggish diffusion

Sluggish diffusion was the third of Yeh's *four core effects* in multicomponent materials.[104,105] The notion of *sluggish diffusion* was first proposed in his seminal paper contributing to initiating the field of (and defining the term) high-entropy alloys,[8] in order to explain surprisingly slow observed rates of spinodal decomposition and precipitation of ordered bcc B2 nanoparticles from an initial fcc supersaturated solid-solution matrix in multicomponent Al_x(CrFeCoNiCu) alloys at intermediate Al contents of 8–11at%. The first direct study of diffusion itself in a high-entropy alloy was by Tsai et al.[13] using three pseudo-binary diffusion couples between alloys with compositions based on the original equiatomic 5-component single-phase fcc Cantor alloy CrMnFeCoNi, but with composition variations of two components in each couple. Tsai et al.[13] used two methods to try and mitigate the problems described above of analysing complex diffusion profiles in multicomponent alloys. First, they kept three component compositions constant in each of the three diffusion couples (a different three components in each couple), so that in each case there was effectively a binary diffusion couple between the other two components. Tsai et al. argued that the diffusion profiles could then be analysed to a first approximation using Darken's equations.[90,92] Second, they used alloys with relatively small composition differences, so that the motion of the Matano interface was negligible. Tsai et al. argued that the composition dependence of interdiffusion could then, again to a first approximation, be ignored. In other words, the directly measured interdiffusion coefficients were taken to be equal to the intrinsic diffusion coefficients of the diffusing components.

Figure 4.34 shows Tsai et al.'s results[13] for the intrinsic diffusion coefficients D_i of the five different components in the original Cantor alloy, plotted on Arrhenius plots as a function of inverse homologous temperature T_m/T where T_m is the melting point, compared with similar intrinsic diffusion coefficients in four Fe-Cr-Ni steels, as well as self-diffusion coefficients in pure Fe, Co and Ni. In all cases, diffusion coefficients in the Cantor alloy are between half and one order of magnitude lower than comparable diffusion coefficients in the steels and pure elements. Tsai et al.'s work[13] has been influential in supporting and promoting the concept of *sluggish diffusion* as one of the most important properties of multicomponent high-entropy solid-solution materials.

Vaidya et al.[106] made the first measurements of tracer diffusion coefficients for diffusion of the radioactive isotope ^{63}Ni in the original equiatomic 5-component single-phase fcc Cantor alloy CrMnFeCoNi and the equiatomic 4-component single-phase fcc modified Cantor alloy CrFeCoNi (i.e. without Mn), and have followed this up[80] with similar measurements of tracer diffusion coefficients for the radioactive isotopes ^{51}Cr, ^{53}Mn, ^{59}Fe and ^{56}Co in the same two single-phase fcc Cantor alloys CrMnFeCoNi and CrFeCoNi. As mentioned above, tracer diffusion experiments are more difficult to perform than interdiffusion experiments between dissimilar alloys, but the resulting tracer diffusion profiles

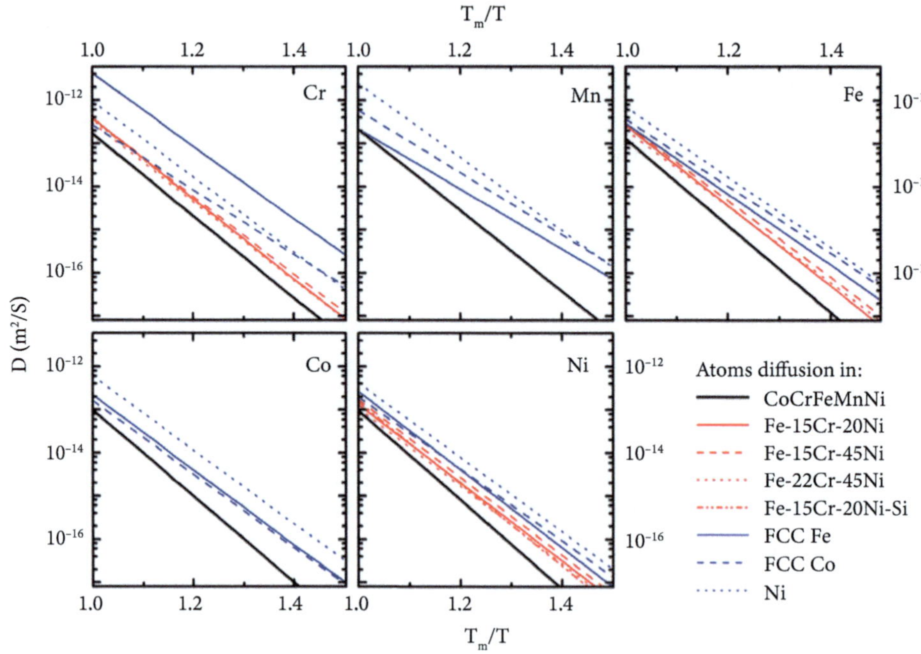

Figure 4.34 Intrinsic diffusion coefficients D_i for i = Cr, Mn, Fe, Co and Ni in the multicomponent single-phase fcc Cantor alloy CrMnFeCoNi on Arrhenius plots versus inverse homologous temperature T_m/T, compared with similar diffusion coefficients in four stainless steels Fe-15Cr-20Ni, Fe-15Cr-45Ni, Fe-22Cr-45Ni and Fe-15Cr-20Ni-Si, as well as self-diffusion in pure Fe, Co and Ni.

Reproduced from Tsai et al. (2013). Sluggish diffusion in CoCrFeMnNi high-entropy alloy. *Acta Materialia 61*: 4887–97.

are much more straightforward to analyse compared with the difficulties associated with deconvoluting the complex interdiffusion profiles that are found in multicomponent alloys. The resulting tracer diffusion coefficients D_i^* are not, therefore, prone to the uncertainties associated with complex deconvolution algorithms. In their conclusions, Vaidya et al.[102,106] throw doubt on the concept of sluggish diffusion in multicomponent solid solutions. Despite this, their results show unequivocally that, although the component tracer diffusion coefficients in the multicomponent alloys are not significantly slower than in similar conventional alloys and pure metals when compared at the same absolute temperature, they are in line with Tsai et al.'s results[13] in being approximately half to one order of magnitude slower when compared on a homologous temperature basis, as shown for Cr, Mn, Fe and Co in Figure 4.35. Comparing the two multicomponent alloys themselves shows that diffusion coefficients in the 5-component original Cantor alloy CrMnFeCoNi are not always slower than in the 4-component modified Cantor alloy CrFeCoNi, i.e. that diffusion rates and diffusion coefficients are (not surprisingly) not simply proportional to the number of components.

Table 4.15 shows Vaidya et al.'s results for the activation energies for diffusion for the different components in the two Cantor alloys. Once again, the values of activation energy for diffusion obtained by Tsai et al.[13] and Vaidya et al.[102,106] are of broadly similar magnitude

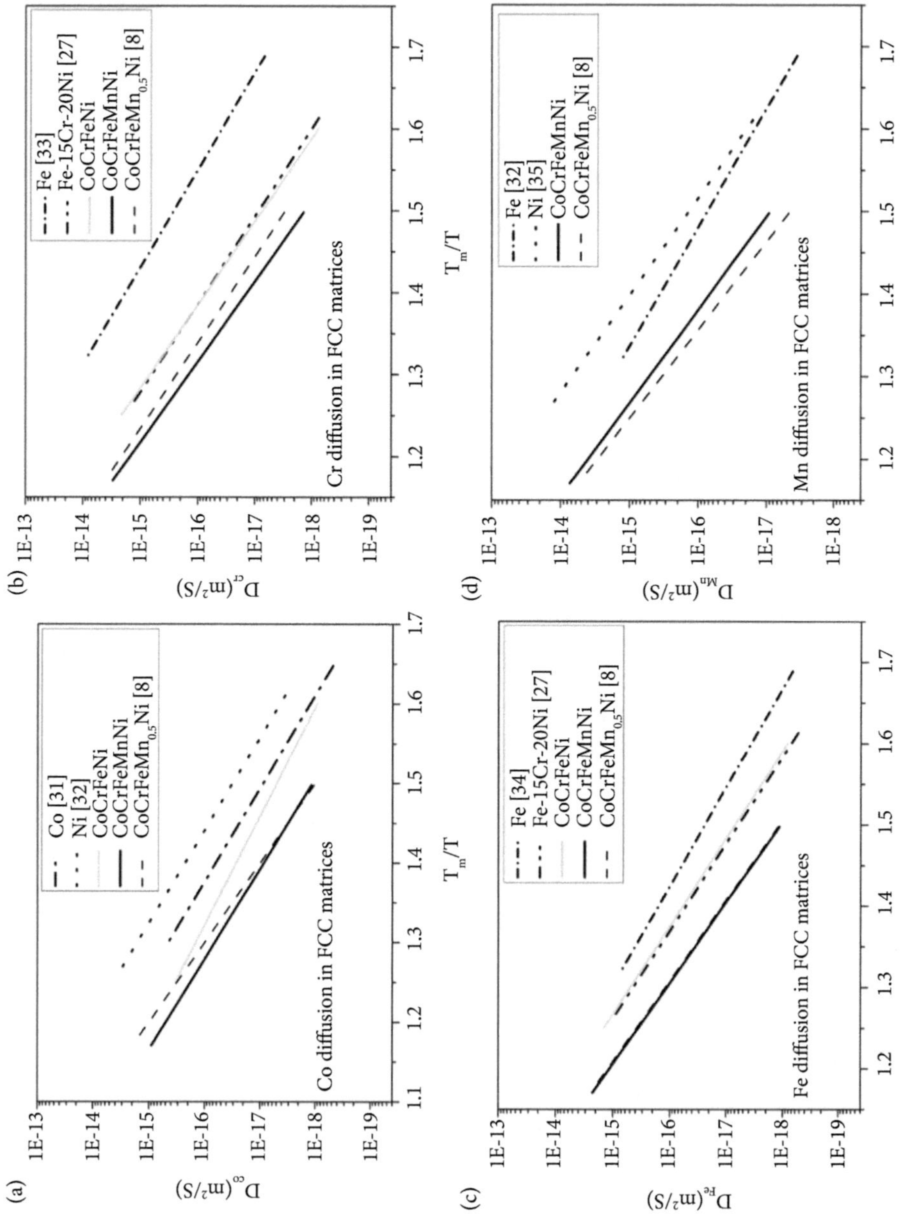

Figure 4.35 Tracer diffusion coefficients D_i^* for i = Cr, Mn, Fe and Co in the multicomponent single-phase fcc Cantor alloy CrMnFeCoNi and the modified single-phase fcc Cantor alloys $CrCoMn_{0.5}FeNi$ and CrCoFeNi on Arrhenius plots versus inverse homologous temperature T_m/T, compared with similar diffusion coefficients in the stainless steel Fe-15Cr-20Ni, as well as self-diffusion in pure Fe, Co and Ni.

Reproduced from Vaidya et al. (2018). Bulk tracer diffusion in CoCrFeNi and CoCrFeMnNi high-entropy alloys. *Acta Materialia 146*: 211–24.

Table 4.15 Activation energies for diffusion of the different components in the original equiatomic 5-component Cantor alloy CrMnFeCoNi and a modified equiatomic 4-component Cantor alloy CrFeCoNi (after Vaidya et al.[102]).

tracer	activation energy Q (kJ/mol) in CrMnFeCoNi	activation energy Q (kJ/mol) in CrFeCoNi
^{51}Cr	313 ± 13	323 ± 5
^{53}Mn	272 ± 13	-
^{59}Fe	309 ± 11	303 ± 3
^{56}Co	270 ± 22	240 ± 20
^{62}Ni	304 ± 9	253 ± 8

to those for self-diffusion in fcc Fe and Ni as can be seen either by direct inspection of Figures 4.34 and 4.35 or, more accurately, by comparing the results in Table 4.15 with those from Table 4.12 of $Q = 2.66$ eV (256 kJ/mol) and 3.1 eV (299.0 kJ/mol) for Fe and Ni respectively. This implies that the sluggishness of half to one order of magnitude in diffusion coefficient D in the Cantor alloy stems primarily from changes in the pre-exponential factor D_o, with similar overall activation energies for diffusion Q, but with a wide range in well energies E_w causing solute atoms, as discussed previously in this sub-section, to become trapped in regions of particularly low well energy.

There have been a number of other studies of interdiffusion in multicomponent materials[107–115] and a detailed review of all the results obtained so far,[116] with substantial divergence between studies claiming to support[107–112] and oppose[113–115] the concept of sluggish diffusion as a core property of high-entropy alloys. As a good example, the study by Dabrowa et al.[113] investigated a total of twenty-one diffusion couples between dissimilar alloys close to the original single-phase fcc Cantor alloy CrMnFeCoNi, the single-phase fcc modified Cantor alloys CrFeCoNi, CrMnCoNi and MnFeCoNi and single-phase fcc $Al_{0.5}$(CrFeCoNi), and used (heroically) a complex deconvolution algorithm involving many detailed assumptions and approximations, fitted thermodynamic parameters and a learning-optimisation process, to analyse the interdiffusion profiles and extract intrinsic diffusion coefficients for all the components in the five alloys. Like Vaidya et al.,[102,106] Dabrowa et al.[113] concluded that diffusion in multicomponent alloys is not sluggish, but their results again show quite clearly that component intrinsic diffusion coefficients in the original 5-component Cantor alloy and some of the 4-component modified Cantor alloys are slower than comparable conventional alloys. To make the point, in their conclusion Dabrowa et al.[113] say '*sluggish diffusion in HEAs* [high-entropy alloys] *is just an artifact*', but nevertheless towards the end of their discussion they say for each of the components one at a time that the intrinsic diffusion coefficient in CrMnFeCoNi is '*significantly slower than most of the other systems*' [including the modified Cantor alloys with fewer components, the fcc conventional alloys and the fcc pure elements] on a homologous temperature basis. They ascribe slower diffusion in some of the multicomponent alloys to the presence of manganese.

Different investigators have undoubtedly, therefore, reached different conclusions concerning the sluggishness of diffusion in multicomponent materials. But these differences are not as stark as they seem. Much of the difference arises from different definitions of *sluggish diffusion*: different *thresholds* at which diffusion can be regarded as sluggish;

whether all multicomponent materials or only some are expected to show sluggish diffusion; whether sluggish diffusion means a lower diffusion coefficient or a lower activation energy; and whether diffusion coefficients are inversely proportional to the number of components. Early investigators perhaps over-emphasised somewhat the potential effect of reduced diffusion rates in multicomponent materials, and similarly later investigators have perhaps been over-enthusiastic in pointing this out. It is quite clear from all the studies that:

1. Diffusion is slower in many multicomponent alloys compared with conventional binary alloys and/or pure metals because of the range of vacancy structures, local lattice distortions, and corresponding vacancy formation and migration energies.[13,102,106,113]
2. The effect is significant, but not enormous, given the scale of the lattice distortions associated with different vacancy structures (very large lattice distortions would lead to a breakdown of the solid-solution structure).[13,102,106,113]
3. The effect can be seen most clearly when comparisons are made on a homologous rather than absolute temperature basis.[13,102,106,113]
4. Diffusion coefficients vary from one multicomponent material to another because of differences in the range of vacancy structures, local lattice distortions, and corresponding vacancy formation and migration energies.
5. Diffusion coefficients are not proportional to the number of components, since different components lead to different levels of lattice distortion.[102,106,113]
6. However the differences between different alloys are not that large, with surprisingly similar diffusion rates when, for instance, Ni is removed,[102] Mn is replaced by Al,[117] or C is added[118,119] to the original Cantor alloy CrMnFeCoNi.
7. Activation energies are also broadly similar across different multicomponent and conventional alloys, in the range 200–300 kJ/mol.[13,106–115]

In summary, there is a clear and not insignificant propensity for sluggish diffusion in multicomponent solid-solution materials such as the Cantor alloys, caused by a wide range of different vacancy structures, local lattice distortions, and corresponding vacancy formation and migration energies, arising from the multiplicity of components. Diffusion coefficients in multicomponent materials are best measured by tracer diffusion experiments, are best compared on a homologous temperature basis, and are typically slower than comparable conventional binary alloys and pure materials by between a half and one order of magnitude, depending on the particular components in the material.

Dislocations and slip

Dislocations

A *dislocation* is one of a number of different kinds of *line defect* in the crystal structure of a material.[3,120,121] The crystal structure is displaced by an amount called the *Burgers vector* $\boldsymbol{b}$ at all points along the dislocation line. An *edge dislocation* is when the Burgers vector $\boldsymbol{b}$ is perpendicular to the dislocation line; a screw dislocation is when it is parallel to the dislocation line; and a mixed dislocation is when it is at some general angle midway between the two:

$$\text{edge}: \boldsymbol{b} \cdot \boldsymbol{l} = 0$$

$$\text{screw}: \boldsymbol{b} \cdot \boldsymbol{l} = 1$$

$$\text{mixed}: 0 < \boldsymbol{b} \cdot \boldsymbol{l} < 1$$

where $\boldsymbol{l}$ is the dislocation line vector, a unit vector parallel to the dislocation line. The dislocation character (whether it is edge, screw or mixed) is determined, therefore, by the dislocation orientation, i.e. the angle $\theta = \cos^{-1}(\boldsymbol{b} \cdot \boldsymbol{l})$. Dislocations are topological defects with Burgers vectors $\boldsymbol{b}$ that are conserved: this means that dislocations cannot end in the middle of the crystal lattice, and must either form loops surrounding the crystal displacement, end at displacement steps at a free surface or a grain boundary, or be part of a network of dislocations with a balance of Burgers vectors at each node of intersecting dislocations.

There is an excess energy per unit length of a dislocation ΔE_d, partly associated with the substantial displacement of atoms along the dislocation line itself, i.e. in the *dislocation core,* and partly associated with the strain energy associated with the somewhat smaller elastic atomic displacements induced throughout the crystal lattice outside the core. The *excess energy per unit length* or *line tension* of a dislocation ΔE_d is given by:[3,120,121]

$$\Delta E_d = E_c + \frac{\alpha G b^2}{4\pi} \ln \frac{R}{r_c} \approx G b^2$$

where E_c and r_c are the energy and radius of the highly distorted dislocation core region, G is the shear modulus, R is the dislocation spacing, the factor $\alpha = 1$ or $1 - \nu$ for screw or edge dislocations respectively, and ν is Poisson's ratio. The largest atomic displacements are in the core of a dislocation, but most of its excess energy is in the large number of smallish atomic displacements in the mid-range elastic strain field.

The most common dislocations are those with the least overall displacement of atoms and, therefore, the lowest excess energy, which is when the Burgers vector displacement is equal to the smallest perfect lattice vector that repeats the crystal structure,[3,120,121] such as $\frac{1}{2}$<110> in an fcc crystal, $\frac{1}{2}$<111> in a bcc crystal or $\frac{1}{3}$<1$\bar{1}$0> in an hcp crystal.[xxiii] According to the third law of thermodynamics, all crystals are perfect at equilibrium at absolute zero and there are, therefore, no equilibrium dislocations (or any kind of defect) at absolute zero. At any realistic temperature, however, the excess energy of creating a dislocation can be balanced by the entropy associated with dislocations being present throughout the crystal, so there is an equilibrium number of dislocations in any crystal that increases with increasing temperature. The equilibrium number of dislocations is rather small, and most materials contain a much higher, non-equilibrium number, usually generated as a consequence of external stresses and strains during manufacturing, handling and use at relatively low temperatures.

In most crystals the dislocation energy is high and dislocations are both difficult to create and immobile. In close-packed metal crystals, however, particularly fcc (and to a lesser extent hcp and bcc), the dislocation energy is much lower and dislocations are relatively easy both to create and to move.[3,120,121] It is because of the relatively easy creation and motion

[xxiii] By convention, Burgers vectors in cubic crystals are described as $\boldsymbol{b} = a/n$<uvw> or sometimes just $\boldsymbol{b} = 1/n$<uvw>, with a direction <uvw> in the crystal lattice, and a magnitude a/n, equal to the distance a/n in the <uvw> direction across the unit cell.

of dislocations that fcc metals (and to a lesser extent hcp and bcc metals) are *soft, ductile* and *tough*. Mobile dislocations move in response to shear stresses across (usually) the close-packed planes of the crystal lattice, i.e. {111} in fcc, {110} in bcc and {0001} in hcp crystals respectively, and as they move they carry the displacement of the Burgers vector with them, so the close-packed planes of atoms effectively roll over each other. This process of *dislocation slip* or *dislocation glide* produces *plastic deformation* or *plastic flow* and results in a permanent set in the crystal. Under an applied load, dislocations in an fcc metal (and to a lesser extent hcp and bcc metals) are created in large numbers and move easily and rapidly throughout the material, so the material is *soft* (deforms at a low applied stress), *ductile* (deforms extensively), and *tough* (resistant to fracture), with local dislocation creation and motion preventing the build-up of local high stresses and also preventing, therefore, the formation and growth of cracks.[3,120–123]

The detailed dynamics of dislocation creation, motion and interaction are quite complex and vary enormously depending on the material and its microstructure.[120–123] Dislocations are only *glissile*, i.e. can only *slip* or *glide*, along *slip planes* containing both their line and Burgers vectors. Screw dislocations have $\boldsymbol{b} \parallel \boldsymbol{l}$ and can slip in principle on any plane that contains the dislocation line, whereas edge and mixed dislocations have $\boldsymbol{b} \nparallel \boldsymbol{l}$, and are restricted to slip on the single plane that contains both, i.e. the plane defined by $\boldsymbol{b} \times \boldsymbol{l}$. In fcc materials whole or *perfect dislocations* with $\boldsymbol{b} = \frac{1}{2}$ <110> can often dissociate into two *Shockley partial dislocations* each with $\boldsymbol{b} = \frac{1}{6}$ <112> separated by a *stacking fault*, and this restricts both edge and screw dislocations to a single {111} *glide plane*. Because of this, most fcc materials exhibit *planar slip*, with arrays of dislocations moving along close-packed {111} planes when they are subjected to a sufficiently large shear stress known as the *critical resolved shear stress* (*CRSS*) or just the *shear flow stress*. In bcc materials, *perfect dislocations* with $\boldsymbol{b} = \frac{1}{2} < 111 >$ do not dissociate into partials, so screw dislocations can indeed slip on several planes and, because of this, bcc materials exhibit *wavy slip*, with arrays of screw dislocations moving on a variety of close-packed {110} (and sometimes other) planes when they are subjected to the critical resolved shear stress or flow stress. In all cases dislocations can become *sessile*, i.e. immobile and incapable of slip or glide, when they interact with microstructural features, such as other dislocations, grain boundaries or precipitate particles. Under continued application of an applied stress, therefore, arrays of dislocations are created, move along suitably oriented slip planes and interact with each other and with other microstructural features, partially or wholly blocking their continued motion, leading to the creation of more dislocations and increased resistance to further plastic flow.

The flow stress

Soft pure metals and conventional alloys can be *hardened* or *strengthened* by a variety of different methods, most of which, as just described, restrict the mobility of the dislocations that would otherwise allow plastic flow to take place easily. In other words, these hardening methods increase the strength and hardness of the material, but at the same time usually reduce its ductility and toughness.[122–124] *Work hardening* is when the mobility of dislocations is restricted by their mutual interaction, as more and more are created by continued plastic flow; *solution hardening* is when the mobility of dislocations is restricted by their interaction with solute atoms in a solid-solution alloy; *grain boundary hardening* is when the mobility of dislocations is restricted by their interaction with grain boundaries in a polycrystalline material; and *precipitation hardening* is when the mobility of dislocations is restricted by their interaction with small precipitate particles formed during heat treatment of an alloy with limited solubility. It is something of a simplification, but the overall *shear flow stress* τ in the material, i.e. the overall shear stress on the close packed planes required to move dislocations, can usually be treated as a straightforward sum of the different contributions to the material strength:[120–124]

$$\tau = \tau_o + \tau_{wh} + \tau_{ss} + \tau_{gb} + \tau_{ph}$$

where τ_o is the *lattice friction stress*, the intrinsic stress needed to move a dislocation in a perfect piece of the crystal, and τ_{wh}, τ_{ss}, τ_{gb} and τ_{ph} are strengthening contributions from work hardening, solution hardening, grain boundaries and precipitates respectively. Each of these contributions to the flow stress in conventional pure metals and alloys is described briefly below, one at a time.

In a perfect crystal, the energy of a dislocation rises and falls periodically as it moves across the close-packed planes from one lattice site to the next. In a pure close-packed metal or a dilute binary alloy, the *lattice friction stress* τ_o is also called the *Peierls–Nabarro stress* and is the shear stress required to overcome the maximum energy of a dislocation at positions midway between adjacent lattice sites. The Peierls–Nabarro stress is low in fcc crystals, so dislocations can move easily; it is somewhat higher in hcp and bcc crystals, so dislocations are somewhat less mobile; and it is much higher in most non-close-packed crystals, so dislocations are largely immobile.[120–124]

The *dislocation density* in a material ρ is defined as the total length of dislocation lines per unit volume or (equivalently) the number of dislocation lines intersecting unit area. Because the dislocation density is the number of dislocations crossing unit *area*, the average dislocation spacing R is inversely proportional to the square root of the dislocation density $R \propto \rho^{-\frac{1}{2}} = \alpha\rho^{-\frac{1}{2}}$ where α is a constant. The dislocation density increases as dislocations are created during plastic flow, typically from about $\rho \approx 10^{10}$–10^{11} lines/m^2 in a very well annealed soft material to $\rho \approx 10^{15}$– 10^{16} lines/m^2 in a highly deformed hard material. The dislocations become pinned and immobilised at points where they interact, with the pinning points separated by the dislocation spacing R. The shear stress at which the dislocations can move τ_{wh} is given then by a balance between the shear force $\tau_{wh}bR$ pushing them between the pinning points, and the dislocation line tension Gb^2 that opposes their bowing motion as they try to move beyond the pinning points:[120–124]

$$\tau_{wh}bR = 2Gb^2$$

$$\therefore \tau_{wh} = 2Gb/R = 2\alpha Gb\rho^{\frac{1}{2}}$$

This is called *parabolic hardening* or *Taylor hardening*, and the increase in flow stress caused by *work hardening* is proportional to the square root of the dislocation density ρ.

In a dilute solid solution binary alloy, individual solute atoms are fairly widely separated. Local strain fields around the solute atoms act as barriers to dislocation motion, with the shear stress required to overcome these barriers depending on the strength of the strain field as determined by the mismatch δ between the sizes of the solute and solvent atoms, and the precipitate spacing as determined[120–124] by the solute concentration c:

$$\tau_{ss} = G\delta^{\frac{3}{2}}bc^{\frac{1}{2}}$$

In other words, the increase in flow stress caused by *solution hardening* is proportional to the square root of the solute concentration c.

In a polycrystalline material the grain boundaries act as barriers to dislocation motion. Dislocations move across each grain and pile up at the surrounding grain boundaries until they generate sufficient force to initiate dislocation motion in adjacent grains. The force exerted per unit length of each dislocation f depends on both the applied shear stress and the Burgers vector, $f = \tau b$, and the number of dislocations n that pile up at the grain boundaries depends on the applied shear stress, the Burgers vector and the grain size, $n \propto \tau d/Gb$. Dislocations can move through the grain boundaries (or a new dislocation is created in the grain on the other side of the grain boundary) when the total force per unit length of pile up reaches a critical value F_{crit} given by:[120–124]

$$F_{crit} = nf \propto \tau_{gb}^2 d/G$$
$$\therefore \tau_{gb} = \beta d^{-\frac{1}{2}}$$

where β is a constant. This is called the *Hall–Petch equation*, and the increase in flow stress caused by grain boundary hardening is proportional to the inverse square root of the grain size d.

When a dilute binary alloy with limited solubility is quenched and then annealed, small precipitate particles are formed throughout the material. Dislocations become pinned and immobilised at points where they interact with the precipitate particles, with the pinning points separated by the precipitate spacing L. In a similar way to the pinning effect of dislocation interactions during work hardening, the shear stress at which the dislocations can move τ_{ph} is given by a balance between the shear force pushing them between the precipitate pinning points $\tau_{ph} bL$ and the dislocation line tension Gb^2 that opposes their bowing motion as they try to move beyond the precipitates:[120–124]

$$\tau_{ph} bL = 2Gb^2$$
$$\therefore \tau_{ph} = 2Gb/L$$

In other words, the increase in flow stress caused by *precipitation hardening* is proportional to the inverse of the precipitate spacing L.

Overall, therefore, the flow stress in a dilute binary polycrystalline precipitation-hardened alloy is given by the sum of these different contributions:

$$\tau = \tau_o + \tau_{wh} + \tau_{ss} + \tau_{gb} + \tau_{ph}$$
$$= \tau_o + 2\alpha Gb\rho^{\frac{1}{2}} + G\delta^{\frac{3}{2}} bc^{\frac{1}{2}} + \beta d^{\frac{1}{2}} + 2Gb/L$$

In pure fcc metals and simple fcc binary alloys, the Peierls–Nabarro stress is low so the initial flow stress is also low and deformation takes place readily under the action of external applied stresses by the easy creation and planar slip on {111} planes of $\frac{1}{2}$ <110> dislocations, that then interact with each other and with obstacles such as solute atoms, grain boundaries and precipitate particles, leading to dense dislocation *tangles* or *forests*.[120–124] In pure fcc metals and simple fcc binary alloys with a low or medium stacking fault energy, such as pure copper and gold and their dilute binary alloys, the dislocations are separated into Shockley partials making it difficult for them to cross slip around obstacles, so the dislocation tangles become increasingly intense, gradually increasing the flow stress until cracks are nucleated and grow, leading to *necking* and *ductile failure*. In pure fcc metals and simple fcc binary alloys with a high stacking fault energy, such as pure aluminium and nickel and their dilute binary alloys, the dislocations are not separated into partials making it easy for many of them to cross slip around obstacles, so the dislocation tangles re-align themselves into a network of well-defined low-angle boundaries surrounding relatively dislocation-free cells, which then continue to deform until cracks are finally nucleated, again leading to necking and *ductile failure*. In pure bcc metals and simple bcc binary alloys, the Peierls–Nabarro stress is higher, so the initial flow stress is also somewhat higher, and deformation under the action of external applied stresses is initially dominated by the difficulty of moving linear $\frac{1}{2}$ <111> screw dislocations along {110} and sometimes other slip planes from one <111> Peierls–Nabarro valley to the next, which takes place by nucleation and lateral growth of kinks in the screw dislocation lines.[120–124] The dislocations again begin to interact with each other and obstacles such as solute atoms, grain boundaries and precipitate particles, again leading to dense dislocation tangles or forests, and cross slip and the formation of low-angle boundaries and cells, with a gradual increase in flow stress again followed by crack nucleation and growth, necking and ductile failure.

Dislocations in fcc Cantor alloys

Although some features are similar, the structure and properties of dislocations are, not surprisingly, more complicated in multicomponent materials such as the fcc Cantor and bcc Senkov alloys. In a multicomponent single-phase solid solution material, there is an enormous number of different atomic structures around different atoms along the length of the core of a dislocation, in the same way as discussed previously for the number of different atomic clusters within the structure generally, as shown in Tables 4.3 and 4.4 and Figures 4.7 and 4.8, and the number of different atomic clusters surrounding vacancies, as shown in Tables 4.13 and 4.14 and Figures 4.26 and 4.27. This means that the atomic structure varies widely from point to point along any dislocation line, and also from time to time at any given point on a dislocation line as it moves from lattice site to lattice site under the action of an applied stress.

From Table 4.3, there are almost twenty trillion (1.9×10^{13}) different oriented first and second near-neighbour atomic clusters and, therefore, different local dislocation structures surrounding each individual atom at the core of a dislocation in the original equiatomic 5-component fcc Cantor alloy CrMnFeCoNi; and from Table 4.4, there are over thirty billion (3.1×10^{10}) in the original equiatomic 5-component Senkov alloy VNbMoTaW. Including first near neighbours only, the numbers are somewhat smaller but still significant, just over a billion (1.2×10^9) and two million (2×10^6) respectively; but including third near neighbours the numbers are much larger again, 1.1×10^{30} and 7.5×10^{18} respectively. Taking the atomic size as ~0.3 nm, the total dislocation length needed to sample all possible first and second near-neighbour dislocation structures is 5.7 km or 9.3 m for the Cantor and Senkov alloys respectively, corresponding to a volume of 0.57 cm^3 or 0.93 mm^3 and a grain size of 8 mm or 1 mm respectively in a well-annealed material with 10^{10} dislocation lines/m^2, and a volume of 5.7×10^6 μm^3 or 9.3×10^3 μm^3 and a grain size of 178 μm or 21 μm respectively in a highly deformed material with 10^{15} dislocation lines/m^2. In other words, there is an enormously extensive variation in local atomic structure along the line of any dislocation in an equiatomic 5-component fcc Cantor or bcc Senkov alloy, which persists over very large volumes of material.[xxiv]

Okamoto et al.[125] used transmission electron microscopy (TEM) techniques to study dislocation structures in the original equiatomic 5-component single-phase fcc Cantor alloy CrMnFeCoNi after compression testing at room temperature and 77 K (−196 °C). Figure 4.36 shows two bright-field (BF) images of the dislocation structure after a few percent strain at 77 K in two different grains: one in Figure 4.36(a) oriented with the (111) slip planes parallel to the thin foil specimen and perpendicular to the electron beam, i.e. with the electron beam direction $\boldsymbol{B}$ = [111]; and the other in Figure 4.28(b) oriented with the slip planes at about 30° to the thin foil specimen, i.e. with the electron beam direction $\boldsymbol{B}$ = [131]. Figure 4.36(a) shows large numbers of $\boldsymbol{b} = \frac{1}{2}[\bar{1}01]$ dislocations, all moving within the (111) plane of the foil in a direction roughly from bottom left to top right, with the dislocation lines mostly in a screw or 60° orientation (i.e. $\theta = \cos^{-1}\boldsymbol{b} \cdot \boldsymbol{l} = 0°$ or 60°), and with generalised planar slip confined to the (111) plane of the thin foil and little or no cross slip into different directions out of the plane. Figure 4.36(b) shows the dislocations lined up into *trains* or *pileups*, with individual dislocation lines running from top to bottom of the thin foil, on a number of different (but parallel) inclined (111) slip planes. These dislocation structures in the Cantor alloy are in general very similar to those found in pure fcc metals and dilute fcc binary alloys with a relatively low *stacking fault energy*, i.e. without easy *cross slip*.

Okamoto et al.[125] confirmed that the stacking fault energy in the Cantor alloy was fairly low, as can be seen from the weak-beam TEM micrograph in Figure 4.37(a) of a dislocation with a Burgers vector $\boldsymbol{b} = \frac{1}{2}[\bar{1}01]$ that has separated into two *Shockley partials* with Burgers vectors $\boldsymbol{b} = \frac{1}{6}$ < 112 >. The high-resolution weak-beam dislocation image in Figure 4.37(a) shows some important special features of dislocations in multicomponent materials such as the Cantor alloy that have since been confirmed by other investigators. First, the dislocations are not straight and are wavy on a near-atomic scale, because of the wide range of different local atomic structures along the dislocation line. Second, the two Shockley partials are separated on average by a distance of about d = 5.7 nm, but this separation distance is not constant along the dislocation line, varying by ± 1 nm over the short segment shown in Figure 4.37(a), again because of the wide range of different local atomic structures in the partial dislocation cores and across the plane of the stacking fault. Figure 4.37(b) shows the results of a large number of similar measurements of the separation of Shockley partials d as a function of dislocation orientation $\theta = \cos^{-1}(\boldsymbol{b} \cdot \boldsymbol{l})$. The separation between the partials is in the range d = 3–5 nm for a screw dislocation with $\theta = 0°$, increasing gradually with increasing θ to d = 5–8 nm for an edge dislocation with $\theta = 90°$. Third, the separation between two Shockley partials represents a balance between a repulsive force proportional to Gb^2 resulting from the overlap of their two stress fields, and an attractive force proportional to the stacking fault energy γ_{sf} trying to reduce the size of the stacking fault between them, so that:[120,121]

$$d = \alpha Gb^2/8\pi\gamma_{sf}$$

[xxiv] It is not, perhaps, entirely fanciful to compare and contrast the total information content coded in an array of twenty trillion different local atomic clusters along a 5.7-km length of a dislocation line coiled up in half a cubic centimetre of the Cantor alloy, with a typical 100 μm diameter human cell containing forty-six chromosomes each of which is a DNA molecule with 50–200 million base pairs, making a total of approximately three billion letters in the genetic script describing and controlling all aspects of human life.[3]

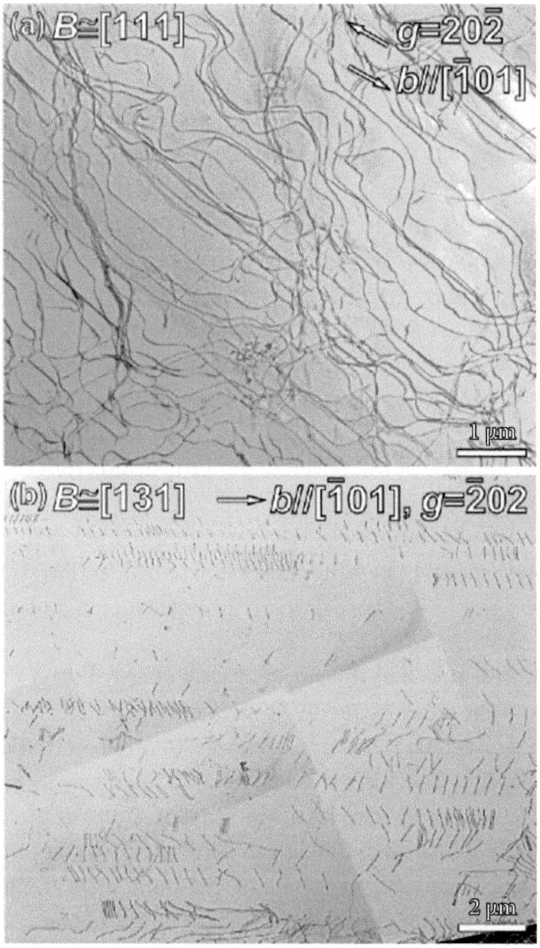

Figure 4.36 TEM BF image of $\frac{1}{2}$ <110> dislocations in the original Cantor alloy CrMnFeCoNi after a few percent compression at 77 K: (a) in a grain with the (111) slip plane oriented parallel to the thin foil specimen and perpendicular to the electron beam B = [111]; and (b) in a grain with the (111) slip plane oriented at about 30° to the thin foil with the electron beam B = [131].
Reproduced from Okamoto et al. (2016). Size effect, critical resolved shear stress, stacking fault energy, and solid solution strengthening in the CrMnFeCoNi high-entropy alloy. *Nature Scientific Reports* 6: Article 35863.

where $\alpha \approx 0.3$–1.6 is a factor depending on the Poisson's ratio and the dislocation character (screw, edge or mixed). The average partial separation of d = 5.7 nm for the dislocation segment in Figure 4.37(a) corresponds to an average stacking fault energy of γ_{sf} = 30 mJ/m^2, but the range of partial separation distances indicates that the stacking fault energy is also not constant along the dislocation line, again because of the wide range of different local atomic structures along the dislocation line, varying by ± 5 mJ/m^2 for the range of dislocations and associated partial separations shown in Figure 4.37(b). Fourth, the flow stress required to move the dislocations is considerably higher than that usually found in pure fcc metals and simple dilute fcc binary alloys (the significance of this is considered in more detail later in this section when discussing the flow stress in multicomponent alloys).

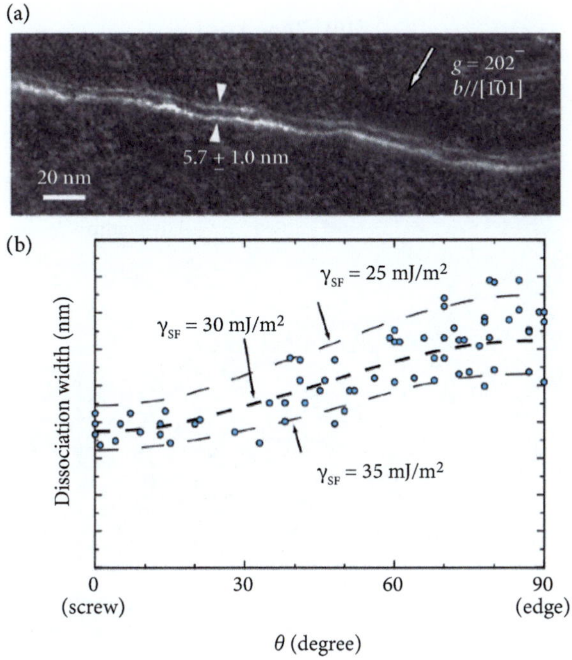

Figure 4.37 (a) High-resolution weak-beam image of a $\frac{1}{2}$ <110> dislocation in the original Cantor alloy CrMnFeCoNi showing a wiggly dislocation line and separation into two $\frac{1}{6}$ <112> Shockley partials; and (b) partial separation distance *d* in a large number of dislocations as a function of dislocation orientation θ, ranging from pure screw (θ = 0°) to pure edge (θ = 90°).
Reproduced from Okamoto et al. (2016). Size effect, critical resolved shear stress, stacking fault energy, and solid solution strengthening in the CrMnFeCoNi high-entropy alloy. *Nature Scientific Reports* 6: Article 35863.

As mentioned above, the general structure of dislocations observed by Okamoto et al.[125] in Cantor alloys has been confirmed by a number of other studies. Figure 4.38 shows another example from Smith et al.[126] of planar slip of dislocations with Burgers vectors $\boldsymbol{b} = \frac{1}{2}$ <110> aligned in pile-ups on inclined {111} slip planes in the original Cantor alloy CrMnFeCoNi after 5% strain, this time in tension and at room temperature. Figure 4.39 shows end-on high-resolution images using high-angle annular dark-field scanning transmission electron microscopy (HAADF-STEM) from two of the dislocations with sharply different partial separation distances of d = 3.2 and 7.8 nm. Smith et al.[126] used this end-on imaging technique to measure the separation distances of thirty dislocations in the strained alloy, all with Burgers vectors $\boldsymbol{b} = \frac{1}{2}$ <110> and orientations θ = 60°. The resulting distribution of separation distances d is shown in Figure 4.40, with an average d = 4.8 nm similar to that found by Okamoto et al.,[125] and with a somewhat wider spread of values, ranging from d = 1–9 nm. Smith et al.[126] quote the results from similar studies in pure fcc metals, showing much less variability (lower by about an order of magnitude) of no more than approximately ± 0.5 nm in partial separation distances and, therefore, much less variability (again lower by about an order of magnitude) in the corresponding stacking fault energy along the length of the dislocation lines. Smith et al.[126] also used high-resolution energy-dispersive (EDS) X-ray mapping to show that the variation in partial separation distance in the Cantor alloy is not caused by a varying nm-scale segregation of different components along the partial dislocation cores.

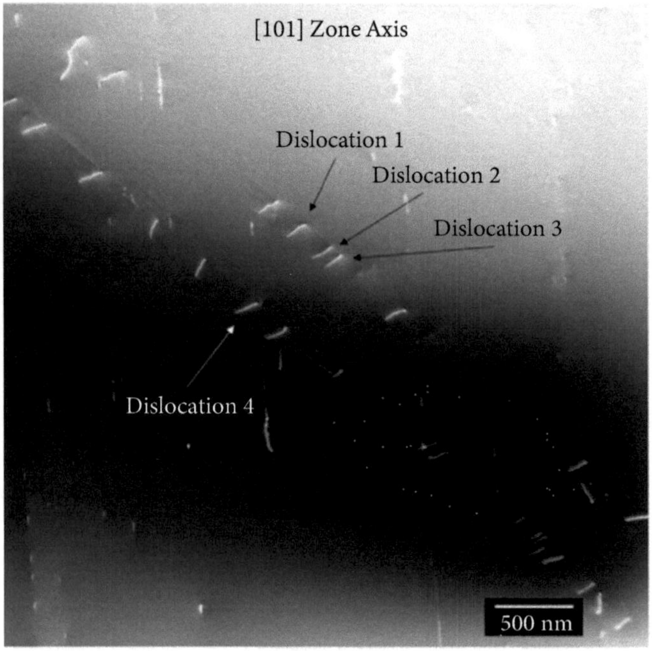

Figure 4.38 TEM BF image of $\frac{1}{2}$ <110> dislocations piled up on inclined {111} slip planes in the original Cantor alloy CrMnFeCoNi after 5% tensile strain.
Reproduced from Smith et al. (2016). Atomic-scale characterization and modelling of 60° dislocations in a high-entropy alloy. *Acta Materialia 110*: 352–63.

Laplanche et al.[127] and Xu et al.[128] employed similar TEM techniques to those used by Okamoto et al.[125] and Smith et al.[126] to investigate dislocation structures in the modified single-phase fcc Cantor alloys CrCoNi and $Al_{0.1}$CrFeCoNi respectively. Figures 4.41 and 4.42 show their results obtained in each case from a large number of dislocations with Burgers vectors $\boldsymbol{b} = \frac{1}{2} < 110 >$ for the separation of Shockley partials d as a function of dislocation orientation $\theta = \cos^{-1}(\boldsymbol{b} \cdot \boldsymbol{l})$. The 3-component modified Cantor alloy CrCoNi and the Al-containing modified Cantor alloy $Al_{0.1}$CrFeCoNi both have lower stacking fault energies than the original 5-component Cantor alloy CrMnFeCoNi and, therefore, correspondingly greater separation of the Shockley partials. As in the original Cantor alloy, the separation between the partials is again spread broadly in both of the modified Cantor alloys: in CrCoNi, in the range d = 4–7 nm for a screw dislocation with θ = 0°, increasing gradually with increasing θ to d = 8–14 nm for an edge dislocation with θ = 90°; and, in $Al_{0.1}$CrFeCoNi, in the range d = 5–18 nm for a screw dislocation with θ = 0°, increasing gradually with increasing θ to d = 9–33 nm for an edge dislocation with θ = 90°. The corresponding stacking fault energies are also spread broadly: in the range γ_{sf} =18–26 mJ/m^2, with an average of γ_{sf} = 22 mJ/m^2 for CrCoNi; and in the range γ_{sf} =6–21 mJ/m^2, with an average value of γ_{sf} = 12 mJ/m^2 for $Al_{0.1}$CrFeCoNi.

Table 4.16 shows measured stacking fault energies for Cantor alloys together with similar results from some typical pure fcc metals and conventional fcc alloys for comparison. The multicomponent Cantor alloys have low stacking fault energies in the range $\gamma_{sf} \approx$ 12–30 mJ/m^2: similar to low stacking fault energy fcc metals such as Ag and Au with $\gamma_{sf} \approx$ 22–32 mJ/m^2; a bit higher than very low stacking fault energy fcc alloys such as α–brass Cu–30%Zn

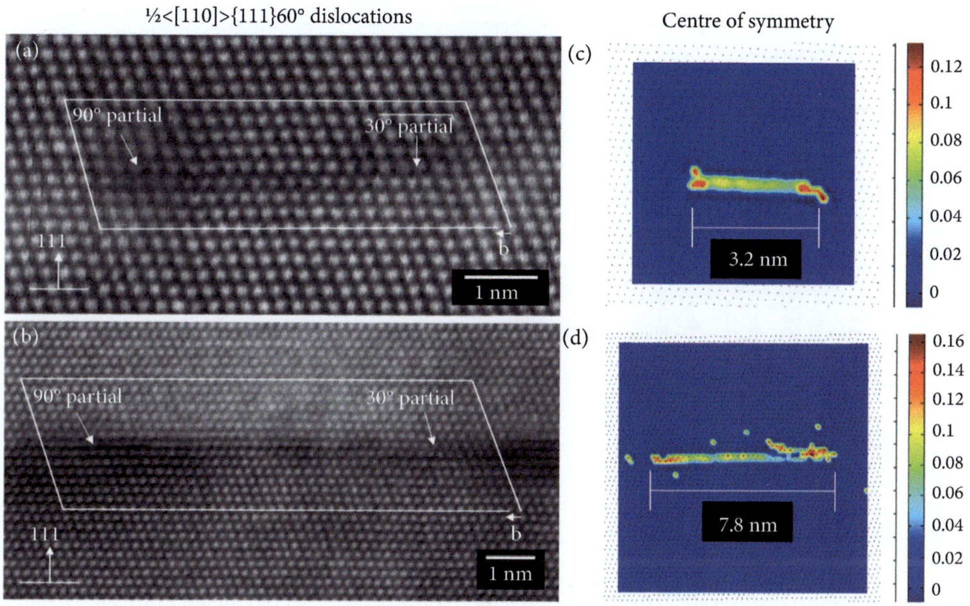

Figure 4.39 HAADF-STEM end-on images of two dislocations in the original Cantor alloy CrMnFeCoNi showing different partial separation distances of (a) $d = 3.2$ nm and (b) $d = 7.8$ nm.

Reproduced from Smith et al. (2016). Atomic-scale characterization and modelling of 60° dislocations in a high-entropy alloy. *Acta Materialia 110*: 352–63.

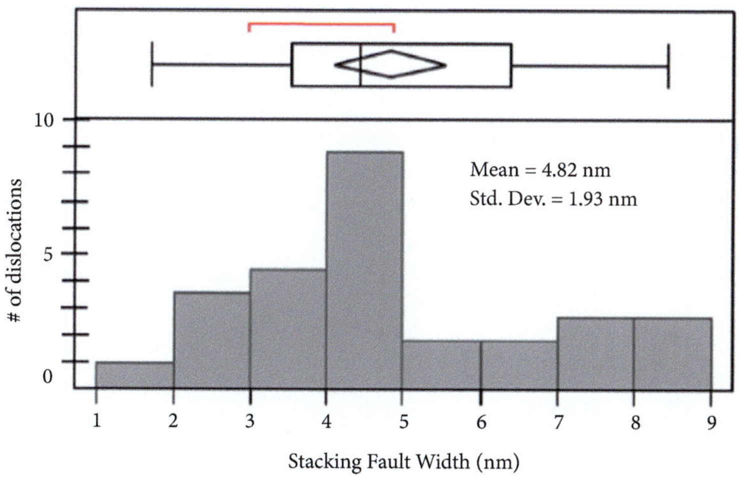

Figure 4.40 Distribution of partial separation distances d for thirty $\frac{1}{2}$ <110> 60° dislocations in the original Cantor alloy CrMnFeCoNi.

Reproduced from Smith et al. (2016). Atomic-scale characterization and modelling of 60° dislocations in a high-entropy alloy. *Acta Materialia 110*: 352–63.

and aluminium bronze Cu–5%Al with $\gamma_{sf} < 12$ mJ/m^2, exhibiting very extended stacking fault ribbons between their Shockley partials, but well below typical high stacking fault energy fcc metals such as pure aluminium and nickel with $\gamma_{sf} \gg 100$ mJ/m^2, exhibiting

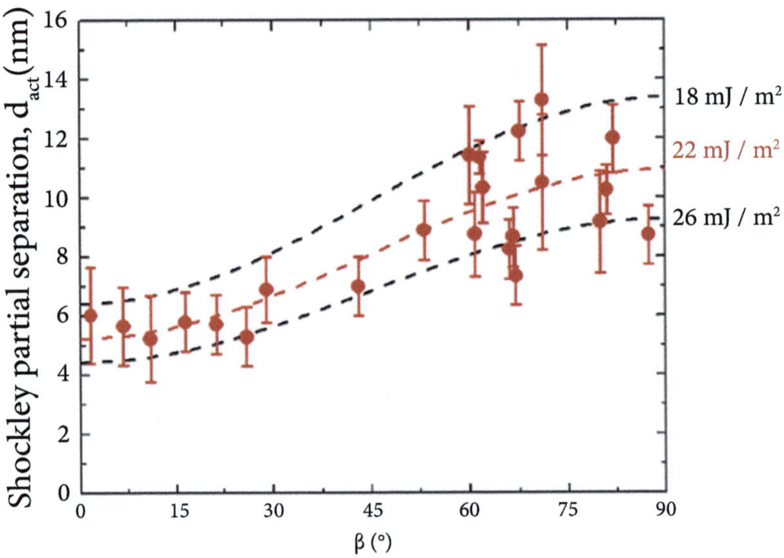

Figure 4.41 Partial separation distance *d* in a large number of $\frac{1}{2}$ <110> dislocations in modified Cantor alloy CrCoNi as a function of dislocation orientation $\theta (= \beta)$, ranging from pure screw ($\theta = 0°$) to pure edge ($\theta = 90°$).

Reproduced from Laplanche et al. (2017). Reasons for the superior mechanical properties of medium-entropy CrCoNi compared to high-entropy CrMnFeCoNi. *Acta Materialia 128*: 292–303.

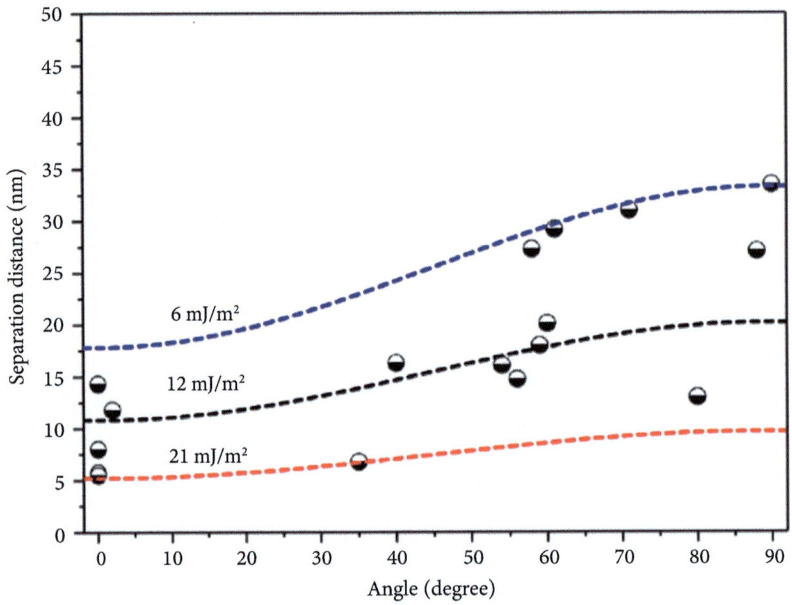

Figure 4.42 Partial separation distance *d* in a large number of $\frac{1}{2}$ <110> dislocations in modified Cantor alloy $Al_{0.1}CrFeCoNi$ as a function of dislocation orientation θ, ranging from pure screw ($\theta = 0°$) to pure edge ($\theta = 90°$).

Reproduced from Xu et al. (2018). Transmission electron microscopy characterization of dislocation structure in a face-centred cubic high-entropy alloy $Al_{0.1}CoCrFeNi$. *Acta Materialia 144*: 107–15.

Table 4.16 Measured stacking fault energies in Cantor alloys γ_{sf} compared with some pure fcc metals (after Cai and Nix[120]) and some conventional fcc alloys (after Lu et al.,[129] Otte and Adler[130] and Rohatgi et al.[131]).

	material	stacking fault energy γ_{sf} (mJ/m^2)
Cantor alloys	CrMnFeCoNi	30
	CrCoNi	22
	$Al_{0.1}$CrFeCoNi	12
pure fcc metals	Al	166
	Ni	128
	Cu	78
	Au	32
	Ag	22
fcc alloys	Fe-20Cr-10Ni	21
	Cu-30Zn	12
	Cu-8Al	<5

effectively no separation of Shockley partials and very easy and extensive cross slip during plastic flow.

Pasianot and Farkas[132] used Sandia's LAMMPS molecular-dynamics simulator, with binary interatomic potentials derived by averaging from approximate embedded-atom (EAM) potentials for each of the individual components to investigate the variation in local atomic structure along the length of dislocations in a 5-component equiatomic modified single-phase fcc Cantor alloy CrFeCoNiCu (i.e. with Cu replacing Mn in the original Cantor alloy). Figure 4.43 shows an example of one of their results for the atomic structure along a length of approximately 10 nm of a screw dislocation with Burgers vector $\boldsymbol{b} = \frac{1}{2}$ <110>, lying on a {111} plane and separated into two Shockley partials with Burgers vectors $\boldsymbol{b} = \frac{1}{6}$ <112>, for a particular initial random configuration of the five components. Overall, Pasianot and Farkas's modelling results confirmed many of the experimental results described previously. As seen in Figure 4.43, the dislocation lines were wavy on an atomic scale, and the partial separation distance varied along the dislocation lines, with a corresponding inverse variation in stacking fault energy. The partial separation distance was lowest for a screw orientation and highest for an edge orientation; and (of course) the corresponding stacking fault energy was highest for a screw orientation and lowest for an edge orientation. As well as varying along the length of any dislocation line, the waviness and the partial separation distance also both varied at any given point on the dislocation as it moved along the {111} slip plane, as can be seen in Figure 4.44. The Peierls–Nabarro stress and the corresponding lattice friction stress were found to be larger than for simple pure fcc metals and dilute binary fcc alloys.

Similar modelling results to those obtained by Pasianot and Farkas[132] for CrFeCoNiCu have also been obtained by Rao et al.[133] for a different 4-component modified single-phase fcc Cantor alloy $Ti_{16.7}Fe_{16.7}Co_{30}Ni_{36.7}$ and by Liu et al.[134] for the original Cantor alloy CrMnFeCoNi. As mentioned previously in the context of modelling electronic band structures and vacancy formation and migration energies in multicomponent alloys, detailed calculations of this type should be treated with considerable caution, since the interaction potentials

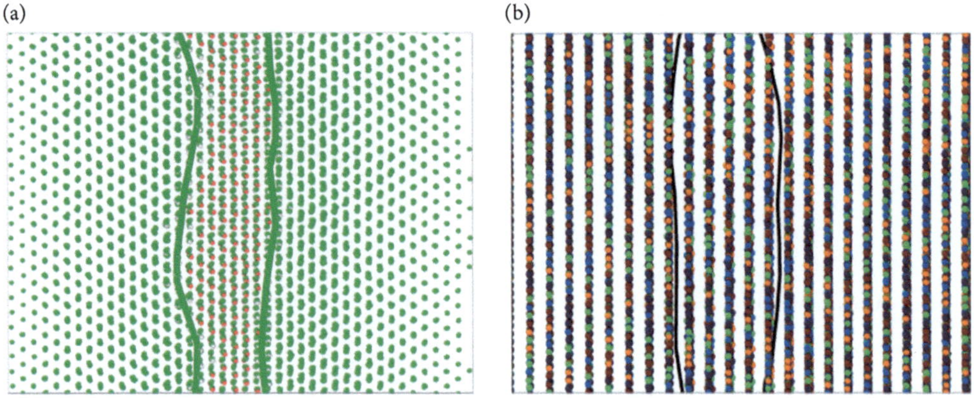

Figure 4.43 MD-modelled local atomic structure along ~10 nm of a $\frac{1}{2}$ <110> screw dislocation lying on a {111} slip plane and separated into two $\frac{1}{6}$ <112> Shockley partials in a modified single-phase fcc Cantor alloy CrFeCoNiCu: (a) atoms coloured according to near-neighbour structure with green = fcc and red = hcp; and (b) atoms coloured according to the five different components in the alloy.

Reproduced from Pasianot and Farkas. (2020). Atomistic modeling of dislocations in a random quinary high-entropy alloy. *Computational Materials Science 173*: Article 109366.

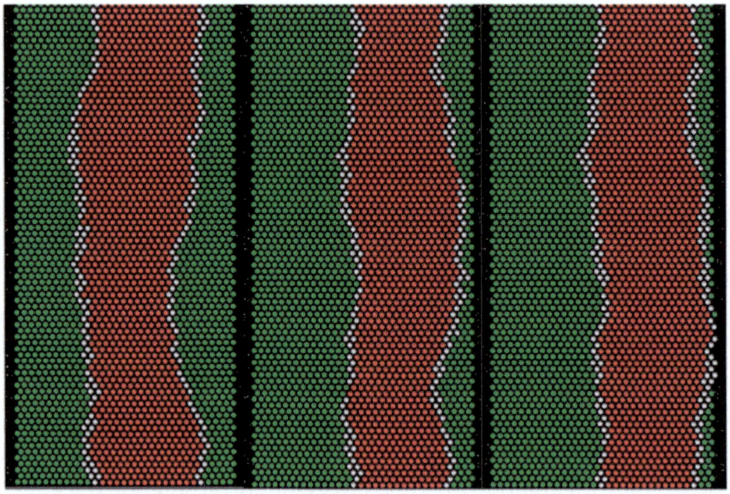

Figure 4.44 MD-modelled local atomic structure along ~25 nm of a $\frac{1}{2}$ <110> edge dislocation lying on a {111} slip plane and separated into two $\frac{1}{6}$<112> Shockley partials in a modified single-phase fcc Cantor alloy CrFeCoNiCu, showing the variation in partial separation distances along the dislocation line at three different times during dislocation glide.

Reproduced from Pasianot and Farkas. (2020). Atomistic modeling of dislocations in a random quinary high-entropy alloy. *Computational Materials Science 173*: Article 109366.

are only approximate and not too realistic, and since the total number of local atomic structures investigated along the cores of the dislocation lines is no more than several thousand at best, well below the total number of almost twenty trillion different local atomic structures (including first and second near neighbours). In other words, such models use only rather approximate calculations and fail to capture anywhere near the full range of different core structures, local lattice distortions and corresponding partial separation distances, which strictly speaking should all be sampled in order to obtain accurate results.

Dislocations in bcc Senkov alloys

Couzinie et al.[135] and Lilensten et al.[136] both used transmission electron microscopy (TEM) techniques to study dislocation structures in the equiatomic 5-component single-phase bcc modified Senkov alloy TiZrNbHfTa after tensile testing at room temperature. Figures 4.45(a)–(c) show three TEM images of the dislocation structures generated in the material by the application of a tensile stress up to plastic strains of approximately 2, 3 and 10% respectively. Large numbers of dislocations are created during deformation and plastic flow, mostly screw dislocations with Burgers vectors $\boldsymbol{b} = \frac{1}{2}$ <111>, moving in trains and pile-ups on a variety of {110}, {112} and other slip planes, distributed in thin bands that gradually increase in number and intensity as the plastic strain builds up in the material. As with the Cantor alloys, these dislocation structures in the modified bcc Senkov alloy are in general very similar to those found in pure bcc metals and dilute binary bcc alloys, although the flow stress required to move the dislocations is again relatively high compared to pure bcc metals and simple dilute bcc binary alloys.

More recent studies by Wang et al.[137] of the modified equiatomic ternary bcc Senkov alloy TiNbMo and by Lee et al.[138] of the equiatomic modified quaternary bcc Senkov alloys TiVNbMo and VCrNbMo have shown, however, that the situation is more complicated, with a range of screw, edge and mixed dislocations gliding on a variety of {110}, {112}, {123} and other slip planes. Figure 4.46 shows an example of a variety of different dislocation types moving on different slip planes during deformation of ternary TiMoNb under the action of a nanoindentation stress field. In pure bcc metals and simple dilute bcc binary alloys, slip is relatively easy except for screw dislocations; deformation is controlled, therefore, by the need to nucleate kinks in screw dislocations lying along one <111> Peierls–Nabarro energy well so as to move them on to the next one; and the resulting dislocation structures are, therefore, dominated by pile-ups of straight screw dislocations, usually predominantly on {110} slip planes. In multicomponent bcc Senkov alloys however, there is a wide range of local atomic structures with associated variation in lattice distortion along the cores of the dislocation lines, making dislocation slip much harder generally and leading to a much wider variety of different slip modes, with more corrugated or wavy dislocations of different orientations on different slip planes, all playing an important role during deformation.

As well as the direct experimental results discussed above, the structure and motion of dislocations of different types moving on different slip planes have also been investigated quite intensively in multicomponent bcc Senkov alloys using a variety of computer modelling techniques, including *ab initio* density functional theory (DFT),[139,140] molecular dynamics (MD) with embedded-atom (EAM) interatomic potentials,[139] phase-field dislocation dynamics (PFDD)[141] and Monte Carlo (MC) simulations,[142] applied to a variety of multicomponent bcc Senkov alloys such as ternary TiZrNb,[139] TiMoNb[141] and VNbTa,[142] and quaternary NbMoTaW.[140] There are advantages and disadvantages of the different modelling schemes, with more fundamental methods such as DFT being the most difficult to apply to large atomic assemblies and, vice versa, methods that cope with large numbers of atoms such as MC being less fundamental. They all suffer, however, from two major problems: first, they all have considerable difficulty in getting suitable and accurate input

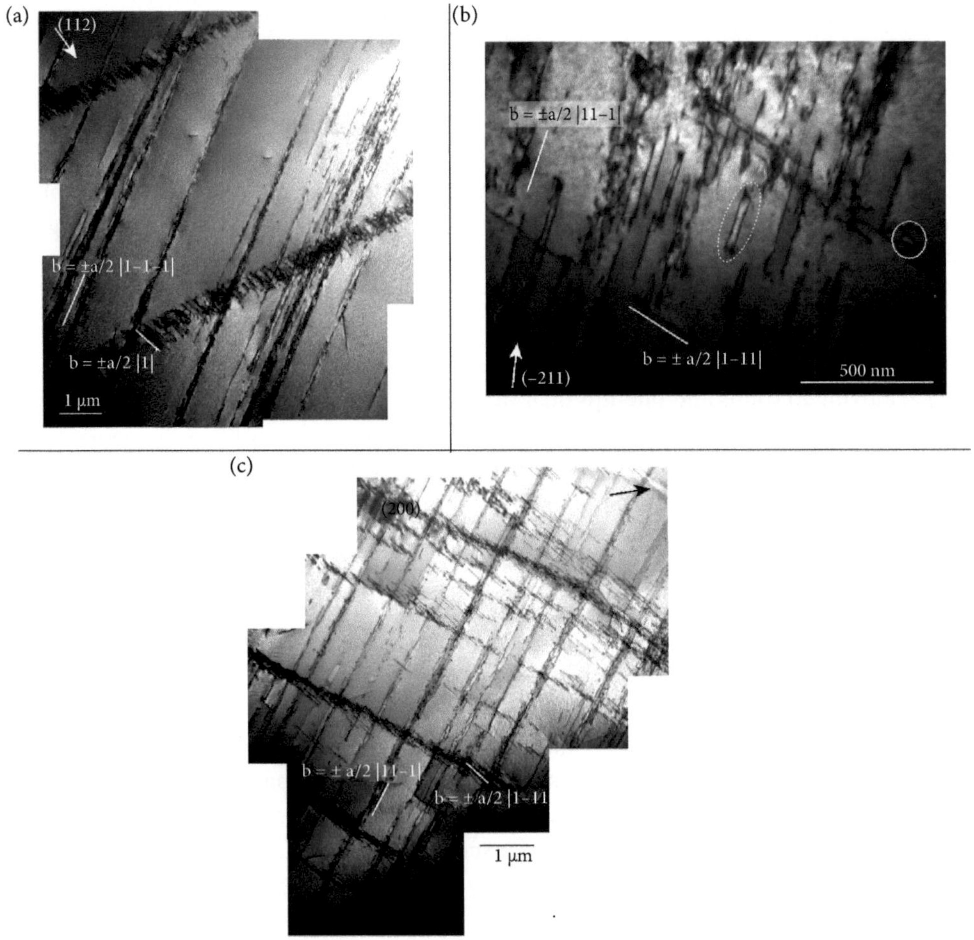

Figure 4.45 Dislocation structures in equiatomic TiZrNbHfTa after: (a) 2.2%, (b) 2.9% and (c) 10% plastic strain in tension at room temperature.

Reproduced from Lilensten et al. (2018). Guillot I. Study of a bcc multi-principal element alloy: Tensile and simple shear properties and underlying deformation mechanism. *Acta Materialia 142*: 131–41.

data for any particular multicomponent alloy, usually requiring, therefore, a number of approximations and estimates with various degrees of certainty; and second that none of them come anywhere near modelling the enormous number of different local atomic structures that are strictly needed to obtain a reasonable sample of even a 3-component material (from Table 4.3, there are fourteen million, just over a billion and just over thirty billion different oriented first and second near-neighbour local atomic structures in equiatomic 3-, 4- and 5-component bcc materials respectively). Nevertheless and perhaps therefore somewhat surprisingly, all the modelling studies[139–142] so far come up with the same key features of the basic structures and properties of dislocations in multicomponent bcc Senkov alloys, which agree well with the experimental results for bcc Senkov alloys[135–138] just discussed:

1. Dislocations in multicomponent bcc Senkov alloys are wavy on an atomic scale because of the large number of different local atomic clusters and corresponding wide

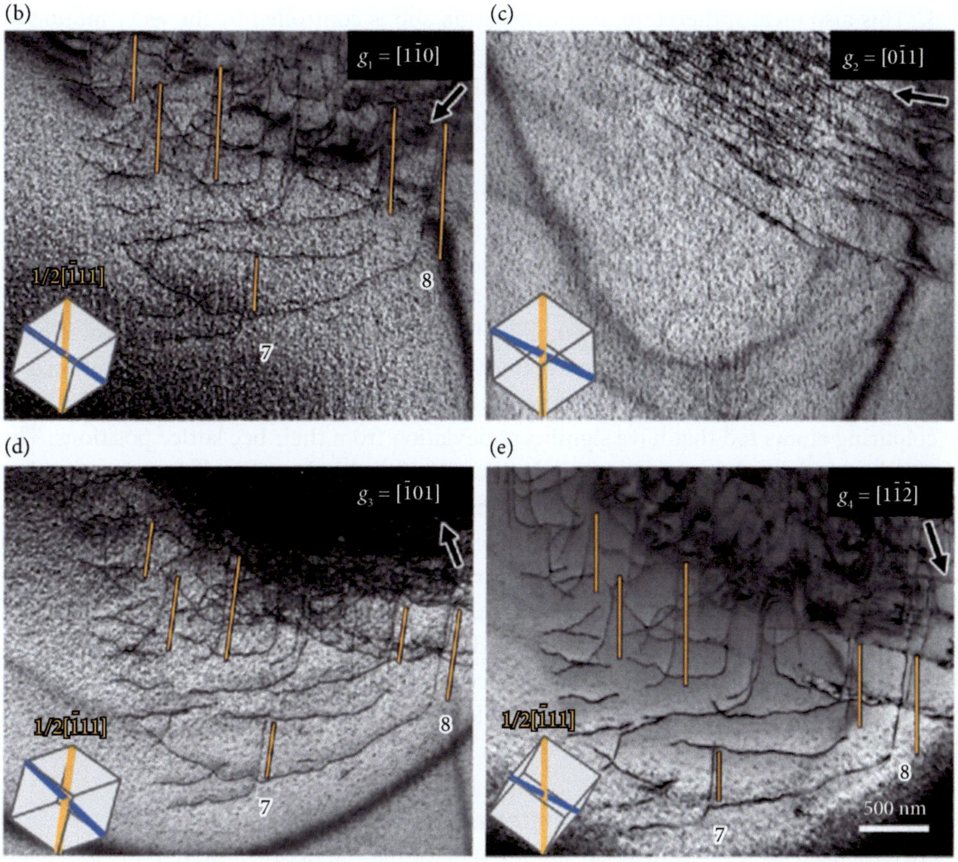

Figure 4.46 Two-beam TEM images of dislocation structures in modified ternary Senkov alloy TiNbMo under four different two-beam conditions, showing a mixture of screw, edge and mixed dislocations slipping on a variety of slip planes. The blue and orange lines show projections of edge and screw orientations respectively.

Reproduced from Wang et al. (2020). Multiplicity of dislocation pathways in a refractory multiprincipal element alloy. *Science 370*: 95–101.

variation in local lattice strain caused by the distribution of different component atoms along the core of the dislocation lines.

2. The large number of different local atomic clusters and corresponding wide variation in local lattice strain caused by the distribution of different component atoms along the core of the dislocation lines produces an increase as well as a variability in the local Peierls–Nabarro stress.
3. This means that dislocation slip is considerably harder for all dislocation types (edge, screw and mixed) moving on all possible slip planes ({110}, {112} and {123}), because the different local atomic clusters act effectively as atomic-scale barriers to dislocation motion, with the resulting flow stress considerably higher than in pure bcc metals and dilute bcc binary alloys.
4. The screw/edge Peierls–Nabarro stress ratio in multicomponent bcc Senkov alloys is much smaller, by up to two orders of magnitude, compared with pure bcc metals and dilute bcc binary alloys

5. This also means, therefore, that dislocation slip is controlled by the jerky motion of wavy dislocations of *all* types on *all* slip planes rather than the dominant mechanism of nucleation of kinks in <111> screw dislocations on {110} planes as usually seen in pure bcc metals and dilute bcc binary alloys.

Figure 4.47 gives some of the results of molecular-dynamics (MD) calculations obtained by Rao et al.[139] using the LAMMPS software system with Zhou–Johnson embedded-atom (EAM) interatomic potentials,[143] showing the structure of two dislocations with Burgers vectors $\boldsymbol{b} = \frac{1}{2}[111]$ after 60 ps of slip under a stress of 0.1G (where G is the shear modulus) at a temperature of 5 K in the ternary modified bcc Senkov alloy TiZrNb. One dislocation is in a pure edge orientation with an average line vector $\boldsymbol{l} = [\bar{1}\bar{1}2]$, and the other dislocation is in a screw orientation with an average line vector $\boldsymbol{l} = [111]$. The dislocation cores are identified by colouring atoms red that have significant deviation from their bcc lattice positions. The edge dislocation core is somewhat spread out compared with the screw dislocation core, but both have wavy lines and slip in a jerky manner. The critical resolved shear stress to move an edge or mixed dislocation was found to be approximately half that required to move a screw dislocation, compared with a much larger edge/screw ratio of about a hundred in pure bcc metals. Figure 4.48 gives another example of the results of modelling dislocation slip, obtained by Fey et al.[138] using phase-field dislocation dynamics (PFDD) calculations, showing very similar jerky motion of wavy screw dislocations with Burgers vectors $\boldsymbol{b} = \frac{1}{2}$ <111> in the ternary modified bcc Senkov alloy TiMoNb, when slipping on four different slip planes, $(1\bar{1}0), (11\bar{2}), (12\bar{3})$ and $(13\bar{4})$.

Plastic flow in multicomponent materials

There have been quite a few detailed studies of deformation, plastic flow, fracture and overall mechanical properties in the original Cantor alloy CrMnFeCoNi, and a number of other modified Cantor alloys, notably 4-component and 3-component single-phase fcc alloys such as CrFeCoNi and CrCoNi, and Al-containing single-phase fcc alloys in the system Al_xCrMnFeCoNi[14,125,144–156]. There are good reviews of this work by Li et al.[157] and George et al.[158]

Otto et al.[144] manufactured billets of the original 5-component single-phase fcc Cantor alloy CrMnFeCoNi by arc melting, drop casting under argon, solution treating for forty-eight hours at 1,200 °C (1,473 K), cold rolling by 87%, and annealing for one hour at 800, 1,000 and 1,200 °C (1,073, 1,273 and 1,423 K respectively), to develop homogeneous equiaxed microstructures with grain sizes of 4.4, 50 and 155 μm respectively as shown in Figure 4.49, in order to study the mechanical properties of the alloy under tensile loading conditions. Figure 4.50 shows typical tensile test curves from specimens with 4.4 and 155 μm grain sizes at temperatures ranging from −196 to 800 °C (77 to 1,073 K). The yield strength, ultimate tensile strength and ductility of the large-grain-size Cantor alloy all increase, from ~100 to 400 MPa, ~0.15 to 1 GPa, and ~30 to 100% respectively, as the testing temperature is decreased progressively from 800 to −196 °C (1,073 to 77 K), in all cases showing prolonged and substantial work hardening, leading to extended ductility and a large increment of the ultimate tensile strength over the yield strength.[144] Similarly, the yield strength, ultimate tensile strength and ductility of the small-grain-size Cantor alloy also all increase, in this case at somewhat higher levels, from ~250 to 600 MPa, ~0.5 to 1.5 GPa, and ~40 to 100% respectively, as the testing temperature is decreased progressively

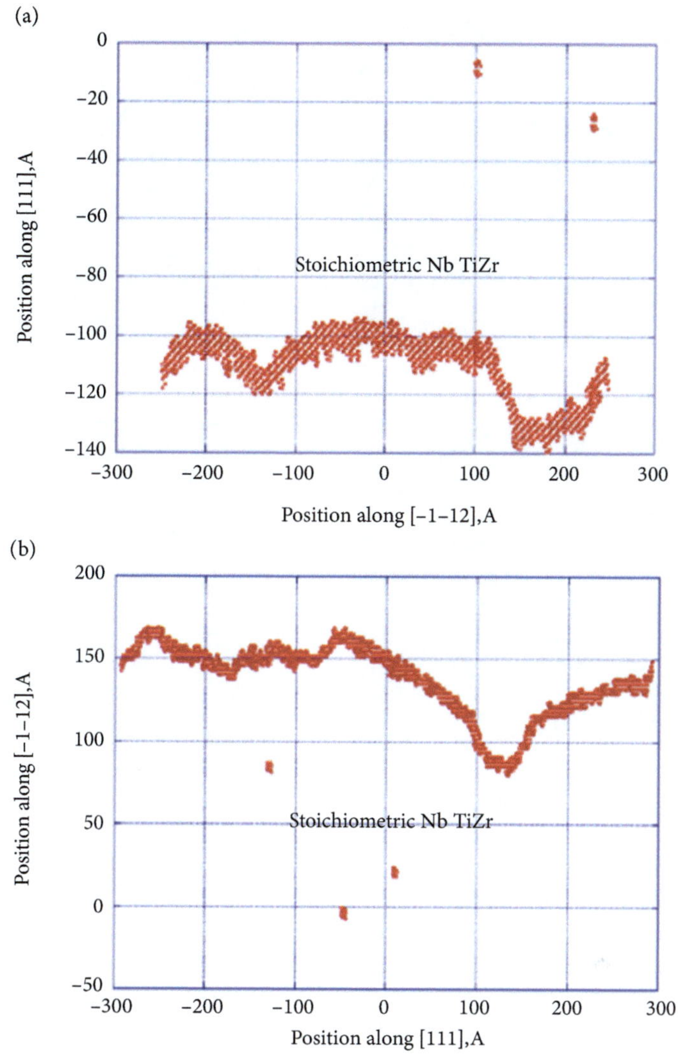

Figure 4.47 MD modelling of structure of $\frac{1}{2}[111]$ dislocations in ternary bcc Senkov alloy TiZrNb: (a) edge orientation with average $\boldsymbol{l} = [\bar{1}\bar{1}2]$; (b) screw orientation with average $\boldsymbol{l} = [111]$.
Reproduced from Rao et al. (2019). Modeling solution hardening in bcc refractory complex concentrated alloys: NbTiZr, $Nb_{1.5}TiZr_{0.5}$ and $Nb_{0.5}TiZr_{1.5}$. *Acta Materialia 168*: 222–36.

from 600 to −196 °C (873 to 77 K), again in all cases showing prolonged and substantial work hardening, leading to extended ductility and a large increment of the ultimate tensile strength over the yield strength.[144] The only exception to this general behaviour is the alloy with 4.4 μm grain size tested at 800 °C (1,073 K), which shows no work hardening, with an ultimate tensile strength equal to the yield strength, and slow deformation and tearing from the yield point to final failure, caused by extensive grain boundary deformation, as might be expected for this special case of particularly small grain size and high testing temperature.

Figure 4.51 shows transmission electron micrographs of typical dislocation structures generated at 20 and −196 °C (293 and 77 K) at strains of between ~4 and 20%, from another

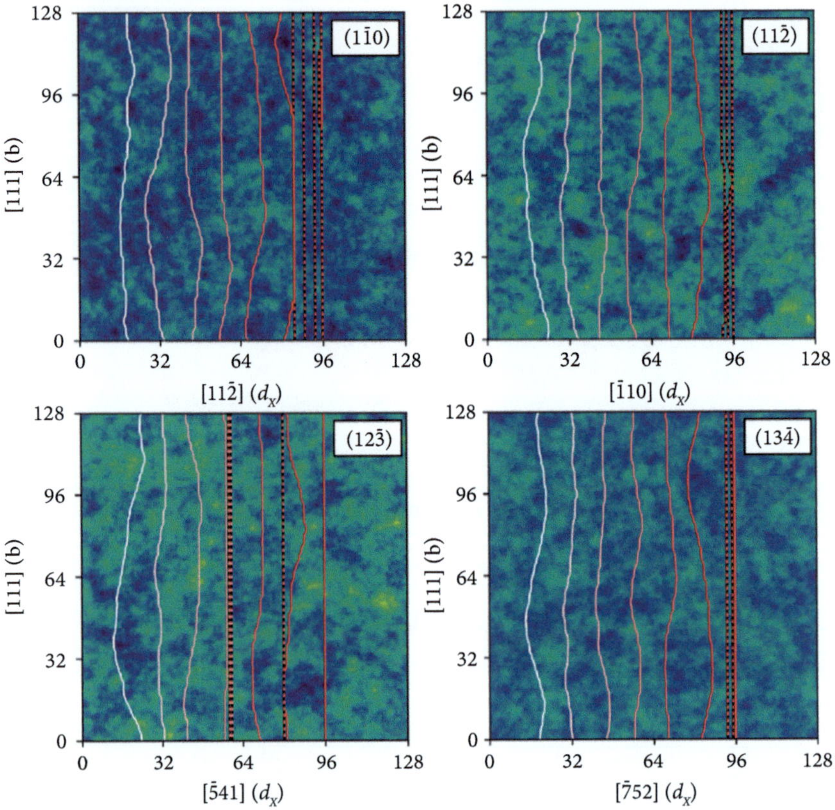

Figure 4.48 PFDD modelling of the jerky motion of four wiggly $\frac{1}{2}$ <111> screw dislocations on $(1\bar{1}0)$, $(11\bar{2})$, $(12\bar{3})$ and $(13\bar{4})$ slip planes in ternary bcc Senkov alloy TiNbMo. Coloured lines represent sequential positions of dislocations as they move under an applied stress, with lighter colours later in time.

Reproduced from Fey et al. (2022). Transitions in the morphology and critical stresses of gliding dislocations in multiprincipal element alloys. *Physical Review Materials 6*: Article 013605.

investigation of tensile deformation in the 5-component single-phase fcc Cantor alloy CrMnFeCoNi by Laplanche et al.[145] As already discussed, deformation in the Cantor alloy takes place similarly in many ways to that seen in low stacking fault energy (SFE) conventional fcc alloys and pure fcc metals,[124,157,158] i.e.:

1. Initial yield takes place by simple planar slip of $\frac{1}{2}$<110> dislocations, mostly separated into Shockley partials, on {111} slip planes.[144,145,157,158]
2. Dislocations pile up at grain boundaries and other obstacles, initiating secondary slip, creating dislocation intersections and increasing the number of obstacles to slip, leading to the formation of extensive dislocation tangles and forest dislocations.[144,145,157,158]
3. Dislocation multiplication and interactions lead to classical parabolic Taylor work hardening, as shown in Figure 4.52, with a square root dependence of flow stress on dislocation density, and a roughly inverse parabolic dependence on plastic strain.[157,158]

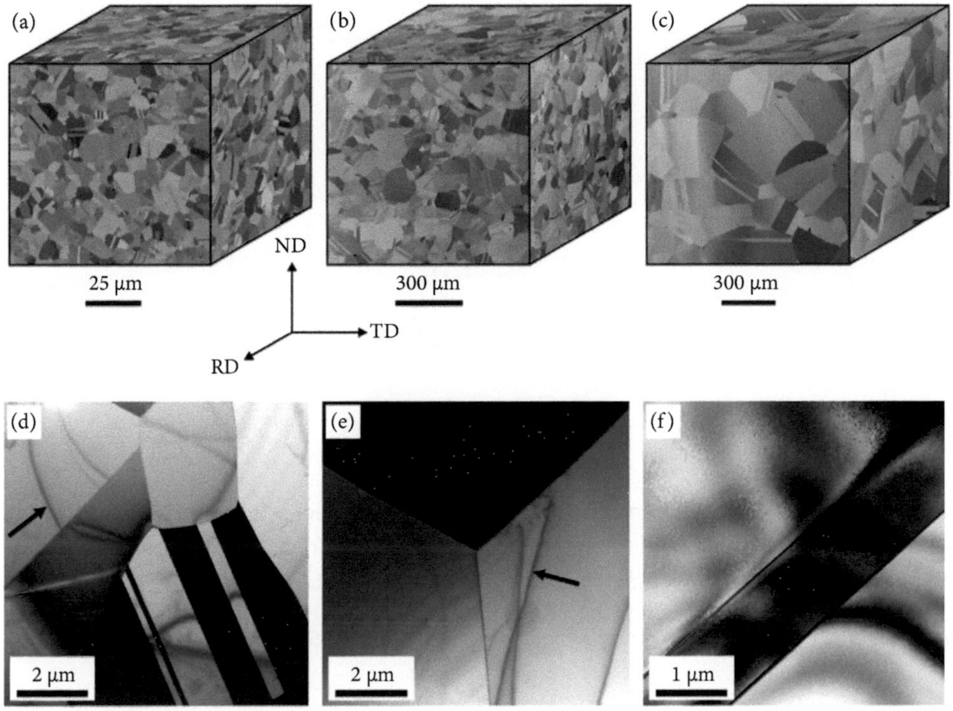

Figure 4.49 Optical and transmission electron micrographs of equiaxed grain structures in the original Cantor alloy CrMnFeCoNi manufactured by casting, rolling and annealing for one hour at 800 °C, 1000 °C and 1200 °C, with grain sizes of (a) 4.4 μm, (b) 50 μm and (c) 155 μm.

Reproduced from Otto et al. (2013). The influence of temperature and microstructure on the tensile properties of a CoCrFeNiMn high-entropy alloy. *Acta Materialia 61*: 5743–55.

4. There is a significant, largely temperature-independent Hall-Petch effect for the flow stress at different grain sizes,[144,145,157,158] as shown in Figure 4.53, with a Hall-Petch coefficient of $k \approx 500\ \mathrm{MPa}/\mu\mathrm{m}^{\frac{1}{2}}$.
5. There is relatively insignificant strain-rate dependence of deformation and flow stress at normal low strain rates,[147,158] but a stronger effect at high strain rates.[145,146]

As discussed previously, there are also, however, important differences compared with conventional fcc alloys and pure fcc metals:

1. Initial yield strengths are higher, because of the presence of different local atomic structures and associated lattice distortions along the length of the dislocation cores, which act as pinning centres of strain, increasing the local Peierls–Nabarro stress and impeding dislocation motion.[144,145,158]
2. Dislocation lines are wavy on a near-atomic scale, again because of the presence of different local atomic structures and associated lattice distortions along the length of the dislocations, as they bow out between the pinning centres of strain under the action of applied stress.[144,145,158]

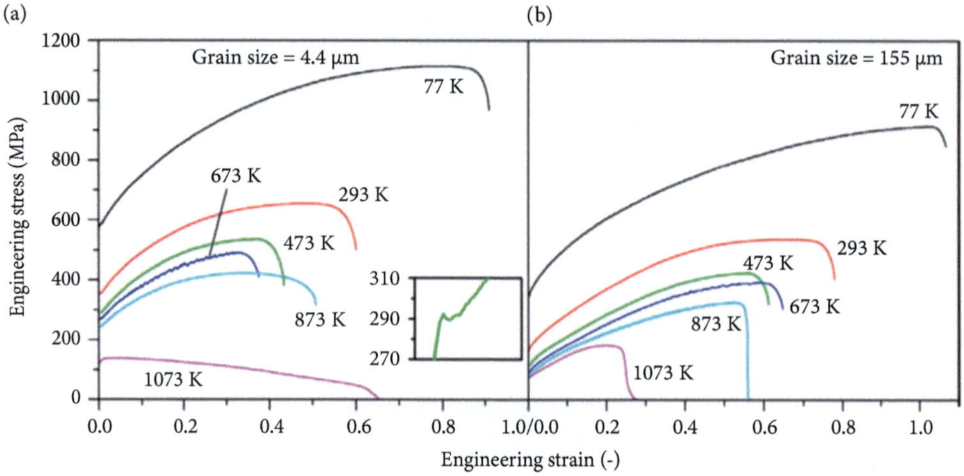

Figure 4.50 Tensile stress-strain curves at temperatures between −196 and 800 °C (77 and 1073K) for the multicomponent single-phase fcc Cantor alloy CrMnFeCoNi as-homogenised, rolled and recrystallised with grain sizes of (a) 4.4 μm and (b) 155 μm.
Reproduced from Otto et al. (2013). The influence of temperature and microstructure on the tensile properties of a CoCrFeNiMn high-entropy alloy. *Acta Materialia 61*: 5743–55.

3. The separation of partials and associated stacking fault energy are both variable, fluctuating from point to point as measured by XRD[150] and TEM,[125] again because of the presence of different local atomic structures and associated lattice distortions along the length of the dislocation cores.
4. Work hardening is high and extends to high levels of plastic strain, again partly because of the presence of different local atomic structures and associated lattice distortions along the length of the dislocations and their associated pinning effect impeding dislocation flow, with an additional contribution from deformation twinning at high strains, high strain rates and/or low temperatures.[144,145,157,158]
5. Necking is delayed and overall ductility enhanced, because of the combination of high initial yield strength and high and sustained work hardening rate, since necking instability sets in according to the Considère criterion when the work-hardening rate falls below the flow stress $d\sigma/d\varepsilon < \sigma$.[158]
6. Strength and ductility both increase on cooling to sub-ambient temperatures, because of the onset of twinning as a major secondary deformation mode,[157] discussed in more detail later in this chapter.

The general features of deformation and plastic flow described above have been demonstrated quite widely for a variety of multicomponent single-phase fcc Cantor alloys.[14,124,144–156]

The shear flow stress in conventional pure metals and dilute binary alloys has been shown previously to be given by the sum of a number of different contributions:

$$\begin{aligned}\tau &= \tau_o + \tau_{wh} + \tau_{ss} + \tau_{gb} + \tau_{ph}\\ &= \tau_o + 2\alpha Gb\rho^{\frac{1}{2}} + G\delta^{\frac{3}{2}}\, bc^{\frac{1}{2}} + \beta d^{-\frac{1}{2}} + 2Gb/L\end{aligned}$$

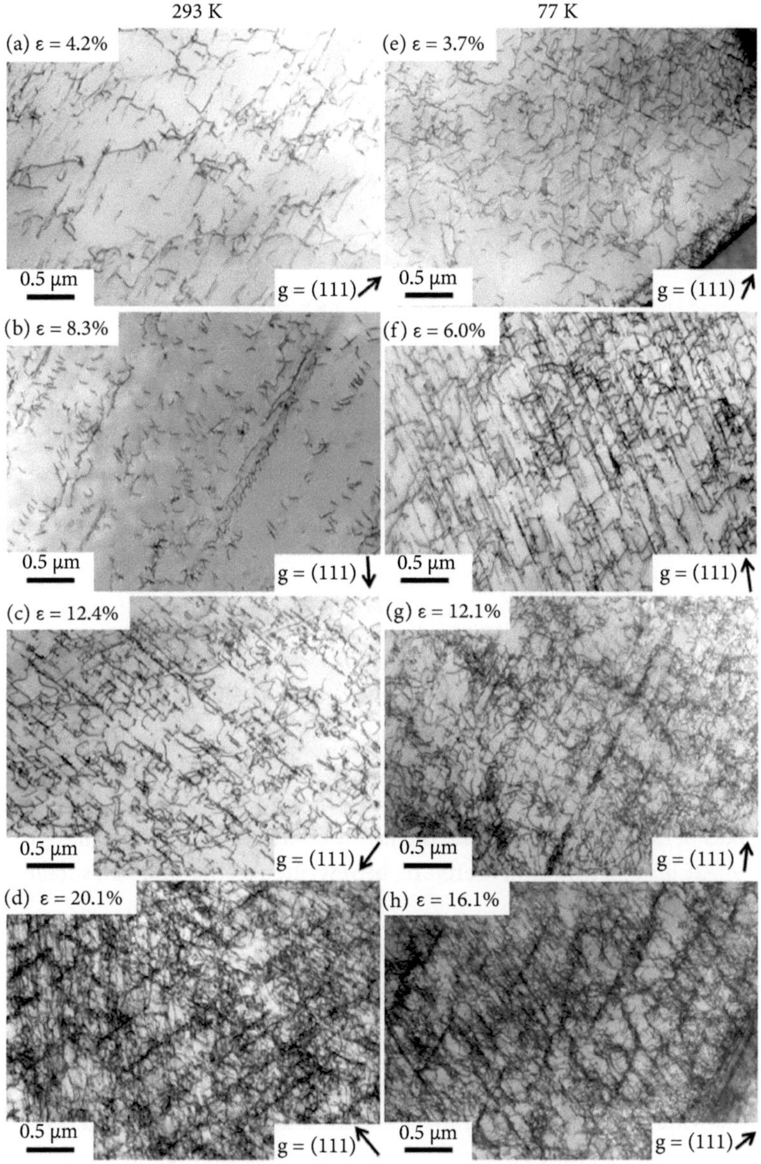

Figure 4.51 Transmission electron micrographs showing dislocation structures in the multicomponent single-phase fcc Cantor alloy CrMnFeCoNi at 20 and −196 °C (273 and 77K) at tensile strains ε between ~4 and 20%.

Reproduced from Laplanche et al. (2016). Microstructure evolution and critical stress for twinning in the CrMnFeCoNi high-entropy alloy. *Acta Materialia 118*: 152–63.

where τ_o is the *Peierls–Nabarro* or *lattice friction stress,* the intrinsic stress needed to move a dislocation in a perfect piece of the crystal, and τ_{wh}, τ_{ss}, τ_{gb} and τ_{ph} are strengthening contributions from *Taylor work hardening, solution hardening, Hall–Petch grain boundary hardening* and *precipitation hardening* respectively. In multicomponent alloys, however, it does not make much sense to separate lattice-friction and solution-hardening effects, since there is an enormous number of different near-neighbour atomic configurations, all

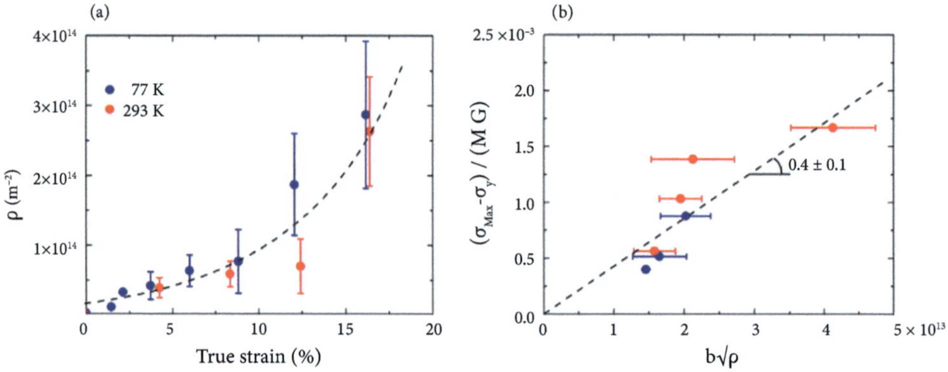

Figure 4.52 Parabolic Taylor work hardening in the original Cantor alloy CrMnFeCoNi during tensile testing at 77 and 293K, with (a) dislocation density ρ increasing proportional to the square of the true strain ϵ; and (b) linear variation of the normalised increase in shear stress τ over the yield shear stress, $(\tau - \tau_y)/bG = (\sigma - \sigma_y)/bMG$, with the square root of the dislocation density $\sqrt{\rho}$, where τ_y, σ_y, b, G and M are the shear and tensile yield stress, Burgers vector, Taylor factor and shear modulus respectively.

Reproduced from Laplanche et al. (2016). Microstructure evolution and critical stress for twinning in the CrMnFeCoNi high-entropy alloy. *Acta Materialia 118*: 152–63.

contributing to different local Peierls–Nabarro stresses as well as different local atomic-level solute stresses and distortions acting as pinning points for the dislocations. Solution hardening is no longer caused by dislocation bowing between widely separated centres of strain around dispersed solute atoms, since every atom is now a centre of local lattice distortion and strain. It makes better sense to combine the lattice friction and solution hardening stresses. The overall shear flow stress τ can, therefore, be rewritten as the sum of a smaller number of contributions:[157,158]

$$\tau = \tau_o + \tau_{wh} + \tau_{gb} + \tau_{ph}$$

$$= \tau_o + 2\alpha G b \rho^{\frac{1}{2}} + \beta d^{-\frac{1}{2}} + 2Gb/L$$

where the multicomponent lattice friction stress τ_o now involves some suitable form of averaging over the wide range of different local Peierls–Nabarro stresses, and also includes the pinning effects of the wide range of different local atomic clusters along the dislocation lines. As already indicated, the effects of work hardening, grain boundaries and precipitates are not dissimilar in multicomponent and simple binary alloys, the first two continuing to obey Taylor parabolic hardening and Hall–Petch strengthening laws, as shown in Figures 4.52 and 4.53.

Varvenne et al.[151,152] developed a detailed Labusch model[159,160] for the intrinsic shear flow stress τ_o and corresponding initial yield strength $\sigma_y = M\tau_o$ (where the Taylor factor M is the ratio of the applied tensile stress to the resulting shear stress on the slip plane), by treating all atoms as solute atoms acting as dislocation pinning centres in a matrix represented by a suitably averaged structure over the variety of different components. The ramifications and applications of this model are discussed further by George et al.[158] Varvenne et al.[151,152] use quite a number of rough and ready assumptions and approximations in order to make the model tractable:

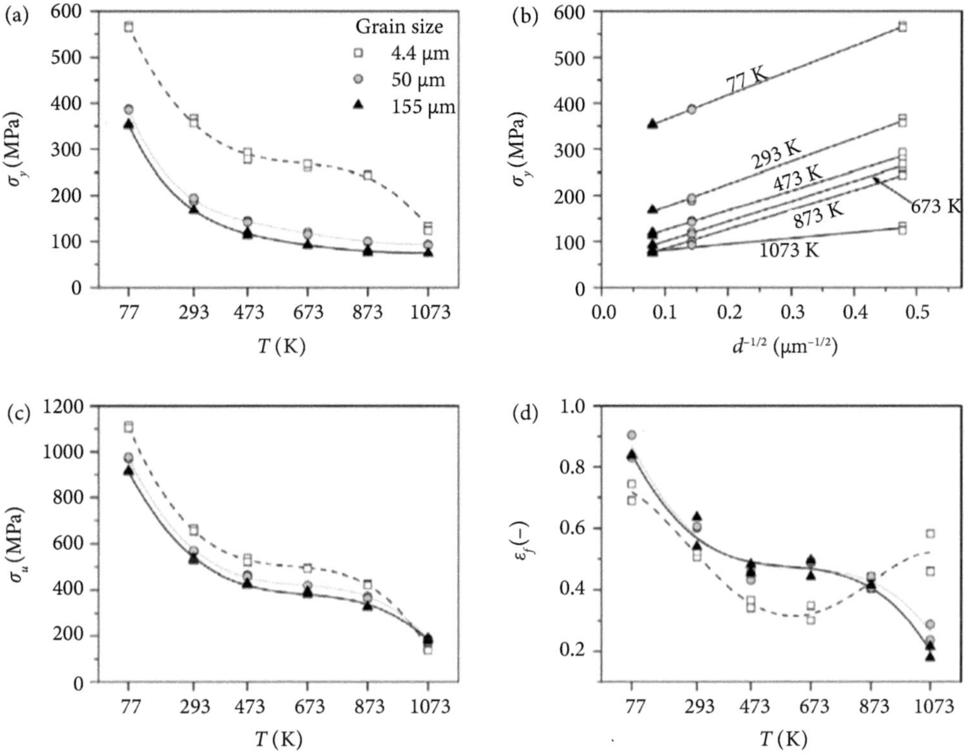

Figure 4.53 Variation of yield strength σ_y with (a) temperature T and (b) grain size d, and variation of (c) ultimate tensile strength σ_u and (d) fracture strain ε_f with temperature T in the original Cantor alloy CrMnFeCoNi.

Reproduced from Otto et al. (2013). The influence of temperature and microstructure on the tensile properties of a CoCrFeNiMn high-entropy alloy. *Acta Materialia 61*: 5743–55.

1. The average matrix structure is taken as that of the pure random concentrated solid solution.
2. The increase in Peierls–Nabarro force itself is ignored as relatively small compared to the dislocation-pinning effect of the wide variety of local atomic clusters.
3. Solute effects on the shear modulus are also ignored as relatively limited compared to the dislocation-pinning effect of the wide variety of local atomic clusters.
4. Variations in Peierls–Nabarro stress and shear modulus from point to point along the dislocation lines are also ignored.
5. Solute strengthening arises, therefore, from the average effect of different atomic clusters acting as centres of strain and barriers to dislocation motion, effectively pinning the wavy shape of the dislocations at all points along their length.
6. The pinning strength of the atomic clusters is calculated very roughly as a root mean square average of the local lattice distortions assessed from the misfits δ_i of the individual i components.
7. The energy barrier to dislocation motion is also averaged very approximately as the amplitude of a simple sinusoidal variation from an averaged dislocation energy well to an averaged dislocation energy peak.[151,152,158]

With these assumptions, there is then a single average energy barrier to dislocation motion ΔE_m, which must be overcome in order to unpin dislocations from the complex arrangement of different atomic clusters and associated lattice distortions along their wavy length. This energy barrier depends in a generalised way on the average dislocation energy or line tension $\Delta E_d \approx Gb^2$ and the average pinning strength as represented by the root mean square of individual atomic misfits normalised to the Burgers vector $\delta_{av}/b = \left(\sum_i x_i(\delta_i/b)^2\right)^{\frac{1}{2}}$, where x_i is the concentration of the ith component and $\delta_i = (r_i - r_{av})/r_{av}$. At absolute zero, the shear stress required to overcome the energy barrier, i.e. the intrinsic lattice friction stress τ_o, is the force acting on a dislocation across an area $b \times \zeta$ required to move it a distance w, where ζ and w are the average wavelength and average amplitude respectively of the wavy dislocation lines. In other words, $\tau_o = \Delta E_m/b\zeta w$ at $T = 0$. At a higher temperature T, the intrinsic lattice friction stress τ_o includes an additional Arrhenius term for the thermal probability of acquiring the barrier energy, given by $\dot{\varepsilon} = \dot{\varepsilon}_o \exp(-\Delta E_m/kT)$ where $\dot{\varepsilon}$ and $\dot{\varepsilon}_o$ are the strain rate and a reference strain rate respectively, and k is Boltzmann's constant. Overall, therefore, the energy barrier ΔE_m, intrinsic lattice stress τ_o and initial yield strength $\sigma_y = M\tau_o$ are given by:

$$\Delta E_m = f(\Delta E_d, \delta_{av}/b) = f\left(Gb^2, \delta_{av}/b\right)$$

$$\tau_o = \sigma_y/M = \frac{\Delta E_m}{b\zeta w}\left[1 - \left(\frac{kT}{\Delta E_m}\ln\frac{\dot{\varepsilon}_o}{\dot{\varepsilon}}\right)^{2/3}\right]$$

where the first term on the right-hand side of the last equation gives the intrinsic lattice flow stress and initial yield strength at absolute zero and the second term gives their temperature and strain rate dependence at more realistic, higher temperatures. Varvenne et al.[151,152] derive an approximate equation for ΔE_m as a function of ζ and w, which is then minimised to determine their optimum values, and thus obtain more accurate results for the energy barrier ΔE_m, lattice friction shear stress τ_o and initial yield strength σ_y:

$$\Delta E_m = KG(\delta_{av}/b)^2$$

$$\tau_o = \sigma_y/M = K'G(\delta_{av}/b)^4\left[1 - \left(\frac{kT}{\Delta E_m}\ln\frac{\dot{\varepsilon}_o}{\dot{\varepsilon}}\right)^{2/3}\right]$$

where K and K' are constants depending on the elastic constants and Poisson's ratio of the particular alloy, and $\delta_{av}/b = \left(\sum_i x_i(\delta_i/b)^6\right)^{\frac{1}{6}}$ is now summed over individual misfits to the sixth power because Varvenne et al.[151,152] average the volumetric rather than the linear local strain. Including other forms of hardening, the overall expression for the flow stress τ is then:

$$\tau = \sigma_y/M = K'G(\delta_{av}/b)^4\left[1 - \left(\frac{kT}{\Delta E_m}\ln\frac{\dot{\varepsilon}_o}{\dot{\varepsilon}}\right)^{2/3}\right] + 2\alpha Gb\rho^{\frac{1}{2}} + \beta d^{-\frac{1}{2}} + 2Gb/L$$

Figure 4.54 shows reasonably good agreement between experimentally measured values of initial yield strength σ_y and predicted values using Varvenne et al.'s model[1561,152] for the multicomponent single-phase fcc Cantor alloy CrMnFeCoNi and the quaternary single-phase fcc modified Cantor alloy CrFeCoNi at different temperatures, as well as for a range of

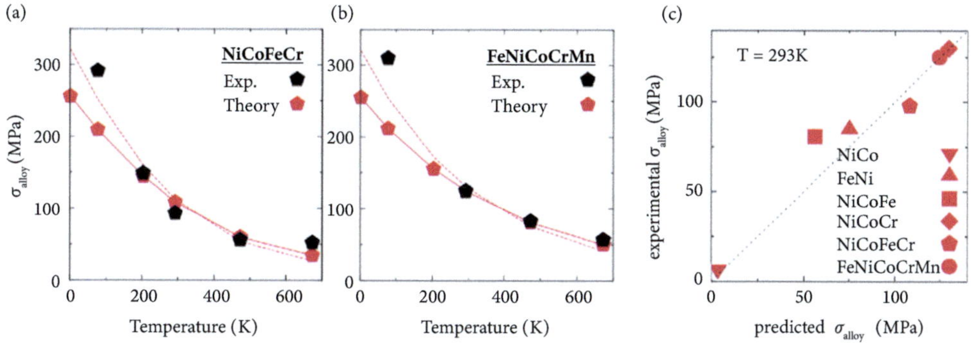

Figure 4.54 Predicted values and experimental measurements of the initial large-grain-size dislocation-free yield stress σ_y (= σ_{alloy}) for: (a) the quaternary single-phase fcc modified Cantor alloy CrFeCoNi and (b) the multicomponent single-phase fcc Cantor alloy CrMnFeCoNi as a function of temperature T; and (c) a range of other binary, ternary and quaternary equiatomic single-phase fcc alloys at room temperature.

Reproduced from Varvenne et al. (2016). Theory of strengthening in fcc high entropy alloys. *Acta Materialia 118*: 164–76.

different quaternary, ternary and binary equiatomic modified Cantor alloys at room temperature.[151,152,158] The experimental results for the initial yield stress σ_y in Figure 4.54 were obtained by removing the effects of work hardening (by using well-annealed, i.e. essentially dislocation-free) materials, grain-boundary strengthening (by using large-grained, i.e. essentially grain-boundary-free) materials and precipitation hardening (by using fully solutionised, i.e. essentially precipitate-free) materials. The good agreement between experiment and theory is impressive since the model has no adjustable parameters. It is also somewhat surprising given the extent of the assumptions and approximations made, essentially ignoring much of the complexity of the energy landscape created by the different atomic clusters and lattice distortions both along the length of the wavy dislocations and in front of them, impeding their lines of motion.

Similar agreement between experiment and Varvenne et al.'s model[151,152] has also been achieved subsequently for other high- and medium-entropy single-phase fcc alloys such as $(CrMnFeCo)_xNi_{1-x}$,[158] $Al_{0.1}CrMnFeCoNi$,[153] CuNiRhPdIrPt[154,155] and CoNiV.[158] Recent experiments have also shown reasonable agreement between the model's predictions for σ_y and nanoindentation hardness measurements obtained from a total of twenty-four modified Cantor alloys, varying the concentration of each of the components one at a time.[14] The predicted activation volume from Varvenne et al.'s model is approximately $300b^3$ at room temperature, with a linear size of $\sqrt[3]{300b^3} \approx 6.7b$, which is: 1) smaller than typical activation volumes in conventional fcc alloys and pure metals, which usually correspond to the separation of dislocation intersections with forest dislocations along the slip plane; but 2) in good agreement with experimental values for Cantor alloys;[134] 3) in reasonable agreement with the expected separation of strong local atomic clusters acting as the main pinning centres; and 4) also in reasonable agreement with the observed wavelength and amplitude of the waviness of the dislocation lines.

Even more surprisingly than the good agreement between experimental results and Varvenne et al.'s model[151,152] shown in Figure 4.54 for the flow stress in multicomponent

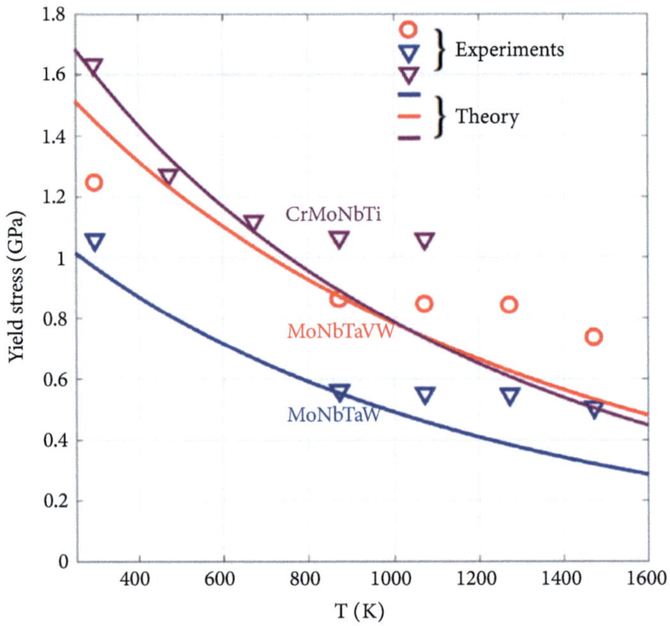

Figure 4.55 Predicted values and experimental measurements of the initial large-grain-size dislocation-free yield stress σ_y as a function of temperature T for three single-phase bcc modified Senkov alloys: 5-component VNbMoTaW and 4-component TiCrNbMo and NbMoTaW. Reproduced from Maresca and Curtin. (2020). Theory of screw dislocation strengthening in random bcc alloys from dilute to high-entropy alloys. *Acta Materialia 182*: 144–62.

single-phase fcc Cantor alloys, essentially the same model but with different constants K and K' also works quite well for the flow stress in multicomponent single-phase bcc Senkov alloys,[161] as shown in Figure 4.55 for 5-component VNbMoTaW and 4-component TiCrNbMo and NbMoTaW, despite the more complex dislocation flow behaviour in these materials discussed previously. Not surprisingly however, as shown in Figure 4.55, the model fails to predict the plateau in flow stress and strength that appears at high temperatures in the Senkov alloys.

In summary, deformation and plastic flow in multicomponent single-phase fcc Cantor and bcc Senkov alloys is in many ways similar to conventional fcc and bcc alloys and pure metals, but with higher yield strengths, ultimate tensile strengths, work-hardening rates and ductilities, all caused essentially by the enormous variation in atomic clusters and associated local lattice distortions and strains along the length of the dislocation lines, acting to restrict but not prevent dislocation generation and subsequent motion.

Grain boundaries

Grain boundary structure

Most materials are *polycrystalline*, i.e. they consist of a densely packed aggregate of differently oriented individual *crystals* or *grains*.[122,123] Within each grain, the atoms or molecules are arranged on the regular crystal lattice of the material, but the crystal orientation varies from one grain to another. Within any polycrystalline material, adjacent grains are separated by planar regions called *grain boundaries*, where the two grains meet and the regular crystalline atomic or molecular arrangement is (inevitably) somewhat disrupted. In fact, *grain boundaries* are the main kind of *planar defect* in crystalline materials, and there is a wide variety of different grain-boundary atomic or molecular structures with different amounts of atomic or molecular disorganisation, depending on the relative orientation of the two grains either side of the grain boundary, and the orientation of the grain boundary plane itself.[29,120] There is, of course, an excess *grain-boundary energy* associated with the disorganisation of the atoms or molecules at any grain boundary which, perhaps not surprisingly, varies even in pure materials and conventional binary alloys in a fairly complex way with the orientation of the two grains either side of the grain boundary and the orientation of the grain boundary plane.[29,120]

The *orientation relationship* or *misorientation* between the two grains either side of a grain boundary is defined by a common rotation axis $\boldsymbol{r}$ and a rotation angle θ. Rotating either grain by the angle θ about $\boldsymbol{r}$ brings it into crystallographic registry with the other grain.[29,120] Once an orientation relationship has been specified, the *grain-boundary plane* is defined by its normal $\boldsymbol{n}$, with the two grains then being positioned either side of this plane.[29,120] In a *symmetric grain boundary* the grain-boundary plane normal $\boldsymbol{n}$ is crystallographically identical in the two grains, i.e. $\boldsymbol{n} = (hkl)$ where h, k and l are the same in both grains, and the crystal lattices of the two grains exhibit a mirror relationship across the grain-boundary plane; in an *asymmetric grain boundary*, on the other hand, the grain-boundary plane is crystallographically different in the two grains, i.e. $\boldsymbol{n} = (hkl)$ where h, k and l are not the same in both grains, and there is no simple mirror relationship between the two grains across the grain boundary. The rotation axis and grain-boundary normal are both unit vectors, each defined by two angles relative to any given coordinate system, i.e. $\boldsymbol{r} = (\phi_r, \psi_r)$ and $\boldsymbol{n} = (\phi_n, \psi_n)$. In other words, grain boundaries in any given material have five independent geometric or crystallographic and, therefore, structural degrees of freedom, with any particular grain boundary defined by five independent parameters θ, ϕ_r, ψ_r, ϕ_n and ψ_n.

Low-angle grain boundaries have misorientation angles θ less than approximately 15°, and consist of well-defined arrays of separated dislocations.[29,120] In a *low-angle tilt grain boundary*, the rotation axis is parallel to the grain-boundary plane and perpendicular to the grain-boundary normal, $\boldsymbol{r} \perp \boldsymbol{n}$, and the atomic structure consists of one or more arrays of edge dislocations that effectively insert a wedge between the two grains, corresponding to a tilt orientation relationship between them; in a *low-angle twist grain boundary*, the rotation axis is perpendicular to the grain-boundary plane and parallel to the grain-boundary normal, $\boldsymbol{r} \parallel \boldsymbol{n}$, and the atomic structure consists of two or more arrays of screw dislocations, corresponding to a twist orientation relationship between the two grains; and in a *low-angle mixed grain boundary* the rotation axis $\boldsymbol{r}$ is midway between being parallel or perpendicular to the grain-boundary plane and the grain-boundary normal, and the atomic structure consists of arrays of both edge and screw dislocations, corresponding to a combined tilt and twist orientation relationship between the two grains. In other words, low-angle grain boundaries can be classified as:[29,120]

$$\text{tilt}: \boldsymbol{r} \cdot \boldsymbol{n} = 0$$

$$\text{twist}: \boldsymbol{r} \cdot \boldsymbol{n} = 1$$

$$\text{mixed}: 0 < \boldsymbol{r} \cdot \boldsymbol{n} < 1$$

With increasing misorientation angle θ, the dislocation arrays in the grain-boundary plane become more and more closely spaced until the individual dislocations can no longer be recognised above about $\theta = 15°$. The *grain-boundary energy* γ_{gb} of a low-angle grain boundary is equal to the energy of the corresponding dislocation arrays, increasing from $\gamma_{gb} = 0$ at $\theta = 0°$ (i.e. when there is no grain boundary) and reaching a maximum at about $\theta = 15°$, as given by the *Read–Shockley equation*:

$$\gamma_{gb} = A\theta\,(B - \ln\theta)$$

where $A \propto Gb$ and $B \propto E_c/Gb^2$ are constants depending on the elastic constants of the material and the exact nature of the grain boundary and its particular dislocation arrays, and G, b and E_c are the shear modulus of the material and the Burgers vector and core energy of the array dislocations respectively.[29,120]

High-angle grain boundaries have misorientation angles θ above approximately 15°, usually with substantial atomic or molecular disorder tightly confined to a width of no more than one or two crystal planes, and, in almost all cases therefore, a correspondingly high and roughly constant grain-boundary energy.[29,120] In other words, the plot of grain-boundary energy γ_{gb} versus rotation angle θ shows a plateau of constant high grain-boundary energy, typically in the range $\frac{1}{2}$–1 J/m^2, for almost all rotation angles θ above approximately 15°. There are, however, some special high-angle grain boundaries at particular orientation relationships where the two grains share a relatively high number of coincident lattice sites. These particular orientation relationships are said to exhibit a *coincident site lattice* or *CSL*, and the corresponding special high-angle grain boundaries are called Σn boundaries, where n is the inverse ratio of the number of coincident sites to the number of normal lattice sites, which geometrically must always be an odd integer.[29,120] Depending on the detailed atomic or molecular structure of the material, some (though not all) of these special CSL orientation relationships and corresponding Σn high-angle boundaries exhibit somewhat reduced atomic or molecular disorganisation, with a correspondingly reduced grain-boundary energy, as manifested by the presence of low-energy dips or *cusps* at these particular orientation angles on the high-angle grain boundary energy plateau of the plot of grain-boundary energy γ_{gb} versus rotation angle θ. Typical examples of special grain boundaries with CSL orientation relationships in fcc materials include: the $\Sigma 3$ twin boundary, formed by a rotation of $\theta = 70.53°$ about $[01\bar{1}]$ with a symmetric (111) grain-boundary plane; the $\Sigma 5$ symmetric twist boundary, formed by a rotation of $\theta = 36.87°$ about [100] with a symmetric (100) grain-boundary plane; and the $\Sigma 11$ symmetric tilt boundary, formed by a rotation of $\theta = 50.48°$ about $[01\bar{1}]$ with a symmetric (311) grain-boundary plane.[120]

Figure 4.56 shows schematically the resulting typical overall variation of grain-boundary energy γ_{gb} with rotation angle θ for symmetric tilt grain boundaries about $r = [011]$ in aluminium, showing the following features: increasing low-angle grain-boundary energy with increasing tilt angle for misorientations $\theta < 15°$; a high-angle grain-boundary energy plateau for misorientations $\theta > 15°$; and low-energy cusps at particular misorientations θ corresponding to special $\Sigma 3$ and $\Sigma 11$ CSL boundaries.[120,162] Overall the grain-boundary structure and energy landscape is clearly complex even in simple pure materials and binary alloys, with many plots similar to Figure 4.56 for each material because of the large number of different sets of grain boundaries with different values of the five independent defining parameters θ, ϕ_r, ψ_r, ϕ_n and ψ_n.

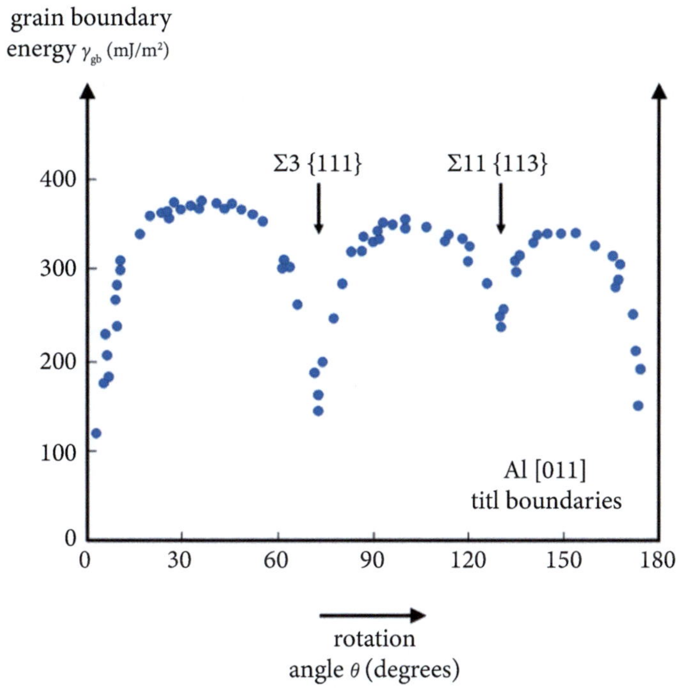

Figure 4.56 Measurements of the variation of grain boundary energy γ_{gb} with rotation angle θ for symmetric [011] tilt boundaries in Al.

Reproduced from Sutton and Baluffi. (1987). On geometric criteria for low interfacial energy. *Acta Metallurgica 35*: 2177–2201.

Grain boundaries in multicomponent solid solutions

As just described, there is a wide range of grain boundaries with a correspondingly wide range of complex structures and energies even in pure materials and simple alloys.[29,120] Not surprisingly, the complexity of grain-boundary structures is much greater still in multi-component high-entropy materials. The topological complexity of accommodating atomic and molecular sites midway between two differently oriented crystals is compounded in a multicomponent solid solution by the additional chemical complexity of accommodating a large number of different component atoms or molecules across the various grain-boundary sites. Table 4.3 shows that there are almost twenty trillion different oriented first and second near-neighbour local atomic clusters in a 5-component equiatomic single-phase fcc Cantor alloy, and Table 4.4 shows that there are just over thirty billion in a 5-component equiatomic single-phase bcc Senkov alloy. In any polycrystalline Cantor or Senkov alloy, this enormous variety of different local atomic clusters has to be distributed across the many different grain-boundary sites all along the many different grain-boundary planes. The complexity and multiplicity of grain boundary structures in multicomponent high-entropy materials arising from such a vast range of different local atomic clusters is, of course, similar to the complexity and multiplicity discussed previously, arising in the same way for electronic, vacancy and dislocation structures in multicomponent high-entropy materials.

Direct observation of grain-boundary structures is not easy, even in pure materials and simple alloys, because of the complex imaging conditions created by the atomic or molecular disorganisation in grain-boundary regions, complicating the use of standard high-resolution imaging techniques such as transmission electron microscopy (TEM) and atom-probe (AP) tomography. And the problems associated with the direct observation of grain-boundary structures are only made worse by the chemical complexity found in multicomponent high-entropy materials. Not surprisingly, therefore, there have been relatively few attempts to investigate multicomponent grain-boundary structures by direct observation. Perhaps more surprisingly, there have also been relatively few attempts to use atomistic modelling techniques, which are more straightforward technically, but always suffer, as remarked previously, from not being able to handle anywhere near the staggeringly large number of different local atomic clusters that are present in multicomponent high-entropy materials.

Lin et al.[163] used high-resolution transmission electron microscopy (HRTEM) to study the atomic-level structure of grain boundaries in the original equiatomic 5-component single-phase fcc Cantor alloy CrMnFeCoNi manufactured by vacuum induction melting high-purity constituents under argon, casting into a water-cooled copper mould, remelting and recasting five times to ensure homogeneity, cutting into 20 mm-diameter, 1.5 mm-thick discs, and then mechanical working by high-pressure torsion (HPT) with a strain of up to ten quasi-constrained revolutions under a pressure of 5 GPa at room temperature, to produce a final nanocrystalline material with a grain size of ~55 nm. Lin et al.[163] reported the detailed structure of only two grain boundaries in the resulting nanocrystalline material. One was a low-angle symmetric tilt boundary, as shown in Figure 4.57, with a $[\bar{1}10]$ rotation axis, a rotation angle of $\theta \approx 6°$ and a grain-boundary plane ±3° from (111) in the two grains, consisting somewhat surprisingly of a single array of sessile $\frac{1}{3}[11\bar{1}]$ Frank dislocations, rather than the expected, more common, glissile $\frac{1}{2}$<110> full or $\frac{1}{6}$<112> Shockley partial dislocations, with a regular dislocation spacing of just under 2 nm corresponding to nine $(11\bar{1})$ planes. The other was a Σ3 twin boundary, as shown in Figure 4.58, with a $[\bar{1}10]$ rotation axis, a rotation angle of $\theta \approx 70.53°$and a (111) grain-boundary mirror plane. Lin et al.[163] also conducted in-situ heating experiments in the high-resolution transmission electron microscope to demonstrate how both boundaries can act effectively as sources and sinks for emitting and absorbing matrix dislocations and stacking faults into and out of the crystal lattices of the two surrounding grains.

There have been several investigations of the detailed grain-boundary structure in multicomponent high-entropy materials using atomistic computer modelling methods.[164–166] Lee et al.[166] used embedded-atom (EAM) interatomic potentials with the LAMMPS molecular-dynamics (MD) modelling package to build computer models of twenty different bicrystals of the original equiatomic single-phase fcc Cantor alloy CrMnFeCoNi, all with symmetric twist grain boundaries with [110] and [111] rotation axes and corresponding [110] and [111] grain-boundary planes, including low-angle boundaries with rotation angles of $\theta \approx 8 - 9°$, high-angle boundaries with high-density CSLs such as Σ3, Σ7 and Σ9, and high-angle boundaries with low density CSLs such as Σ51, Σ57 and Σ93. Between half a million and a million atoms were modelled in each bicrystal, with grain-boundary dimensions of 10 × 10 nm, i.e. there were approximately one or two thousand atoms in the grain-boundary plane, and most of the atoms were in the body of the two grains either side of the grain boundary. Moreover, the modelling was restricted to the topological structure of the atomic positions at the grain boundary, with no attempt to investigate the complex chemical effects of the variety of different atomic species at different sites. Not surprisingly therefore, the resulting grain-boundary

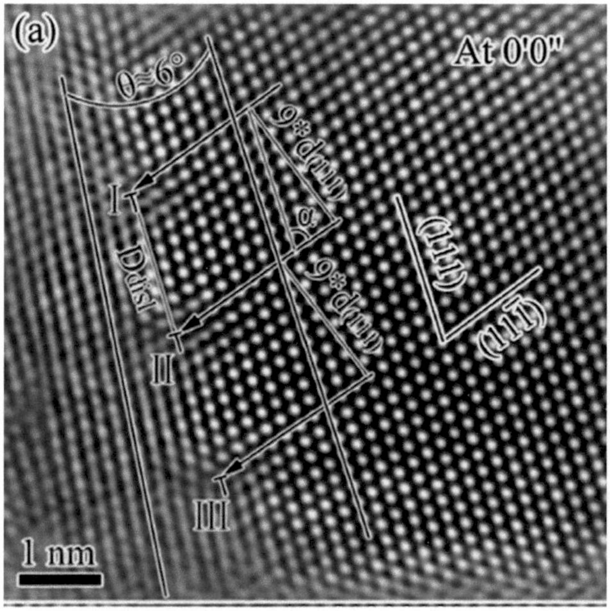

Figure 4.57 HRTEM image of the regular dislocation array at a low-angle symmetric tilt grain boundary in the Cantor alloy CrMnFeCoNi.

Reproduced from Lin et al. (2017). In-situ high-resolution transmission electron microscopy investigation of grain boundary dislocation activities in a nanocrystalline CrMnFeCoNi high-entropy alloy. *Journal of Alloys and Compounds 709*: 802–7.

atomic structures, some of which are shown in Figure 4.59, are essentially similar to those found in pure fcc metals and simple binary fcc alloys.

Wynblatt and Chatain[164] also used embedded-atom (EAM) interatomic potentials with the LAMMPS molecular-dynamics (MD) modelling package to build a tricrystal of the original equiatomic single-phase fcc Cantor alloy CrMnFeCoNi, containing two pseudo-Σ13 high-angle symmetric twist grain boundaries with an [001] rotation axis, a rotation angle of $\theta = 22.62°$ and an [001] grain-boundary plane.[xxv] The tricrystal contained 8,320 atoms of which approximately 250 were located on each grain-boundary plane, with the five different component atoms initially distributed randomly across the grain and grain-boundary sites, followed by 45,000 random MC atom swops, and a final thermal MD relaxation, in order to investigate chemical segregation of the different component atoms to and from the grain boundary. Wynblatt and Chatain[164] found significant positive segregation of Cr atoms and slight positive segregation of Mn atoms into the two grain boundaries from the bulk of the grains, with a corresponding negative segregation of the other three components Fe, Co and Ni away from the two grain boundaries into the bulk of the grains, as shown in Figure 4.60. Clearly, these results should be treated with considerable caution, since the number of

[xxv] The [001] planes in an fcc crystal consist of alternating A and B configurations, corresponding to alternating arrays of cube corner and cube face atoms respectively. A standard Σ13 [001] CSL is an [001] twist boundary with both grains terminating with A configurations at the [001] grain-boundary plane; a pseudo-Σ13 [001] CSL is an [001] twist boundary with one grain terminating with an A configuration and the other grain terminating with a B configuration at the [001] grain-boundary plane.

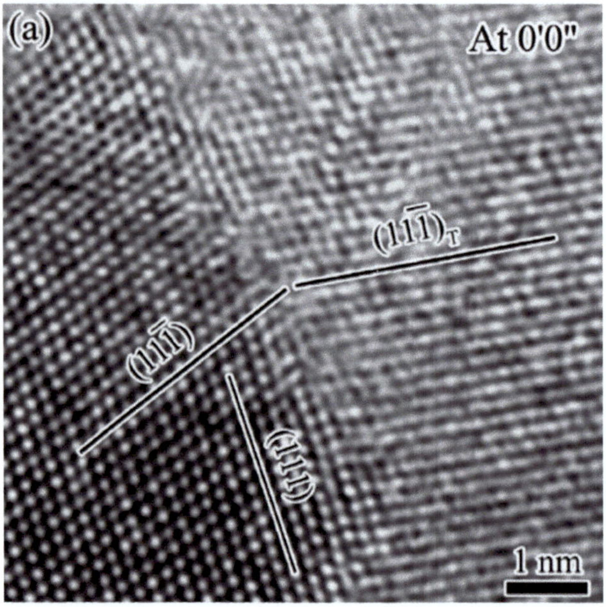

Figure 4.58 HRTEM image of a twin boundary in the Cantor alloy CrMnFeCoNi.

Reproduced from Lin et al. (2017). In-situ high-resolution transmission electron microscopy investigation of grain boundary dislocation activities in a nanocrystalline CrMnFeCoNi high-entropy alloy. *Journal of Alloys and Compounds 709*: 802–7.

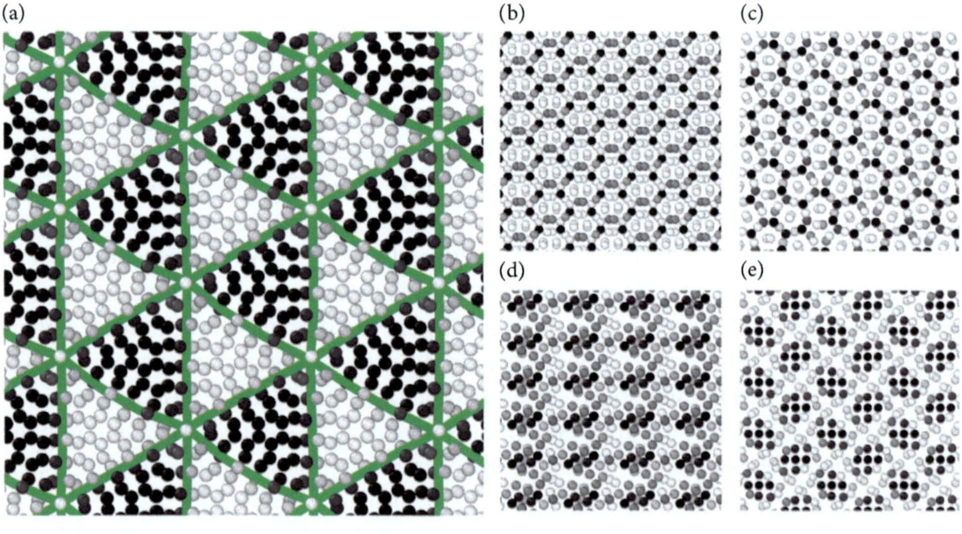

Figure 4.59 MD models of high-angle symmetric twist grain boundaries in the Cantor alloy CrMnFeCoNi: (a)–(c) rotation axis $r = [110]$; (d)–(e) rotation axis $r = [111]$: (a) $\Sigma 111$, (b) $\Sigma 7$, (c) $\Sigma 39$, (d) $\Sigma 9$ and (e) $\Sigma 17$.

Lighter-shaded atoms represent progressively greater deviation from regular fcc crystal lattice sites; green lines = dislocations.

Reproduced from Lee et al. (2021). Tensile deformation behavior of twist grain boundaries in CoCrFeMnNi high entropy bicrystals. *Nature Scientific Reports 11*: Article 428.

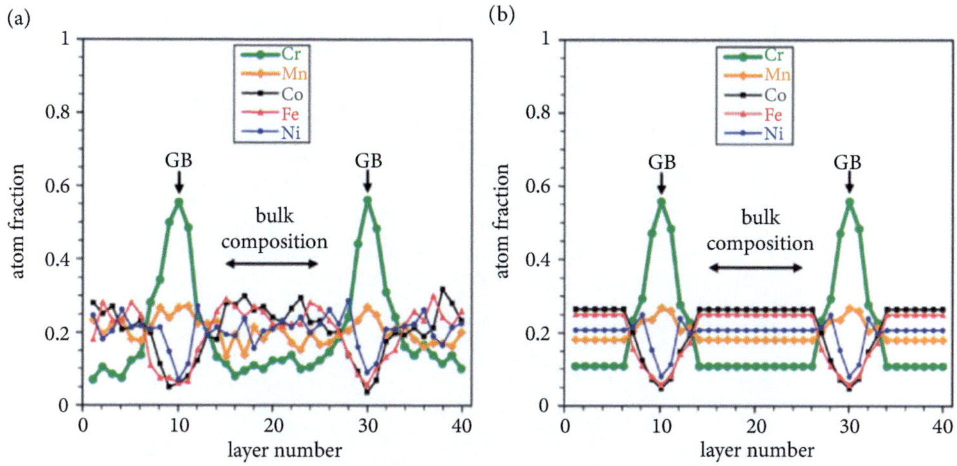

Figure 4.60 MD/MC modelling of segregation of the different component atoms into (Cr and, to a lesser extent, Mn) and out of (Fe, Co and Ni) two pseudo-Σ13 high-angle symmetric twist grain boundaries in the Cantor alloy CrMnFeCoNi.

Reproduced from Wynblatt and Chatain. (2019). Modeling grain boundary and surface segregation in multicomponent high-entropy alloys. *Physical Review Materials* 3: Article 054004.

grain-boundary sites considered is well below the nearly twenty trillion different local atomic structures expected to be found distributed across the grain-boundary sites in an equiatomic 5-component fcc alloy, as shown in Table 4.3.

The most detailed computer modelling study so far of grain boundaries in multicomponent high-entropy materials was conducted by Farkas,[165] who again used LAMMPS MD modelling with a convenient but somewhat unrealistic interatomic potential to build a nanocrystalline computer sample of the Cantor alloy CrMnFeCoNi containing about a million atoms. The average grain size was 40 nm, with nine differently oriented grains separated by twenty-five different pure tilt grain boundaries, all with a common [110] rotation axis, but with random misorientations that included six low-angle boundaries with rotation angles $\theta < 15°$, four high-angle, high-density CSL boundaries (four Σ3 and one Σ11), and fifteen random high-angle grain boundaries with general rotation angles $\theta > 15°$. The grain boundary structures were visualised with OVITO, DXA and CNA software algorithms, and were compared with the same set of boundaries in an equivalent single-component fcc nanocrystalline material, built using a somewhat artificial average CrMnFeCoNi interatomic potential. Figure 4.61 shows an example of one of the random high-angle grain boundaries in the Cantor alloy visualised with a range of different techniques. There were only relatively minor differences between the grain-boundary structures in the multicomponent Cantor alloy and the artificial single-component alloy comparison.

In summary, there has been relatively little detailed atomic-level experimental or computer modelling investigation of the structure of grain boundaries in multicomponent high-entropy solid solutions. Results obtained to date show grain boundary structures in multicomponent high-entropy solid solutions that in general are not very different from those found in pure metals and simple binary alloys. A much wider range of local atomic structures in grain boundaries in multicomponent high-entropy solid solutions is expected to lead to significantly different behaviour in grain-boundary properties such as segregation,

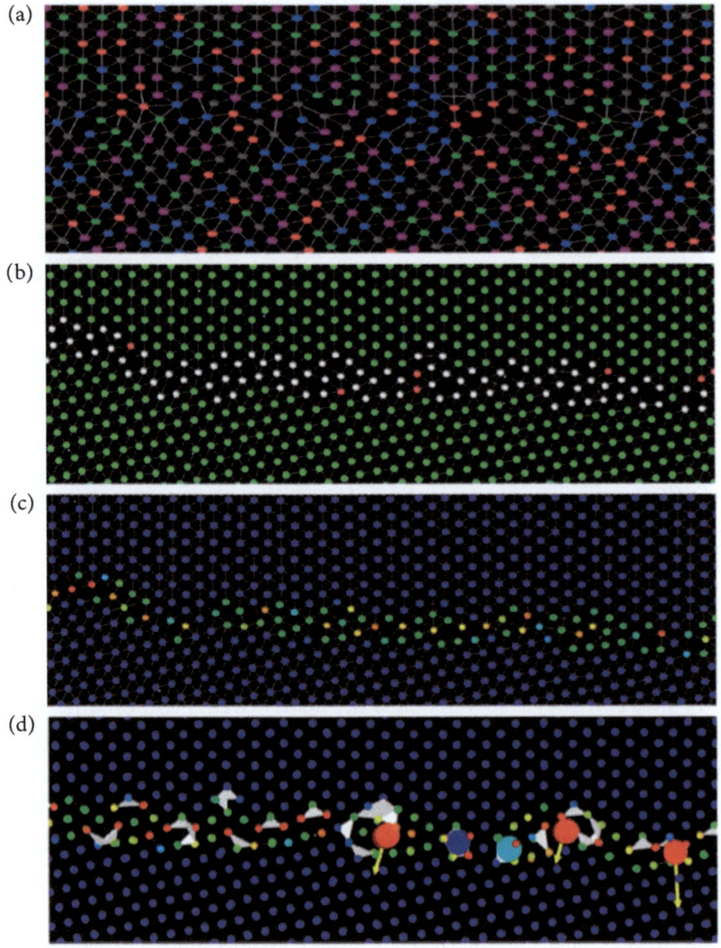

Figure 4.61 MD modelling of a random high-angle grain boundary in the Cantor alloy CrMnFeCoNi, visualised by: (a) different component atoms coloured differently; (b) different local structures coloured differently (green-fcc; red-hcp; white-other); (c) different atoms coloured according to their local centro-symmetry (from blue to red on a ten-point scale); and (d) different local structures coloured differently according to their dislocation character.
Reproduced from Farkas. (2020). Grain boundary structure in high-entropy alloys. *Journal of Materials Science* *55*: 9173–83.

grain-boundary diffusion, grain-boundary/dislocation interactions and grain-boundary migration, but the associated structures and mechanisms are yet to be identified in any detail.

Stacking faults and twins

Twins can form in different ways. *Deformation twins* form as regions where the crystal lattice is sheared under the action of an applied stress, as an alternative to dislocation slip as a mechanism of plastic flow; *growth twins* form during phase transformations such as solidification from a liquid or thin film deposition from a vapour, as a consequence of faults generated while atoms or molecules are adding on to the growing crystalline solid/liquid or solid/vapour interface; and *annealing twins* form during recrystallisation and grain growth, again as a consequence of faults generated while atoms are adding on to the growing dislocation-free grains.[29,120,121] There has been quite a lot of scientific interest in the structure and development of deformation twins in multicomponent single-phase fcc Cantor alloys since, as remarked previously, twinning can become an important additional source of plastic flow in these materials at low temperatures, leading to outstanding combinations of strength and ductility for cryogenic applications, as will be discussed in more detail in Chapter 6.

Twins and stacking faults are closely related in fcc materials.[29,120,121] In an fcc crystal the close-packed {111} planes are stacked in the sequence ABCABCABCABC. A normal *twin* in fcc is equivalent to a Σ3 CSL between the two adjacent twinned crystals, with a {111} mirror plane for the twin boundary; this sounds complicated, but is more easily described as an ABCABCBACBA stacking sequence of {111} planes (with the twin boundary at the middle C plane). A normal or *intrinsic stacking fault*[xxvi] in fcc is equivalent to a single missing {111} plane of atoms, i.e. an ABCABCBCABCA stacking sequence of {111} planes (with a missing A plane between the second C and third B planes from the left). It can be seen by inspection of the stacking sequences that the intrinsic stacking fault is equivalent to two twin boundaries on two adjacent {111} planes (the second C and third B planes from the left), so the twin boundary energy is approximately half the stacking fault energy: $\gamma_t \approx \frac{1}{2}\gamma_{sf}$. Since the stacking sequence of close-packed {0001} planes in an hcp crystal (identical in structure to the close-packed {111} planes in fcc) is ABABABABAB, upon inspection we see that the twin in fcc is equivalent to three adjacent planes of hcp (the middle BCB sequence of planes).[xxvii] As discussed in a previous section, the stacking fault energy is low in multicomponent single-phase fcc Cantor alloys so, not surprisingly given the structural similarities, the twin boundary energy and the fcc/hcp interface boundary energy are also low.[29,120,121] During deformation therefore, there is a close competition between shearing by three different modes: slip of <110> dislocations separated into two Shockley <112> partials with a stacking fault between them; deformation twinning; and fcc → hcp martensite transformation. And depending upon the exact alloy composition, temperature and configuration of applied stresses, deformation twinning and martensite transformation can both, as mentioned above, provide additional mechanisms for plastic flow, and this is important in determining the mechanical properties of Cantor alloys, as is discussed in more detail in Chapter 6.

There have been quite a few electron back-scatter diffraction (EBSD) pattern and TEM studies of the formation and structure of deformation twins under the action of tensile and compressive stresses at low temperatures in the original and modified multicomponent single-phase fcc Cantor alloys.[167–171] Figure 4.62 shows one example of bright-field (BF) transmission electron micrographs (BF-TEM) of twins formed in the original Cantor alloy CrMnFeCoNi during tensile deformation at 77K (–196 °C) after strains of 5, 18 and

[xxvi] The normal stacking fault in an fcc crystal is technically called an *intrinsic stacking fault* with an ABCABCBCABCA stacking sequence of {111} planes. In an fcc crystal, there is also an *extrinsic stacking fault*, which is equivalent to the insertion of a single extra {111} plane of atoms, i.e. an ABCABCBABCABC stacking sequence of {111} planes. It can be seen by inspection that an extrinsic stacking fault is equivalent to two intrinsic stacking faults on two adjacent {111} planes (the third B and third A planes from the left), so it has a higher energy and is, therefore, less common.

[xxvii] It is also perhaps worth noticing that a twin in fcc is equivalent to inserting an intrinsic stacking fault on every {111} plane on one side of the twin boundary; and it is also perhaps worth noticing that hcp is equivalent to fcc after inserting an intrinsic stacking fault on every other {111} plane.

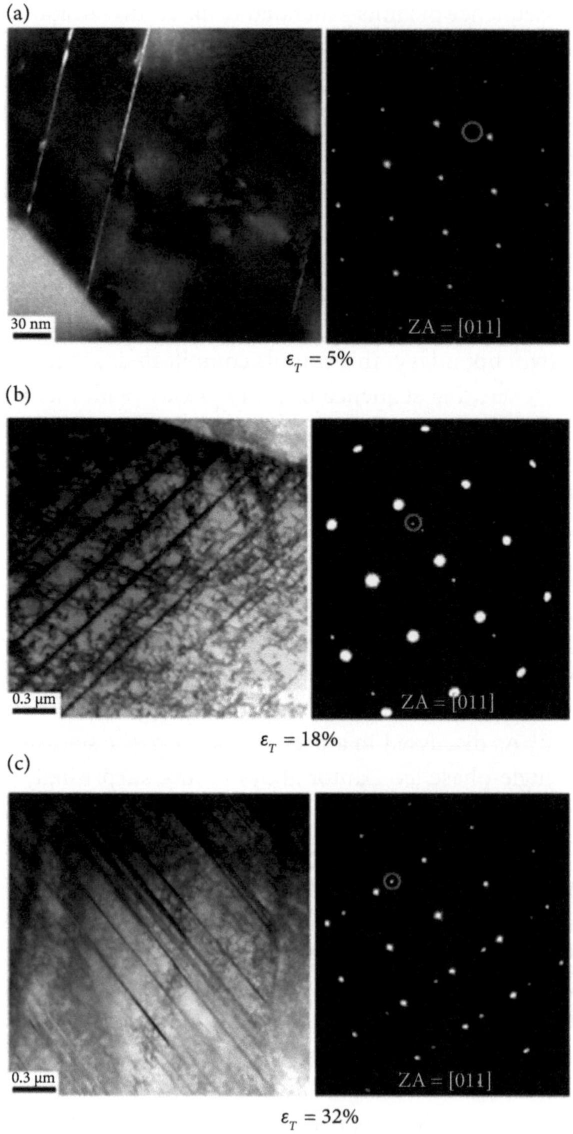

Figure 4.62 TEM micrographs of deformation twins in the single-phase fcc Cantor alloy CrMnFeCoNi after tensile strains of (a) 5%, (b) 18% and (c) 32%.
Reproduced from Moon et al. (2017). On the strain rate-dependent deformation mechanism of CoCrFeMnNi high-entropy alloy at liquid nitrogen temperature. *Materials Research Letters* 5: 472–7.

32%.[145] Twinning becomes a major deformation mechanism as well as dislocation flow at 77K in the original Cantor alloy and, as shown in Figure 4.62, the twin density increases with increasing tensile strain, from a volume fraction of v_f = 1% at 5% strain to v_f = 17% at 32% strain.[167] Deformation twinning in Cantor alloys has been shown to be very effective as a method of work hardening,[172] in initiating shear bands,[170] and in deflecting crack growth.[168] In general, twinning is found to be more significant as a deformation mechanism at lower temperatures, higher strain rates and larger grain sizes, and is also dependent on alloy composition and grain orientation,[167–171] with in all cases a subtle balance between on the one hand the CRSS and on the other hand the stacking fault, twin boundary and fcc/hcp interface energies.

There have been many attempts to use *ab initio* and other atomistic modelling methods to try to explore the mechanisms and resulting structures of deformation twinning in multicomponent high-entropy materials,[173–176] similar to the use of these methods, as described previously, in trying to study atomic diffusion, dislocation slip and grain-boundary structure. The results are again somewhat unsatisfactory given the difficulty of constructing a large number of suitable interatomic potentials, as well as the near impossibility of modelling enough atoms to include all of the gigantic number of different local atomic configurations. Nevertheless, they broadly support the direct experimental results discussed briefly above.

Grain size

There has been a lot of interest in controlling the grain size of multicomponent high-entropy materials as one of the methods of tuning their properties for potential applications, as is discussed in more detail in Chapter 6. Single-phase fcc Cantor alloys and bcc Senkov alloys have been manufactured with grain sizes ranging from large single crystals down to very fine-scale nanocrystalline materials. *Single crystals* of a variety of single-phase fcc Cantor alloys have been manufactured, as discussed in more detail in Chapter 5, by *Bridgman solidification*, including $Al_{0.3}CrFeCoNi$,[177] CrMnFeCoNi,[178–180] CrFeCoNi,[181] and CrCoNi,[180] in order to measure properties such as tracer diffusion,[178] thermal expansion,[182] elastic stiffness,[182] CRSS[177,181] and creep rate.[179,180]

Multicomponent *nanocrystalline* materials can be manufactured by a variety of different methods, including *sputter deposition*,[183–185] *high pressure torsion* (HPT)[186–189] and *mechanical alloying* (MA),[190–194] as also discussed in more detail in Chapter 5. A wide variety of different nanocrystalline multicomponent materials have been studied, including: fcc Cantor alloys such as CrFeCoNi,[191] CrMnFeCoNi,[188,189,197] $Al_{0.1}CrFeCoNi$[184] and $(AlTi)_x(FeCoNiCu)_{1-x}$; bcc Senkov alloys such as TiNbTa,[194] TiZrNbTa,[194] VZrNbMoW,[192] VNbMoTaW[193] and TiVCrZnNb;[190] and many other multiphase nanocrystalline and nanocomposite materials. Further details can be found in the reviews by Koch,[195] Haché et al.[196] and Sathiyamoorthy and Kim.[197] Sputter-deposited and mechanically alloyed (MA) nanocrystalline powder grain sizes can often be as low as 5–10 nm;[184,185,190,194] high-pressure torsion (HPT) produces somewhat larger grain sizes of 30–100 nm;[187–189] and consolidated MA powders have typical grain sizes of 100 nm–1 μm.[194–198] Figure 4.63 shows an example of bright-field and high-resolution transmission electron microscopy (BF-TEM and HRTEM) images from a sintered MA nanocrystalline equiatomic bcc VNbMoTaW Senkov alloy powder with a grain size of ~6 nm;[193] and Figure 4.64 shows an example of electron back-scatter diffraction (EBSD) maps from a sintered MA nanocrystalline equiatomic CrMnFeCoNi fcc Cantor alloy powder with grain sizes between 200 nm and 2 μm depending on the sintering conditions.[198] Multicomponent nanocrystalline materials are found to have extremely high hardnesses and strengths, up to 15–16 GPa, together with considerable ductility and fracture toughness, as will be discussed in more detail in Chapter 6.

The most common method of controlling grain size in metallurgical materials is by *thermomechanical processing* or *TMP*, using a mixture of deformation and heat treatment

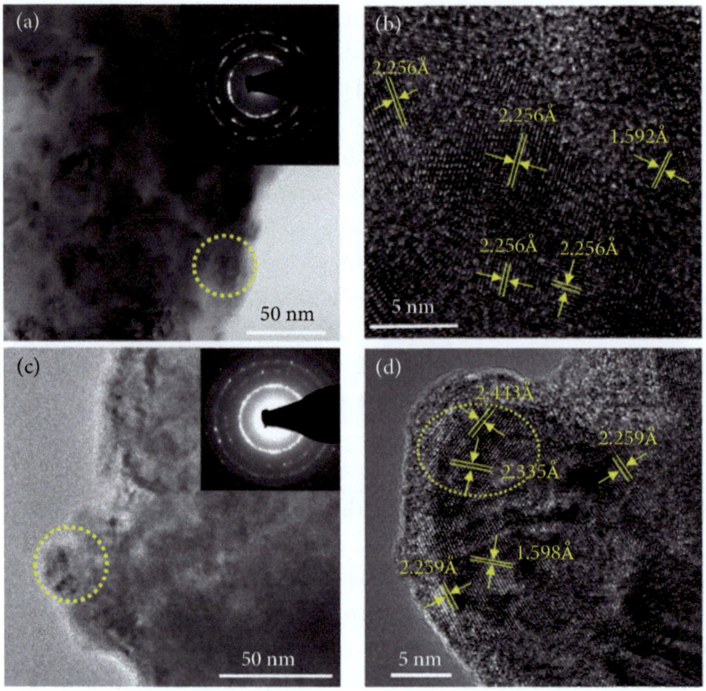

Figure 4.63 Sintered MA bcc VNbMoTaW Senkov alloy powder: (a) BF-TEM and (b) HRTEM images after sintering at 1000 °C; and (c) BF-TEM and (d) HRTEM images after sintering at 1150 °C.

Reproduced from Xin et al. (2018). Ultrahard bulk nanocrystalline VNbMoTaW high-entropy alloy. *Journal of Alloys and Compounds 769*: 597–604.

to *homogenise* and then *recrystallise* the material,[199–201] so as to produce a homogeneous composition and an equiaxed grain structure. The range of thermomechanical processing treatments that can be used is enormous: with a wide range of different types of deformation available, such as rolling, swaging, stretching, bending, torsion, etc.; a wide range of different heat treatments available, using different annealing temperatures, annealing times and heating and cooling rates; and a wide range of different possible multiple deformation and annealing sequences. In general, recrystallisation takes place during heat treatment after deformation in several stages: *recovery*, when the dislocations generated by deformation are reorganised into low-angle grain boundaries by a combination of cross-slip and climb; *recrystallisation* by the nucleation of new undeformed, dislocation-free grains followed by growth of the new low-energy undeformed grains to consume the old high-energy deformed grains; and finally *grain growth* as the larger dislocation-free grains continue to grow and consume the smaller dislocation-free grains, gradually increasing the grain size and decreasing the overall grain-boundary energy.[199–201] The resulting *grain size*, *texture* (i.e. the range of crystal orientations) and *grain-boundary character* (i.e. the spread of different types of grain boundary) depend in a complex way on the exact details of the thermomechanical processing that is used, providing considerable scope for detailed metallurgical engineering to fine tune the final microstructure of the material and, therefore, its resulting material properties.[199–201]

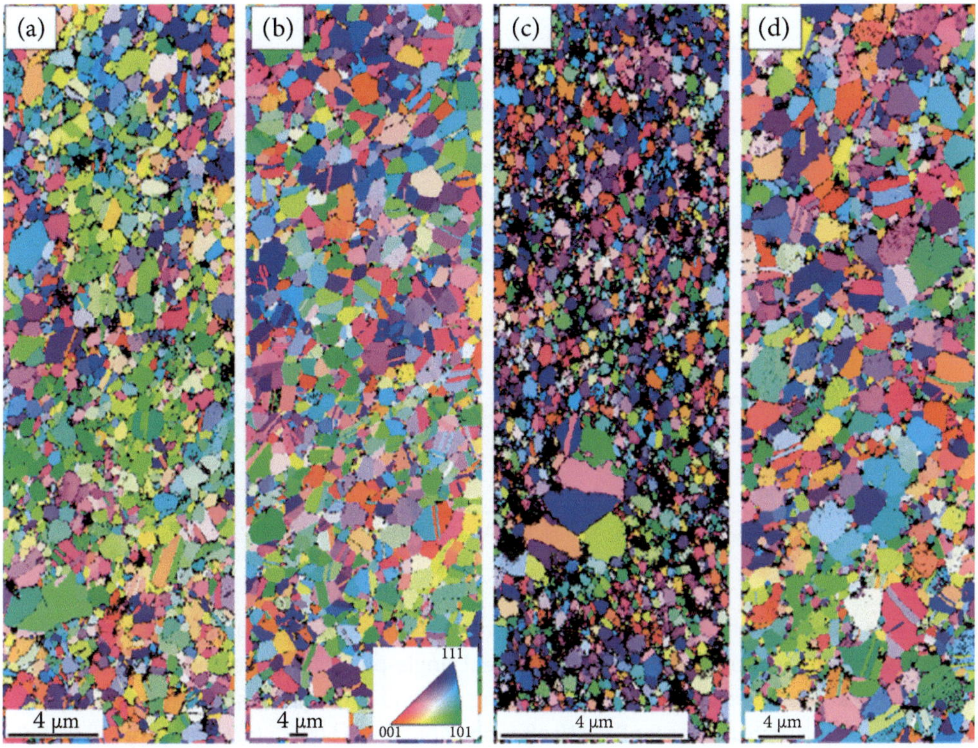

Figure 4.64 EBSD images of sintered MA fcc CrMnFeCoNi Cantor alloy powder: (a) twenty mins MA, sintered at 900 °C; (b) twenty mins MA, sintered at 1100 °C; (c) sixty mins MA, sintered at 900 °C; and (d) sixty mins MA, sintered at 1100 °C.

Reproduced from Joo et al. (2017). Structure and properties of ultrafine-grained CoCrFeMnNi high-entropy alloys produced by mechanical alloying and spark plasma sintering. *Journal of Alloys and Compounds 698*: 591–604.

There have been quite a few studies of recrystallisation, grain growth and texture in the original equiatomic 5-component single-phase fcc Cantor alloy CrMnFeCoNi[202–208] and its equiatomic single-phase fcc quaternary (CrFeCoNi, MnFeCoNi and CrMnCoNi) and ternary (CrFeNi, CrCoNi, MnFeNi, MnCoNi and FeCoNi) subsystems.[207,209–211] Recrystallisation and grain growth in these multicomponent materials are found to be considerably slower than in fcc pure metals and binary alloys, essentially caused by the wide range of local atomic structures and corresponding lattice strains, slowing the rates of atomic diffusion during heat treatment.[202–204,207,210] Recrystallisation temperatures in multicomponent equiatomic single-phase fcc Cantor alloys are, therefore, in the range T_r = 700–900 °C (973–1,173 K) for one-hour anneals, as shown by the variation of hardness versus annealing temperature for several different alloys[210] in Figure 4.65. These correspond to homologous temperatures relative to the melting point T_r/T_m = 0.6–0.75, depending on the particular alloy composition and the extent of the initial deformation treatment and subsequent heat treatment,[210] considerably higher than more typical values in conventional pure metals and binary alloys of T_r/T_m = 0.3–0.5. The resulting final recrystallised grain sizes are, as a consequence, very small and slow to coarsen, often no larger than 1–5 µm after one hour at the lower annealing temperatures of 700–800 °C (973–1,073 K)[202–204,207,210] and much smaller

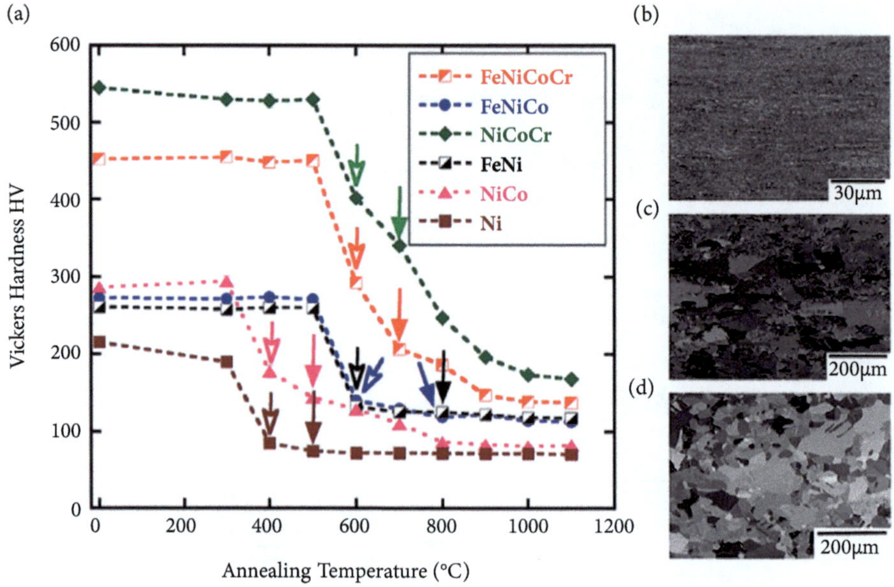

Figure 4.65 (a) Vickers hardness versus temperature for one hour anneals after 92% reduction in thickness by rolling for various multicomponent single-phase fcc Cantor alloys and pure fcc nickel (open and solid arrows show the beginning and end of recrystallisation respectively); (b)–(d) typical microstructures for FeCoNi after one hour at (b) 500 °C; (c) 600 °C; and (d) 800 °C, corresponding to (a) no recrystallisation; (b) partial recrystallisation; and (c) full recrystallisation.

Reproduced from Wu et al. (2014). Recovery, recrystallization, grain growth and phase stability of a family of fcc-structured multicomponent equiatomic solid solution alloys. *Intermetallics 46*: 131–40.

than found in fcc nickel,[210] as shown again for several different alloys in Figure 4.66, largely independent of the extent of the initial deformation treatment,[202–204,207–210] though larger grain sizes are achieved by annealing at higher temperatures, as also shown in Figure 4.66.

As shown for example in Figure 4.67, grain growth in multicomponent single-phase fcc Cantor alloys obeys classical power-law kinetics:[202,204,210]

$$d^n - d_o^n = kt$$

where d_o and d are respectively the initial grain size and the grain size after annealing for time t at a temperature T, n is the grain-growth kinetic exponent, and k is the temperature-dependent grain-growth kinetic rate constant. The grain-growth exponent for multicomponent single-phase fcc Cantor alloys is found to be relatively low,[202,204,210] in the range $n = 3–4$ as shown in Figure 4.67, and the grain-growth kinetic constant obeys an Arrhenius variation with temperature:

$$k = k_o \exp\ (-Q/RT)$$

where k_o is a constant, R is the gas constant, and Q is the activation energy for grain growth.[202,204,210] The activation energy $Q\ \approx$ 325 kJ/mol in the original Cantor alloy

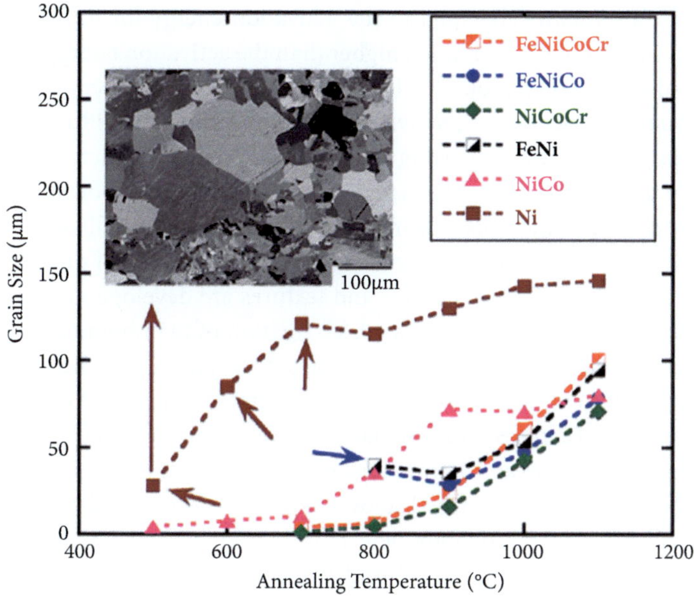

Figure 4.66 Grain size versus temperature for one hour anneals after 92% cold rolling for various multicomponent single-phase fcc Cantor alloys and pure fcc Ni, with inset showing abnormal grain growth in fcc Ni.

Reproduced from Wu et al. (2014). Recovery, recrystallization, grain growth and phase stability of a family of fcc-structured multicomponent equiatomic solid solution alloys. *Intermetallics 46*: 131–40.

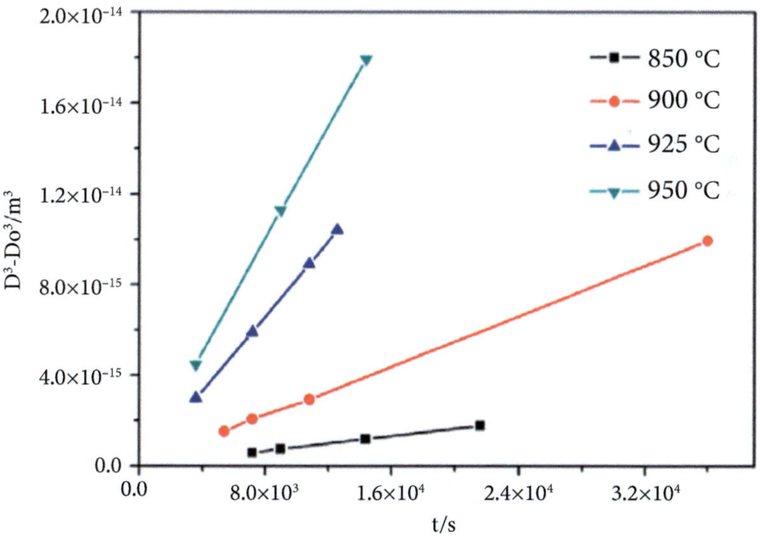

Figure 4.67 Grain size D versus time t at different annealing temperatures T for the original equiatomic single-phase fcc Cantor alloy CrMnFeCoNi after 70% cold rolling.

Reproduced from Liu et al. (2013). Grain growth and the Hall–Petch relationship in a high-entropy FeCrNiCoMn alloy. *Scripta Materialia 68*: 526–9.

CrMnFeCoNi,[204] approximately equal to the activation energy for diffusion in CrMnFe-CoNi, as shown in Table 4.15, and much higher than the activation energy for grain growth in pure nickel of $Q \approx 150$ kJ/mol.[178] A low activation energy for recrystallisation in pure fcc metals is usually ascribed to a relative ease of grain-boundary diffusion in pure-metal high-angle grain boundaries, suggesting that it is the complex structure of multicomponent high-angle grain boundaries discussed previously that is responsible for restricting grain-boundary diffusion and thus slowing down the rate of recrystallisation.

As a result of the various effects described above, a wide range of different, often quite complex, recrystallisation microstructures and textures are developed in multicomponent single-phase fcc Cantor alloys, often with a high fraction of twin boundaries, as shown for example in Figure 4.68, with the exact details again depending on the extent of the initial deformation treatment and subsequent heat treatment, and again caused by the wide range of local atomic structures and associated lattice strains promoting the nucleation of different crystal orientations and twins during grain-boundary migration.[202–211] In general, similar results to those described above have also been found for recrystallisation and grain growth in other multicomponent high-entropy materials, such as single-phase fcc $Al_{0.1}CrFeCoNi$[212] and $CrFeCoNi_2$[211] and two-phase fcc + bcc $Al_{0.3}CrFeCoNi$[213] and $Al_{0.5}CrFeCoNi$[214].

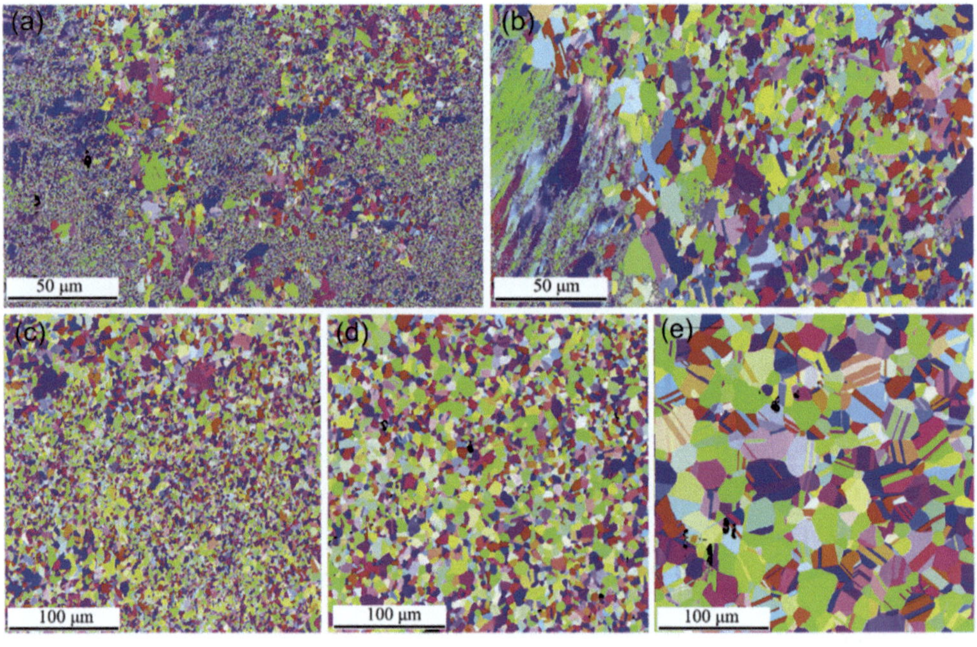

Figure 4.68 EBSD maps of recrystallised structures in the original single-phase fcc Cantor alloy CrMnFeCoNi annealed for different times and temperatures after deformation by 90% cold rolling: (a)–(b) partially recrystallised after annealing for one hour at 600 and 650 °C respectively; (c)–(e) fully recrystallised after annealing for one hour at 700, 800 and 900 °C respectively.

Reproduced from Gu et al. (2019). Regulating the strength and ductility of a cold-rolled FeCrCoMnNi high-entropy alloy via annealing treatment. *Materials Science and Engineering* A755: 289–94.

5 Summary

Multicomponent phase space is populated by vast multicomponent high-entropy single-phase regions: regions of simple close-packed multicomponent disordered fcc, bcc and hcp solid solutions, as well as regions of multicomponent intermetallic and ceramic compounds such as aluminides, oxides and borides, with disordered solid solutions on their anion and/or cation sublattices. It seems clear that, when the alloying content in a material is increased to higher concentrations and larger numbers of components, the entropy of mixing promotes the formation of random solid solutions on one or more sublattices of existing crystalline structures rather than the formation of completely new fixed-stoichiometry compounds. The resulting vast solid-solution regions in multicomponent phase space contain materials that are rightly known as high-entropy materials, separated by similarly vast regions of multiphase materials, so far much less studied, that do not always have such high entropy.

Multicomponent high-entropy single-phase solid solutions, whether fully disordered as in fcc, bcc and hcp, or disordered on one or more sublattices as in multicomponent intermetallic or ceramic compounds, have enormous numbers (typically in the billions or trillions) of different local atomic or molecular structures with associated variations in local properties extending over very large spatial regions within each material. In particular, the substantial and extensive variations in local atomic and molecular structures in multicomponent high-entropy solid-solution materials create associated substantial and extensive variations in local mechanical and electronic lattice strain, typically on the order of approximately one or two percent on average and several percent at maximum. These mechanical and electronic strains and their widespread variation from point-to-point throughout individual crystals and grains influence many of the most fundamental properties of multicomponent high-entropy solid-solution materials, notably reducing electrical and thermal conductivity, slowing down rates of vacancy hopping and atomic diffusion, restricting dislocation slip and thus increasing material strength without loss of ductility, and slowing down rates of grain-boundary migration and recrystallisation.

References

1. Gaskell DR, Laughlin DE. (2017). *Introduction to the Thermodynamics of Materials*, 6th edn. Boca Raton: CRC Press.
2. Hillert M. (2008). *Phase Equilibria, Phase Diagrams and Phase Transformations*, 2nd edn. Cambridge: Cambridge University Press.
3. Cantor B. (2020). *The Equations of Materials.* Oxford: Oxford University Press.
4. Cantor B. (2021). Multicomponent high-entropy Cantor alloys. *Progress in Materials Science* 120: Article 100754.
5. Cantor B. (2021). A personal perspective on the discovery and significance of multicomponent high-entropy alloys. In J. Brechtl, PK Liaw, eds., *Fundamental Studies of High-Entropy Materials: Theory, Experiments and Applications.* New York: Springer, 1–36.
6. Andersson J, Guillermet AF, Hillert M. (1975). A compound energy model of ordering in a phase with sites of different coordination numbers. *Acta Materialia* 34: 437–45.

7. Muggianu YM, Gambino M, Brosj JP. (1975). Enthalpies of formation of liquid alloys Bi-Ga-Sn at 273K—choice of analytical representation of integral and partial thermodynamic functions of mixing for this ternary system. *Journal de Chimie Physique et de Physico-Chimie Biologique* 2: 83–8.
8. Yeh JW, Chen SK, Lin SJ, Gan JY, Chin TS, Shun TT, Tsau CH, Chang SY. (2004). Nanostructured high entropy alloys with multiple principal elements: Novel alloy design concepts and outcomes. *Advanced Engineering Materials* 6: 299–303.
9. Miracle DB, Senkov ON. (2017). A critical review of high-entropy alloys and related concepts. *Acta Materialia* 122: 448–511.
10. Murty BS, Yeh JW, Ranganathan S, Bhattacharjee PP. (2019). *High-Entropy Alloys*. Amsterdam: Elsevier.
11. Brechtl J, Liaw PK. (eds). (2021). *Fundamental Studies of High-Entropy Materials: Theory, Experiments and Applications*. New York: Springer.
12. Laurent-Brocq M, Akhatova A, Perrière L, Chebini S, Sauvage X, Leroy E. (2015). Insights into the phase diagram of the CrMnFeCoNi high entropy alloy. *Acta Materialia* 88: 355–65.
13. Tsai KY, Tsai MH, Yeh JW. (2013). Sluggish diffusion in CoCrFeMnNi high-entropy alloy. *Acta Materialia* 61: 4887–97.
14. Broomfield ME, Christofidou KA, Jones NG. (2019). Effect of Co on the phase stability of CrMnFeCoxNi high entropy alloys following long-duration exposures at intermediate temperatures. *Intermetallics* 114: Article 106582.
15. Bracq G, Laurent-Brocq M, Varvenne C, Perrière, L., Curtin WA, Joubert J-M, GuillotI. (2019). Combining experiments and modelling to explore the solid solution strengthening of high and medium entropy alloys. *Acta Materialia* 177: 266–79.
16. Pickering EJ, Muñoz-Moreno R, Stone HJ, Jones NG. (2016). Precipitation in the equiatomic high-entropy alloy CrMnFeCoNi. *Scripta Materialia* 113: 106–9.
17. Otto F, Diouhy A, Pradeep KG, Kubenova, Raabe D, Eggeler G, George EP. (2016). Decomposition of the single-phase high-entropy alloy CrMnFeCoNi after prolonged anneals at intermediate temperatures. *Acta Materialia* 112:40–52.
18. Schuh B, Mendez-Martin F, Völker B, George EP, Clemens H, Pippan R, Hohenwarter A. (2015). Mechanical properties, microstructure and thermal stability of a nanocrystalline CoCrFeMnNi high-entropy alloy after severe plastic deformation. *Acta Materialia* 96: 258–68.
19. Zhou W, Fu LM, Liu P, Xu XD, Chen B, Zhu GZ, Wang XD, Shan AD, Chen MW. (2017). Deformation stimulated precipitation of a single-phase CoCrFeMnNi high entropy alloy. *Intermetallics* 85: 90–7.
20. Lee KH, Hong S-K, Hong SI. (2019). Precipitation and decomposition in CoCrFeMnNi high entropy alloy at intermediate temperatures under creep conditions. *Materialia* 8: Article 100445.
21. Lee U, Straumal B, Park N. (2021). Dynamic precipitation of σ-phase and element partitioning in equiatomic CoCrFeMnNi high-entropy alloy. *Materials Science and Engineering* A804: Article 140739.
22. Laplanche G, Berglund S, Reinhart C, Kostka A, Fox F, George EP. (2018). Phase stability and kinetics of σ-phase precipitation in CrMnFeCoNi high-entropy alloys. *Acta Materialia* 161: 338–51.

23. Hultgren R. (1973). *Selected Values of the Thermodynamic Properties of Binary Alloys.* Cleveland: ASM International.
24. Otto F, Yang Y, Bei H, George EP. (2013). Relative effects of enthalpy and entropy on the phase stability of equiatomic high-entropy alloys. *Acta Materialia* 61: 2628–38.
25. Gao MC, Yeh JW, Liaw PK, Zhang Y. (eds). (2016). *High-Entropy Alloys: Fundamentals and Applications.* Basel: Springer.
26. Tsai MH, Yeh JW. (2014). High-entropy alloys: A critical review. *Materials Research Letters* 2: 107–23.
27. Koh KM, Chen CC. (1992). *Principles and Techniques in Combinatorics.* Singapore: World Scientific, 1–50.
28. Philips FC. (1973). *An Introduction to Crystallography.* New York: Wiley.
29. Kelly A, Knowles KM. (2012). *Crystallography and Crystal Defects.* New York: Wiley.
30. Vegard JL. (1921). Die konstitution der mischkristalle und die raumfüllung der atome. *Zeitschrift für Physik* 5: 17–26.
31. Alonso JC, Alonso OC. (2021). Derivation of unit cell volume and lattice parameter of cubic high entropy alloys from volume size factors. *Intermetallics* 137: Article 107299.
32. Greenwood NN, Earnshaw A. (1997). *Chemistry of the Elements,* 2nd edn. Oxford: Butterworth-Heinemann.
33. Cantor B, Chang ITH, Knight P, Vincent AJB. (2004). Microstructural development in equiatomic multicomponent alloys. *Materials Science and Engineering* A375: 213–18.
34. He JY, Liu WH, Wang H, Wu Y, Liu XJ, Nieh TJ, Lu ZP. (2014). Effects of Al addition on structural evolution and tensile properties of the FeCoNiCrMn high-entropy alloy system. *Acta Materialia* 62: 105–13.
35. Okamoto NL, Yuge K, Tanaka K, Inui H, George EP. (2016). Atomic displacement in the CrMnFeCoNi high entropy alloy: A scaling factor to predict solid solution strengthening. *American Institute of Physics* 6: Article 125008.
36. Oh HS, Ma D, Leyson GP, Grabowski B, Park ES, Körmann F, Raabe D. (2016). Lattice distortions in the FeCoNiCrMn high entropy alloy studied by theory and experiment. *Entropy* 18: 321–9
37. Song H, Tian F, Hu QM, Vitos L, Wang Y, Shen J, Chen N. (2018). Local lattice distortions in high-entropy alloys. *Physical Review Materials* 1: 023404.
38. Yen CC, Huang GR, Huang YC, Yeh HW, Luo DJ, Hsieh KT, Huang EW, Yeh JW, Lin SJ, Wang CC, Kuo CL, Chang SY, Lo YC. (2020). Lattice distortion effect on elastic anisotropy of high-entropy alloys. *Journal of Alloys and Compounds* 818: 152876.
39. Owen LR, Pickering EJ, Playford HY, Stone HJ, Tucker MG, Jones NG. (2017). An assessment of the lattice strain in the CrMnFeCoNi high-entropy alloy. *Acta Materialia* 122: 11–18.
40. Tan YY, Su MY, Xie ZC, Chen ZJ, Gong Y, Zheng LR, Shi Z, Mo G, Li Y, Li LW, Wang HY, Dai LH. (2021). Chemical composition dependent local lattice distortions and magnetism in high entropy alloys. *Intermetallics* 129: Article 107050.
41. Tong Y, Jin K, Bei H, Ko JYP, Pagan DC, Zhang Y, Zhang FX. (2018). Local lattice distortion in NiCoCr, FeCoNiCr and FeCoNiCrMn concentrated alloys investigated by synchrotron X-ray diffraction. *Materials and Design* 155: 1–7.

42. Tong Y, Zhao S, Jin K, Bei H, Ko JPY, Zhang Y, Zhang FX. (2018). A comparison study of local lattice distortion in $Ni_{80}Pd_{20}$ binary alloy and FeCoNiCrPd high entropy alloy. *Scripta Materialia* 156: 14–18.
43. Ge H, Tian F. (2019). A review of ab initio calculation on lattice distortion in high-entropy alloys. *Journal of Metals* 71: 4225–37.
44. Fernandez-Caballero A, Wróbel JS, Mummery PM, Nguyen-Manh D. (2017). Short-range order in high-entropy alloys: theoretical formulation and application to MoNbTaVW system. *Journal of Phase Equilibria and Diffusion* 38: 391–403.
45. Senkov ON, Miracle DB, Chaput KJ. (2018). Development and exploration of refractory high entropy alloys—A review. *Journal of Materials Research* 33: 3092–3218.
46. Miracle DB, Tsai MH, Senkov ON, Soni V, Banerjee R. (2020). Refractory high entropy superalloys. *Scripta Materialia* 187: 445–52.
47. Zou Y, Maiti S, Steurer W, Spolenak R. (2014). Size-dependent plasticity in an $Nb_{25}Mo_{25}TA_{25}W_{25}$ refractory high-entropy alloy. *Acta Materialia* 65: 85–97.
48. Guo W, Dmowski W, Noh JY, Rack P, Liaw PK, Egami T. (2013). Local atomic structure of a high-entropy alloy: An X-ray and neutron scattering study. *Metallurgical and Materials Transactions* 44A: 1994–7.
49. Maiti S, Steurer W. (2016). Structural disorder and its effect on mechanical properties in single-phase TaNbHfZr high-entropy alloy. *Acta Materialia* 106: 87–97.
50. Shannon RD. (1976). Revised effective ionic radii and systematic studies of interatomic distances in halides and chalcogenides. *Acta Crystallographia* 32A: 751–67.
51. Rost CM, Sachet E, Borman T, Moballegh A, Dickey EC, Hou D, Jones JL, Curtarolo S, Maria JP. (2015). Entropy-stabilized oxides. *Nature Communications* 6: Article 8485.
52. Owen LR, Jones NG. (2018). Lattice distortions in high-entropy alloys. *Journal of Materials Research* 33: 2954–69.
53. Cowley JM. (1950). An approximate theory of order in alloys. *Physical Review* 77: 669–75.
54. Li JH, Guo HB, Lai WS, Liu BX. (2004). Microchemical inhomogeneity of multicomponent systems and its evaluation from interatomic potentials. *Physical Review B* 69: Article 172201.
55. Ceguerra AV, Powles RC, Moody MP, Ringer SP. (2010). Quantitative description of atomic architecture in solid solutions: A generalized theory for multicomponent short-range order. *Physical Review B* 82: Article 132201.
56. De Fontaine D. (1971). The number of independent pair-correlation functions in multicomponent systems. *Journal of Applied Crystallography* 4: 15–19.
57. Muniandy Y, He M, Eizadjou M, George EP, Kruzic JJ, Ringer SP, Gludovatz B. (2021). Compositional variations in equiatomic CrMnFeCoNi high-entropy alloys. *Materials Characterization* 180: Article 111437.
58. Yao MJ, Pradeep KJ, Tasan CC, Raabe D. (2014). A novel, single phase, non-equiatomic FeMnNiCoCr high-entropy alloy with exceptional phase stability and tensile ductility. *Scripta Materialia* 72–73: 5–8.
59. Zhang FX, Zhao S, Jin K, Xue H, Velisa G, Bei H, Huang R, Ko JYP, Pagan DC, Neuefeind JC, Weber WJ, Zhang Y. (2017). Local structure and short-range order in a NiCoCr solid solution alloy. *Physical Review Letters* 118: Article 205501.

60. Zhang R, Zhao S, Ding J, Chong Y, Jia T, Ophus C, Asta M, Ritchie RO, Minor AM. (2020). Short-range order and its impact on the CoCrNi medium-entropy alloy. *Nature* 581: 283–7.
61. Chaudhary V, Soni V, Gwalani B, Ramanujan RV, Banerjee R. (2020). Influence of non-magnetic Cu on enhancing the low temperature magnetic properties and Curie temperature of $FeCoNiCrCu_x$ high entropy alloys. *Scripta Materialia* 182: 99–103.
62. Yuan H, Tsai M-H, Sha G, Liu F, Horita Z, Zhu Y, Wang JT. (2016). Atomic-scale homogenization in an fcc-based high-entropy alloy via severe plastic deformation. *Journal of Alloys and Compounds* 686: 15–23.
63. Gwalani B, Choudhuri D, Soni V, Ren Y, Styles M, Hwang JY, Nam SJ, Ryu H, Hong SH, Banerjee R. (2017). Cu assisted stabilization and nucleation of L12 precipitates in Al0.3CuFeCrNi2 fcc-based high-entropy alloy. *Acta Materialia* 129: 170–82.
64. Shinde D, Fritze S, Thuvander M, Malinovskis P, Riekehr L, Jansson U, Stiller K. (2019). Elemental distribution in CrNbTaTiW-C high-entropy alloy thin films. *Microscopy and Microanalysis* 25: 489–500.
65. Kostiuchenko T, Körmann F, Neugebauer J, Shapeev A. (2019). Impact of lattice relaxations on phase transitions in a high-entropy alloy studied by machine learning potentials. *NPJ Computing Materials* 5: 1–7.
66. Qiu Y, Hu YJ, Taylor A, Styles MJ, Marceau RKW, Ceguerra AV, Gibson MA, Liu ZK, Fraser HL, Birbilis N. (2017). A lightweight single-phase AlTiVCr compositionally complex alloy. *Acta Materialia* 123: 115–24.
67. Hu R, Jin S, Sha G. (2022). Application of atom probe tomography in understanding high entropy alloys: 3D local chemical compositions in atomic scale analysis. *Progress in Materials Science* 123: Article 100854.
68. Ferrari A, Dutta B, Gubaev K, Ikeda Y, Srinivasan P, Grabowski B, Körmann FHW. (2020). Frontiers in atomistic simulations of high entropy alloys. *Journal of Applied Physics* 128: Article 150901.
69. Kittel C. (1957). *Introduction to Solid State Physics*, 2nd edn. New York: Wiley, 234–346.
70. Sutton AP. (1993). *Electronic Structure of Materials*. Oxford: Oxford University Press.
71. Barber ZH (ed). (2005). *Introduction to Materials Modelling*. London: Maney.
72. Giustino F. (2014). *Materials Modelling Using Density Functional Theory*. Oxford: Oxford University Press.
73. Car R, Parrinello M. (1985). Unified approach for molecular dynamics and density functional theory. *Physical Review Letters* 55: 2471–4.
74. Born M, Oppenheimer R. (1927). Zur quantentheorie der molekeln [On the quantum theory of molecules]. *Annalen der Physik* 84: 457–84.
75. Mu Y, Liu H, Liu Y, Zhang X, Jiang Y, Dong T. (2017). An ab initio and experimental studies of the structure, mechanical parameters and state density on the refractory high-entropy alloy systems. *Journal of Alloys and Compounds* 714: 668–80.
76. Chang C, Zhang H. (2022). First-principles calculations to investigate elastic and thermodynamic properties of $FeAlNi_xCrMn$ quinternary alloys. *Journal of Materials Research and Technology* 18: 1322–32.
77. Tong Y, Bai L, Liang X, Chen Y, Zhang Z, Liu J, Li Y, Hu Y. (2020). Influence of alloying elements on mechanical and electronic properties of NbMoTaWX (X = Cr, Zr, V, Hf and Re) refractory high entropy alloys. *Intermetallics* 126: Article 106928.

78. Karabin M, Mondal WR, Östlin A, Ho WGD, Dobrosavljevic V, Tam KM, Terletska H, Chioncel L, Wang Y, Eisenbach M. (2022). Ab initio approaches to high-entropy alloys: a comparison of CPA, SQS and supercell methods. *Journal of Materials Science* 57: 10677–90.
79. Wencka M, Krnel M, Jelen A, Vrtnik S, Luzar J, Kozelj P, Gacnik D, Meden A, Hu Q, Wang C, Guo S, Dolinsek J. (2022). Electronic transport properties of the Al0.5TiZrPdCuNi alloy in the high-entropy alloy and metallic glass forms. *Scientific Reports* 12: Article 2271.
80. Wu D, Kusada K, Nanba Y, Koyama M, Yamamoto T, Toriyama T, Matsumura S, Seo O, Gueye I, Kim J, Singgapulige L, Kumara R, Sakata O, Kawaguchi S, Kobota Y, Kitagawa H. (2022). Noble-metal high-entropy-alloy nanoparticles: atomic-level insight into the electronic structure. *Journal of the American Chemical Society* 144: 3365–9.
81. Uporov SA, Ryltsev RE, Sidorov VA, Estemirova SK, Sterkhov EV, Balyakin IA, Chtchelkatchev NM. (2022). Pressure effects on electronic structure and electrical conductivity of TiZrHfNb high-entropy alloy. *Intermetallics* 140: Article 107394.
82. Bag P, Su YC, Kuo YK, Lai YC, Wu SK. (2021). Physical properties of face-centred cubic structured high-entropy alloys: Effects of NiCo, NiFe and NiCoFe alloying with Mn, Cr and Pd. *Physical Review Materials* 5: Article 085003.
83. Samolyuk GD, Mu S, May AF, Sales BC, Wimmer S, Mankovsky S, Ebert H, Stocks GM. (2018). Temperature-dependent electronic transport in concentrated solid solutions of the 3d-transition metals Ni, Fe, Co and Cr from first principles. *Physical Review B* 98: Article 165141.
84. Mu S, Samolyuk GD, Wimmer S, Troparevsky MC, Khan S, Mankovsky S, Ebert H, Stocks GM. (2019). Uncovering electron scattering mechanisms in NiFeCoCrMn derived concentrated solid solution and high entropy alloys. *NPJ Computational Materials* 5: 1–8.
85. Jin K, Sales BC, Stocks GM, Samolyuk GD, Daene M, Weber WJ, Zhang Y, Bei H. (2016). Tailoring the physical properties of Ni-based single-phase equiatomic alloys by modifying the chemical complexity. *Scientific Reports* 6: Article 20159.
86. Zhang Y, Stocks GM, Jin K, Lu C, Bei H, Sales BC, Wang L, Béland LK, Stoller RE, Samolyuk GD, Caro M, Caro A, Weber W. (2015). Influence of chemical disorder on energy dissipation and defect evolution in concentrated solid solution alloys. *Nature Communications* 6: Article 8736.
87. Uporov SA, Ryltsev RE, Estemirova SK, Sterkhov EV, Chtchelkatchev NM. (2021). Stable high-entropy TiZrHfNbVCrMoMnFeCoNiAl Laves phase. *Scripta Materialia* 193: 108–11.
88. Ryltsev R, Gaviko V, Estemirova S, Sterkhov E, Cherepanova L, Yagodin D, Chtchelkatchev N, Dubinin N, Uporov S. (2021). Laves phase formation in high entropy alloys. *Metals* 11: Article 1962.
89. Ryltsev RE, Estemirova SK, Gaviko VS, Yagodin DA, Bykov VA, Sterkhov EV, Cherepanova LA, Sipatov IS, Balyakin IA, Uporov SA. (2022). Structural evolution in TiZrHfNb high-entropy alloy. *Materialia* 21: Article 101311.
90. Uporov S, Bykov V, Pryanichnikov S, Shubin A, Uporova N. (2017). Effect of synthesis route on structure and properties of AlCoCrFeNi high-entropy alloy. *Intermetallics* 83: 1–8.
91. Abbaschian R, Abbaschian L, Reed-Hill RE. (2009). *Physical Metallurgy Principles*, 4th edn. Boston: Cengage Learning, chs 7 and 12, 194–215; 348–88.
92. Shewmon P. (2016). *Diffusion in Solids*, 2nd edn. New York: Springer.

93. Thomas SL, Patala S. (2020). Vacancy diffusion in multi-principal element alloys: The role of chemical disorder in the ordered lattice. *Acta Materialia* 196: 144–53.
94. Kottke J, Utt D, Laurent-Brocq M, Fareed A, Gaertner D, Perrière L, Rogal L, Stukowski A, Albe K, Divinski SV. (2020). Experimental and theoretical study of tracer diffusion in a series of $(CoCrFeMn)_{100-x}Ni_x$ alloys. *Acta Materialia* 194: 236–48.
95. Choi WM, Jo YH, Sohn SS, Lee S, Lee B. (2018). Understanding the physical metallurgy of the CoCrFeMnNi high-entropy alloy: An atomistic simulation study. *NPJ Computing Materials* 4: 1.
96. Li Y, Li R, Peng Q, Ogata S. (2020). Reduction of dislocation, mean free path, and migration barriers using high entropy alloy: Insights from the atomistic study of irradiation damage of CoNiCrFeMn. *Nanotechnology* 31: Article 425761.
97. Manara D, Seibert A, Gouder T, Benes O, Martel L, Colle J-Y, Griveau J-C, Walter O, Canbriani A, Blanco OD, Staicu D, Wiss T, Vigier J-F. (2020). Experimental methods. In MHA Piro, ed., *Advances in Nuclear Fuel Chemistry*. Oxford: Woodhead-Elsevier, ch 2, 89–158.
98. Somoza A, Dupasquier A. (2011). Vacancies in aluminium and solute-vacancy interactions in aluminium alloys. In R Lumley, ed., *Fundamentals of Aluminium Metallurgy: Production, Processing and Applications*. Oxford: Woodhead-Elsevier, ch 14, 386–421.
99. Sugita K, Matsuoka N, Mizuno M, Araki H. (2020). Vacancy formation enthalpy in CoCrFeMnNi high-entropy alloy. *Scripta Materialia* 176: 32–5.
100. Sugita K, Ogawa R, Mizuno M, Araki H, Yabuuchi A. (2022). Vacancy migration energies in CrMnFeCoNi, CrFeCoNi and CrFeNi alloys and their effect on atomic diffusion. *Scripta Materialia* 208: Article 114339.
101. Schulte CW, Campbell JL, Jackman JA. (1978). Temperature-dependence of positron-annihilation in iron and vanadium. *Applied Physics* 16: 29–34.
102. Vaidya M, Pradeep KG, Murty BS, Wilde G, Divinsky SV. (2018). Bulk tracer diffusion in CoCrFeNi and CoCrFeMnNi high entropy alloys. *Acta Materialia* 146: 211–24.
103. Jodi DE, Listyawan TA, Hruska P, Cizek J, Park N, Lee U. (2021). Study of vacancies in Fex(CoCrMnNi)100-x medium- and high-entropy alloys by positron annihilation spectroscopy. *Scripta Materialia* 194: Article 113654.
104. Yeh JW. (2006). Recent progress in high entropy alloys. *Annale de Chimie Science des Matériaux* 31: 633–48.
105. Yeh JW. (2013). Alloy design strategies and future trends in high-entropy alloys. *Journal of Metals* 65: 1759–71.
106. Vaidya M, Trubel S, Murty BS, Wilde G, Divinsky S. (2016). Ni tracer diffusion in CoCrFeNi and CoCrFeMnNi high entropy alloys. *Journal of Alloys and Compounds* 88: 994–1001.
107. Dabrowa J, Kucza W, Cieslak G, Kulak T, Daniele ski M, Yeh JW. (2016). Interdiffusion in the FCC-structured Al-Co-Cr-Fe-Ni high entropy alloys: Experimental studies and numerical simulations. *Journal of Alloys and Compounds* 674: 455–62.
108. Bake DL, Redeye G. On the diffusion in high-entropy alloys. *Materials Letters*, 2016, vole 164, pp. 111–113.
109. Kulkarni K, Chatham GPS. (2015). Investigations of quaternary interdiffusion in a constituent system of high entropy alloys. *American Institute of Physics Advances* 5: Article 097162.

110. Chen W, Zhang L. (2017). High-throughput determination of interdiffusion coefficients for Co-Cr-Fe-Mn-Ni high-entropy alloys. *Journal of Phase Equilibria and Diffusion* 38: 457–65.
111. Li Q, Chen W, Hong J, Zhang L, Chen Q, Liu Z. (2018). On sluggish diffusion in fcc AlCoCrFeNi high-entropy alloys: an experimental and numerical study. *Metals* 8: Article 16.
112. Wang R, Chen W, Phuong J, Zhang L. (2018). Experimental and numerical studies on the sluggish diffusion in face cantered cubic Co-Cr-Cu-Fe-Ni high-entropy alloys. *Journal of Materials Science and Technology* 34: 1791–8.
113. Dabrowa J, Zeus M, Kucza W, Cieslak G, Berent K, Crepe T, Kulik T, Danielewski M. (2019). Demystifying the sluggish diffusion effect in high entropy alloys. *Journal of Alloys and Compounds* 783: 193–207.
114. Zhang C, Zhang F, Jin K, Bei H, Chen S., Cao W, Zhu J, Lv D. (2017). Understanding of the elemental diffusion behavior in concentrated solid solution alloys. *Journal of Phase Equilibria and Diffusion* 38: 434–44.
115. Vaidya M, Pradeep KG, Murty BS, Wilde G, Divinski S. (2017). Radioactive isotopes reveal a non sluggish kinetics of grain boundary diffusion in high entropy alloys. *Science Reports* 7: Article 12293.
116. Dabrowa J, Danielewsky M. (2020). State-of-the-art diffusion studies in the high entropy alloys. *Metals* 10: 347–91.
117. Hashimoto N, Ono Y. (2021). Mobility of point defects in CoCrFeNi-base high entropy alloys. *Intermetallics* 133: Article 107182.
118. Lukianova OA, Rao Z, Kulitckii V, Li Z, Wilde G, Divinski SV. (2020). Impact of interstitial carbon on self-diffusion in CoCrFeMnNi high entropy alloys. *Scripta Materialia* 188: 264–8.
119. Lu E, Zhao J, Makkonen I, Mizohata K, Li Z, Hua M, Djurabekova F, Tuomisto F. (2021). Enhancement of vacancy diffusion by C and N interstitials in the equiatomic FeMnNiCoCr high entropy alloy. *Acta Materialia* 215: Article 117093.
120. Cai W, Nix WD. (2016). *Imperfections in Crystalline Solids.* Cambridge: Cambridge University Press,209–432.
121. Hull D, Bacon DJ. (2011). *Introduction to Dislocations*, 5th edn. Oxford: Butterworth-Heinemann/Elsevier, 157–250.
122. Ashby MF, Jones DRH. (2012). *Engineering Materials: An Introduction to Properties, Application and Design*, 4th edn. Oxford: Butterworth-Heinemann/Elsevier chapters 9 and 10, 135–156.
123. Martin JW. (1996). *Materials for Engineering.* London: Institute of Materials, 69–78.
124. Honeycombe RWK. (1984). *The plastic deformation of metals.* London: Edwin Arnold, chs 4 and 5, 80–143.
125. Okamoto NL, Fujimoto S, Kambara Y, Kawamura M, Chen ZMT, Matsunoshita H, Tanaka K, Inui H, George EP. (2016). Size effect, critical resolved shear stress, stacking fault energy, and solid solution strengthening in the CrMnFeCoNi high-entropy alloy. *Nature Scientific Reports* 6: Article 35863.

126. Smith TM, Hooshmand MS, Esser BD, Otto F, McComb DW, George EP, Ghazisaeidi M, Mills MJ. Atomic-scale characterization and modelling of 60° dislocations in a high-entropy alloy. *Acta Materialia*, 2016, vol 110, pp 352–363.
127. Laplanche G, Kostka A, Reinhart C, Hunfeld J, Eggeler G, George EP. (2017). Reasons for the superior mechanical properties of medium-entropy CrCoNi compared to high-entropy CrMnFeCoNi. *Acta Materialia* 128: 292–303.
128. Xu XD, Liu P, Tang Z, Hirata A, Song SX, Nieh TG, Liaw PK, Liu CT, Chen MW. (2018). Transmission electron microscopy characterization of dislocation structure in a face-centred cubic high-entropy alloy $Al_{0.1}CoCrFeNi$. *Acta Materialia* 144: 107–15.
129. Lu J, Hultman L, Holmström E, Antonsson KH, Grehk M, Li W, Vitos L, Golpayegani A. (2015). Stacking fault energies in stainless steels. *Acta Materialia* 97: 305–15.
130. Otte HM, Adler RPI. (1966). Formation of stacking faults in polycrystalline brass during tensile deformation. *Philosophical Magazine* 16: 239–52.
131. Rohatgi A, Vecchio K, Gray G. (2001). The influence of stacking fault energy on the mechanical behaviour of Cu and Cu-Al alloy: Deformation twinning, work hardening and dynamic recovery. *Metallurgical Transactions* A32: 135–45.
132. Pasianot R, Farkas D. (2020). Atomistic modeling of dislocations in a random quinary high-entropy alloy. *Computational Materials Science* 173: Article 109366.
133. Rao SI, Woodward C, Parthasarathy TA, Senkov O. (2017). Atomistic simulations of dislocation behaviour in a model fcc multicomponent concentrated solid solution alloy. *Acta Materialia* 134: 188–94.
134. Liu X, Pei Z, Eisenbach M. (2019). Dislocation core structures and Peierls stresses of the high entropy alloy NiCoFeCrMn and its subsystems. *Materials and Design* 180: Article 107955.
135. Couzinie JP, Lilensten L, Champion Y, Dirras G, Perrière L, Guillot I. (2015). On the room temperature deformation mechanisms of a TiZrHfNmMo refractory high-entropy alloy. *Materials Science and Engineering* A645: 255–63.
136. Lilensten L, Couzinie JP, Perrière L, Hocini A, Keller C, Dirras G, Guillot I. (2018). Study of a bcc multi-principal element alloy: Tensile and simple shear properties and underlying deformation mechanism. *Acta Materialia* 142: 131–41.
137. Wang F, Balbus GH, Xu S, Su Y, Shin J, Rottmann PF, Knipling KE, Stinville JC, Mills LH, Senkov ON, Beyerlein IJ, Pollock TM, Gianola DS. (2020). Multiplicity of dislocation pathways in a refractory multiprincipal element alloy. *Science* 370: 95–101.
138. Lee C, Maresca F, Feng R, Chou Y, Ungar T, Widom M, An K, Poplawsky JD, Chou YC, Liaw PK, Curtin WA. (2021). Strength can be controlled by edge dislocations in refractory high-entropy alloys. *Nature Communications* 12: Article 5474.
139. Rao SI, Akdim B, Antillon E, Woodward C, Parthasarathy TA, Senkov ON. (2019). Modeling solution hardening in bcc refractory complex concentrated alloys: NbTiZr, $Nb_{1.5}TiZr_{0.5}$ and $Nb_{0.5}TiZr_{1.5}$. *Acta Materialia* 168: 222–36.
140. Yin S, Ding J, Asta M, Ritchie RO. (2020). Ab initio modeling of the energy landscape for screw dislocations in body-centred cubic high-entropy alloys. *NPJ Computational Materials* 6: Article 110.

141. Fey LTW, Xu S, Su Y, Hunter A, Beyerlein IJ. (2022). Transitions in the morphology and critical stresses of gliding dislocations in multiprincipal element alloys. *Physical Review Materials* 6: Article 013605.
142. Zhou X, He S, Marian J. (2021). Cross-kinks control screw dislocation strength in equiatomic bcc refractory alloys. *Acta Materialia* 211: Article 116875.
143. Lin DY, Wang SS, Peng DL, Li M, Hui XD. (2013). An n-body potential for a Zr-Nb system based on the embedded-atom method. *Journal of Physics: Condensed Matter* 25: Article 105404.
144. Otto F, Diouhý A, Somsen C, Bei H, Eggeler G, George EP. (2013). The influence of temperature and microstructure on the tensile properties of a CoCrFeNiMn high-entropy alloy. *Acta Materialia* 61: 5743–55.
145. Laplanche G, Kostka A, Horst OM, Eggeler G, George EP. (2016). Microstructure evolution and critical stress for twinning in the CrMnFeCoNi high-entropy alloy. *Acta Materialia* 118: 152–63.
146. Liu W, Wu Y, He J, Nieh T, Lu Z. (2013). Grain growth and the Hall–Petch relationship in a high-entropy FeCrNiCoMn alloy. *Scripta Materialia* 68: 526–9.
147. Gali A, George EP. Tensile properties of high- and medium-entropy alloys. (2013). *Intermetallics* 39: 74–8.
148. Kang M, Won JW, Kwon JB, Na YS. (2017). Intermediate strain rate deformation behavior of a CoCrFeMnNi high-entropy alloy. *Materials Science and Engineering* A707: 16–21.
149. Park JM, Moon J, Bae JW, Jang MJ, Park J, Lee S, Kim HS. (2018). Strain rate effects of dynamic compressive deformation on mechanical properties and microstructure of CoCrFeMnNi high-entropy alloy. *Materials Science and Engineering* A719: 155–63.
150. Zaddach AJ, Niu C, Koch CC, Irving DL. (2013). Mechanical properties and stacking fault energies of NiFeCrCoMn high-entropy alloy. *Journal of Metals* 65: 1780–9.
151. Varvenne C, Luque A, Curtin WA. (2016). Theory of strengthening in fcc high entropy alloys. *Acta Materialia* 118: 164–76.
152. Varvenne C, Leyson GPM, Ghazisaeidi M, Curtin WA. (2017). Solute strengthening in random alloys. *Acta Materialia* 124: 660–83.
153. Varvenne C, Curtin WA. (2017). Strengthening of high entropy alloys by dilute solute additions: CoCrFeNiAlx and CoCrFeNiMnAlx alloys. *Scripta Materialia* 138: 92–5.
154. Sohn S, Liu Y, Liu J, Gong P, Prades-Rodel S, Blatter A, Scanlon BE, Broadbridge CC, Schroers J. (2017). Noble metal high entropy alloys. *Scripta Materialia* 126: 29–32.
155. Yin B, Curtin WA. (2019). First-principles-based prediction of yield strength in the RhIrPdPtNiCu high-entropy alloy. *NPJ Computational Materials* 5: Article 14.
156. Laplanche G, Bonneville J, Varvenne C, Curtin WA, George EP. (2018). Thermal activation parameters of plastic flow reveal deformation mechanisms in the CrMnFeCoNi high-entropy alloy. *Acta Materialia* 143: 257–64.
157. Li Z, Zhao S, Ritchie RO, Meyers MA. (2019). Mechanical properties of high-entropy alloys with emphasis on face-centered cubic alloys. *Progress in Materials Science* 102: 296–345.
158. George EP, Curtin WA, Tasan CC. (2020). High entropy alloys: A focused review of mechanical properties and deformation mechanisms. *Acta Materialia* 188: 435–74.

159. Labusch R, Grange G, Ahearn J, Haasen P. (1975). Rate processes in plastic deformation of materials. In JCM Li, AK Mukherjee, eds, *Proceedings from the John E. Dorn Symposium*, Metals Park: ASM, 26.
160. Leyson GPM, Curtin WA. (2016). Solute strengthening at high temperatures. *Modelling and Simulation in Materials Science and Engineering* 24: Article 065005.
161. Maresca F, Curtin WA. (2020). Theory of screw dislocation strengthening in random bcc alloys from dilute to high-entropy alloys. *Acta Materialia* 182: 144–62.
162. Sutton AP, Baluffi RW. (1987). On geometric criteria for low interfacial energy. *Acta Metallurgica* 35: 2177–2201.
163. Lin Q, An X, Liu H, Tang Q, Dai P, Liao X. (2017). In-situ high-resolution transmission electron microscopy investigation of grain boundary dislocation activities in a nanocrystalline CrMnFeCoNi high-entropy alloy. *Journal of Alloys and Compounds* 709: 802–7.
164. Wynblatt P, Chatain D. (2019). Modeling grain boundary and surface segregation in multicomponent high-entropy alloys. *Physical Review Materials* 3: Article 054004.
165. Farkas D. (2020). Grain boundary structure in high-entropy alloys. *Journal of Materials Science* 55: 9173–83.
166. Lee H, Shabani M, Pataky GJ, Abdeljawad F. (2021). Tensile deformation behavior of twist grain boundaries in CoCrFeMnNi high entropy bicrystals. *Nature Scientific Reports* 11: Article 428.
167. Moon J, Hong SI, Bae JW, Jang MJ, Yim D, Kim HS. (2017). On the strain rate-dependent deformation mechanism of CoCrFeMnNi high-entropy alloy at liquid nitrogen temperature. *Materials Research Letters* 5: 472–7.
168. Kim SW, Kim JH. (2018). In-situ observations of deformation twins and crack propagation in a CoCrFeMnNi high-entropy alloy. *Materials Science and Engineering* A718: 321–5.
169. Fu W, Gan K, Huang Y, Ning Z, Sun J, Cao F. (2021). Elucidating the transition of cryogenic deformation mechanism of CrMnFeCoNi high entropy alloy. *Journal of Alloys and Compounds* 872: Article 159606.
170. Li X, Sun S, Zou Y, Zhu Q, Tian Y, Wang J. (2022). Twin-coupled shear bands in an ultrafine-grained CoCrFeMnNi high-entropy alloy deformed at 77K. *Materials Research Letters* 10: 385–91.
171. Wang M, Li Z, Raabe D. (2018). *In-situ* SEM observation of phase transformation and twinning mechanisms in an interstitial high-entropy alloy. *Acta Materialia* 47: 236–46.
172. Wu Y, Bönisch M, Alkan S, Abuzaid W, Sehitoglu H. (2018). Experimental determination of latent hardening coefficients in FeMnNiCoCr. *International Journal of Plasticity* 105: 239–60.
173. Zhang YH, Zhuang Y, Hu A, Kai JJ, Liu CT. (2017). The origin of negative stacking fault energies and nano-twin formation in fcc high-entropy alloys. *Scripta Materialia* 130: 96–9.
174. Huang S, Huang H, Li W, Kim D, Lu S, Li X, Holmström E, Kwon SK, Vitos L. (2018). Twinning in metastable alloys. *Nature Communications* 9: Article 2381.
175. Aitken ZH, Zhang YW. (2019). Revealing the deformation twinning nucleation mechanism of bcc HEAs. *MRS Communications* 9: 406–12.
176. Yang Z, Lu S, Tian Y, Gu Z, Sun J, Vitos L. (2022). Theoretical and experimental study of phase transformation and twinning behaviour in metastable high-entropy alloys. *Journal of Materials Science and Technology* 99: 161–8.

177. Ma SG, Zhang SF, Qiao JW, Wang ZH, Gao MC, Jiao ZM, Yang HJ, Zhang Y. (2014). Superior high tensile elongation of a single-crystal $CoCrFeNiAl_{0.3}$ high-entropy alloy by Bridgman solidification. *Intermetallics* 54: 104–9.
178. Gaertner D, Kottke J, Wilde G, Divinski SV, Chumlyakov Y. (2018). Tracer diffusion in single crystalline CoCrFeNi and CoCrFeMnNi high-entropy alloys. *Journal of Materials Research* 33: 3184–91.
179. Gadelmeier C, Haas S, Lienig T, Manzoni A, Feuerbacher M, Glatzel U. (2020). Temperature dependent solid solution strengthening in the high entropy alloy CrMnFeCoNiin single crystalline state. *Metals* 10: Article 1412.
180. Glatzel U, Schleifer F, Gadelmeier C, Krieg F, Müller M, Mosbacher M, Völkl R. (2021). Quantification of solid solution strengthening and internal stresses through creep testing of Ni-containing single crystals at 980C. *Metals* 11: Article 1130.
181. Wu Z, Gao YF, Bei H. (2015). Single crystal plastic behaviour of a single-phase, fcc equiatomic FeNiCrCo alloy. *Scripta Materialia* 109: 108–12.
182. Laplanche G, Schneider M, Scholz F, Frenzel J, Eggeler G, Schreuer J. (2020). Processing of a single-crystalline CrCoNi medium-entropy alloy and evolution of its thermal expansion and elastic stiffness coefficients with temperature. *Scripta Materialia* 177: 44–8.
183. Feng X, Zhang J, Xia Z, Fu W, Wu K, Liu G, Sun J. (2018). Stable nanocrystalline NbMoTaW high entropy alloy thin films with excellent mechanical and electrical properties. *Materials Letters* 210: 84–7.
184. Feng X, Surjada JU, Li X, Lu Y. (2020). Size dependency in stacking fault-mediated ultrahard high-entropy alloy thin films. *Journal of Alloys and Compounds* 844: Article 156187.
185. Xiao LL, Zheng ZQ, Guo SW, Huang P, Wang F. (2020). Ultra-strong nanostructures CrMnFeCoNi high-entropy alloys. *Materials and Design* 194: Article 108895.
186. Lee DH, Choi IC, Seok MY, He J, Lu Z, Suh JY, Kawasaki M, Langdon TG. Jang JI. (2015). Nanomechanical behavior and structural stability of a nanocrystalline CoCrFeNiMn high-entropy alloy processed by high-pressure torsion. *Journal of Materials Research* 30: 2804–15.
187. Zheng R, Chen JC, Xiao W, Ma C. (2016). Microstructure and tensile properties of nanocrystalline $(FeNiCoCu)_{1-x}Ti_xAl_x$ high-entropy alloys processed by high pressure torsion. *Intermetallics* 74: 38–45.
188. Rusakova HV, Fomenko LS, Smirnov SN, Podolskiy AV, Shapovalov YO, Tabachnikova ED, Tikhonovsky MA, Levenets AV, Zehetbauer MJ, Schafler E. (2021). Low temperature micromechanical properties of nanocrystalline CoCrFeNiMn high-entropy alloy. *Materials Science and Engineering* 828: Article 142116.
189. Keil T, Bruder E, Laurent-Brocq M, Durst K. (2021). From diluted solid solutions to high entropy alloys: Saturation grain size and mechanical properties after high pressure torsion. *Scripta Materialia* 192: 43–8.
190. Dwivedi A, Koch CC, Rajulapati KV. (2016). On the single phase fcc solid solution in nanocrystalline CRNbTiVZn high-entropy alloy. *Materials Letters* 183: 44–7.
191. Praveen S, Basu J, Kashyap S, Kottada RS. (2016). Exceptional resistance to grain growth in nanocrystalline CoCrFeNi high entropy alloy at high homologous temperatures. *Journal of Alloys and Compounds* 662: 361–7.

192. Oleszak D, Anatolak-Dudka A, Kulik T. (2018). High entropy multicomponent WMoNbZrV alloy processed by mechanical alloying. *Materials Letters* 232: 160–2.

193. Xin SW, Zhang M, Yang TT, Zhao YY, Sun BR, Shen TD. (2018). Ultrahard bulk nanocrystalline VNbMoTaW high-entropy alloy. *Journal of Alloys and Compounds* 769: 597–604.

194. Shkodich NF, Kuskov KV, Sedegov AS, Kovalev ID, Pantaleeva AV, Vergunova YS, ScheckYB, Panina E, Stepanov N, Serhiienko I, Moskovskikh D. (2022). Refractory TaTiNb, TaTiNbZr and TaTiNbZrX (X=Mo,W) high entropy alloys by combined use of high energy ball milling and spark plasma sintering: structural characterization, mechanical properties, electrical resistivity and thermal conductivity. *Journal of Alloys and Compounds* 893: Article 162030.

195. Koch CC. (2017). Nanocrystalline high-entropy alloys. *Journal of Materials Research* 32: 3435–44.

196. Haché MJR, Cheng C, Zou Y. (2020). Nanostructured high-entropy materials. *Journal of Materials Research* 35: 1051–75.

197. Sathiyamoorthi P, Kim HS. (2022). Nanocrystalline high entropy alloys: Processing and properties. In FG Caballero, ed., *Encyclopedia of Materials: Metals and Alloys*, vol 2. Oxford: Elsevier, 372–80.

198. Joo SH, Kato H, Jang MJ, Moon J, Kim EB, Hong SJ, Kim HS. (2017). Structure and properties of ultrafine-grained CoCrFeMnNi high-entropy alloys produced by mechanical alloying and spark plasma sintering. *Journal of Alloys and Compounds* 698: 591–604.

199. Doherty RD, Hughes DA, Humphreys FJ, Jonas JJ, Jensen DJ, Kassner ME, King WE, McNelley TR, McQueen HJ, Rollett AD. (1997). Current issues in recrystallization: A review. *Materials Science and Engineering* A238: 219–74.

200. Gottstein G, Shvindlerman LS. (2009). *Grain Boundary Migration in Metals*, 2nd edn. Boca Raton: CRC Press.

201. Humphreys J, Rohrer GS, Rollett A. (2017). *Recrystallization and Related Annealing Phenomena*, 3rd edn. Amsterdam: Elsevier.

202. Liu WH, Wu Y, He JY, Nieh TG, Lu ZP. (2013). Grain growth and the Hall-Petch relationship in a high-entropy FeCrNiCoMn alloy. *Scripta Materialia* 68: 526–9.

203. Bhattacharjee PP, Sathiaraj GD, Zaid M, Gatti JR, Lee C, Tsai CW, Yeh JY. (2014). Microstructure and texture evolution during annealing of equiatomic CoCrFeMnNi high-entropy alloy. *Journal of Alloys and Compounds* 587: 544–52.

204. Otto F, Hanold NL, George EP. (2014). Microstructural evolution after thermomechanical processing in an equiatomic single-phase CoCrFeMnNi high-entropy alloy with special focus on twin boundaries. *Intermetallics* 54: 39–48.

205. Laplanche G, Horst O, Eggeler G, George EP. (2015). Microstructural evolution of a CoCrFeMnNi alloy after swaging and annealing. *Journal of Alloys and Compounds* 647: 548–57.

206. Sathiaraj GD, Bhattacharjee PP. (2015). Analysis of microstructure and microtexture during grain growth in low stacking fault energy equiatomic CoCrFeMnNi high entropy and Ni-60wt%Co alloys. *Journal of Alloys and Compounds* 637: 267–76.

207. Chen BR, Yeh AC, Yeh JW. (2016). Effect of one-step recrystallization on the grain boundary evolution of CoCrFeMnNi high entropy alloy and its subsystems. *Nature Scientific Reports* 6: Article 22306.

208. Gu J, Ni S, Liu Y, Song M. (2019). Regulating the strength and ductility of a cold rolled FeCrCoMnNi high-entropy alloy via annealing treatment. *Materials Science and Engineering* A755: 289–94.

209. Shiyamoorthi P, Asghari-Rad P, Bae JW, Kim HS. (2019). Fine tuning of tensile properties in CrCoNi medium entropy alloy through cold rolling and annealing. *Intermetallics* 113: Article 106578.

210. Wu Z, Bei H, Otto F, Pharr GM, George EP. (2014). Recovery, recrystallization, grain growth and phase stability of a family of fcc-structured multicomponent equiatomic solid solution alloys. *Intermetallics* 46: 131–40.

211. Yang Z, He F, Wu Q, Zhang K, Cui D, Guo B, Han B, Li J, Wang J, Wang Z. (2021). Distinct recrystallization kinetics in NiCoCrFe-based single-phase high-entropy alloys. *Metallurgical and Materials Transactions* A52: 3799–3810.

212. Wu SW, Xu L, Ma XD, Jia YF, Mu YK, Jia YD, Wang G, Liu CT. (2021). Effect of annealing temperatures on microstructures and deformation behavior of Al0.1CrFeCoNi high-entropy alloy. *Materials Science and Engineering* 805: Article 140523.

213. Wang X, Zhang Z, Wang Z, Ren X. (2022). Microstructural evolution and tensile properties of Al0.3CoCrFeNi high-entropy alloy associated with B2 precipitates. *Materials* 15: Article 1215.

214. Guo T, Li J, Wang J, Wang WY, Liu Y, Luo X, Kou H, Beaugnon E. (2018). Microstructure and properties of bulk $Al_{0.5}CoCrFeNi$ high-entropy alloy by cold rolling and subsequent annealing. *Materials Science and Engineering* A729: 141–8.

Further Reading

Cai W, Nix WD. (2016). *Imperfections in Crystalline Solids*. Cambridge: Cambridge University Press.

Cantor B. (2021). Multicomponent high-entropy Cantor alloys. *Progress in Materials Science* 120: Article 100754.

Cantor B. (2020). *The Equations of Materials*. Oxford: Oxford University Press.

Hillert M. (2008). *Phase Equilibria, Phase Diagrams and Phase Transformations*, 2nd edn. Cambridge: Cambridge University Press.

Hull D, Bacon DJ. (2011). *Introduction to Dislocations*, 5th edn. Oxford: Butterworth-Heinemann/Elsevier.

Kittel C. (1957). Semiconductor crystals; Fermi surfaces and metals; Superconductivity. *Introduction to Solid State Physics*, 2nd edn. New York: Wiley, chs 8–10.

Miracle DB, Senkov ON. (2017). A critical review of high-entropy alloys and related concepts. *Acta Materialia* 122: 448–511.

Murty BS, Yeh JW, Ranganathan S, Bhattacharjee PP. (2019). *High-Entropy Alloys*. Amsterdam: Elsevier.

Humphreys J, Rohrer GS, Rollett A. (2017). *Recrystallization and Related Annealing Phenomena*, 3rd edn. Amsterdam: Elsevier.

Shewmon P. (2016). *Diffusion in Solids*, 2nd edn. New York: Springer.

Sutton AP. (1993). *Electronic Structure of Materials.* Oxford: Oxford University Press.

Sutton AP, Baluffi RW. (1995). *Interfaces in Crystalline Materials.* Oxford: Oxford University Press.

5
Making Multicomponent Materials

1 Introduction

As discussed in previous chapters, we have discovered in just the last decade or so that multicomponent phase space is truly enormous, containing literally trillions of different materials, almost all so far completely uninvestigated or even conceived. This discovery immediately raises the critical question of how best to chart such a vast range of possible new multicomponent materials with potentially exciting new and valuable properties. This chapter discusses two aspects of this problem: first it considers the different strategies that have been developed for trying to map parts of multicomponent phase space in a scientifically organised way, using methods such as multicomponent substitution, multicomponent Hume-Rothery rules, thermodynamic calculations, atomistic modelling, machine learning and high-throughput measurements; and second, it then describes some of the manufacturing methods that are available and have been used for making multicomponent materials in a controlled and reliable way.

2 Exploring Multicomponent Phase Space

Trial and error

We saw in Chapter 2 that the first discoveries of multicomponent high-entropy materials were made essentially by *trial and error*. In the late 1970s and early 1980s, Alain Vincent and I discovered the first *multicomponent high-entropy materials*—several multicomponent single-phase fcc (now called) Cantor alloys, several multicomponent two-phase fcc + bcc alloys, and several multicomponent multiphase alloys—by melting and casting a range of different equiatomic component mixtures selected largely on a trial-and-error basis.[1–3] Similarly, Jien-Wei Yeh has described how later, but independently, in the 1990s, he asked his research students to make a range of multicomponent materials based on copper, aluminium and molybdenum, again selected largely on a trial-and-error basis, leading to the discovery of a variety of interesting new multicomponent single-phase, two-phase and multiphase alloys, and to his subsequent development of the concept of *high-entropy materials*, and their *four core properties*.[4–6] The phrase trial and error sounds, and clearly is, a bit too haphazard a process for it to be treated as an ideal scientific approach to the problems of how to explore multicomponent phase space and discover new multicomponent materials. But it would be wrong in general to disparage the use of trial and error in science: the role of *discovery* was essential to the initial development of scientific methods in the fifteenth and sixteenth centuries,[7,8] and trial-and-error experimentation and exploration has been the basis of many of our most important scientific discoveries, as described

Fundamentals of Multicomponent High-Entropy Materials. Brian Cantor, Oxford University Press. © Brian Cantor (2024).
DOI: 10.1093/9780191986710.003.0005

in more detail in Chapter 1. In fact, trial and error is sometimes given the somewhat grander name of the *Edisonian approach*, reflecting the importance placed by the famous nineteenth-century scientist, inventor and businessman Thomas Edison on using trial-and-error methods to make progress with important scientific and engineering opportunities and developments.[9]

In a memorable definition of the uncertainty of doing scientific research,[10] Werner von Braun once said, 'research is what I am doing when I don't know what I am doing'. And many of the most important scientific discoveries (including multicomponent high-entropy materials) could not initially have been made in any other way than a pure, unknowing, exploratory trial-and-error approach. It is certainly true that, in some cases, scientific phenomena are predicted before they are demonstrated. But in many other cases (including multicomponent high-entropy materials), they have to be discovered first because they are just too surprising and unexpected to be predicted in advance. We just don't know about them so we *couldn't* predict them. Nevertheless, it is not really satisfactory to continue scientific investigation of particular phenomena with only pure trial-and-error methods and serendipitous discovery. Following initial discovery, we make better progress by trying to develop underlying theories or at least organised methodologies to help guide more effectively further discoveries and deeper understanding. The rest of this section considers some of the more structured methods that have been developed to try and navigate the vast undiscovered regions of multicomponent phase space and the vast variety of different multicomponent materials. These methods have had some, albeit somewhat limited, success, as described in the following sub-sections, but the scale and complexity of both multicomponent phase space itself and the different multicomponent materials that occupy it means that there continues to be (and indeed will almost certainly continue to be for a long time) an important place for trial-and-error experimentation in the overall advancement of the field.

Multicomponent substitution

The first more organised method of exploring multicomponent phase space that was developed was *equiatomic substitution*, in which one of the components in a known alloy or material of interest is replaced or substituted with an equiatomic mixture of chemically similar components. This method was first employed by Kim et al.,[11] who replaced nickel and zirconium with equiatomic multicomponent mixtures of, respectively, late and early transition metals in the well-known bulk metallic-glass-forming (BMG) alloy $Ni_{50}Zr_{40}Al_{10}$. Kim et al.[12–15] also varied the relative amounts of the early and late transition metals and the aluminium content, thus developing a wide range of *multicomponent amorphous alloys* and (crystallised) *multicomponent intermetallic compounds* with compositions of the form $LM_xEM_yAl_z$ where LM and EM are, respectively, mixtures of late and early transition metals, $x = 10–60\%$, $y = 10–60\%$ and $z = 1–x–y = 0–30\%$. Similar methods were used to manufacture the first *multicomponent ceramic compounds*: equimolar mixtures of mono-oxides such as MgO, CuO, NiO, etc., forming rock-salt-structured multicomponent mono-oxide materials with generic compositions MO, where M is an equiatomic mixture of bivalent metallic cations.[16] And similar substitution methods have since been used to develop many other *multicomponent amorphous alloys*,[17] *multicomponent intermetallic compounds*[18] and *multicomponent ceramic compounds*,[19–21] as discussed in more detail in Chapter 4.

It is, in fact, better to describe this method more generically as *multicomponent substitution*, since the *equiatomic* nature of the replacement of one component with several others is not really critical, and *non-equiatomic substitution* is just as useful. The key feature is the substitution of one component by many (or at least several) others, not whether they are added in equiatomic or non-equiatomic proportions.

As already explained, the multicomponent substitution method of exploring multicomponent phase space starts with a particular material of interest, with one or more particularly desirable properties. The final properties of the resulting multicomponent-substituted materials will then, of course, vary depending on the exact components chosen for substitution and their composition. Putting this another way, multicomponent substitution can be regarded as a method for *multicomponent tuning* of the properties of the starting material, using its composition as the tuning variable, both to enhance its main properties of interest and to add secondary properties. This line of thinking and approach (i.e. multicomponent tuning) together with variations in initial manufacturing and downstream thermomechanical processing (TMP) will almost certainly underpin future work to improve gradually the mechanical, electrical and other functional properties of particular target multicomponent high-entropy materials of potential engineering interest and application.

The number of phases

The Gibbs phase rule

As derived and discussed in more detail in the last chapter, the Gibbs phase rule[22,23] gives a relationship between the number of components c, the number of phases p, and the number of thermodynamic degrees of freedom f in any material at equilibrium:

$$p + f = c + 2$$

For a given temperature and pressure, e.g. at room temperature and atmospheric pressure, the number of degrees of freedom is reduced by two, and the Gibbs phase rule becomes:[22,23]

$$p + f = c$$

The number of degrees of freedom cannot be less than zero, so the number of phases in any material at equilibrium and at room temperature and atmospheric pressure is given by:[22,23]

$$p = c - f \leq c$$

In other words, for a material at equilibrium, at room temperature, and at atmospheric pressure, the number of phases can have any value between one and the number of components, or $c \geq p \geq 1$. And for a material not at equilibrium, the number of phases can be higher, or $p \geq 1$ (including $p > c$).

As expected from and as allowed by the Gibbs phase rule, the number of phases found in equilibrated binary materials is usually one or two,[24] and under non-equilibrium conditions is sometimes higher, three or occasionally four.[25–27] Similarly, the number of phases found in equilibrated ternary materials is usually one, two or three,[28] and under non-equilibrium conditions is sometimes higher, four or occasionally five.[25–27] For small c in binary and ternary materials, p is in general, therefore, spread fairly evenly across the range 1 to c and is occasionally somewhat higher (because of a non-equilibrium structure). For large c in multicomponent materials, however, it is surprisingly common to find only one or two phases, with sometimes three or four phases, and only very rarely more than three or four phases. We have already seen this in previous chapters and it is also well discussed in previous reviews of multicomponent high-entropy materials.[3,18,29–35] The recent book by Murty et al.,[18] for instance, lists over 200 separate reports of different single-phase fcc multicomponent materials and over 150 separate reports of different single-phase bcc multicomponent materials. These results may partly reflect a degree of selection bias, given the scientific interest there has been in multicomponent single-phase solid solutions. Nevertheless, for large values of c in multicomponent materials, it is surprisingly common to find no more than one or two phases and, unlike in binary and ternary materials, p is not in general spread evenly across the range 1 to c.

The number of single-phase materials

Compilations of standard binary and ternary phase diagrams[24,28] show that the number of single-phase materials n_s in a given alloy system can vary from 0 to 100%, with an average $\overline{n}_s$ of, very roughly, 10–20%, i.e. $0 \leq n_s \leq 100\%$ and $\overline{n}_s \approx 10 - 20\%$ in binary and ternary systems. The factors that affect the number of single-phase materials in a particular binary or ternary system are: 1) the extent of terminal solid solubility; 2) the presence or otherwise of compound phases; and 3) the stoichiometric or non-stoichiometric nature of any compound phases. The number of single-phase materials is small when the terminal solid solubility is low, when there are one or more binary or ternary compounds restricting the terminal solubility range, and when these compounds are highly stoichiometric, i.e. when they form line or near-line compounds with fixed compositions and narrow single-phase fields. In contrast, the number of single-phase materials is high when there is substantial terminal solid solubility, when there are no or few compounds, and when any compounds are non-stoichiometric, with a considerable spread in composition and correspondingly wide single-phase fields.

The number of single-phase materials in a multicomponent system depends, therefore, firstly on whether solid solubility (both in the terminal solid solutions and in any stoichiometric compounds) is enhanced or diminished with multiple solutes, and secondly on whether the number of stoichiometric compounds increases or decreases in a multicomponent material. It turns out that both of these effects are important: as discussed in some detail in previous chapters, there are many large single-phase regions of multicomponent versions of well-known binary and ternary phases in multicomponent phase space (single phases with close-packed crystal structures such as fcc, bcc and hcp, as well as single phases with compound structures such as borides and aluminides, Laves phases, spinels and perovskites), indicating fairly clearly that solubility (both terminal solid solubility and the stoichiometric range of compounds) increases with multiple solutes, but that the number of stoichiometric compounds does not.

The reasons for relatively few phases in multicomponent materials

Underlying these phase diagram effects, there are two basic reasons for finding relatively few phases in multicomponent materials: 1) a high configurational entropy of mixing; and 2) a paucity of new multicomponent intermetallic or ceramic compound structures. The first of these is well known and is of course also the reason for the name of the field being *high-entropy materials*; the second is less well known and has hardly been commented on previously. We discuss both of these effects briefly one at a time.

The structure of a material depends on its *Gibbs free energy* G, as given at any temperature T by:[3,22,23,36]

$$G = \sum_i x_i \mu_i^{ss} = \sum_i x_i \mu_i + \Delta G_{mix} = \sum_i x_i \mu_i + \Delta H_{mix} - T\Delta S_{mix}$$

where x_i is the molar fraction of the ith component, μ_i and μ_i^{ss} are, respectively, the *chemical potential* (or *molar free energy*) of the ith component before and after mixing, and ΔG_{mix}, ΔH_{mix} and ΔS_{mix} are, respectively, the *free energy of mixing, heat (or enthalpy) of mixing* and *entropy of mixing*. The first term on the right-hand side is the free energy of the unmixed components, and the last two terms are the free energy change caused by mixing them. We have already discussed in considerable detail in Chapter 4 the general implications of this well-known and important expression in determining the structure of multicomponent materials, and later in this chapter we discuss in even more detail its use in thermodynamic modelling of multicomponent material structures and phase diagrams. Here we only need to point out simply that the structure of any material depends on its Gibbs free energy G, and that this in turn depends on two effects caused by mixing the different components: on the one hand, the heat of mixing ΔH_{mix} as determined mainly by the interaction energies of the component atoms and molecules; and on the other hand, the entropy of mixing ΔS_{mix} as determined mainly by the configurational entropy of mixing the different component atoms or molecules. In other words, the structure of the material depends on a balance between 1) the tendency of the component atoms or molecules to reduce their energy by interacting to form stoichiometric compounds (as represented by ΔH_{mix}) and 2) the tendency of the component atoms or molecules to increase their entropy by mixing randomly to form extended solid solutions (as represented by ΔS_{mix}). In binary and ternary alloys, the configurational entropy of mixing ΔS_{mix} is relatively low because there are only two or three components, and beyond anything other than very dilute alloys the structure is, therefore, often dominated by the heat of mixing ΔH_{mix}, i.e. by atomic or molecular interactions reducing the energy and enthalpy of the material by forming fixed-composition (or near-fixed-composition) stoichiometric intermetallic or ceramic compounds. In multicomponent materials, on the other hand, the configurational entropy of mixing ΔS_{mix} is relatively high because there are many components, and the energy reduction associated with atomic or molecular interactions can often then be suppressed by the increased entropy caused by random mixing of the different components on a suitable crystal lattice. This is the first reason why there are often relatively few phases found in multicomponent materials and, as already mentioned, it is also the reason why they have come to be known as *high-entropy materials*, as discussed in more detail in previous chapters.

There is a rich array of different binary and ternary compounds with different crystal structures, different point and lattice symmetries (i.e. different point groups and space groups), and different sub-lattices.[37,38] The compilation by Westbrook and Fleischer,[38] for instance, reports over two hundred distinct compound crystal structures. However, there have been few, if any, reports of *new multicomponent compounds* with unique crystal structures and multiple sub-lattices, different and distinct from existing well-known binary and ternary compounds.[3,18,29–35] This is quite a striking observation, which has not been much remarked on previously. After all, a variety of new binary and ternary compounds was discovered each time alloy structures were extended by new techniques such as high-speed solidification, vapour deposition and mechanical alloying, but similar behaviour has not been replicated by the extension of alloy structures into multicomponent phase space. The paucity of new and distinctive multicomponent compounds seems to be something to do with the fact that we live in a three-dimensional world.[39] The different crystal structures of different binary and ternary compounds represent the various different ways that two or three differently sized atoms and molecules can be packed into three-dimensional space, allowing for any spatial effects of directionality in bonding.[i] Given the relatively restricted size range of component atoms, there is only a limited number of ways that spatial packing can be achieved, with only a limited number of different sub-lattices for the different atoms and molecules. Most binary and ternary compounds have no more than two or three sub-lattices in their crystal structure. Increasing the number of components to make an intrinsically different multicomponent compound would require increasing the number of different sub-lattices to accommodate the different atomic species, and this is just not possible or, at least, is fairly difficult in our limited three-dimensional space.[ii] In any case and for whatever reason, the number of different compounds does not scale with the number of components, i.e. the number of different compounds does not keep on increasing as we increase the number of components above two or three in a multicomponent material. Instead, the additional components are accommodated as substitutional solutes on the most appropriate lattice, sub-lattice or sub-lattices out of the available binary or ternary disordered solid solutions and/or intermetallic or ceramic compound structures, increasing their range of non-stoichiometry. This combination of limited or no new multicomponent compounds and increased solubility and non-stoichiometry of existing binary and ternary compounds is the second, so far largely unremarked, reason why there are often relatively few phases found in multicomponent materials.

As discussed in Chapter 4 and as shown in Figure 4.1 and Table 4.1, the ideal (i.e. random) configurational entropy of mixing ΔS_{mix} in a multicomponent solid solution and, therefore, its effectiveness in suppressing the compound formation associated with the heat of mixing ΔH_{mix} increases with increasing number of components c, at first rapidly for up to five or ten components, and then more slowly as the number of components rises above about ten. Moreover, the size of the ideal configurational entropy of mixing ΔS_{mix} is substantial but not enormous, larger than the binary interaction energies between chemically fairly similar

[i] And strong ionic or covalent bonds are themselves intrinsically one-dimensional in nature, with each bond resulting from the electrostatic attraction between individual atoms in a molecule.

[ii] This comment excludes the rather special case of large bio-organic molecules, which are not usually crystalline.

components, but not that large compared with the binary interaction energies between more strongly reacting, chemically fairly dissimilar components. Whether or not the configurational entropy of mixing ΔS_{mix} is larger than the heat of mixing ΔH_{mix} and thus suppresses compound formation in a given multicomponent material depends, therefore, on the number and range of different components in the material, their chemical variety and their associated propensity to react and form compounds. This means that the likelihood of forming a single-phase multicomponent solid solution is expected to increase with increasing number of components up to five or ten, as long as the components are reasonably similar chemically (for instance, all late transition metals or all early transition metals, as in most of the Cantor and Senkov alloys respectively, or all metals with bivalent cations in the multicomponent mono-oxides), but then to decrease with increasing number of components beyond about ten, as the chemical range of such a large number of components inevitably increases. There just aren't enough chemically similar components to prevent strong binary chemical interaction energies, strong compound formation and a large heat of mixing ΔH_{mix} with more than about ten components. The probability of forming a single-phase multicomponent solid solution should, therefore, first increase and then decrease with increasing number of components c. We have already seen very considerable evidence for the first of these effects, i.e. increased formation of single-phase multicomponent solid solutions in materials with four, five and six components. There is some, albeit only very limited, evidence[40,41] for the second effect, i.e. decreased formation of single-phase multicomponent solid solutions as the number of components rises further, mainly because has been almost no investigation of materials with ten or more components.

Multicomponent Hume-Rothery rules

Terminal solid-solubility levels in binary systems are usually described by the *Hume-Rothery rules*,[42] which say that solubility increases when:

1. Solvent and solute atoms have similar atomic size.
2. Solvent and solute atoms have similar valence.
3. Solvent and solute atoms have similar electronegativity.

Conversely, the propensity for compound formation in binary systems is described by the inverse of the Hume-Rothery rules, i.e. compound formation is promoted when:

1. Solvent and solute atoms have different atomic size.
2. Solvent and solute atoms have different valence.
3. Solvent and solute atoms have different electronegativity.

In other words, when a solvent and solute in a binary alloy obey the Hume-Rothery rules, they exhibit extended or complete terminal solid solubility, and limited or no formation of compounds; and, conversely, when a solvent and solute in a binary alloy do not obey the Hume-Rothery rules, they exhibit limited or no terminal solid solubility, and the formation of one or more compounds, which itself further restricts the extent of terminal solid solubility.[42]

Notice that the first Hume-Rothery rule is associated with the mechanical effect of different solvent and solute atomic size on *lattice strain*, causing an increase in the internal energy and enthalpy of the solution, thus limiting the solubility; and the second and third Hume-Rothery rules are associated with the electronic effect of different solvent and solute atoms leading to *chemical interactions* between them, causing compound formation, thus again limiting the solubility. In some formulations, the Hume-Rothery rules are concatenated to two rather than three rules:[3]

1. Solvent and solute atoms have similar atomic size.
2. Solvent and solute atoms have similar electron/atom ratio.

since valence and electronegativity both influence chemical interactions between solvent and solute atoms via differences in electron/atom ratio.[3,42]

There have been many attempts to develop *extended Hume-Rothery rules* suitable for application to multicomponent materials,[43–57] and detailed discussions of these attempts are provided in the book by Murty et al.[18] and the reviews by Miracle and Senkov[58] and Soni et al.[59] The basic objective is to develop a predictive model, or at least predictive criteria, for the formation of extended solid solutions in multicomponent materials. As with the normal Hume-Rothery rules in binary alloys, these attempts rely on *parameterising* two different basic effects, both of which act to limit intersolubility between the different components: first the *lattice strain* caused by different atomic sizes; and second the propensity for compound formation associated with *chemical interactions* between the different atoms. In very general terms, and again as with binary alloys, large differences in atomic size and valency more or less preclude extensive intersolubility and promote strongly the formation of compounds. However, in multicomponent systems many different methods can be used and indeed have been proposed for measuring the effect of differences in atomic size and chemical activity, and there are considerable difficulties in finding a suitable way of parameterising the key effects simply and effectively.

The most common parameter used to determine the atomic size difference between the different components is the atomic misfit;[18,36] and the most common parameter used to determine the chemical interaction between the different components is the heat of mixing ΔH_{mix} (though it is worth noticing that the heat of mixing strictly speaking includes a combination of both size and chemical interaction effects).[18,58] Zhang et al.[43] were the first to propose using atomic misfit δ and heat of mixing ΔH_{mix} to identify multicomponent chemical compositions that promote the formation of extended solid solutions. They calculated the atomic misfit as a weighted root mean square average δ_{av} given by:

$$\delta_{av} = \frac{\sqrt{\sum_i x_i (r_i - \bar{r})^2}}{\bar{r}}$$

where x_i and r_i are the molar fraction and atomic radius of the ith component, $\bar{r} = \sum_i x_i r_i$ is the mean atomic radius,[iii] and individual values of atomic radius r_i were taken from tabulated values[60] of the Goldschmidt metallic radii of the pure elemental components i, effectively

[iii] We use different notations to emphasise the different kinds of average: a bar notation for a simple weighted mean and an *av* subscript for a root mean square (RMS) average. Thus for atomic radii, the mean and RMS values

ignoring any relaxation of atomic and electronic structure of the different components (and, therefore, ignoring any change of atomic size) when dissolved in the multicomponent solid solution. They calculated the heat of mixing ΔH_{mix} using the well-known Miedema method:[61,62]

$$\Delta H_{mix} = \sum_{ij} x_i x_j \omega_{ij} \approx \sum_{ij} \alpha x_i x_j \Delta H_{ij}$$

where α is a crystallographic parameter for the number of near neighbours, and ω_{ij} and ΔH_{ij} are interaction energies and heats of solution for binary mixtures of the ith and jth components, effectively assuming that the multicomponent material is a regular solution with only limited interaction energies ω_{ij} between the different components, ignoring three-way, four-way and higher interactions, assuming that binary interaction energies can be added linearly on a pairwise basis, and approximating binary interaction energies as given by binary heats of solution in the different i–j subsystems, which are usually measured (and are, therefore, only accurate) in relatively dilute alloys.[3,22,23,36]

Calculations of this type have been performed by a number of investigators, and Figure 5.1 shows a typical Cartesian plot of the heat of mixing ΔH_{mix} versus the atomic misfit δ_{av}, onto which is mapped a range of different multicomponent alloys and their structures.[46] For 10 kJ/mol $\gtrsim \Delta H_{mix} \gtrsim -10$ kJ/mol and $\delta_{av} < 6-7\%$, multicomponent materials tend to form single phase solid solutions; for $\Delta H_{mix} \lesssim -10$ kJ/mol and $\delta_{av} > 6 - 7\%$, multicomponent materials tend to form amorphous alloys; and for $\Delta H_{mix} \approx -10$ kJ/mol and $\delta_{av} \approx 6 - 7\%$, there is an intermediate zone where multicomponent materials tend to form either an intermetallic compound or a multiphase mixture of a solid solution and one or more intermetallic compounds.[46]

Some of the difficulties of this parameterisation approach to developing extended multicomponent Hume-Rothery rules are immediately apparent. First, accurate values of parameters such as atomic misfit and heat of mixing can only really be obtained by direct (and often difficult) measurements on the multicomponent materials themselves, preventing accurate *ab initio* prediction (since the multicomponent materials have to be made first) and only allowing accurate parameterisation to be used, therefore, as an *ex post facto* explanation. Second, for almost all multicomponent materials there is, therefore, no real alternative to using less accurate proxy parameters taken from tabulated data for the pure components and/or simple binary mixtures, effectively assuming (almost certainly incorrectly in many cases) that multicomponent properties can be approximated by linear superposition of dilute alloy data. And third, the structure of a material is, as pointed out previously, determined by its Gibbs free energy, so conventional Hume-Rothery size and valency parameters that only affect the heat of mixing need to be further tensioned in multicomponent materials against the relatively large entropy of mixing (this is not necessary in binary and ternary materials with much lower entropies of mixing). Zhang et al.[43] recognised this last difficulty and extended their two-dimensional (δ_{av}, ΔH_{mix}) mapping of different material structures, as shown in Figure 5.1, into a three-dimensional (δ_{av}, ΔH_{mix}, ΔS_{mix}) mapping,

are $\bar{r} = \sum_i x_i r_i$ and $r_{av} = \sqrt{\sum_i x_i r_i^2}$; and for atomic misfit, the mean and RMS values are $\bar{\delta} = \sum_i x_i (r_i - \bar{r})/\bar{r}$ and $\delta_{av} = \sqrt{\sum_i x_i (r_i - \bar{r})^2}/\bar{r}$.

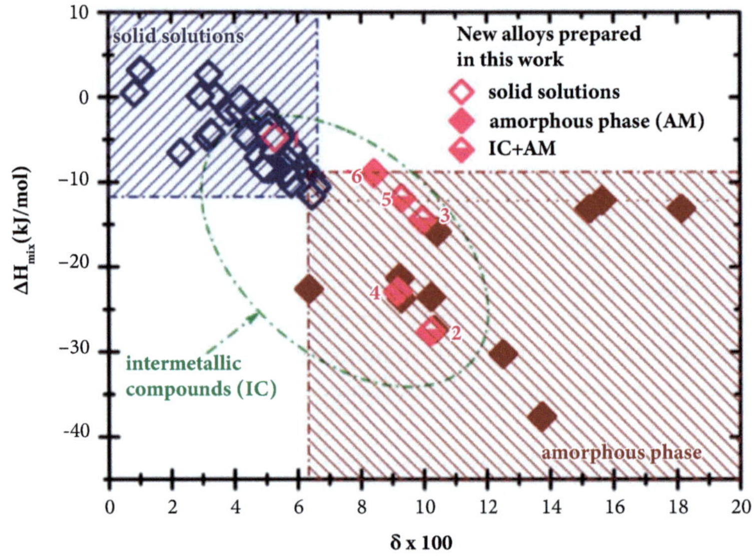

Figure 5.1 Two-dimensional map of heat of mixing ΔH_{mix} versus average atomic misfit δ_{av} showing regions of multicomponent single-phase solid solutions, multicomponent amorphous alloys and multicomponent intermetallic compounds.

Reproduced from Guo et al. (2013). More than entropy in high-entropy alloys: forming solid solutions or amorphous phase. *Intermetallics 41*: 96–103.

as shown in Figure 5.2. Although scientifically more satisfying, this results in considerably greater complexity and inconvenience without really providing much greater clarity or insight. Not surprisingly of course, extended multicomponent solid solutions are found in materials with relatively high values of ΔS_{mix} as well as relatively low values of δ_{av} and ΔH_{mix}.

Since Zhang et al.'s initial work,[43] there have been very many attempts to develop improved multicomponent Hume-Rothery rules in the same sort of way as described above, all with the objective of trying to predict the composition ranges of extended multicomponent solid solutions as against amorphous alloys, intermetallic compounds or multiphase materials. A variety of parameterisation and mapping schemes have been proposed in order to try to separate more effectively the different composition regions with different structures in a manner similar to that shown in Figures 5.1 and 5.2, including modified parameters, modified estimation methods, different combinations of parameters, and different mapping techniques. A wide range of other parameters have been tried, including for instance the mean valence electron concentration $\overline{VEC} = \sum_i x_i VEC_i$, the electronegativity range $\Delta\chi = \sqrt{\sum_i x_i(\chi_i - \overline{\chi})^2}$ and the ratio of entropy and heat of mixing $\Omega = T_m \Delta S_{mix}/\Delta H_{mix}$, where VEC_i and χ_i are the number of valence electrons and electronegativity of the ith component respectively, and T_m is the melting point of the material. This work is listed somewhat exhaustively in an extended chapter in the book by Murty et al.[18] and in a review by Soni et al.,[59] though without much attempt in either case to evaluate the different methods used. In the end though, Murty et al. conclude[18] that 'there are still no universal criteria to predict the formation of different phase types', and Soni et al.[59] conclude that there are still 'barriers that impede their [i.e. different

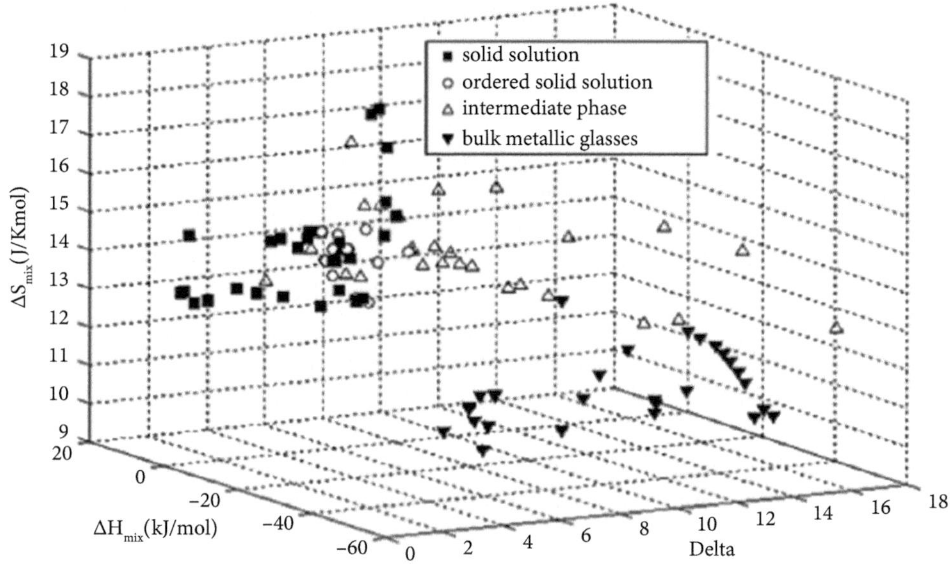

Figure 5.2 Three-dimensional map of entropy of mixing ΔS_{mix} versus heat of mixing ΔH_{mix} and atomic misfit δ_{av} (labelled delta) showing regions of multicomponent single-phase solid solutions, multicomponent amorphous alloys and multicomponent intermetallic compounds.
Reproduced from Zhang et al. (2008). Solid-solution phase formation rules for multi-component alloys. *Advanced Engineering Materials 10*: 534–8.

parameters'] ability to forecast'. These conclusions are almost certainly correct and are almost certainly the inevitable outcome of a methodology that contains the difficulties described above, of needing to use relatively inaccurate proxy parameters from pure components and dilute alloys that can only be combined using unjustified assumptions of linear superposition that are largely inappropriate in complex multicomponent high-entropy materials. In general, the different proposed methods all end up demonstrating what we already know, namely that extended solid solutions are preferred in multicomponent materials with small differences in atomic size, small differences in chemical reactivity and large numbers of components (and, therefore, a correspondingly large configurational entropy of mixing).

Thermodynamic modelling

As described in the last few sub-sections, *trial and error*, *multicomponent substitution* and *multicomponent Hume-Rothery rules* (or *parameterisation*) have proved to be useful, if somewhat limited, as initial methods for charting the complexities of multicomponent phase space. These methods are essentially heuristic and semi-empirical in approach and unable, therefore, to make clear *ab initio* predictions of new multicomponent material structures and properties. There have been two main ways of trying to use more fundamental methods to explore multicomponent phase space and the range of potential multicomponent materials, namely *thermodynamic modelling* and *atomistic modelling*.[18,58] These methods both require extensive computer calculations based on fundamental scientific theories of, respectively, thermodynamics and quantum mechanics. They are, therefore, much more satisfactory scientifically, but they both suffer from the same two key difficulties: firstly, they both rely on theoretical expressions that have largely been derived for pure components and dilute binary or ternary solutions; and secondly, they both rely on material parameters that have largely been measured only for pure components and dilute binary or ternary solutions. This means that, although in principle more suitable for making *ab initio* predictions of multicomponent material structures and properties, in practice they usually require either extended (and often inaccurate) extrapolation from our existing knowledge of pure components and dilute alloys, or manufacture of the multicomponent materials in the first place, removing the possibility of real *ab initio* prediction. We discuss *thermodynamic modelling* and *atomistic modelling* techniques and their application one at a time, in this sub-section and the next one respectively.

Over the last few decades, *thermodynamic modelling* has become one of the most important and most regularly used of scientific methods supporting the development of improved microstructures and properties of new materials.[62–65] In general, and for the reasons mentioned above, thermodynamic modelling techniques are never really fully guaranteed in predicting the structure and properties of a newly proposed material, even in relatively simple dilute binary and ternary systems. But they have become sufficiently reliable when applied in favourable circumstances (i.e. for already well-researched materials, requiring interpolation within a known composition range rather than extrapolation beyond it) to provide valuable underpinning and guidance to experimental investigations of new dilute binary and ternary materials, often thereby shortening the time and effort required to deliver important improvements in material structure and properties. Not surprisingly, however, their usefulness is considerably more circumscribed in more complex and less well-known multicomponent materials.[3,18,58]

Thermodynamic modelling relies essentially on the well-known fundamental equation we have already discussed to some extent for the *Gibbs free energy G* of a multicomponent high-entropy solid-solution material at any temperature T:[3,22,23,36]

$$G = \sum_i x_i \mu_i^{ss} = \sum_i x_i \mu_i + \Delta G_{mix} = \sum_i x_i \mu_i + \Delta H_{mix} - T\Delta S_{mix}$$

where x_i is the molar fraction of the ith component, μ_i and μ_i^{ss} are, respectively, the *chemical potential* (or *molar free energy*) of the ith component before and after mixing, and ΔG_{mix}, ΔH_{mix} and ΔS_{mix} are, respectively, the *free energy of mixing, heat (or enthalpy) of mixing* and *entropy of mixing*. The first term on the right-hand side is the free energy of the unmixed components, and the last two terms are the free energy change caused by mixing them.

As described in Chapter 4, if the solid solution is *ideal*, there is no *interaction energy* between the different atoms, which act in the solid solution, therefore, effectively like a packing of non-interacting billiard balls. The heat of mixing is then zero, i.e. $\Delta H_{mix} = 0$, and the entropy of mixing is given by the configurational entropy of a random distribution of the different atoms within the solid solution:[iv]

$$\Delta S_{mix} = -R\sum_i x_i \ln x_i$$

where R is the gas constant. The overall ideal free energy of the material is then[3,22,23,36]:

$$G = \sum_i x_i\mu_i^{ss} = \sum_i x_i\mu_i + RT\sum_i x_i \ln x_i$$

If the solid solution is not ideal but instead is *near-ideal* or *regular*, there are small pair-wise, three-way, four-way, etc., interaction energies ω_{ij}, ψ_{ijk}, χ_{ijkl}, etc., between the different atoms, so the heat of mixing is not zero, i.e. $\Delta H_{mix} = f(\omega_{ij}, \psi_{ijk}, \chi_{ijkl} \ldots) \neq 0$[v] but the atomic interactions are insufficiently strong to affect the entropy of mixing significantly, which is still, therefore, given by the configurational entropy of a random distribution of the different atoms within the solid solution. The overall regular free energy of the material is then:[3,22,23,36]

$$G = \sum_i x_i\mu_i^{ss} = \sum_i x_i\mu_i + \Delta H_{mix}\left(\omega_{ij}, \psi_{ijk}, \chi_{ijkl} \ldots\right) + RT\sum_i x_i \ln x_i$$

If the solid solution deviates further from ideality and is *non-regular*, there are larger interaction energies ω_{ij}, ψ_{ijk}, χ_{ijkl} etc. between the different atoms, so the entropy of mixing is no longer given only by the configurational entropy of a random distribution of the different atoms within the solid solution. The overall non-regular free energy of the material is then:[3,22,23,36]

$$G = \sum_i x_i\mu_i^{ss} = \sum_i x_i\mu_i + \Delta H_{mix}\left(\omega_{ij}, \psi_{ijk}, \chi_{ijkl} \ldots\right) + RT\sum_i x_i \ln x_i - T\Delta S_{mix}^{xs}\left(\omega_{ij}, \psi_{ijk}, \chi_{ijkl} \ldots\right)$$

where $\Delta S_{mi}^{xs} = f'\left(\omega_{ij}, \psi_{ijk}, \chi_{ijkl} \ldots\right) \neq 0$ is the excess entropy of mixing over and above the random configurational entropy. Notice that there are several contributions to the excess entropy ΔS_{mix}^{xs}:

1. excess *configurational* entropy of mixing arising from a non-random distribution of the different atoms, caused by their interactions;
2. excess *vibrational* entropy of mixing arising from changes in the vibrational spectra of the different atoms caused by their interactions; and

[iv] Boltzmann's equation gives the entropy of a particular atomic arrangement as $k \ln w$ where w is its probability as given by the number of equivalent atomic arrangements and k is Boltzmann's constant. For one mole of an ideal (i.e. random) multicomponent solid solution, $\ln w = -N\sum_i x_i \ln x_i$ where N is Avogadro's number, so the configurational entropy of mixing is $\Delta S_{mix} = -R\sum_i x_i \ln x_i$, where $R = kN$ is the gas constant.[1,2,3,4,5]

[v] For the avoidance of doubt, notice that the heat of mixing is a function of *all* the interaction energies, i.e. the functional dependence of $\Delta H_{mix} = f(\omega_{ij}, \omega_{ijk}, \chi_{ijkl})$ is a shorthand notation for its dependence on $\omega_{AB}, \omega_{BC}, \omega_{AC} \ldots$, $\psi_{ABC}, \psi_{ABD}, \psi_{ACD} \ldots, \chi_{ABCD}, \chi_{ABCE}, \chi_{ACDE} \ldots$, etc., where A, B, C, D, E, etc. are the different individual components. The same notation is also used in the next paragraph for the functional dependence of the excess entropy of mixing ΔS_{mix}^{xs}.

3. excess *magnetic* and *electronic* entropy of mixing arising from changes in the magnetic and electronic interactions between the different atoms.

Notice also that the excess entropy of mixing ΔS_{mix}^{xs} can be positive or negative, since the entropy of mixing in the non-regular solution can be lower or higher than the ideal random configurational entropy.

Beginning with the seminal contribution by Kaufmann and Bernstein,[63] *thermodynamic modelling* represents a collective scientific programme conducted by a large number of scientists over the last several decades to try to find ways of calculating and then minimising the values of the free energy of mixing ΔG_{mix} and, thus, the total free energy G for all possible phases in all possible materials in order to be able to predict their structures.[64,65] This is a truly heroic scientific enterprise, requiring the gradual development of methods to estimate in a semi-empirical and self-consistent way the interaction energies $\omega_{ij}, \psi_{ijk}, \chi_{ijkl} \ldots$, etc. and thus the non-regular heats and excess entropies of mixing, $\Delta H_{mix}\left(\omega_{ij}, \psi_{ijk}, \chi_{ijkl} \ldots\right)$ and $\Delta S_{mix}^{xs}\left(\omega_{ij}, \psi_{ijk}, \chi_{ijkl} \ldots\right)$, for all possible phases and all possible phase combinations for any given set of starting components i, j, k, etc. and compositions x_i, x_j, x_k etc. This is equivalent to being able to predict all possible stable and metastable multicomponent phase diagrams. The resulting approximations, calculation methods and underlying material parameter datasets are often, therefore, referred to as *Calphad* (short for *calculation of phase diagrams*), and a variety of Calphad software programs and datasets are now widely available and widely used to perform such thermodynamic calculations and predictions, but as mentioned above they are most effective when applied to relatively simple and well-known dilute binary and ternary systems, and become increasingly unreliable and often unusable for more complex and concentrated multicomponent systems. To be fair to the Calphad enterprise, it was begun and largely developed when almost all significant materials were manufactured by traditional alloying methods, i.e. were composed of one or at most two major components with usually no more than one or two dilute alloying additions. It was never initially expected that it might have to cope with multicomponent materials consisting of four, five, six and more major components, conditions which, not surprisingly, put the calculation techniques and underlying databases under considerable strain.

In the Calphad methodology, the chemical potentials (or molar free energies) of the pure components μ_i are mostly obtained from experimental measurements of properties such as specific heat, thermal expansion, bulk modulus and electrochemical potential, often stored using semi-empirical fitting functions. Equilibria between different phases are mostly obtained from phase diagrams determined experimentally by techniques such as calorimetry, resistivity and metallography, again often stored using semi-empirical fitting functions. Free energies of metastable phases are obtained by extrapolation from corresponding stable phases at different compositions, temperatures and pressures. Simplifying assumptions are used to make the thermodynamic calculations tractable:

1. The non-ideal (or excess) component of the free energy of mixing $\Delta G_{mix}^{xs} = \Delta H_{mix}\left(\omega_{ij}, \psi_{ijk}, \chi_{ijkl} \ldots\right) - T\Delta S_{mix}^{xs}\left(\omega_{ij}, \psi_{ijk}, \chi_{ijkl} \ldots\right)$ is assumed to be a linear sum of independent binary, ternary, quaternary etc. interactions between the different components i, j, k, l, etc.:

$$\Delta G_{mix}^{xs} \approx \Delta G_{mix,\ bin}^{xs}\left(\omega_{ij}\right) + \Delta G_{mix,\ tern}^{xs}\left(\psi_{ijk}\right) + \Delta G_{mix,\ quat}^{xs}\left(\chi_{ijkl}\right) + \ldots$$

2. The independent binary, ternary, quaternary, etc. components of the non-ideal (or excess) free energy of mixing are assumed to be linear sums of individual binary *i–j*, ternary *i–j–k*, quaternary *i–j–k–l*, etc. subsystem interactions, in each case proportional to the compositions of the individual components x_i, x_j, x_k, x_l, etc. (the *Muggianu approximation*):

$$\Delta G^{xs}_{mix,\ bin}\left(\omega_{ij}\right) \approx \sum_{i,j} x_i x_j \omega_{ij}$$

$$\Delta G^{xs}_{mix,\ tern}\left(\psi_{ijk}\right) \approx \sum_{i,j,k} x_i x_j x_k \psi_{ijk}$$

$$\Delta G^{xs}_{mix,\ quat}\left(\chi_{ijkl}\right) \approx \sum_{i,j,k,l} x_i x_j x_k x_l \chi_{ijkl}$$

3. Higher-order terms in the non-ideal (excess) free energy of mixing ΔG^{xs}_{mix} are often ignored, and only binary and (occasionally) ternary contributions included.
4. In other words, the free energy of each multicomponent phase is usually approximated by a truncated linear series expansion with non-linear and higher-order terms ignored:

$$\begin{aligned} G = \sum_i x_i \mu_i^{ss} &= \sum_i x_i \mu_i + \Delta H_{mix}\left(\omega_{ij}, \psi_{ijk}, \chi_{ijkl} \ldots\right) + RT \sum_i x_i \ln x_i \\ &\quad - T\Delta S^{xs}_{mix}\left(\omega_{ij}, \psi_{ijk}, \chi_{ijkl} \ldots\right) \\ &\approx \sum_i x_i \mu_i + RT \sum_i x_i \ln x_i + \sum_{i,j} x_i x_j \omega_{ij} + \sum_{i,j,k} x_i x_j x_k \psi_{ijk} \end{aligned}$$

The resulting very large datasets are regularly assessed by Calphad scientists (at regular workshops and in the literature) in order to increase gradually the consistency between multiple different measurements by different experimenters using different experimental techniques often with different reference points. Notice that the free energies of the various phases in a multicomponent system are built up essentially by evaluating the pure components, then the binary subsystems, then the ternary subsystems, ignoring (usually) any higher-order subsystems, with each interaction treated as independent and summed linearly. Even with all these simplifying assumptions, a 5-component material requires ideally the full evaluation of a total of five pure components and ten binary and ten ternary subsystems, with an extensive range of underlying materials parameters needed for the calculations. To take a more extreme example, a 10-component material requires ideally the full evaluation of a staggering total of ten pure components, and 45 binary and 120 ternary subsystems, with an even more extensive range of underlying materials parameters needed for the calculations.[vi]

Despite the obvious limitations described above (ignoring higher-order and non-linear interactions and requiring large numbers of often unknown underlying materials parameters), thermodynamic modelling has proved to be surprisingly successful, with a

[vi] For a material with *c* components, the number *n* of *k*-level subsystems is the number of different ways of selecting *k* objects from a total of *c* objects and is given by the law of combinations without repetition as the binomial coefficient $n = \binom{c}{k} = c!/k!\,(c-k)!$, which can be taken from Pascal's triangle.

number of cases of predicted multicomponent phase structures that are in good agreement with experimental observations. Unfortunately, there is also a similar number of other cases of predicted multicomponent phase structures that are not in particularly good agreement with experimental observations. In other words, thermodynamic modelling cannot be used reliably as a guaranteed predictor of multicomponent phase structures without experimental verification, but it has, nevertheless, proved to be valuable as a technique to be used alongside experimental investigations, supporting and guiding them.

One of the first attempts to use Calphad thermodynamic modelling techniques to predict multicomponent high-entropy material microstructures was a study by Hsieh et al.[66] of tungsten-inert-gas (TIG) welded hard-facing 5-component $Al_{0.3}MnCrFe_{1.5}Ni_{0.5}$ and $Al_{0.3}MnCrFe_{1.5}Ni_{0.5}$ overlays, which were found to consist of a mixture of disordered bcc, ordered B2 bcc and disordered fcc phases. These experimental results agreed reasonably well with Hsieh et al.'s[66] Calphad-calculated equilibrium quinary phase diagram, as shown in Figure 5.3 on a vertical section for $Al_xMnCrFe_{1.5}Ni_{0.5}$ over the composition range $x = 0–1$, indicating a disordered bcc phase for temperatures above approximately $T = 700$ °C (973K), decomposing to form a two-phase mixture of an ordered B2 bcc phase and a disordered fcc phase at lower temperatures. All three phases were present in the experimental samples, in varying proportions depending on the details of subsequent post-deposition heat treatment, and Hsieh et al.[66] speculated that the decomposition reaction failed to go to completion because of sluggish diffusion in the multicomponent high-temperature bcc solid solution.

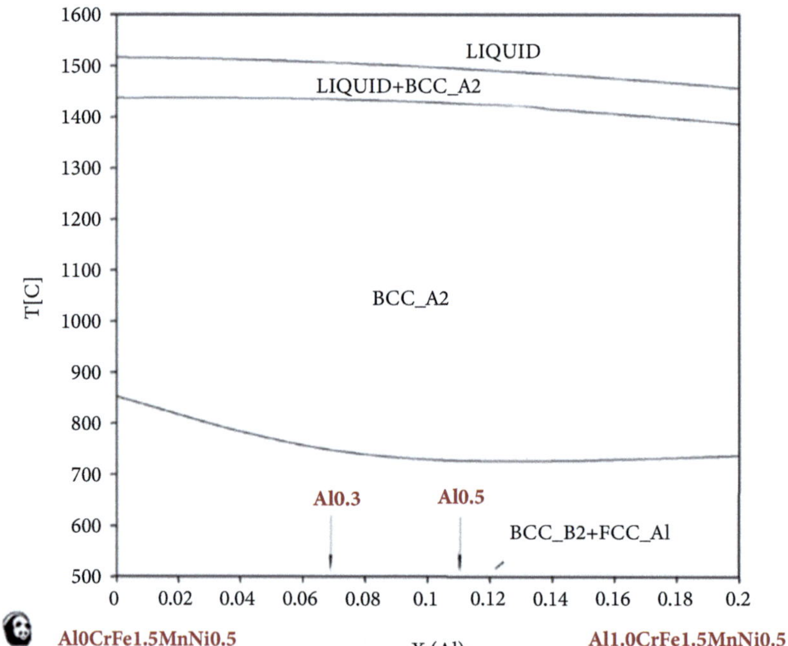

Figure 5.3 Calphad thermodynamic modelling of a vertical section of the phase diagram for 5-component $Al_xCrMnFe_{0.5}Ni_{1.5}$ alloys with compositions of $x = 0–1$.

Reproduced from Hsieh et al. (2009). The microstructure and phase equilibrium of new high performance high-entropy alloys. *Journal of Alloys and Compounds 483*: 209–12.

Another early attempt to use Calphad thermodynamic modelling techniques to predict multicomponent high-entropy material microstructures was by Huang et al.,[67] who studied laser-clad 5-component AlSiTiVCr coatings that were found experimentally to consist of a mixture of disordered bcc, $(TiV)_5Si_3$ silicide and $(VCr)_5Al_8$ aluminide phases, again in proportions depending on the details of subsequent post-deposition heat treatment. These experimental results again agreed reasonably well with Huang et al.'s[67] Calphad-calculated equilibrium quinary phase diagram, as shown in Figure 5.4 on a vertical section for $Ti_x(AlSiVCr)_{1-x}$ over the composition range $x = 0.26$–0.65 (corresponding to the composition range marked by Huang et al. on Figure 5.4 between $Ti_{0.08}(AlSiVCr)_{0.23}$ and $Ti_{0.32}(AlSiVCr)_{0.17}$), indicating a two-phase mixture of disordered bcc + $(TiV)_5Si_3$ silicide for the equiatomic composition AlSiTiVCr at temperatures above approximately $T = 900$ °C (1,173K), decomposing to form a three-phase mixture of disordered bcc phase + $(TiV)_5Si_3$ silicide + $(VCr)_5Al_8$ aluminide at lower temperatures.

In both these studies, Calphad calculations gave reasonable agreement with measured phase structures and compositions. But other early attempts to use Calphad thermodynamic modelling techniques to predict multicomponent high-entropy material microstructures gave less accurate results. For instance, Senkov et al.[68] studied a 6-component vacuum-arc melted $TiCrZrNbMo_{0.5}Ta_{0.5}$ alloy. The as-solidified structure of the multicomponent alloy consisted of a dendritic disordered bcc phase enriched in the heavier elements Nb,

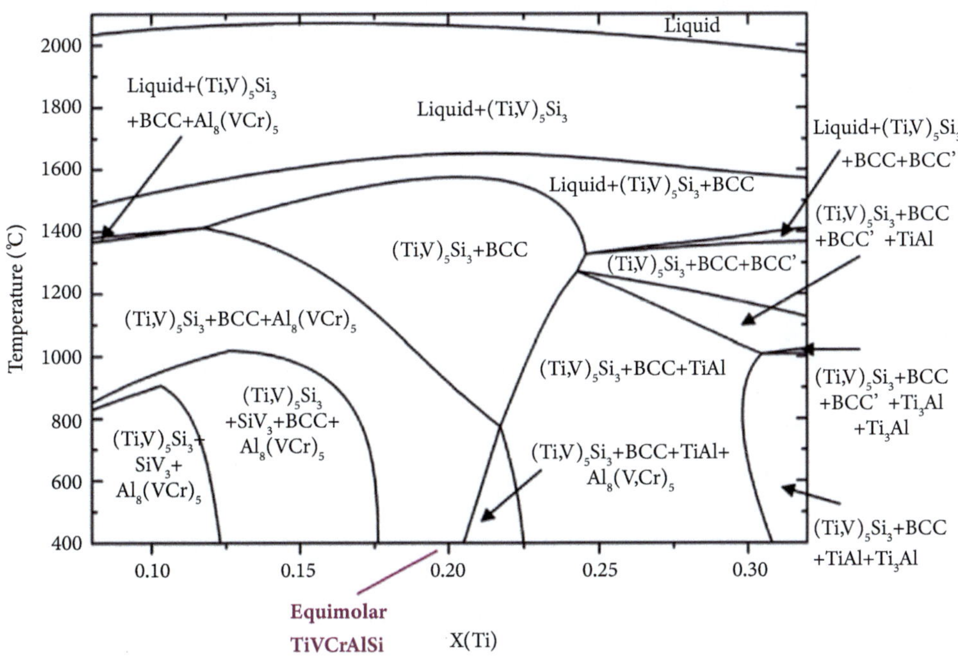

Figure 5.4 Calphad thermodynamic modelling of a vertical section of the phase diagram for 5-component $Ti_x(AlSiVCr)_{1-x}$ alloys with compositions between $x = 0.26$–0.65, corresponding to the composition range between $Ti_{0.08}(AlSiVCr)_{0.23}$ and $Ti_{0.32}(AlSiVCr)_{0.17}$.

Reproduced from Huang et al. (2011). Thermal stability and oxidation resistance of laser-clad TiVCrAlSi high-entropy alloy coatings on Ti-6Al-4V alloy. *Surface and Coatings Technology 206*: 1389–95.

Mo and Ta, and a two-phase interdendritic eutectic mixture of a second disordered bcc phase enriched in the lighter elements Ti and Zr and a Cr-rich C15 $MgCu_2$-type cubic Laves phase. The same three phases were maintained in the alloy after hot isostatic pressing (HIP) for three hours at 1,450 °C (1,723K), as well as after HIP followed by heat treatment for one hundred hours at 1,000 °C (1,273K), but (not surprisingly) with different proportions and compositions of the three phases in each case and with substantially modified microstructures. These results did not agree particularly well with Senkov et al.'s[68] Calphad-calculated equilibrium phase diagram for the 6-component alloy as shown in Figure 5.5, indicating no eutectic reaction, a single disordered bcc phase at high temperatures between approximately T = 1,200 °C (1,473K) and the solidus and liquidus temperatures of T_s ≈ 1,600 °C (1,873K) and T_l ≈ 2,500 °C (2,773K), and the formation of a second disordered bcc phase and the C15 $MgCu_2$-type cubic Laves phase only by precipitation at lower temperatures. In other words, the single disordered bcc phase predicted to form after annealing at high temperature was not seen experimentally, and the second disordered bcc phase was found to form principally by a eutectic reaction rather than the predicted precipitation reaction. Similar detailed discrepancies were found between the experimental results and Calphad-calculated predictions of the non-equilibrium Scheil-solidified multicomponent alloy structure. In another early study, Gao and Alman[69] investigated Calphad-calculated phase diagrams for a variety of multicomponent high-entropy alloys, including the original 5-component equiatomic single-phase fcc Cantor alloy CrMnFeCoNi, other modified equiatomic single-phase fcc Cantor alloys such as MnFeCoNi, NiCuPdPt and NiCuPdPtRh, single-phase hcp CoOsReRu, two-phase fcc AlCrFeCoNiCu and the single-phase bcc Senkov alloy TiZrNbHfTa, with the required assessment of the different

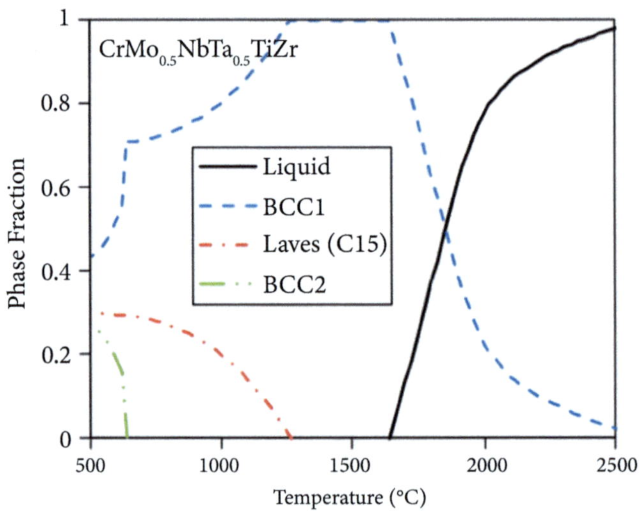

Figure 5.5 Calphad thermodynamic modelling of the equilibrium phase structure of a 6-component $TiCrZrNbMo_{0.5}Ta_{0.5}$ alloy as a function of temperature T.

Reproduced from Senkov et al. (2013). Searching for next single-phase high-entropy alloy compositions. *Entropy 15*: 4504–19.

binary, ternary, quaternary, etc. subsystems proving to be quite difficult, leading to the conclusion[69] that 'thermodynamic databases are crude for most HEA [high-entropy alloy] systems'.

There have since been many other studies using a mixture of direct measurements and Calphad predictions of multicomponent material phases and microstructures,[18,58,59] which have generally followed a similar pattern of giving sometimes quite good and sometimes less good agreement between the calculations and the observed structures. This is exemplified well by a series of studies[59,70–75] of the range of 5-component alloys $Al_xCrFeCoNi$ with x = 0.1–1.7 subjected to a variety of different heat treatments for various times at temperatures in the range 700–1,250 °C (973–1,523K), with variable results for the agreement between Calphad predictions and experimental observations, as shown in Table 5.1. Details of Calphad calculations of the phase structures of a range of other multicomponent high-entropy materials are given in the reviews by Murty et al.,[18] Soni et al.[59] and Zhang and Gao,[76] and some examples are given in Table 5.2, again showing somewhat variable results for the agreement between prediction and observation.

In summary, as mentioned above, the results of extensive Calphad calculations on a wide variety of different multicomponent alloys do not overall justify the use of thermodynamic modelling as a guaranteed method for prediction of multicomponent material structures. But the method is, nevertheless, a helpful aid to experimental investigations, particularly, as discussed in a later sub-section, as a useful sift for a strategy of high-throughput examination of particular regions of multicomponent phase space.

Table 5.1 Calphad-calculated and experimentally observed equilibrium phase structures from different studies of $Al_xCrFeCoNi$ alloys.[59,70–75]

$Al_xCrFeCoNi$ composition x	heat treatment	equilibrium phases	
		observed	predicted
0.3	72h at 700 °C	fcc + $L1_2$	fcc + B2
0.3	72h at 900 °C	fcc + B2	fcc + B2
0.25	24h at 900 °C	fcc + B2	fcc + B2
0.25	24h at 1100 °C	fcc	fcc
0.375–1.0	24h at 100 °C	fcc + bcc	fcc + bcc + B2
1.25–2.0	24h at 1100 °C	bcc	bcc + B2
0.3	2h at 900–1100°C	fcc	fcc
0.5	2h at 900–1100°C	fcc + bcc	fcc + B2
0.7–0.9	2h at 900–1100°C	fcc + B2	fcc + bcc + B2
1.5	2h at 900–1100°C	bcc + B2	bcc + B2
0.3	1h at 850 °C	fcc + B2	fcc + B2
0.3	50h at 1250 °C	fcc	fcc
0.5	50h at 1250 °C	fcc	fcc + B2
0.7	50h at 1250 °C	fcc + B2	fcc + bcc + B2
0.1	800h at 750 °C	fcc + fcc + Cr-rich phase	fcc + B2

Table 5.2 Calphad-calculated and experimentally observed equilibrium phase structures from different multicomponent alloys.[70,77–82]

alloy	heat treatment	equilibrium phases	
		observed	predicted
$Al_{1.7}CrFeCoNi_{0.3}$	100h at 1000 °C	σ + B2	bcc + B2
$Al_{0.3}CrFeCoNi_{1.7}$	100h at 1000 °C	fcc + $L1_2$	fcc
CrMnFeCoNi	12000h at 500 °C	$L1_0$ + bcc + B2	fcc + σ
CrMnFeCoNi	12000h at 700 °C	fcc + σ	fcc + σ
CrMnFeCoNi	12000h at 900 °C	fcc	fcc
CrMnFeCoCu	72h at 850 °C	fcc + bcc	fcc + σ
MnFeCoNiMo	72h at 1000 °C	fcc + μ	fcc + μ
VMnFeCoNi	72h at 1000 °C	fcc	fcc + σ
VCrMnCoNi	72h at 1000 °C	fcc + bcc	fcc + σ
TiVNbMoTaW	as-solidified	bcc	bcc

Atomistic modelling

Over the last few decades, *atomistic modelling*, like *thermodynamic modelling* discussed in the last sub-section, has also become one of the most important and most regularly used scientific methods supporting the development of improved microstructures and properties of new materials.[83,84] In general, atomistic modelling techniques, for similar reasons to thermodynamic modelling techniques, are never really fully guaranteed in predicting the structure and properties of a newly proposed material, even in relatively simple dilute binary and ternary systems. But they have also become sufficiently reliable when applied in favourable circumstances (i.e. for some of the simpler properties of materials consisting of well-defined arrangements of relatively small groups of atoms and/or molecules) to provide valuable underpinning and guidance to experimental investigations of new dilute binary and ternary materials, often thereby shortening the time and effort required to deliver important improvements in material structure and properties.[83,84] Not surprisingly, however, their usefulness is again considerably more circumscribed in more complex multicomponent materials.[3,18,58]

As described in the last sub-section, *thermodynamic modelling* rests on the well-established fundamental theories of *thermodynamics* that seek to explain materials based on the statistical mechanics of large assemblies of atoms and/or molecules, concentrating essentially on calculating their *collective* behaviour, and thus their resulting structures and properties.[22,36,83] *Atomistic modelling*, which we describe and discuss in this sub-section, rests on similarly well-established fundamental theories that seek to explain materials based on the *quantum mechanics* and (where appropriate) the *classical mechanics* of small groups or clusters of atoms and/or molecules, concentrating essentially on calculating their *individual* behaviour, and thus again their resulting structures and properties.[83,84] The ideal in atomistic modelling is *ab initio modelling*, i.e. describing the structure and properties of a material using calculations derived exclusively and directly from first principles. But in truth, *ab initio* modelling can never be achieved for any realistic material based fully on first principles, because the calculations are too difficult, and progress can only be made by making assumptions and approximations which, to some extent, compromise the purity of the *ab initio* first-principles ideal.

There are three basic atomistic modelling approaches, namely *density functional theory (DFT)*, *molecular dynamics (MD)*, and *Monte Carlo methods (MC)*, each of which has its own particular advantages and disadvantages.[83] In this sub-section, we briefly describe each of these three approaches and then discuss their application to multicomponent high-entropy materials.

Density functional theory

A material is a collection of different atoms and molecules, each of which is composed in turn of a collection of more fundamental particles, namely its constituent electrons, protons and neutrons. *Density functional theory (DFT)* uses *quantum mechanics* to describe the structure and properties of a material from the collective behaviour of the set of all the constituent electrons, protons and neutrons in all the different atoms and molecules of the material.[84,85] In quantum mechanics, the behaviour of any set of fundamental particles is described by the *many-body time-dependent wave function* $\Psi(\{\boldsymbol{r}_i\}, t)$, which gives the probability of finding the set of different fundamental particles $\{i\} = i, j, k \ldots$, etc., at positions $\{\boldsymbol{r}_i\} = \boldsymbol{r}_i, \boldsymbol{r}_j, \boldsymbol{r}_k \ldots$ etc. respectively at any given time t.[vii] The value of any individual property p is given by integrating the wave function:[84,85]

$$p = \langle \Psi | \mathcal{P} | \Psi \rangle = \int \Psi^* \mathcal{P} \Psi dV$$

where Ψ^* is the complex conjugate of Ψ, $\mathcal{P}$ is an operator[viii] describing the property p as a function of Ψ, the integration is over the volume V occupied by the material, and the expression in angular brackets is the compact Dirac notation for the full integral on the right hand side.[ix] Determining any property p of the material corresponds, therefore, to finding suitable expressions for the wave function Ψ and the operator $\mathcal{P}$, and then performing the integration above over the volume V. There is always a series of solutions to the integration corresponding to the ground state and a series of excited states of the material within the volume V.

As a particularly important example, the energy of the material E is given by the *Hamiltonian operator* (or just the *Hamiltonian*) $\mathcal{H}$:

$$\mathcal{H} = \mathcal{T} + \mathcal{V} = -\sum_i \frac{\hbar^2}{2m_i} \nabla_i^2 + \frac{1}{2} \sum_{i \neq j} \frac{e^2}{4\pi\epsilon_o} \frac{z_i z_j}{(\boldsymbol{r}_i - \boldsymbol{r}_j)}$$

[vii] More accurately, the probability of finding the set of fundamental particles $\{i\}$ at positions $\{\boldsymbol{r}_i\}$ at time t is given by the square of the wave function $\Psi\Psi^*$ where Ψ^* is the complex conjugate of Ψ. The particles are all somewhere in the atom at time t, so $\int \Psi\Psi^* dV = 1$ when the integral is over the volume V occupied by the material, i.e. by its collection of fundamental particles.

[viii] An operator (denoted here by curly capitals) is a function of a function, its domain and range are both sets of functions, and it maps any given function in its domain to another function in its range. In *quantum mechanics*, an operator $\mathcal{P}$ represents a property p in the sense that it maps any given wave function Ψ defined over some region of space and time $(\boldsymbol{r}, t)$ to give the value of the property as a function p over the same region of space and time $(\boldsymbol{r}, t)$.

[ix] We will not use the Dirac notation any further in what follows. It is given here simply in the interest of completeness.

where $\mathcal{T}$ and $\mathcal{V}$ are kinetic and potential energy operators given respectively on the right hand side by the sum of the individual kinetic energies of all the constituent fundamental particles i and the sum of the individual electrostatic attractions or repulsions of all pairs of constituent fundamental particles i and j, and where $\hbar = h/2\pi$, h is Planck's constant, m_i and $\boldsymbol{r}_i$ are the mass and position of the ith particle, e is the charge on the electron, $z_i = -1, +1$ and 0 for electrons, protons and neutrons respectively, and $\nabla_i^2 = \partial/\partial x_i^2 + \partial/\partial y_i^2 + \partial/\partial z_i^2$ is the Laplacian operator for the ith particle. Notice that there are three terms in the potential operator $\mathcal{V} = V_{nn} + V_{ee} + V_{en}$ for the three cases when i and j are both nuclear particles, both electrons and mixed respectively. The wave function is obtained by solving the *time-dependent Schrödinger equation*:[84,85]

$$i\hbar\frac{\partial\Psi}{\partial t} = \mathcal{H}\Psi$$

and the ground and excited state energies of the material are then obtained by its integration:

$$E = \langle\Psi|\mathcal{H}|\Psi\rangle = \int \Psi^*\mathcal{H}\Psi dV$$

Solving these equations is fiendishly difficult even for just one or two atoms, because of the large number (except for hydrogen atoms) of constituent electrons, protons and neutrons, and also because of the complex structure of the equations themselves. *Density functional theory* represents a concerted effort to overcome these difficulties, another truly heroic scientific enterprise (like Calphad, as discussed in the last sub-section) made collectively by many scientists over many years, gradually developing a series of approximations to try and make this problem tractable, as follows:[84,85]

1. *Time independence*: we concentrate exclusively on stationary (i.e. equilibrium) states of the material, by ignoring the time dependence of the wave function, i.e. we replace $\Psi(\{\boldsymbol{r}_i\}, t)$ with $\Psi(\{\boldsymbol{r}_i\})$, which now has to obey the simpler time-independent (rather than time-dependent) Schrödinger equation $E\Psi = \mathcal{H}\Psi$.
2. *Separation of variables* (the *Car–Parrinello method*[86]): the wave function is split into two separate wave functions, one for all the electrons and one for all the protons and neutrons in the atomic nuclei, i.e. we replace $\Psi(\{\boldsymbol{r}_i\})$ with $\Psi_e(\{\boldsymbol{r}_i\})\Psi_n(\{\boldsymbol{r}_{i'}\})$ where Ψ_e and Ψ_n are many-body wave functions for the collections of electrons and nuclear particles respectively, and $\{i\}$ and $\{i'\}$ are now the separate sets of all electrons and all nuclear particles at positions $\{\boldsymbol{r}_i\}$ and $\{\boldsymbol{r}_{i'}\}$ respectively.
3. *Clamped nuclei* (the *Born–Oppenheimer approximation*[87]): we concentrate exclusively on the electrons (which determine, of course, the chemical and electronic properties of the material), treating the atomic nuclei as fixed centres of electronic charge, and ignoring the motion of the protons and neutrons in the nuclei, since they are much heavier than the electrons, i.e. we ignore $\Psi_n(\{\boldsymbol{r}_{i'}\})$ and replace $\Psi_e(\{\boldsymbol{r}_i\})\Psi_n(\{\boldsymbol{r}_{i'}\})$ with just $\Psi_e(\{\boldsymbol{r}_i\})$, reducing the total number of variables dramatically.
4. With the *Born–Oppenheimer approximation*[87] we can also ignore the electrostatic potential between pairs of nuclear particles V_{nn} and rewrite the potential energy

operator $\mathcal{V} = V_{ee} + V_{en}$ as a sum of two terms only, one for the electrostatic repulsion between the different electrons V_e and one for the electrostatic attraction between the electrons and the fixed nuclei V_n.

5. *Electron density* (the *Hohenberg–Kohn theorem*[88]): we concentrate exclusively on the ground state of the collection of electrons, and the energy can then be shown to be a function of just the local electron density $\rho(\boldsymbol{r})$ rather than the set of all electron positions $\{\boldsymbol{r}_i\}$, again reducing the total number of variables dramatically. This is why the overall method is known as density functional theory.
6. *Independent electrons*: we replace the many-body wave function with a set of independent *single-electron wave functions*, i.e. we replace $\Psi_e(\{\boldsymbol{r}_i\})$ with the product $\prod_i \psi_i(\boldsymbol{r}_i)$, where $\{\psi_i(\boldsymbol{r}_i)\}$ is a set of single-body electron wave functions, one for each electron moving separately within the nuclear potential of the material, each of which now has to obey the *single-body Schrödinger equation* $E\psi = \mathcal{H}\psi$. This corresponds to the assumption that the *collective* probability $\Psi_e(\{\boldsymbol{r}_i\})$ of finding electrons i, j, k, etc. at $\boldsymbol{r}_i$, $\boldsymbol{r}_j$, $\boldsymbol{r}_k$ etc. is equal to the product of the *independent* probabilities of finding them there $\{\psi_i(\boldsymbol{r}_i)\}$, and reduces the problem from solving one complex differential equation in many dimensions to solving many times a simpler differential equation in three dimensions.
7. *Mean-field*: the electrostatic repulsion between the different electrons V_e is approximated as the *Hartree potential* V_H, the classical Coulomb repulsion that each electron experiences in the *mean field* of all the others,[89] i.e. the potential operator is rewritten as $\mathcal{V} = V_{en} + V_H$ where $V_H = \int e^2\rho(\boldsymbol{r})\,d\boldsymbol{r}/4\pi\varepsilon_o(\boldsymbol{r} - \boldsymbol{r}')$ and ε_o is the permittivity of free space.
8. *Exchange-correlation* (the *Kohn–Sham equation*[90]): taken together the *independent electron* and *mean-field approximations* violate Pauli's exclusion principle and ignore correlations between the motions of different electrons, and these effects are corrected for by incorporating in the Hamiltonian an extra so-called *exchange-correlation* potential term V_{xc}, i.e. the potential operator is further rewritten as $\mathcal{V} = V_{en} + V_H + V_{xc}$.
9. *Local-density* (*LDA*) and *generalised gradient* (*GGA*) *approximations*: we estimate V_{xc} semi-empirically in each specific case using either the *local-density approximation* $V_{xc} = \int \varepsilon_{xc}\rho(\boldsymbol{r})\,\boldsymbol{dr}$ or the *generalised gradient approximation* $V_{xc} = \int \varepsilon_{xc}(r)\,\rho(r)\,\nabla\rho(r)\,dr$, where ε_{xc} is the exchange-correlation energy for homogeneous or inhomogeneous electron distributions respectively.[83,84]
10. *Crystalline material* (*Bloch's theorem*[91]): we can use the translational symmetry of a crystalline material to represent its atomic or molecular structure by a relatively small number of atoms in a single unit cell, with the wave functions $\{\psi_i(\boldsymbol{r})\}$ given by *Bloch waves* $\psi_i(r) = \phi_i(\boldsymbol{r})\,u_i(\boldsymbol{r})$ where $\phi_i(\boldsymbol{r}) = e^{i\boldsymbol{kr}}$ are plane waves modulated by a periodic function $u_i(\boldsymbol{r}) = u_i(\boldsymbol{r} + \boldsymbol{R})$ with the periodicity of the crystal lattice, $\boldsymbol{k}$ is the wave vector and $\boldsymbol{R}$ is any symmetry operation in the crystal lattice.
11. *Non-crystalline material*: For non-crystalline or complex molecular materials, the same method can be used by constructing a *supercell* of a limited number of atoms representative of the material structure surrounded by periodic boundary conditions.

12. *Pseudopotentials*:[92] we can remove high-frequency oscillations of the wavefunctions $\{\psi_i(\mathbf{r})\}$ near the nuclei that make it difficult to truncate the plane-waves $\phi_i(\mathbf{r})$, by replacing the electron-nuclear potential V_{en} with a matching pseudopotential V_{en}' smoothed in the nuclear regions.
13. *Valence electrons*: we can concentrate on the chemically important outer *valence electrons* by absorbing into the *pseudopotentials* the nuclear screening effect of the otherwise inert inner *core electrons*, effectively treating the material as a mixture of ions and valence electrons only and thus reducing the number of wave functions $\{\psi_i(\mathbf{r})\}$ to just those of the valence electrons.[83]

With all these approximations, the equations defining the wave functions of the fundamental particles in the material and operators such as the Hamiltonian for the overall material energy become tractable but still not straightforward to solve. Finding solutions still requires a complex set of iterative computer calculations: starting with a trial set of wave functions $\{\psi_i(\mathbf{r})\}$; determining an initial electron charge density distribution $\rho(\mathbf{r})$ and a corresponding initial Hamiltonian $\mathcal{H}$; applying the latter to the trial set of wave functions to determine the energy E and its gradient ∇E; using the gradient to calculate an improved set of wave functions with lower energy; and then repeating the process iteratively until the calculations converge on a *self-consistent* set of wave functions with minimum energy.[83,84] DFT computer calculations of this type are quite intensive and are usually used for no more than several tens or a small number of hundreds of atoms, though larger numbers are possible in special cases with additional approximations and long computing times.

Molecular dynamics

In DFT we try to predict the structure and properties of a material consisting of an assembly of atoms and molecules from the behaviour of their constituent fundamental particles, i.e. from the collective behaviour of the set of electrons, protons and neutrons that comprise the atoms and molecules of the material.[83,84] Solving the resulting *many-body time-dependent Schrödinger equation* is so fearsomely difficult that we have to introduce, as explained above, a series of simplifying assumptions and approximations. Notably this simplification process includes:[83–92] the *Car–Parrinello approximation* to separate out the electrons from the nuclear protons and neutrons; the *Born–Oppenheimer approximation* so we can ignore the nuclei as heavy, clamped in position, and effectively immobile compared to the light, mobile electrons; the *ground-state* assumption so we can ignore the particle dynamics and concentrate on the material at equilibrium; the *Kohn–Sham equation* treating the behaviour of the set of electrons as the sum of the behaviours of the individual electrons one at a time; the *Hohenberg–Kohn theorem* so we can treat the kinetic and potential energies of each electron as functions of the local electron density rather than the set of all electron positions; the use of a *periodic structure* to reduce the complex wave function to a simpler set of orthonormal plane waves in the first Brillouin zone for a small number of atoms in the unit cell; and a *pseudopotential assumption* so as to ignore the *core electrons* and treat them as part of clamped *ions* rather than just clamped *nuclei*.[83–92] With these assumptions, the problem is reduced to solving the much simpler *single-body time-dependent Schrödinger equation* for the individual valence electrons one at a time.

This is a tremendous step forward towards achieving near–*ab initio* calculation of the structure and properties of a number of materials. But the simplification obtained by making these assumptions and approximations comes at quite a high price, essentially restricting our solutions to materials consisting of rather few atoms with fixed (pre-determined and unchanging) positions of their nuclei, and also requiring a certain amount of semi-empirical estimation on a case-by-case basis, somewhat undermining the *ab initio* nature of the calculations.

We use *molecular simulation methods*, namely molecular dynamics (*MD*) and Monte-Carlo (*MC*) *methods* to step back and take a longer view, mostly removing our attention from the electrons and, instead, re-focusing our attention on the behaviour of the atoms and molecules themselves, in order to deal with the structure of larger, more representative, and freer (i.e. dynamic) groups of atoms or molecules. The behaviour of a material composed of an assembly of n atoms or molecules is described classically by the application of Newton's laws of motion, i.e. by n partial differential equations arising from the application of Newton's second law to each of the n atoms:[93,94]

$$m_i \frac{d^2 \boldsymbol{r_i}}{dt^2} = m_i \boldsymbol{a}_i = \boldsymbol{F_i}\left(\{\boldsymbol{r_i}\}\right) = -\nabla V_i\left(\{\boldsymbol{r_i}\}\right)$$

where m_i, $\boldsymbol{r_i}$ and $\boldsymbol{a}_i = d^2\boldsymbol{r}_i/dt^2$ are the mass, position and acceleration of the ith atom, $\boldsymbol{F}_i$ is the force exerted on the ith atom by the gradient vector $\nabla = (\boldsymbol{a}\partial/\partial x + \boldsymbol{b}\partial/\partial y + \boldsymbol{c}\partial/\partial z)$ of the potential energy field V_i at position $\boldsymbol{r}_i$, with $\boldsymbol{a}$, $\boldsymbol{b}$ and $\boldsymbol{c}$ unit vectors in the x, y, and z directions respectively, and with $\boldsymbol{F}_i$, V_i and ∇V_i all functions of the set of all n atom positions $\{\boldsymbol{r}_i\} = \boldsymbol{r}_i, \boldsymbol{r}_j, \boldsymbol{r}_k \ldots$ etc. for atoms i, j, k... etc. respectively.

The resulting set of n partial differential equations of motion is very difficult to solve analytically for any significant number of atoms or molecules. In MD simulations of a material and its constituent atoms or molecules, the set of partial differential equations is solved instead, therefore, by using a series of approximate finite difference calculations: i.e. by selecting an initial set of atomic or molecular positions $\{\boldsymbol{r}_i\}$ and velocities $\{\boldsymbol{v}_i\} = \{d\boldsymbol{r}_i/dt\}$; imposing a small time increment τ; using a potential energy field $V_i\left(\{\boldsymbol{r}_i\}\right)$ to calculate the forces on the atoms or molecules $\{\boldsymbol{F}_i\}$ and, therefore, their accelerations $\{\boldsymbol{a}_i\}$, and a new set of positions $\{\boldsymbol{r}_i\}$ and velocities $\{\boldsymbol{v}_i\}$; imposing an additional time increment τ; recalculating again a new set of positions $\{\boldsymbol{r}_i\}$ and velocities $\{\boldsymbol{v}_i\}$; and so on, effectively letting the material, its constituent atoms and molecules, and their structure and properties evolve gradually under the imposed potential energy field in a stepwise way with time.[93,94] This is achieved by using a Taylor expansion for the time variation of the atomic and molecular positions $\{\boldsymbol{r}_i\}$:

$$\boldsymbol{r_i}(t+\tau) = \boldsymbol{r_i}(t) + \tau \frac{d\boldsymbol{r_i}}{dt}(t) + \frac{\tau^2}{2}\frac{d^2\boldsymbol{r_i}}{dt^2}(t) + \ldots = \boldsymbol{r_i}(t) + \tau \boldsymbol{v_i}(t) + \frac{\tau^2}{2}\boldsymbol{a_i}(t) + \ \ldots$$

and ignoring higher-order terms (i.e. terms of order $\geq \tau^3$), so that the set of n partial differential equations is converted into a set of n approximate algebraic difference equations. This is equivalent to approximating the three-dimensional curves describing the evolution of the positions of all the atoms and molecules in the material with piecewise linear segments.[93,94]

There are various ways of incrementing the positions, velocities and accelerations of the atoms and molecules, the *Euler*, *Verlet*, *velocity Verlet* and *leapfrog* methods, each of which have different advantages and disadvantages of accuracy and speed of calculation.[93] As an example, the Verlet algorithm eliminates the velocity term by summing Taylor expansions at $(t + \tau)$ and $(t - \tau)$:

$$\boldsymbol{r}_i(t+\tau) + \boldsymbol{r}_i(t-\tau) = 2\boldsymbol{r}_i(t) + \tau^2 \boldsymbol{a}_i(t)$$

so the position of atom i at $(t + \tau)$ can be calculated from its previous positions at t and $(t - \tau)$, using an acceleration $\boldsymbol{a}_i$ obtained from Newton's second law with the force on the atom $\boldsymbol{F}_i$ given by differentiating the potential energy V_i. The velocity $\boldsymbol{v}_i$ and, therefore, the kinetic energy T_i of each atom can then be calculated using the *central time difference* method:

$$\boldsymbol{v}_i(t) = \frac{d\boldsymbol{r}_i}{dt}(t) = \frac{\boldsymbol{r}_i(t+\tau) - \boldsymbol{r}_i(t-\tau)}{2\tau}$$

$$T_i(t) = \tfrac{1}{2} m_i \boldsymbol{v}_i^2(t) = \frac{\{\boldsymbol{r}_i(t+\tau) - \boldsymbol{r}_i(t-\tau)\}^2}{8\tau^2}$$

There are various boundary conditions that can be used to contain the atoms or molecules within the material, the so-called *microcanonical* (*NVE*), *canonical* (*NVT*), *grand canonical* (*μVT*), or *isothermal–isobaric* (*NPT*) *ensembles*, with different combinations of fixed numbers of atoms or molecules N, volume V, total energy E, chemical potential μ and pressure P, corresponding respectively to *adiabatic*, *isothermal*, *isopotential* and *isobaric* ambient conditions for the material.[83] And there are various ways of describing the potential energies V_i from which the forces on the atoms and molecules and, therefore, their accelerations are calculated, corresponding to different combinations of *ionic*, *covalent*, *metallic* and *van der Waals bonding*, and closely related to the pairwise, three-way, four-way, etc., *interaction energies* ω_{ij}, ψ_{ijk}, χ_{ijkl} used for the non-ideal heat of mixing in thermodynamic modelling, as described in the last sub-section. In fact, one of the main difficulties in molecular dynamics calculations is in defining appropriate functions for the potential energies representing the interactions between all the different atoms and/or molecules in the material. The potential energies V_i can, of course, include some or all of the translational, vibrational and rotational kinetic energies of atomic motion as appropriate to the material of interest and can also include additional potential energies arising for instance from external applied magnetic, electrical, mechanical or gravitational fields, again as appropriate to the material of interest.[83,94]

Overall, molecular dynamics algorithms are available which allow fairly readily for the calculation of the behaviour of up to a few million atoms or molecules in a given material, with a typical time step of no more than a few fs, and with the cumulative error of approximating a curve by a series of stepwise linear segments building up gradually and preventing useful results beyond some tens or hundreds of ns. *Classical molecular dynamics* (*CMD*) as described above can sometimes be replaced by *ab initio molecular dynamics* (*AIMD*), using density functional theory to calculate some or all of the interatomic forces, interaction energies and potentials. This is clearly more fundamental and more satisfying scientifically, but of course, as described previously, comes at the price of restricting again the number of atoms and molecules and, therefore, the materials that can be investigated.[83,93,94]

Monte Carlo methods

Unlike density functional theory, which uses quantum mechanics to achieve near-*ab initio* calculations of the structure of a material from the behaviour of its fundamental constituent electrons, protons and neutrons, molecular dynamics uses (mostly) classical Newtonian mechanics to calculate the structure of a material from its constituent atoms and molecules. This enables the investigation of larger atomic and molecular assemblies in a material of interest, and also their dynamical as well as static behaviour. *Monte Carlo methods* allow even larger assemblies of atoms and molecules to be studied, by ignoring the specific dynamics of the atomic and molecular motion and concentrating instead on the instantaneous static material structure.

In Monte Carlo (MC) simulations, like molecular dynamics (MD) simulations, the structure and properties of a material are determined from the potential energy field $V_i(\{\boldsymbol{r}_i\})$ describing the potential energy of each atom or molecule in the material i at position $\boldsymbol{r}_i$ arising from the interaction energies between the set of all atoms or molecules in the material $\{i\} = i, j, k \ldots$ at positions $\{\boldsymbol{r}_i\} = \boldsymbol{r}_i, \boldsymbol{r}_j, \boldsymbol{r}_k \ldots$ In Monte Carlo simulations however, unlike molecular dynamics simulations, this is not done by explicitly solving Newton's laws of motion to calculate the force on each atom and, therefore, its acceleration and motion, but is achieved instead by an iterative quasi-static stochastic process.[83,94] We calculate the total potential energy of all the atoms and molecules in the material $E = \sum_i V_i(\{\boldsymbol{r}_i\})$ for an initial set of atomic and molecular positions $\{\boldsymbol{r}_i\}$; change the position $\boldsymbol{r}_i$ of one or other of the atoms or molecules i selected at random, by a randomly selected distance and direction to a new position $\boldsymbol{r}'_i$; re-calculate the new total potential energy $E' = \sum_i V_i$; adopt the new structure or stay with the initial structure according to whether respectively $E' < E$ or $E' > E$; change the position of another atom or molecule again selected at random; again re-calculate the new total potential energy; again adopt the new structure or stay with the initial structure according to whether respectively $E' < E$ or $E' > E$; and then continue this process iteratively until the material structure stabilises at a steady-state with an equilibrium minimum energy E_{min}. Essentially, the material reduces its energy from its initial randomly selected structure to its final equilibrium structure not, as in MD, by a series of physical steps calculated from Newton's laws of motion, but instead by a series of quasi-static random steps. The different structures through which the material progresses randomly from its initial to its equilibrium structure are called *microstates* of the material.[83,94] The process is clearly speeded up by selecting a plausible near-equilibrium initial structure, and by selecting each random step, i.e. each possible microstate, from a plausible range of changes in structure reasonably likely to reduce or at least not increase too much the overall energy.

The MC method described above is essentially a probabilistic way of determining the equilibrium structure and minimum energy E_{min} of a material at absolute zero, i.e. at temperature $T = 0$. One of the major strengths of the Monte Carlo method is the simplicity with which it can be modified[83,94] to determine the structure of a material at realistic temperatures $T > 0$, using a technique known as the *Metropolis method*. We again calculate the total potential energy of all the atoms and molecules in the material $E = \sum_i V_i(\{\boldsymbol{r}_i\})$ for an initial set of atomic and molecular positions $\{\boldsymbol{r}_i\}$; again change the position $\boldsymbol{r}_i$ to a new position r'_i for one or other of the atoms or molecules i selected at random, by a randomly selected distance and direction; and again re-calculate the new total potential energy $E' = \sum_i V_i$; but we then use different probabilities to determine whether to adopt or reject the new structure. At absolute zero $T > 0$, we effectively adopted the new structure with a probability $p = 1$ if $E' < E$ and $p = 0$ if $E' > E$; but at realistic temperatures $T > 0$ we instead adopt the new structure with a probability $p = 1$ if $E' < E$ and $p = \exp\{-(E' - E)/kT\}$ if $E' > E$.[x] After continuing the calculation iteratively through a series of microstates, the material no longer stabilises at a fixed equilibrium structure and minimum energy E_{min}, because at realistic temperatures its entropy allows it to occupy microstates with energy E greater than the minimum energy E_{min} with a probability given by the Boltzmann factor $\exp\{-(E - E_{min})/kT\}$. Instead of its structure stabilising, the material progresses iteratively through a range of different microstates with different energies and probabilities given by different values of the Boltzmann factor $\exp\{-(E - E_{min})/kT\}$. Adopting the new structure in each step with probability $p = 1$ if $E' < E$ and $p = 0$ if $E' > E$ corresponds[83,94] to minimising the overall internal energy of the material E at absolute zero $T = 0$; whereas adopting the new structure in each step with probability $p = 1$ if $E' < E$ and $p = \exp\{-(E' - E)/kT\}$ if $E' > E$ corresponds[83,94] to minimising the free energy of the material $E - TS$ at a realistic temperature $T > 0$, where S is the entropy of the material.[xi] In other words, the *Metropolis method* allows explicitly for the effect of entropy at a realistic temperature $T > 0$.

A set of microstates generated stochastically by selecting from a fixed set of possibilities according to fixed probabilistic rules is called a *Markov chain* in probabilistic theory.[83,94] The set of microstates generated in this way for a given material corresponds to a *microcanonical* (*NVE*), *canonical* (*NVT*), *grand canonical* (μVT), or *isothermal–isobaric* (*NPT*) *ensemble* of the atoms and molecules of the material, with different combinations respectively of fixed numbers of atoms or molecules N, volume V, total energy E, chemical potential μ and pressure P, describing the statistical mechanics of the material structure under respectively *adiabatic, isothermal, isopotential* and *isobaric* ambient conditions.[83,94] In each case, the properties of the material can be obtained by taking an ensemble average across the range of microstates generated. The iterative random steps and the potential energies V_i can, as with molecular dynamics modelling, be chosen to include some or all of the translational, vibrational and rotational kinetic energies of atomic and molecular motion as appropriate to the material of interest; and can also include additional potential energies arising for instance from external applied magnetic, electrical, mechanical or gravitational fields, again as appropriate to the material of interest. Chemical effects can also be readily included by using random steps involving swopping atoms of different types.

Overall, Monte Carlo like molecular dynamics algorithms are available which allow for the calculation of the behaviour of up to perhaps a few million atoms or molecules in a given material, with the advantage of incorporating entropy effects at realistic temperatures. As

[x] This is usually achieved using a random number generator to select a value for p between 0 and 1, with the new structure adopted if $p < \exp\{-(E' - E)/kT\}$ and discarded if $p > \exp\{-(E' - E)/kT\}$.

[xi] Strictly speaking, the *Monte Carlo* method described at realistic temperatures $T > 0$ minimises the Helmholtz free energy $F = E - TS$ for a *canonical ensemble* of material microstates under conditions of constant number of atoms or molecules N, volume V and temperature T (since $dF = -PdV - SdT$ where P is the pressure); and minimises the Gibbs free energy $G = H - TS = E + PV - TS$ for an *isobaric–isothermal ensemble* of material microstates under conditions of constant number of atoms or molecules N, pressure P and temperature T (since $dG = VdP - SdT$).

with molecular dynamics simulations, classical Monte Carlo methods can be replaced by *quantum Monte Carlo* (*QMC*) methods, using density functional theory to calculate some or all of the interatomic forces, interaction energies and potentials, but again only at the expense of restricting the number of atoms and molecules and, therefore, the materials that can be investigated.[83,93,94]

Atomistic modelling of multicomponent high-entropy materials

As mentioned above, atomistic modelling has proved in recent years to be a valuable method of studying materials consisting of relatively small groups of atoms or molecules with relatively simple structures, but there are considerable difficulties with extending its use to more complex multicomponent high-entropy materials. Chapter 4 described the use of atomistic modelling to investigate the behaviour of various defects in multicomponent high-entropy materials, including vacancies, dislocations and grain boundaries. And we have already noted, therefore, the two main overarching difficulties with using atomistic modelling methods:

1. The maximum number of atoms or molecules that can be studied by atomistic modelling stretches with modern high-performance parallel computers to a million or two in very favourable circumstances (though it is often much lower in less favourable cases). But this is still only a minuscule fraction of the total number of local atomic environments to be found in multicomponent high-entropy solid-solution materials, even for fairly low numbers of components, typically billions or trillions of different local atomic environments in 4- or 5-component materials, as discussed in some detail in Chapter 4. In other words, despite the advances in atomistic modelling methods, they still struggle to provide results that are anywhere near representative of the complexity of the atomic structures in multicomponent high-entropy materials.
2. As with thermodynamic modelling, a large number of parameters is needed to perform atomistic modelling calculations in multicomponent high-entropy materials, and they mostly have to be estimated using simplistic linear addition and/or extrapolation from values for pure components and/or dilute binary or ternary materials, methods that are likely to be unreliable in many cases. In other words, the promise of using atomistic modelling to predict the properties of multicomponent high-entropy materials *ab initio* from first principles, without first conducting initial trial-and-error experiments followed by reasonably accurate measurement of multicomponent material properties, is in general somewhat illusory.

There are two basic ways that have been used to try to overcome the limited number of atoms and molecules that can be handled by atomistic modelling techniques when applied to multicomponent high-entropy materials with billions and trillions of different local atomic configurations.[95–97] These are the *coherent potential approximation* (*CPA*) and the *special quasi-random structure* (*SQS*) methods. Both approaches have had limited though far from complete success, and neither is really *ab initio* since in both cases it is still necessary to know or guess the basic crystal structure in advance.

In the coherent potential approximation (CPA) approach,[95,96] the atoms or molecules of the multicomponent high-entropy material are initially all replaced by an *effective medium*

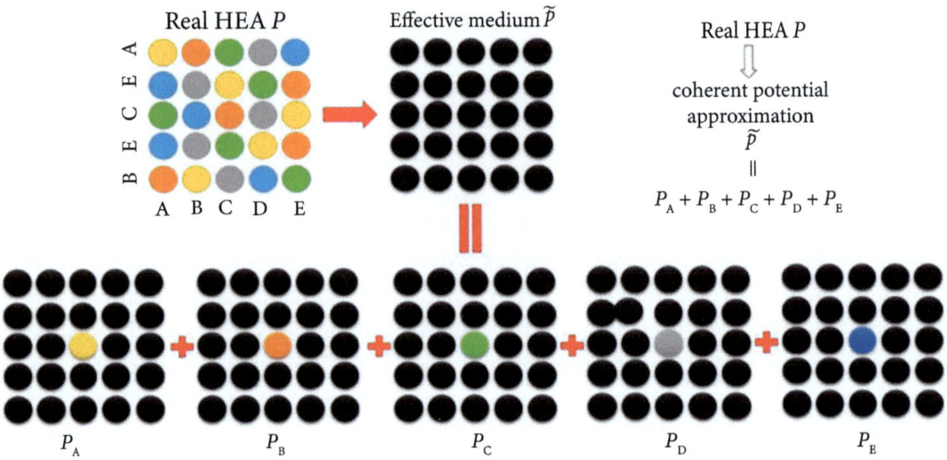

Figure 5.6 Schematic of coherent potential approximation (CPA) in an equiatomic 5-component high-entropy alloy ABCDE on a square two-dimensional lattice, with the real potential P approximated first by all atoms having the same potential $\tilde{P}$, and then by individual atoms having their own potentials P_A, P_B, P_C, P_D and P_E as single-atom solutes embedded in the effective medium potential $\tilde{P}$.

Reproduced from Tian. (2017). A review of solid-solution models of high-entropy alloys based on *ab initio* calculations. *Frontiers in Materials 4*: Article 36.

with an average *coherent potential*, with each of the component atoms then treated one at a time as single atoms embedded in the surrounding coherent (and constant) potential,[xii] as shown schematically in Figure 5.6. In other words, the multicomponent high-entropy material is first treated as if every atom on the pre-selected crystal structure is identical, with a set of properties given by a weighted average of the individual properties of all the component atoms in the material, and then each different type of individual component atom is treated as a (very) dilute-solution single atom in the mean field of all the other averaged atoms. The CPA method is computationally efficient (fairly obviously) since all atoms are treated (mostly) the same; but it is not physically very accurate, since this assumption only works for properties that depend rather simply on the average behaviour of the component atoms in the material, such as the mean density and crystal lattice parameter. More complex physical behaviours and properties such as vacancy hopping or dislocation slip depend explicitly on the local atomic configurations and are not, therefore, well represented by CPA methods.

Another commonly used technique[95,97] is to develop a *special quasi-random structure* (*SQS*), which is a supercell representing the structure of the multicomponent material with a (relatively smallish) number of unit cells of the pre-selected crystal structure and a random patterning of the different component atoms across the crystal lattice sites within these few unit cells, as shown in Figure 5.7, the exact pattern of atoms selected to have a *radial distribution function* (*RDF*) as near as possible similar to the measured radial distribution function of the material itself. The number of atoms that can be treated in this way is usually

[xii] A common method is to use *muffin-tin* potentials for the embedded single atoms, with a spherically symmetrical potential field within a muffin-tin shaped region surrounding the single embedded atom and a constant mean-field potential for the average of all the other atoms outside the muffin-tin region surrounding the single embedded atom.

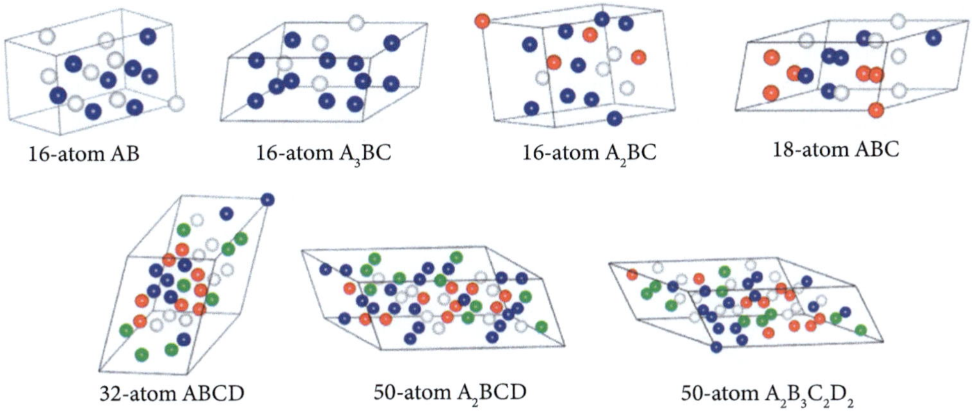

Figure 5.7 Monte Carlo-optimised special quasi-random structure (MC-SQS) supercells used to model fcc binary, ternary and quaternary alloys in the 4-component CrFeCoNi alloy system. Reproduced from Yang and Zhong. (2021). *Ab initio* modeling of fcc FeCoCrNi high-entropy alloys with full composition range. *Journal of Phase Equilibria and Diffusion 42*: 656–72.

on the order of several tens or maybe one or two hundred at most, and even then, it is not usually possible to consider rigorously all possible random atomic structures (a *general-SQS* or *gen-SQS*) to get the optimum radial distribution function. As an alternative, therefore, a starting random atomic structure is selected and then improved gradually using a limited number of Monte Carlo atom swaps (an *MC-SQS*). The problems of the SQS method are fairly obvious. Firstly, as just mentioned, it is often not feasible rigorously to optimise the radial distribution function of the SQS supercell; secondly, the radial distribution function is in any case a relatively crude measure of atomic structure, which only describes the first and, sometimes, second near-neighbour environment of atoms in the material, and does so only in an averaged, smeared-out way; and thirdly (and partly for these reasons) an SQS with a radial distribution function similar to that measured for a given multicomponent material is no real guarantee of a faithful representation of the billions and trillions of different local atomic structures and their associated complex (in many cases) atomic behaviour.

There has been quite a large number of atomistic modelling studies of various equiatomic and non-equiatomic multicomponent materials, including single-phase fcc Cantor alloys, single phase bcc Senkov alloys, some multiphase (mostly fcc + bcc) alloys and one or two multicomponent compounds.[95–116] One of the first detailed investigations was by Ma et al.,[98] studying the original equiatomic single-phase fcc Cantor alloy composition CrMnFeCoNi, using exact muffin-tin orbital (EMTO) potentials for each of the individual component solute atoms embedded in an effective medium of all the other atoms within the coherent potential approximation (CPA) in a single unit cell. Detailed self-consistent calculations were undertaken for fcc, hcp and bcc CrMnFeCoNi crystal structures in non-magnetic, paramagnetic and ferromagnetic configurations, corresponding respectively to having the electron spins either ignored, distributed randomly, or aligned. As Ma et al.[98] comment, the main difficulty with using the CPA approach is that only a small number of atoms are used for the model with no consideration of local atomic relaxations, so the wide range of local atomic environments that are present within the alloy structure, as discussed in detail in the last chapter, are effectively ignored. Not surprisingly, therefore, while Ma et al.'s calculations correctly predict a paramagnetic fcc structure above a temperature of about 500 K, they incorrectly predict a transformation to paramagnetic hcp at lower temperatures and ferromagnetic hcp close to absolute zero, as shown by the free energies of formation in Figure 5.8. The calculations give reasonable predictions for the temperature variations of thermal expansion coefficient and bulk modulus, but not their absolute values, as shown in Figure 5.9.

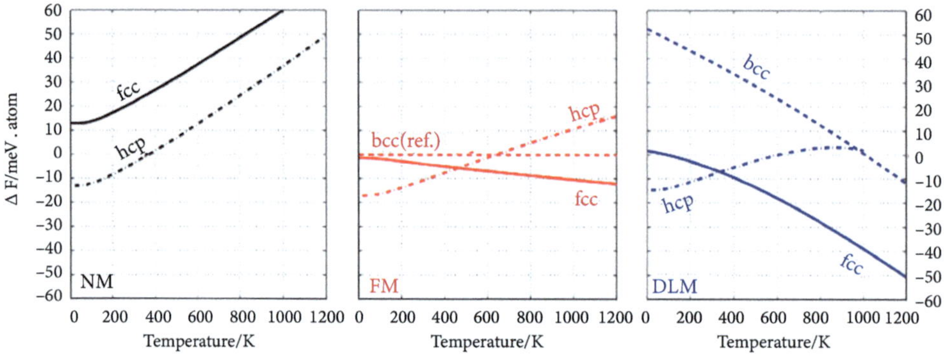

Figure 5.8 EMTO-CPA atomistic modelling of free energy of formation ΔF versus temperature T for non-magnetic (NM), ferromagnetic (FM) and paramagnetic (DLM) configurations of the 5-component equiatomic Cantor alloy CrMnFeCoNi with fcc, bcc and hcp crystal structures.
Reproduced from Ma et al. (2015). *Ab initio* thermodynamics of the CoCrFeMnNi high entropy alloy: Importance of entropy contributions beyond the configurational one. *Acta Materialia 100*: 90–7.

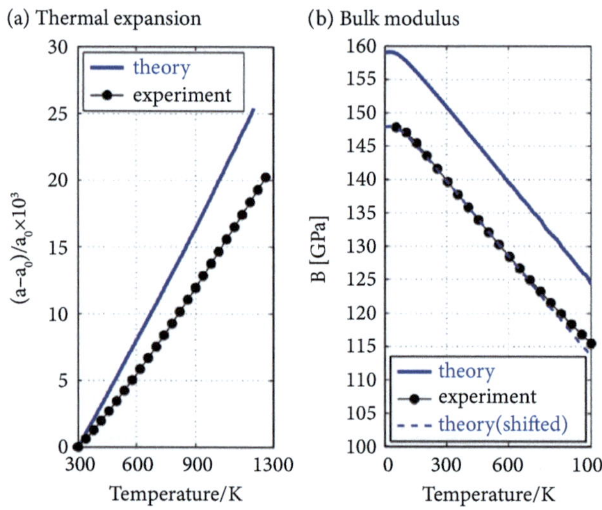

Figure 5.9 EMTO-CPA atomistic modelling of thermal expansion coefficient $(a - a_o)/a_o$ and bulk modulus B versus temperature T for the 5-component equiatomic single-phase paramagnetic fcc Cantor alloy CrMnFeCoNi, compared with experimental values.
Reproduced from Ma et al. (2015). *Ab initio* thermodynamics of the CoCrFeMnNi high entropy alloy: Importance of entropy contributions beyond the configurational one. *Acta Materialia 100*: 90–7.

Sun et al.[99] used similar atomistic modelling techniques to investigate the effect of composition on the relative stability of paramagnetic single-phase fcc and bcc crystal structures (ignoring the possibility of hcp) in modified Cantor alloys with additions of aluminium, i.e. 6-component $Al_xCrMnFeCoNi$ alloys in the range $x = 0–5$. They again used exact muffin-tin orbital (EMTO) potentials for each of the individual component solute atoms embedded in an effective medium of all the other atoms within the coherent potential approximation, and the effect of temperature was included by calculating configurational and excess magnetic entropy contributions (though the latter were small). Overall, Sun et al.'s[99] atomistic modelling calculations predicted, as observed experimentally, that the stable alloy structure (i.e. the alloy structure with the lowest Gibbs free energy) changes with increasing aluminium content from single-phase fcc to two-phase fcc + bcc and then to single-phase bcc. Figure 5.10 shows Sun et al.'s predicted phase boundaries[99] compared with He et al.'s experimental results[100] for the same range of alloys, and also with similar atomistic modelling predictions[101] for 5-component $Al_xCrFeCoNi$ alloys (i.e. without Mn). Predicted and measured fcc solvus compositions for the 6-component alloys are in good agreement, in both cases roughly constant with temperature at an aluminium content of approximately $x = 0.5$; but predicted and measured bcc solvus lines for the 6-component alloys are not in such good agreement, the predicted bcc solvus decreasing with increasing temperature from approximately $x = 1.6$ at 0 K to $x = 1$ at 1,000 K (i.e. with the fcc + bcc region narrowing), whereas the measured bcc solvus was found to remain constant at about $x = 1$ (and, therefore, the width of the fcc + bcc region also remained approximately constant).

Yang and Zhong[102] used density functional theory (DFT) calculations obtained with the Vienna *ab initio* simulation software package (VASP) using a generalised gradient approximation (GGA) for the exchange-correlation function to investigate a wide range of different compositions of multicomponent 4-component single-phase fcc Cantor alloys across the whole of the quaternary CrFeCoNi system, using a Monte Carlo optimised SQS with 16-atom AB and A_3B supercells for the binary alloy compositions, 16-atom A_2BC and 18-atom ABC supercells for the ternary alloy compositions, and 32-atom ABCD and 50-atom A_2BCD and $A_3B_3C_2D_2$ supercells for the quaternary alloy compositions, as shown in Figure 5.7. Simple inspection of Figure 5.7 shows how far removed the MC-SQS supercells are, irrespective of their similar RDFs, from representing accurately the enormous number and complexity of local environments that are actually present in the various multicomponent alloys. Yang and Zhong[102] studied the effect of composition on physical properties such as the mean fcc lattice parameter, local lattice distortion and heat of mixing, as well as mechanical properties such as the bulk and shear moduli. As an example of their results, Figure 5.11 shows colour maps of the variation of predicted fcc enthalpy of formation versus composition across a series of ternary and quaternary sections through the phase diagram. As discussed in a previous sub-section of this chapter, the stability of the fcc phase is expected to correlate with a relatively low heat of mixing and, therefore, enthalpy of formation. As can be seen in Figure 5.11, the quaternary alloys are predicted, therefore, in rather general agreement with experimental observations, to be fcc over most of the quaternary phase diagram, except for alloys close to the chromium-rich corner that have the highest predicted fcc enthalpies of formation and correspondingly exhibit bcc rather than fcc crystal structures. Figure 5.11 exemplifies the extensive range of predicted properties that can in principle be achieved by such computer-based atomistic modelling methods. Figure 5.12 shows a comparison of the predicted and experimentally measured values of fcc lattice parameter,

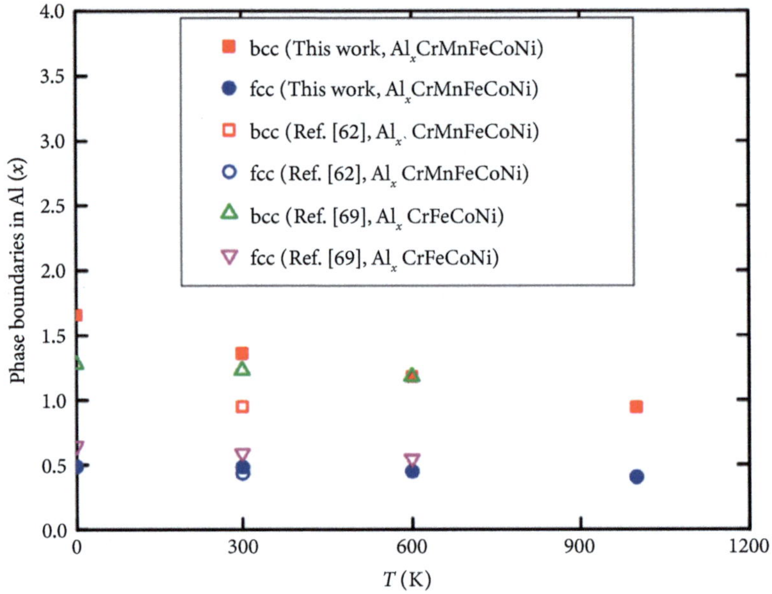

Figure 5.10 EMTO-CPA atomistic modelling of fcc and bcc solvus compositions in 6-component Al_xCrMnFeCoNi alloys compared with He et al.'s measurements[100] (quoted on graph as ref 62) and EMTO-CPA modelling of 5-component Al_xCrFeCoNi alloys[101] (quoted on graph as ref 69).

Data from Sun et al. (2017). Phase selection rule for Al-doped CrMnFeCoNi high-entropy alloys from first principles. *Acta Materialia 140*: 366–74.

enthalpy of formation, and bulk modulus, for a series of selected compositions within the quaternary system. As can be seen by comparing Figures 5.9 and 5.12, Yang and Zhong's[102] atomistic modelling calculations are similar to those of Ma et al.,[98] in the sense that they again predict trend behaviours reasonably well, but do not in general predict accurately the absolute values of the various physical and mechanical properties.

Yang et al.[103] have also made similar density functional theory (DFT) calculations, again with VASP and with similar MC-SQS supercells, to explore the relative formation enthalpies of fcc and bcc crystal structures for a wide variety of alloys with different compositions spread across the 5-component AlCrFeCoNi system. Some of their calculated results for the difference between the enthalpies of formation of fcc and bcc crystal structures $\Delta H_f^{fcc} - \Delta H_f^{bcc}$ are shown in Figure 5.13. Experimentally,[104] the structure of alloys of this type is found to depend on the valence electron concentration (VEC), and is single-phase fcc for VEC > 8, two-phase fcc + bcc for 6.67 < VEC < 8, and single-phase bcc for VEC < 6.67. As can be seen in Figure 5.13, atomistic modelling of a large number of different alloys replicates this observed behaviour to some extent, but only in the most generalised way, i.e. with a tendency to positive and negative values of $\Delta H_f^{fcc} - \Delta H_f^{bcc}$ with respectively lower and higher values of VEC. But there are also very sharp high and low variations in predicted enthalpies of formation at very similar VEC values, above and below the line corresponding to $\Delta H_f^{fcc} - \Delta H_f^{bcc} = 0$ or $\Delta H_f^{fcc} = \Delta H_f^{bcc}$, indicating that the calculations are not a good guide to the structure of individual alloys.

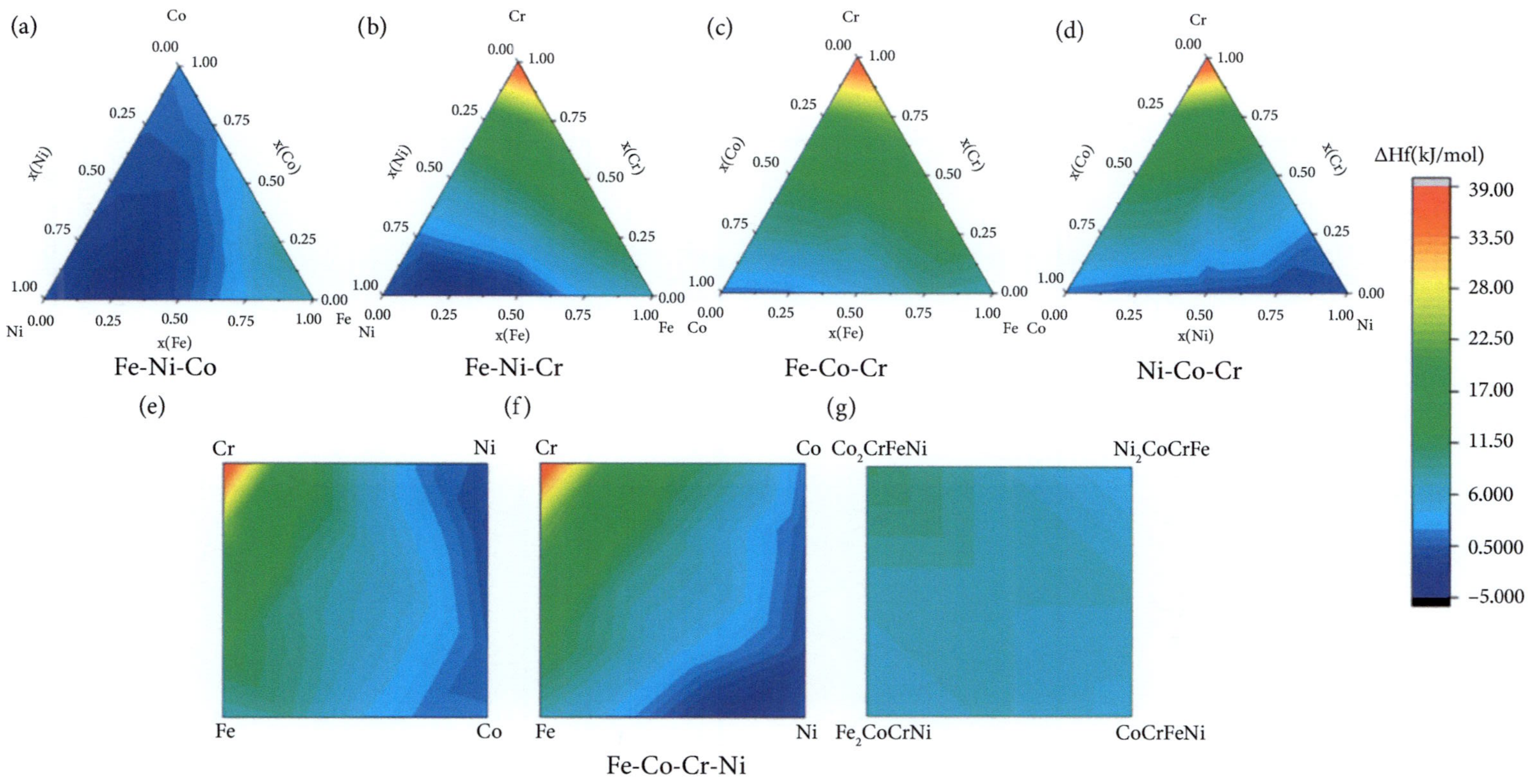

Figure 5.11 Colour maps from MC-SQS atomistic modelling of the fcc heat of formation ΔH_f versus composition on a variety of ternary and quaternary sections in the 4-component CrFeCoNi alloy system.

Reproduced from Yang and Zhong. (2021). *Ab initio* modeling of fcc FeCoCrNi high-entropy alloys with full composition range. *Journal of Phase Equilibria and Diffusion 42*: 656–72.

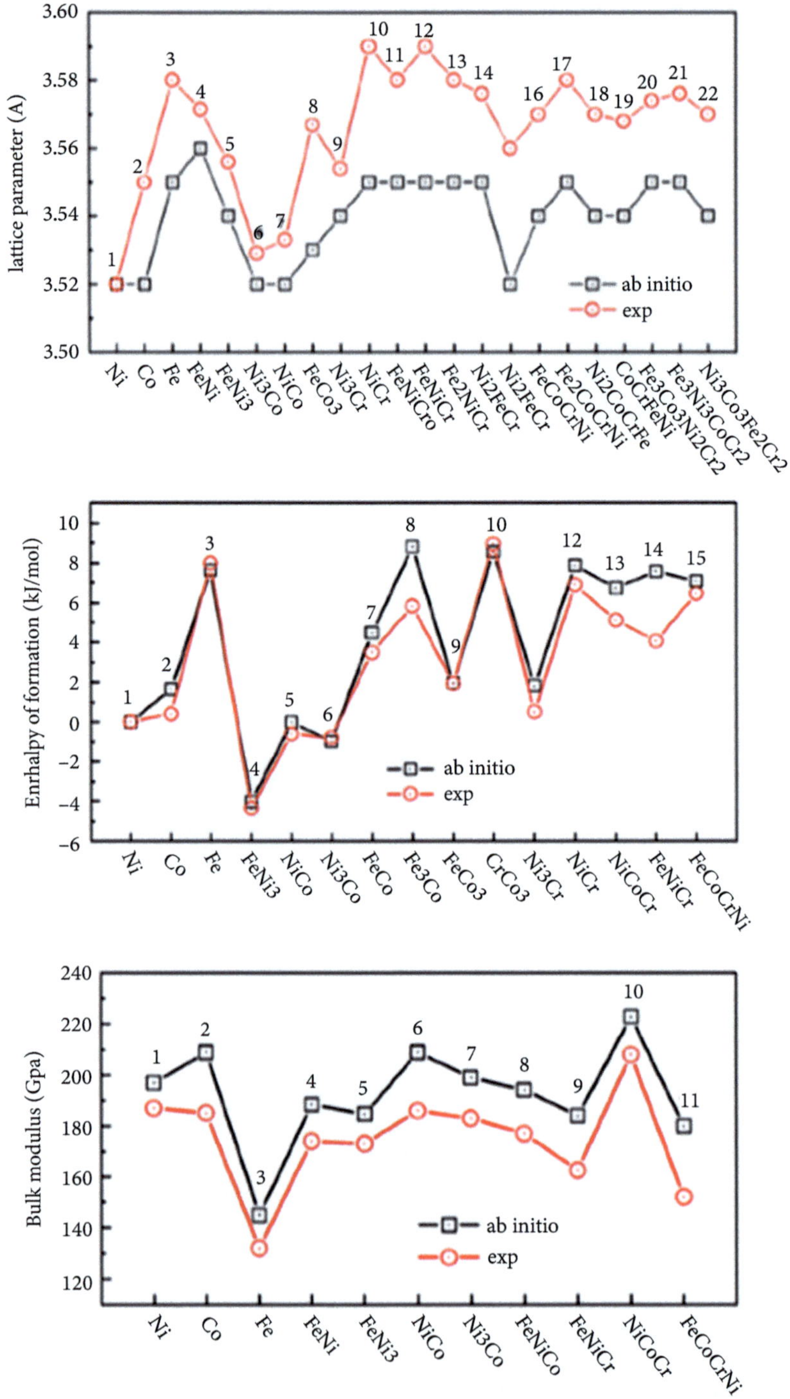

Figure 5.12 MC-SQS atomistic modelling compared with measurements of lattice parameters, heats of formation and bulk moduli for alloys in the four-component CrFeCoNi system.

Reproduced from Yang and Zhong. (2021). *Ab initio* modeling of fcc FeCoCrNi high-entropy alloys with full composition range. *Journal of Phase Equilibria and Diffusion 42*: 656–72.

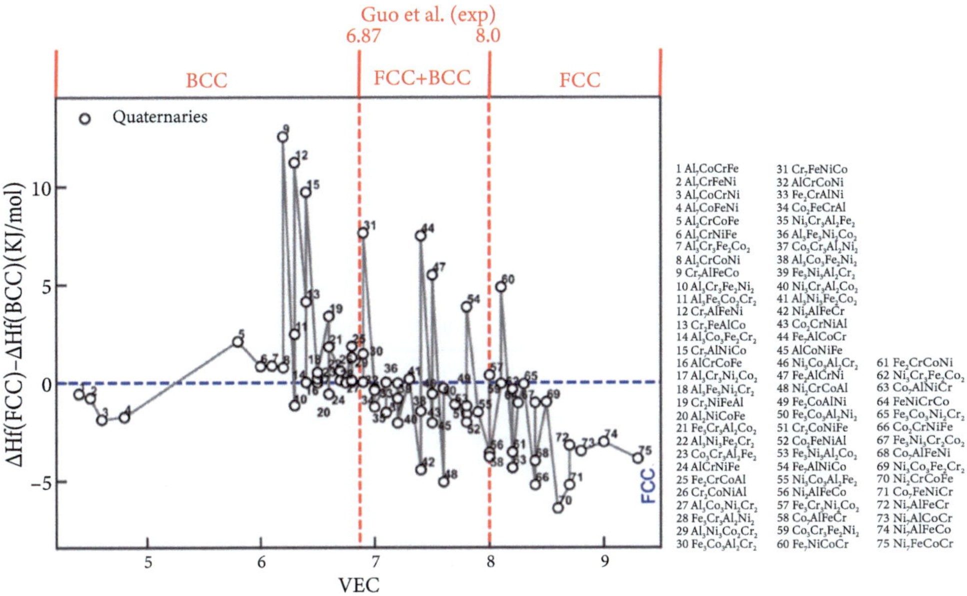

Figure 5.13 MC-SQS atomistic modelling of the difference between fcc and bcc heats of formation $\Delta H_f^{fcc} - \Delta H_f^{bcc}$ versus valence electron concentration (VEC) for quaternary alloys in the AlCrFeCoNi system, compared with measured fcc and bcc ranges.

From Guo et al. (2011). Effect of valence electron concentration on stability of fcc or bcc phase in high entropy alloys. *Journal of Applied Physics 109*, Article 103505. Data from Yang and Zhong. (2021). *Ab initio* modeling of fcc FeCoCrNi high-entropy alloys with full composition range. *Journal of Phase Equilibria and Diffusion 42*: 656–72.

Tian et al.[95,105] used embedded muffin-tin orbitals for individual component atoms as solutes within the effective-medium coherent potential approximation (EMTO-CPA), similar to Ma et al.[98] and Sun et al.,[99] to try to predict the structures and properties of a number of multicomponent high-entropy materials, including the effect of vanadium concentration in 5-component $TiV_xZrNbMo$ Senkov alloys. Tian et al.'s results[95,105] predicted correctly a single-phase bcc structure for the Senkov alloys, with calculated bcc heats of formation lower than corresponding calculated fcc and hcp values, as shown in Table 5.3. And Figure 5.14 shows calculated electronic density of states curves for some of the different Senkov alloys, again exemplifying the extensive range of predicted properties that can in principle be achieved by such computer-based atomistic modelling methods. Unfortunately, however, the same methods again predicted correctly only trends rather than accurate absolute values of properties such as lattice parameter, elastic constants and bulk and shear moduli. Gao et al.[97] used density functional theory (DFT) calculations with VASP and MC-SQS supercells, similar to Yang et al.[102,103] also to investigate the structure and properties of a variety of multicomponent high-entropy alloys. Gao et al.'s results[97] again predicted correctly the structure of a range of different alloys and also gave reasonable estimates of their lattice parameters, as shown by the comparison of calculated and measured values in Table 5.4, though predictions varied considerably with different selected MC-SQS supercells.

Table 5.3 EMTO-CPA atomistic modelling of fcc, bcc and hcp heats of formation in $TiV_xZrNbMo$ alloys (after Tian et al.[95,105]).

alloy	ΔH_f^{fcc} (kJ/mol)	ΔH_f^{bcc} (kJ/mol)	ΔH_f^{hcp} (kJ/mol)
$TiV_{0.25}ZrNbMo$	−13.0	9.4	11.1
$TiV_{0.50}ZrNbMo$	−1.0	21.7	23.4
$TiV_{0.75}ZrNbMo$	−2.1	20.6	22.5
$TiV_{1.00}ZrNbMo$	5.7	28.1	29.6
$TiV_{1.25}ZrNbMo$	12.7	35.5	37.5
$TiV_{1.50}ZrNbMo$	23.4	46.3	48.4

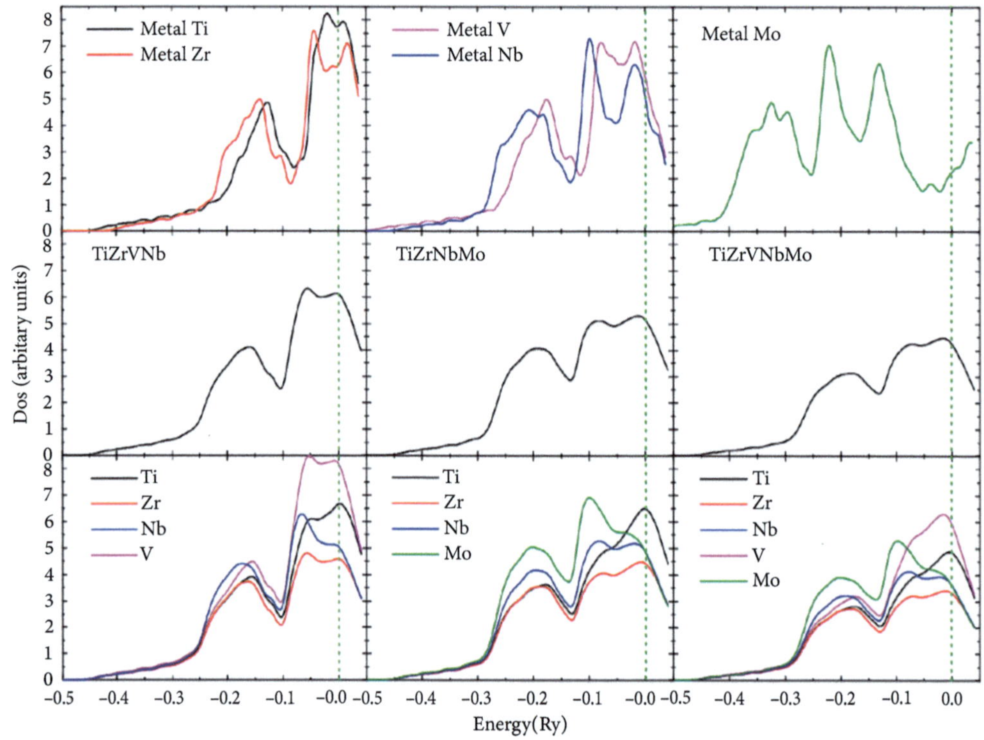

Figure 5.14 EMTO-CPA atomistic modelling of electronic density of states (DOS) curves for Ti, V, Zr, Nb and Mo pure metals (top row), and TiVZrNb, TiZrNbMo and TiVZrNbMo multicomponent single-phase bcc Senkov alloys: total DOS in second row and partial DOS for each component in third row.

Data from Tian. (2017). A review of solid-solution models of high-entropy alloys based on *ab initio* calculations. *Frontiers in Materials 4*: Article 36.

There have been a number of other atomistic modelling investigations of multicomponent high-entropy alloys[106–113] and compounds,[114–116] many of them discussed in the reviews by Tian,[95] Tian et al.[96] and Gao et al.,[97] all showing that valuable information can be obtained by atomistic modelling of multicomponent materials, but that the results are somewhat variable and not in general sufficiently accurate for reliable *ab initio* predictions of their structure and

Table 5.4 MC-SQS atomistic modelling of lattice parameters of fcc Cantor and bcc Senkov alloys, compared with experimental measurements (after Gao et al.[97]).

	alloy	predicted lattice parameter a_o (nm)	measured lattice parameter a_o (nm)
fcc	CrFeCoNi	0.3548	0.3575
	CrMnFeCoNi	0.3559	0.3597
bcc	NbMoTaW	0.3237	0.3213
	VNbMoTaW	0.3188	0.3183
	TiZrNbHfTa	0.3411	0.3404

properties. As with computer-based Calphad thermodynamic modelling methods, we can conclude that computer-based atomistic modelling methods can provide important guidance and support for investigations of multicomponent high-entropy materials, but that they do not give guaranteed certainty in their predictions and do not, therefore, replace the need for substantial experimental programmes of exploration.

Machine learning

Machine-learning algorithms

As we have seen in the last two sub-sections, methods based on fundamental physical theories of *thermodynamics*, *quantum mechanics* and *classical mechanics* have been used fairly extensively to try to explore the complexity of multicomponent phase space and thus predict the structure and properties of multicomponent high-entropy materials. These methods have had some, but only limited, success. Much of the difficulty of making progress lies in the sheer vastness of multicomponent phase space, the extent and complexity of multicomponent atomic and molecular structures, and our starting position of deep (though improving) theoretical and experimental ignorance. Recent work[117–119] has, therefore, begun to extend into using the modern fundamental mathematical techniques of *data science*, *machine learning (ML)*, and *artificial intelligence (AI)*, a group of essentially *statistical* methods designed to identify underlying classification and organisational patterns in large and complex data sets obtained in situations of substantial theoretical uncertainty and randomness.[120–122] These methods seem at first sight to be ideally suited to exploring multicomponent phase space and finding new multicomponent materials with desirable structures and properties. We shall see, however, that although machine learning methods can indeed, as with thermodynamic and atomistic modelling, be very helpful and insightful, they are still limited by the fact that they essentially work by discovering patterns in existing data, and are, therefore, successful at *interpolating* to predict new information within a known range of materials, but not really successful at *extrapolating* to predict new information in previously unexplored regions of multicomponent phase space.[117–122]

Machine learning (ML)[117–122] operates on a *data set* of *objects*, in our case different materials $\{x_{ij}\}$, each with a set of known *features*, in our case different material properties, with the label $i = 1$–N representing the set of N different materials and the label $j = 1$–M representing the set of M different known properties for each of these materials. The objective of machine learning is to develop a function f that predicts a set of *targets* or, in our case, additional new properties $\{y_{ik}\}$ from the known properties $\{x_{ij}\}$, with the label $k = 1$–D representing the set of D different new properties to be predicted. In *supervised learning* we already know the values of the target properties $\{y_{ik}\}$, and in *unsupervised learning* we don't. In either case, we develop a *statistical model* in the form of a function or algorithm of the form:

$$\{\dot{y}_{ik}\} = f(\{x_{ij}\})$$

where the dot notation $\{\dot{y}_{ik}\}$ indicates here that the values of the target properties are those predicted by operation of the function f on the known properties $\{x_{ij}\}$, which are not in general exactly the same as the undotted actual (known or unknown) target properties $\{y_{ik}\}$. In general, f depends on a set of initially unknown parameters $\{p_l\}$, with the label $l = 1$–E representing the set of E unknown parameters. Alongside the prediction function f, we also develop an optimisation function $g(\{\dot{y}_{ik} - y_{ik}\})$ (or just $g(\{\dot{y}_{ik}\})$ for the case of unsupervised learning) to assess how good f is at predicting the target properties. The overall machine-learning process consists then of the following steps:[117–122] apply a proposed prediction function f with initial assumed parameters $\{p_l\}$ to the known properties $\{x_{ij}\}$ so as to predict a set of target new properties $\{\dot{y}_{ik}\}$; evaluate the optimisation function g; use g to help adjust the parameters $\{p_l\}$ so as to modify and improve the prediction function f; re-apply the modified prediction function f with modified parameters $\{p_l\}$ to the known properties $\{x_{ij}\}$ so as to predict a modified set of target new properties $\{\dot{y}_{ik}\}$; repeat this process iteratively until the optimisation function g has been minimised and the predicted target properties are within some desired range of the actual target properties. The final resulting prediction and optimisation functions f and g are the final *machine-learning algorithm*, which can then be used in a *feed-forward* way to predict target properties for other proposed materials, or in a *feed-backward* way to predict possible new materials with desirable target properties.[117–122]

The initial data set $\{x_{ij}\}$ is often divided into a *training subset* $\{x'_{ij}\}$, typically at least 75–80% of the full data set, which is used to develop the final machine-learning algorithm with its optimised function f and optimised parameters $\{p_l\}$ by the iterative process described above, and a *testing subset* $\{x''_{ij}\}$, typically no more than 20–25% of the full data set, which is then used to check and score the quality of the final learning algorithm with its optimised function f and optimised parameters $\{p_l\}$. There is usually an initial careful *data cleansing* process to remove obvious outliers and improve consistency in the initial data set; and there is also usually an initial careful *feature selection* process (in our case material property selection process) to identify the key predictive properties and remove redundant ones. When there is a limited data set, the statistics of the learning process are often improved by *bootstrapping* and *cross-validation*, dividing the data randomly into n equal subsets, and progressively using each as the testing subset to optimise the selection of property features $\{x_{ij}\}$, learning algorithm f and g, and parameters $\{p_l\}$. The known and target material properties $\{x_{ij}\}$ and $\{y_{ik}\}$ can include physical properties with ranges of continuous numerical values, such as composition, processing conditions, melting point, density, strength, conductivity, etc., as well as non-continuous and/or non-numeric data, such as crystal structure, type of microstructure, number of phases, etc. When there are only continuous numerical target properties, the learning problem is essentially a multiple *regression* to determine the functional dependence of $\{\boldsymbol{y}_{ik}\}$ on $\{\boldsymbol{x}_{ij}\}$; and when there are only non-continuous and/or non-numerical target properties, the learning problem is essentially a multiple *classification* of the materials into different y_k–categories.

A wide range of different types of machine-learning algorithm has been developed,[117–122] with different prediction and optimisation functions *f* and *g*, and different associated advantages and disadvantages that are more or less useful depending on the type of problem to be solved, i.e. depending on factors such as the size and quality of the data set, the number and complexity of the known and target properties, the balance desired between precision and simplicity of the final learning algorithm, the speed of computer optimisation, and so on. Some of the different types of learning algorithm that have been developed are as follows:[117–122]

1. *Multiple linear regression* (*MLR*): In a multiple linear regression or MLR learning algorithm, as shown schematically in Figure 5.15, the function *f* predicts each of the target properties $\dot{\boldsymbol{y}}_{ik}$ as a linear sum of terms proportional to each of the known properties $\boldsymbol{x}_{ij}$:

$$f:\ \dot{\boldsymbol{y}}_{ik} = \sum_j w_{ijk}\boldsymbol{x}_{ij} + b_{ijk}$$

   There are $N \times D$ linear sums of this type, one for each target property *k* of each material *i*, and there are $N \times M \times D$ regression coefficients or *weights* w_{ijk} and constants or *biases* b_{ijk}, which together are the $E = 2(N \times M \times D)$ parameters $\{p_l\}$ of the prediction function *f* to be optimised by the learning algorithm. More complex quadratic, cubic and other non-linear regression equations can also be used. A commonly used

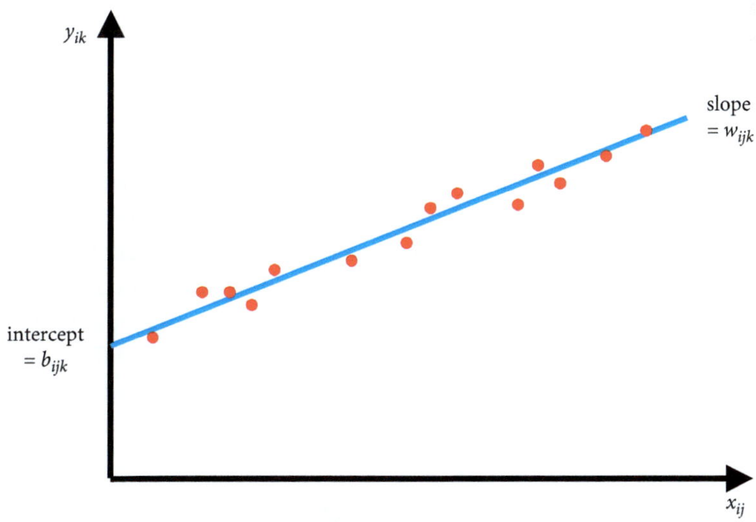

Figure 5.15 Multiple linear regression (MLR), minimising deviation of output target properties $\boldsymbol{y}_{ik}$ (red) versus input known properties $\boldsymbol{x}_{ij}$ with best fit lines (blue).

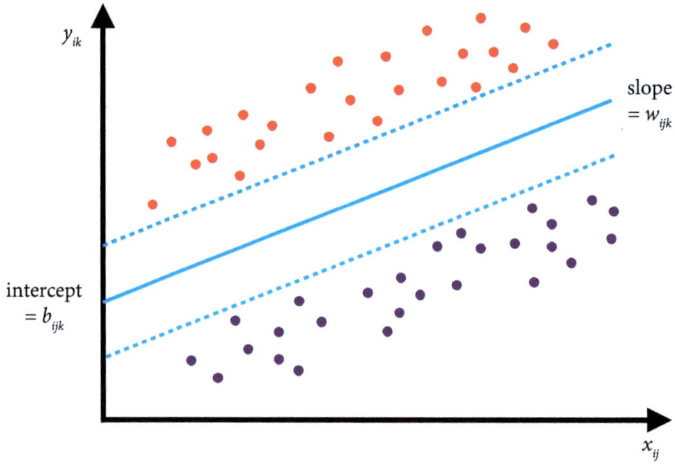

Figure 5.16 Linear support vector machine (SVM), maximising separation of output target classes $\boldsymbol{y}_{ij}$ (red and green) versus input known properties $\boldsymbol{x}_{ij}$ with best fit lines (blue).

optimisation function for regression problems is the least squares sum of differences between predicted and actual target properties:

$$g: \min \sum_{i,k} (\dot{\boldsymbol{y}}_{ik} - \boldsymbol{y}_{ik})^2$$

though alternative optimisation functions can also be used.

2. *Support vector machine* (*SVM*): In a linear support vector machine or SVM learning algorithm, as shown schematically in Figure 5.16, the function f classifies the target properties $\dot{\boldsymbol{y}}_{ik}$ into different sets according to a similar linear sum of terms proportional to each of the known properties $\boldsymbol{x}_{ij}$:

$$f: \dot{\boldsymbol{y}}_{ik} = \text{sign} \sum_{j} w_{ijk}\boldsymbol{x}_{ij} + b_{ijk}$$

In other words, each target property is assigned a value ±1 corresponding to having or not having the property, and there are again $N \times M \times D$ weights w_{ijk} and biases b_{ijk}, which are the $E = 2(N \times M \times D)$ parameters $\{p_l\}$ of the prediction function f to be optimised by the learning algorithm. Once again, more complex quadratic, cubic, and other non-linear equations can also be used. A commonly used optimisation function for classification problems is the maximum margin between the different classes of the target properties, as shown in Figure 5.16, which can be achieved by minimising the Euclidean norm of the linear coefficients w_{ijk}:

$$g: \min \sum_{i,j,k} {w_{ijk}}^2$$

though alternative optimisation functions can also be used.

3. *Decision tree learning*: In a decision tree learning algorithm, as shown schematically in Figure 5.17, the function f describes a structured set of decision steps with a tree like structure, starting with the known properties $\boldsymbol{x}_{ij}$ and narrowing down to a final decision about the predicted values of the target properties $\dot{\boldsymbol{y}}_{ik}$. This corresponds to a nested set of functions:

$$f : \dot{\boldsymbol{y}}_{ik} = f'''\left(f''\left(f'\left(\boldsymbol{x}_{ij}\right)\right)\right)$$

where f', f'', f''', etc., are the decision steps at each point of the tree. The decision process can be represented as a branching graph with a series of nodes, one of the known properties at each node, and the left or right branch followed according to the value of the known property, as shown in Figure 5.17. Least-squares and maximum-margin optimisation functions g can again be used for regression or classification problems respectively. There is considerable flexibility in the choice of the structure of a decision tree graph, and a *random forest* (*RF*) learning algorithm takes, therefore, the average of a set of different decision tree graphs used on randomly determined subsets of the data set.
4. *Neural network* (*NN*): In a neural network or NN learning algorithm, as shown schematically in Figure 5.18, the function f describes a structured set of *nodes* or *neurons* that combine the known properties $\boldsymbol{x}_{ij}$ in a series of so-called *hidden layers* of intermediate composite properties $\boldsymbol{x}'_{ij}$, $\boldsymbol{x}''_{ij}$ etc, until finally predicting the target properties $\dot{\boldsymbol{y}}_{ik}$. This corresponds again to a nested set of functions:

$$f : \dot{\boldsymbol{y}}_{ik} = f'''\left(f''\left(f'\left(\boldsymbol{x}_{ij}\right)\right)\right)$$

where f', f'', f''', etc., are the combination steps at each layer of the network. There are often no more than one or two hidden layers, as shown in Figure 5.18. Each node takes outputs from the previous layer and combines them to provide an output to nodes in the next layer using, for instance, linear regression or classification functions:

$$f' : \boldsymbol{x}'_{im} = \sum_j w_{ijm}\boldsymbol{x}_{ij} + b_{ijm}$$

$$\text{or } f' : \boldsymbol{x}'_{im} = \text{sign}\sum_j w_{ijm}\boldsymbol{x}_{ij} + b_{ijm}$$

where m, n, etc., are the number of nodes in each layer, and w_{ijm}, w'_{ijm}, etc. and b_{ijm}, b'_{ijm}, etc. are weights and biases for each node in each layer, though alternative functions can again be used. The weights and biases as well as the number of hidden layers and number of nodes in each layer are all parameters $\{p_l\}$ of the network. Least-squares and maximum-margin optimisation functions g can again be used for regression or classification problems respectively, though alternative optimisation functions can again be used.

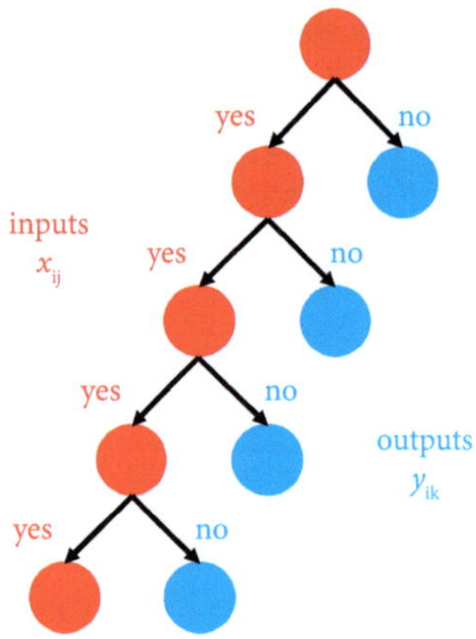

Figure 5.17 Decision tree, with sequential yes/no decision on input known properties $\boldsymbol{x_{ij}}$ leading to output target classes $\boldsymbol{y_{ij}}$ (red and blue).

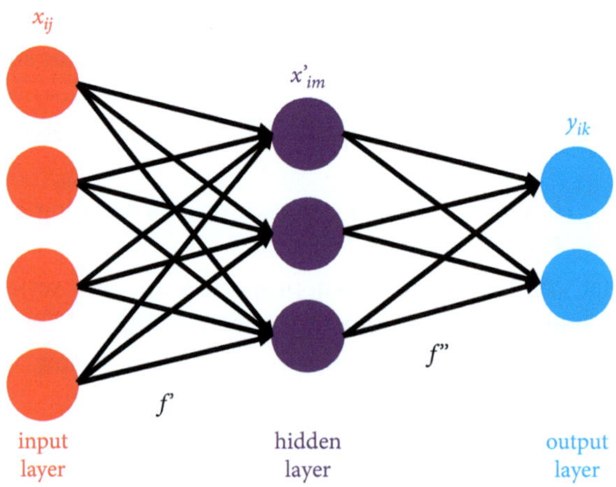

Figure 5.18 Neural network with input layer of known properties $\boldsymbol{x_{ij}}$ (red) determining hidden layer of intermediate composite properties $\boldsymbol{x'_{im}}$ (green) and then output layer of target properties $\boldsymbol{y_{ik}}$.

5. *k-nearest neighbours* (*kNN*): In a k-nearest neighbour or kNN learning algorithm, as shown schematically in Figure 5.19, the function f predicts the target properties $\dot{\boldsymbol{y}}_{\boldsymbol{ik}}$ for any new material i from the average of the known or optimised target properties $\dot{\boldsymbol{y}}_{\boldsymbol{ik}}$ of the k materials with known properties $\boldsymbol{x_{ij}}$ that are nearest neighbours to those of i:

$$f:\ \dot{\boldsymbol{y}}_{\boldsymbol{ik}} = k^{-1} \sum_{kNN} \boldsymbol{y_{ik}}$$

Least-squares and maximum-margin optimisation functions g can again be used for regression or classification problems respectively, though alternative optimisation functions can again be used.

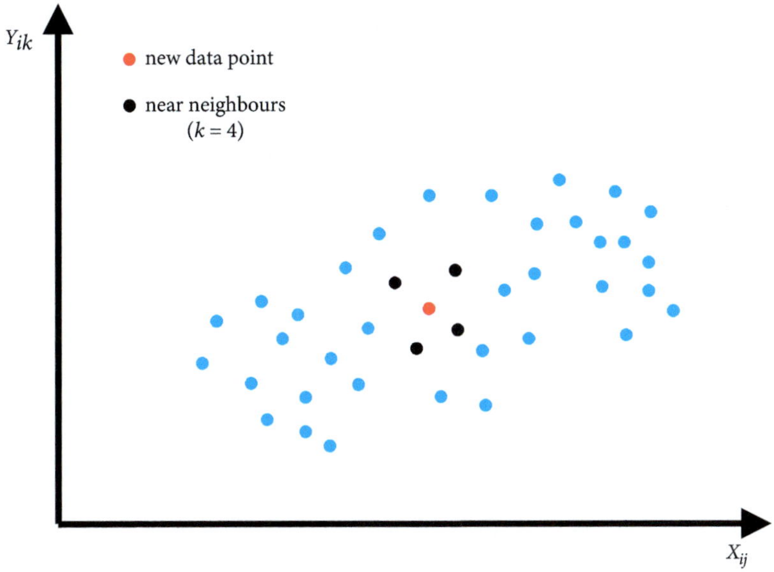

Figure 5.19 *k*-nearest neighbours (*k*NN) with output target properties $\boldsymbol{y_{ik}}$ (red) determined from input known properties $\boldsymbol{x_{ij}}$ (green and black) using only nearest neighbour data (black).

Machine learning for multicomponent materials

Machine-learning software packages are widely available and easy to use. Unsurprisingly therefore, there have now been quite a large number of studies of multicomponent materials using a wide range of different machine-learning methods and algorithms, and these studies have been reviewed by Qiao et al.,[117] Hart et al.[118] and Liu et al.,[119] albeit fairly briefly in each case. Machine learning has been used in two different ways to investigate multicomponent materials:[117–119] 1) to try to predict the occurrence of different phases and crystal structures, essentially providing a statistically enhanced form of the multicomponent Hume-Rothery rules discussed in a previous sub-section; and 2) to try to predict specific desirable multicomponent material properties (so far almost always mechanical properties) directly rather than via their structures. In the rest of this sub-section, we first describe some examples of these two slightly different approaches, and then make some general observations about the advantages and disadvantages of applying machine learning to multicomponent materials.

One of the first applications of machine learning to investigate the structure of multicomponent materials was by Islam et al.[123] They used a data set of 118 multicomponent alloys, each with five known properties, namely the mean valence electron concentration ($\overline{VEC}$), root mean square average Pauling electronegativity ($\Delta\chi$), root mean square average atomic misfit calculated from pure component Goldschmidt atomic radii (δ_{av}), heat of mixing estimated by Miedema's method (ΔH_{mix}) and ideal configurational entropy of mixing (ΔS_{mix}):

$$\overline{VEC} = \sum_i x_i VEC_i$$

$$\Delta\chi = \sqrt{\sum_i x_i(\chi_i - \overline{\chi})^2}$$

$$\delta_{av} = \frac{\sqrt{\sum_i x_i(r_i - \overline{r})^2}}{\overline{r}}$$

$$\Delta H_{mix} = \sum_{ij} 4x_i x_j \Delta H_{ij}$$

$$\Delta S_{mix} = -R\sum_i x_i \ln x_i$$

where x_i, χ_i and r_i are respectively the composition, Pauling electronegativity and Goldschmidt atomic radius of the ith component of each alloy, $\overline{\chi} = \sum_i x_i\chi_i$ and $\overline{r} = \sum_i x_i r_i$ are respectively the mean Pauling electronegativity and Goldschmidt radius of the alloy, and ΔH_{ij} are the heats of mixing of the alloy's binary i–j subsystems. Islam et al.[123] used a neural network with an input layer of five nodes for each of the five known material properties, two hidden layers each of ten self-learning nodes with outputs given by linear combination of inputs from the previous layer, and a final output layer of three yes/no classification nodes for each of three possible target material single-phase structure predictions (amorphous, solid solution and intermetallic compound). The actual structures of the multicomponent alloys were already known as part of the data set, so it was a supervised neural network, learning through a series of iterations to predict the alloy structures with decreasing training loss and increasing accuracy, as measured respectively by the percentage of incorrect and correct classifications into the three different single-phase structures.

Figure 5.20 shows a 5 × 5 matrix of binary cross-correlation plots for Islam et al.'s[123] five known material properties, with corresponding cross-correlation coefficients in Table 5.5. The different material properties showed, as expected, a variety of positive, negative and zero correlations; and the histograms on the matrix diagonal showed, also as expected, that all five properties were relevant to the determination of the material structure. Figure 5.21 shows the gradual decrease in training loss and corresponding increase in prediction accuracy when the neural-network learning algorithm was applied to the full data set with an increasing number of iterations (epochs), reaching 75% after 10^3 iterations and >99% after 10^4 iterations. But, when the optimised neural network learning algorithm was subsequently applied *de novo* via a fourfold cross-validation process to each of four randomly selected quarters of the data set, the accuracy dropped to lower values in the range 76–87%. The final weights w_{ij} of the linear functions at the different nodes of the neural network showed that the known material properties had varying influence on its predictions, in the decreasing order $\overline{VEC} > \Delta\chi > \delta_{av} > \Delta H_{mix} > \Delta S_{mix}$.

There has been a number of other, similar applications of machine learning to try to predict the phases and crystal structures of multicomponent metallic alloys[124–130] and ceramic compounds, mainly monocarbides and diborides,[131–137] using a variety of different material data sets, material property inputs, learning algorithms, prediction and optimisation functions, and testing and assessment procedures.[124–137] These have all used very similar

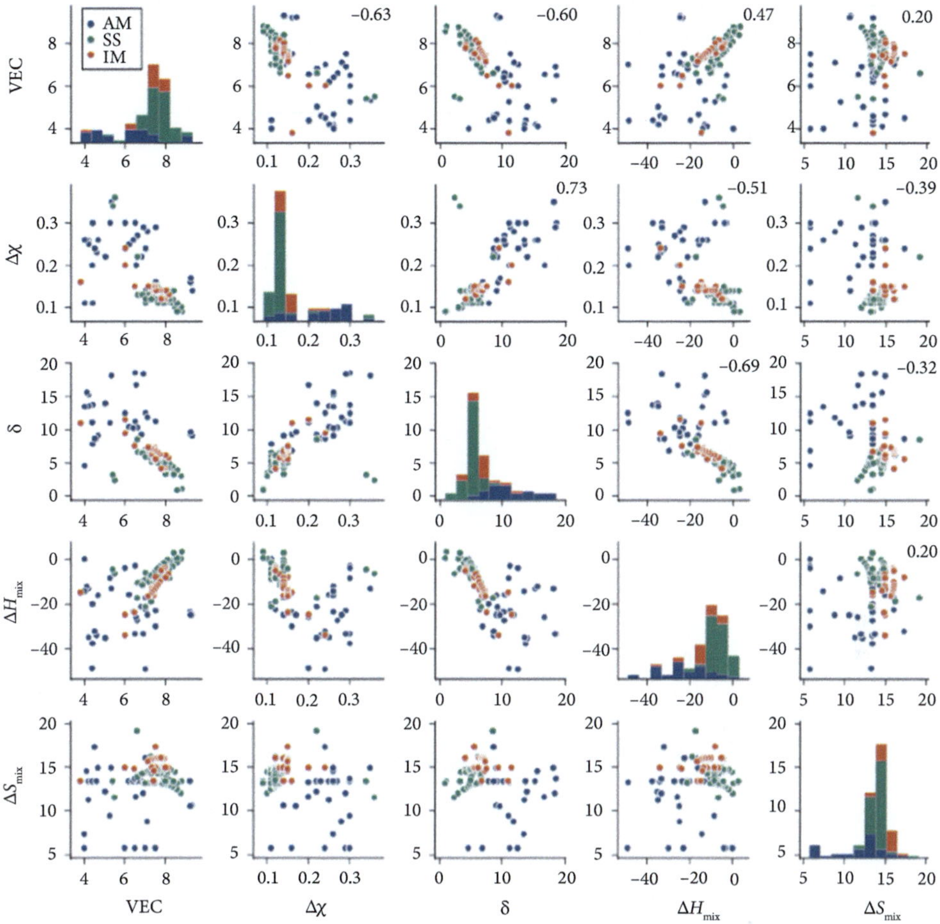

Figure 5.20 Cross-correlation plots for average valency (VEC), electronegativity ($\Delta\chi$), atomic misfit (δ_{av}), and heat (ΔH_{mix}) and entropy (ΔS_{mix}) of mixing for 118 multicomponent alloys.

Data from Islam et al. (2018). Machine learning for phase selection in multi-principal element alloys. *Computational Materials Science 150*: 230–5.

combinations of material properties as inputs, typically alloy composition (x_i) as well as valency (VEC), electronegativity ($\Delta\chi$), atomic misfit (δ_{av}), and heat (ΔH_{mix}) and entropy (ΔS_{mix}) of mixing, but have employed a variety of different learning algorithms such as neural networks, random forest decision trees, support vector machines, and k-near neighbours, to try to classify the alloys into different types of phase (usually amorphous, solid solution or intermetallic compound), crystal structure (usually fcc, bcc or hcp) and/or number of phases (single or multiple).

The different studies all exhibit binary cross-correlation plots for the material properties with somewhat variable correlation coefficients, similar to those shown in Figure 5.20 and Table 5.5 (not surprisingly since properties such as VEC and electronegativity are likely to be at least partially correlated, whereas properties such as heat and entropy

Table 5.5 Cross-correlation coefficients for average valency (VEC), electronegativity ($\Delta\chi$), atomic misfit (δ_{av}), and heat (ΔH_{mix}) and entropy (ΔS_{mix}) of mixing for 118 multicomponent alloys (after Islam et al.[123]).

	VEC	$\Delta\chi$	δ_{av}	ΔH_{mix}	ΔS_{mix}
VEC	1	−0.63	−0.60	0.47	0.20
$\Delta\chi$	−0.63	1	0.73	−0.51	−0.39
δ_{av}	−0.60	0.73	1	−0.69	−0.32
ΔH_{mix}	0.47	−0.51	−0.69	1	0.20
ΔS_{mix}	0.20	−0.39	−0.32	0.20	1

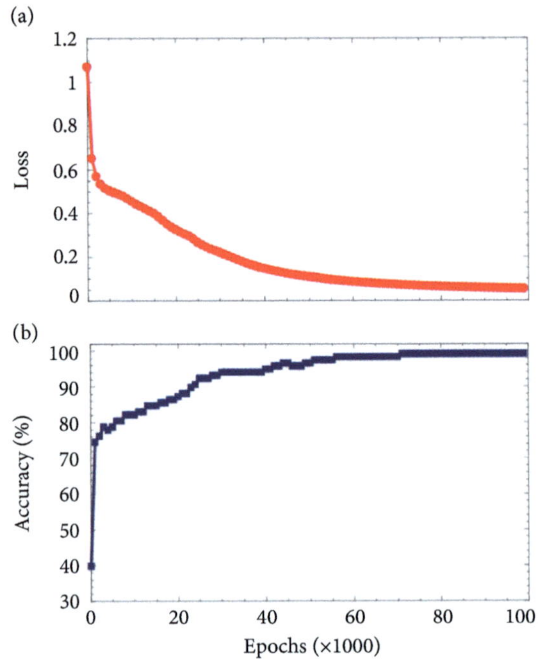

Figure 5.21 Variation of training loss and accuracy with increasing number of iterations (epochs) for a neural network predicting the phases of 118 multicomponent alloys.
Data from Islam et al. (2018). Machine learning for phase selection in multi-principal element alloys. *Computational Materials Science 150*: 230–5.

of mixing are not).[123–137] The different studies also all demonstrate significant effects for all the selected material properties (again not surprisingly, since as discussed in an earlier sub-section these material properties were all identified previously as important parameters in the context of trying to develop multicomponent Hume-Rothery rules to predict multicomponent material structures).[123–137] In many studies, the limited amount of available experimental data for testing and model building is supplemented with predictions from Hume-Rothery parameterisation and/or thermodynamic and atomistic modelling, leading to a certain amount of circularity in the apparent accuracy of the machine-learning predictions.[123–137] There is no agreement between different studies about the optimum machine-learning algorithm for the most accurate prediction of properties, some preferring neural networks, some support vector machines, some random forests, etc.; and there is also no agreement between different studies about the relative importance of the different material properties, some finding VEC most important in predicting properties, some atomic misfit, some electronegativity, some alloy composition, etc.[123–137] The final accuracy of the learning algorithms are all high after repeated iteration with the initial training sets, well above 90%, but accuracy decreases sharply, typically to 50–80%, when the learning algorithms are applied *de novo* to a separate testing set or via cross-validation to a random subset of the main data set.[123–137] This reduction in accuracy is partly caused by the intrinsic nature of machine learning methods, which are good at *interpolation* within a known range (the initial training set), but less good at *extrapolation* into an unknown range (a new testing set or newly randomised subset); and it is partly a reflection of previous work using the same material properties to develop multicomponent Hume-Rothery rules, which as commented on in an earlier sub-section gives reasonable *ex-post facto* categorisation of multicomponent materials but is insufficiently precise to give confident prediction of their structures. More details are given in the reviews by Qiao et al.,[117] Hart et al.[118] and Liu et al.[119]

One of the first applications of machine learning to the considerably more ambitious problem of trying to predict directly the mechanical properties of multicomponent materials was by Chang et al.[138] They used a data set of ninety-one alloys in the 8-component AlCrMnFeCoNiCoMo system, including a range of binary, ternary, quaternary, and 5-, 6- and 7-component alloys, each with twenty-four known properties, namely the eight composition values x_i and corresponding weighted component hardness, density and atomic mass values, H_i, ρ_i and am_i respectively. Like Islam et al.,[123] Chang et al.[138] also used a neural network, but with an input layer of twenty-four nodes for each of the twenty-four known material properties, one hidden layer of three self-learning nodes with outputs given by linear combination of inputs from the previous layer, and a final output layer predicting one target material property of alloy hardness H. Figure 5.22 shows the resulting correlation between final predicted and previously measured known hardness values of the different alloys, which agreed reasonably well for the whole data set, with a correlation coefficient of 0.97 corresponding to a mean hardness error of $\Delta H = 35$ kg/mm^2. The neural network was then used to try to predict *de novo* five high-hardness alloys in the 6-component AlCrMnFeCoNi subsystem for subsequent manufacture and measurement, with some success in finding high-hardness alloys but with considerably less accuracy, as shown by the results in Table 5.6.

A more recent detailed example of an attempt to predict the hardness and, thereby, the yield strength of multicomponent materials is by Wen et al.,[139,140] who used a data set of 155 different quaternary and 5- and 6-component alloys in the AlCrMnFeCoNi system; a total of fifty-four different material-property inputs, including alloy composition (x_i) and crystal structure (fcc, bcc or mixed) as well as component and alloy physical, chemical and mechanical properties such as valency (VEC), electronegativity ($\Delta\chi$), atomic misfit (δ_{av}), heat of formation (H_f), heat of mixing (ΔH_{mix}), entropy of mixing (ΔS_{mix}), bulk, shear and Young's moduli (B, G and E), Poisson's ratio (ν), and cohesive energy (E_c); a total of twelve different learning algorithms including neural networks, support vector machines, random forest decision trees and k-near neighbours; and a variety of bootstrap cross-validation assessment techniques. The optimised learning model was then applied to a virtual data set of

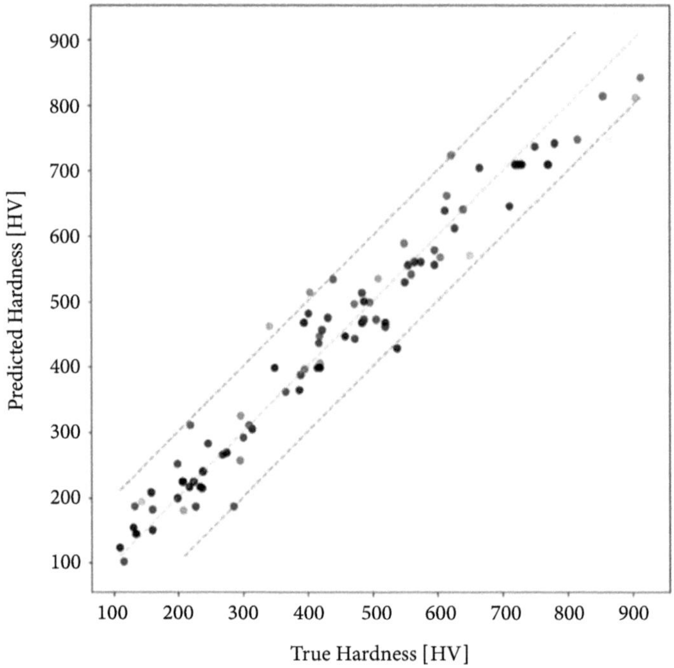

Figure 5.22 Predicted versus measured hardness for a neural network applied to 91 multicomponent alloys.

Data from Chang et al. (2018). Prediction of the composition and hardness of high-entropy alloys by machine learning. *Journal of Metals 71*: 3433–42.

Table 5.6 Predicted versus measured hardness for new high-hardness AlCrMnFeCoNi alloys predicted by a neural network (after Chang et al.[138]).

alloy	predicted hardness (kg/mm^2)	measured hardness (kg/mm^2)
$Al_{11}Cr_{22}Mn_5Fe_{22}Co_{18}Ni_{22}$	303	198
$Al_{30.5}Cr_{18.5}Mn_5Fe_{16.5}Co_{16}Ni_{13.5}$	505	522
$Al_{30}Cr_{35}Mn_{18}Fe_6Co_6Ni_5$	670	605
$Al_{25.5}Cr_{35}Mn_{15.5}Fe_{10}Co_9Ni_5$	670	628
$Al_{24}Cr_{35}Mn_{7.5}Fe_{10}Co_{18}Ni_{5.5}$	670	650

1.9 million trial compositions across the AlCrMnFeCoNi system to try to predict *de novo* twenty-one new high-hardness alloys for subsequent manufacture and measurement. The whole learning and prediction process was repeated three times using subsets of the composition data alone, some material properties alone, and a mixture of both, with rather little difference between the three sets of results, demonstrating fairly effectively that just a few material properties or alloy composition alone can be used with surprisingly good results. Figure 5.23 shows a comparison between predicted and measured alloy hardnesses for the full training data set and for the new predicted alloys. As can be seen in Figure 5.23, Wen et al.[139,140] were successful, like Chang et al.,[138] in finding new alloys with considerably

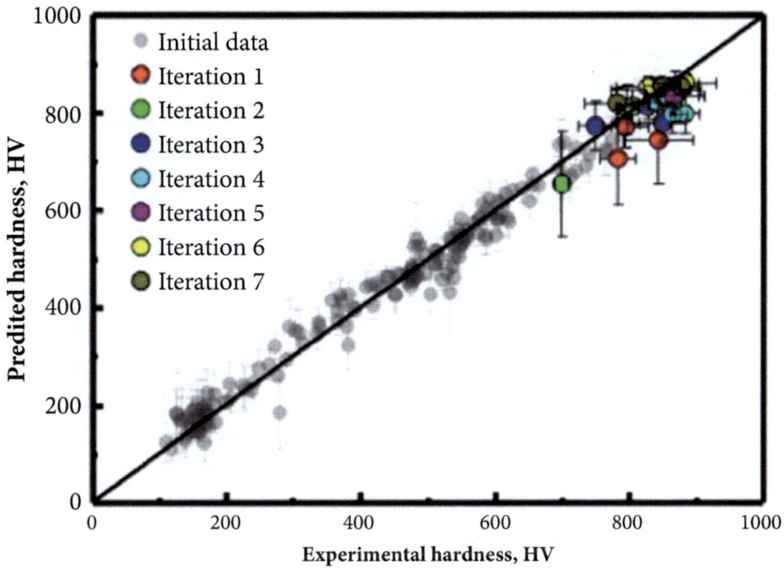

Figure 5.23 Predicted versus measured hardness for a neural network applied to 155 multicomponent alloys (initial training data) together with twenty-one predicted new high-hardness alloys (three in each of seven iterations).
Data from Wen et al. (2019). Machine learning assisted design of high-entropy alloys with desired property. *Acta Materialia 170*: 109–17.

higher hardness than any in the initial training data set, but with noticeably less accuracy in the predicted hardness values. There has been a number of other investigations[141–143] of machine-learning methods to try to predict directly the mechanical properties of multicomponent materials, with broadly similar results, i.e. training is effective in developing learning models which work well for the initial training data set, and are able to predict new materials with improved properties, but only with considerably reduced accuracy. More details are given in the reviews by Qiao et al.,[117] Hart et al.[118] and Liu et al.[119]

Overall, we can conclude that machine-learning methods provide a valuable tool for exploring multicomponent materials, with considerable success in organising and regularising the wide range of multicomponent material structures and properties.[117–119,123–143] However, they only work well at predicting new material structures and properties when trained initially with a large amount of previous experimental data, and when applied within the same known range of materials. This means that they are not really suitable for prediction of completely new materials with new structures and properties in previously unknown regions of multicomponent phase space but are ideally suited to guiding and supporting experimental investigations.

High throughput studies

We have seen how trial and error, parameterisation via extended multicomponent Hume-Rothery rules, thermodynamic modelling, atomistic modelling and machine learning can all provide insight into the formation of different structures and properties in multicomponent materials, but that they all fail to achieve fully reliable predictions. The underlying problem is, of course, the vastness and complexity of multicomponent phase space. Recent approaches to discovering new multicomponent materials have, therefore, focused onto using high-throughput scanning of large numbers of alloy compositions to try to cope with the enormous scale of the enterprise, and a belt-and-braces combination of both prediction and experimentation to try to cope with its enormous complexity.

The high-throughput approach was first described by Miracle et al.[144] in 2014. The overall methodology is as follows: select a region of multicomponent phase space of potential interest; determine a set of desired material properties and a corresponding set of component properties likely to underpin them; apply a sequence of high-throughput predictive and experimental assessment stages to a large number of alloy compositions in the selected region of multicomponent phase space; sift out at each stage compositions unable or unlikely to meet the desired component and/or material properties; and thereby narrow down the large number of initial materials to a few candidate compositions that can be subjected to a full experimental testing programme. Miracle et al.[144] proposed the following more detailed strategy: a first stage of applying high-throughput (i.e. relatively simple, approximate and, therefore, rapid) Calphad thermodynamic modelling to a large initial set of materials so as to determine a predicted range of simple properties such as melting point, freezing range, density, number of phases and crystal structure, selecting on that basis a somewhat smaller but still large number of promising candidate materials; then a second stage of applying high-throughput experiments to sample actual material properties, selecting on that basis an even smaller number, now just a handful, of even more promising candidates; and a final stage of full experimental trials for the few remaining candidates for the desired final engineering application. In implementing this kind of programme subsequently, others have used atomistic modelling and machine learning instead of, or as well as, thermodynamic modelling, but in all cases concentrating on speed rather than accuracy so as to enable rapid predictive sifting of a large number of initial materials; and have followed this up with a variety of novel high-throughput experimental methods for preparing a still quite large number of candidate materials, such as multiple diffusion couples, co-sputtering from multiple targets, and graded-powder additive manufacturing, with properties measured by rapid automated surface microprobe methods such as energy-dispersive (EDS) X-ray composition measurements, X-ray diffraction (XRD) phase analysis, nanoindentation mechanical testing and nano-probe chemical and electronic testing.

In recent years, there have been quite a few high-throughput studies of this type,[145–156] though the detailed methodologies have varied quite widely (it has also become somewhat fashionable to use the term ‘high throughput’ to describe incorrectly fairly conventional experimental studies). Following the initial description by Miracle et al.,[144] the high-throughput experimental methods that have since been used and the results that have been obtained by applying them to multicomponent materials have been reviewed by Li et al.[157] and Li et al.[158] Early high-throughput studies[40,145] concentrated on assessing large numbers of phase diagrams using Calphad thermodynamic modelling to try to identify solid solution compositions, largely without associated additional high-throughput experiments, and with moderate results when compared to previous experimental measurements. For instance, Senkov et al.[40] used rapid Calphad modelling to investigate over 130,000 alloys

Table 5.7 High-throughput Calphad predictions of phases in some equiatomic alloys compared with previous measurements (after Senkov et al.[40]).

alloy	phases measured	phases predicted at T_m	phases predicted at 600 °C
AlCrFeCoNiCu	bcc + fcc	fcc + B2 + bcc	fcc + B2 + bcc + bcc
AlCrMnFeNiCu	bcc	bcc + B2 + fcc + fcc	bcc + B2 + fcc + fcc
AlTiVCrMn	B2	B2	B2
AlTiVCrNb	B2	bcc	bcc + B2
TiZrNbHfTa	bcc	bcc	bcc + hcp
CrFeNiCuZr	bcc	bcc + Fe_2Zr + fcc + $Ni_{10}Zr_7$	bcc + Fe_2Zr + fcc + $Ni_{10}Zr_7$ + $Ni_{21}Zr_8$
CrMnFeCoNi	fcc	fcc	fcc + bcc
VFeCoNiCu	fcc	fcc + fcc	fcc + fcc
VnbMoTaW	bcc	bcc	bcc + bcc

from a starting palette of twenty-six different pure elemental components, and predicted about 5,000 new equiatomic 3-, 4-, 5- and 6-component solid-solution alloys. Some of Senkov et al.'s[40] predicted structures are compared with previous experimental measurements in Table 5.7, showing reasonable agreement in some cases and poor agreement in others. Lederer et al.[138] also used Calphad modelling methods to investigate a large number of equiatomic 2-, 3-, 4- and 5-component alloys, with about 87% accuracy when comparing their predictions with previous results for 117 of the binaries (and a lower accuracy expected for higher-order multicomponent systems).

As a more typical example of the high-throughput approach, Feng et al.[146] used rapid Calphad modelling backed up with *ab initio* molecular-dynamics (AIMD) and Monte-Carlo (MC) methods to try to predict lightweight high-temperature alloys in the 5-component AlTiCrMnFe system, sifting through 3,200 different non-equiatomic alloy compositions and finding just eight that met their criteria of a melting point $T_m > 1{,}250$ °C (1,523 K), a single-phase disordered bcc A2 structure at T_m, and a two-phase disordered bcc A2 + ordered bcc $L2_1$ structure with a volume fraction of $L2_1$ between 0.05–0.5 at% at both 0.5 T_m and 0.8 T_m. These criteria were intended to produce alloys that could be solution treated in a single-phase disordered bcc A2 region at high temperature, followed by quenching and annealing or controlled cooling to form an $L2_1$ precipitation-hardened microstructure. The eight candidate alloys were then manufactured by arc melting and casting for conventional microstructural and mechanical property characterisation by scanning and transmission electron microscopy (SEM and TEM), atom probe microscopy (AP), synchrotron X-ray and neutron diffraction (XRD and ND) and compression testing. Only two of the eight alloys had the predicted microstructure of disordered bcc A2 + ordered bcc $L2_1$ precipitates; the other six had instead a microstructure of equiaxed ordered bcc $L2_1$ grains with disordered bcc A2 at the grain and antiphase domain boundaries. Both types of alloy microstructure exhibited good room- and high-temperature mechanical properties, as shown in the plot of yield strength versus temperature in Figure 5.24, comparing Feng et al.'s[146] alloy 1 (A2 matrix + $L2_1$ precipitates) and alloy 8 ($L2_1$ grains + A2 boundaries) with a variety of other conventional and multicomponent alloys.

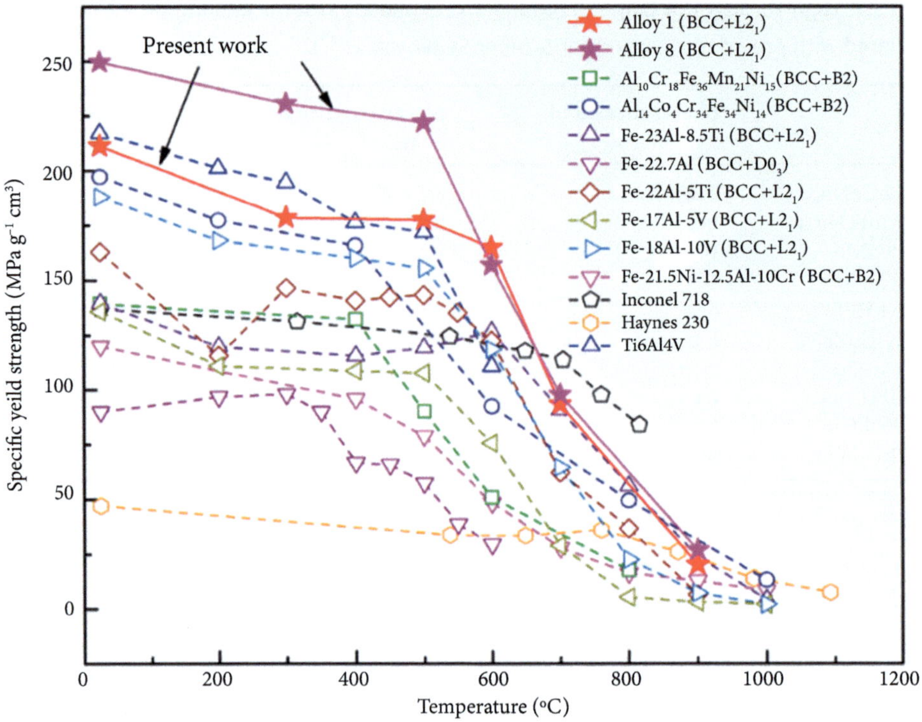

Figure 5.24 Specific yield strength versus temperature for high-throughput Calphad-predicted alloys $Al_{20}Ti_5Cr_5Fe_{50}Mn_{20}$ (alloy 1: B2 matrix + L2$_1$ precipitates) and $Al_{15}Ti_5Cr_5Fe_{50}Mn_{25}$ (alloy 8: L2$_1$ grains + B2 boundaries) compared with other conventional and multicomponent alloys.

Data from Feng et al. (2021). High-throughput design of high-performance lightweight high-entropy alloys. *Nature Communications 12*: Article 4329.

The well-known 5-component AlCrFeCoNi system has been subjected to a number of high-throughput investigations, using diffusion couple,[147] co-sputtering,[148] and graded laser-deposition melting[149] methods, with the results confirming, as noted in previous chapters, the sensitivity of structure and properties in this range of materials to minor variations in alloy composition and manufacturing method. Shang et al.[147] studied the microstructure and mechanical properties as a function of Al content in an $Al_{1.1}$CrFeCoNi–CrFeCoNi diffusion couple, forming a series of Al_xCrFeCoNi alloys with a composition gradient from $x = 0$–1.1, using scanning (SEM) and transmission (TEM) electron microscopy, energy dispersive X-ray spectrometry (EDS), and nanoindentation tests. The diffusion couple exhibited three composition regions, as shown in Figure 5.25, with a single-phase fcc structure in the low Al region, a two-phase structure of (Cr, Fe, Co)-rich disordered bcc A2 matrix + (Al, Ni)-rich ordered bcc B2 nanoparticle precipitates in the intermediate Al region, and an inverted two-phase structure of (Al, Ni)-rich ordered bcc B2 matrix + (Cr, Fe, Co)-rich disordered bcc A2 nanoparticle precipitates in the high Al region, as shown in Figure 5.26 for the latter two microstructures by transmission electron micrographs from the interface between the intermediate and high Al regions. The regions of bcc A2 matrix + B2 precipitate and B2 matrix + A2 precipitate both exhibited

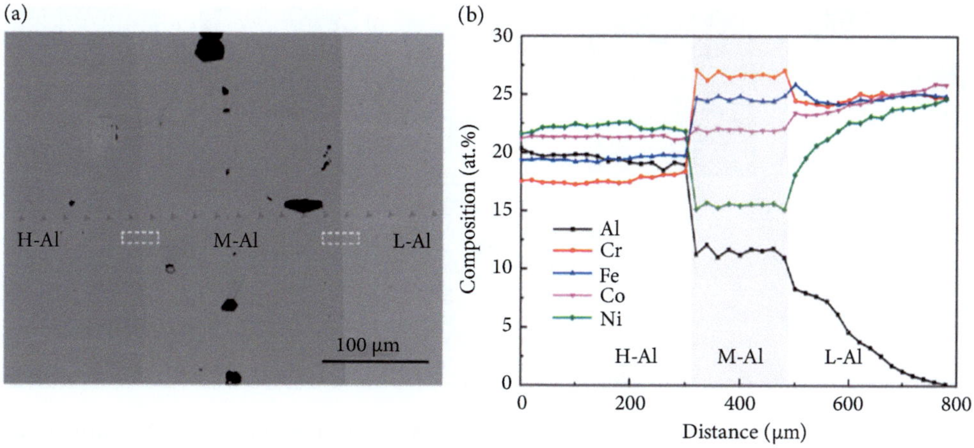

Figure 5.25 SEM image and EDS profiles showing low- (L), medium- (M) and high- (H) Al regions in an $Al_{1.1}CrFeCoNi$—CrFeCoNi diffusion couple.

Reproduced from Shang et al. (2022). High-throughput experimental study on the microstructural and compositional variations of mechanical properties for AlCoCrFeNi high-entropy alloys. *Journal of Alloys and Compounds 917*: Article 165513.

high values of indentation hardness and Young's modulus, as shown in Figure 5.27. Shi et al.[148] studied the microstructure and corrosion properties as a function of Al content in a co-sputter-deposited film manufactured from two separate sputter targets of CrFeCoNi and Al, forming a series of graded $Al_xCrFeCoNi$ alloys with a composition gradient from $x = 0–2.7$, using scanning (SEM) and transmission (TEM) electron microscopy, energy dispersive X-ray spectrometry (EDS), and potentiodynamic polarisation tests. The co-sputtered film exhibited a single-phase disordered fcc structure at low Al contents and a sharp transition at a composition of approximately $x = 1$ to a single-phase disordered bcc structure at high Al contents, as shown in the X-ray diffraction (XRD) traces in Figure 5.28. The films showed excellent corrosion protection, with low corrosion current densities and high pitting potentials compared to a range of other high-entropy and conventional materials, as shown in Figure 5.29. Li and Flores[149] studied the microstructure and mechanical properties in a series of $Al_xCrFeCoNi$ alloys with a composition gradient from $x = 0–1$, manufactured by laser surface melting a track of deposited Al powder on a CrFeCoNi substrate, varying the composition and the cooling rate by varying the Al powder feed rate, and the laser power and scanning speed, using scanning (SEM) and transmission (TEM) electron microscopy, energy dispersive X-ray spectrometry (EDS), and nanoindentation tests. The laser-melted surface track exhibited a two-phase disordered fcc + disordered bcc A2 structure at all cooling rates with an Al content of $x = 0.56$; a two-phase disordered fcc + disordered bcc A2 structure at low cooling rates and a two-phase disordered A2 bcc + ordered bcc B2 structure at high cooling rates with an Al content of $x = 0.6$; and a two-phase disordered A2 bcc + ordered bcc B2 structure at all cooling rates with an Al content of $x = 1$. The bcc A2 + B2 structures exhibited very high hardness values as shown in Figure 5.30.

Melia et al.[150] manufactured through-thickness graded single-phase bcc Senkov alloys in the NbMoTaW system by varying the feed rates from four independent Nb, Mo, Ta and W powder hoppers during laser surface deposition and melting. They manufactured

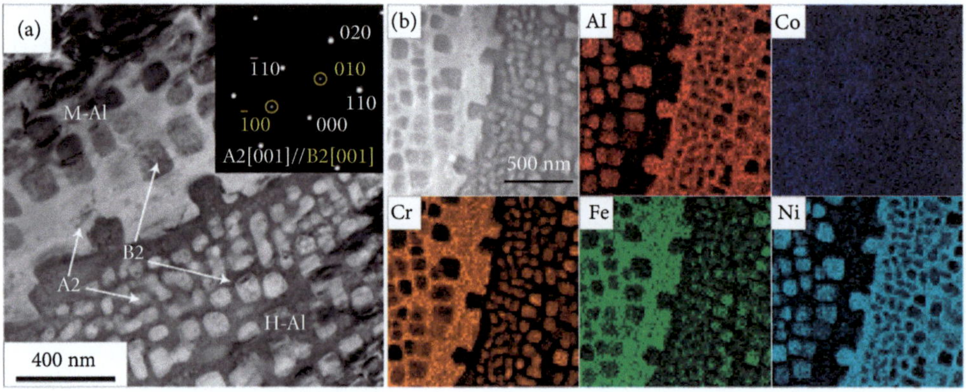

Figure 5.26 TEM image and EDS maps showing A2 matrix + B2 precipitate and B2 matrix + A2 precipitate microstructures in medium- (M) and high- (H) Al regions respectively in an $Al_{1.1}$CrFeCoNi—CrFeCoNi diffusion couple.

Reproduced from Shang et al. (2022). High-throughput experimental study on the microstructural and compositional variations of mechanical properties for AlCoCrFeNi high-entropy alloys. *Journal of Alloys and Compounds 917*: Article 165513.

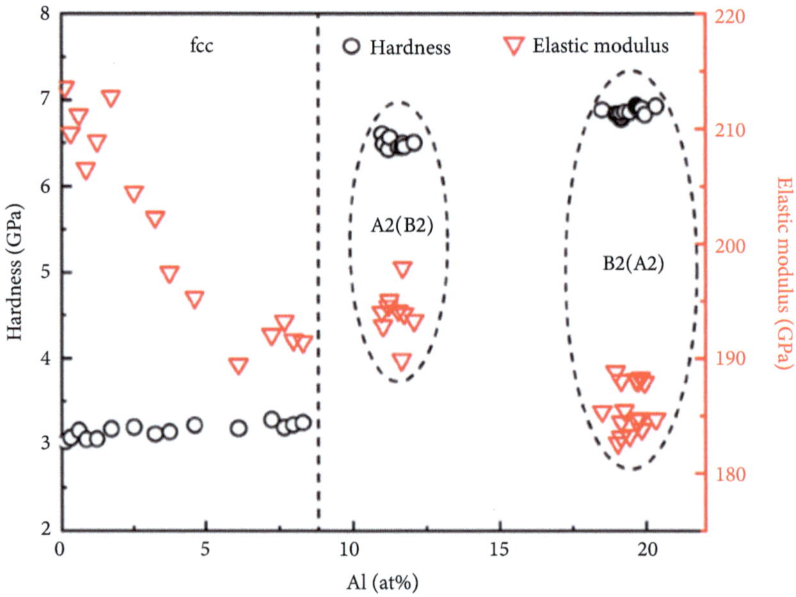

Figure 5.27 Nanoindentation hardness and Young's modulus versus Al content in an $Al_{1.1}$CrFeCoNi—CrFeCoNi diffusion couple.

Data from Shang et al. (2022). High-throughput experimental study on the microstructural and compositional variations of mechanical properties for AlCoCrFeNi high-entropy alloys. *Journal of Alloys and Compounds 917*: Article 165513.

a series of ~12 mm-thick plate-shaped deposits, each on a Mo build plate, and each made up of fifty-seven deposition layers melted continuously during deposition with a 2 mm laser spot scanning at 5 mm/s. The deposits were examined by a combination of optical and scanning electron (SEM) microscopy, electron back-scattering diffraction (EBSD)

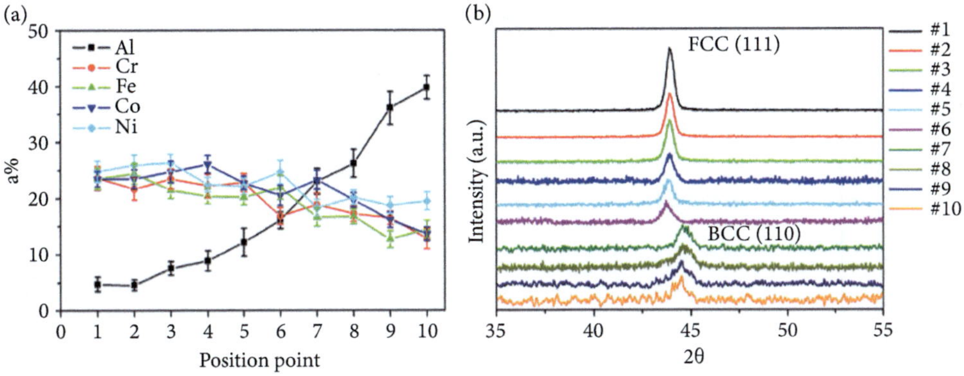

Figure 5.28 XRD traces of transition from fcc to bcc with increasing Al content in co-sputtered $Al_xCrFeCoNi$.

Data from Shi et al. (2020). High-throughput synthesis and corrosion behaviour of sputter-deposited nanocrystalline $Al_x(CoCrFeNi)_{100-x}$ combinatorial high-entropy alloys. *Materials and Design 195*: Article 109018.

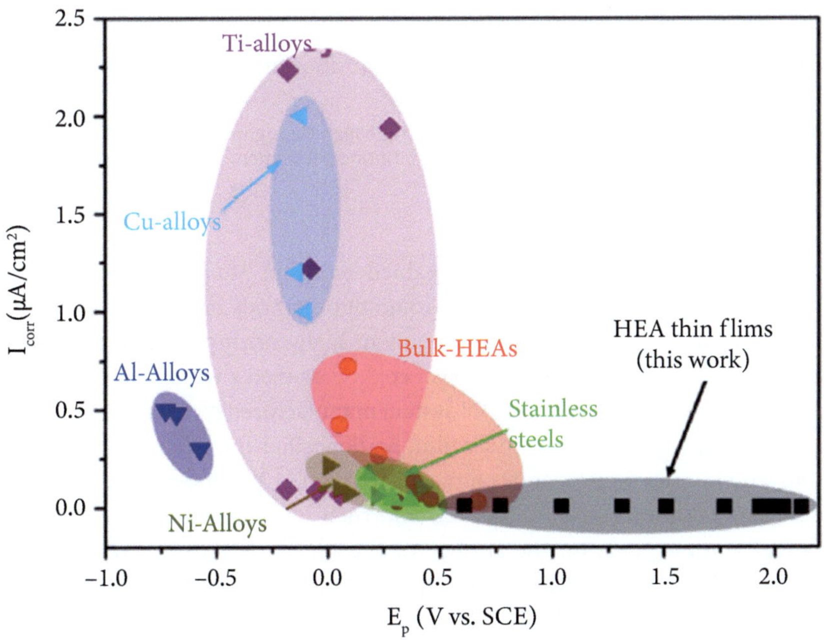

Figure 5.29 Corrosion current I_{corr} and pitting potential E_p in co-sputtered $Al_xCrFeCoNi$ and other alloys.

Data from Shi et al. (2020). High-throughput synthesis and corrosion behaviour of sputter-deposited nanocrystalline $Al_x(CoCrFeNi)_{100-x}$ combinatorial high-entropy alloys. *Materials and Design 195*: Article 109018.

patterns and micro-indentation scratch tests. Figure 5.31 shows a typical set of results for a $(MoTaW)_xNb_{1-x}$ deposit with a through-thickness graded composition varying from $x = 0$–1, with a fairly equiaxed single-phase bcc grain structure, an overall {100} texture, a somewhat non-linear composition variation, and an increasing hardness with decreasing

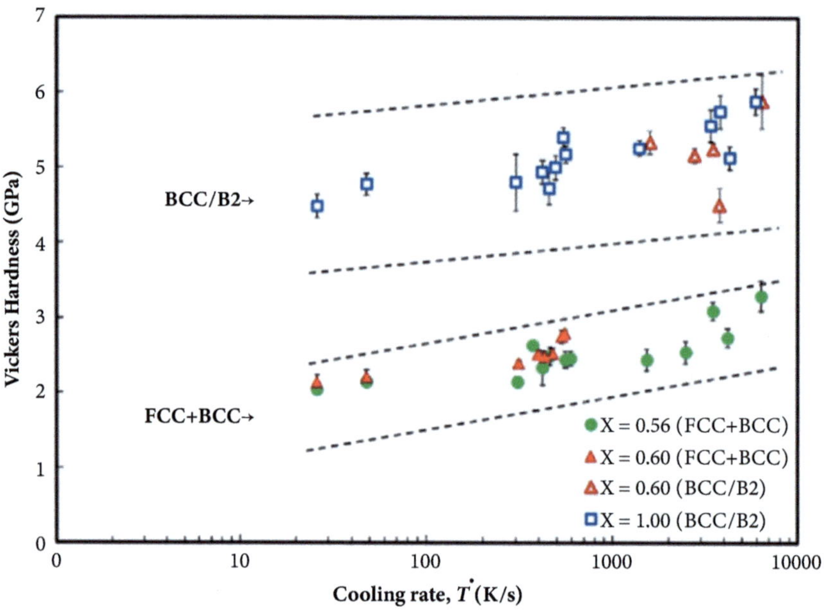

Figure 5.30 Nanoindentation hardness versus composition and cooling rate in laser melted $Al_xCrFeCoNi$ alloys.

Reproduced from Li and Flores. (2020). Laser processing as a high-throughput method to investigate microstructure-processing-property relationships in multiprincipal element alloys. *Journal of Alloys and Compounds 825*: Article 154025.

Nb content. Similar results have been obtained by other studies of a variety of laser-melted graded Senkov alloys.[151,152] High-throughput methods have also recently begun to be applied to the investigation of multicomponent high-entropy ceramic compounds and functional materials.[153–156] For instance, Hsu et al.[153] studied a wide range of multicomponent mononitrides in the (TiVZrNbHf)N system manufactured by reactive radio-frequency (RF) co-sputtering a graded 1 μm thick film, co-deposited from four separate targets (V, Hf, Nb and TiZr) under a nitrogen N_2 atmosphere onto (100) single-crystal silicon Si substrates at a temperature of 400 °C (673K), and then investigated at 396 separate points with different compositions on the surface of the film by electron probe microanalysis (EPMA), X-ray diffraction (XRD), scanning transmission electron microscopy (STEM) and nanoindentation. Figure 5.32 shows a series of electron probe microanalysis (EPMA) composition maps of the graded film, and Figure 5.33 shows corresponding hardness contours.

Other examples and further details of high-throughput studies of multicomponent high-entropy materials can be found in the reviews by Li et al.[157] and Li et al.[158] Overall we can conclude that a series of stages of high-throughput thermodynamic, atomistic and machine-learning calculations combined with high-throughput experiments such as graded diffusion couples, sputter-deposited films and additive-manufactured deposits can provide quite an effective way of identifying potentially valuable new multicomponent high-entropy materials. However, the necessarily rough-and-ready nature of the calculations required to scan at high speed through a large number of candidate materials, and the difficulties of accurate probe-based micro- or nano-scale surface measurements needed to scan through

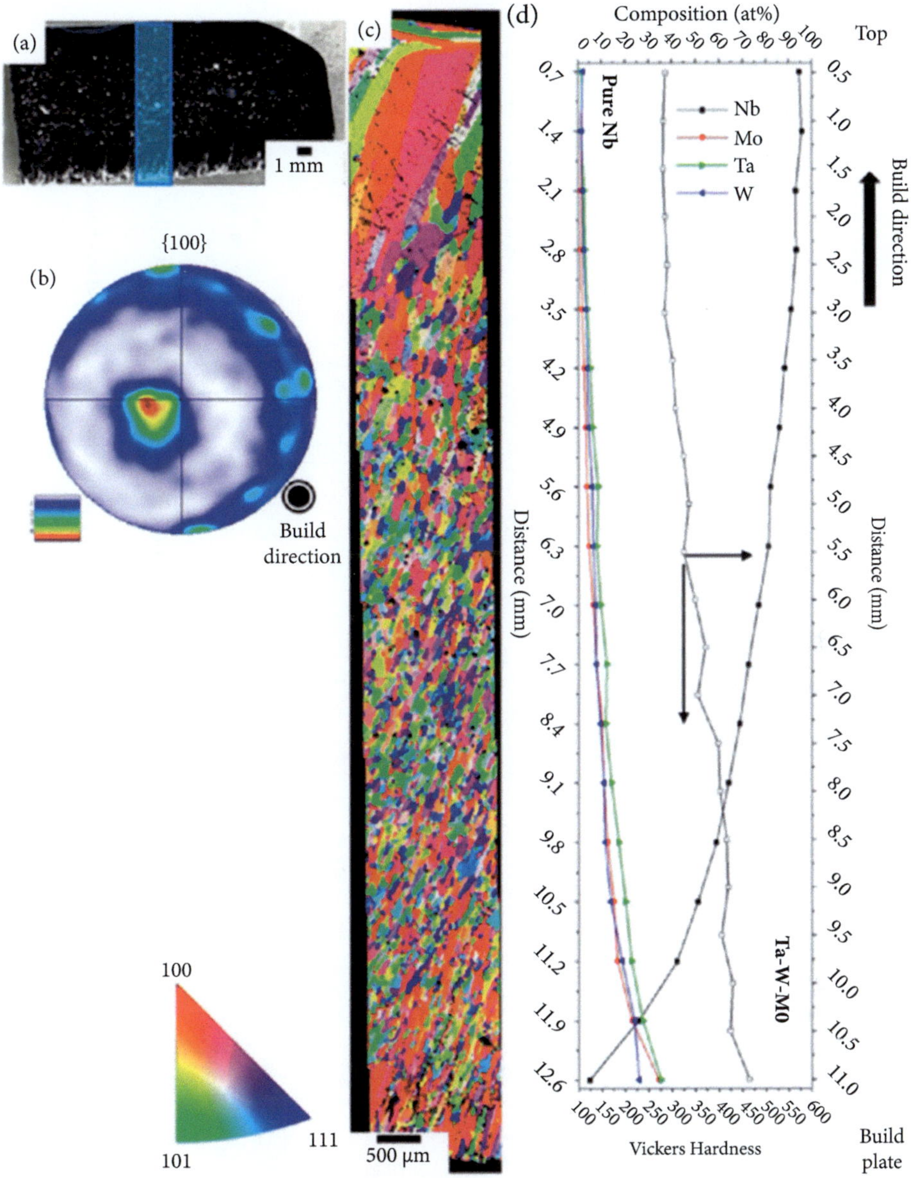

Figure 5.31 Laser melted $(MoTaW)_xNb_{1-x}$ deposit with through-thickness graded composition varying from $x = 0–1$: (a) optical image of the deposit; (b) (100) pole figure; (c) through-thickness microstructure; (d) through-thickness variation of composition and hardness.

Reproduced from Melia et al. (2020). High-throughput additive manufacturing and characterization of refractory high-entropy alloys. *Applied Materials Today 19*: Article 100560.

graded materials rather than conventional bulk experimental measurements, means that good results are not guaranteed and that many potentially valuable materials may be missed and/or discarded when candidate materials are sifted in the different high-throughput theoretical and experimental stages.

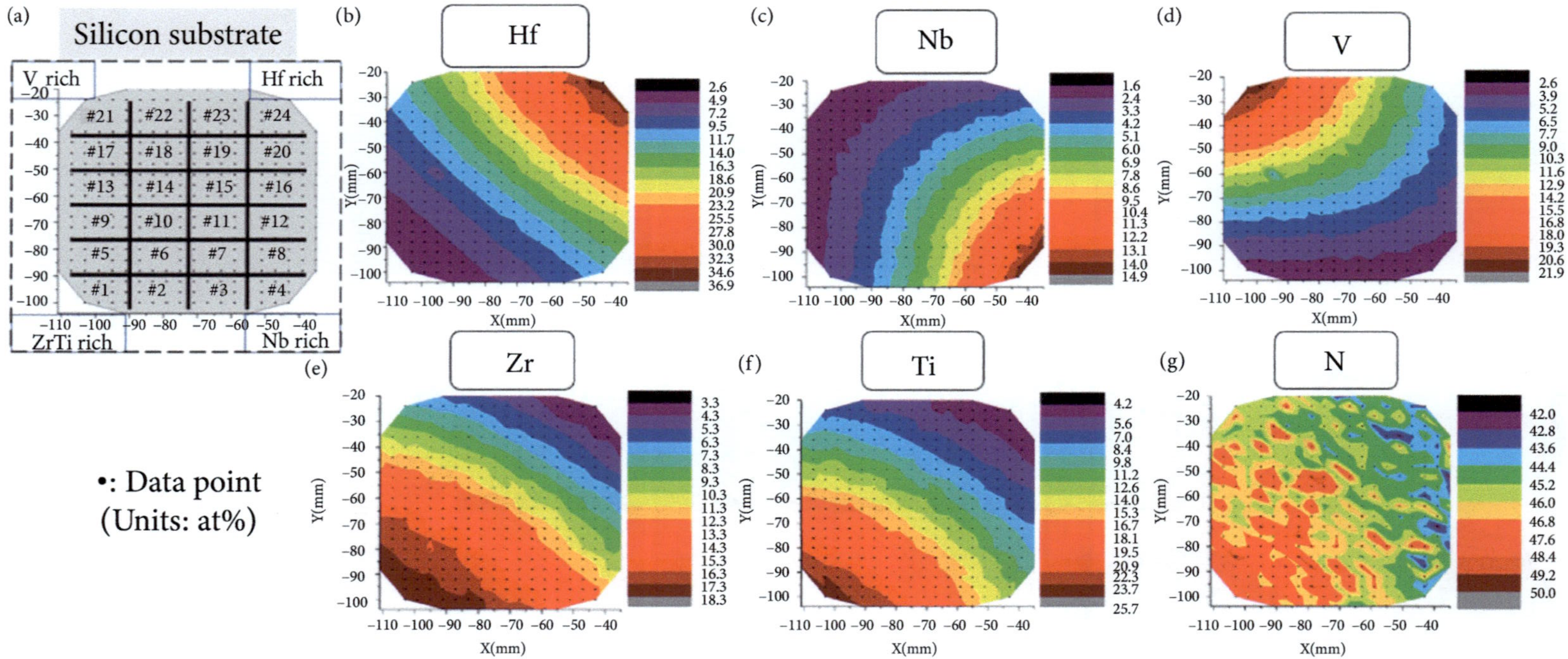

Figure 5.32 Graded co-sputtered (TiVZrNbHf)N film: relative geometry of sputter targets, substrate and 396 characterisation points; (b)–(g) EPMA composition maps of the six components.

Reproduced from Hsu et al. (2022). Combinatorial synthesis of reactively co-sputtered high entropy nitride (HfNbTiVZr) coatings: Microstructure and mechanical properties. *Surface and Coatings Technology 442*: Article 128564.

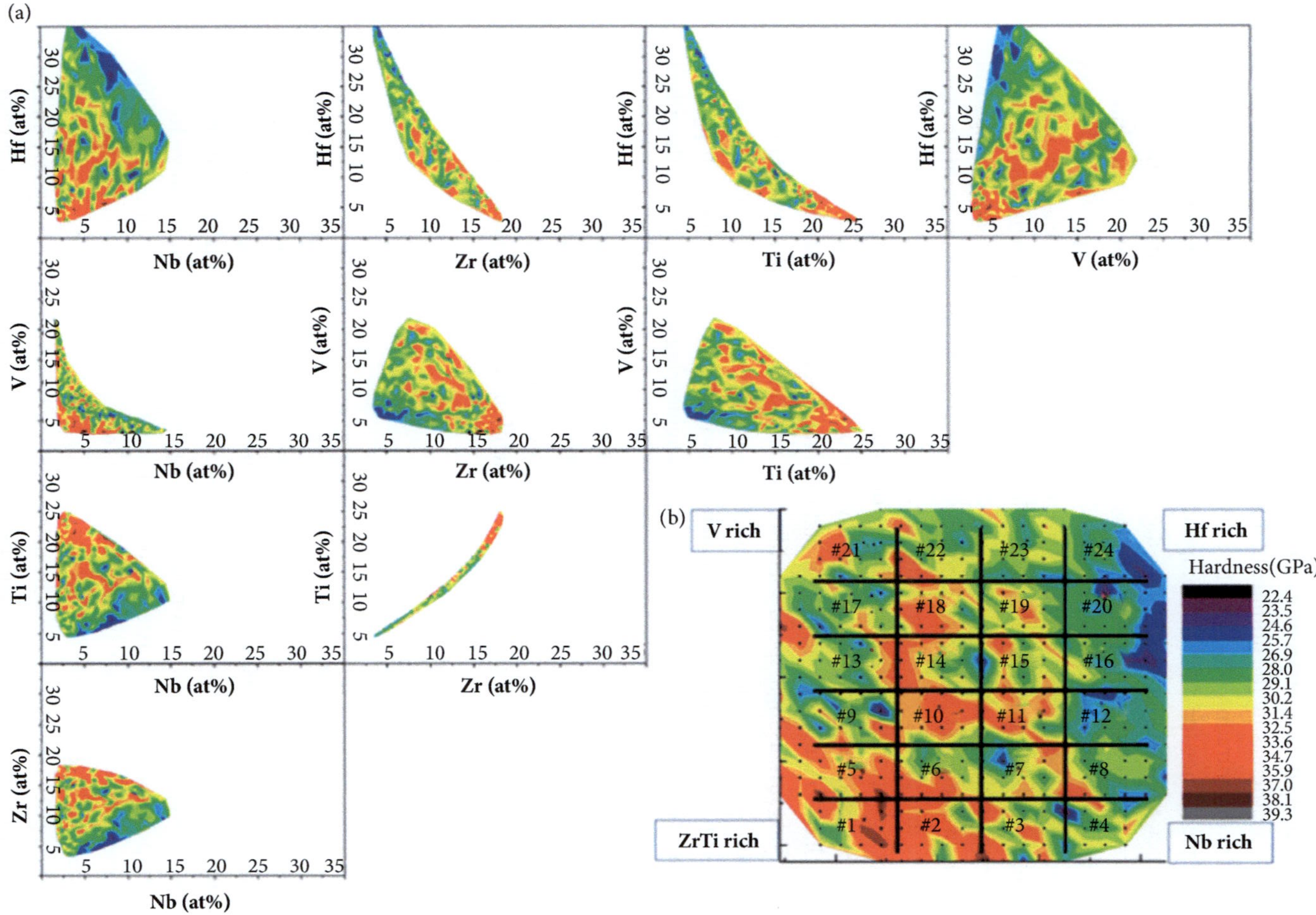

Figure 5.33 Hardness variation with composition in graded co-sputtered (TiVZrNbHf)N film: (a) binary sections through multicomponent composition variation; (b) composition variation across the film surface.

Reproduced from Hsu et al. (2022). Combinatorial synthesis of reactively co-sputtered high entropy nitride (HfNbTiVZr) coatings: Microstructure and mechanical properties. *Surface and Coatings Technology 442*: Article 128564.

3 Manufacturing Multicomponent Materials

As we have seen in previous chapters, human development has been based for millennia on a conventional alloying strategy of selecting a single principal component for a material and then tweaking its properties by adding one or two alloying components at relatively low concentrations. Increasing the number of different alloying elements much beyond one or two and their concentration beyond a few per cent was generally avoided because it led to an increasing tendency for different components to react chemically and form multiple compound phases, and an associated increasing difficulty of processing and manufacturing the materials. A strong conventional wisdom developed gradually over many years, almost a taboo, against the use of highly alloyed materials. As we have also seen in previous chapters and in the previous section of this chapter, however, we have discovered relatively recently that, with sufficiently large numbers of alloying elements added at sufficiently large concentrations, the configurational entropy of mixing often acts to suppress the various chemical reactions between the different components, preventing the formation of multiple compound phases, so that processing and manufacturing the materials become much easier than might otherwise be expected. In fact, multicomponent high-entropy materials can in many cases be made in a relatively straightforward way by most conventional material processing methods.[18,159] Unlike many other advanced materials that require complex and specialised processing methods such as high-speed solidification, thin-film vapour deposition or severe mechanical deformation, multicomponent and high-entropy materials are not difficult to make in bulk form, with considerable fundamental advantages for measuring their properties as well as practical advantages for using them in engineering applications, as described in more detail in Chapter 6.

Multicomponent high-entropy materials can, therefore, be manufactured by most conventional methods, using liquid, vapour and/or solid processing routes. The next few sub-sections give a brief overview of the different processes that have been used, most of which are quite well known and have already been mentioned in previous chapters.

Casting

Many multicomponent and high-entropy metallic and ceramic alloy materials can be manufactured relatively easily by conventional *melting and casting* of the starting components.[18,159] Complex shaped objects can be made either by *shape casting* or by casting *billets* or *preforms* followed by *working* and *machining* to shape. Not surprisingly, there are typical problems of evaporation, oxidation and decomposition when melting alloys from a mixture of components that are unstable and reactive or that have very different melting points, but these problems are not in general worse than those that are found with most well-known binary and ternary dilute alloys, and can usually be overcome by mixing master alloys rather than pure components and by using a carefully controlled atmosphere and/or suitable fluxes during melting and casting. Not surprisingly, there are also typical casting defects such as segregation, porosity and inclusions, but these are again not significantly worse than those that are found with most well-known dilute binary and ternary alloys and can usually be ameliorated if not fully removed by conventional casting control of molten metal temperature, atmosphere, feeding and cooling conditions.

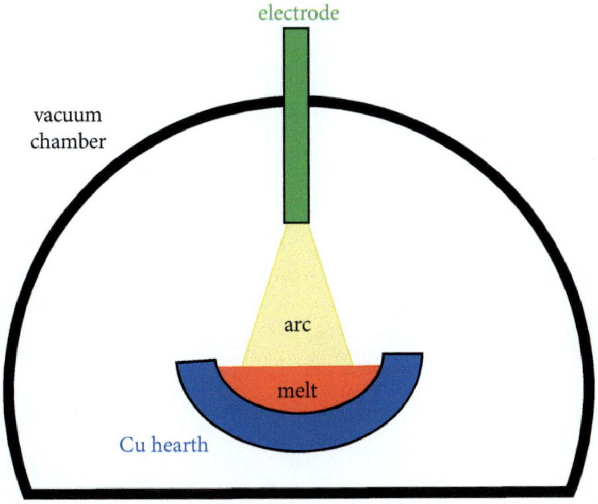

Figure 5.34 Schematic of arc melting in a water-cooled Cu hearth.

A very common and convenient processing method for making multicomponent materials for initial assessment is by *arc melting*, usually in a vacuum or under argon, followed by solidification after powering down to allow the molten material to be quenched by the water-cooled melting hearth.[18,159] The typical arrangement is shown schematically in Figure 5.34. Multiple remelting and resolidification is frequently used to promote chemical homogeneity, and shaped products can easily be made by simple bottom pouring from the hearth into a suitably shaped graphite or copper mould. Arc melting and casting is a very flexible process and is widely applicable to many different multicomponent materials. It works well for anything ranging from small test buttons of a few tens of grams up to larger shaped castings of a few tens of kilograms, and it has been used very widely in many studies of multicomponent high-entropy materials over the last decade or so.

As already shown in Figures 3.8, 3.14, 3.20 and 3.41, as-cast multicomponent high-entropy material microstructures are typically dendritic, and are often fully single phase with interdendritic segregation, or composed of single-phase dendrites with interdendritic binary or ternary eutectic phases, with the detailed structure depending on the composition of the material and the cooling rate during casting.[160–164] Figure 5.35 shows examples of fcc dendrites and fcc + bcc interdendritic eutectic in as-cast non-equiatomic 5-component $CrFeCoNiNb_x$ hypoeutectic, eutectic and hypereutectic alloys. As discussed previously, the probability of forming a fully single-phase solid-solution microstructure increases when the alloy components are reasonably similar chemically, i.e. when they have similar atomic sizes and electronegativities, and when the cooling rate during casting is relatively high. As with conventional materials, the size and spacing of as-cast multicomponent dendrites and the size and spacing of any interdendritic secondary and tertiary eutectic phases are reduced significantly when the cooling rate during casting is high. Conventional melting and chill casting leads to micron-sized dendritic and/or dendritic/eutectic as-cast microstructures,[160–163] as shown in Figures 3.8, 3.14, 3.20, 3.41 and 5.35.

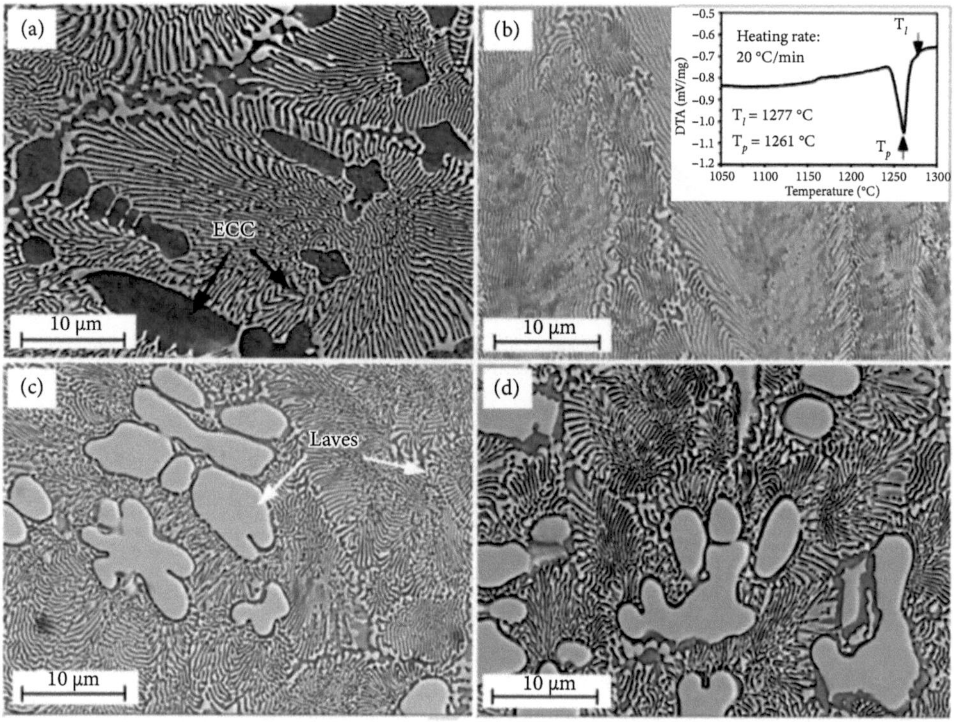

Figure 5.35 Dendritic solidification in arc melted and cast Cr_3Fe_5CoNi: fcc dendrites (A) and interdendritic fcc + bcc eutectic (B).

Reproduced from Zhang et al. (2016). Fabrication routes. In Gao et al., eds., *High-entropy alloys: Fundamentals and applications.* Basel: Springer.

Rapid solidification

Rapid solidification processes such as *melt spinning* or *powder atomisation*, shown schematically in Figures 5.36 and 5.37 can be used to increase the likelihood of an as-cast fully single-phase solid-solution microstructure, and to form much finer nanoscale dendritic and/or dendritic/eutectic microstructures,[18,159] but they do require relatively complex subsequent consolidation processing to make bulk products. Table 5.8 gives some examples of melt spun multicomponent high-entropy materials[2,32,165–170] with others given in the book by Murty et al.[18] Figure 5.38 shows a typical SEM image of the irregular columnar crystal structure found on cross sections through melt spun ribbons, in this case for the equiatomic 5-component single-phase fcc Cantor alloy CrMnFeCoNi.[165]

Single crystals

Single crystals of multicomponent high-entropy single-phase solid solutions have been made in a number of cases by conventional *Bridgman crystal growth*, as shown schematically in Figure 5.39, and the same method has also been used to make *directionally solidified* (*DS*) dendritic and dendritic/eutectic multicomponent multiphase microstructures.[171–178]

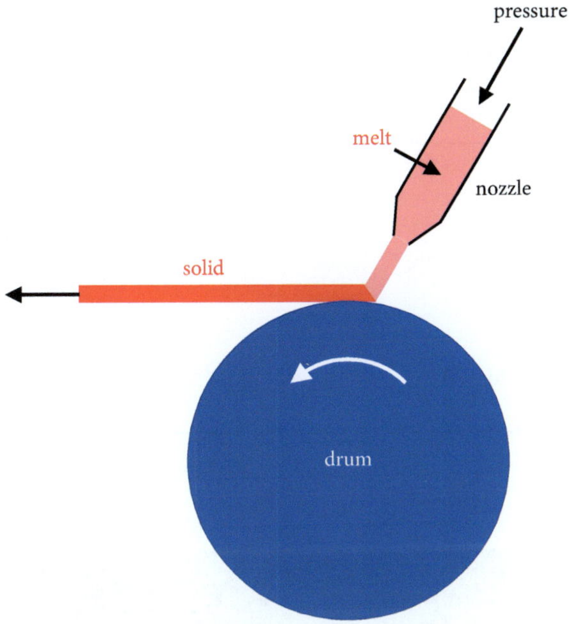

Figure 5.36 Schematic of melt spinning.

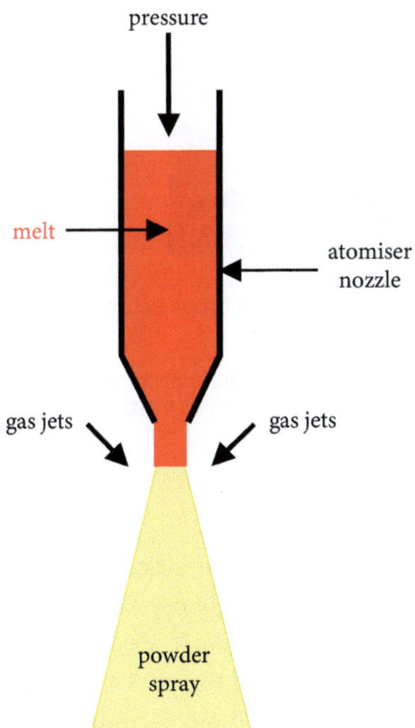

Figure 5.37 Schematic of powder atomisation.

Table 5.8 Some examples of multicomponent high-entropy materials that have been processed by melt spinning.[2,18,32,165–170]

alloy	structure
CrMnFeCoNi	fcc
CrMnFeCoNiCu	fcc
CrMnFeCoNiGe	fcc + bcc
TiCrMnFeCoNi	fcc + bcc
$Si_{0.2}FeCoNiCu_{0.2}$	fcc
$AlFeCoNi_3Mo_{0.2}$	$L1_2$ + bcc
AlCrFeCoNiCu	bcc + B2 + fcc
$(Cr_{0.5}FeCoNiMo_{0.5})B_x$	amorphous
TiVCrNiZr	Laves phase

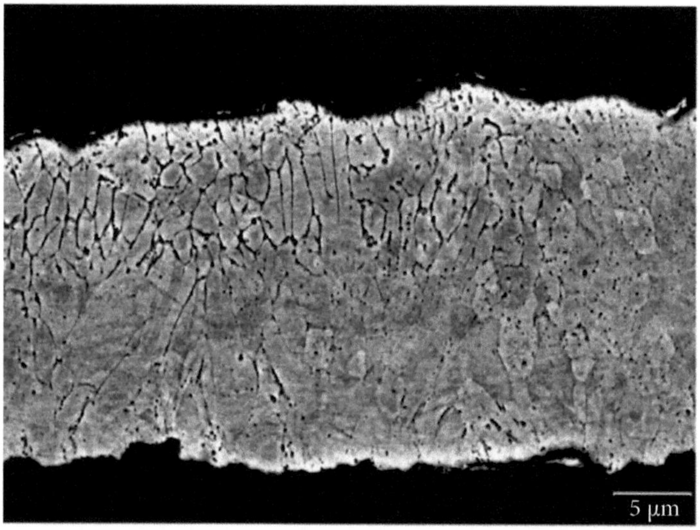

Figure 5.38 SEM image of irregular columnar single-phase fcc crystals on a cross section through the melt-spun Cantor alloy CrMnFeCoNi.

Reproduced from Cech et al. (2020). A novel ultrafine grained high-entropy alloy with excellent combination of mechanical and soft magnetic properties. *Journal of Magnetism and Magnetic Materials 502*: Article 166513.

Table 5.9 shows a selection of single crystal and directionally solidified multicomponent materials that have been manufactured by the Bridgman and other single-crystal growth methods. Figure 5.40 shows an example of a Bridgman-grown near-single crystal of a 5-component equiatomic two-phase bcc + B2 AlCrMnFeCo alloy,[176] and Figure 5.41 shows a typical microstructure of a directionally solidified dendritic equiatomic 5-component single-phase fcc Cantor alloy CrMnFeCoNi.[178]

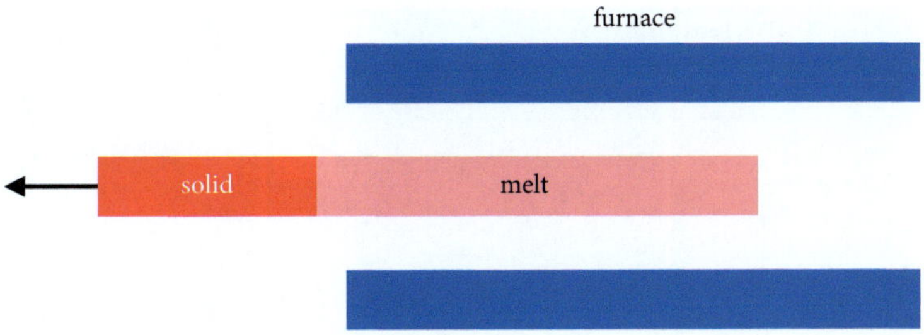

Figure 5.39 Schematic of Bridgman solidification.

Table 5.9 Multicomponent high-entropy alloys made by Bridgman and other single-crystal growth methods.[171–178]

alloy	processing method
CrCoNi	Bridgman
CrFeCoNi	Bridgman
CrFeCoNi	floating zone
CrMnFeCoNi	Bridgman
$Al_{0.3}CrFeCoNi$	Bridgman
AlCrMnFeCo	Bridgman

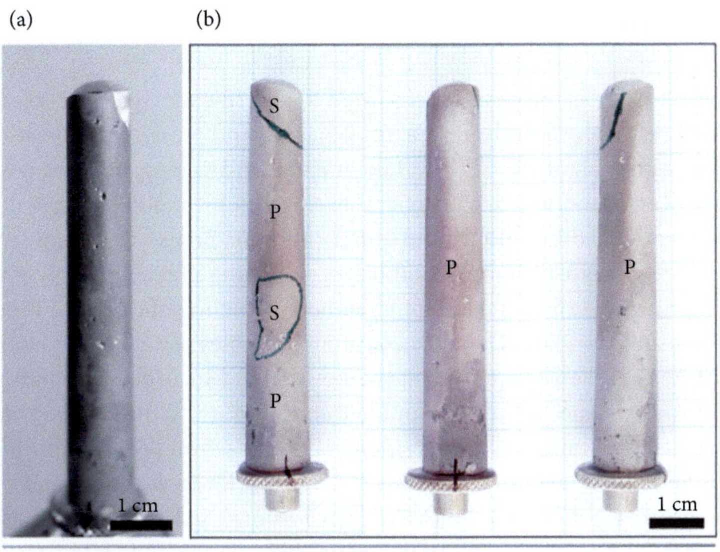

Figure 5.40 Bridgman-grown 5-component equiatomic AlCrMnFeCo alloy consisting of one main single crystal (P) and two small secondary crystals (S).

Reproduced from Feuerbacher et al. (2014). Superior high-tensile elongation of a single-crystal $CoCrFeNiAl_{0.3}$ high-entropy alloy by Bridgman solidification. *Intermetallics 54*: 104–9.

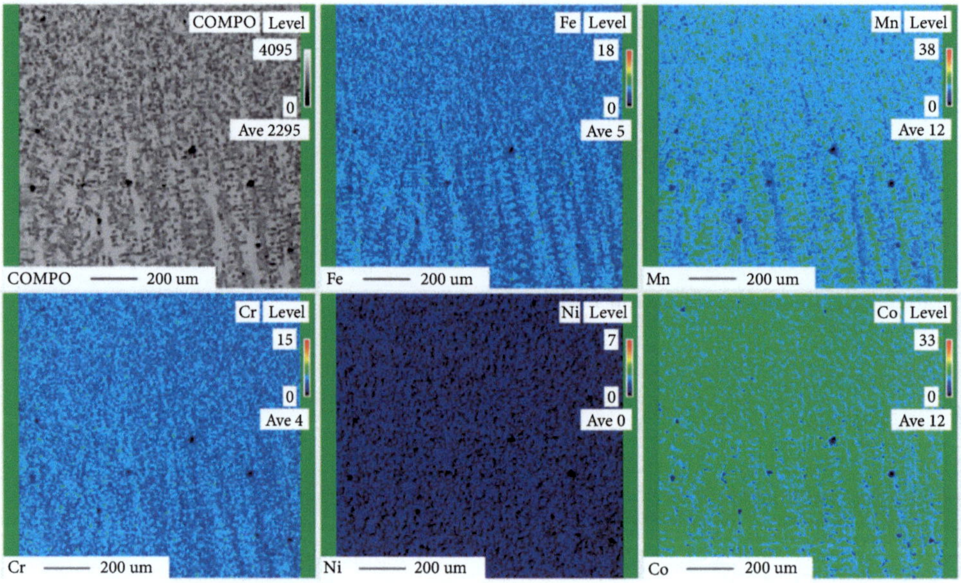

Figure 5.41 SEM image and EPMA composition maps of Bridgman-grown directionally solidified (DS) dendritic 5-component equiatomic single-phase fcc Cantor alloy CrMnFeCoNi alloy.

Reproduced from Zheng et al. (2020). Microstructure evolution and mechanical property of directionally solidified CoCrFeMnNi high-entropy alloy. *Intermetallics 119*: Article 106723.

Additive manufacturing

Additive manufacturing is the generic term used to describe a range of processes that have become popular relatively recently for making objects by progressively depositing and melting powders on a growing surface.[179–181] Like melting and casting, additive manufacturing processes have been found to work in general just as well for multicomponent materials as they do for conventional dilute binary and ternary alloy materials.[18,159,181] Different additive manufacturing processes use different methods of delivering a regular supply of new powder material to the growing surface and different methods of melting the powder to consolidate it into the growing surface.[179–181] The most common processes use a focused laser or electron beam scanning across the surface to melt the powder and incorporate it into the growing object. The powdered material is either supplied continuously to the surface as it is scanned, and the process is then called *laser deposition* (*LD*) or *electron beam deposition* (*EBD*), or the powder is laid down layer by layer on the surface prior to scanning and melting each layer, and the process is then called *selective laser melting* (*SLM*) or *selective electron beam melting* (*EBSM*)[179–181] (scanning is controlled so that melting and solidification only take place selectively in particular places to enable the manufacture of complex shaped objects). A typical laser deposition arrangement is shown schematically in Figure 5.42.

Additive manufacturing processes are very flexible, and can be used for a wide variety of often complex-shaped objects made from a wide variety of metal, ceramic, polymer and other powdered materials, usually using *3D printing* with a *computer-aided design* (*CAD*)

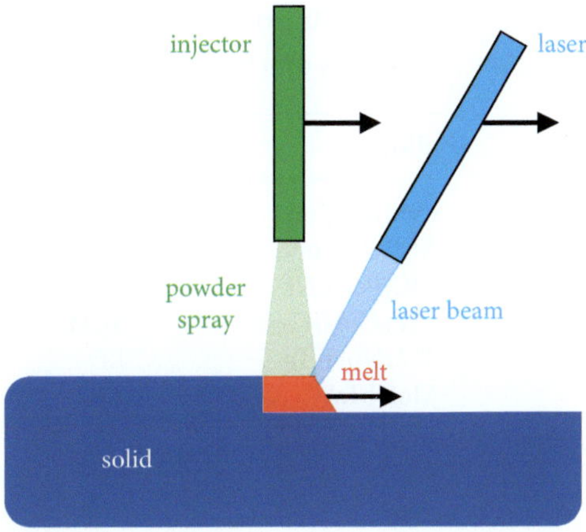

Figure 5.42 Schematic of laser deposition.

program to control the delivery of the powder and the motion of the laser or electron beam as it scans over the growing surface.[179–181] This allows the manufacture of complex shaped objects with minimal loss of material compared to more traditional processing methods such as shape casting or machining. It is important in additive manufacturing, however, to control carefully the rate of delivery of the powder, the intensity of the laser or electron beam and the scanning speed across the surface so as to ensure good melting and consolidation of each surface layer and thus prevent the development of excessive porosity and/or oxide inclusions during processing.[179–181] Best results often require further post-processing consolidation such as *hot isostatic pressing* (*HIP*) to ensure full integrity of the final material.

Table 5.10 gives some typical examples of multicomponent high-entropy materials that have been processed by additive manufacturing,[181–188] with more given in the review by Torralba et al.[181] The as-processed microstructures are usually irregular columnar dendritic grains roughly perpendicular to the growth direction, i.e. normal to and often growing through the sequential surface layers, often with an admixture of some more equiaxed dendritic grains.[181–188] Figure 5.43 shows a typical example of a laser-deposited (LD) equiatomic 5-component single-phase fcc Cantor alloy CrMnFeCoNi, with optical and SEM images obtained using electron back-scatter patterns (EBSP), indicating irregular columnar grains growing epitaxially through the irregular layer boundaries.[182]

Vapour deposition

Thin and thick film surface coatings can be made by a variety of methods involving *vapour deposition*, i.e. by the creation of a vapour followed by its deposition onto a suitable substrate.[159] Different processing methods differ in how the vapour is created and how it is directed onto the substrate, and include techniques such as: *thermal evaporation*, in which a small amount of material is heated to boil off a stream of vaporised atoms or molecules; *laser melting* or *electron-beam melting*, in which a small amount of material is similarly evaporated by a pulsed laser or electron beam; *filament heating* in which a high voltage is applied

Table 5.10 Some examples of multicomponent high-entropy materials that have been processed by additive manufacturing.[181–188]

alloy	processing method	structure
CrMnFeCoNi	LD	fcc
Al_xCrFeCoNi	LD	bcc + fcc + B2 + σ
AlCrFeCoNi	SEBM	bcc + fcc + B2
$CrFeCoNiC_{0.05}$	SLM	fcc + $Cr_{23}C_6$
TiZrNbHfTa	LD	bcc
MoNbTaW	LD	—

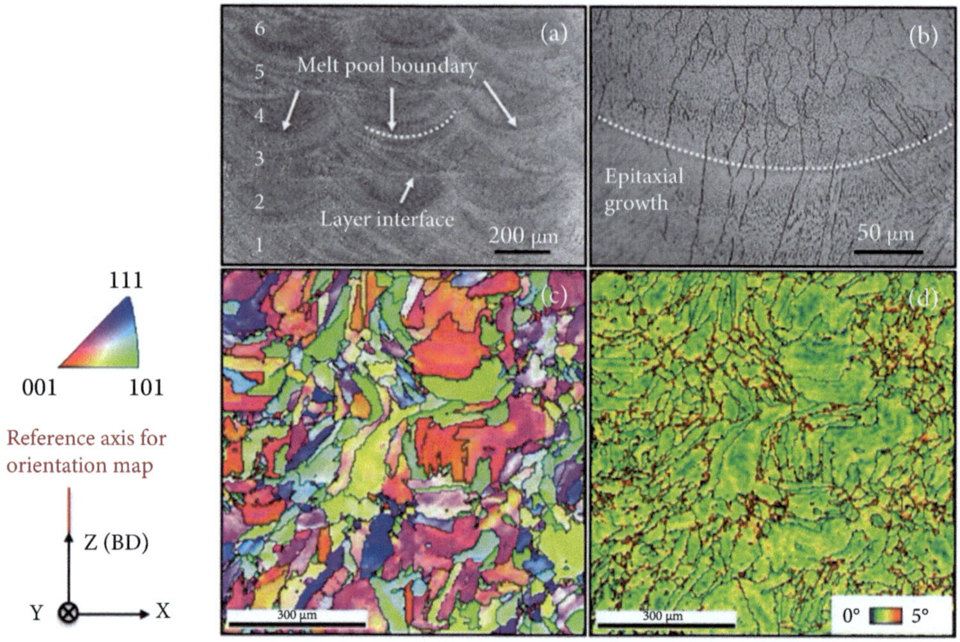

Figure 5.43 (a) and (b): optical micrographs and (c) and (d): scanning electron micrographs (SEM) using electron back-scattered patterns (EBSP) of laser-deposited Cantor alloy CrMnFeCoNi.

*Layers in x–y plane; vertical z-axis = build direction BD.
Reproduced from Guan et al. (2019). Microstructure and enhanced strength of laser additive manufactured CoCrFeNiMn high-entropy alloy. *Materials Science and Engineering* A744: 137–44.

to a fine wire to emit a stream of ions; and *sputtering*, in which an ionised plasma is used to remove atoms or molecules mechanically from the surface of a target cathode plate.[159] The stream of atoms, molecules or ions is usually created in a vacuum chamber or in a controlled atmosphere in order to prevent oxidation, but also sometimes in order to promote particular reactions with the gas atmosphere before deposition of the reaction products onto the substrate.

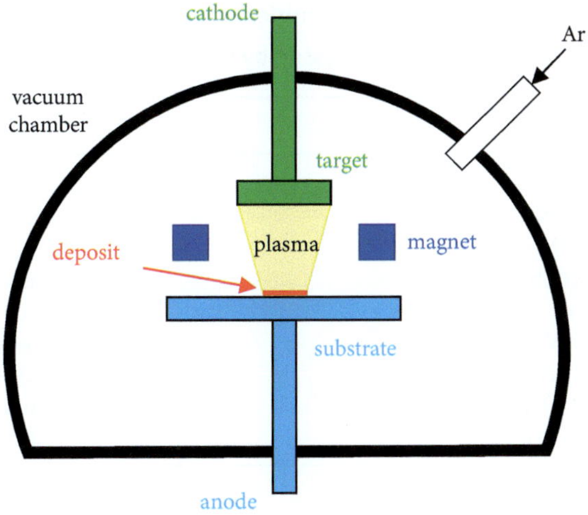

Figure 5.44 Schematic of magnetron sputtering.

Under most circumstances, the resulting material is a thin film ranging from a few nm to a few tens of nm thick, with a microstructure that can range from being an epitaxial single crystal at low deposition rates and high substrate temperatures and with a single crystal substrate to being nanocrystalline or amorphous at high deposition rates and low substrate temperatures and with a polycrystalline or amorphous substrate.[159] In some cases, however, higher rates of deposition can be achieved and the resulting materials are much thicker, up to several hundred nm or even several μm or tens of μm thick. A well-known example is *magnetron sputtering*, in which controlled magnetic fields intensify the ionising plasma and thus increase the sputtering and deposition rates. Magnetron sputtering has been used quite widely, therefore, to manufacture multicomponent high-entropy materials, and a typical magnetron sputtering arrangement is shown schematically in Figure 5.44.

Table 5.11 gives some examples of multicomponent high-entropy materials that have been manufactured by *sputter deposition*,[189–195] with others given in the review by Zhang et al.[159] The ceramic nitrides and oxides in Table 5.11 were manufactured by *reactive sputter deposition*, using respectively Ar/N_2 or Ar/O_2 gas mixtures instead of pure Ar to create the sputtering plasma. Figure 5.45 shows SEM cross-section images[192] of the typical columnar grain structure in sputter-deposited 7-component ($AlSi_xTiCrNbTa$)N nitride films with $x = 0–15$ at%.

Mechanical alloying

A convenient and effective method for alloying different starting materials and ensuring good mixing at the atomic or molecular level is by *ball milling*.[18,159] Powder mixtures of the starting components with the desired final composition are subjected to continual mechanical attrition by hard inert balls in a controlled atmosphere in a vibrating or rotating sealed chamber. A typical ball mill arrangement is shown schematically in Figure 5.46. When the main objective of the process is to alloy powders of different starting materials the process is

Table 5.11 Some examples of multicomponent high-entropy materials that have been processed by sputter deposition.[159,189–195]

alloy	structure
CrFeCoNiCu	fcc
$Al_xCrFeCoNiCu$	fcc + bcc
$CrFeCoNiNb_x$	fcc
AlSiTiVZrNbMoTa	amorphous
NbMoTaW	bcc
$(NbMoTaW)N_{0.5}$	rock salt
$(AlSi_xTiCrNbTa)N$	rock salt
$(BAlSiTiCr)N_x$	amorphous
$(AlTiCrZrTa)O_x$	amorphous

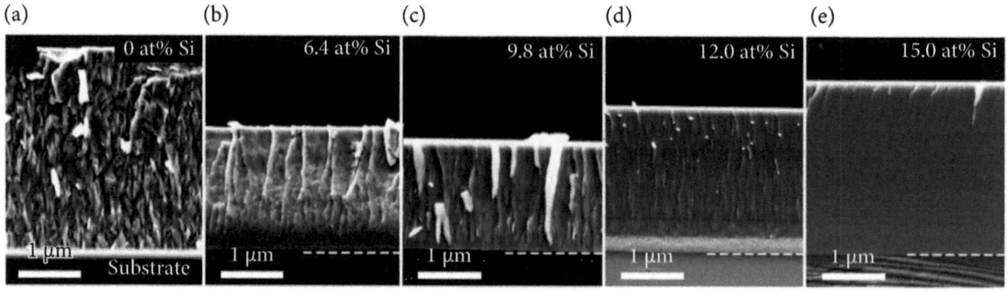

Figure 5.45 SEM cross-sectional images[185] of the columnar grain structure of fracture surfaces in sputter-deposited ($AlSi_xTiCrNbTa$)N films with $x = 0$–15 at%.

Reproduced from Kretschmer et al. (2021). Improving phase stability, hardness and oxidation resistance of reactively magnetron-sputtered (AlCrNbTaTi)N thin films by Si alloying. *Surface and Coatings Technology 416*: Article 127162.

called *mechanical alloying*; when the main objective of the process is to ensure atomic-level mixing in a single, already pre-alloyed powder the process is called *mechanical milling*. After sufficient milling time, typically several hours or even days, the starting components become very well mixed at an atomic or molecular level, and the resulting fine-scale powder material is often (though not always) single-phase nanocrystalline or amorphous, depending on its composition, the milling time and temperature, and the surrounding atmosphere.

Mechanical alloying and mechanical milling are very flexible and can be used to manufacture a wide variety of different multicomponent metallic, ceramic and composite materials, but processing is rather slow, energy intensive and only really effective for relatively small amounts of material.[196–207] As with rapidly solidified materials discussed in a previous sub-section, mechanically alloyed powders require relatively complex subsequent consolidation processing in order to make bulk products, often by *hot isostatic* pressing *(HIP)* or *spark plasma sintering (SPS)*. Table 5.12 shows some typical multicomponent high-entropy materials that have been made by mechanical alloying.[196–207]

Figure 5.47 shows a typical example of a series of X-ray diffraction (XRD) traces taken after different milling times from 0–60 hours during the manufacture of an equiatomic

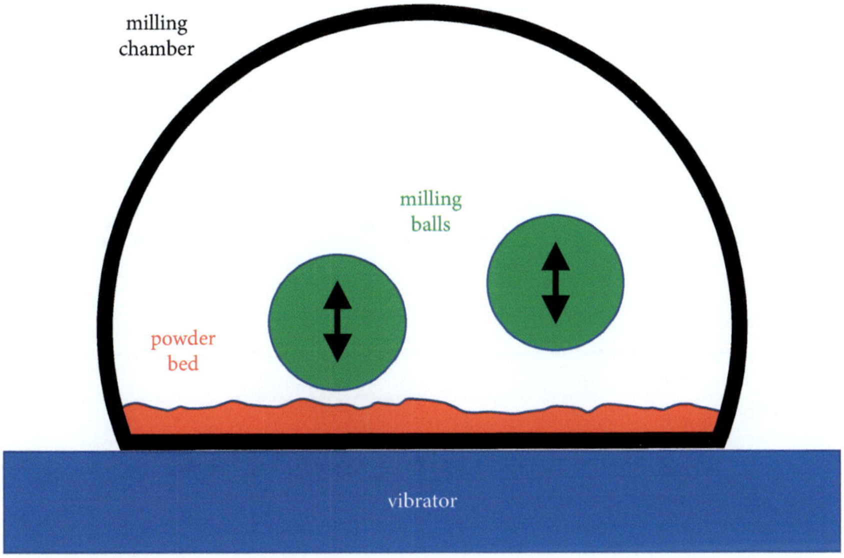

Figure 5.46 Schematic of ball milling.

Table 5.12 Some examples of multicomponent high-entropy materials that have been processed by mechanical alloying.[196–207]

alloy	structure
CrFeCoNiCu	fcc
AlCrFeCoNi	bcc
AlCrMnFeCoNi	bcc + fcc
AlTiCrFeCuZn	bcc
AlTiCrFe	bcc
CoNiCuZn	fcc
VNbMoTaW	bcc
VCrFeMo	bcc
VCrMnFeMo	bcc + σ
SiCrFeCoNi	amorphous
BAlSiFeNiNb	amorphous
(TiZrNbTaHf)(BCN)	rock salt
$(FeNbMoTaW)_3N$	amorphous + bcc

5-component single-phase bcc AlCrFeCoNi alloy from its pure components by mechanical alloying in a planetary ball mill.[197] Figure 5.48(a) shows a scanning electron microscope (SEM) image of the resulting fully alloyed powder after milling for sixty hours at room temperature; Figure 5.48(b) shows a corresponding energy-dispersive X-ray spectrum (EDS) of the fully alloyed powder composition; and Figures 5.49(a) and (b) show corresponding high-magnification bright-field transmission electron microscope (TEM) images of the fully alloyed nanocrystalline microstructure, with an inset selected area electron diffraction pattern (SADP) of the bcc structure.[197]

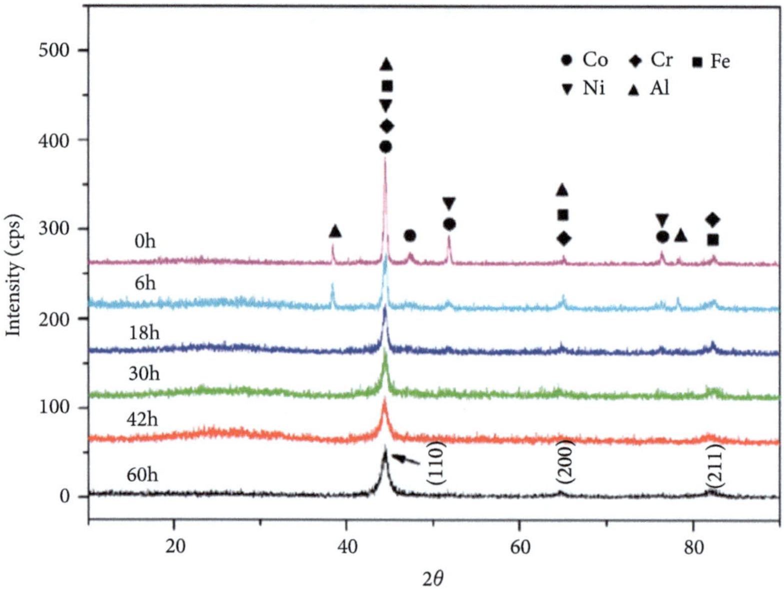

Figure 5.47 XRD traces from equiatomic 5-component single-phase bcc AlCrFeCoNi at different times during manufacture by mechanical alloying from pure starting components.
Data from Ji et al. (2008). Synthesis and characterization of nanocrystalline AlFeTiCrZnCu high-entropy solid solution by mechanical alloying. *Journal of Alloys and Compounds 460*: 253–7.

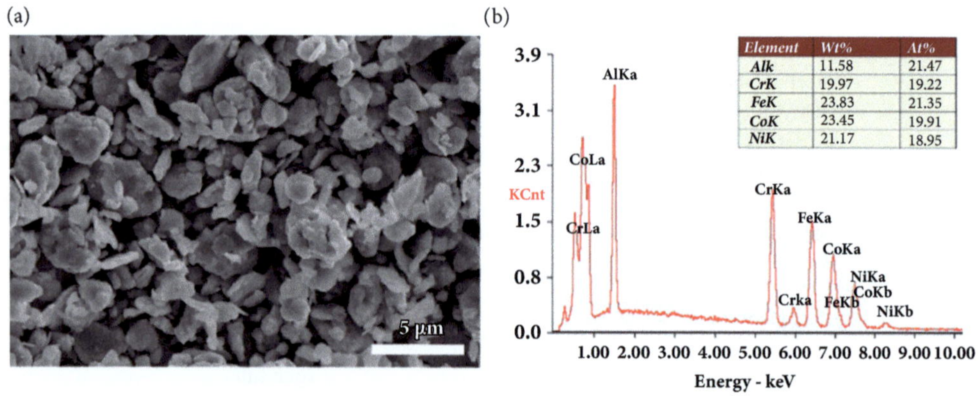

Figure 5.48 (a) SEM image and (b) EDS composition from equiatomic 5-component single-phase bcc AlCrFeCoNi after sixty hours ball milling.
Reproduced from Ji et al. (2008). Synthesis and characterization of nanocrystalline AlFeTiCrZnCu high-entropy solid solution by mechanical alloying. *Journal of Alloys and Compounds 460*: 253–7.

Chemical pyrolysis

One of the most common methods of manufacturing multicomponent metallic alloys and ceramic compounds is *chemical pyrolysis*, in which a mixed solution of different precursor salts is heated in a suitable inert or reactive atmosphere, causing their decomposition and

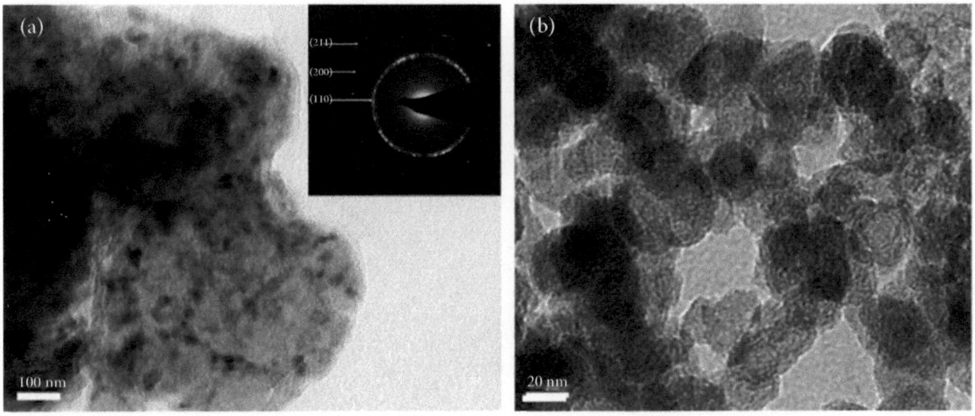

Figure 5.49 (a) and (b): TEM images and inset SADP of bcc nanocrystalline microstructure in equiatomic 5-component single-phase bcc AlCrFeCoNi after sixty hours ball milling.
Reproduced from Ji et al. (2008). Synthesis and characterization of nanocrystalline AlFeTiCrZnCu high-entropy solid solution by mechanical alloying. *Journal of Alloys and Compounds 460*: 253–7.

Table 5.13 Some examples of multicomponent high-entropy materials that have been processed by chemical pyrolysis.[211–217]

alloy	structure
FeCoNiPdPt	fcc
FeCoNiCuPdSnPtAu	fcc
RuRhReOsIr	hcp
RuRhPdOsIrPt	fcc
CoNiCuPdSnPtAu	fcc
MnCoNiCuRhPdSnIrPtAu	fcc
$(MnFeCoNiCu)_3O_4$	spinel
$La(CrMnFeCoNi)O_3$	perovskite
$(YLaNdSmGd)(CrMnFeCoNi)O_3$	perovskite
$(Ca_{0.5}Ba_{0.5}LaPrNdSm)(Fe_3CoNiCu_{0.5})O_3$	perovskite

inter-reaction to produce a friable or powdered deposit with the desired metallic or ceramic composition.[18,159,207–210]

Different starting mixtures of multiple salts and different gaseous atmospheres of multiple reactive gases can be used to synthesise a wide variety of multicomponent metallic alloys and ceramic compounds, in many cases readily forming high-entropy materials with multiple cations and multiple anions in a single crystal structure, the details depending on the composition and reaction time and temperature. Chemical pyrolysis has begun to be used quite widely in the last few years to manufacture a range of different multicomponent materials for a range of different potential functional applications. Table 5.13 gives some examples of multicomponent materials made by chemical pyrolysis, and more are given in the book by Murty et al.[18] and the reviews by Zhang et al.,[159] Zhang et al.,[207] Kusada et al.,[208] Li et al.[209] and Wang et al.[210]

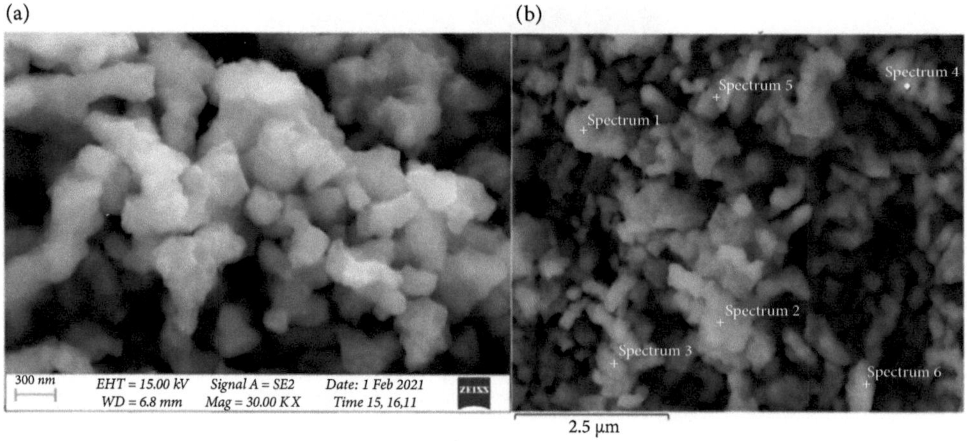

Figure 5.50 SEM images of calcined 11-component $(Ca_{0.5}Ba_{0.5}LaPrNdSm)(Fe_3CoNiCu_{0.5})O_3$ perovskite powder.

Reproduced from Yang et al. (2021). A high-entropy perovskite cathode for solid oxide fuel cells. *Journal of Alloys and Compounds 872*: Article 159633.

Figure 5.50 shows a typical scanning electron microscope (SEM) image of 50–100 nm sized 11-component single-phase cubic perovskite $(Ca_{0.5}Ba_{0.5}LaPrNdSm)$ $(Fe_3CoNiCu_{0.5})O_3$ powder manufactured by calcining mixed precursor nitrates dissolved in water and then heated for three hours at 1,000 °C (1,273 K).[217] Corresponding X-ray diffraction (XRD) traces, energy-dispersive (EDS) X-ray composition maps, and high-resolution transmission electron microscope (HRTEM) images demonstrated that all the ten cation components were distributed randomly across the A and B lattice sites of a single-phase ABO_3 cubic perovskite structure, as shown for instance in the EDS composition maps in Figure 5.51.

Thermomechanical processing

As-cast or powder-consolidated billets of material are often subjected to subsequent *thermomechanical processing (TMP)*, i.e. combinations of working, heat treatment and machining, in order to control the final shape of the product as well as the final material microstructure.[218,219] In general, multicomponent high-entropy alloys respond to the range of different thermomechanical processes in broadly the same way as conventional dilute binary and ternary alloys, though the detailed behaviour will, of course, depend on the precise material composition and applied processing conditions. Thus multicomponent metallic alloys can often readily be worked by processes such as *rolling*, *forging* and *extrusion* to manufacture respectively sheet, forgings and bar products; multicomponent metallic and ceramic powder preforms are often further consolidated and shaped by hot isostatic pressing and spark plasma sintering; homogeneity and grain size are often controlled by suitable combinations of deformation and heat treatment to produce *recrystallisation* and *strain relief*; and surface treatments such as *carburisation*, *anodisation*, *coating* and *shot peening* are used to harden product surfaces and provide protection from cracking and excessive oxidation or corrosion.[218,219]

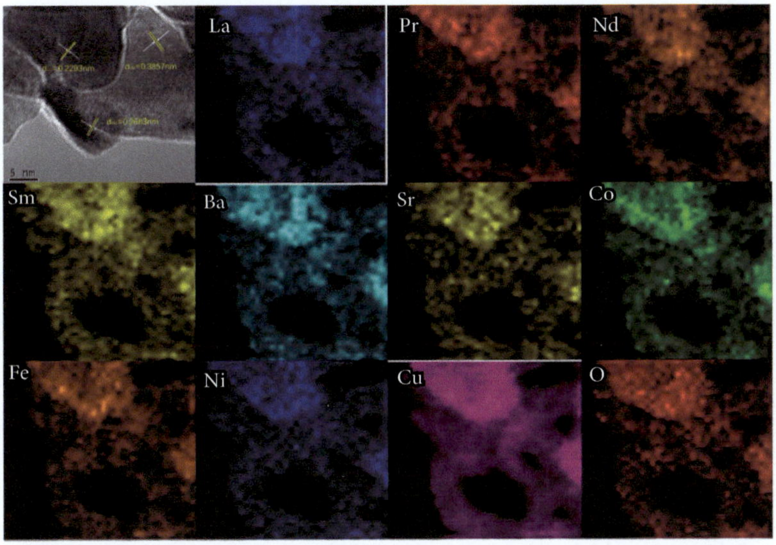

Figure 5.51 TEM image and EDS composition maps of 11-component single-phase cubic perovskite $(Ca_{0.5}Ba_{0.5}LaPrNdSm)(Fe_3CoNiCu_{0.5})O_3$ powder calcined from nitrate precursors for three hours at 1,000 °C.

Reproduced from Yang et al. (2021). A high-entropy perovskite cathode for solid oxide fuel cells. *Journal of Alloys and Compounds 872*: Article 159633.

A number of novel methods have been developed that use *severe plastic deformation* (*SPD*) to manufacture nanocrystalline materials with improved hardness and strength.[220,221] Severe plastic deformation can be achieved by a variety of different processes using different deformation modes, including *high-pressure torsion* (*HPT*), in which a bar of material is deformed by several hundred per cent by continuous axial torsion; *equal-channel angular pressing*, (*ECAP*), in which a bar of material is sheared through a right-angle channel; and *accumulative roll-bonding* (*ARB*), in which a sheet of material is alternately rolled, folded, re-rolled, re-folded and so on, many times over.[220,221] High-pressure torsion and equal-channel angular pressing are shown schematically in Figures 5.52 and 5.53, and Table 5.14 gives examples of multicomponent high-entropy materials that have been manufactured by severe plastic deformation,[222–231] with other examples given in the reviews by Koch[220] and Zhilyaev and Langdon.[221] Figure 5.54 shows typical high-resolution transmission electron microscope (HRTEM) images and energy-dispersive (EDS) X-ray composition maps of a typical nanocrystalline microstructure in an equiatomic 4-component single-phase bcc TiZrNbHf Senkov alloy manufactured by high-pressure torsion.[230]

4 Summary

A variety of different strategies have been employed to try to explore the vast regions of multicomponent phase space and thus identify materials with particular structures and desirable properties. These include experimental trial and error (sometimes called the Edisonian approach), multicomponent substitution of particular components with a number

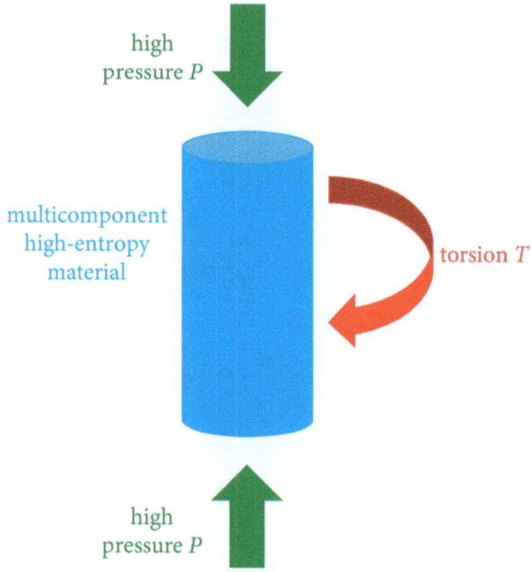

Figure 5.52 Schematic of high-pressure torsion (HPT).

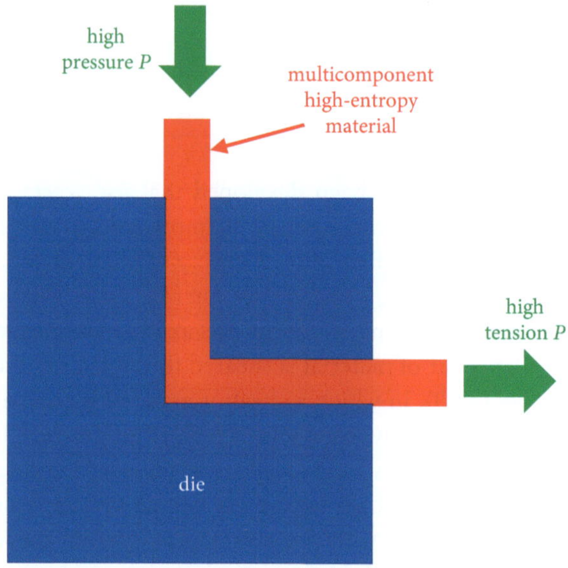

Figure 5.53 Schematic of equal-channel angular pressing (ECAP).

of chemically similar ones, parameterisation to develop an extended set of multicomponent Hume-Rothery rules, *ab initio* predictions based on fundamental thermodynamic and atomistic modelling, and data-driven statistical prediction using machine-learning and artificial-intelligence algorithms. None of these strategies has proved to be fully satisfactory. Trial-and-error experimentation and multicomponent substitution are, fairly obviously, somewhat hit and miss; parameterisation into multicomponent Hume-Rothery rules do not get much beyond binary Hume-Rothery rules, demonstrating not much more than that single phases tend rather broadly to be preferred when components have similar atomic sizes and electronic structures; computer-based thermodynamic Calphad and atomistic density

Table 5.14 Nanocrystalline multicomponent high-entropy alloys made by severe plastic deformation methods.[222–231]

alloy	processing method
CrMnFeCoNi	HPT
CrMnFeCoNi	ECAP
CrMnFeCoNi	rolling
$Cr_2Mn_{25}Fe_{41}Co_8Ni_{24}$	rolling
$Cr_{12.5}Fe_{55}Co_{17.5}Ni_{10}Mo_5$	HPT
$Al_{0.1}CrFeCoNi$	ECAP
$(CrFeCoNi)_{94}Al_4Ti_2$	HPT
TiZrNbHfTa	HPT
TiZrNbHf	HPT
MgAlTiFeZn	ECAP

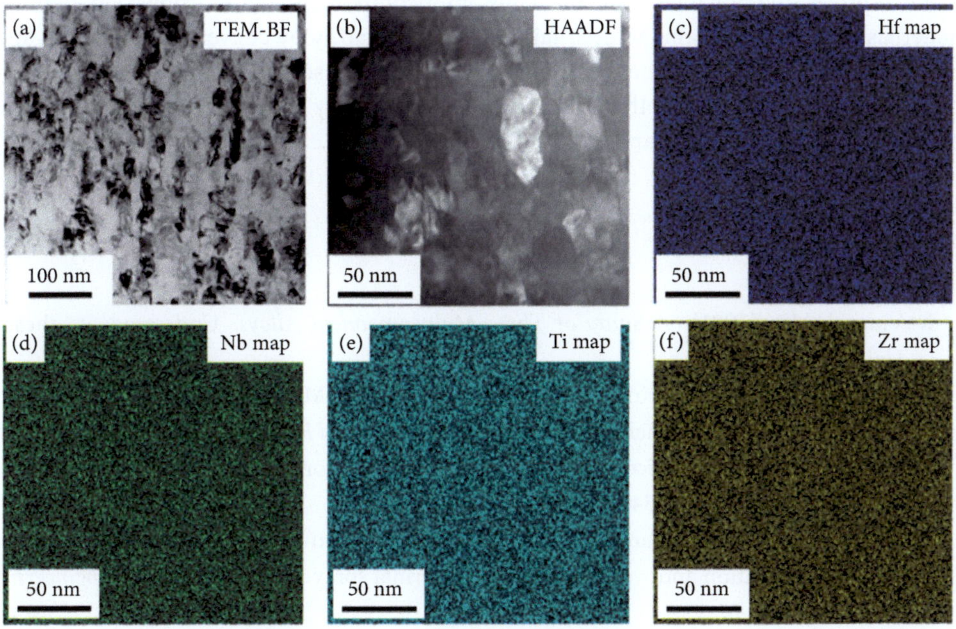

Figure 5.54 (a) bright-field (BF-TEM) and (b) high-angle annular dark-field (HAADF-TEM) images and (c)–(f) EDS composition maps of nanocrystalline equiatomic 4-component single-phase bcc TiZrNbHf manufactured by HPT.

Reproduced from Hung et al. (2014). Equal-channel angular extrusion of a low-density high-entropy alloy produced by high-energy cryogenic mechanical alloying. *JOM 66*: 2021–9.

functional theory (DFT), molecular dynamics (MD) and Monte Carlo (MC) calculations require many simplifying assumptions and many uncertain input material parameters, and still cannot handle the billions of different local atomic structures found in multicomponent high-entropy materials; and machine-learning (ML) algorithms only work well when interpolating within a set of known materials rather than being able to extrapolate into unknown regions of phase space. In the end, there is still no real substitute for a more traditional bootstrapping approach, using a combination of (hopefully) inspired experimentation supported by theoretical predictions as a useful but by no means guaranteed guide. This is now being done at scale by using high-throughput methods, i.e. by selecting an initial (large) range of candidate materials in a particular region of multicomponent phase space, using rapid (somewhat simplified) theoretical methods to sift out a smaller (but still quite large) number of candidate materials, followed by rapid (again somewhat simplified) experiments to sift out a (small) handful of best candidates, and only then applying a full-blown testing programme of experimental measurements and theoretical calculations.

For many millennia, we have restricted our materials to alloys with relatively few components in relatively dilute proportions. This has been because, as the number of components and their compositions begin to increase, the corresponding increase in levels of compound formation begins to make material processing difficult. We now know, however, that, as the number of components and their compositions increase further, the configurational entropy of mixing can often suppress compound formation, promoting single phases and making material processing much less difficult. Thus, multicomponent high-entropy solid-solution materials can often be manufactured fairly straightforwardly by conventional melting and casting, vapour deposition, chemical synthesis, powder consolidation and solid-state thermomechanical processing methods, in a broadly similar way to those used to manufacture more conventional dilute binary and ternary alloy materials.

References

1. Vincent AJB. (1981). *A Study of Three Multicomponent Alloys.* Undergraduate thesis. University of Sussex.
2. Cantor B, Chang ITH, Knight P, Vincent AJB. (2004). Microstructural development in equiatomic multicomponent alloys. *Materials Science and Engineering A375–377*: 213–18.
3. Cantor B. (2021). Multicomponent high-entropy Cantor alloys. *Progress in Materials Science 120*, Article 100754.
4. Huang PK, Yeh JW, Shun TT, Chen SK. (2004). Multi-principal-element alloys with improved oxidation and wear resistance for thermal spray coating. *Advanced Engineering Materials* 6: 74–8.
5. Yeh JW, Chen SK, Lin SJ, Gan JY, Chin TS, Shun TT, Tsau CH, Chang SY. (2004). Nanostructured high entropy alloys with multiple principal elements: Novel alloy design concepts and outcomes. *Advanced Engineering Materials* 6: 299–303.
6. Yeh JW. (2021). My trip from physics to high entropy materials. In J Brechtl, PK Liaw, eds., *Fundamental Studies of High-Entropy Materials: Theory, Experiments and Applications.* New York: Springer, chapter 2.
7. Cohen IB. (1976). The eighteenth-century origins of the concept of scientific revolution. *Journal of the History of Ideas 37*: 257–88.
8. Wootton D. (2016). *The Invention of Science: A New History of the Scientific Revolution.* London: Penguin Books.
9. Wills I. (2019). *Thomas Edison: Success and Innovation through Failure.* Studies in history and Philosophy of Science series, vol 52. Cham: Springer.

10. Weber RL. (1973). *A Random Walk in Science.* Bristol: Institute of Physics.
11. Kim KB. (2004). *Formation of Amorphous and Nanocrystalline Phases in Novel Multicomponent Alloys.* Doctorate thesis. University of Oxford.
12. Kim KB, Warren PJ, Cantor B. (2003). Formation of metallic glasses in novel (Ti33Zr33Hf33)((100-x-y)(Ni50Cu50)(x)Al(y) alloys. *Materials Transactions 44*: 411–13.
13. Kim KB, Warren PJ, Cantor B. (2003). Metallic glass formation in multicomponent (Ti, Zr, Hf, Nb)-(Ni, Cu, Ag)-Al alloys. *Journal of Non-Crystalline Solids 317*: 17–22.
14. Kim KB, Warren, Cantor B. (2005). Crystallization behaviour of novel $(TiZrHf)_{100-x-y}$ $(NiCu)_x$ Al_y with $x = 48–55$. *Journal of Metastable and Nanocrystalline Materials 24–25*: 657–60
15. Kim KB, Warren PJ, Cantor B. (2004). Glass-forming ability of novel multicomponent (Ti33Zr33Hf33)-(Ni50Cu50)-Al alloys developed by equiatomic substitution. *Materials Science and Engineering A375*: 317–21.
16. Rost CM, Sachet E, Borman T, Moballegh A, Dickey EC, Hou D, Jones JL, Curtarolo S, Maria JP. (2015). Entropy-stabilized oxides. *Nature Communications 6*: Article 8485.
17. Suryanarayana C, Inoue A. (2018). *Bulk Metallic Glasses.* Boca Raton: CRC Press, 1–170.
18. Murty BS, Yeh JW, Ranganathan S, Bhattacharjee PP. (2019). *High-Entropy Alloys.* Amsterdam: Elsevier.
19. Zhang RZ, Reece MJ. (2019). Review of high entropy ceramics: Design, synthesis, structure and properties. *Journal of Materials Chemistry A 7*: 22148–62.
20. Sarkar A, Breitung B, Hahn H. (2020). High entropy oxides: The role of entropy, enthalpy and synergy. *Scripta Materialia 187*: 43–8.
21. Xiang H, Xing Y, Dai FZ, Wang H, Su L, Miao L, Zhang G, Wang Y, Qi X, Yao L, Wang H, Zhao B, Li J, Zhou Y. (2021). High-entropy ceramics: Present status, challenges, and a look forward. *Journal of Advanced Ceramics 10*: 385–441.
22. Gaskell DR, Laughlin DE. (2017). *Introduction to the Thermodynamic of Materials*, 6th edn. Boca Raton: CRC Press.
23. Cantor B. (2020). *The Equations of Materials.* Oxford: Oxford University Press.
24. Massalski TB. (ed). (1990). *Binary Alloy Phase Diagrams.* Detroit: ASM International.
25. Reed RC. (2006). *The Superalloys: Fundamentals and Applications.* Cambridge: Cambridge University Press.
26. Farrar JCM. (2004). *The Alloy Tree: A Guide to Low-Alloy Steels, Stainless Steels and Nickel-Base Alloys.* Cambridge: Woodhead.
27. Davis JR (ed). (1993). *Aluminium and Aluminium Alloys.* Metals Park, OH: ASM International.
28. Villars P, Prince A, Okamoto H. (1995). *Handbook of Ternary Alloy Phase Diagrams.* Metals Park, OH: ASM International.
29. Yeh JW. Recent progress in high entropy alloys. (2006). *Annale de Chimie Science des Matériaux 31*: 633–48.
30. Yeh JW. (2013). Alloy design strategies and future trends in high-entropy alloys. *Journal of Metals 65*: 1759–71.
31. Gao MC, Yeh JW, Liaw PK, Zhang Y. (eds). (2016). *High-Entropy Alloys: Fundamentals and Applications.* Basel: Springer.
32. Cantor B. (2014). Multicomponent and high entropy alloys. *Entropy 16*: 4749–68.

33. Tsai MH, Yeh JW. (2014). High-entropy alloys: A critical review. *Materials Research Letters 2*: 107–23.
34. Zhang Y, Zuo TT, Tang Z, Gao, MC, Dahmen K, Liaw PK, Lu ZP. (2014). Microstructures and properties of high entropy alloys. *Progress in Materials Science 61*: 1–93.
35. Cantor B. (2021). A personal perspective on the discovery and significance of multicomponent high-entropy alloys. In J. Brechtl, PK Liaw, eds., *Fundamental Studies of High-Entropy Materials: Theory, Experiments and Applications.* New York: Springer, 1–36.
36. Hillert M. (2008). *Phase Equilibria, Phase Diagrams and Phase Transformations*, 2nd edn. Cambridge: Cambridge University Press.
37. Villars P, Calvert LD. (1991). *Pearson's Handbook of Crystallographic Data for Intermetallic Phases.* Metals Park, OH: ASM International.
38. Westbrook JH, Fleischer RL (eds). (2000). *Intermetallic Compounds: Vol 1: Crystal Structures of Intermetallic Compounds.* Chichester: Wiley-Blackwell.
39. Mackay AL. (2001). On complexity. *Crystallographic Reports 46*: 524–6.
40. Senkov ON, Miller JD, Miracle DB, Woodward C. (2015). Accelerated exploration of multi-principal element alloys with solid solution phases. *Nature Communications 6*: Article 6529.
41. Cao Q, Fan Z. Private communication.
42. Mizutani U. (2019). *Hume-Rothery Rules for Structurally Complex Alloy Phases.* Boca Raton: CRC Press.
43. Zhang Y, Zhou YJ, Lin JP, Chen GL, Liaw PK. (2008). Solid-solution phase formation rules for multi-component alloys. *Advanced Engineering Materials 10*: 534–8.
44. Yang X, Zhang Y. (2012). Prediction of high-entropy stabilized solid-solution in multi-component alloys. *Materials Chemistry and Physics 132*: 233–8.
45. Zhang Y, Yang X, Liaw PK. (2012). Alloy design and properties optimization of high-entropy alloys. *Journal of Metals l64*: 830–8.
46. Guo S, Hu Q, Ng C, Liu CT. (2013). More than entropy in high-entropy alloys: forming solid solutions or amorphous phase. *Intermetallics 41*: 96–103.
47. Takeuchi A, Amiya K, Wada T, Yubuta K, Zhanb W, Makino A. (2013). Entropies in alloy design for high entropy and bulk glassy alloys. *Entropy 15*: 3810–21.
48. Ren MX, Li BX, Fu HZ. (2013). Formation condition of solid solution type high-entropy alloy. *Transactions of the Nonferrous Metals Society of China (English Edition) 23*: 991–5.
49. Poletti MG, Battezzati L. (2014). Electronic and thermodynamic criteria for the occurrence of high entropy alloys in metallic systems. *Acta Materialia 75*: 297–306.
50. Salishchev GA, Tikhonovsky MA, Shaysultanov DG, Stepanov ND, Kuznetsov AV, Kolodiy IV, Tortika AS, Senkov ON. (2014). Effect of Mn and V on structure and mechanical properties of high-entropy alloys based on CoCr-FeNi system. *Journal of Alloys and Compounds 591*: 11–21.
51. Nong ZS, Zhu JC, Cao Y, Yang XW, Lai ZH, Liu Y. (2014). Stability and structure prediction of cubic phase in as cast high entropy alloys. *Materials Science and Technology 30*: 363–9.
52. Guo S. (2015). Phase selection rules for cast high entropy alloys: An overview. *Materials Science and Technology 31*: 1223–30.

53. Wang Z, Huang Y, Yang Y, Wang J, Liu CT. (2015). Atomic size effect and solid solubility of multicomponent alloys. *Scripta Materialia 94*: 28–31.
54. Ye YF, Wang Q, Lu J, Liu CT, Yang Y. (2015). The generalized thermodynamic rule for phase selection in multicomponent alloys. *Intermetallics 59*: 75–80.
55. King DJM, Middleburgh SC, Mc Gregor AG, Cortie MB. (2016). Predicting the formation and stability of single phase high-entropy alloys. *Acta Materialia 104*: 172–9.
56. Senkov ON, Miracle DB. (2016). A new thermodynamic parameter to predict formation of solid solutions or intermetallic phases in high entropy alloys. *Journal of Alloys and Compounds 658*: 603–7.
57. Caraballo IT, Rivera-Diaz-del-Castillo PEJ. (2016). A criterion for the formation of high entropy alloys based on lattice distortion. *Intermetallics 71*: 76–87.
58. Miracle DB, Senkov ON. (2017). A critical review of high-entropy alloys and related concepts. *Acta Materialia 122*: 448–511.
59. Soni VK, Sanyal S, Raja Rao K, Sinha SK. (2021). A review on phase prediction in high entropy alloys. *Journal of Mechanical Engineering Science 235*: 6268–86.
60. Greenwood NN, Earnshaw A. (1997). *Chemistry of the Elements*, 2nd edn. Oxford: Butterworth-Heinemann.
61. Weeber AW. (1987). Application of the Miedema model to formation enthalpies and crystallisation temperatures of amorphous alloys. *Journal of Physics F 17*: 809–13.
62. Li C, Yuan Y, Li F, Wei Q, Huang Y. (2022). Modification and verification of Miedema model for predicting thermodynamic properties of binary precipitates in multi-element alloys. *Physica B: Condensed Matter 627*: Article 413540.
63. Kaufman L, Bernstein H. (1970). *Computer Calculations of Phase Diagrams.* New York: Academic Press.
64. Saunders N, Miodownik AP. (1998). *Calphad: Calculation of Phase Diagrams: A Comprehensive Guide.* Oxford: Pergamon Elsevier.
65. Lukas HL, Fries SG, Sundman B. (2007). *Computational Thermodynamics.* Cambridge: Cambridge University Press.
66. Hsieh KC, Yu CF, Hsieh WT, Chiang WR, Ku JS, Lai JH, Tu CP, Yang CC. (2009). The microstructure and phase equilibrium of new high performance high-entropy alloys. *Journal of Alloys and Compounds 483*: 209–12.
67. Huang C, Zhang Y, Shen J, Vilar R. (2011). Thermal stability and oxidation resistance of laser clad TiVCrAlSi high entropy alloy coatings on Ti-6Al-4V alloy. *Surface and Coatings Technology 206*: 1389–95.
68. Gao MC, Alman DE. (2013). Searching for next single-phase high-entropy alloy compositions. *Entropy 15*: 4504–19.
69. Senkov ON, Zhang F, Miller JD. (2013). Phase composition of a CrMoNbTaTiZr high entropy alloy: Comparison of experimental and simulated data. *Entropy 15*: 3796–3809.
70. Saal JE, Berglund IS, Sebastian JT, Liaw PK, Olson GB. (2018). Equilibrium high entropy alloy phase stability from experiments and thermodynamic modeling. *Scripta Materialia 146*: 5–8.

71. Shun TT, Du YC. (2009). Microstructure and tensile behaviors of fcc $Al_{0.3}CoCrFeNi$ high entropy alloy. *Journal of Alloys and Compounds 479*: 157–60.
72. Kao YF, Chen SK, Chen TJ, Chu PC, Yeh JW, Lin SJ. (2011). Electrical, magnetic and Hall properties of $Al_xCoCrFeNi$ high-entropy alloys. *Journal of Alloys and Compounds 509*: 1607–14.
73. Wang WR, Wang WL, Yeh JW. (2014). Phases, microstructure and mechanical properties of $Al_xCoCrFeNi$ high-entropy alloys at elevated temperatures. *Journal of Alloys and Compounds 589*: 143–52.
74. Tang Q, Huang Y, Cheng H, Liao X, Langdon TG, Dai P. (2016). The effect of grain size on the annealing-induced phase transformation in an $Al_{0.3}CoCrFeNi$ high entropy alloy. *Materials Design 105*: 381–5.
75. He F, Wang Z, Wu Q, Li J, Wang J, Liu CT. (2017). Phase separation of metastable CoCrFeNi high entropy alloy at intermediate temperatures. *Scripta Materialia* 126: 15–19.
76. Zhang C, Gao MC. (2016). CALPHAD modelling of high-entropy alloys. In MC Gao, JW Yeh, PK Liaw, Y Zhang (eds.), *High Entropy Alloys: Fundamentals and Applications.* Cham: Springer International, 399–444.
77. Zhang C, Zhang F, Diao H, Gao MC, Tang Z, Poplawsky JD, Liaw PK. (2016). Understanding phase stability of AlCoCrFeNi high entropy alloys. *Materials and Design 109*: 425–33.
78. Sistla HR, Newkirk JW, Liou F. (2015). Effect of Al/Ni ratio and heat treatment on phase transformations and microstructure of $Al_xFeCoCrNi_{2x}$ high entropy alloys. *Materials Design 81*: 113–21.
79. Otto F, Dlouhý, Pradeep KG, Kuběnovà, Raabe D, Eggeler G, George EP. (2016). Decomposition of the single-phase high-entropy alloy CrMnFeCoNi after prolonged anneals at intermediate temperatures. *Acta Materialia 112*: 40–52.
80. Ma D, Yao M, Pradeep KG, Tasan CC, Springer H, Raabe D. (2015). Phase stability of non-equiatomic CoCrFeMnNi high entropy alloys. *Acta Materialia 98*: 288–96.
81. Otto F, Yang Y, Bei H, George E. (2013). Relative effects of enthalpy and entropy on the phase stability of equiatomic high-entropy alloys. *Acta Materialia 61*: 2628–38.
82. Zhang B, Gao MC, Zhang Y, Yang S, Guo SM. (2015). Senary refractory high entropy alloy MoNbTaTiVW. *Materials Science and Technology 31*: 1207–13.
83. Barber ZH (ed.). (2005). *Introduction to Materials Modelling.* London: Maney.
84. Giustino F. (2014). *Materials Modelling Using Density Functional Theory.* Oxford: Oxford University Press.
85. Hasnip PJ. (2005). Advanced atomistic modelling. In: ZH Barber (ed.), *Introduction to Materials Modelling.* London: Maney, 29–44.
86. Car R, Parrinello M. (1985). Unified approach for molecular dynamics and density functional theory. *Physical Review Letters 55*: 2471–4.
87. Born M, Oppenheimer R. (1927). Zur Quantentheorie der Molekeln [On the quantum theory of molecules]. *Annalen der Physik 84*: 457–84.
88. Hohenberg P, Kohn W. (1964). Inhomogeneous electron gas. *Physical Review B 136*: 864–71.

89. Hartree D. (1928). The wave mechanics of an atom with a non-Coulomb central field. *Proceedings of the Cambridge Philosophical Society 24*: 89–110.
90. Kohn W, Sham L. (1965). Self-consistent equations including exchange and correlation effects. *Physical Review A 140*: 1133–8.
91. Bloch F. (1928). Über die Quantenmechanik der Elektronen in Kristallgittern. *Zeitschrifft für Physik 52*: 555–600.
92. Heine V. (1970). The pseudopotential concept. *Solid State Physics 24*: 1–36.
93. Alavi S. (2020). *Molecular Simulations.* Weinheim: Wiley, 5–33
94. Satoh A. (2011). *Introduction to the Practice of Molecular Simulation: Molecular Dynamics, Monte Carlo, Brownian Dynamics, Dissipative Particle Dynamics.* Burlington: Elsevier Science.
95. Tian F. (2017). A review of solid-solution models of high-entropy alloys based on *ab initio* calculations. *Frontiers in Materials 4*: Article 36.
96. Tian F, Wang Y, Irving DL, Vitos L. (2016). Applications of coherent potential approximations to HEAs. In MC Gao, JW Yeh, PK Liaw, Y Zhang (eds.), *High-Entropy Alloys: Fundamentals and Applications.* Cham: Springer International, 299–332.
97. Gao MC, Niu C, Jiang C, Irving DL. Applications of special quasi-random structures to high-entropy alloys. In MC Gao, JW Yeh, PK Liaw, Y Zhang (eds.), *High-Entropy Alloys: Fundamentals and Applications.* Cham: Springer International, 333–368.
98. Ma D, Grabowski B, Körmann F, Neugebauer J, Raabe D. (2015). *Ab initio* thermodynamics of the CoCrFeMnNi high entropy alloy: Importance of entropy contributions beyond the configurational one. *Acta Materialia 100*: 90–7.
99. Sun X, Zhang H, Lu S, Ding X, Wang Y, Vitos L. (2017). Phase selection rule for Al-doped CrMnFeCoNi high-entropy alloys from first principles. *Acta Materialia 140*: 366–74.
100. He JY, Liu WH, Wang H, Wu Y, Liu XJ, Nieh TG, Lu ZP. (2014). Effects of Al addition on structural evolution and tensile properties of the FeCoNiCrMn high-entropy alloy system. *Acta Materialia 62*: 105–13.
101. Tian FY, Delczeg L, Chen NX, Varga LK, Shen J, Vitos L. (2013). Structural stability of $NiCoFeCrAl_x$ high-entropy alloy from *ab initio* theory. *Physical Review B 88*: Article 085128.
102. Yang S, Zhong Y. (2021). Ab initio modeling of fcc FeCoCrNi high entropy alloys with full composition range. *Journal of Phase Equilibria and Diffusion 42*: 656–72.
103. Yang S, Liu G, Zhong Y. (2022). Revisit the VEC criterion in high entropy alloys (HEAs) with high throughput ab initio calculations: A case study with AlCoCrFeNi system. *Journal of Alloys and Compounds 916*: Article 165477.
104. Guo S, Ng C, Lu J, Liu C. (2011). Effect of valence electron concentration on stability of fcc or bcc phase in high entropy alloys. *Journal of Applied Physics 109*: Article 103505.
105. Tian F, Varga LK, Chen N, Shen J, Vitos L. (2014). Ab initio design of elastically isotropic $TiZrNbMoV_x$ high-entropy alloys. *Journal of Alloys and Compounds 599*: 19–25.
106. McAlpine SW, Logan JV, Short MP. (2021). Predicting single phase stability and segregation in the NbMoTaTi-(W,V) high entropy alloy system with the vacancy exchange potential. *Scripta Materialia 191*: 29–33.

107. Zhao Q, Li J, Fang Q, Feng H. (2019). Effect of Al solute concentration on mechanical properties of Al_xFeCuCrNi high-entropy alloys: A first principles study. *Physica B:Condensed Matter 566*: 30–7.
108. Niu C, Zaddach AJ, Koch CC, Irving DL. (2016). First principles exploration of near-equiatomic NiFeCrCo high entropy alloys. *Journal of Alloys and Compounds 672*: 510–20.
109. Song H, Tian F, Hu QM, Vitos L, Wang Y, Shen J, Chen N. (2017). Local lattice distortion in high-entropy alloys. *Physical Review Materials 1*: Article 023404.
110. Karabin M, Mondal WR, Östlin A, Ho WGD, Dobrosavljevic V, Tam KM, Terletska H, Chioncel L, Wang Y, Eisenbach M. (2022). Ab initio approaches to high-entropy alloys: A comparison of CPA, SQS and supercell methods. *Journal of Materials Science 57*: 10677–90.
111. Chen SM, Ma ZJ, Qiu S, Zhang LJ, Zhang SZ, Yang R, Hu QM. (2022). Phase decomposition and strengthening in HfNbTaTiZr high entropy alloy from first-principles calculations. *Acta Materialia 225*: Article 117582.
112. Biermair F, Razumovskiy VI, Ressel G. (2022). Influence of alloying on thermodynamic properties of AlCoCrFeNiTi high entropy alloys from DFT calculations. *Computational Materials Science 202*: Article 110952.
113. San S, Tong Y, Bei H, Kombaiah B, Zhang Y, Ching WY. (2021). First-principles calculation of lattice distortion in four single phase high entropy alloys with experimental validation. *Materials and Design 209*: Article 110071.
114. Liu S, Cao P, Lin DY, Tian F. (2018). Stability of $L2_1$ $(NiM)_2TiAl$ (M=Co, Fe) in high-entropy alloys. *Journal of Alloys and Compounds 764*: 650–5.
115. Wang S, Chen S, Jia Y, Hu Z, Huang H, Yang Z, Dong A, Zhu G, Wang D, Shu D, Tian F, Dai Y, Sun B. (2019). Fcc-$L1_2$ ordering transformation in equimolar FeCoNiV multi-principal element alloy. *Materials and Design 168*: Article 107648.
116. Ryitsev, R, Gaviko V, Estemirova S, Sterkhov E, Cherepanova L, Yagodin D, Chtchelkatchev N, Dubinin N, Uporov S. (2021). Laves formation in high entropy alloys. *Metals 11*: Article 1962.
117. Qiao L, Liu Y, Zhu J. (2021). A focused review on machine learning aided high-throughput methods in high entropy alloy. *Journal of Alloys and Compounds 877*: Article 160295.
118. Hart GLW, Mueller T, Toher C, Curtarolo S. (2021). Machine learning for alloys. *Nature Reviews Materials* 6: 730–55.
119. Liu X, Xu P, Zhao J, Lu W, Li M, Wang G. (2022). Material machine learning for alloys: Applications, challenges and perspectives. *Journal of Alloys and Compounds 921*:Article 165984.
120. Burkov A. (2019). *The Hundred-Page Machine Learning Book.*. Quebec City: True Positive Incorporated.
121. Flach P. (2012). *Machine Learning: The Art and Science of Algorithms That Make Sense of Data.* Cambridge: Cambridge University Press.
122. Russell S, Norvig P. (2022). *Artificial Intelligence: A Modern Approach*, 4th edn. Harlow: Pearson Education, 669–873.
123. Islam N, Huang W, Zhuang HL. (2018). Machine learning for phase selection in multi-principal element alloys. *Computational Materials Science 150*: 230–5.

124. Bobbili R, Ramakrishna B, Madhu V. (2022). Development of machine learning based models for design of high entropy alloys. *Materials Technology 37*: Article 2046930.
125. Huang W, Martin P, Zhuang HL. (2019). Machine learning phase prediction of high-entropy alloys. *Acta Materialia 169*: 225–36.
126. Mandal P, Choudhury A, Mallick AB, Ghosh M. (2022). Phase prediction in high entropy alloys by various machine learning modules using thermodynamic and configurational parameters. *Metals and Materials International 29*: 38–52.
127. Yan Y, Lu D, Wang K. (2021). Accelerated discovery of single-phase refractory high entropy alloys assisted by machine learning. *Computational Materials Science 199*: Article 110723.
128. Nassar AE, Mullis AM. (2021). Rapid screening of high entropy alloys using neural networks and constituent elements. *Computational Materials Science 199*: Article 110755.
129. Kaufmann K, Vecchio KS. (2020). Searching for high entropy alloys: A machine learning approach. *Acta Materialia 198*: 178–222.
130. Jaiswal UK, Krishna YV, Rahul MR, Phanikumar G. (2021). Machine learning-enabled identification of new medium- to high-entropy alloys with solid solution phases. *Computational Materials Science 197*: Article 110623.
131. Wen T, Ye B, Liu H, Ning S, Wang CZ, Chu Y. (2019). Formation criteria for binary metal diboride solid solutions established through combinatorial methods. *Journal of the American Ceramic Society 103*: 3338–48.
132. Kaufmann K, Maryanovsky D, Mellor WM, Zhu C, Rosengarten AS, Harrington TJ, Oses C, Toher C, Curtarolo S, Vecchio KS. (2020). Discovery of high-entropy ceramics via machine learning. *NPJ Computational Materials 6*: Article 42.
133. Dai FZ, Wen B, Sun Y, Xiang H, Zhou Y. (2020). Theoretical prediction on thermal and mechanical properties of high entropy (ZrHfTiNbTa)C by deep learning potential. *Journal of Materials Science and Technology 43*: 168–74.
134. Dai FZ, Sun Y, Wen B, Xiang H, Zhou Y. (2021). Temperature dependent thermal and elastic properties of high entropy $(TiZrHfNbTa)B_2$: Molecular dynamics simulation by deep learning potential. *Journal of Materials Science and Technology 72*: 8–15.
135. Mellor WM, Kaufmann K, Dippo OF, Figuera SD, Schrader GD, Vecchio KS. (2021). Development of ultrahigh-entropy ceramics with tailored oxidation behavior. *Journal of the European Ceramic Society 41*: 5791–5800.
136. Zhang J, Xu B, Xiong Y, Ma S, Wang Z, Wu Z, Zhao S. (2022). Design high-entropy carbide ceramics from machine learning. *NPJ Computational Materials 8*: Article 5.
137. Mitra R, Bajpai A, Biswas K. (2022). Machine learning based approach for phase prediction in high entropy borides. *Ceramics International 48*: 16695–706.
138. Chang YJ, Jui CY, Lee WJ, Yeh AC. (2018). Prediction of the composition and hardness of high-entropy alloys by machine learning. *Journal of Metals 71*: 3433–42.
139. Wen CW, Zhang Y, Wang C, Xue D, Bai Y, Antonav S, Dai L, Lookman T, Su Y. (2019). Machine learning assisted design of high entropy alloys with desired property. *Acta Materialia 170*: 109–17.

140. Wen C, Wang C, Zhang Y, Antonov S, Xue D, Lookman T, Su Y. (2021). Modeling solid solution strengthening in high entropy alloys using machine learning. *Acta Materialia 212*: Article 116917.

141. Bhandari U, Rafi, MR, Zhang C, Yang S. (2021). Yield strength prediction of high entropy alloys using machine learning. *Materials Today Communications 26*: Article 101871.

142. Huang X, Jin C, Zhang C, Zhang H, Fu H. (2021). Machine learning assisted modelling and design of solid solution hardened high entropy alloys. *Materials and Design 211*: Article 110177.

143. Bundela AS, Rahul MR. (2022). Machine learning-enabled framework for the prediction of mechanical properties in new high entropy alloys. *Journal of Alloys and Compounds 908*: Article 164578.

144. Miracle DB, Miller JD, Senkov ON, Woodward C, Uchie MD, Tiley J. (2014). Exploration and development of high entropy alloys for structural applications. *Entropy 16*:494–525.

145. Lederer Y, Toher C, Vecchio KS, Curtarolo S. (2018). The search for high entropy alloys: a high-throughput *ab initio* approach. *Acta Materialia 159*: 364–83.

146. Feng R, Zhang C, Gao MC, Pei Z, Zhang F, Chen Y, Ma D, An K, Poplawsky JD, Ouyang L, Ren Y, Hawk JA, Widom M, Liaw PK. (2021). High-throughput design of high-performance lightweight high-entropy alloys. *Nature Communications 12*: Article 4329.

147. Shang G, Jiang L, Liu ZZ, Lu XG. (2022). High-throughput experimental study on the microstructural and compositional variations of mechanical properties for AlCoCrFeNi high entropy alloys. *Journal of Alloys and Compounds 917*: Article 165513.

148. Shi Y, Yang B, Rack PD, Guo S, Liaw PK, Zhao Y. (2020). High-throughput synthesis and corrosion behaviour of sputter-deposited nanocrystalline $Al_x(CoCrFeNi)_{100-x}$ combinatorial high-entropy alloys. *Materials and Design 195*: Article 109018.

149. Li M, Flores KM. (2020). Laser processing as a high-throughput method to investigate microstructure-processing-property relationships in multiprincipal element alloys. *Journal of Alloys and Compounds 825*: Article 154025.

150. Melia MA, Whetten SR, Puckett R, Jones M, Heiden MJ, Argibay N, Kustas AB. (2020). High-throughput additive manufacturing and characterization of refractory high entropy alloys. *Applied Materials Today 19*: Article 100560.

151. Moorehead M, Bertsch K, Niezgoda M, Parkin C, Elbakshwan M, Sridharan K, Zhang C, Thoma D, Couet A. (2020). High-throughput synthesis of Mo-Nb-Ta-W high-entropy alloys via additive manufacturing. *Materials and Design 187*: Article 108358.

152. Dobbelstein H, George EP, Gurevich EL, Kostka A, Ostendorf A, Laplanche G. (2021). Laser metal deposition of refractory high-entropy alloys for high-throughput synthesis and structure-property characterization. *International Journal of Extreme Manufacturing 3*: Article 015201.

153. Hsu SY, Lai YT, Chang SY, Tsai SY, Duh JG. (2022). Combinatorial synthesis of reactively co-sputtered high entropy nitride (HfNbTiVZr) coatings: Microstructure and mechanical properties. *Surface and Coatings Technology 442*: Article 128564.

154. Liu B, Zhao J, Liu Y, Xi J, Li Q, Xiang H, Zhou Y. (2021). Application of high-throughput first-principles calculations in ceramic innovation. *Journal of Materials Science and Technology 88*: 143–57.

155. Wen et al., Formation.
156. Zhang et al., Design.
157. Li Z, Ludwig A, Savan A, Springer H, Raabe D. (2018). Combinatorial metallurgical synthesis and processing of high-entropy alloys. *Journal of Materials Research 33*: 3156–69.
158. Li R, Xie L, Wang WY, Liaw PK, Zhang Y. (2020). High-throughput calculations for high-entropy alloys: A brief review. *Frontiers in Materials 7*: Article 290.
159. Zhang Y, Koch CC, Ma SG, Zhang H, Pan Y. (2016). Fabrication routes. In MC Gao, JW Yeh, PK Liaw, Y Zhang (eds.), *High-Entropy Alloys: Fundamentals and Applications*. Cham: Springer International, 151–79.
160. Laurent-Brocq M, Akhatova A, Perrière L, Chebini S, Sauvage X, Leroy E, Champion Y. (2015). Insights into the phase diagram of the CrMnFeCoNi high entropy alloy. *Acta Materialia 88*: 355–65.
161. Senkov ON, Wilks GB, Miracle DB, Chuang CP, Liaw PK. (2010). Refractory high-entropy alloys. *Intermetallics 18*: 1758–65.
162. Muralikrishna GM, Esther ACM, Guruvidyathri K, Watermeyer P, Liebscher CH, Kulkarni KN, Wilde G, Divinski SV, Murty BS. (2020). Novel multicomponent B2-ordered aluminides: Compositional design, synthesis, characterization and thermal stability. *Metals 10*: Article 1411.
163. Chanda B, Das J. (2018). Composition dependence on the evolution of nanoeutectic in $CoCrFeNiNb_x$ ($0.45<x<0,65$) high entropy alloys. *Advanced Engineering Materials 20*: Article 1700908.
164. Cech J, Hausild P, Karlik M, Capek J, Prusa F. (2021). Indentation size effect in CoCrFeMnNi HEA prepared by various techniques. *Materials 14*: Article 237246.
165. Chen C, Zhang H, Fan Y, Zhang W, Wei R, Wang T, Zhang T, Li F. (2020). A novel ultrafine grained high entropy alloy with excellent combination of mechanical and soft magnetic properties. *Journal of Magnetism and Magnetic Materials 502*: Article 166513.
166. Wang Q, Li Y, Yang Y, Zhang X, Guo Y, Jia Z, Shen B. (2022). Self-supported AlFeNiCoMo high-entropy alloy with micropillar array structure for efficient oxygen evolution reaction. *APL Materials 10*: Article 101106.
167. Zhang Y, Chen Z, Cao D, Zhang J, Zhang P, Tao Q, Yang X. (2019). Concurrence of spinodal decomposition and nano-phase precipitation in a multi-component AlCoCrCuFeNi high-entropy alloy. *Journal of Materials Research and Technology 8*: 726–36.
168. Koga GY, Travessa D, Zepon G, Coimbrão DD, Jorge Jr AM, Berger JE, Roche V, Lepretre JC, Bolfarino C, Kiminami CS, Wang F, Zhu SL, Inoue A, Botta WJ. (2021). Corrosion resistance of pseudo-high entropy Fe-containing amorphous alloys in chloride-rich media. *Journal of Alloys and Compounds 884*: Article 161090.
169. Yadav TP, Mukhopadhyay S, Mishra SS, Mulhopadhyay NK, Srivastava ON. (2017). Synthesis of a single phase of high entropy Laves intermetallics in the TiZrVCrNi equiatomic alloy. *Philosophical Magazine Letters 97*: 494–503.
170. Laplanche G, Schneider M, Scholz F, Frenzel J, Eggeler G, Schreuer J. (2020). Processing of a single-crystalline CrCoNi medium entropy alloy and evolution of its thermal expansion and elastic stiffness coefficients with temperature. *Scripta Materialia 177*: 44–8.
171. Glatzel U, Schleifer F, Gadelmeier C, Krieg F, Müller M, Mosbacher M, Völkl R. (2021). Quantification of solid solution strengthening and internal stresses

through creep testing of Ni-containing single crystals at 980 °C. *Metals 11*: Article 1071130.

172. Wu Z, Gao YF, Bei H. (2015). Single crystal plastic behavior of a single-phase, fcc equiatomic FeNiCrCo alloy. *Scripta Materialia 109*: 108–12.
173. Gaertner D, Kottke J, Wilde G, Divinski SV, Chumlyakov Y. (2018). Tracer diffusion in single crystalline CoCrFeNi and CoCrFeMnNi high entropy alloys. *Journal of Materials Research 33*: 3184–91.
174. Gadelmeier C, Haas S, Lienig T, Manzoni A, Feuerbacher M, Glatzel U. (2020). Temperature dependent solid solution strengthening in the high entropy alloy CrMnFeCoNi in single crystalline state. *Metals 10*: Article 111412.
175. Feuerbacher M, Würtz, Kovács A, Thomas C. (2017). Single-crystal growth of a FeCoCrMnAl high-entropy alloy. *Materials Research Letters 5*: 128–34.
176. Ma SG, Zhang SF, Qiao JW, Wang ZH, Gao MC, Jiao ZM, Yang HJ, Zhang Y. (2014). Superior high tensile elongation of a single-crystal $CoCrFeNiAl_{0.3}$ high-entropy alloy by Bridgman solidification. *Intermetallics 54*: 104–9.
177. Zheng H, Xu Q, Chen R, Qin G, Li X, Su Y, Guo J, Fu H. (2020). Microstructure evolution and mechanical property of directionally solidified CoCrFeMnNi high entropy alloy. *Intermetallics 119*: Article 106723.
178. Singh S, Ramakrishna S, Singh R. (2017). Material issues in additive manufacturing: A review. *Journal of Manufacturing Processes 25*: 185–200.
179. Herzog D, Seyda V, Wycisk E, Emmelmann C. (2016). Additive manufacturing of metals. *Acta Materialia 117*: 371–92.
180. Torralba JM, Campos M. (2020). High entropy alloys manufactured by additive manufacturing. *Metals 10*: Article 050639.
181. Guan S, Wan D, Solberg K, Berto F, Welo T, Yue TM, Chan KC. (2019). Additive manufacturing of fine-grained and dislocation-populated CrMnFeCoNi high entropy alloy by laser engineered net shaping. *Materials Science and Engineering A761*: Article 138056.
182. Chew Y, Bi GJ, Zhu ZG, Ng FL, Weng F, Liu SB, Nai SML, Lee BY. (2019). Microstructure and enhanced strength of laser additive manufactured CoCrFeNiMn high entropy alloy. *Materials Science and Engineering A744*: 137–44.
183. Xiang S, Li J, Luan H, Amar A, Lu S, Li K, Zhang L, Liu X, Le G, Wang X, Qu F, Zhang W, Wang D, Li Q. (2019). Effects of process parameters on microstructures and tensile properties of laser melting deposited CrMnFeCoNi high entropy alloys. *Materials Science and Engineering A743*: 412–17.
184. Joseph J, Hodgson P, Jarvis T, Wu X, Stanford N, Fabijanic DM. (2018). Effect of hot isostatic pressing on the microstructure and mechanical properties of additive manufactured $Al_xCoCrFeNi$ high entropy alloys. *Materials Science and Engineering 733*: 59–70.
185. Wu W, Zhou R, Wei B, Ni S, Liu Y, Song M. (2018). Nanosized precipitates and dislocation networks reinforced C-containing CoCrFeNi high entropy alloy fabricated by selective laser melting. *Materials Characterization 144*: 605–10.
186. Dobbelstein H, Gurevich EL, George EP, Ostendorf A, Laplanche G. (2018). Laser metal deposition of a refractory TiZrNbHfTa high-entropy alloy. *Additive Manufacturing 24*: 386–90.

187. Dobbelstein H, Thiele M, Gurevich EL, George EP, Ostendorf A. (2016). Direct metal deposition of refractory high entropy alloy MoNbTaW. *Physics Procedia 83*: 624–33.
188. Braekman BR, Depla D. (2015). Structure formation and properties of sputter deposited $Nb_xCoCrCuFeNi$ high entropy alloy thin films. *Journal of Alloys and Compounds 646*:810–15.
189. Chen TK, Wong MS. (2007). Structure and properties of reactively sputtered $Al_xCoCrCuFeNi$ oxide films. *Thin Solid Films 516*: 141–6.
190. Tsai MH, Yeh JW, Gan JW. (2008). Diffusion barrier properties of AlMoNbSiTaTiVZr high-entropy alloy layer between copper and silicon. *Thin Solid Films 516*: 5527–30.
191. Lee JW, Chen CY, Chen YJ, Tzeng CH, Chen YI. (2022). Effects of nitrogen flow rate on the structural, mechanical and anti-corrosive properties of co-sputtered $(NbTaMoW)N_x$ films. *Journal of Materials Research and Technology 21*: 1890–1902.
192. Kretschmer A, Kirnbauer A, Moraes V, Primetzhofer D, Yalamanchili K, Rudiger H, Mayrhoher PH. (2021). Improving phase stability, hardness and oxidation resistance of reactively magnetron sputtered (AlCrNbTaTi)N thin films by Si alloying. *Surface and Coatings Technology 416*: Article 127162.
193. Tsai CW, Lai SW, Cheng KH, Tsai MH, Davison A, Tsau CH, Yeh JW. (2012). Strong amorphization of high-entropy AlBSiCrTi nitride film. *Thin Solid Films 520*: 2613–18.
194. Lin MI, Tsai MH, Shen WJ, Yeh JW. (2010). Evolution of structure and properties of multi-component $(AlCrTaTiZr)O_x$ films. *Thin Solid Films 518*: 2732–7.
195. Ji W, Wang W, Wang H, Zhang J, Wang Y, Zhang F, Fu Z. (2015). Alloying behavior and novel properties of CoCrFeNiMn high-entropy alloy fabricated by mechanical alloying and spark plasma sintering. *Intermetallics 56*: 24–7.
196. Ji W, Fu Z, Wang W, Wang H, Zhang J, Wang Y, Zhang F. (2014). Mechanical alloying synthesis and spark plasma sintering consolidation of CoCrFeNiAl high-entropy alloy. *Journal of Alloys and Compounds 589*: 61–6.
197. Varalakshmi S, Kamaraj M, Murty BS. (2008). Synthesis and characterization of nanocrystalline AlFeTiCrZnCu high entropy solid solution by mechanical alloying. *Journal of Alloys and Compounds 460*: 253–7.
198. Varalakshmi S, Kamaraj M, Murty BS. (2010). Formation and stability of equiatomic and nonequiatomic nanocrystalline CuNiCoZnAlTi high entropy alloys by mechanical alloying. *Metallurgical and Materials Transactions A41*: 2703–9.
199. Wang C, Ji W, Fu Z. (2014). Mechanical alloying and spark plasma sintering of CoCrFeNiMnAl high-entropy alloy. *Advanced Powder Technology 25*: 1334–8.
200. Kang B, Lee J, Ryu HJ, Hong SH. (2018). Ultra-high strength WNbMoTaV high-entropy alloys with fine grain structure fabricated by powder metallurgical process. *Materials Science and Engineering A712*: 616–24.
201. Raza A, Ryu HJ, Hong SH. (2018). Strength enhancement and density reduction by the addition of Al in CrFeMoV based high-entropy alloy fabricated through powder metallurgy. *Materials and Design 157*: 97–104.

202. Raza A, Kang B, Lee J, Ryu HJ, Hong SH. (2018). Transition in microstructural and mechanical behavior by reduction of sigma-forming element content in a novel high entropy alloy. *Materials and Design 145*: 11–18.

203. Vaidya M, Armugam S, Kashyap S, Murty BS. (2015). Amorphization in equiatomic high entropy alloys. *Journal of Non-Crystalline Solids 413*: 8–14.

204. Wang J, Zheng Z, Xu J, Wang Y. (2014). Microstructure and magnetic properties of mechanically alloyed FeSiBAlNi(Nb) high entropy alloys. *Journal of Magnetism and Magnetic Materials 355*: 58–64.

205. Guan J, Li D, Yang Z, Wang B, Cai D, Duan X, He P, Jia D, Zhou Y. (2020). Synthesis and thermal stability of novel high-entropy metal boron carbonitride ceramic powders. *Ceramics International 46*: 26581–9.

206. Smelzer JA, Burton MT, Hornbuckle BC, Giri AK, Darling KA, Harmer MP, Marvel CJ. (2021). Optimization of cryogenic mechanical allying parameters to synthesize ultrahard refractory high entropy materials. *Materials and Design 210*: Article 110070.

207. Zhang Y, Wang D, Wang S. (2022). High-entropy alloys for electrocatalysis: Design, characterization and applications. *Small 18*: Article 2104339.

208. Kusada K, Mukoyoshi M, Wu D, Kitagawa H. (2022). Chemical synthesis, characterization and properties of multi-element nanoparticles. *Angewandte Chemie 61*: Article 09616.

209. Li H, Lai J, Li Z, Wang L. (2021). Multi-sites electrocatalysis in high-entropy alloys. *Advanced Functional Materials 31*: Article 2106715.

210. Wang B, Yao Y, Yu X, Wang C, Wu C, Zou Z. (2021). High entropy alloys: From theory to experiment. *Journal of Materials Chemistry A 9*: 19410–38.

211. Yao Y, Huang Z, Xie P, Lacey SD, Jacob RJ, Xie H, Chen F, Nie A, Pu T, Rehwoldt M, Yu D, Zachariah MR, Wang C, Yassar RS, Li J, Hu L. (2018). Carbothermal shock synthesis of high-entropy alloy nanoparticles. *Science 359*: 1489–94.

212. Yusenko KV, Riva S, Carvalho PA, Yusenko MV, Arnaboldi S, Sukhikh AS, Hanfland M, Gromilov SA. (2017). First hexagonal close-packed high-entropy alloy with outstanding stability under extreme conditions and electrocatalytic activity for methanol oxidation. *Scripta Materialia 138*: 22–7.

213. Gao S, Hao S, Huang Z, Yuan Y, Han S, Lei L, Zhang X, Shahbazian-Yassar R, Lu J. (2020). Synthesis of high-entropy alloy nanoparticles on supports by the fast-moving bed pyrolysis. *Nature Communications 11*: Article 2016.

214. Wang D, Liu Z, Du S, Zhang Y, Li H, Xiao Z, Chen W, Chen R, Wang Y, Zou Y, Wang S. (2019). Low-temperature synthesis of small-sized high-entropy oxides for water oxidation. *Journal of Materials Chemistry A 7*: 24211–16.

215. Wu D, Kusada K, Yamamoto T, Toriyama T, Matsumura S, Kawaguchi S, Kubota Y, Kitigawa H. (2020). Platinum-group-metal high-entropy-alloy nanoparticles. *Journal of the American Ceramic Society 142*:13833–8.

216. Witte R, Sarkar A, Kruk R, Eggert B, Brand A, Wende H, Hahn H. (2019). High entropy oxides: An emerging prospect for magnetic rare earth–transition metal perovskites. *Physical Review Materials* 3: Article 034406.

217. Yang Q, Wang G, Wu H, Beshiwork BA, Tian D, Zhu S, Yang Y, Lu X, Ding Y, Ling Y, Chen Y, Lin B. (2021). A high-entropy perovskite cathode for solid oxide fuel cells. *Journal of Alloys and Compounds 872*: Article 159633.
218. Martin JW. (1996). *Materials for Engineering*. London: Institute of Materials.
219. Ashby MF, Jones DRH. (2012). *Engineering Materials: An Introduction to Properties, Application and Design*, 4th edn. Oxford: Butterworth-Heinemann.
220. Koch CC. (2017). Nanocrystalline high-entropy alloys. *Journal of Materials Research 32*:3435–44.
221. Zhilyaev AP, Langdon TG. (2008). Using high-pressure torsion for metal processing: Fundamentals and applications. *Progress in Materials Science 53*: 893–979.
222. Shahmir H, Assghari-Rad P, Mehranpour S, Forghani F, Kim HS, Nili-Ahmadabadi M. (2021). Evidence of fcc to hcs and bcc martensite transformation in a CoCrFeNiMn high-entropy alloy by severe plastic deformation. *Materials Science and Engineering A807*: Article 140875.
223. Picak S, Yilmaz HC, Karaman I. (2021). Simultaneous deformation twinning and martensitic transformation on CoCrFeMnNi high entropy alloy at high temperatures. *Scripta Materialia 202*: Article 113995.
224. Won JW, Lee S, Park SH, Kang M, Lim KR, Park CH, Na YS. (2018). Ultrafine-grained CoCrFeMnNi high-entropy alloy produced by cryogenic multi-pass caliber rolling. *Journal of Alloys and Compounds 742*: 290–5.
225. Jeong HT, Kim WJ. (2022). Effect of roll speed ratio on the texture and microstructural evolution of an fcc high-entropy alloy during differential speed rolling. *Journal of Materials Science and Technology 111*: 152–66
226. Kwon H, Asghari-Rad PA, Park JM, Sathiyamoorthi P, Bae JW, Moon J, Zargaran A, Choi YT, Son S, Kim HS. (2021). Synergistic strengthening from refinement and nano-scale precipitates in non-equiatomic CoCrFeNiMo medium-entropy alloy. *Intermetallics 135*: Article 107212.
227. Bian YL, Feng ZD, Zhang NB, Li YX, Wang XF, Zhang BB, Cai Y, Lu L. (2022). Ultrafast severe plastic deformation in high-entropy alloy $Al_{0.1}$CoCrFeNi via equal channel angular pressing. *Materials Science and Engineering A847*: Article 143221.
228. Lee DH, Park JM, Yang G, He J, Lu Z, Suh JY, Kawasaki M, Ramamurty U, Jang J. (2019). Nano-graining a particle-strengthened high-entropy alloy. *Scripta Materialia 163*: 24–8.
229. Cizek J, Hausild P, Cieslar M, Melikhova O, Vlasák T, Janeçek M, Král R, Harcuba P, Lukác F, Zýka J, Málek J, Moon J, Kim HS. (2018). Strength enhancement of high entropy alloy HfNbTaTiZr by severe plastic deformation. *Journal of Alloys and Compounds 768*: 924–37.
230. Hung PT, Kawasaki M, Han JK, Lábár JL, Gubicza J. (2020). Thermal stability of a nanocrystalline HfNbTiZr multi-principal element alloy processed by high-pressure torsion. *Materials Characterization 168*: Article 110550.
231. Hammond VH, Atwater MA, Darling KA, Nguyen HQ, Kecskes LJ. (2014). Equal-channel angular extrusion of a low-density high-entropy alloy produced by high-energy cryogenic mechanical alloying. *JOM 66*: 2021–9.

Further Reading

Barber ZH (ed.). (2005). *Introduction to Materials Modelling*. London: Maney.

Burkov A. (2019). *The Hundred-Page Machine Learning Book*. Quebec City: True Positive Incorporated.

Cantor B. (2020). *The Equations of Materials*. Oxford: Oxford University Press.

Flach P. (2012). *Machine Learning: The Art and Science of Algorithms That Make Sense of Data*. Cambridge: Cambridge University Press.

Gaskell DR, Laughlin DE. (2017). *Introduction to the Thermodynamic of Materials*, 6th edn. Boca Raton: CRC Press.

Hillert M. (2008). *Phase Equilibria, Phase Diagrams and Phase Transformations*, 2nd edn. Cambridge: Cambridge University Press.

Martin JW. (1996). *Materials for Engineering*. London: Institute of Materials.

Mizutani U. (2019). *Hume-Rothery Rules for Structurally Complex Alloy Phases*. Boca Raton: CRC Press.

Murty BS, Yeh JW, Ranganathan S, Bhattacharjee PP. (2019). *High-Entropy Alloys*. Amsterdam: Elsevier.

Satoh A. (2011). *Introduction to the Practice of Molecular Simulation: Molecular Dynamics, Monte Carlo, Brownian Dynamics, Dissipative Particle Dynamics*. Burlington: Elsevier Science.

Saunders N, Miodownik AP. (1998). *Calphad: Calculation of Phase Diagrams: A Comprehensive Guide*. Oxford: Pergamon Elsevier.

6

Properties of Multicomponent Materials

1 Introduction

We have seen how multicomponent phase space is enormous in size, with an unimaginably large number of multicomponent materials that have never been made or even thought of. We have also seen how this wide range of new multicomponent and high-entropy materials has a wide range of complex atomic and molecular structures, nanostructures and microstructures, and, therefore, a correspondingly wide range of novel physical, mechanical, electrical, magnetic, chemical and other properties, with great potential for engineering applications.

As described in Chapter 1, humankind moves forward by discovering and utilising new and valuable materials, but so far we have only scratched the surface at the edges of multicomponent phase space, avoiding almost completely the gigantic range of possible multicomponent materials in its centre. And this vast central field of multicomponent high-entropy materials undoubtedly contains many exciting new materials with wonderful new properties yet to be found. This chapter describes some of the properties and potential applications of multicomponent high-entropy materials that have begun to be investigated during the intensive research of the last decade or so. Several thousand papers have now been published on multicomponent high-entropy materials, and we have already discovered many new materials and properties, some of which are discussed in the rest of this chapter. We do not try to cover in detail all the work that has been done, but instead we concentrate on a few examples, on those properties and applications discovered so far that are particularly innovative and potentially valuable. To be more specific, we discuss some aspects of the mechanical properties of multicomponent metallic alloys that have been investigated in most detail so far, as well as some functional properties of both multicomponent metallic alloys and multicomponent ceramic compounds that have begun to be studied in detail more recently. Despite the substantial work that has been done, however, it is important and perhaps salutary to realise that we have still hardly made much of an inroad into multicomponent phase space and that there are still many millions of wonderful materials yet to be found.

The famous *law of mixtures* or *rule of mixtures* for a conventional macroscopic composite material gives many of its properties as a weighted linear sum of the properties of its individual separated component phases:[1,2]

$$\text{composite materials}: P = \sum_i x_i p_i$$

where P is the value of the property of the composite material and x_i and p_i are respectively the composition and property value of the ith component phase. This is a very convenient relationship that works well and is used quite widely to estimate many of the properties of conventional macroscopic composite materials in which the individual phase regions are

Fundamentals of Multicomponent High-Entropy Materials. Brian Cantor, Oxford University Press. © Brian Cantor (2024).
DOI: 10.1093/9780191986710.003.0006

usually mm, µm or at least nm in size. In a macroscopic composite, the interface regions between the different component phases constitute a relatively small part of the material and this is why many of the properties can be expressed by the law of mixtures as a weighted sum of the properties of the component phases. In a multicomponent high-entropy material on the other hand, the components are mixed on a much finer atomic or molecular scale, and the properties depend on the much more complex details of interactions between individual atoms and their surrounding near neighbours, and thus on the resulting wide variety of local atomic configurations discussed in detail in Chapter 4. Not surprisingly the properties of multicomponent high-entropy materials are often, therefore, non-linear and not in general given by the composite law of mixtures:

$$\text{multicomponent materials} : P \neq \sum_i x_i p_i$$

Nevertheless, for some of the simpler physical properties, such as density, elastic modulus and thermal expansion coefficient, the effects of local interactions between neighbouring atoms and/or molecules are not too strong, and in these cases multicomponent material properties, while not given accurately by the law of mixtures, are still fairly close to the properties of their components, $P \approx \sum_i x_i p_i$. As we have already seen in previous chapters, there are other properties, such as diffusion and dislocation slip, that are much more strongly influenced by local interactions between neighbouring atoms and/or molecules, and in these cases there is a considerable degree of non-linearity, so that multicomponent material properties deviate substantially from the properties of their components, $P \not\approx \sum_i x_i p_i$. In this chapter, we concentrate on the latter properties, since they provide greater potential for new materials development and novel engineering applications.

2 Mechanical Properties of Multicomponent Materials

Cryogenic Cantor alloys

We have already seen in Chapter 4 that the original equiatomic single-phase fcc Cantor alloy CrMnFeCoNi exhibits remarkable mechanical properties,[3–7] with its ultimate tensile strength (UTS) and ductility reaching as high as approximately 1 GPa and 100% respectively when the temperature is reduced to −196 °C (77K) for material manufactured with a small equiaxed grain size of 4.4 µm, as shown by the results of Otto et al.[3] in Figure 6.1. Such a significant increase in both strength and ductility as the temperature is reduced below room temperature is particularly unusual and surprising behaviour, and as a consequence the Cantor alloy has been described as having '*outstanding cryogenic mechanical properties*'.[4] Moreover, the fracture toughness of the alloy has been measured[5] at room temperature as ~220 MPa/√m with a grain size of 6 µm, and its high-cycle fatigue strength has also been measured[6] at room temperature as approaching 300 MPa after 10^7 cycles with a somewhat smaller grain size of 0.65 µm. These excellent mechanical property results have led to a substantial body of research[4,7,8] investigating: 1) other mechanical properties of the original Cantor alloy CrMnFeCoNi, such as compression,

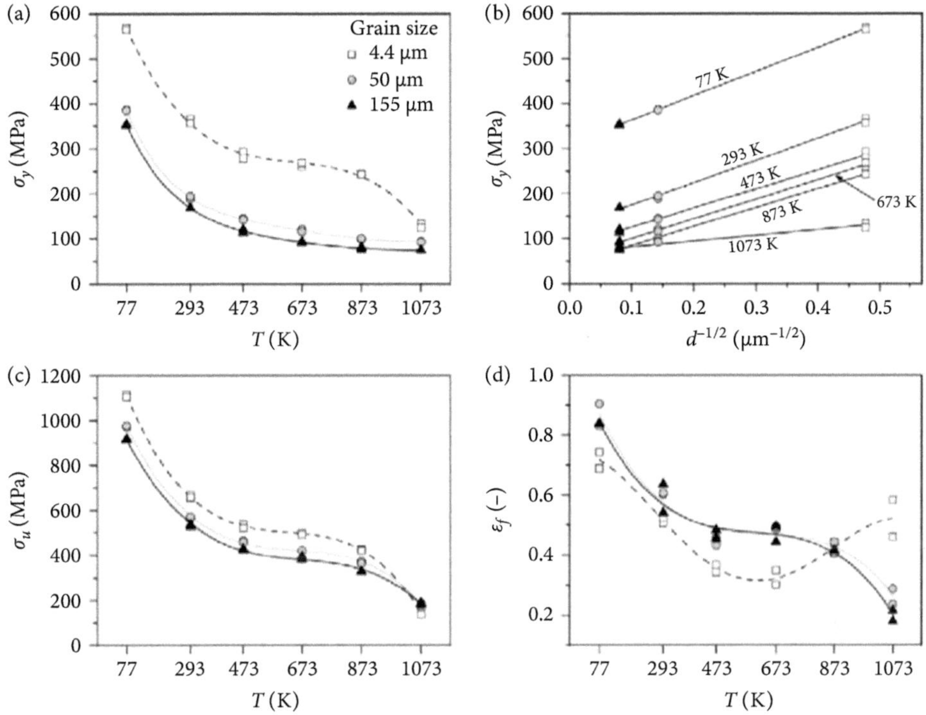

Figure 6.1 Variation of tensile yield strength σ_y with (a) temperature *T* and (b) grain size *d*, and variation of (c) tensile ultimate tensile strength σ_u and (d) tensile fracture strain ϵ_f with temperature *T* in the original single-phase fcc Cantor alloy CrMnFeCoNi.
Reproduced from Otto et al. (2013). The influence of temperature and microstructure on the tensile properties of a CoCrFeNiMn high-entropy alloy. *Acta Materialia 61*: 5743–55.

torsion, wear and impact resistance; as well as 2) extensions to other single-phase fcc modified Cantor alloys, both with fewer and more components as well as with non-equiatomic compositions; and 3) alloying additions and thermomechanical treatments designed to enhance strength and other mechanical properties using, for instance, precipitation hardening, grain boundary hardening, enhanced twinning response, transformation hardening, and surface treatment.[4,7,8] For instance, Figure 6.2 shows the effect on room temperature yield strength, ultimate tensile strength (UTS) and ductility of additional components in a variety of different single-phase fcc Cantor alloys.[8] Much of this research is described in detail in the reviews by Li et al.,[4] George et al.[7] and Li et al.[8] In what follows, we do not try to provide an overview of all this work, and instead concentrate on describing and discussing the mechanical properties of one or two particularly interesting materials.

Gludovatz et al.[9] studied the mechanical properties of the equiatomic ternary single-phase fcc Cantor alloy CrCoNi, manufactured by arc melting, drop casting into rectangular cross-section copper moulds, followed by cold forging and cross rolling into 10-mm-thick sheet and then recrystallisation for one hour at 800 °C (1,073K) to produce a material with an equiaxed, albeit somewhat variable, grain size in the range 5–50 μm. At room temperature, CrCoNi has a yield strength and fracture toughness comparable to the original Cantor alloy

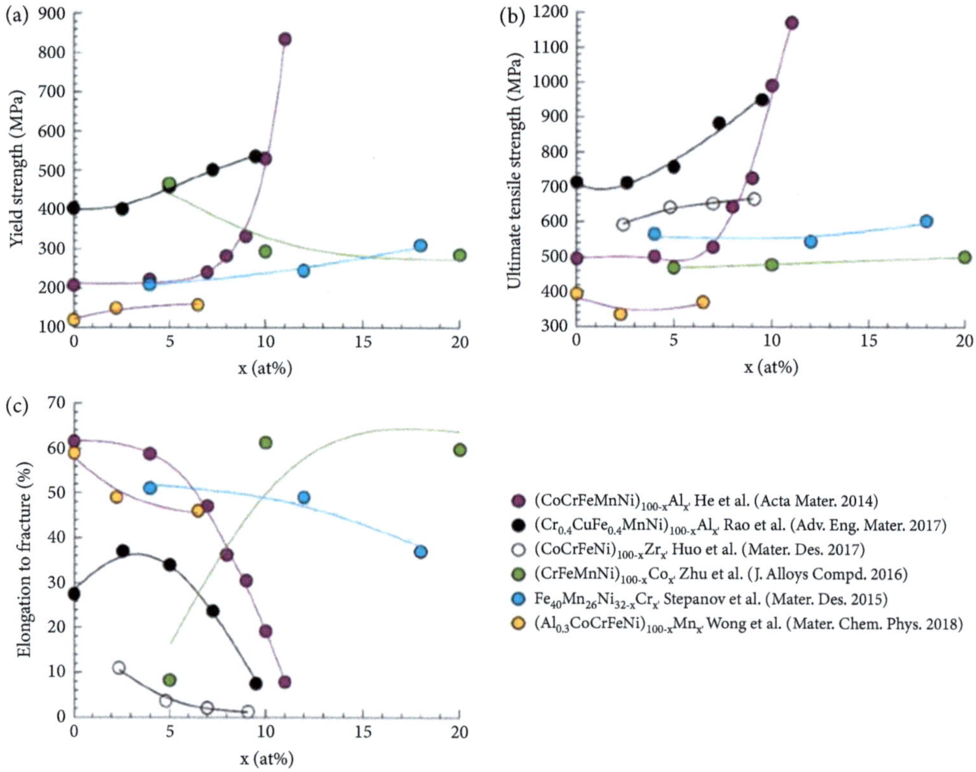

Figure 6.2 Effect of composition on (a) yield strength, (b) ultimate tensile strength and (c) ductility for a variety of single-phase fcc Cantor alloys.

Reproduced from Li et al. (2021). Mechanical behaviour of high-entropy alloys. *Progress in Materials Science 118*: Article 100777.

CrMnFeCoNi shown in Figure 6.1, but exhibits significantly higher ultimate tensile strength and ductility, as shown in Figure 6.3. At cryogenic temperatures the mechanical properties of CrCoNi are even more impressive, with a yield strength, ultimate tensile strength, ductility and fracture toughness of ~650 MPa, 1.3 GPa, 90% and 270 MPa/√m respectively[9] as also shown in Figure 6.3. The excellent toughness at low temperatures of CrCoNi, and indeed of CrMnFeCoNi, is caused by a combination of intense nanoscale deformation twinning as well as extensive dislocation slip, as shown in Figures 6.4 and 6.5 for CrCoNi at −196 °C (77K) and CrMnFeCoNi at −150 °C (123K) respectively,[10,11] a mechanism known as twinning-induced plasticity (TWIP),[8] with further resistance to crack growth also coming from bridging ligaments of material behind any crack tip.[12] Deformation twinning is a well-known mechanism used to increase work hardening and fracture toughness in conventional TWIP steels and other alloys.[13,14] Overall, Gludovatz et al. comment that 'in terms of . . . crack-initiation and crack-growth toughnesses, the NiCoCr medium-entropy alloy represents one of the toughest materials in any materials class ever reported'.[9]

It is now well established, as discussed in some detail in Chapter 4, that Cantor alloys have low stacking fault energies, typically 15–30 mJ/m^2 at room temperature, which decrease even further on cooling to cryogenic temperatures.[23–25] Lower stacking fault energy corresponds to lower twinning energy, which is why Cantor alloys such as CrMnFeCoNi and CrCoNi

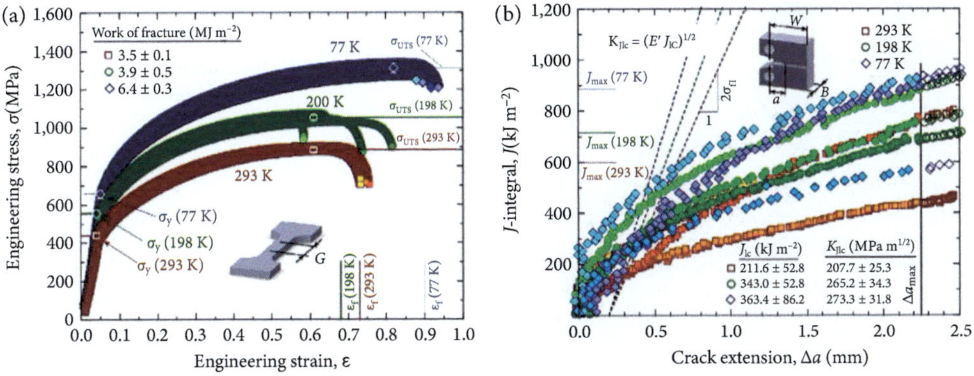

Figure 6.3 Variation of (a) tensile stress–strain curves and (b) fracture-toughness crack-extension curves with testing temperature for single-phase fcc ternary CrCoNi.
Reproduced from Gludovatz et al. (2016). Exceptional damage-tolerance of a medium-entropy alloy NiCoCr at cryogenic temperatures. *Nature Communications* 7: Article 10602.

are also able to enhance their ductility, work hardening and resistance to necking and fracture by adding deformation twinning plasticity (TWIP) to conventional dislocation slip and multiplication.[7,8,9,26] Laplanche et al.[17,26] have investigated the details of dislocation slip and deformation twinning behaviour and their effect on work hardening, ductility and toughness during tensile testing at room temperature and −196 °C (77K) in the 5-component CrMnCoFeNi and ternary CrCoNi Cantor alloys manufactured by vacuum-induction melting, chill casting in a steel mould, homogenisation for forty-eight hours at 1,473K (1,200 °C), progressive step-swaging by over 60% reduction in cross-section and recrystallisation for one hour at 1,173K (900 °C), to produce an equiaxed grain size of about 16 μm. Typical true stress–true strain ($\sigma - \epsilon$) curves and corresponding work-hardening ($d\sigma/d\epsilon - \sigma$) curves for the two alloys at room temperature, 20 °C (293K), and low temperature, −196 °C (77K), are shown in Figure 6.6. In all cases, deformation takes place in three stages,[17,26] as shown in the variation of work-hardening rate $d\sigma/d\epsilon$ with increasing stress and strain: 1) an initial stage of dislocation slip, with a high work hardening rate that decreases rapidly with increasing stress and strain; 2) a second stage beginning with the onset of deformation twinning at a point of inflection in the work-hardening curve, followed by enhanced ductility and a slower decrease (at room temperature) or a plateau (at low temperature) in the work-hardening rate; and 3) final necking and ductile fracture when Considère's criterion is met with $d\sigma/d\epsilon \leq \sigma$. The extent of deformation twinning is considerably greater in ternary CrCoNi compared with 5-component CrMnFeCoNi, and is considerably greater at low temperature in both alloys, as can be seen clearly in Figure 6.6. The onset of deformation twinning takes place at a true strain of approximately 5% at −196 °C (77K) and 10% at 20 °C (293K), corresponding to a roughly temperature-independent critical true stress value of about 800 MPa, and a critical shear stress for twinning of 230–260 MPa in both alloys.[17,26]

Similar results showing good strength and ductility at low and high temperatures associated with a combination of dislocation slip and deformation twinning have been obtained in other studies of the original 5-component single-phase fcc Cantor alloy CrMnFeCoNi[15–17] and the reduced ternary single-phase fcc Cantor alloy CrCoNi,[13–19] and have also been obtained in other single-phase fcc modified Cantor alloys, for example, $Al_{0.1}CrFeCoNi$,

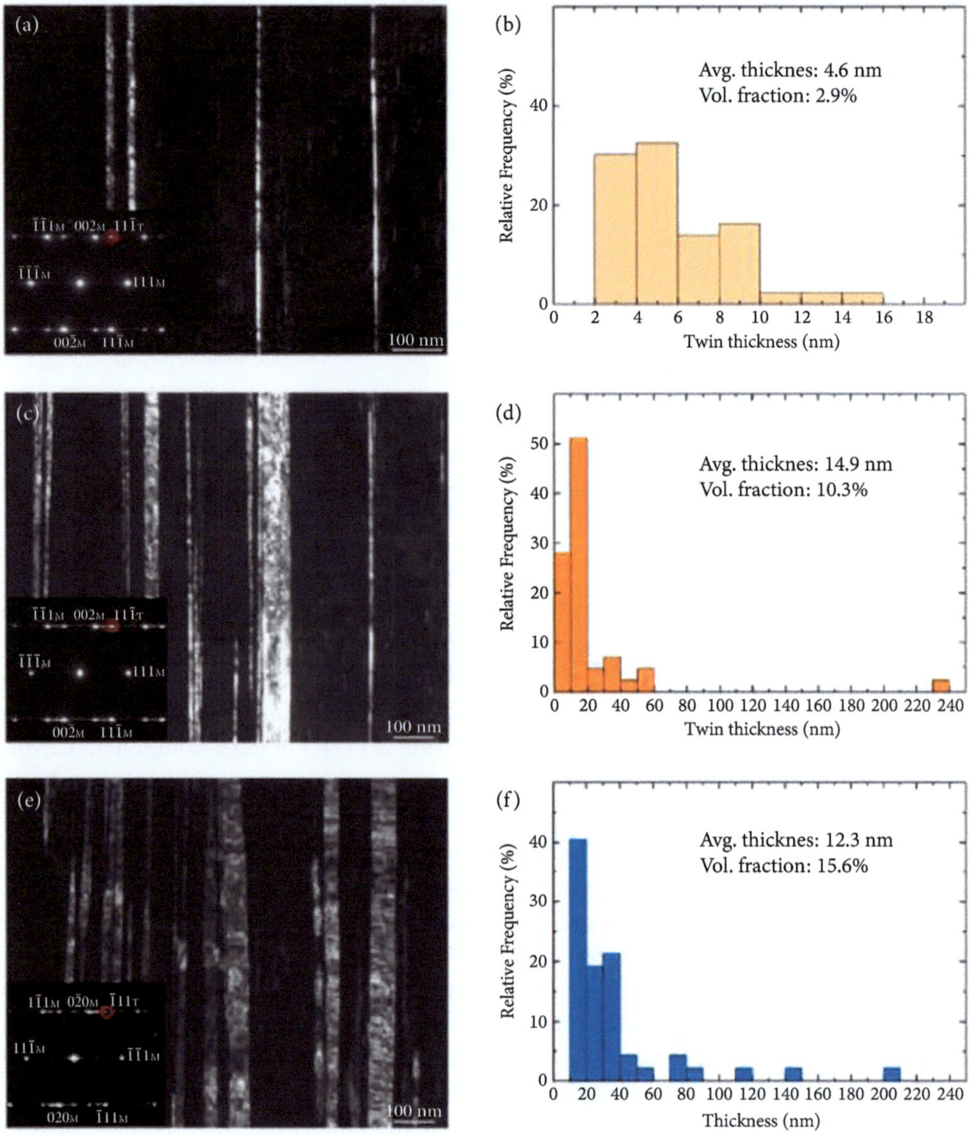

Figure 6.4 TEM images of deformation twinning in compression-tested single crystal CrCoNi: (a) and (b) 100% strain at room temperature; (c) and (d) 150% strain at room temperature; and (e) and (f) 150% strain at −196 °C (77K).

Reproduced from Li et al. (2022). Tensile and compressive plastic deformation behavior of medium entropy CrCoNi single crystals from cryogenic to elevated temperatures. *International Journal of Plasticity 148*: Article 103144.

$Al_{0.2}CrCoNiTa_{0.07}$, CrFeCoNi and CrFeNi.[20–23] Table 6.1 lists of some of the multicomponent high-entropy single-phase fcc alloys in which twinning is an important additional deformation mode as well as dislocation slip, with others listed and discussed in the review by Li et al.[8]

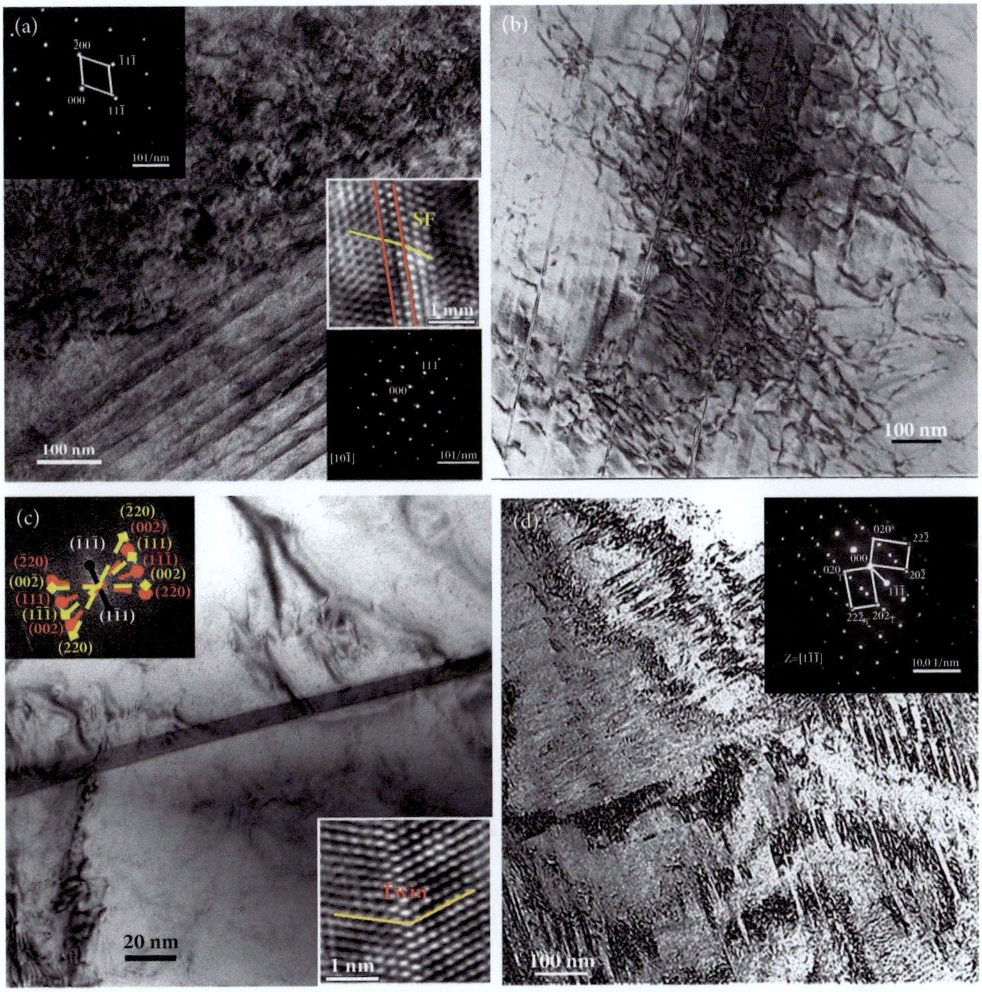

Figure 6.5 TEM images of dislocation slip and deformation twinning in compression-tested CrMnFeCoNi at −150 °C (123K): (a) 5.4% strain; (b) and (c) 9.3% strain; and (d) 11.2% strain.
Reproduced from Fu et al. (2021). Elucidating the transition of cryogenic deformation mechanism of CrMnFeCoNi high-entropy alloy. *Journal of Alloys and Compounds 872*: Article 159606.

A lower stacking fault energy also corresponds to a lower energy barrier for the transformation from fcc to hcp ϵ-martensite, since stacking faults, twins and hcp ϵ-martensite represent different but related changes to the stacking sequence in the underlying fcc structure, as discussed previously in Chapter 4. This provides the possibility of another deformation mode known as transformation-induced plasticity (TRIP).[3,8,9,26] Li et al.[27–29] studied modified Cantor alloys without Ni, with lower Cr and Co, and with varying Mn and Fe concentrations, in the system $Cr_xMn_yFe_{80-2x-y}Co_x$, to develop materials with a mixed fcc/hcp structure, so that deformation could be further enhanced by utilising TRIP via the fcc → hcp ϵ-martensite phase transformation, as shown in Figure 6.7. An alloy of composition $Cr_{10}Mn_{30}Fe_{50}Co_{10}$ with a grain size of 4.5 μm was shown to have a room temperature yield strength, UTS and ductility of 350 MPa, 900 MPa and 80% respectively.[27]

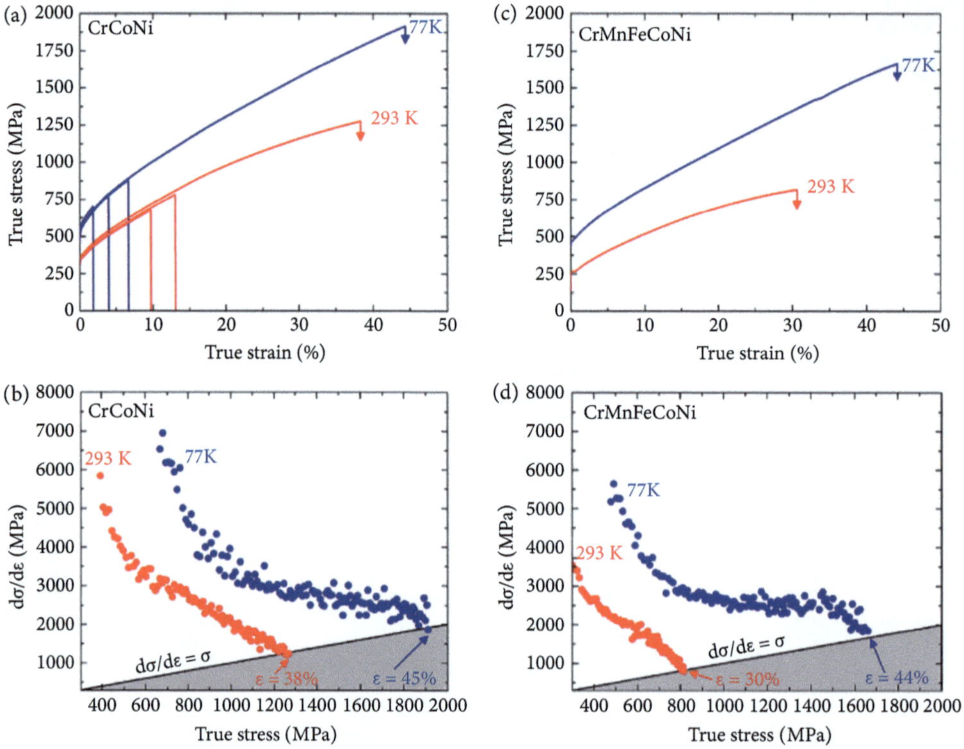

Figure 6.6 Tensile true stress–strain curves and work-hardening curves for single-phase ternary CrCoNi and 5-component CrMnFeCoNi single-phase fcc Cantor alloys at 20 °C (293K) and −196 °C (77K).

Reproduced from Laplanche et al. (2017). Reasons for the superior mechanical properties of medium-entropy CrCoNi compared to high-entropy CrMnFeCoNi. *Acta Materialia 128*: 292–303.

Once again, similar results showing good strength and ductility at low and high temperatures associated with a combination of dislocation slip, deformation twinning and martensitic transformation have been obtained in a variety of other multicomponent single-phase and multiphase fcc Cantor alloys. Some examples are listed in Table 6.2, with more cases listed and discussed in the review by Li et al.[8]

This type of alloy design has been taken further, by adding interstitial carbon and using thermomechanical treatment to develop bimodal and trimodal grain-size structures.[30,31] A cast alloy of composition $Cr_{10}Mn_{30}Fe_{49.5}Co_{10}C_{0.5}$ was hot rolled 50% at 900 °C (1,173K), homogenised for two hours at 1,200 °C (1,473K), water-quenched, cold rolled 67%, tempered for three to ten minutes at 400 °C (673K), and finally annealed for three to ten minutes at 650–750 °C (923–1,023K). The resulting alloy microstructures consisted of complex bimodal and trimodal mixtures of highly deformed μm- and nm-sized grains and twins. The yield strength, UTS and ductility at room temperature were 1.3 MPa, 1.5 MPa and 14% after annealing at 400 °C (623K), and 600 MPa, 900 MPa and 60% after annealing at 650–750 °C (923–1,023K),[30] as shown in the tensile stress–strain curves in Figure 6.8. Deformation took place, as expected, by a complex mixture of dislocation slip, twinning and martensite transformation, all contributing to the resulting excellent combinations of

Table 6.1 Some single-phase fcc Cantor alloys that exhibit TWIP, i.e. twinning as well as dislocation slip as important deformation modes.[8]

alloy	temperature (K)
CrCoNi	77; 298
CrFeCoNi	298
$Al_{0.1}CrFeCoNi$	298
CrMnFeCoNi	77; 298; 673; 773
$Al_{0.6}Ti_{0.4}CrMnFeCoNi$	298
VCrMnFeCoNi	298

Table 6.2 Some single-phase fcc and multiphase fcc + hcp Cantor alloys that exhibit TRIP, i.e. martensitic transformation as well as dislocation slip and twinning as important deformation modes.[8]

alloy
CrCoNi
CrFeCoNi
$Cr_{10}Mn_{30}Fe_{50}Co_{10}$
$Cr_{20}Mn_{20}Fe_{34}Co_{20}Ni_6$
$Cr_{20}Mn_{20}Fe_{20}Co_{30}Ni_{10}$
$Al_{7.5}Fe_{25}Co_{25}Ni_{25}Cu_{17.5}$
$Al_1Si_5Cr_{15}Mn_{20}Fe_{39}Co_{20}$
$Si_5Cr_{15}Mn_{28}Fe_{42}Co_{10}$

strength and ductility.[30] Like TWIP, TRIP has also been used in conventional steels and alloys.[3,7,32]

More recently, attempts to improve the mechanical properties of multicomponent fcc Cantor alloys have concentrated on using additional interstitial and/or substitutional solute elements and thermomechanical processing to develop additional grain boundary, precipitation and grain-size hardening effects. Seol et al.[33] doped both the original equiatomic single-phase fcc Cantor alloy CrMnFeCoNi and a modified single-phase fcc alloy with lower Mn and Co content $(CrFe)_{0.8}(MnCo)_{0.2}$ in each case with boron in order to strengthen the grain boundaries. The modified alloy was selected because of its particularly low stacking fault energy and, therefore, a deformation mechanism dominated by twinning rather than dislocation slip.[33] Boron is known to segregate to grain boundaries in conventional steels and other alloys, leading to two major effects, both of which improve the mechanical properties: 1) reduced grain boundary mobility and, therefore, lower final grain size after homogenisation and recrystallisation; and 2) increased grain boundary cohesion and strength.[34–36] Seol et al.[33] found that the addition of as little as 30 ppm boron led to considerable increments in yield strength and ultimate tensile strength, without detriment to ductility. The original Cantor alloy CrMnFeCoNi with 30-ppm boron had a room temperature yield strength, ultimate tensile strength and ductility of 850 MPa, 1.04 GPa and almost 60% respectively. Despite these substantial improvements in mechanical properties, the overall deformation

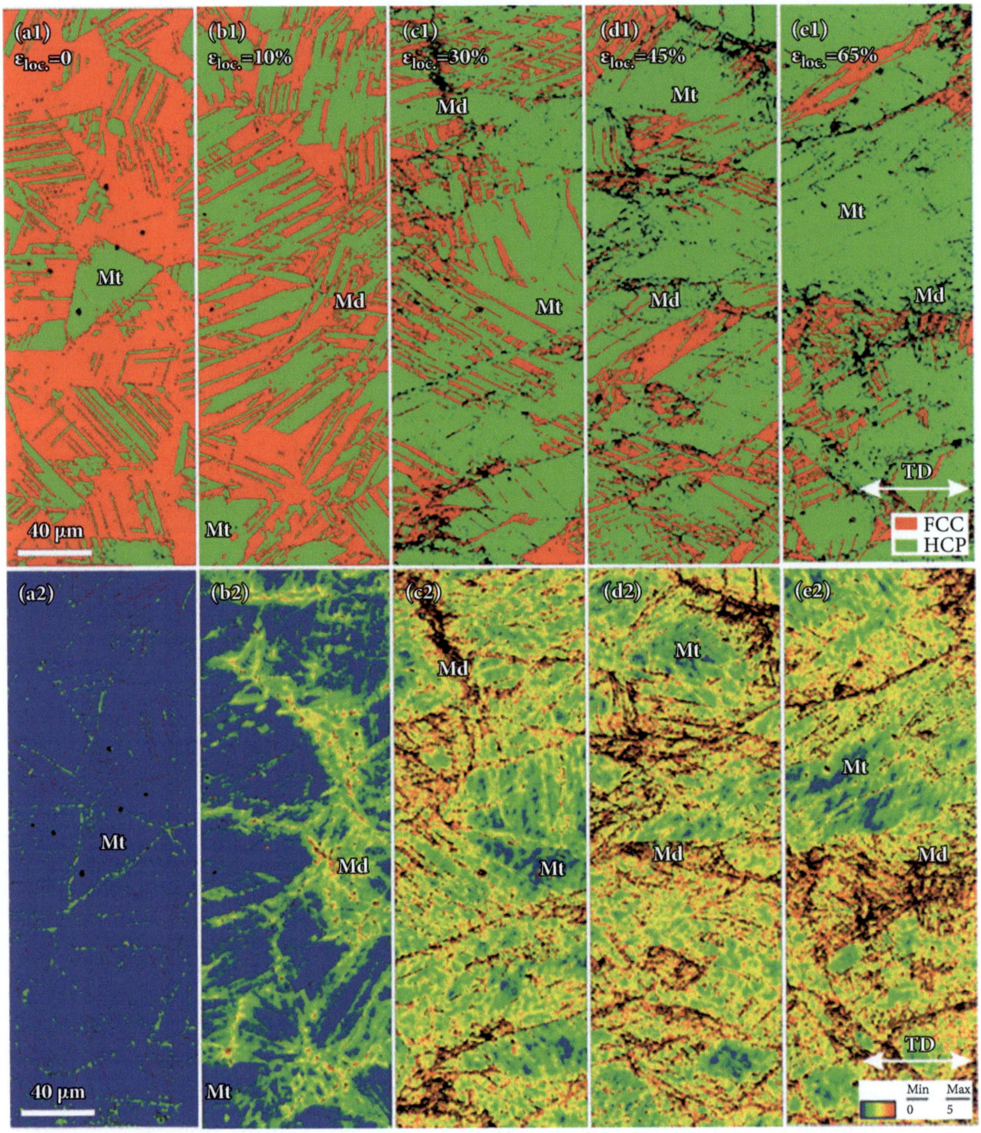

Figure 6.7 Electron back-scatter diffraction (EBSD) patterns (a1–e1) and kernel average misorientation (KAM) maps (a2–e2) showing thermal (Mt) and deformation (Md) hcp ϵ-martensite in quaternary $Cr_{11}Mn_{28}Fe_{50}Co_{11}$ after tensile straining at room temperature to (a) 0%; (b) 10%; (c) 30%; (d) 45%; and (e) 65%.

Reproduced from Li et al. (2016). Microstructure evolution and critical stress for twinning in the CrMnFeCoNi high-entropy alloy. *Acta Materialia 118*: 152–63.

behaviour was, however, largely unchanged, indicating that the effect of boron was caused directly by the expected grain boundary and grain size hardening effects rather than by a change in deformation mode.[33]

Many nickel superalloys and high-strength stainless steels have structures consisting of an fcc matrix, which are hardened by a combination of suitable additional alloying components and thermomechanical processing so as to develop a microstructure of coherent $L1_2$

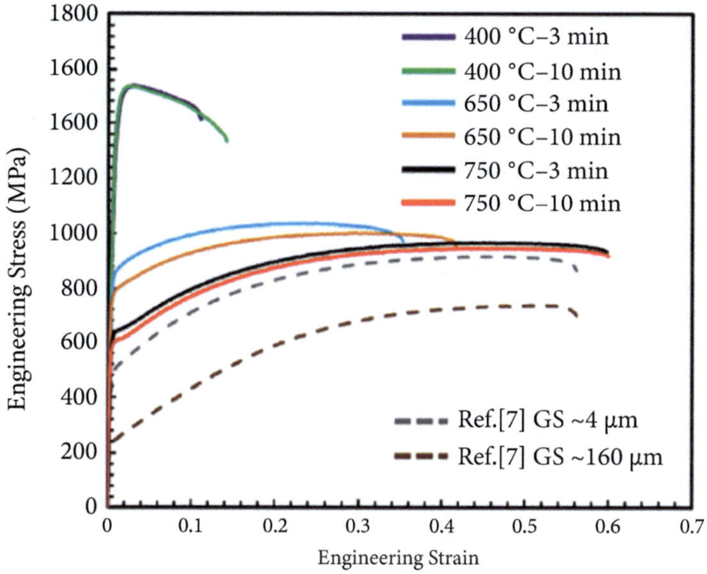

Figure 6.8 Room temperature tensile stress–strain curves for a 5-component $Cr_{10}Mn_{30}Fe_{49.5}Co_{10}C_{0.5}$ alloy after annealing for 3–10 minutes at 400, 650 and 750 °C.
Reproduced from Su et al. (2019). Hierarchical microstructure design to tune the mechanical behavior of an interstitial TRIP–TWIP high-entropy alloy. *Acta Materialia 163*: 40–54.

γ'-Ni_3(Al, Ti) precipitates within the fcc matrix, which confer enhanced room temperature as well as high temperature mechanical properties.[32,43] The same alloying strategy has recently begun to be used for a variety of modified Cantor alloys.[38–43] Table 6.3 gives details of alloy compositions, thermomechanical treatments and resulting room temperature yield strengths, ultimate tensile strengths and ductilities. Careful homogenisation, deformation, recrystallisation and ageing treatments can be used to develop room temperature yield strengths above 1 GPa and ultimate tensile strengths above 1.5 GPa, allied with ductilities above 50%. Microstructures typically show an fcc matrix with dense arrays of coherent $L1_2$ γ'-Ni_3(Al, Ti) and sometimes other precipitates in the size range 10–40 nm, which are clearly very effective in restricting easy dislocation flow, producing strong precipitation hardening, and preventing the onset of premature fracture.[33–42] There have been fewer mechanical property measurements at elevated temperatures, but Li et al.[43] quote yield and ultimate tensile strengths and ductilities in the range 550–620 MPa, 700–950 MPa and 28–40% respectively for testing temperatures in the range 600–700 °C (873–973K).

Li et al.[4] have followed Gludavatz et al.[5] and plotted the range of mechanical properties that have been measured for multicomponent and high entropy materials, including single-phase fcc and multiphase fcc-based Cantor alloys, on an Ashby map of fracture toughness versus yield strength, in order to compare them with a wide range of other materials such as high-strength steels, nickel superalloys, hard ceramics, and metallic and oxide glasses, as shown in Figure 6.9. The range of mechanical properties of the high-entropy alloys, notably the single-phase and multiphase fcc-based Cantor alloys, exceed those of all the other materials, in particular in exhibiting exceptionally high fracture toughnesses at very high yield strengths. Li et al.[4] state that 'the CrMnFeCoNi alloy represents one of the toughest materials reported to date, with a plane-strain fracture toughness, K_{1c}, that exceeds 200 MPa/$\sqrt{}$m and

Table 6.3 Mechanical properties of Cantor alloys[38–43] precipitation hardened with coherent $L1_2^{\gamma'}$-$Ni_3(Al, Ti)$ precipitates.

alloy	condition	YS (MPa)	UTS (MPa)	ϵ (%)
$(CrFeCoNi)_{94}Al_4Ti_2$	H	~200	503	~69
$(CrFeCoNi)_{94}Al_4Ti_2$	H/CR/RX/aged 1073K	645	1094	39
$(CrFeCoNi)_{94}Al_4Ti_2$	H/CR/aged 923K	1005	1273	17
$CrFeCoNi_2\,Cu_{0.2}Al_{0.2}$	H/CR/aged 973K	720	1050	31
$CrFeCoNi_2\,Cu_{0.2}Al_{0.2}$	H/CR/aged 1073K	460	700	32
$(Cr_{10}Fe_9Co_{17.5}Ni_{49})Al_{7.5}Ti_7$	as-cast	859	1044	14.7
$(Cr_{10}Fe_9Co_{17.5}Ni_{48})Al_{7.5}Ti_7Mo_1$	as-cast	899	1064	14.3
$(FeCoNi)_{86}Al_7Ti_7$	not given	1100	1500	50
$(FeCoNi)_{86}Al_8Ti_6$	not given	1000	1250	35
$(Cr_{15}Fe_{21}Co_{28}Ni_{28})Al_4Ti_4$	H/CR/RX/aged 1073K	855	1031	27
$(Cr_9Fe_{27}Co_{27}Ni_{27})Al_8Ti_1Nb_1$	H/CR/RX/aged 1168K	863	1286	28.2
	H/CR/RX/aged 1168K then 993K			

*H = homogenised, CR = cold rolled, RX = recrystallised, YS = yield strength, UTS = ultimate tensile strength, ϵ = ductility.

with an outstanding tensile strength exceeding 1 GPa'. George et al.[7] have also plotted the range of mechanical properties that have been measured for multicomponent and high-entropy materials, including single-phase fcc and multiphase fcc-based Cantor alloys, on a slightly different Ashby map of tensile strength versus elongation, again to compare them with the highest strength steels and other metallic alloys, as shown in Figure 6.10. George et al.[7] say only that: 'The observed mechanical properties are well within those observed in martensitic steels, advanced high strength steels, nickel-based alloys, etc.' However, as can be seen on Figure 6.10, the very highest combinations of strength and ductility are once again shown by some of the multicomponent high-entropy fcc-based Cantor alloys.

In summary, multicomponent single-phase and near-single-phase fcc Cantor alloys have mechanical properties that are comparable to and, in some carefully engineered cases, exceeding those of the very strongest and toughest conventional materials, including high strength alloy steels, nickel superalloys and advanced engineering ceramics. In particular, many of the multicomponent fcc-based Cantor alloys have quite exceptional mechanical properties at cryogenic temperatures with: 1) high strength caused by the complexity of local atomic configurations and associated lattice strains, restricting but not preventing dislocation flow; and 2) extensive ductility and remarkable fracture toughness caused by the onset at low temperatures, as well as dislocation slip, of additional nanoscale twinning (TWIP) and martensitic transformation (TRIP) deformation modes. The multicomponent fcc-based Cantor alloys also have excellent room and elevated temperature mechanical properties, and intensive research is underway to improve strength levels by using different alloy compositions, different manufacturing methods and thermomechanical processing to enhance grain-size, solution and precipitation hardening effects. Overall, these results are very encouraging given that we have only just begun, in the last few years, to explore such conventional materials engineering approaches. The mechanical properties of

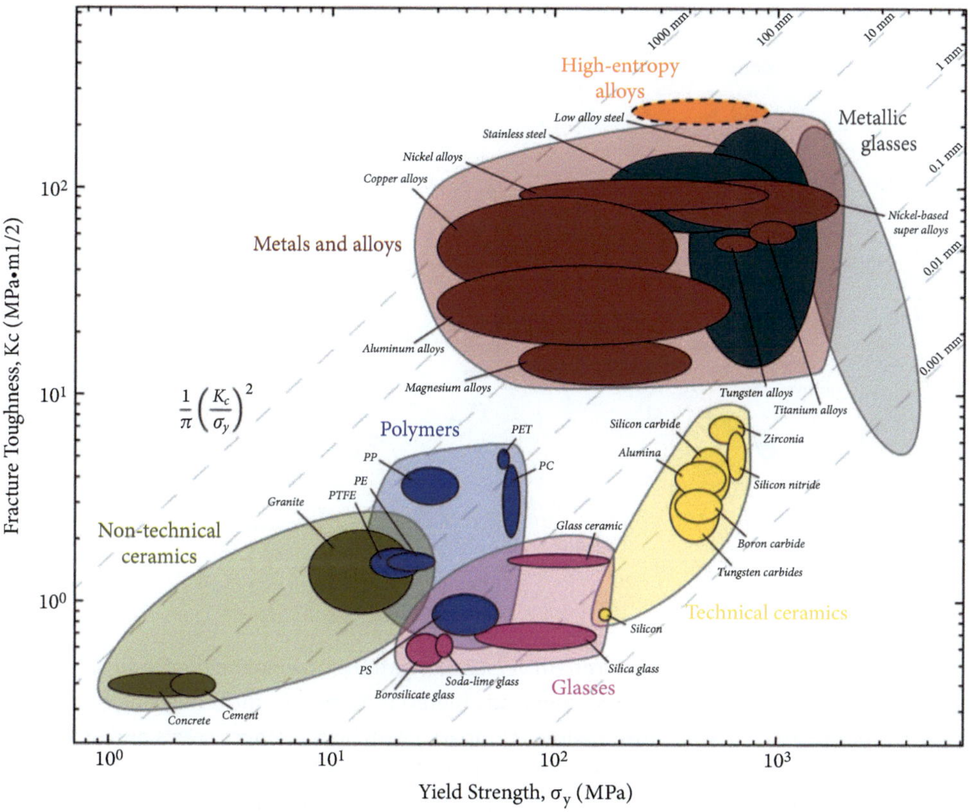

Figure 6.9 Ashby map of fracture toughness K_c versus yield strength σ_y showing the mechanical properties of multicomponent and high-entropy materials including single-phase fcc and multiphase fcc-based Cantor alloys, compared with high-strength steels, nickel superalloys, hard ceramics and metallic and oxide glasses.

Reproduced from Gludovatz et al. (2014). A fracture-resistant high-entropy alloy for cryogenic applications. *Science 345*: 1153–8 with permission from AAAS.

multicomponent fcc Cantor alloys are clearly still far from optimised, with an enormous range of potential compositions, treatments and resulting microstructures yet to be explored.

High-temperature Senkov alloys

As discussed in the last sub-section, multicomponent high-entropy single-phase and multiphase fcc Cantor alloys have outstanding mechanical properties at cryogenic temperatures, with exceptional combinations of strength and toughness. At room temperature and at elevated temperatures they continue to have good ductility, formability and toughness, with current research concentrating on using variations in composition and thermomechanical treatment to try to improve their strength and hardness, particularly at higher temperatures.

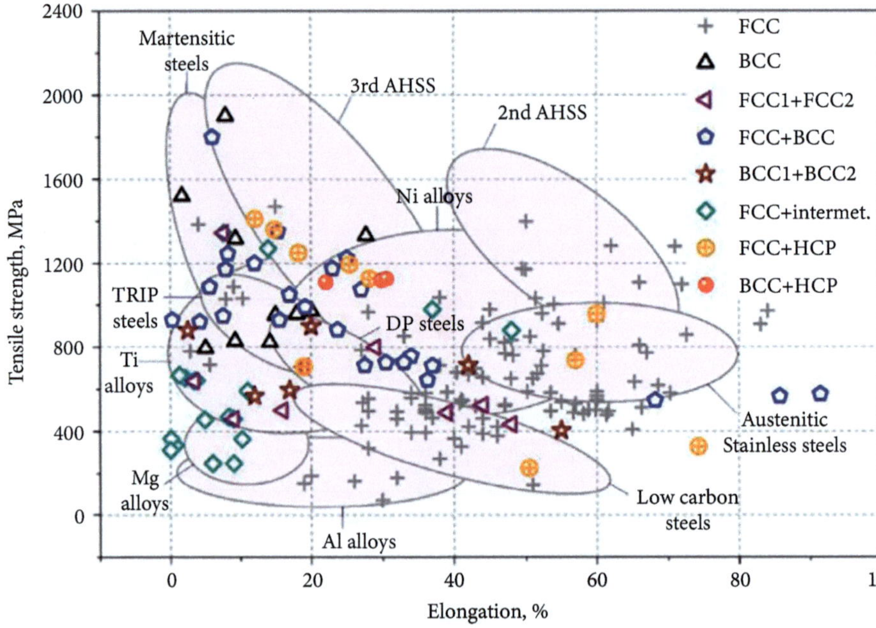

Figure 6.10 Ashby map of tensile strength versus elongation showing the mechanical properties of multicomponent and high-entropy materials (symbols) including single-phase fcc and multiphase fcc-based Cantor alloys, compared with high-strength steels and other metallic alloys (shaded regions) (second and third AHSS, DP and TRIP are second- and third-generation high-strength steels, dual-phase steels and transformation-induced plasticity steels respectively).

Reproduced from George et al. (2020). High-entropy alloys: A focused review of mechanical properties and deformation mechanisms. *Acta Materialia 188*: 435–74.

The situation is, in many ways, reversed for multicomponent high-entropy single-phase and multiphase bcc Senkov alloys (also often called *refractory high-entropy alloys* or *RHEAs*), which have good strength and hardness at room and elevated temperatures, with current research concentrating on variations in composition and thermomechanical treatment to try to improve their ductility, formability and toughness, particularly at room temperature.

We have already seen in Chapter 4 that the two original equiatomic single-phase bcc Senkov alloys,[43] quaternary NbMoTaW and quinary VNbMoTaW, both have excellent strength at room and elevated temperatures, with yield stresses of approximately 1.0 GPa and 1.3 GPa at room temperature and 0.5 GPa and 1.0 GPa at 1,000 °C (1,273K) respectively, as shown in Senkov et al.'s original compressive stress-strain curves[43] in Figure 6.11. As can also be seen in Figure 6.11, the two original equiatomic single-phase bcc Senkov alloys have extensive ductility at high temperatures, typically 15–30% throughout the temperature range 600–1,600 °C (873–1,873K), but are relatively brittle at room temperature with ductility limited to no more than 2–3%. These excellent high-temperature mechanical property results have led, as with the fcc Cantor alloys, to a substantial body of research[44–48] investigating: 1) other mechanical properties of the original Senkov alloys NbMoTaW and VNbMoTaW, such as fatigue, creep, wear and impact resistance; as well as 2) extensions to other single-phase bcc modified Senkov alloys, both with fewer and more components as

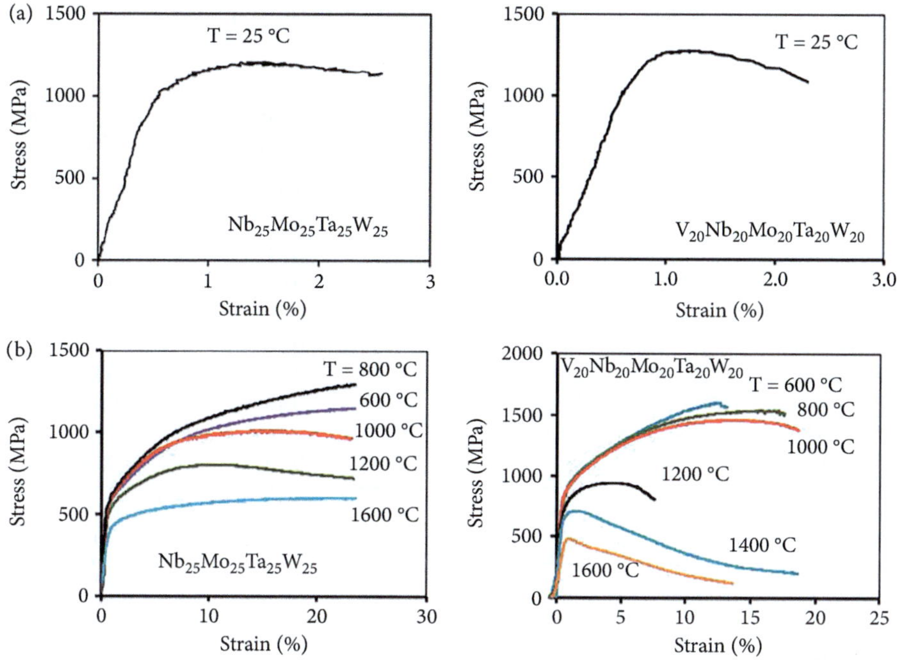

Figure 6.11 Compressive stress-strain curves at (a) room temperature and (b) elevated temperatures for the two original equiatomic single-phase bcc Senkov alloys NbMoTaW and VNbMoTaW.

Reproduced from Senkov et al. (2018). Development and exploration of refractory high entropy alloys: A review. *Journal of Materials Research 33*: 3092–3128.

well as with non-equiatomic compositions; and 3) alloying additions and thermomechanical treatments designed to enhance strength and ductility using, for instance, precipitation and grain-size effects.[44–48] Once again, we do not try to cover here all this research, which is discussed in detail in several substantial review articles,[44–48] but instead concentrate on a number of particularly interesting alloys and effects.

Senkov et al.[44,49] collected data for strength versus temperature for a total of fifteen single-phase bcc and eleven multiphase bcc-based equiatomic and non-equiatomic Senkov alloys, with compressive yield strengths ranging from 0.9–2.1 GPa at room temperature, 0.1–0.9 GPa at 1,000 °C (1,273K) and 0.02–0.7 GPa at 1,200 °C (1,473K). The alloys had a mixture of as-cast and recrystallised microstructures, all with a relatively large grain size, and the multiphase alloys were nearly all just two phase, consisting of bcc plus one or other of either a second bcc phase, a multicomponent Laves phase or a multicomponent silicide. Typical phase structures and yield strengths are shown in Table 6.4 for a selection of these Senkov alloys. The alloy yield strengths show little or no correlation with alloy melting point T_m or atomic radius misfit δ, but for the two-phase alloys increase very approximately proportional to the volume fraction V of the second phase independent of its nature, as shown in Figure 6.12. As shown in Figure 6.13, yield strengths remain high and ductilities remain low at relatively low temperatures, $T < 0.6T_m$, with deformation taking place by dislocation slip as discussed in Chapter 4; but yield strengths decrease and ductilities increase at higher temperatures, $T > 0.6T_m$, when atomic diffusion begins to be significant, leading

Table 6.4 Yield strength versus temperature for a variety of single-phase bcc and two-phase bcc-based Senkov alloys.

alloy	phases	σ_y at 23 °C (GPa)	σ_y at 1000 °C (GPa)	σ_y at 1200 °C (GPa)
$Al_{0.4}TiZrNbHf_{0.6}Ta$	bcc	1.8	0.30	0.09
AlTiCrMo	bcc	1.1	0.37	0.10
TiZrNbMoHf	bcc	1.5	0.81	0.56
VNbMoTaW	bcc	1.2	0.84	0.73
VTiNbMoTaW	bcc	1.5	0.75	0.66
$Al_{0.24}TiVZrNb$	bcc + Laves	1.2	0.08	0.02
AlTiCrNbMo	bcc + Laves	1.0	0.55	0.10
$Si_{0.3}TiV_{0.5}NbMo_{0.5}Hf$	bcc + M_5Si_3	1.6	0.40	0.16
$TiCr_{0.3}ZrNb$	bcc + Laves	1.6	0.14	0.03
TiVZrNbTa	bcc + bcc	1.4	0.19	0.08

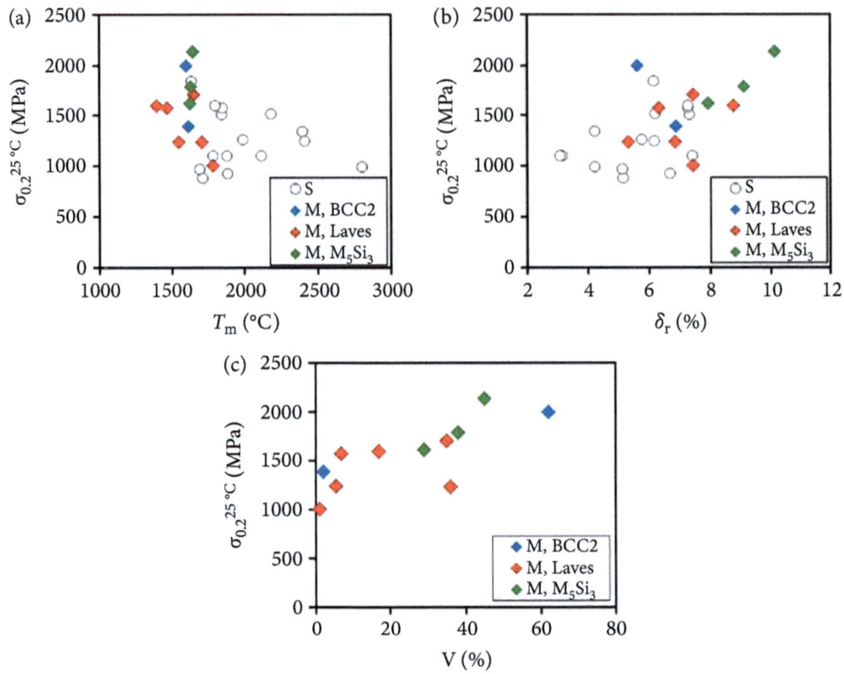

Figure 6.12 Variation of room-temperature yield strength $\sigma_{0.2}$ with (a) alloy melting point T_m, (b) atomic radius misfit δ, and (c) volume fraction V of second phase for a variety of multicomponent single-phase (S) and multiphase (M) bcc Senkov alloys.

Reproduced from Senkov et al. (2019). High-temperature strength of refractory complex concentrated alloys. *Acta Materialia 175*: 394–405.

ultimately to the onset of dynamic recovery and diffusion creep. Strength levels in multicomponent Senkov alloys are generally higher across the temperature range than comparable refractory-based commercial alloys, as indicated in Figure 6.13.

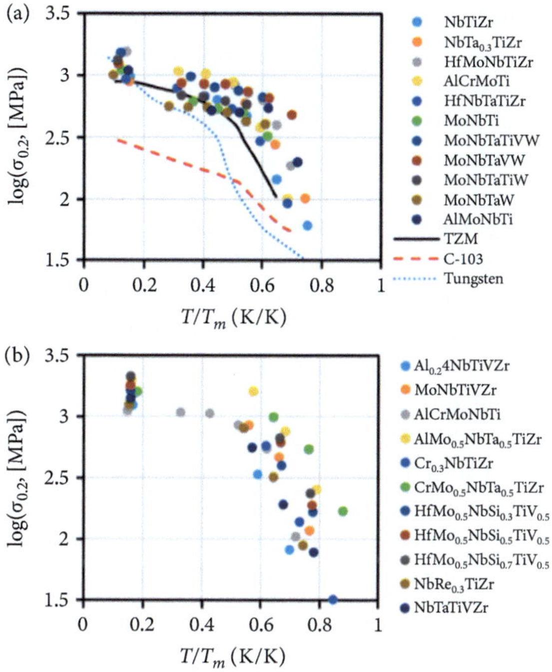

Figure 6.13 Variation of yield strength $\sigma_{0.2}$ with homologous temperature T/T_m for a variety of multicomponent (a) single-phase and (b) multiphase bcc Senkov alloys, compared with pure tungsten W, a Mo-based commercial alloy TMZ and a Nb-based commercial alloy C-103.
Reproduced from Senkov et al. (2019). High-temperature strength of refractory complex concentrated alloys. *Acta Materialia 175*: 394–405.

Given their excellent high-temperature combinations of strength, ductility and fracture toughness, there has been considerable effort to improve the room temperature brittleness of multicomponent single-phase bcc and multiphase bcc-based Senkov alloys.[44–48] It is well established that increasing the concentration of group IV elements such as Ti, Zr and Hf enhances room temperature ductility, whereas increasing the concentration of group V and VI elements such as Nb, Mo, Ta and W impairs room temperature ductility and causes excessive brittleness.[44–48] Essentially there is a ductile-brittle transition in multicomponent bcc Senkov alloys, with a transition temperature that decreases when there are fewer valence electrons, leading in some cases to ductilisation at room temperature. The basic brittleness of these alloys is thought to be a consequence of the fundamental instability of the bcc structure causing cleavage when it is subjected to tensile stresses, which can be overcome at higher temperatures by a Jahn–Teller electronic distortion of the lattice, made easier when there are fewer valence electrons, leading to shear instead of tensile deformation and fracture.[44,50]

The most well-known example of this ductilisation effect in multicomponent bcc Senkov alloys is the substantial room-temperature ductility of the equiatomic quinary alloy TiZrNbHfTa,[51–53] which as discussed in Chapter 3 is unlike the original Senkov alloys NbMoTaW and VNbMoTaW, both of which are quite brittle at room temperature, as shown in Figure 6.11. Senkov and Semiatin[51] manufactured multicomponent TiZrNbHfTa sheet by vacuum-arc melting followed by hot isostatic pressing at 207 MPa for two hours at 1,200 °C (1,473K) under argon, annealing for twenty-four hours at 1,200 °C (1,473K) also under argon, cold rolling by 86.4% with a series of passes at a strain rate of ~0.84/s, and finally recrystallisation by annealing for two hours at 800, 1,000 and 1,200 °C (1,073, 1,273 and 1,473K) to produce an equiaxed grain size of 20–30 μm. The alloy was single-phase bcc as-cast, as-rolled, and after annealing at 1,000 or 1,200 °C (1,273 or 1,473K), but was two-phase after annealing at 800 °C (1,073K) with a second bcc phase precipitating at the lower annealing temperature. As shown in the room-temperature true stress–true strain curves in Figure 6.14: with a deformed single-phase bcc elongated grain structure in the as-rolled sheet, the alloy had a yield strength of 1.2 GPa, with limited ductility of just under 5%; with a recrystallised two-phase bcc + bcc equiaxed grain structure after annealing at 800 °C (1,073K), it had a somewhat higher yield strength of 1.3 GPa, but with even poorer ductility of less than 2%; but with a recrystallised single-phase bcc equiaxed grain structure after annealing at 1,000 °C (1,273K), it had a slightly lower yield strength of 1.1 GPa, but much greater ductility of approximately 10%. Couzinié et al.[52] and Lilensten et al.[53] have studied the tensile and compressive deformation behaviour of equiatomic TiZrNbHfTa in considerable detail, as discussed in Chapter 4, showing that deformation is dominated by dislocation slip, initially on one and then on several slip systems, with relatively low rates of work hardening, no evidence of twinning, and very similar dislocation structures and a similar activation volume to those found in pure bcc metals, albeit at much higher flow stresses.

Similarly, Han et al.[54], Zhang et al.[55] and Wang et al.[56] have shown that progressive additions of Ti, V or Ni to the original equiatomic quaternary NbMoTaW Senkov alloy or the simpler equiatomic NbMoTa ternary subsystem gradually improves the room temperature ductility at the same time as increasing the yield and ultimate tensile or compressive strength, as shown for $Ti_xNbMoTaW$ in Figure 6.15. These alloys continue to have excellent high-temperature strengths and ductilities, as shown in the compressive stress–strain curves and work-hardening curves for $V_xNbMoTa$ at 800 and 1,000 °C (1,073 and 1,273K) in Figure 6.16, with strengths in the range 0.8–1.2 GPa and ductilities greater

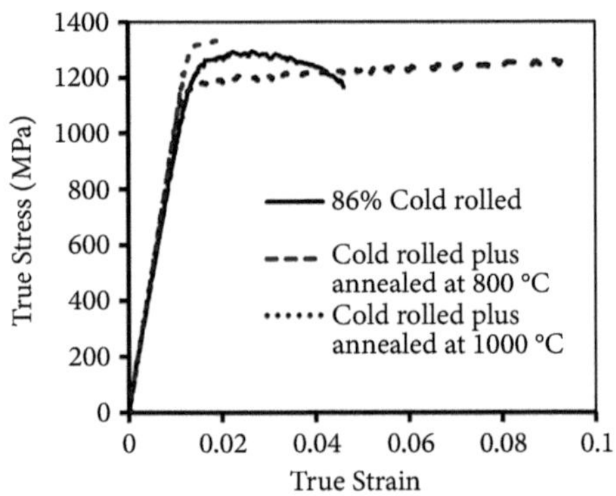

Figure 6.14 Tensile stress-strain curves for multicomponent TiZrNbHfTa as-cold rolled 86.4% and after recrystallisation by annealing for two hours at 800 and 1000 °C.

Reproduced from Senkov and Semiantin. (2015). Microstructure and properties of a refractory high-entropy alloy after cold working. *Journal of Alloys and Compounds 649*: 1110–23.

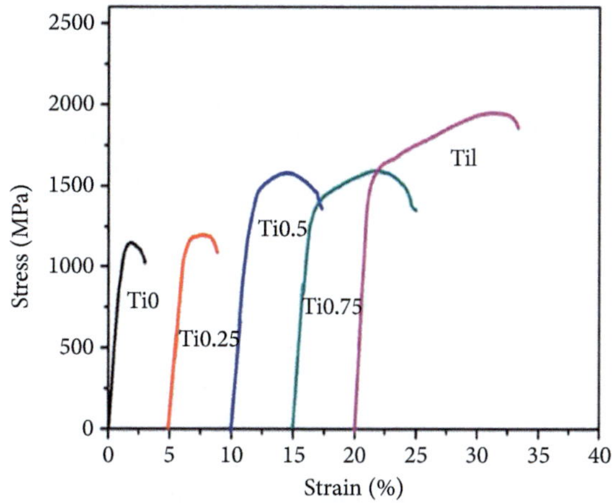

Figure 6.15 Room-temperature compressive stress–strain curves for multicomponent $Ti_xNbMoTaW$ Senkov alloys with $x = 0–1$.

Reproduced from Han et al. (2018). Microstructures and mechanical properties of $Ti_xNbMoTaW$ refractory high-entropy alloys. *Materials Science and Engineering A712*: 380–5.

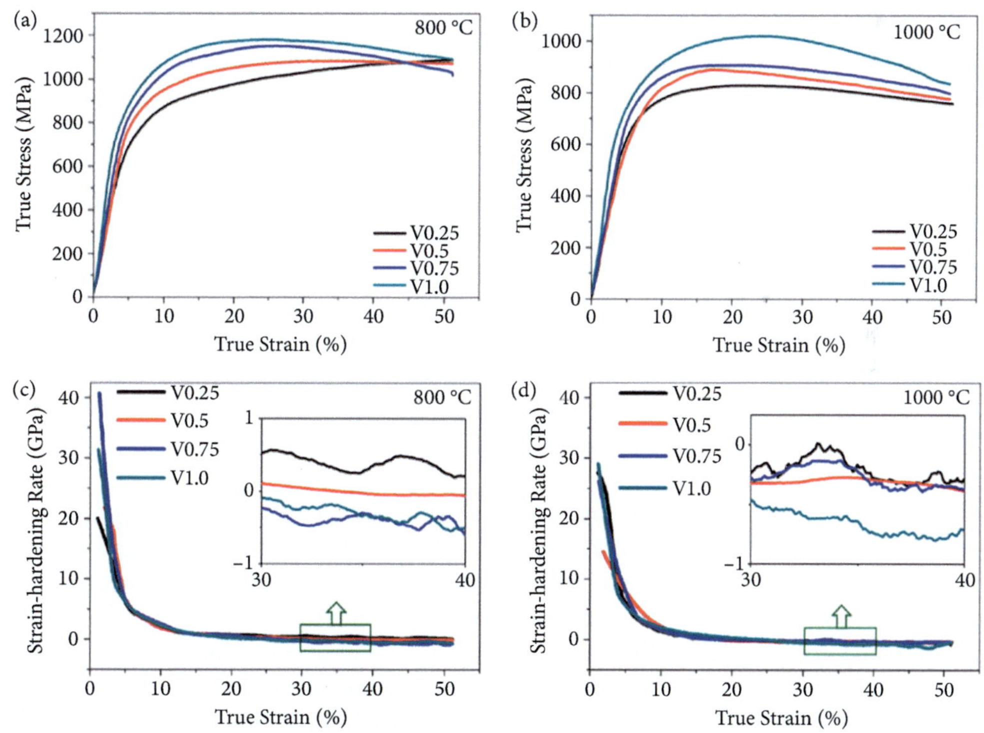

Figure 6.16 High-temperature compressive stress–strain curves and corresponding work-hardening curves at 800 and 1000 °C for multicomponent $V_xNbMoTa$ Senkov alloys with $x = 0–1$.

Reproduced from Wang et al. (2021). Effects of vanadium concentration on mechanical properties of $V_xNbMoTa$ refractory high-entropy alloys. *Materials Science and Engineering A808*: Article 140848.

Table 6.5 Yield strength σ_y, maximum compressive strength σ_m and fracture strain ε_f for some equiatomic single-phase bcc Senkov alloys.

alloy	temperature T (°C)	yield strength σ_y(GPa)	maximum strength σ_m (GPa)	fracture strain ε_f (%)
NbMoTa	25	1.2	1.2	15
TiNbMoTa	25	1.2	1.4	23
VNbMoTa	25	1.2	1.8	35
VNbMoTa	800	0.9	1.2	>50
VNbMoTa	1000	0.8	1.0	>50

than 50%. Some typical data for yield strengths, ultimate compressive strengths and fracture strains are collected together in Table 6.5. Similar results have been obtained in other alloys.[57–60]

Various strategies are now being pursued to try to improve further the mechanical properties of bcc-based Senkov alloys, particularly their high temperature strength and room temperature ductility, and also particularly their tensile rather than compressive properties.[44–48] These different strategies include:[44–48]

1. Addition of interstitial elements such as B, O and N to enhance grain boundary cohesion, at levels low enough to prevent grain boundary embrittlement and formation of hard borides, oxides or nitrides.[61,62]
2. Addition of elements such as Al, C and Si to develop two-phase microstructures, aiming to use aluminide, carbide or silicide precipitation hardening to improve high-temperature strength, again at low enough levels to avoid reductions in room-temperature ductility and toughness.[63–65]
3. Modification of alloy composition to increase room-temperature and low-temperature ductility by destabilising the bcc structure and thus promoting martensitic transformation and associated TRIP.[66]
4. Severe plastic deformation to reduce grain size and use the Hall–Petch effect to increase strength without loss of ductility.[67]

As an example, Gao et al.[61] investigated the effect of adding boron to bcc Senkov alloys by manufacturing a series of $B_xTiZrNbMo_{0.5}Hf_{0.5}$ alloys with boron contents ranging from x = 0.1–0.9 at% by arc melting the pure metals and TiB_2 under argon, re-melting and casting five times to ensure homogeneity, followed by microstructural characterisation, hardness and compression testing. The alloys consisted of a multicomponent dendritic bcc matrix, with increasing amounts of MB_2 boride as the boron content was increased from 0.1 to 0.9 at%, as shown by the X-ray diffraction (XRD) patterns in Figure 6.17. The MB_2 boride was in the form of fine-scale interdendritic bcc + MB_2 eutectic for low boron contents up to about 0.5 at%, and then coarser primary MB_2 particles for higher boron contents >0.5 at%. The corresponding compressive stress-strain curves in Figure 6.18 show: 1) a small 0.1 at% addition of boron increases the yield strength, ultimate strength and ductility from approximately 1.3 GPa, 1.5 GPa and 24% to 1.5 GPa, 2.0 GPa and 30% respectively; and 2) further increases in boron content up to 0.9 at% produce further

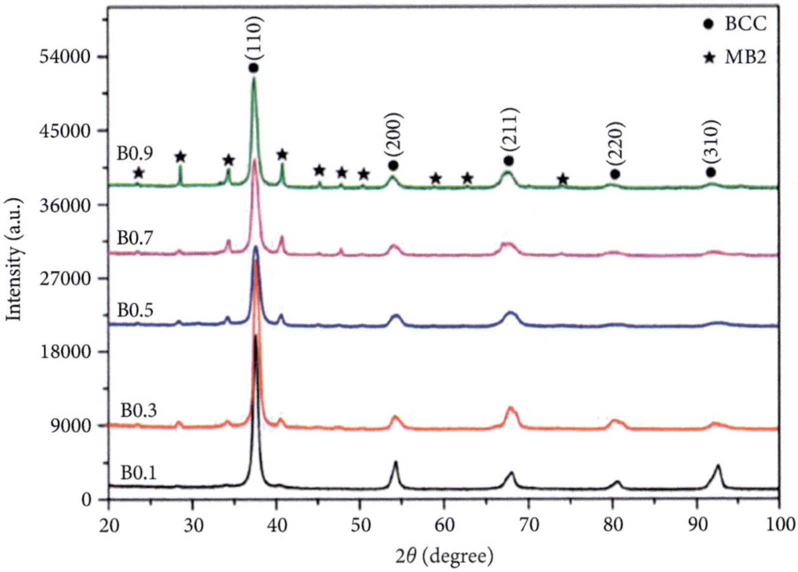

Figure 6.17 XRD traces from multicomponent $B_xTiZrNbMo_{0.5}Hf_{0.5}$ alloys with $x = 0.1–0.9$ at%. Reproduced from Gao et al. (2021). In-situ development of MB_2 and their effect on microstructure and mechanical properties of refractory $Hf_{0.5}Mo_{0.5}NbTiZr$ high-entropy alloy composites. *International Journal of Refractory Metals and Hard Materials 96*: Article 105473.

progressive increases in yield and ultimate strength, up to almost 2.0 and 2.2 GPa respectively, but with sharp reductions in ductility and plastic strains falling to no more than a few per cent.

As another example, Senkov et al.[63] investigated the effect of adding Al to bcc Senkov alloys, manufacturing a variety of alloys with different compositions in the AlTiZrNbMoTa system by vacuum arc-melting in a reduced inert gas atmosphere, followed by hot isostatic pressing (HIP) for two hours at 1,400 °C (1,673K) and 207 MPa to remove as-cast porosity, annealing for six hours at 1,400 °C (1,673K) and furnace cooling at 20 °C/min (0.3K/s) to room temperature. The alloys all consisted of polycrystalline equiaxed grains, with a grain size of approximately 50 μm and some residual evidence of an internal dendritic sub-grain solidification structure. The detailed microstructures were more complex: the matrix grains in some cases were single-phase multicomponent disordered bcc; the matrix grains in other cases were a nanoscale basket-weave mixture of multicomponent disordered bcc and multicomponent ordered bcc B2, as shown in Figure 6.19; and relatively coarse intermetallic particles were formed at the grain boundaries and sub-grain boundaries in some of the alloys. As shown in the compression stress-strain curves in Figure 6.20, the mixed bcc + B2 alloys exhibited yield and ultimate strengths in the range 1.8–2.5 GPa with up to 4% ductility at room temperature, and retained strengths of 0.4–1.0 GPa with >50% ductility at 1,000 °C (1,273K).

Lilensten et al.[66] developed a TRIP alloy by modifying the composition of the equiatomic Senkov alloy TiZrNbHfTa by decreasing the quantity of bcc-stabiliser elements Nb and Ta in favour of non-bcc elements Ti, Zr and Hf. The resulting modified Senkov alloy had a composition of $Ti_{35}Zr_{27.5}Nb_5Hf_{27.5}Ta_5$ and was manufactured by induction-melting and arc-casting, followed by cold-rolling 60–70% and solution treatment for thirty minutes at

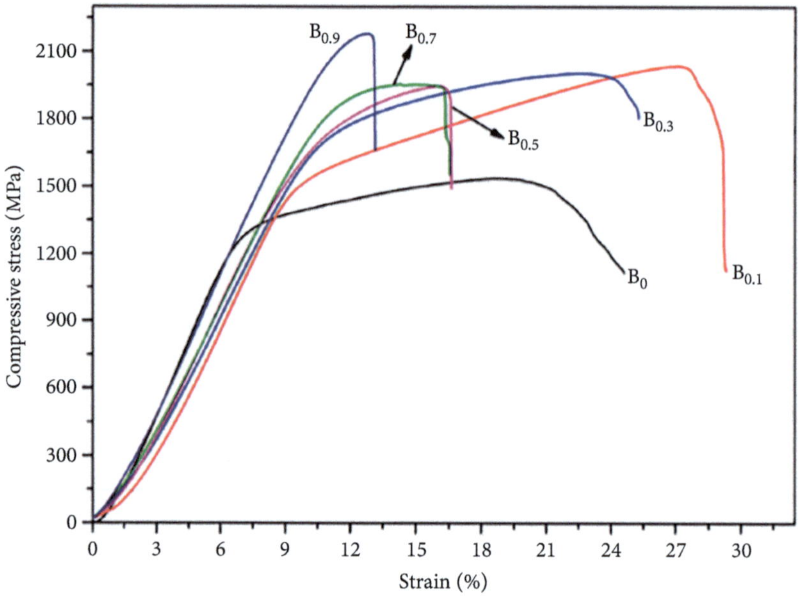

Figure 6.18 Compressive stress–strain curves from multicomponent $B_xTiZrNbMo_{0.5}Hf_{0.5}$ alloys with x = 0–0.9 at%.

Reproduced from Gao et al. (2021). In-situ development of MB_2 and their effect on microstructure and mechanical properties of refractory $Hf_{0.5}Mo_{0.5}NbTiZr$ high-entropy alloy composites. *International Journal of Refractory Metals and Hard Materials 96*: Article 105473.

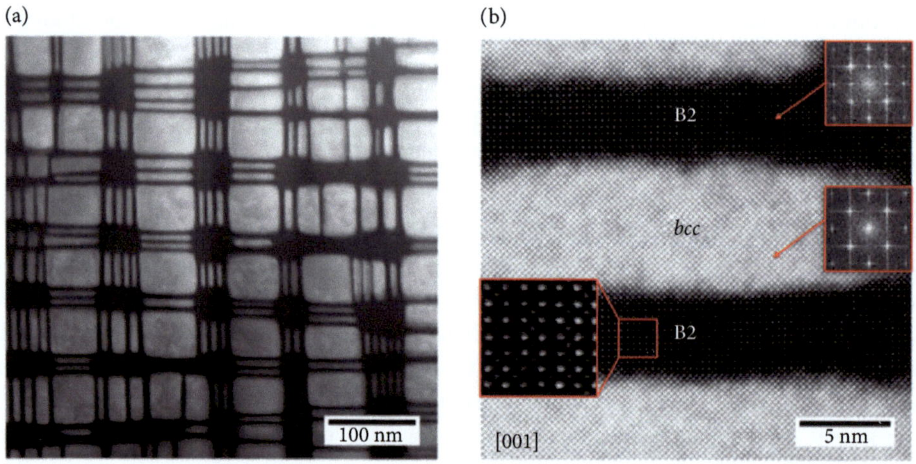

Figure 6.19 TEM images of basket-weave mixture of disordered bcc + ordered B2 structure of grains in $AlTiZrNbMo_{0.5}Ta_{0.5}$ Senkov alloy.

Reproduced from Senkov et al. (2018). Compositional variation effects on the microstructure and properties of a refractory high-entropy superalloy $AlMo_{0.5}$ $NbTa_{0.5}$ TiZr. *Materials and Design 139*: 498–511.

800 and 900 °C (1.073 and 1,273K) to achieve an equiaxed bcc structure. The resulting tensile yield strength, UTS and tensile ductility were approximately 0.6 GPa, 1.0 GPa and 23% respectively, with deformation enhanced by a stress-induced bcc $\beta \rightarrow$ orthorhombic α'' martensitic transformation, as shown by the increasing martensite content with increasing

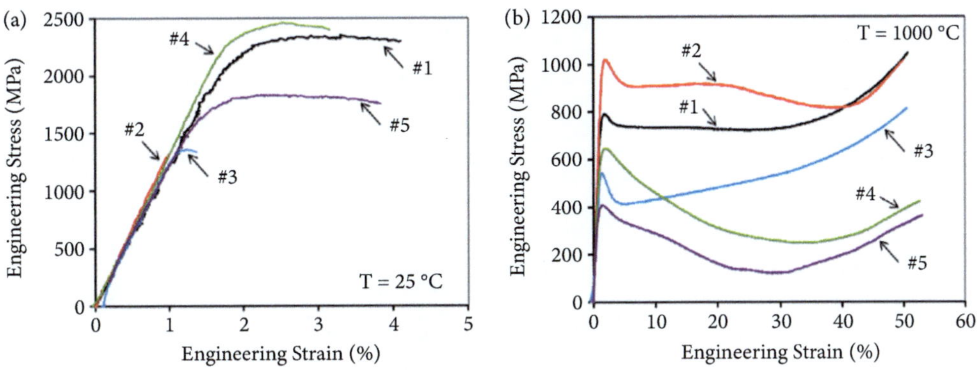

Figure 6.20 Compression stress–strain curves at 25 and 1,000 °C for Senkov alloys $AlTiZrNbMo_{0.5}Ta_{0.5}$ (#1), $AlTiZr_{0.5}NbMo_{0.5}Ta_{0.5}$ (#2), $AlTiZr_{0.5}NbTa_{0.5}$ (#3), $Al_{0.5}TiZrNbMo_{0.5}Ta_{0.5}$ (#4), and $Al_{0.25}TiZrNbTa$.

Reproduced from Senkov et al. (2018). Compositional variation effects on the microstructure and properties of a refractory high-entropy superalloy $AlMo_{0.5}NbTa_{0.5}TiZr$. *Materials and Design 139*: 498–511.

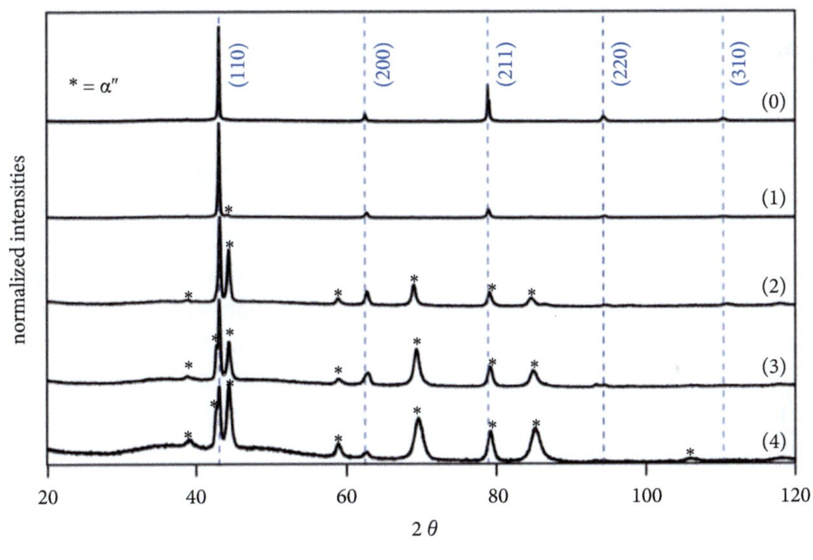

Figure 6.21 XRD patterns from modified Senkov alloy $Ti_{35}Zr_{27.5}Nb_5Hf_{27.5}Ta_5$ after varying amounts of tensile deformation: (0)–(4) = 0, 4, 7, 11 and 17% tensile strain respectively.

*Unmarked peaks are disordered bcc β and asterisked peaks are orthorhombic α'' martensite. Reproduced from Lilensten et al. (2017). Design and tensile properties of a bcc-Ti-rich high-entropy alloy with transformation-induced plasticity. *Materials Research Letters 5*: 110–16.

strain on the X-ray diffraction (XRD) patterns in Figure 6.21, and the deformation-induced martensitic structure shown clearly in the electron back-scattered diffraction (EBSD) maps in Figure 6.22.

Much of the interest in the mechanical properties of multicomponent bcc Senkov alloys arises from their sustained high-temperature strength, which makes them potential candidate materials from which to manufacture some of the critical rotating and other components in the hottest regions of industrial and aeroengine gas turbines. For applications of

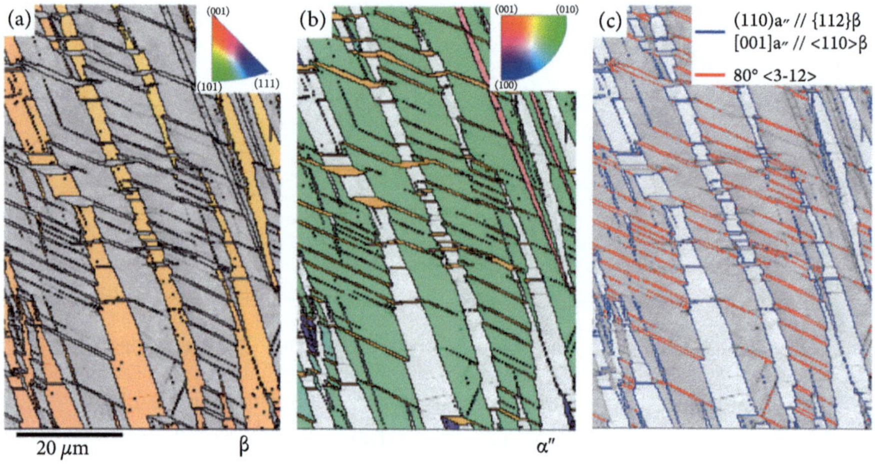

Figure 6.22 EBSD maps of bcc $\beta \rightarrow$ orthorhombic α'' martensitic structure caused by transformation-induced plasticity (TRIP) in modified Senkov alloy $Ti_{35}Zr_{27.5}Nb_5Hf_{27.5}Ta_5$ after 7% tensile strain.

Reproduced from Lilensten et al. (2017). Design and tensile properties of a bcc-Ti-rich high-entropy alloy with transformation-induced plasticity. *Materials Research Letters* 5: 110–16.

this type, reducing alloy weight and thus increasing specific strength is extremely important, leading to the preferred use of lighter elements such Al, Si and Ti.[44–48] Moreover, aggressive gas conditions put a premium on good high-temperature oxidation and corrosion resistance, leading to the preferred use of elements that form resistant oxide coatings, such as Al and Cr.[44–48] Overall, this means that it is far from straightforward to adjust multicomponent Senkov alloy compositions so as to achieve a good balance between high temperature strength, room temperature ductility, low density, and oxidation and gas-corrosion resistance, with much further research still needed, and indeed underway, to try to solve this complex optimisation problem.

Li et al.[8] have compared the room-temperature mechanical properties of multicomponent single-phase fcc Cantor alloys, multicomponent single-phase bcc Senkov alloys and multicomponent multiphase alloys on Ashby maps of tensile yield strength and ultimate tensile strength (UTS) versus failure strain, as shown in Figure 6.23. These maps show clearly: the excellent room-temperature ductility and toughness of the single-phase fcc Cantor alloys, as discussed in the last sub-section; the excellent room-temperature strength but poor ductility of the single-phase bcc alloys, as discussed in this sub-section; and the potential for improving these properties further using thermomechanical treatments to develop multiphase structures via transformations such as precipitation, ordering and martensite formation. As can be seen in Figure 6.23, single-phase bcc Senkov alloys have room-temperature yield strengths, ultimate tensile strengths and fracture strains ranging up to 1.9 MPa, 2.0 MPa and 23% respectively; and two-phase bcc-based Senkov alloys have room-temperature yield strengths, ultimate tensile strengths and fracture strains ranging up to 2.1 MPa, 2.6 MPa and almost 100% respectively. It is, of course, also clear that the number of different alloy compositions and treatments that have been investigated is still a minute fraction of the total range of possibilities, so there is still much room for optimising multicomponent bcc Senkov

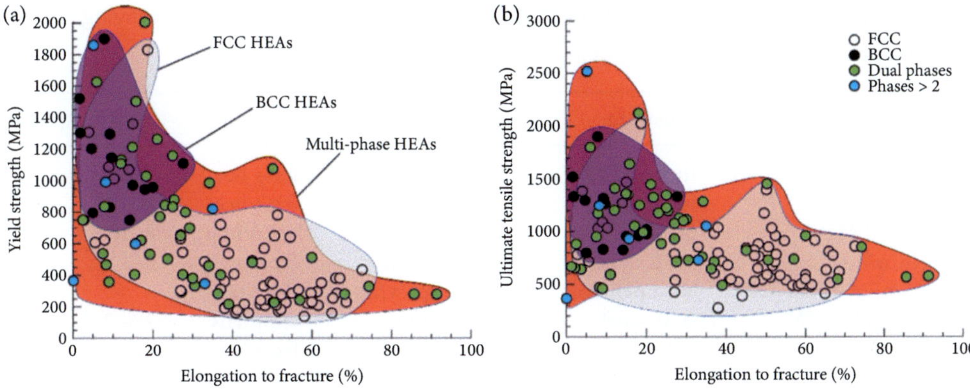

Figure 6.23 Ashby maps of (a) tensile yield strength and (b) ultimate tensile strength versus fracture strain for multicomponent single-phase fcc, single-phase bcc and multiphase materials.

Reproduced from Li et al. (2021). Mechanical behaviour of high-entropy alloys. *Progress in Materials Science 118*: Article 100777.

alloys for future engineering applications. As Li et al. comment, 'continued research and development of HEAs with ultra-high strength and exceptional ductility are still imperative'.[8]

Hard coatings

Multicomponent high-entropy metallic, ceramic and composite materials can be deposited onto a surface to provide hard, wear-resistant and corrosion-resistant coatings, using techniques such as laser and plasma cladding, flame and plasma spraying, thermal evaporation and sputtering. The manufacturing methods and resulting properties of such multicomponent coatings deposited on a range of different substrate materials have begun to be studied in some detail. Table 6.6 gives examples of relatively thick coatings that have been manufactured from the liquid phase by spraying and cladding processes;[68–74] Table 6.7 gives examples of thinner coatings that have been manufactured from the vapour phase by sputtering;[75–79] and more examples of other coating/substrate combinations are given in the reviews by Arshad et al.,[80] Oses et al.,[81] Zhang and Reece[82] and Xiang et al.[83] Many multicomponent high-entropy coatings are found to exhibit low friction coefficients and low levels of degradation by wear and/or corrosion.[68–83]

Ni superalloy and stainless steel components are protected from aggressive and high-temperature gas environments in many engineering uses by the application of a complex thermal-barrier coating (TBC), typically several hundred μm thick, consisting of an underlying sprayed metallic bondcoat (often an MCrAlY alloy where M = Ni or Co), covered by a thermally grown oxide or TGO (usually alumina Al_2O_3), on top of which is a thermally insulating ceramic topcoat (often yttria-stabilised zirconia YSZ). The ceramic topcoat is usually somewhat porous, and it protects the metal component from exposure to high temperatures in the ambient gases. The underlying metallic bondcoat provides oxidation and gas corrosion protection below the porous ceramic, and together with the thermally grown oxide also provides strain relief by preventing a sharp change in thermal expansion

Table 6.6 Examples of multicomponent coatings deposited from the liquid by spraying and cladding processes.[72–78]

coating	substrate	process	structure	hardness(GPa)
$Al_xCrFeNiCu$	Inconel 718	plasma cladding	fcc + bcc	3.5–7
AlTiCrCoNi	Haynes 230	thermal spraying	fcc + bcc	
AlTiCrFeNbMoW	M2 steel	laser cladding	bcc + hcp	8–10.5
CrFeNiCoMo	304L stainless	laser cladding	fcc + σ	6–12
$(LaNdSmEuGd)_2Zr_2O_7$	Ni superalloy	plasma spraying	fluorite	
$Ti_xCr_2FeCoNi/TiN$	904L steel	laser cladding	fcc + TiN	2.5–7
AlTiCrFeCoNi	316 stainless	plasma spraying	fcc + bcc	6–8

Table 6.7 Examples of multicomponent coatings deposited from the vapour by sputtering (am = amorphous).[79–83]

coating	substrate	process	structure	hardness (GPa)
CrFeCoNi	Si	sputtering	fcc	8.5–10.4
TiZrNbHfTa	Ti-6Al-4V	sputtering	am	12.5
TiCrZrNbMo	304 stainless	sputtering	bcc + am	9.7
$(AlTiZrCrHf)N_x$	Si	sputtering	fcc	16–33
$(VNbMoTaW)N_x$	Si	sputtering	bcc or fcc	18–32

at the boundary between the ceramic and the metallic component. However, failures still take place under prolonged exposure when the ambient hot gases and/or oxygen from the ceramic diffuse into and react with the bondcoat material. Thick plasma-sprayed and/or laser-clad multicomponent high-entropy alloy coatings such as those shown in Table 6.6 are being investigated as possible improved bondcoat materials because of their low diffusivity and, therefore, inherent resistance to microstructural degradation and oxidation at high temperature. As an example, Xu et al.[68] manufactured a series of ~250 μm-thick multicomponent $Al_xCrFeNiCu$ coatings with $x = 0.5$, 1, 1.5 and 2 on Inconel 718 substrates by laser cladding, with five simultaneous and independent pure metal powder feeds into a molten pool on the substrate surface created by a focused Ar gas laser beam scanning at 260 mm/s. With increasing Al content the laser-clad $Al_xCrFeNiCu$ coatings changed structure from columnar single-phase fcc to equiaxed single-phase bcc grains, and increased in hardness from ~3 to ~7 GPa, as shown in Figures 6.24 and 6.25 respectively. The coatings were all found to be very resistant to degradation by grain growth, oxidation or corrosion at high temperatures, because of low diffusion rates associated with local lattice strains in the multicomponent high-entropy materials, as discussed in detail in Chapter 4.

Sputtered multicomponent high-entropy alloy thin films such as those shown in Table 6.7 are being investigated more for their potential as ultra-hard and wear-resistant coatings. As an example, Xia et al.[79] manufactured a series of ~1.5 μm-thick multicomponent $(VNbMoTaW)_{1-x}N_x$ coatings with $x = 0–0.45$ on B-doped Si single crystal and polyimide substrates by two sputtering methods: 1) cathode arc deposition (CAD) from a compacted VNbMoTaW cathode, with a background vacuum of 8×10^{-4} Pa, in 5 Pa of flowing Ar and

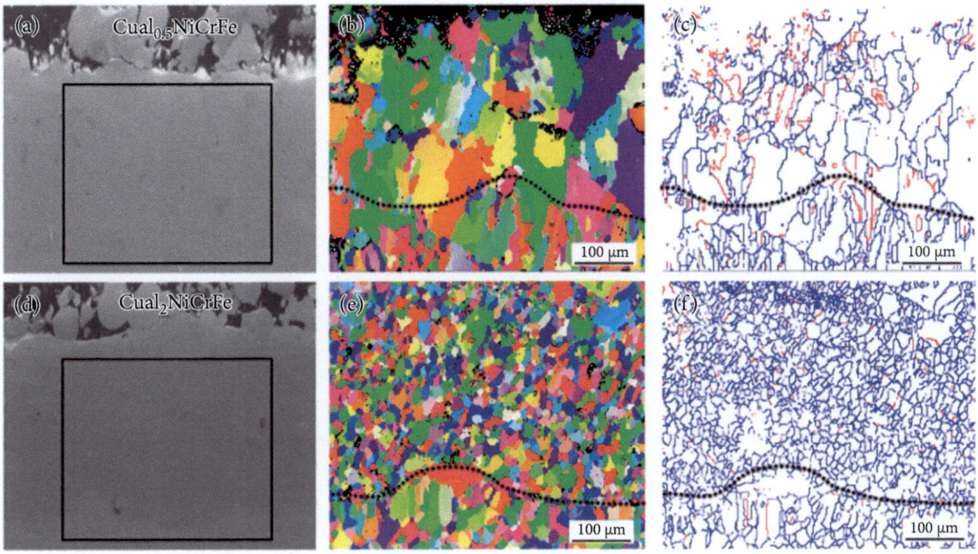

Figure 6.24 SEM images of cross-sections through laser-clad $Al_xCrFeNiCu$ coatings on Inconel 718.

Reproduced from Xu et al. (2021). TGO and Al diffusion behavior of $CuAl_xNiCrFe$ high-entropy alloys fabricated by high-speed laser cladding for TBC bond coats. *Corrosion Science 192*: Article 109781.

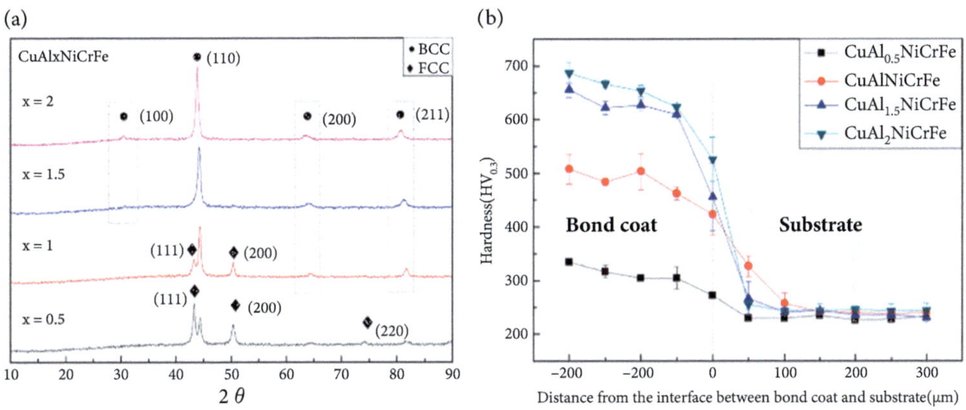

Figure 6.25 (a) XRD patterns and (b) hardnesses profiles from laser-clad $Al_xCrFeNiCu$ coatings on Inconel 718.

Reproduced from Xu et al. (2021). TGO and Al diffusion behavior of $CuAl_xNiCrFe$ high-entropy alloys fabricated by high-speed laser cladding for TBC bond coats. *Corrosion Science 192*: Article 109781.

N_2 gas, with 120 A electric current and a corresponding deposition rate of ~140 nm/min; and 2) dc magnetron sputtering (DCMS) again from a compacted powder VNbMoTaW cathode, with a background vacuum of 4×10^{-4} Pa, in 0.5 Pa of Ar and N_2 gas flow, with a discharge current of 0.15 A, and a corresponding deposition rate of 26–32 nm/min. Coatings with a low nitrogen content, x = 0–0.2, were disordered single-phase bcc, with nitrogen in interstitial solid solution; and coatings with a high nitrogen content, x = 0.2–0.45, were single-phase fcc with an ordered rock-salt transition-metal nitride structure, as shown in the

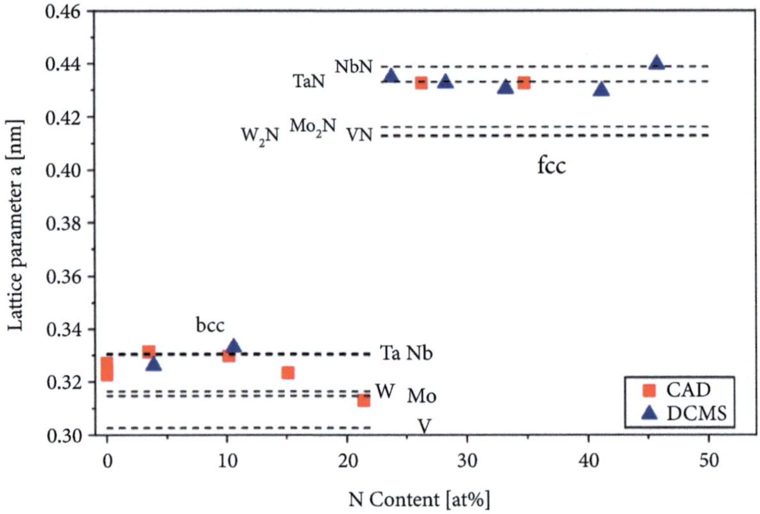

Figure 6.26 XRD measurements of crystal structure in sputtered $(VNbMoTaW)_{1\text{-}x}N_x$ coatings.
Reproduced from Xia et al. (2021). Influence of the nitrogen content on the structure and properties of MoNbTaVW high-entropy alloy thin films. *Journal of Alloys and Compounds 850*: Article 156740.

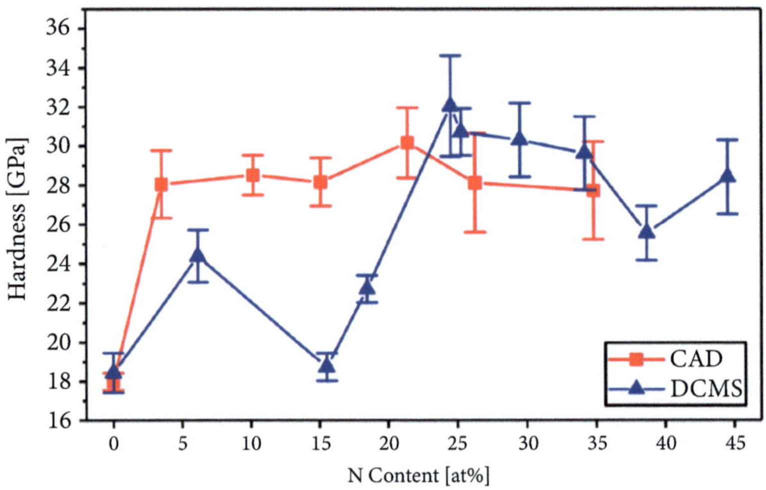

Figure 6.27 Nanoindentation measurements of hardness in sputtered $(VNbMoTaW)_{1\text{-}x}N_x$ coatings.
Reproduced from Xia et al. (2021). Influence of the nitrogen content on the structure and properties of MoNbTaVW high-entropy alloy thin films. *Journal of Alloys and Compounds 850*: Article 156740.

X-ray diffraction (XRD) results in Figure 6.26. The hardnesses of the high-nitrogen coatings were in the range 28–32 GPa, as shown in Figure 6.27.

As a final example, Tüten et al.[76] manufactured 800 nm thick multicomponent amorphous TiZrNbHfTa coatings on a Ti-6wt%Al-4wt%V alloy substrate by radio-frequency (RF) sputtering from an arc-melted TiZrNbHfTa cathode, with a background vacuum of 10^{-5} Pa, in a flowing Ar sputtering gas, with an electric power of 100 W. The resulting dense TiZrNbHfTa

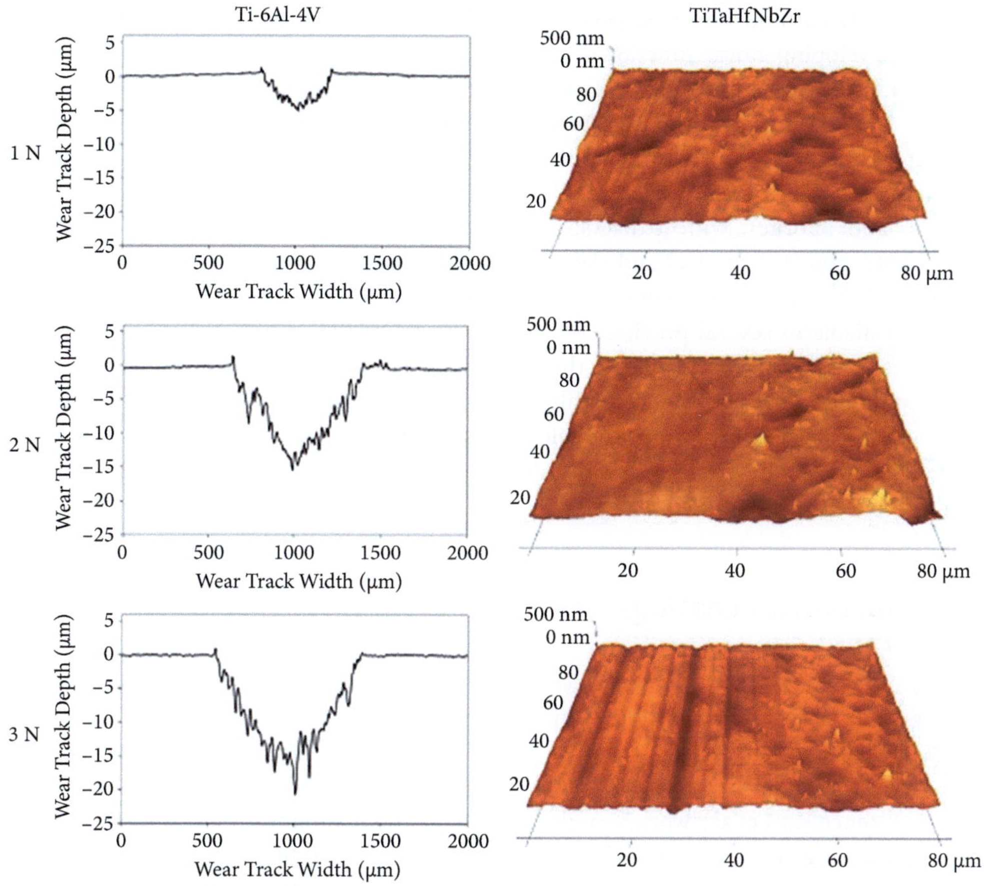

Figure 6.28 Sliding wear damage for different applied forces on Ti-6Al-4V substrate (profilometer measurements on LHS) and sputtered TiZrNbHfTa coating (atomic force microscope AFM images on RHS).

Reproduced from Tüten et al. (2019). Microstructure and tribological properties of TiTaHfNbZr high entropy alloy coatings deposited on Ti-6Al-4V substrates. *Intermetallics 105*: 99–106.

coatings had hardnesses in the range 18 ± 2.5 GPa and were essentially undamaged in ball-on-disc sliding wear tests with a 6-mm-diameter alumina ball sliding at 10 mm/s for a total sliding distance of thirty metres under applied loads of up to 3 N, as shown in Figure 6.28.

3 Functional Properties of Multicomponent Materials

Following on from the initial discovery of multicomponent high-entropy materials,[84,85] there has now been substantial investigation of their mechanical properties and associated potential applications,[4,7,8] in particular for single-phase fcc and related Cantor alloys, single-phase bcc and related Senkov alloys, and metallic, ceramic and composite hard coatings, as discussed in the last section. By comparison there has been much less study of the functional and other properties of multicomponent materials. The great potential for developing novel functional properties in multicomponent materials and tuning their properties by varying the composition has only been recognised much more recently,[80–83,86–88] and in the last few years has begun to be explored in detail, both in multicomponent high-entropy metallic alloys[87,88] and in multicomponent high-entropy ceramic alloys,[80–83] including materials such as borides, silicides, chalcogenides, perovskites and fluorites, with a diverse range of properties such as magnetic, dielectric, electrolytic, catalytic, thermal and electrical conduction, and corrosion and radiation protection.[80–83,87,88] We discuss some of these studies in the following sub-sections, with more details available in several previous reviews of the functional properties of multicomponent materials[80–83,87,88] available in the literature. As with the mechanical properties of multicomponent high-entropy materials discussed in the last section, many of their novel functional properties arise from the wide variation in local atomic structures and associated local mechanical and electronic lattice strains described and discussed in detail in Chapter 4.

Thermal barrier coatings

The investigation of *multicomponent ceramics* and their functional properties and applications was initiated, as discussed in Chapter 4, by Rost et al.'s seminal study[86] of the single-phase high-entropy rock-salt-structured mixed mono-oxide (MgCoNiCuZn)O, although there were a number of earlier studies of multicomponent nitride and boride hard coatings.[89–91] Since Rost et al.'s work,[86] multicomponent and high-entropy ceramics have since been studied much more intensively. In particular, a variety of different single-phase high-entropy oxides, carbides, chalcogenides, silicates and aluminates with multicomponent metallic cations have been shown to exhibit low values of *thermal conductivity*, caused by the variation in cation sizes and resulting lattice strains throughout the different crystal structures, leading to substantial disruption of lattice heat conduction by phonon scattering and attenuation. Table 6.8 shows a selection of these materials, which are being investigated as potential *thermal* and *environmental barrier coating materials* (*TBCs* and *EBCs*) for a range of different applications,[92–98] with other materials and applications given in the reviews by Oses et al.,[81] Zhang and Reece[82] and Xiang et al.[83]

As an example, Braun et al.[92] used pulsed laser deposition to manufacture epitaxial multicomponent ceramic mono-oxide thin films of (MgCoNiCuZn)O and (MgCoNiCuZnM)O alloys where M = Sc, Sb, Sn, Cr and Ge, all grown on (100)-oriented single-crystal MgO substrates. The multicomponent ceramic films were all single-phase single crystals with the characteristic mono-oxide NaCl rock-salt cubic crystal structure, but with an anisotropic tetragonal distortion as-deposited, having a constrained in-plane lattice parameter equal to that of the underlying MgO substrate and a different, unconstrained relaxed through-thickness lattice parameter, slightly greater or slightly smaller than that of MgO depending on the particular combination of multicomponent cations. The thermal conductivity of the 5-component (MgCoNiCuZn)O alloy mono-oxide was already low, $k = 2.9$ W/m/K, but was further reduced, roughly halved, in all the six-component (MgCoNiCuZnM)O alloy mono-oxides, in the range $k = 1.4$–1.7 W/m/K as shown in Figure 6.29. The decrease in thermal conductivity on going from five to six components as shown in Figure 6.29 was caused by

Table 6.8 Some multicomponent high-entropy ceramics with low thermal conductivity for potential use in thermal and environmental barrier coatings (TBCs and EBCs).[92–98]

ceramicalloy	processingmethod	structure	thermal conductivity k (W/m/K)
(MgCoNiCuZn)O	laser deposition	rock salt	2.9
(MgCoNiCuZnSb)O	laser deposition	rock salt	1.4
$(TiZrCeHfSn)O_2$	sintering	fluorite	1.3
$(Ca_{0.5}Y_{0.5}ZrCeHf)O_{2-\delta}$	spark plasma sintering	fluorite	1.1
(TiZrNbHfTa)C	spark plasma sintering	rock salt	6.4
$(Sn_{0.7}Ge_{0.2}Pb_{0.1})Mn_{0.11}Te$	spark plasma sintering	rock salt	2.0
(GeAgBi)Se	vacuum melting	rock salt	0.4
$(SmGd)_2(TiZrHf)_2O_7$	sintering	pyrochlore	1.5
$(LaCeNdSmEu)_2Zr_2O_7$	sintering	pyrochlore	2.1
$(YEuErYbLu)Al_5O_{12}$	sintering	Al garnet	3.6
$(YNdSmEuYb)_4Al_2O_9$	spark plasma sintering	monoclinic	1.5

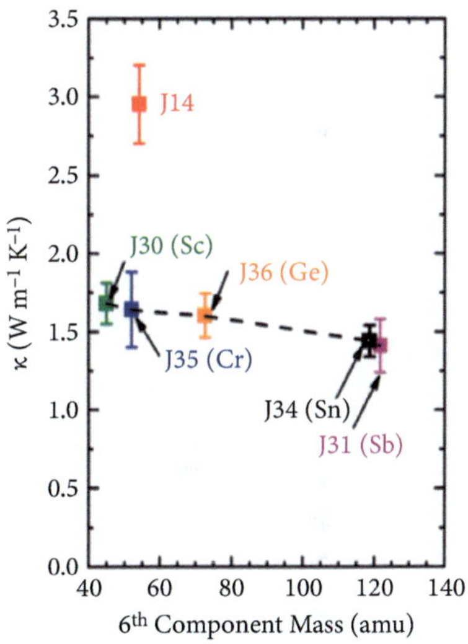

Figure 6.29 Thermal conductivity k of multicomponent high-entropy ceramic thin films of (MgCoNiCuZn)O (J14) and a series of (MgCoNiCuZnM)O alloys where M = Sc, Sb, Sn, Cr and Ge (J30, J31, J34, J35 and J36) respectively.
Reproduced from Braun et al. (2018). Charge-induced disorder controls the thermal conductivity of entropy-stabilized oxides. *Advanced Materials 30*: Article 1805004.

an increase in local lattice distortion at both cation and anion lattice sites, as confirmed by a combination of X-ray diffraction (XRD), extended X-ray absorption fine structure (EXAFS), and density functional theory (DFT) modelling.[92]

High-entropy ceramics have a number of advantages over more conventional single component ceramics as materials for *thermal* and *environmental barrier coatings* (*TBCs* and *EBCs*). As discussed in Chapter 5, multicomponent high-entropy alloys tend to have low levels of atomic diffusivity caused by wide variations in component atomic sizes and associated lattice strains, leading to considerable resistance to high temperature degradation by either microstructural coarsening or gaseous oxidation or corrosion. Moreover, as discussed in the last section, multicomponent high-entropy ceramics also have attractive mechanical properties, including ultra-high hardness, high strength and stiffness, and high fracture toughness. The combination of low thermal conductivity, thermal stability, oxidation resistance, high hardness, high stiffness and resistance to fracture is very attractive in protective TBC and EBC coating materials.[80–83] Figure 6.30 shows typical scanning electron microscope (SEM) images and corresponding energy dispersive (EDS) X-ray composition maps of the cross-sectional microstructures of three multicomponent oxide thermal barrier coatings, $(YZrCeYbHf)O_{2-\delta}$, $(YZrCeGdHf)O_{2-\delta}$ and $(CaYZrCeHf)O_{2-\delta}$ manufactured by high-energy ball milling of the individual metal oxide powders followed by spark plasma sintering at 1,800 °C (2,073K) for five minutes at 50 MPa, the first two oxide alloys with a single-phase solid-solution fluorite crystal structure, and the last with a secondary, precipitated $CaHfO_3$ phase.[93] Figure 6.31 shows an Ashby map of thermal conductivity versus elastic modulus for a large number of different materials, with the mono-oxides studied by Braun et al.[92] (labelled entropy-stabilised oxides or ESOs) having the best combination of stiff, strong and thermally insulating properties. Another advantage of high-entropy ceramics is the ability to tune their thermal expansion coefficient by varying the composition, so as to match the expansion of the underlying substrate and thus reduce problems with spalling when used as thermal or environmental barrier coatings (TBCs and EBCs).[80–83]

The natural silica SiO_2 scale on the surface of high-strength, silicon-based ceramics such as silicon nitride Si_3N_4, silicon carbide SiC, and fibre-reinforced silicon carbide composites SiC_f/SiC is attacked by water vapour at high temperatures, forming silicon hydroxide $Si(OH)_4$ and leading to corrosion and degradation of the underlying material. Silicon-based ceramics are usually, therefore, protected by a multilayer thermal and environmental barrier coating (TEBC), typically consisting of a bottom adhesive bondcoat of silicon Si, a

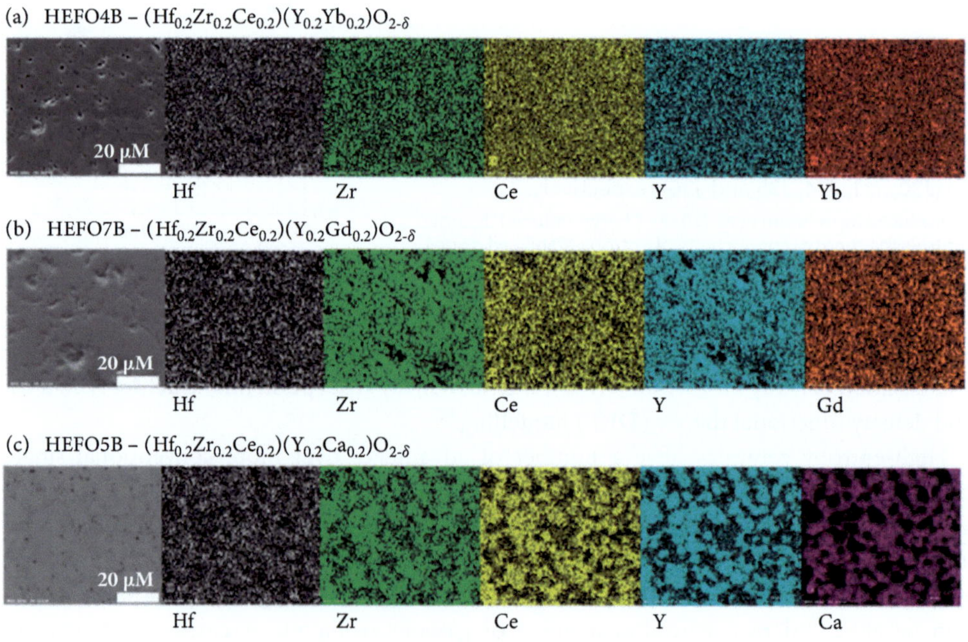

Figure 6.30 SEM images and EDS composition maps of the cross-sectional microstructures of $(YZrCeYbHf)O_{2-\delta}$, $(YZrCeGdHf)O_{2-\delta}$ and $(CaYZrCeHf)O_{2-\delta}$ thermal barrier coatings.
Reproduced from Gild et al. (2018). High-entropy fluorite oxides. *Journal of the European Ceramic Society 38*: 3578–84.

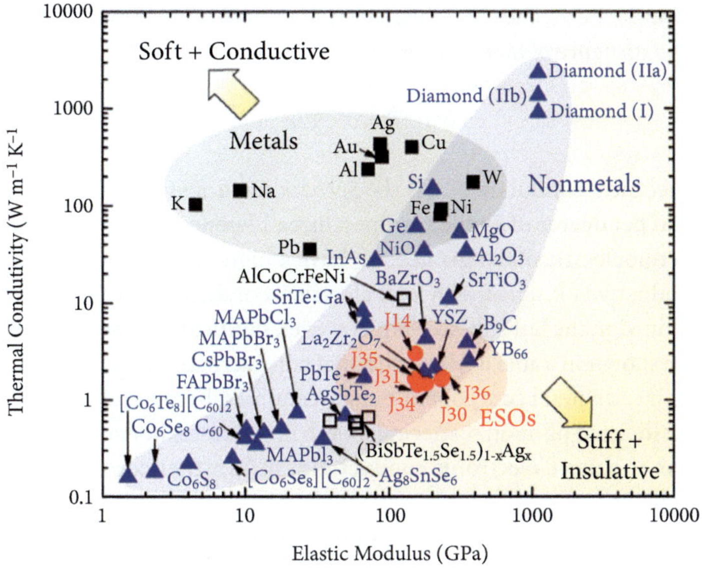

Figure 6.31 Ashby map of thermal conductivity versus elastic modulus for a variety of different materials.

Reproduced from Braun et al. (2018). Charge-induced disorder controls the thermal conductivity of entropy-stabilized oxides. *Advanced Materials 30*: Article 1805004.

stress-relieving interlayer of plasma-sprayed mullite $3Al_2O_3.2SiO_2$, and a water-resistant plasma-sprayed topcoat of ytrria-stabilized zirconia YSZ (ZrO_2–8wt%Y_2O_3) or barium strontium aluminosilicate BSAS ($(1-x)BaO.xSrO.Al_2O_3.2SiO_2$).[98,99] Multicomponent rare-earth silicates and aluminates are being investigated as potential improved TEBC topcoats for high-temperature silicon-based ceramic combustor parts in aeroengines.[98,99] As an example, Zhao et al.[98] manufactured $(YNdSmEuYb)_4Al_2O_9$ coatings by co-precipitation from a multiple solution of individual rare-earth nitrates, followed by filtration, drying and calcining, then ball milling, re-drying and re-screening, and finally spark plasma sintering at 1,923K (1,700 °C) for four minutes under a pressure of 30 MPa. The resulting multicomponent aluminate had an excellent combination of properties: good phase stability, with a single phase monoclinic structure for all temperatures from room temperature up to 1,473K (1,200 °C); with volume and linear thermal expansion coefficients that match well to mullite; and with a particularly low thermal conductivity of 1.5 W/m/K at room temperature, again caused by phonon scattering from the variation in local cation and anion lattice strains.

Thermoelectric materials

The *thermoelectric effect* is either the creation of a voltage in a material by applying a temperature difference across its ends (the *Seebeck effect*), or *vice versa* the creation of a temperature difference by applying a voltage across its ends (the *Peltier effect*). Thermoelectric materials are important in cooling and heating applications and are being used to regenerate electricity from waste heat. The size of the effect for any particular material at a temperature T is measured by its *figure of merit* zT, given by:

$$zT = \sigma S^2 T/k$$

where σ is the electrical conductivity, S is the *Seebeck coefficient* or *thermoelectric power* (the voltage generated per degree of applied temperature difference) and k is the thermal conductivity. Good thermoelectric materials need to have a high figure of merit zT and, therefore, low thermal conductivity k, allied with high electrical conductivity σ, which is quite unusual. As already explained in the last sub-section, however, multicomponent high-entropy ceramics such as those shown in Table 6.8 have thermal conductivities typically two or three times smaller than their individual component binary ceramics, because of variations in local lattice strain and associated phonon scattering, but in many cases they still maintain reasonable or may even have enhanced electronic conductivity and are therefore potentially valuable as thermoelectric materials.

There have, therefore, been a number of studies of multicomponent thermoelectric materials.[81–83] As an example, germanium chalcogenides such as germanium telluride GeTe and selenide GeSe are attractive thermoelectric materials but can usually only be made as p-type semiconductors. Roychowdhury et al.[96] developed an n-type multicomponent high-entropy thermoelectric selenide by alloying GeSe with $AgSbSe_2$ by vacuum melting and casting to form materials with the generic composition $(GeSe)_{1-x}(AgSbSe_2)x$ (equivalent to $Ge_{1-x}Ag_xSb_xSe_{1+x}$). Pure GeSe is orthorhombic, but the as-cast crystal structure changes on alloying with $AgSbSe_2$, to rhombohedral at low levels, $x = 0.1$, cubic at intermediate levels, $x = 0.3$–0.5, and finally hexagonal at high levels, $x = 0.5$–0.9. The cubic alloys with $x = 0.3$–0.5 can be used as n-type semiconductors,[94] with low thermal conductivity, $k \sim 0.45$ W/m/K, relatively high electrical conductivity, $\sigma \sim 25$ S/cm, a maximum thermoelectric efficiency of $zT = 0.45$ and a corresponding maximum thermoelectric power of $S^2\sigma = 3.8$ µW/cm/K^2 at 670 °C (943K) for $x = 0.5$, as shown in Figure 6.32.

Catalytic nanoparticles

A wide variety of different solid *catalysts* is used to accelerate industrial gaseous and solution-based chemical production processes.[100] The key requirement for a successful catalyst is the presence of a large number of activated surface sites at which the relevant chemical species can be absorbed easily so as to facilitate and promote critical steps in the desired chemical reactions of the production process.[100,101] In the last few years, multicomponent metallic and ceramic materials, manufactured in the form of small *nanoparticles* on a variety of supporting substrates have been found to be very effective in a range of catalytic processes.[102–105] The reasons for the success of multicomponent metallic and ceramic nanoparticles as catalytic materials undoubtedly vary from case to case, and in most cases remain to be elucidated in any real detail. However, there are clearly two important underlying factors: first the relative convenience with which most multicomponent high-entropy materials can be manufactured by a number of different methods as very fine-scale nanoparticles, thus presenting a high

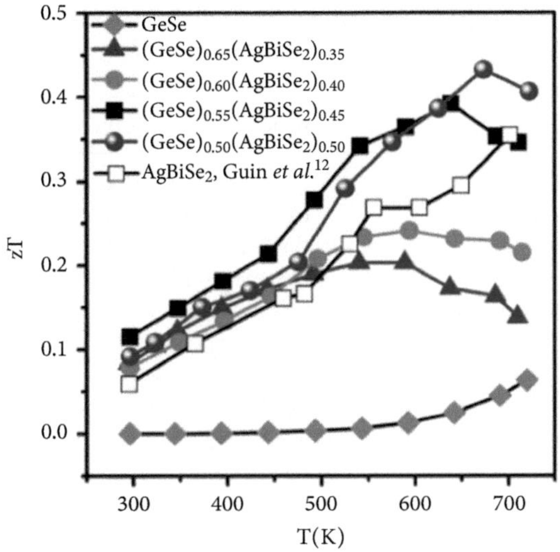

Figure 6.32 Temperature variation of thermoelectric figure of merit *zT* for high-entropy $(GeSe)_{1-x}(AgSbSe_2)_x$ ceramics.

Reproduced from Roychowdhury et al. (2018). Stabilizing n-type cubic GeSe by entropy-driven alloying of $AgBiSe_2$: Ultralow thermal conductivity and promising thermoelectric performance. *Angewandte Chemie 57*: 15167–71.

specific surface area on which catalysis can potentially take place; and second, the wide variety of local lattice site chemistry, mechanical strain and electronic structure, providing a wealth of atomically distributed activated surface sites at which adsorption of reactant species and thus catalysis of the relevant reactions can take place effectively.[102–105]

A wide variety of different multicomponent high-entropy metallic and ceramic alloy nanoparticles has been manufactured using a similarly wide variety of different methods of synthesis, including processes such as pyrolysis, electrodeposition, colloidal co-precipitation, mechanical milling, liquid metal dealloying (etching), reactive co-sputtering and pulsed laser ablation. A selection[106–114] of typical multicomponent metallic and ceramic nanoparticles and their processing methods is shown in Table 6.9, with more examples given in the recent reviews by Zhang et al.,[102] Kusada et al.,[103] Li et al.[104] and Wang et al.[105]

Yao et al.[106] were among the first to recognise the potential of multicomponent high-entropy nanoscale particles as highly efficient catalyst materials offering a combination of high specific surface area, atomic distribution of activated sites, and tunable chemical selectivity. Yao et al.[106] developed a carbothermal shock (CTS) synthesis process, in which a large number of individual micron-sized metallic salt particles were initially deposited on a supporting carbon nanofibre bed and were then electrodischarge-heated to ultra-high temperatures of 2,000–3,000K for 50–60 ms, leading to rapid melting, alloying and droplet ablation, yielding final populations of supported alloy nanoparticles, 10–20 nm in size, with complete fcc solid solution structures for up to eight or more chemical components. Figure 6.33 shows EDS X-ray composition maps and a high-angle annular dark-field scanning transmission electron microscope (HAADF-STEM) image of the typical CTS-manufactured

Table 6.9 Multicomponent high-entropy metallic and ceramic nanoparticles for potential use as catalysts.[106–114]

alloy	processing method	structure
FeCoNiPdPt	carbothermal shock	fcc
FeCoNiCuPdSnPtAu	carbothermal shock	fcc
RuRhReOsIr	thermal decomposition	hcp
CoNiCuPdSnPtAu	fast-moving bed pyrolysis	fcc
MnCoNiCuRhPdSnIrPtAu	fast-moving bed pyrolysis	fcc
FeCoNiLaPt	electrodeposition	amorphous
VCrMnCoNiCuInGd	electrodeposition	amorphous
AlNiCuPdPtAu	dealloying	fcc
AlFeCoNiCuPdPtAu	dealloying	fcc
CrFeCoNiCu	mechanical milling	fcc
$(MnFeCoNiCu)_3O_4$	solvothermal	spinel
$Ru_{0.13}(BaSrBi)(ZrHfTiFe)O_3$	sonication	perovskite
$(Co_{0.22}Cu_{0.28}Zn_{0.25}Ga_{0.11}In_{0.16})S$	cation exchange	wurzite

structure of 20-nm 8-component FeCoNiCuPdPtAuSn nanoparticles, demonstrating their single-phase solid-solution crystal structure. As with many of the synthesis methods shown in Table 6.9, phase separation was observed in binary and ternary nanoparticles, but solid solutions were easily obtained with four, five or more components, and proved to be very stable even after exposure at high temperatures. Excellent catalytic activity in carbothermal shock (CTS) manufactured nanoparticles resulted from the atomic-scale dispersion of multiple component atoms and the atomic-scale activation from local lattice mechanical and electronic strains. Yao et al.[106] achieved 100% catalytic conversion of ammonia with 99% selectivity to form NO_x at the relatively low temperature of 700 °C (973K) using CTS-synthesised 5-component RuRhPdCePt nanoparticles supported on carbon nanofibres, with no detectable degradation after thirty hours of continuous catalytic operation at 700 °C (973K).

Gao et al.[108] developed a fast-moving bed pyrolysis (FMBP) method for rapid synthesis of a wide variety of different high-entropy nanoparticles as potential catalyst materials. In FMBP, unlike conventional fixed bed pyrolysis (FBP), a substrate coated with multiple mixed metal precursor salts is rapidly drawn through a hot reaction chamber and thus heated rapidly to high temperature, ensuring simultaneous rather than sequential pyrolysis of the different components, leading to the deposition of large numbers of 10–30 nm-sized, high-entropy solid solution particles, again containing up to ten or more components. Figure 6.34 shows energy-dispersive (EDS) X-ray composition maps and a high-angle annular dark-field scanning transmission electron microscope (HAADF-STEM) image of the typical FMBP manufactured structure of a 30-nm-sized 10-component MnCoNiCuRhPdSnIrPtAu nanoparticle on a graphene oxide (GO) substrate, again demonstrating the nanoparticle single-phase solid-solution crystal structure. Many different multicomponent compositions were manufactured by Gao et al.[108] on a variety of substrates, including graphene oxide (GO), alumina γ–Al_2O_3, porous zeolites and carbon black.[108] The detailed size, structure and composition of the resulting nanoparticles could be controlled within quite tight limits

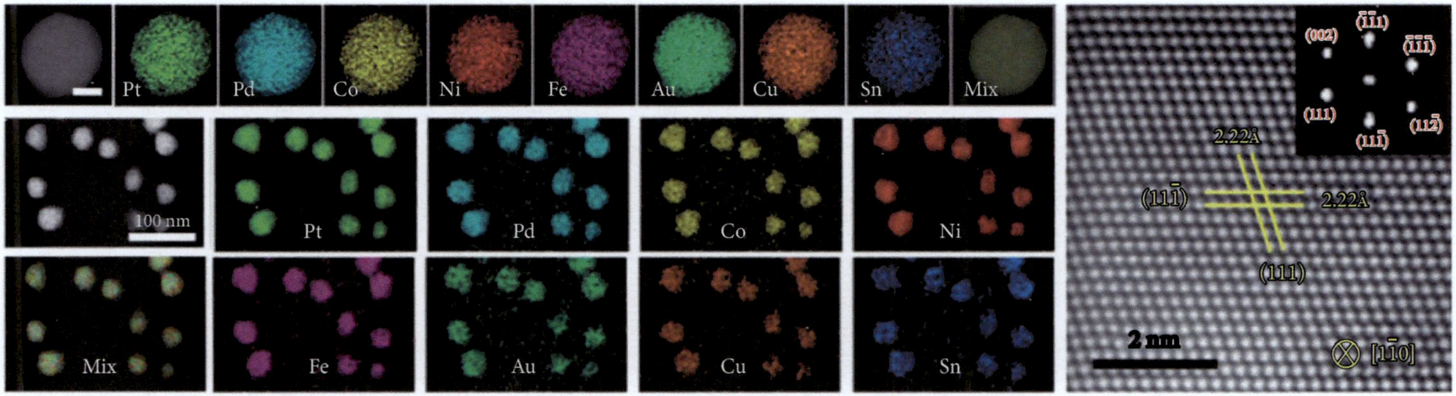

Figure 6.33 EDS composition maps of individual and multiple carbothermal shock (CTS) manufactured 8-component FeCoNiCuPdPtAuSn nanoparticles on a carbon nanofibre substrate (individual scale bar = 10 nm; multiple scale bar = 100 nm), and an HAADF-STEM image of their single-phase fcc structure.

Reproduced from Yao et al. (2018). Carbothermal shock synthesis of high-entropy alloy nanoparticles. *Science 359*: 1489–94.

by varying the nature of the substrate, its loading with different metal precursors, the pyrolysis temperature and the speed of the moving bed. Five-component FMBP FeCoPdIrPt nanoparticles on graphene oxide (FeCoPdIrPt/GO) were shown to provide considerably more efficient catalysis as the working electrode for the hydrogen evolution reaction (HER) in a solution of one molar potassium hydroxide (KOH) than either single, binary or ternary nanoparticles (such as Fe/GO, CoPd/GO or PdIrPt/GO) or conventional platinised graphite Pt/C, with the multicomponent nanoparticles FeCoPdIrPt/GO operating, for instance, at a much lower overvoltage than platinised carbon Pt/C, and with twenty-six times greater mass activity under equivalent conditions.[108]

As an example of high-entropy ceramic catalysts, Okejiri et al.[113] manufactured a variety of ruthenium-containing multicomponent perovskite nanoparticles based on barium zirconate $BaZrO_3$ using a sonication process to precipitate the nanoparticles from an initial solution of multiple precursors. Mixed chloride salts were dissolved in 20 molar sodium hydroxide solution, $NaOH/H_2O$, and then exposed to ultrasonic radiation with an immersed titanium horn for ten minutes at 20 kHz at ambient temperature, followed by centrifugal separation of the resulting precipitated multicomponent 20–40 nm-sized nanoparticles, and washing and drying at 80 °C (353K) on various substrates. Figure 6.35 shows a series of energy-dispersive (EDS) X-ray composition maps and an HAADF-STEM image from a typical sonicated 30 nm-sized 8-component $Ru_{0.13}(BaSrBi)(ZrHfTiFe)O_3$ oxide nanoparticle, showing its single-phase perovskite solid-solution structure. Catalytic oxidation of carbon monoxide CO to carbon dioxide CO_2 is extremely important in decontamination and depollution processes, with ruthenium a very effective catalytic substrate.[113] The multicomponent $Ru_{0.13}(BaSrBi)(ZrHfTiFe)O_3$ nanoparticles were loaded into a CO conversion reactor on a quartz wool substrate and exposed to a constant flow rate of 1% CO in dry air, with the catalytic performance compared, as shown in Figure 6.35, to multicomponent $(BaSrBi)(ZrHfTiFe)O_3$ nanoparticles manufactured in exactly the same way but without ruthenium, as well as simple binary $BaRuO_3$ also manufactured in the same way.[113] As can be seen in Figure 6.36, multicomponent $Ru_{0.13}(BaSrBi)(ZrHfTiFe)O_3$ achieved ~50% conversion of CO to CO_2 after < 1 s at 90 °C (363K) compared with 50% conversion only at a much higher temperature of 255 °C (528K) for $BaRuO_3$ even though it contains a much higher concentration of ruthenium, and much lower conversion rates for $(BaSrBi)(ZrHfTiFe)O_3$ even at a higher temperature of 300 °C (573K). The superior multicomponent nanoparticle catalytic performance shown in Figure 6.36 was attributed to the atomic distribution of distorted and thus activated ruthenium lattice sites on the surface of the nanoparticles.[113]

Multicomponent high-entropy metallic and ceramic nanoparticles have been shown to be effective catalysts for a wide variety of different chemical reactions in many industrial processes, including oxygen evolution and reduction reactions (OER and ORR), hydrogen evolution and reduction reactions (HER and HOR), methanol and ethanol oxidation reactions (MOR and EOR) and nitrogen reduction. Table 6.10 lists a number of examples,[104,105] with others given in the reviews by Zhang et al.,[102] Kusada et al.,[103] Li et al.[104] and Wang et al.[105]

In summary, many multicomponent high-entropy metallic alloys and ceramic compounds have been found to be highly effective as catalysts for a wide variety of different oxidation, reduction and other chemical reactions. Effective catalysis in such multicomponent high-entropy materials is achieved by a combination of several effects: partly because

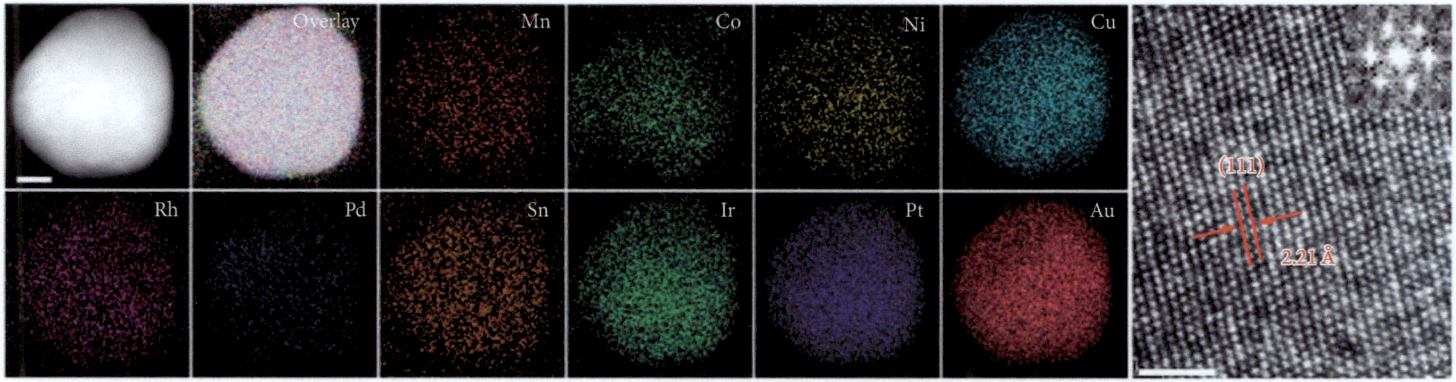

Figure 6.34 EDS composition maps of an individual fast moving bed pyrolysis (FMBP) manufactured ten-component MnCoNiCuRhPdSnIrPtAu nanoparticle on a GO substrate (scale bar = 10 nm), and an HAADF-STEM image of its single-phase fcc structure.

Reproduced from Gao et al. (2020). Synthesis of high-entropy alloy nanoparticles on supports by the fast-moving bed pyrolysis. *Nature Communications 11*: Article 2016.

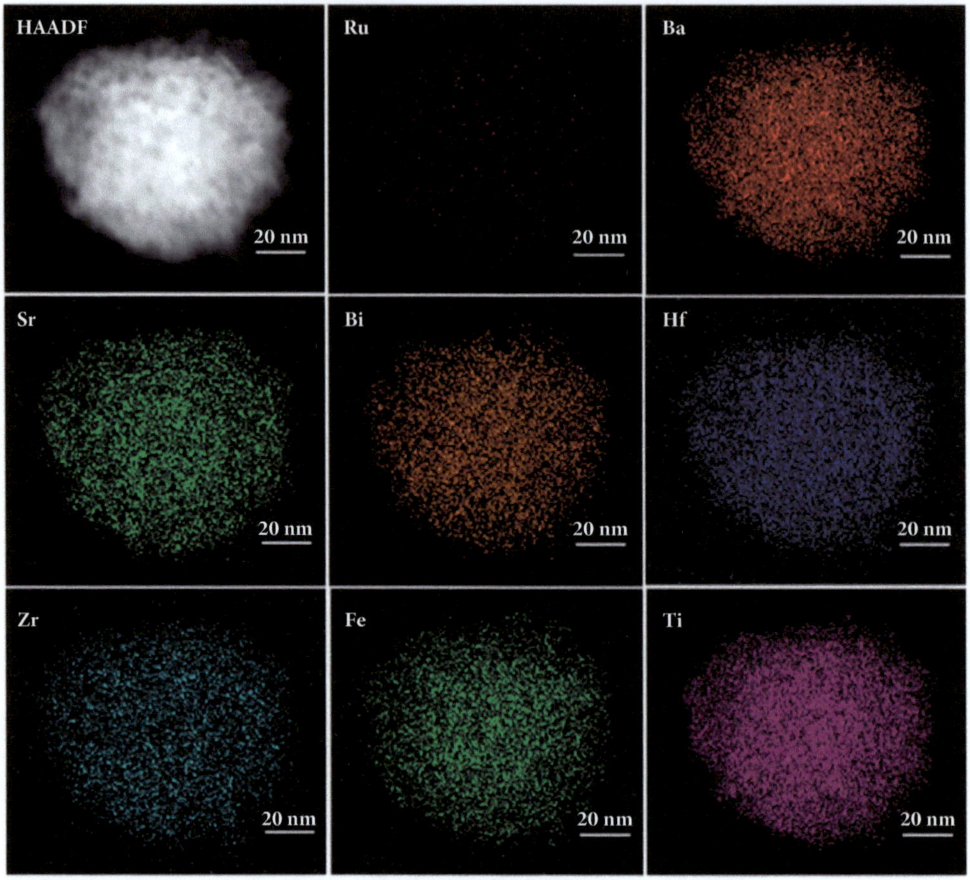

Figure 6.35 EDS composition maps and an HAADF-STEM image of an individual sonicated 8-component $Ru_{0.13}$(BaSrBi)(TiZrFeHf)O_3 perovskite oxide nanoparticle.

Reproduced from Okejiri et al. (2019). Room-temperature synthesis of high-entropy perovskite oxide nanoparticle catalysts through ultrasonication-based method. *ChemSusChem 13*: 111–15.

multicomponent materials can readily be made as very finely divided nanoparticles with a high surface area on which catalysis can take place; partly because multicomponent materials exhibit a wide range of local atomic structures and, therefore, a correspondingly wide range of surface activated states suitable for promoting effective catalysis; and partly because multicomponent alloy compositions can readily be tuned to provide high specificity for particular, desirable chemical reactions.

Radiation resistance

Materials used in the manufacture of *nuclear reactors* need to have good mechanical properties under conditions of high temperature and high pressure, at the same time as being resistant to corrosion and damage caused by radiation emitted from the fissile nuclear fuel.[115] We have already seen in the last section that multicomponent high-entropy fcc Cantor and bcc Senkov alloys have outstanding combinations of strength and ductility at low and high

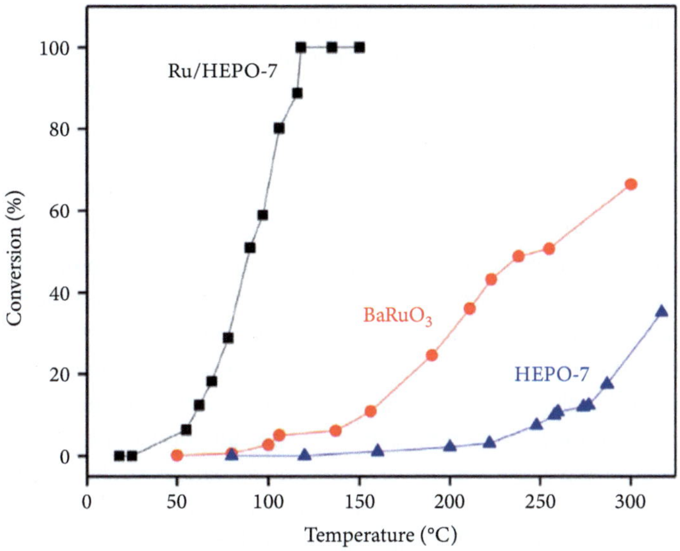

Figure 6.36 Conversion of CO to CO_2 by $Ru_{0.13}(BaSrBi)(ZrHfTiFe)O_3$ (labelled Ru/HEPO-7), $BaRuO_3$ and $(BaSrBi)(ZrHfTiFe)O_3$ (labelled HEPO-7) nanoparticles.

Reproduced from Okejiri et al. (2019). Room-temperature synthesis of high-entropy perovskite oxide nanoparticle catalysts through ultrasonication-based method. *ChemSusChem 13*: 111–15.

Table 6.10 Typical chemical reactions catalysed by multicomponent high-entropy nanoparticles.[104,105]

alloy	processing method	reaction
MnFeCoNiCu	solvothermal	OER
AlCrFeCoNi	mechanical milling	OER
CrMnFeCoNiNb	sputtering	ORR
FeCoNiPdPt	carbothermal shock	ORR
FeCoPdIrPt	fast-moving bed pyrolysis	HER
CrFeCoNiMo	arc melting	HER
FeCoNiCuAgPt	sputtering	MOR
RuRhReOsIr	pyrolysis	MOR
RuRhPdOsIrPt	polyol	EOR
FeCoNiCuMo	carbothermal shock	NH_3 oxidation
CuPdAgPtAu	cryogrinding	CO_2 reduction

temperatures;[4,7,8,44–48] they have also been found to have excellent resistance to a variety of problems caused by *radiation damage* from beams of high energy electrons, heavy ions and neutrons.[116–120] Multicomponent high-entropy single-phase and multiphase fcc Cantor and bcc Senkov alloys have, therefore, been subjected to a considerable amount of investigation, exploring the complex details of their radiation resistance and thus their suitability for potential use in future nuclear reactors (so-called *generation-4 reactors*). The results of these investigations have been reviewed by a number of recent publications.[116–120]

When a material is subjected to radiation in an intense beam of electrons, ions or neutrons, direct collisions by the radiating particles can sometimes displace the constituent atoms or molecules from their positions in the crystal lattice of the material, causing a *displacement cascade*, with considerable local generation of heat (a *thermal spike*) and local disturbance in the atomic structure of the crystal.[115] The intensity of an incident beam of radiation is measured typically by its *fluence* (the number of radiating particles per unit cross-sectional area in the beam) and the *dose* (the fluence integrated over the exposure time); but its overall effect is more helpfully measured by the corresponding number of *displacements per atom* (or *dpa*), which is usually calculated by an *SRIM* (*stopping range of ions in matter*) computer program, and which (fairly obviously) also depends on the temperature, pressure, and composition and microstructure of the material being irradiated.[115] In conventional metallic alloys, atomic displacements create *Frenkel pairs* of vacancies and interstitials, which begin to diffuse away from the displacement region, at the same time as congregating and collapsing to form defect clusters such as intrinsic and extrinsic *dislocation loops* and *stacking fault tetrahedra*.[115] The radiation-induced defects and their migration and clustering cause a series of undesirable changes in the material microstructure, including *segregation*, enhanced *microstructural degradation*, *radiation hardening* and *radiation embrittlement*, *void swelling* and *gas bubble* formation.[115] A considerable number of studies have now shown that these effects can be removed at low radiation doses and ameliorated at high radiation doses in many multicomponent high-entropy single-phase fcc Cantor and bcc Senkov alloys, compared with more conventional stainless steels and nickel superalloys.[116–132]

Most of the initial work on radiation damage in multicomponent high-entropy materials concentrated on the original 5-component equiatomic fcc Cantor alloy CrMnFeCoNi, related 4- and 3-component fcc sub-systems such as CrFeCoNi and CrCoNi, modified 5-component fcc Cantor alloys such as CrFeCoNiCu and CrFeCoNiPd, and the well-known fcc + bcc alloys in the $Al_xCrMnFeCoNi$ and $Al_xCrFeCoNi$ series.[121–127] Table 6.11 gives some examples of radiation studies of multicomponent single-phase fcc and multiphase fcc-based Cantor alloys, with more given in the review by Zhang et al.[119] It is important that materials used in the construction of nuclear reactors have low *activation*, i.e. do not themselves become excessively radioactive under the influence of the primary radiation. This depends on the reactivity of the different constituent atoms present in the materials, with components such as manganese (Mn), cobalt (Co) and niobium (Nb), for instance, having relatively high neutron-capture cross-sections and thus the potential for forming dangerous and long-lived radioactive isotopes. More recent studies have begun to concentrate, therefore, on multicomponent high-entropy bcc Senkov alloys such as equiatomic and non-equiatomic 4-component VCrTaW, TiVNbTa and TiVCrTa, 5-component VCrFeTaW and TiVZrHfTa, and 6-component AlTiNiCuZrPd,[128–134] which also have particularly good mechanical properties at elevated temperatures, as discussed earlier in this chapter. Table 6.12 gives some examples of radiation studies of multicomponent single-phase bcc and multiphase bcc-based Senkov alloys, with more given in the review by Shi et al.[120]

Enhanced resistance to degradation by radiation in multicomponent high-entropy Cantor and Senkov alloys is caused by a variety of different interrelated effects, all essentially linked to the wide variety of different local atomic structures and associated lattice distortions found in these materials, as discussed in detail in Chapter 4. These different effects are discussed in the reviews by Zhang et al.[119] and Shi et al.,[120] and can be summarised as follows:

Table 6.11 Some examples of radiation damage studies in multicomponent high-entropy fcc-based Cantor alloys.[121–127]

alloy	radiation	radiation damage
CrMnFeCoNi	3 MeV Ni^{2+}; 53 dpa; 773K	30 × less swelling than pure Ni
CrMnFeCoNi	3 MeV Ni^{2+}; 3 dpa	similar grain boundary segregation to 404 stainless steel
CrFeCoNi	3 MeV Ni^{2+}; 53 dpa; 773K	20 × less swelling than pure Ni
CrFeCoNi	1.2 MeV e^-; 1.2 dpa; 673K	40 × slower dislocation loop growth than pure Ni
CrCoNi	3 MeV Ni^{2+}; 53 dpa; 773K	6 × less swelling than pure Ni
CrCoNi	1.3 MeV Au^+; 0.57 dpa	3 × less damage than pure Ni
CrFeCoNiCu	2MeV e^-; 68 dpa; 298 + 773K	fcc phase stability
$Al_xCrFeCoNi$	3 MeV Au^+; 65 dpa	15 × less swelling than 316 stainless steel; fcc phase stability

Table 6.12 Some examples of radiation damage studies in multicomponent high-entropy bcc-based Senkov alloys.[128–134]

alloy	radiation	radiation damage
$V_{11}Cr_{15}Ta_{36}W_{38}$	1 MeV Kr^{2+}; 8 dpa; 1073K	no dislocation loops
$V_{11}Cr_{15}Ta_{36}W_{38}$	3 MeV Cu^+; 8 dpa; 1050K	3–5 nm Cr/V rich precipitates
TiVNbTa	1 MeV Kr^{2+}; 2 dpa; 50K	5 × fewer loops than V
TiVZrTa	2 MeV V^+; 3.6 dpa; 773K	no hardening
TiVCrTa	2 MeV V^+; 3.6 dpa; 773K	no hardening
$VCrFeTa_xW_x$	—	low activation
TiVZrHfTa	4.4MeV Ni^{2+}; 40 dpa; 298K	no amorphisation; low hardening
TiZrNbHfTa	0.3 MeV Ni^+; 30 dpa; 373K	no hardening; bcc phase stability; good swelling resistance
$Al_{0.5}TiNiCuZrPd$	2 MeV e^-; 0.04 dpa; 298–673K	stable amorphisation

1. A wide variety of local lattice distortions leads to strong phonon scattering and reduced thermal conductivity, making multicomponent high-entropy materials attractive as thermal barrier coatings, as discussed earlier in this section. The same effect intensifies the thermal spike around displacement cascades during irradiation, enhancing recombination of interstitials and vacancies, and reducing the total number of primary radiation defects.
2. A wide variety of local lattice distortions slows down atomic diffusion rates, as discussed in detail in Chapter 4. Slower diffusion enhances phase stability during exposure to radiation, limits segregation to grain boundaries and reduces the nucleation and growth of secondary precipitate phases.
3. Slower diffusion also prevents clustering of residual unrecombined interstitials and vacancies, reducing the number and size of larger radiation defects such as dislocation loops and stacking fault tetrahedra.

4. Fewer point defects, fewer defect clusters, reduced segregation and reduced precipitation all limit the extent of radiation hardening and embrittlement.
5. Fewer defect clusters and slower diffusion prevent the nucleation and growth of voids and gas bubbles, limiting the extent of radiation swelling and associated macroscopic distortion.

He et al.[123] irradiated single crystal thin films of equiatomic 4-component single-phase fcc CrFeCoNi and observed the evolution of their microstructure *in situ* in a JEM-1300NEF high-voltage electron microscope (HVEM) operating with 0.4, 0.5 and 1.3 MeV electrons at a dose rate of 5×10^{18} electrons/cm^2/s up to a damage level of 1.2 dpa at 400 °C (673K). A mixture of elliptical faulted $\frac{1}{3}$<111> Frank interstitial loops and polygonal $\frac{1}{2}$<110> perfect dislocation loops nucleated within the first minute of irradiation (<0.05 dpa); the ratio of Frank/perfect loops was 3:1; the number density of both types of loop saturated after five minutes of irradiation (~0.2 dpa); and continued irradiation led to no new loops and only relatively slow growth of the existing Frank and perfect loops. Figures 6.37(a) and (b) show respectively high-resolution images of the growth of Frank and perfect loops; and Figures 6.37(c) and (d) show respectively the evolution of the size and aspect ratio of both types of loop with increasing damage level. Individual loops in the multicomponent alloy grew with a time exponent between 0.5 and 1.0, but in all cases very much more slowly, at 45–110 nm/dpa, than the corresponding growth rate, 2,400–4,300 nm/dpa, of loops in fcc pure

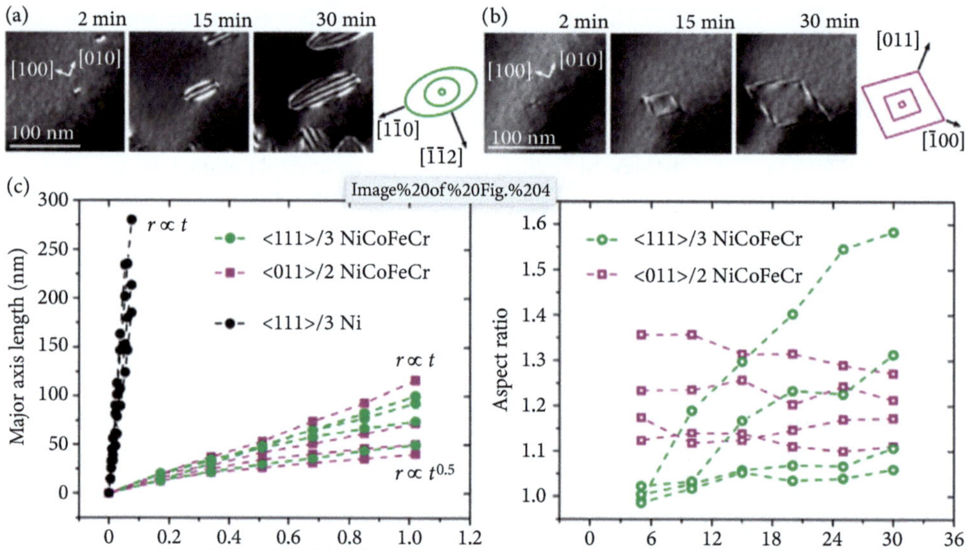

Figure 6.37 Radiation damage in CrFeCoNi after 1.2 dpa by 1.3 MeV electrons: (a) and (b) HVEM images of the growth of $\frac{1}{3}$<111> Frank and $\frac{1}{2}$<110> perfect dislocation loops; (c) and (d) dislocation loop size and aspect ratio versus radiation time and damage level for individual Frank (green) and perfect (purple) loops, compared to pure Ni (black).

Reproduced from He et al. (2016). Enhanced damage resistance and novel defect structure of CrFeCoNi under *in situ* electron irradiation. *Scripta Materialia 125*: 5–9.

Ni under similar irradiation conditions, i.e. more slowly by a factor of 40–60. He et al.[123] ascribe this enormous reduction in dislocation loop growth rate in the multicomponent high-entropy fcc Cantor alloy CrFeCoNi to a lower mobility of both interstitials and vacancies, caused by the compositional complexity of the alloy and the associated wide variation in local atomic structure and wide spread of activation energies for diffusion.

Jin et al.[121] studied the growth of voids and the associated swelling caused by irradiation with 3 MeV Ni^{2+} ions at the University of Tennessee–Oak Ridge Ion Beam Materials Lab (IBML) with a fluence of up to 5×10^{16} ions/cm^2 up to damage levels of 53 dpa at 500 °C (773K) in single crystal specimens of the original 5-component equiatomic single-phase fcc Cantor alloy CrMnFeCoNi and reduced equiatomic 4-, 3- and 2-component single-phase fcc alloys CrFeCoNi, CrCoNi, FeCoNi and NiCo, as well as pure fcc Ni. In each case, radiation-induced swelling was measured from the step height of the irradiated surface region using a Wyko NT9100 3D optical profilometer with sub-nm vertical resolution in phase-shift interferometry (PSI) mode. Figure 6.38(a) shows a comparison of three-dimensional images of the irradiated region step heights in CrMnFeCoNi, CrCoNi, CoNi and pure Ni; Figures 6.38(b) and (c) show the growth rate with increasing fluence in pure Ni and CoNi; and Figure 6.38(d) compares the step heights and corresponding percentage volume swelling in all the different materials. Radiation swelling increased in the order CrMnFeCoNi and FeCoNi (<0.2%) < CrFeCoNi and FeNi (~0.33%) < CrCoNi (~1.1%) < CoNi (~3.5%) < Ni (~6.7%), i.e. was more than 33.5 times lower in the Cantor alloy CrMnFeCoNi than in pure Ni. The same effect is shown even more graphically in corresponding transmission electron

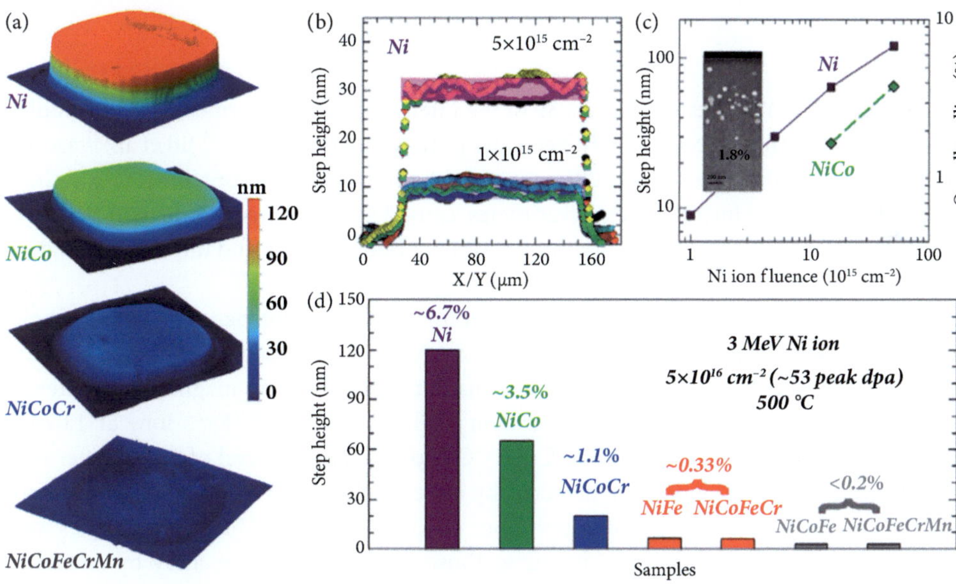

Figure 6.38 Radiation swelling in different fcc materials after 53 dpa by 3 MeV Ni^{2+} ions: (a) three-dimensional step profiles; (b) and (c) step height growth; and (d) final step height and swelling.

Reproduced from Jin et al. (2016). Effects of compositional complexity on the ion-irradiation induced swelling and hardening in Ni-containing equiatomic alloys. *Scripta Materialia 119*: 65–70.

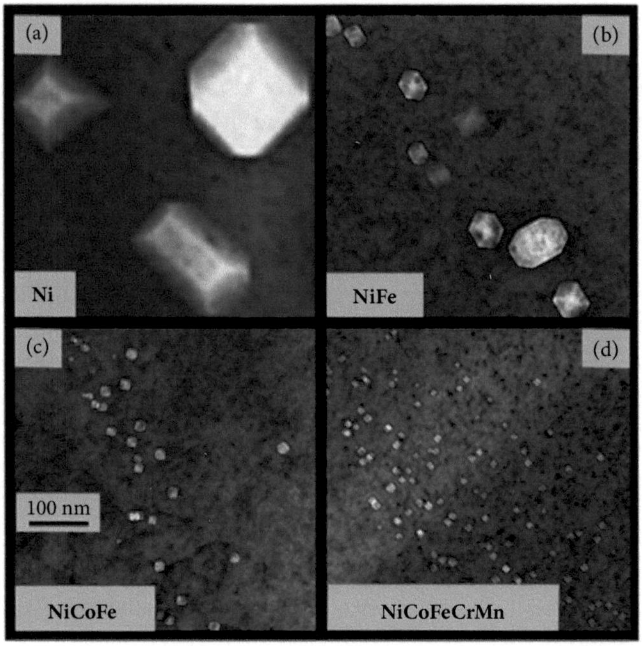

Figure 6.39 HVEM images of radiation-induced voids in different fcc materials after 53 dpa by 3 MeV Ni^{2+} ions.

Reproduced from Jin et al. (2016). Effects of compositional complexity on the ion-irradiation induced swelling and hardening in Ni-containing equiatomic alloys. *Scripta Materialia 119*: 65–70.

microscope (TEM) images in Figure 6.39 of the internal voids that formed and contributed to swelling in CrMnFeCoNi, FeCoNi, FeNi and pure Ni. Like He et al.,[123] Jin et al.[121] ascribe reduced radiation damage in the multicomponent high-entropy Cantor alloy CrMnFeCoNi to a lower mobility of interstitials and vacancies, caused by the compositional complexity of the alloy and an associated wide variation in local atomic structure and diffusion activation energies.

El Atwani et al.[128] developed a non-equiatomic 4-component nanocrystalline single-phase bcc Senkov alloy with a composition $V_{11}Cr_{15}Ta_{36}W_{38}$, designed to have good mechanical properties and low activation, and manufactured the alloy by magnetron sputtering followed by irradiation at the IVEM-Tandem facility with 1 MeV Kr^{2+} ions at 1,073 K (800 °C) and 3 MeV Cu^{+} ions at 1,050K (777 °C) up to a damage level of 8 dpa. After irradiation, no dislocation loops were detected by either high-resolution electron microscopy (HREM) or atom-probe (AP) tomography, but ~3–5-nm-sized Cr/V-rich precipitates were formed, as shown in Figure 6.40, similar to those found in the same alloy by annealing at high temperature without irradiation. Nanoindentation measurements showed that the as-sputtered alloy had a very high hardness of ~14 GPa, which increased to ~15 GPa (7.1%) after plain annealing or irradiation up to 1.6 dpa at 1,073K (800 °C), and increased further to ~16 GPa (14.2%) after stronger irradiation up to 8 dpa at the same temperature, as shown in Figure 6.41. In other words, heat treatment with or without irradiation led to moderate hardening caused by nanoscale Cr and V precipitation.

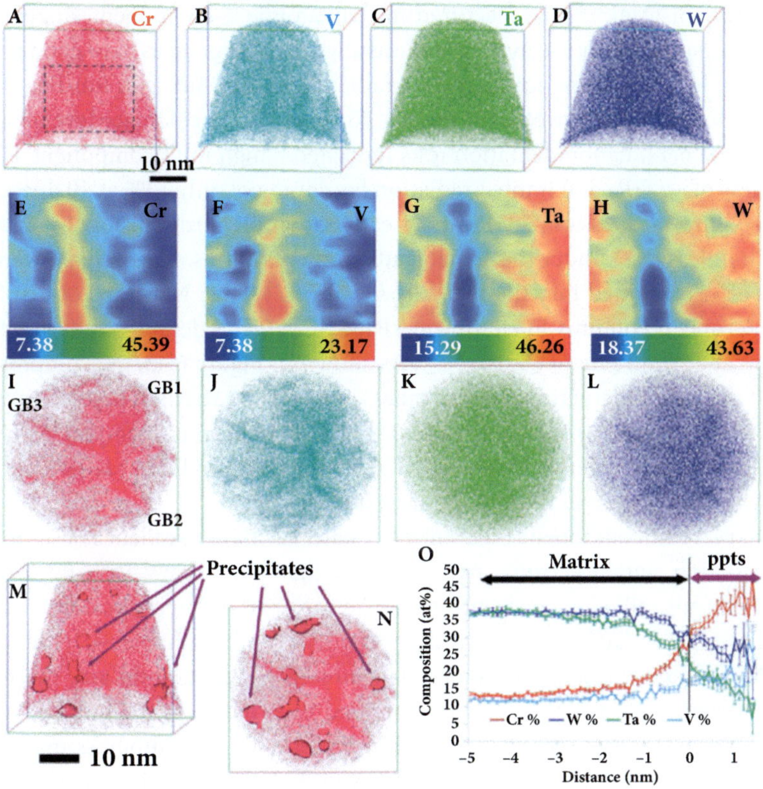

Figure 6.40 AP Cr, V, Ta and W composition maps of radiation-induced Cr/V grain-boundary segregation and precipitation in nanocrystalline bcc $V_{11}Cr_{15}Ta_{36}W_{38}$ after 8 dpa by three MeV Cu^+ ions: (A)–(D) side-view; (E)–(H) a 25 × 20 × 1 nm slice; (I)–(L) top-view; (M) and (N) reconstruction with a 25% Cr isosurface; (O) composition profiles.

Reproduced from Ei Atwani et al. (2019). Outstanding radiation resistance of tungsten-based high-entropy alloys. *Science Advances 5*: Article 2002.

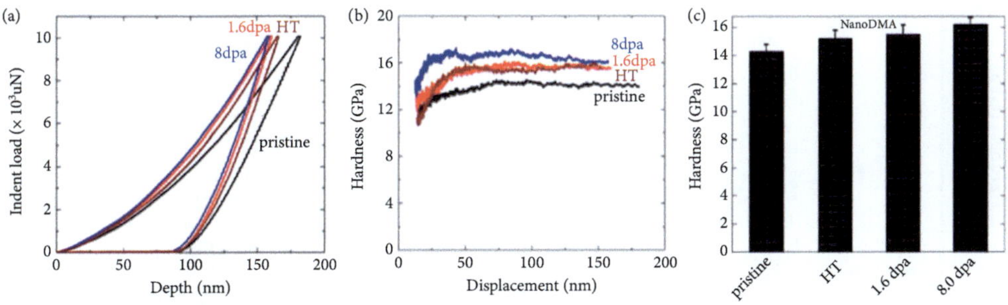

Figure 6.41 Nanoindentation measurements in nanocrystalline bcc $V_{11}Cr_{15}Ta_{36}W_{38}$ as-sputtered (pristine), heat-treated at 1,073K (HT) and irradiated up to 1.6 and 8 dpa by 1 MeV Kr^+ ions at 1,073K: (a) load-displacement curves; (b) hardness-displacement curves; and (c) hardness values.

Reproduced from Ei Atwani et al. (2019). Outstanding radiation resistance of tungsten-based high-entropy alloys. *Science Advances 5*: Article 2002.

Sadeghilaridjani et al.[132] manufactured an equiatomic 5-component single-phase bcc TiVCrHfTa Senkov alloy by arc melting and casting, also designed to have good mechanical properties and low activation, followed by irradiation at the Michigan Ion-Beam Lab with 4.4 MeV Ni^{2+} ions at a fluence of just over 10^{17} ions/cm^2 up to a damage level of 40 dpa. As shown in Figure 6.42, nanoindentation tests indicated a high multicomponent TiVCrHfTa alloy hardness of ~6 GPa and limited radiation hardening of only 13% at 40 dpa, compared with a lower alloy hardness of ~2 GPa and greater radiation hardening of 50% in a conventional 304 stainless steel under the same irradiation conditions. Micropillars were focused ion beam (FIB) machined into the two alloys in an FEI Nova Nanolab 200FIB-SEM, followed by micropillar compression testing in a P188 SEM Picoindenter with a 5 μm flat punch diamond tip at a strain rate of 8×10^{-3} s^{-1}. As shown in Figure 6.43, the compression tests agreed with the nanoindentation measurements, with a yield strength of 1.06–1.09 GPa and limited radiation hardening of 28% up to 1.30–1.48 GPa in the multicomponent TiVCrHfTa alloy, compared with a lower yield strength of 0.39–0.5 GPa and higher radiation hardening of 54% up to 0.7–0.79 GPa in the 304 stainless steel.

In summary, multicomponent high-entropy materials have excellent resistance to damage by irradiation with electrons, neutrons and light and heavy ions for a variety of interlinked reasons, all associated with the wide range of different local atomic environments and associated lattice strains, as discussed in detail in Chapter 4: 1) lattice strains cause phonon scattering and low thermal conductivity, containing any thermal spike', enhancing vacancy-interstitial recombination, and thus reducing the number of primary radiation defects; 2) lattice strains reduce diffusivities and, thus, suppress microstructural degradation by segregation and precipitation; and 3) lower diffusivities also limit defect clustering and, therefore, associated radiation hardening, pore formation and radiation swelling.

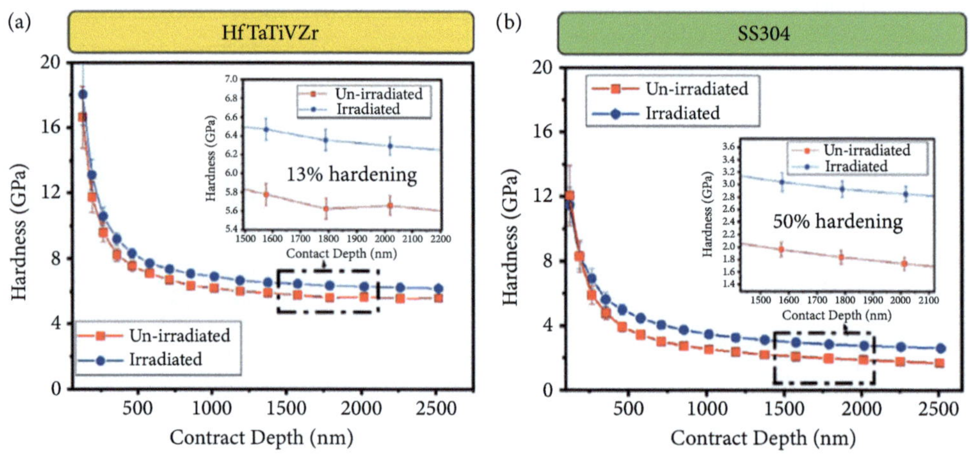

Figure 6.42 Nanoindentation hardness versus depth in (a) bcc TiVZrHfTa and (b) 304 stainless-steel irradiated up to 40 dpa with 4.4 MeV Ni^{2+} ions.

Reproduced from Sadeghilaridjani et al. (2020). Ion irradiation response and mechanical behavior of reduced activity high-entropy alloy. *Journal of Nuclear Materials 529*: Article 151955.

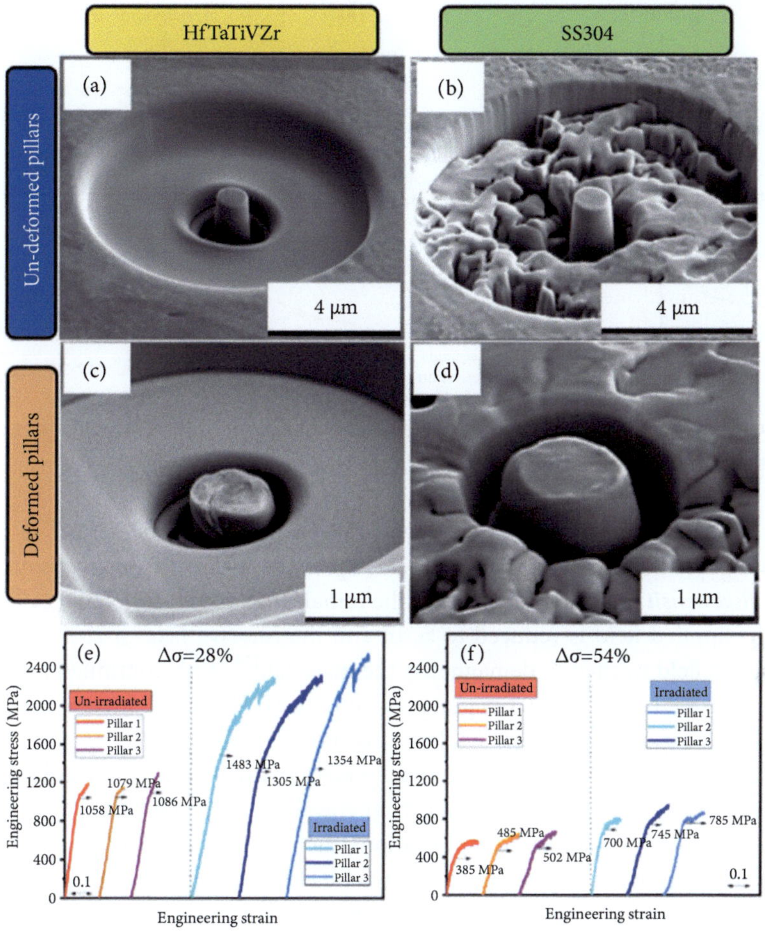

Figure 6.43 Micropillar compression tests in bcc TiVZrHfTa and 304 stainless-steel irradiated up to 40 dpa with 4.4 MeV Ni^{2+} ions: (a) and (b) as-machined micropillars; (c) and (d) compression tested micropillars; and (e) and (f) compression test stress–strain curves.
Reproduced from Sadeghilaridjani et al. (2020). Ion irradiation response and mechanical behavior of reduced activity high-entropy alloy. *Journal of Nuclear Materials 529*: Article 151955.

Magnetic materials

When an external magnetic field H is applied to a ferromagnetic material, it becomes magnetised, with an internal induced magnetic field or *magnetisation* $M = \chi H$ where χ is the *magnetic susceptibility* of the material.[135] The resulting *magnetic flux density B* at any point outside the material is a sum of the effects of both applied and induced magnetic fields, $B = \mu_o (H + M) = \mu_o (1 + \chi) H = \mu H$, where $\mu = \mu_o (1 + \chi)$ is the *magnetic permeability* of the material and μ_o is the magnetic permeability in a vacuum.[135] The induced magnetisation M and, therefore, the magnetic flux density B vary in a complicated way with external applied field H, as shown schematically in Figure 6.44. The magnetisation and flux density

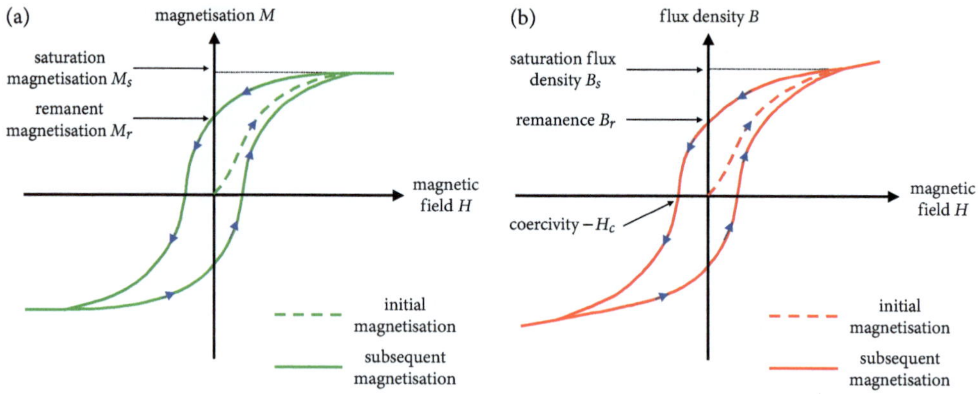

Figure 6.44 Schematic magnetisation and demagnetisation loop for a ferromagnetic material (a) magnetisation *M* and (b) flux density *B*.

increase initially with increasing applied field, up to the *saturation magnetisation* M_s and *saturation flux density* B_s; the *remanence* is the remaining magnetisation M_r or flux density B_r when the applied field is reduced to zero; and the *coercive field* or *coercivity* H_c is the reverse applied field needed to demagnetise the material.[135] With continued magnetisation and demagnetisation the material migrates around its *M–H* or *B–H hysteresis loop* with an energy loss proportional to the area enclosed, as shown in Figure 6.44.

Hard magnetic materials are difficult to magnetise and demagnetise, with large values of remanence M_r and B_r, coercivity H_c, and magnetisation energy or $(BH)_{max}$ product. They are used as permanent magnets in applications such as motors, disc drives and sensors. Important conventional hard magnetic materials include Alnico (Fe alloys with 8–12 wt% Al, 15–25 wt% Ni and 5–25 wt% Co), ferrites (mixed oxides containing ferrite Fe_2O_3) and Fe-Nd-B alloys (based on the boride $Nd_2Fe_{14}B$).[135] Soft magnetic materials, on the other hand, are easy to magnetise and demagnetise, with small values of remanence M_r and B_r, coercivity H_c, and magnetisation energy or $(BH)_{max}$ product. They are used in applications that require easy magnetic reversals such as transformers and magnetic memories. Important conventional soft magnetic materials include silicon–iron (Fe with 3–4 wt% Si), Fe-Ni alloys (with a range of compositions around equiatomic FeNi), and amorphous alloys (Fe with 15–20 at% B, C and Si).[135]

Magnetisation is caused by the spins on the electrons inside a material.[135] Most materials are not ferromagnetic and are only weakly magnetised, because the spins on the different electrons are not aligned and, therefore, largely or completely cancel each other out. *Ferromagnetic materials* (i.e. strongly magnetised materials) represent just a few special cases that contain particular transition metal atoms with large numbers of unpaired electron spins, and that have particular crystal structures with a strong exchange interaction between the electrons in adjacent atoms, causing their spins to become aligned, rather than opposed.[135] In *hard ferromagnetic materials*, the alignment of the different electron spins is fixed by being pinned strongly to the underlying crystal structure of the material; but in *soft ferromagnetic materials* the alignment of the different electron spins is not so strongly pinned and is, therefore, easily broken up and/or reorientated. As we have seen in previous sections

Table 6.13 Examples of multicomponent high-entropy ferromagnetic materials.[139–148]

alloy	processing method	structure
AlMnFeCoNi	arc melting	bcc + B2
MnFeCoNiGa	arc melting	fcc + B2
$AlCr_{1-x}FeCo_xNi$	laser deposition	bcc + B2
AlFeCoNi	arc melting	bcc
$Al_{0.1}FeCoNi$	arc melting	fcc + bcc
SiFeCoNi	arc melting	fcc + M_3Si
FeCoNi	arc melting	fcc
$(AlSi)_xFeCoNi$	arc melting	fcc + bcc
BAlSiFeNi	mechanical alloying	amorphous
BAlSiFeNiNb	mechanical alloying	amorphous
$Cr_{10}Fe_{40}Co_{10}Ni_{40}$	arc melting	fcc
FeCoNiCuPd	arc melting	fcc + fcc
$AlFeCoNi(TiCu)_x$	arc melting	bcc
$Al_{0.5}FeCo_{0.5}LaPrNd$	arc melting	amorphous

and chapters, many multicomponent high-entropy materials contain substantial amounts of ferromagnetic components such as iron and nickel and not surprisingly, therefore, there has been considerable investigation of their magnetic properties, the results of which have been reviewed in a number of previous publications.[136–138] Table 6.13 gives a number of examples of multicomponent high-entropy ferromagnetic materials that have been studied previously,[139–148] with more given in the reviews by Chaudhary et al.,[136] Kumari et al.[137] and Huang et al.[138]

Most of the multicomponent high-entropy materials that have been studied exhibit soft ferromagnetic behaviour. As an example, Zuo et al.[142] manufactured a series of 4-component $Al_xFeCoNi$ and $Si_xFeCoNi$ alloys with $x = 0–1$ by arc melting and casting, re-melting and re-casting several times to ensure homogeneity. The structure of the base 3-component FeCoNi alloy was fcc; it changed to fcc + bcc and then bcc with increasing additions of Al; and it changed to fcc + M_3Si silicide with increasing additions of Si. The alloys were all soft ferromagnets, as can be seen in the narrow $M–H$ hysteresis loops and very low values of remanence and coercivity, as measured by vibrating sample magnetometry and as shown in Figure 6.45. Not surprisingly, the saturation magnetisation M_s decreased gradually with increasing additions of non-magnetic Al or Si, and the coercivity H_c was higher in the two-phase alloys at intermediate Al and at high Si content, as shown in Figure 6.46. As another example, Borkar et al.[141] manufactured a series of 5-component $AlCr_{1-x}FeCo_xNi$ alloys with $x = 0–1$ by graded laser deposition. The alloys had a structure of spinodally decomposed disordered bcc + B2 ordered bcc, with the amount of disordered bcc phase increasing with increasing Co content from $x = 0$ to $x = 1$. The alloys again all exhibited soft ferromagnetic behaviour, with narrow $M–H$ hysteresis loops and very low values of remanence and coercivity, as shown in Figure 6.47(a), and with a saturation magnetisation M_s that increased with

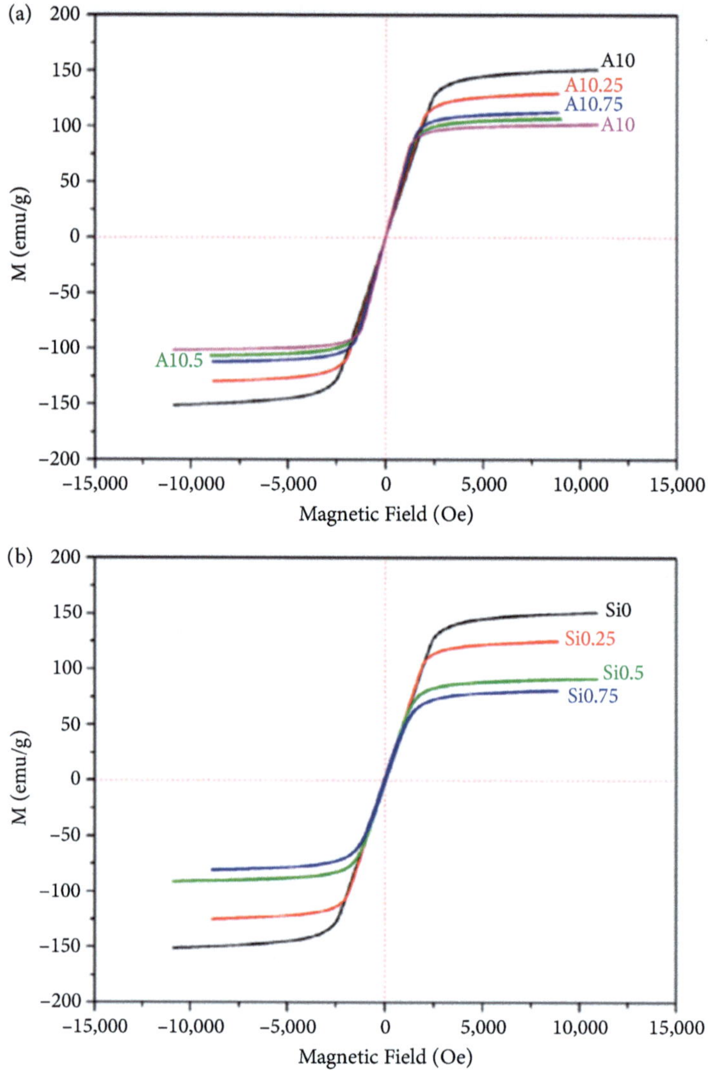

Figure 6.45 Soft ferromagnetic *M–H* hysteresis loops in multicomponent $Al_xFeCoNi$ and $Si_xFeCoNi$ alloys.

Reproduced from Zuo et al. (2013). High-entropy alloys with high saturation magnetization, electrical resistivity and malleability. *Scientific Reports 3*: Article 01455.

increasing additions of Co, and a coercivity H_c that was higher at intermediate compositions, as shown in Figure 6.47(b).

It is not too surprising that most multicomponent high-entropy alloys exhibit soft ferromagnetic behaviour. A wide distribution of different local atomic environments, as discussed in Chapter 4, promotes a correspondingly wide distribution of different exchange interactions between the electron spins on adjacent atoms, effectively preventing the magnetisation from being pinned too strongly to any particular orientation in the underlying crystal structure, and thus allowing for easy magnetisation and demagnetisation. More recently, however, a number of multicomponent high-entropy materials have been studied that

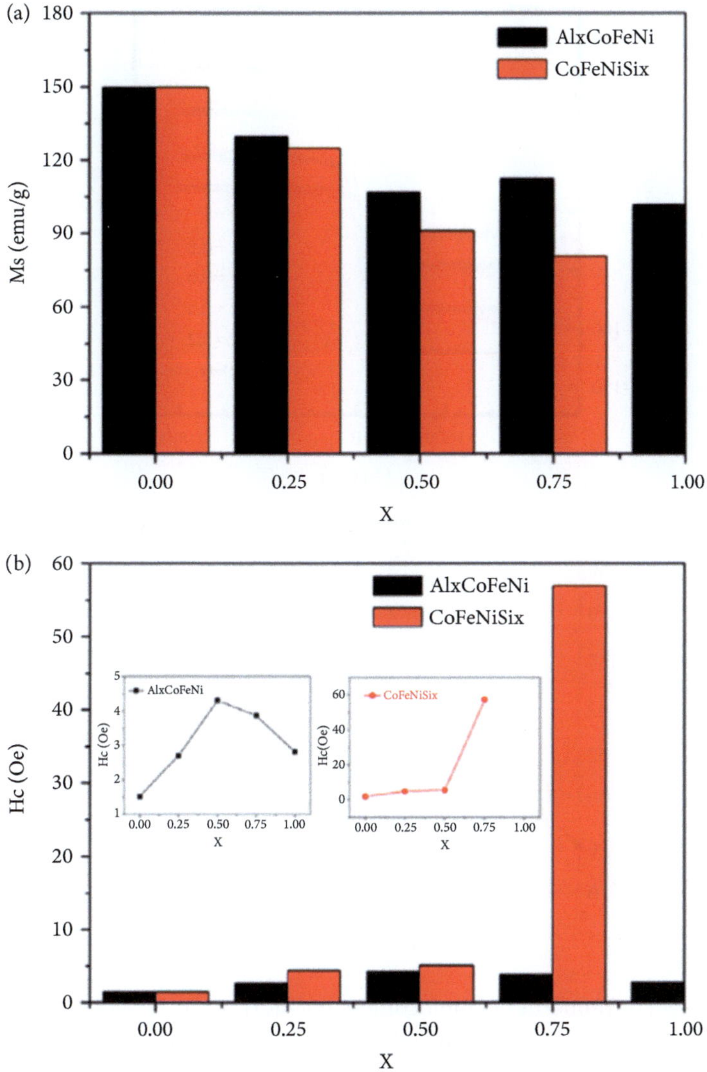

Figure 6.46 Variation of saturation magnetisation M_s and coercivity H_c with Al and Si content x in multicomponent $Al_xFeCoNi$ and $Si_xFeCoNi$ alloys.

Reproduced from Zuo et al. (2014). Effects of Al and Si addition on the structure and properties of CoFeNi equal atomic ratio alloy. *Journal of Magnetism and Magnetic Materials 371*: 60–8.

exhibit hard ferromagnetic behaviour, with a fine-scale two-phase microstructure used to pin the magnetisation in place. As an example, Na et al.[147] manufactured by arc melting and then casting a series of $AlFeCoNiCu_x$ alloys with $x = 0.1–1$, $AlFeCoNi(TiCu)_x$ alloys with $x = 0.2–0.4$, and $AlFe_xCo_yNi(TiCu)_{0.4}$ with x and $y = 1$ or 2. The base 4-component AlFeCoNi alloy was single-phase bcc with a narrow soft-ferromagnetic $M–H$ hysteresis loop, as shown in Figure 6.48(a). The 5- and 6-component $AlFeCoNiCu_x$, $AlFeCoNi(TiCu)_x$, and $AlFe_xCo_yNi(TiCu)_{0.4}$ alloys contained small nanocrystalline fcc Cu precipitates embedded in a bcc matrix that developed an internal fine-scale two-phase spinodal decomposition structure on heat treating for three hours at temperatures between 600–800 °C

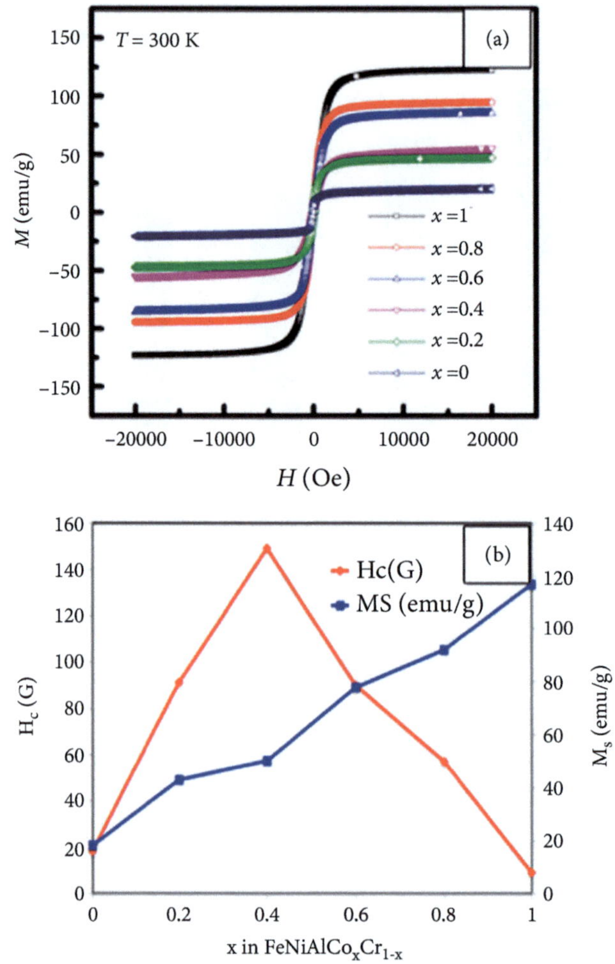

Figure 6.47 Magnetic properties of multicomponent $AlCr_{1-x}FeCo_xNi$ alloys: (a) soft ferromagnetic *M–H* hysteresis loops; and (b) variation of saturation magnetisation M_s and coercivity H_c with Co content *x*.

Reproduced from Borkar et al. (2017). A combinatorial approach for assessing the magnetic properties of high-entropy alloys: Role of Cr in $AlCo_xCr_{1-x}FeNi$. *Advanced Engineering Materials 10*: Article 1700048.

(873–1,073K). The alloys exhibited increasingly hard ferromagnetic behaviour with increasing Cu and Ti content, as shown in Figure 6.48, with increased *M–H* hysteresis loop width and increased coercivity, caused by increasingly effective pinning of electron spins and, therefore, magnetisation by the nanocrystalline Cu precipitates and the fine-scale matrix spinodal structure.

Huang et al.[138] have plotted an Ashby map of remanence versus coercivity, as shown in Figure 6.49, for a wide variety of different conventional and multicomponent high-entropy magnetic materials. The diagonal dashed lines on Figure 6.49 represent lines of equal $(BH)_{max}$ product, i.e. equal magnetic hysteresis and equal magnetisation/demagnetisation energy, with soft ferromagnetic materials having low remanence B_r and coercivity H_c at the lower left-hand side, and hard ferromagnetic materials having high remanence B_r

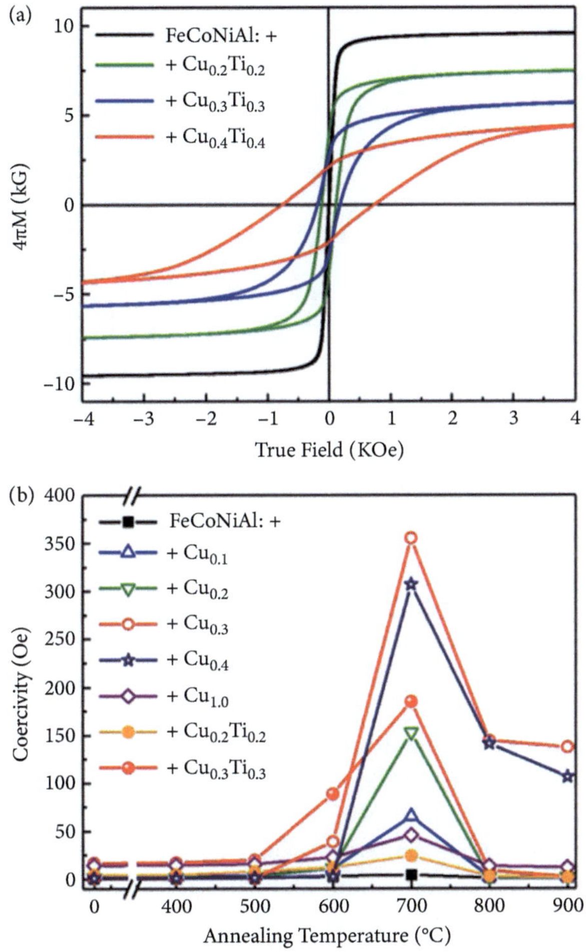

Figure 6.48 Magnetic properties of multicomponent $AlFeCoNiCu_x$ and $AlFeCoNi(TiCu)_x$ alloys: (a) *M–H* hysteresis loops and (b) variation of coercivity H_c with annealing temperature.
Reproduced from Na et al. (2021). Hard magnetic properties of $FeCoNiAlCu_xTi_x$-based high-entropy alloys. *AIP Advances 11*: Article 015210.

and coercivity H_c at the upper right-hand side, as indicated by the red and blue arrows respectively at the top of the figure. Conventional alloys are shown on Figure 6.49 as regions surrounded by solid coloured lines (with a different colour for each different alloy family, as indicated in the legend); and multicomponent high-entropy alloys are shown on Figure 6.49 as regions surrounded by broken coloured lines (again with a different colour for each different alloy family, as indicated in the legend). As can be seen in Figure 6.49, the best multicomponent high-entropy soft and hard ferromagnetic alloys are not as good as, respectively, the best conventional soft and hard ferromagnetic alloys. This is also shown in Table 6.14, which compares typical measured values of saturation magnetisation and coercivity in some multicomponent high-entropy and conventional soft and hard ferromagnetic alloys.[136,146–148] Huang et al.[138] concluded that '. . . the hard magnets made from commercial alloys have $B_r \times H_c$ greater than those

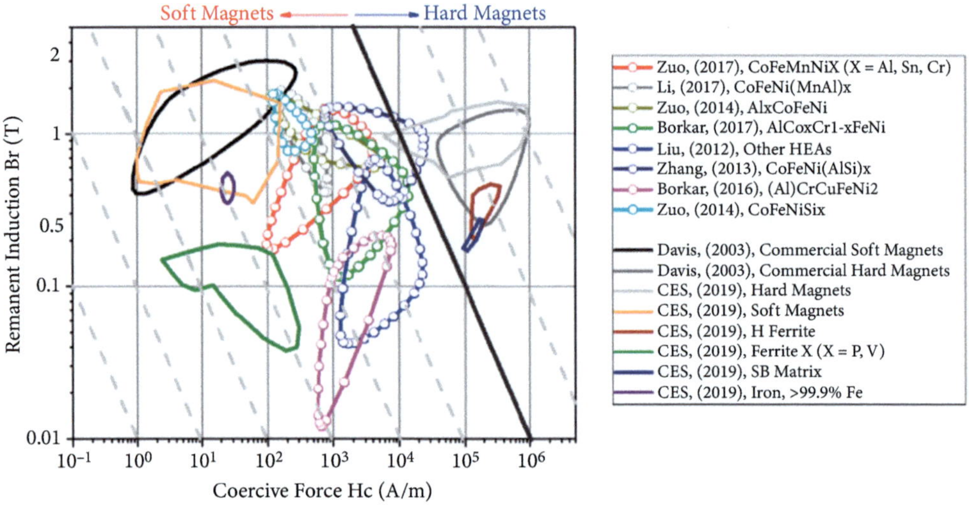

Figure 6.49 Ashby map of remanence B_r versus coercivity H_c for different conventional and multicomponent high-entropy alloys.

Reproduced from Huang et al. (2020). Mechanical and magnetic properties of the high-entropy alloys for combinatorial approaches. *Crystals 10*: Article 10030200.

of the high entropy alloys . . . [and] . . . commercial alloys have a better soft-magnet performance with small products of $B_r \times H_c$'. However, multicomponent alloys are very far from being optimised, and their properties are likely to be improved rapidly in the next few years. Moreover, multicomponent alloys have other magnetic and mechanical properties, such as low magnetostriction, high strength and ductility, and high corrosion and wear resistance, which can readily be tuned by varying the alloy composition. This means that they have considerable potential to be used in many complex engineering applications. Huang et al.[138] also concluded, therefore, that ' . . . when applications require a specific function needing more than the optimization of a single property, high-entropy alloys present combinatorial advantages since the alloys are much more diverse'.

In summary, some multicomponent high-entropy metallic alloys containing ferromagnetic components such as Fe and Ni exhibit excellent soft ferromagnetic properties because of the wide range of different local atomic structures discussed in detail in Chapter 4, with a correspondingly wide range of different magnetic exchange interactions between adjacent atoms not easily aligned with and, therefore, not easily pinned by the underlying crystal structure. On the other hand, other multicomponent high-entropy metallic alloys containing ferromagnetic components such as Fe and Ni exhibit excellent hard ferromagnetic properties when they are heat treated and undergo nanoscale precipitation and/or spinodal decomposition, so the magnetic exchange interactions between adjacent atoms become well pinned by the underlying multiphase nanostructure. The best hard and soft multicomponent ferromagnetic materials discovered so far are not, in either case, as good as the best hard and soft conventional materials already available. However, multicomponent ferromagnetic materials exhibit additional properties such as high strength and ductility, and high wear and corrosion resistance, that make them attractive for use in particular complex and demanding engineering applications. Moreover, multicomponent ferromagnetic materials are far from optimised so far, with literally millions of alternative alloy compositions and microstructures yet to be explored.

Corrosion resistance

The wide variety of different local atomic structures in multicomponent high-entropy single-phase materials suppresses atomic mobility and thus reduces the diffusion rates of the component atoms and molecules, as discussed in detail in Chapter 4. This increases the thermal stability of many multicomponent high-entropy single-phase (and multiphase) materials and makes them resilient to attrition and damage caused by processes such as wear, irradiation and corrosion, all of which rely on atomic diffusion to sustain the various damage mechanisms. The first two of these processes are discussed earlier in this chapter. Not surprisingly, there has also been considerable investigation of the corrosion resistance of multicomponent high-entropy materials, and the results achieved so far have been reviewed in some detail.[149,150]

Corrosion of a metallic material is the electrolytic process of *anodic oxidation* and thus dissolution of the component metal atoms at the surface of the material:[151]

$$M \longrightarrow M^{n+} + ne^{-}$$

The resulting metal ions can be removed into the surrounding liquid or gas, or can react with it to form a compound film, often an oxide, on the material surface. Not surprisingly, the corrosion rate is very dependent on the exact details of the corroding environment, in particular its nature (whether it is gaseous or liquid), its composition (the acidity and the presence or otherwise of reacting oxygen or other anionic species), and the ambient pressure and temperature. In general, anodic oxidation reactions become *polarised*,[151] i.e. the electron current density i and, therefore, the corresponding rate of oxidation of the component metal atoms at the surface (or, in short, the corrosion rate), vary in a complex way with the electric potential at the material surface E, as shown schematically in Figure 6.50. There is no corrosion at zero current density, $i = 0$, and the potential at the material surface is then $E_a^o = -\Delta G_a^o/nF$ where E_a^o and ΔG_a^o are respectively the *standard electrode potential* and the *standard state free energy change* for the anodic oxidation reaction, and F is the Faraday constant (the charge per mole of electrons). There are typically three stages of polarisation when the anode surface potential E_a is increased by applying an anodic *overpotential* $\eta_a = E_a - E_a^o$ to initiate corrosion by drawing a current density i:

1. *Activation polarisation* for small values of η_a, with the current density i and the corresponding corrosion rate increasing as the anodic oxidation reaction is driven faster by an increasing free energy change $\Delta G_a = -nFE_a$;
2. Concentration polarisation leading to passivation for intermediate values of η_a, as the current density i and the corresponding corrosion rate become constant or drop to near zero and the oxidation reaction is stifled, with the oxidised metal ions increasingly unable to diffuse away from the corroding material, which builds up a passivating oxide or other compound film on its surface; and finally
3. Breakdown for large values of η_a, as the current density i and the corresponding corrosion rate again increase, usually dramatically, by severe pitting of the surface oxide film and intense localised corrosion.[151]

Table 6.14 Saturation magnetisation M_s and coercivity H_c in some multicomponent high-entropy and conventional soft and hard ferromagnetic materials.[136,146–148]

alloy	magnetisation M_s(MA/m)	coercivity H_c(Oe)
soft		
FeCoNi	1.13	1.5
$FeCoNiAl_{0.2}$	0.98	0.54
FeCoNiAlMn	0.90	1.2
FeCoNiCuPd	0.82	1.45
silicon-iron $Fe_{97}Si_3$	1.59	0.06
Hypernik FeNi	1.27	0.10
Metglas 2628SC	0.70	0.006
hard		
$AlFeCoNiTi_{0.4}Cu_{0.4}$	0.41	955
$AlFe_2CoNiTi_{0.4}Cu_{0.4}$	0.47	1,078
$Al_{0.5}FeCo_{0.5}NiLaPrNd$	0.09	3,522
Alnico 4	0.94	1,860
$Al_{10}Fe_{30}Nd_{60}$	0.14	3,623
$Nd_2F_{14}B$	1.27	9,433

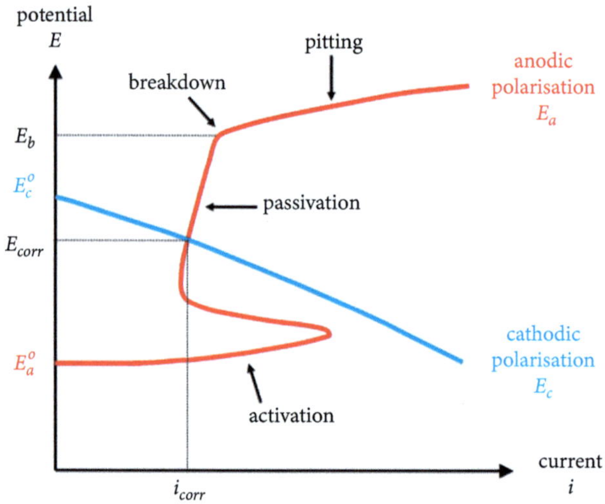

Figure 6.50 Schematic anodic and cathodic polarisation curves showing anodic and cathodic potential E_a and E_c respectively versus current i, with electrical balance at their crossover giving corrosion potential E_{corr} and corrosion current i_{corr}.

Corrosion by anodic oxidation of the metal atoms at the surface of a material as described above cannot proceed without an associated (electrically attached) *cathodic reduction* reaction, often the reduction of oxygen to form oxide or hydroxyl ions:[151]

$$O_2 + 4e^- \longrightarrow 2O^{2-}$$

or

$$O_2 + 2H_2O + 4e^- \longrightarrow 4OH^-$$

The cathodic reduction reaction often takes place at more noble (i.e. less reactive) sites on the surface of the corroding material itself but can also take place on the surface of an adjacent, more noble material attached electrically to (i.e. in contact with) the corroding material. In either case, the cathodic reduction reaction also becomes polarised. At zero current density, $i = 0$, the potential at the surface is again $E_c^o = -\Delta G_c^o/nF$ but E_c^o and ΔG_c^o are now respectively the higher (more noble) standard electrode potential and standard state free energy change for the cathodic reduction reaction; and the *cathodic* current density increases on applying (in this case) a cathodic *underpotential* (negative overpotential) $\eta = E_c^o - E_c$, as the reduction reaction is driven faster by an increasing free energy change $\Delta G_c^o = -nFE_c$. The combination of polarised anodic oxidation and cathodic reduction reactions establishes an electrical circuit, as shown schematically in Figure 6.50, balancing the rate of production and consumption of electrons where the two polarisation curves cross at the *corrosion potential* E_{corr} and *corrosion current density* i_{corr} that determine the overall corrosion rate, i.e. the rate of dissolution of the material. Key corrosion parameters for any given material in any given corroding environment are the corrosion potential and current density E_{corr} and i_{corr}, which need to be as small as possible for good corrosion resistance, and the breakdown or *pitting potential* E_b and breakdown or *pitting overpotential* $\eta_b = E_b - E_a^o$, which need to be as large as possible for good corrosion resistance, as shown in Figure 6.50.

The various studies of corrosion in multicomponent high-entropy materials have been conducted under many different corrosion conditions, so it is not always easy to compare the different results.[149,150] Helpfully however, Shi et al.[149] and Nascimento et al.[150] have compiled data obtained for the corrosion of a number of multicomponent and conventional materials under roughly comparable conditions. Table 6.15 shows some of the multicomponent alloy results, for alkali corrosion of $Al_xCrMnFe_{1.5}Ni_{0.5}$ and $CrFeCoNiCu_x$ alloys in a 0.6 or 1 molar aqueous sodium chloride NaCl solution.[149,152,153] Table 6.16 shows similarly some of the multicomponent alloy results for acidic corrosion of $Al_xCrMnFe_{1.5}Ni_{0.5}$ and $Al_xCrFeCoNi$ alloys in a 0.5 molar aqueous sulphuric acid H_2SO_4 solution.[149,152,154] As can be seen in Table 6.15, increasing additions of Al to $CrMnFe_{1.5}Ni_{0.5}$ and Cu to CrFeCoNi reduce the alkali corrosion resistance as shown by an increasing corrosion current

Table 6.15 Alkali corrosion of $Al_xCrMnFe_{1.5}Ni_{0.5}$ and $CrFeCoNiCu_x$ in 0.6 or 1 molar NaCl in water.[149,152,153]

alloy	solution	corrosion current density i_{corr} ($\mu A/cm^2$)
$CrMnFe_{1.5}Ni_{0.5}$	1 M NaCl	0.45
$Al_{0.3}CrMnFe_{1.5}Ni_{0.5}$	1 M NaCl	0.69
$Al_{0.5}CrMnFe_{1.5}Ni_{0.5}$	1 M NaCl	1.02
CrFeCoNi	0.6 M NaCl	0.04
$CrFeCoNiCu_{0.5}$	0.6 M NaCl	0.72
CrFeCoNiCu	0.6 M NaCl	1.23

Table 6.16 Acid corrosion of AlxCrMnFe$_{1.5}$Ni$_{0.5}$ and Al$_x$CrFeCoNi multicomponent high-entropy alloys in 0.5 molar H_2SO_4 in water.[149,152,154]

alloy	solution	corrosion current density i_{corr} (μ A/cm^2)
CrMnFe$_{1.5}$Ni$_{0.5}$	0.5 M H_2SO_4	31.4
Al$_{0.3}$CrMnFe$_{1.5}$Ni$_{0.5}$	0.5 M H_2SO_4	73.9
Al$_{0.5}$CrMnFe$_{1.5}$Ni$_{0.5}$	0.5 M H_2SO_4	68.2
CrFeCoNi	0.5 M H_2SO_4	15.8
Al$_{0.25}$CrFeCoNi	0.5 M H_2SO_4	16.7
Al$_{0.5}$CrFeCoNi	0.5 M H_2SO_4	13.4
AlCrFeCoNi	0.5 M H_2SO_4	13.1

density. On the other hand, Table 6.16 shows how increasing additions of Al to either Al$_x$CrMnFe$_{1.5}$Ni$_{0.5}$ and Al$_x$CrFeCoNi alloys initially improve the acid corrosion resistance, as shown by a lower corrosion current at low Al content, but then reduce acid corrosion resistance with a higher corrosion current at higher Al content.

Nascimento et al.[150] have plotted Ashby maps for a large number of multicomponent and conventional materials, based on the breakdown pitting potential and overpotential E_b and η_b for alkali corrosion in aqueous sodium chloride NaCl solution as shown in Figure 6.51, and the corrosion potential and current density E_{corr} and i_{corr} for acid corrosion in aqueous sulphuric acid H_2SO_4 solution as shown in Figure 6.52. In alkali NaCl the two best multicomponent alloys with the highest pitting potential and overpotential E_b and η_b are Al$_{0.5}$TiVCrNb$_{0.5}$ and Al$_{0.3}$CrFeCoNi (labelled 67 and 66 respectively and shown in a green circle in Figure 6.51),[150,155,156] with considerably better corrosion resistance than conventional 304 stainless steel or Ti-6Al-4V. In acid H_2SO_4, the two best multicomponent alloys with the lowest corrosion potential and corrosion current E_{corr} and i_{corr} are CrFeCoNiMo$_{0.4}$ and AlTi$_{0.5}$CrFeCoNi (labelled 30 and 58 respectively and shown in a green circle in Figure 6.52),[150,157,158] again with considerably better corrosion resistance than conventional 304 stainless steel or Ti-6Al-4V.

As an example of the excellent *alkali* corrosion resistance of multicomponent high-entropy alloys, Yan et al.[156] manufactured a series of Al$_{0.3}$Cr$_x$FeCoNi alloys with $x = 0$–2 by arc melting and casting into a water-cooled copper mould, followed by immersion mass loss and potentiodynamic electrochemical corrosion measurements in an aqueous 3.5 wt% NaCl solution with an EG&G Princeton Applied Research VersaSTAT III electrochemical workstation using a three-electrode cell with the multicomponent alloy as the working electrode, a Pt foil counter-electrode and a saturated calomel (SCE) reference electrode, scanning at 50 mV/min from 50 mV below the open circuit potential. Immersion tests showed essentially no mass loss, no surface degradation and, therefore, no detected corrosion after thirty days in the NaCl solution for all Cr-containing alloys with $x \geq 0.5$. Potentiodynamic curves of anodic current density i versus applied surface potential E, as shown in Figure 6.53, indicate clearly the rapid progressive increase in breakdown potential required to initiate pitting in the passivating surface oxide film from $E_b = 0.1$ to 0.5 V as the Cr content in the alloys was increased from $x = 0$ to 1, with film breakdown fully suppressed even with applied surface potentials of up to $E = 2$ V when the Cr alloy content was increased even further to $x \geq 1.5$.

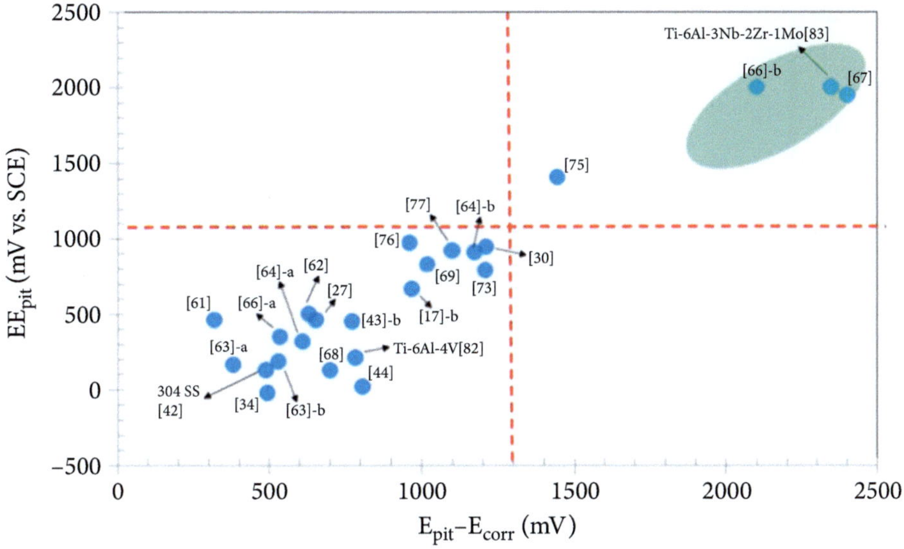

Figure 6.51 Ashby map of breakdown pitting potential E_b $(= E_{pit})$ and overpotential η_b $(= E_{pit} - E_{corr})$ for alkali corrosion of multicomponent high-entropy (numbered) and conventional alloys in NaCl.

Reproduced from Nascimento et al. (2022). A review on corrosion of high-entropy alloys: Exploring the interplay between corrosion properties, alloy composition, passive film stability and materials selection. *Materials Research 25*: Article 20210442.

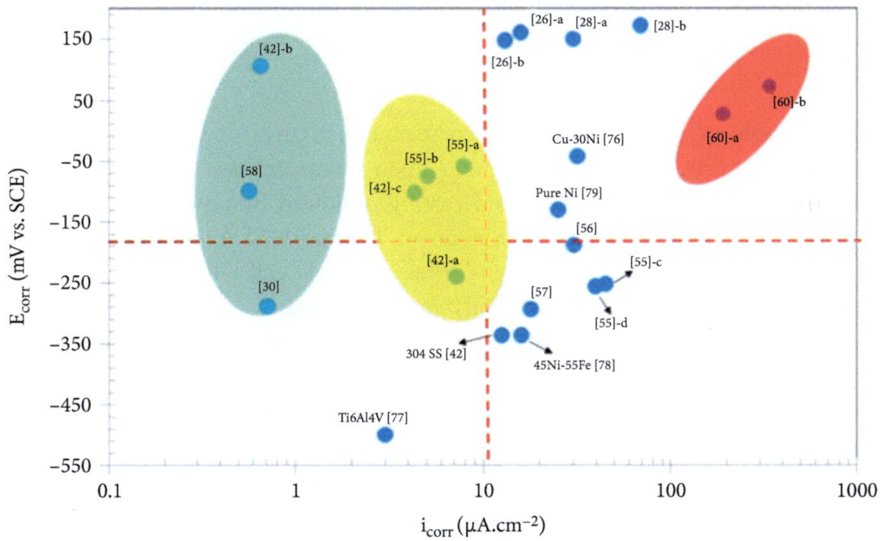

Figure 6.52 Ashby map of corrosion potential E_{corr} and corrosion current i_{corr} for acid corrosion of multicomponent high-entropy (numbered) and conventional alloys in H_2SO_4.

Reproduced from Nascimento et al. (2022). A review on corrosion of high-entropy alloys: Exploring the interplay between corrosion properties, alloy composition, passive film stability and materials selection. *Materials Research 25*: Article 20210442.

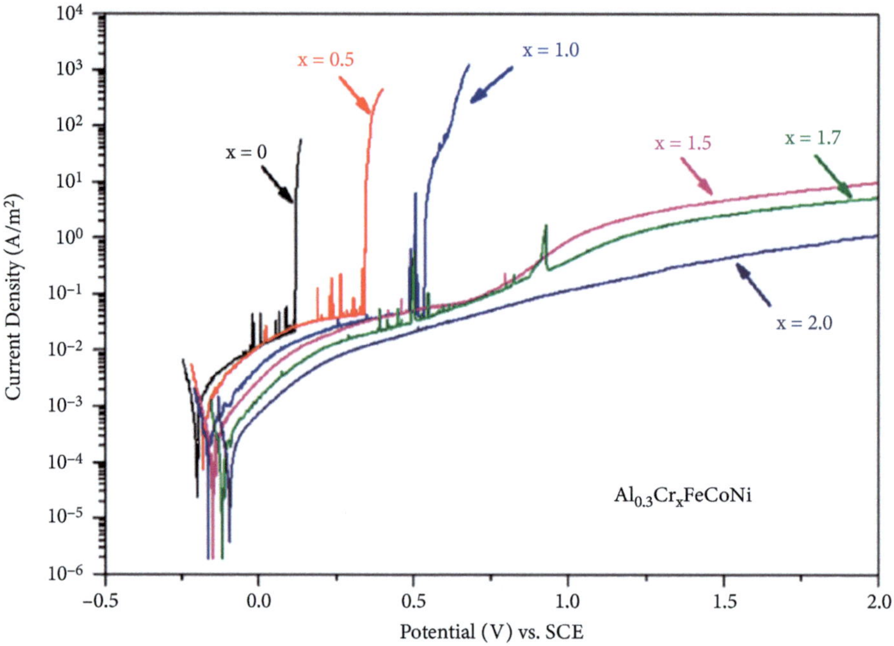

Figure 6.53 Potentiodynamic measurements of anodic polarisation, showing current density *i* versus applied surface potential *E* for multicomponent $Al_{0.3}Cr_xFeCoNi$ alloys with $x = 0–2$ in aqueous 3.5 wt% NaCl.

Reproduced from Yan et al. (2010). Electrochemical passive properties of $Al_xCoCrFeNi$ ($x = 0, 0.25, 0.50, 1.00$) alloys in sulfuric acids. *Corrosion Science 52*: 1026–34.

As an example of the excellent *acid* corrosion resistance of multicomponent high-entropy alloys, Li et al.[155] designed and then manufactured a corrosion-resistant $Al_{0.5}TiVCrNb_{0.5}$ alloy by arc melting and casting, followed by potentiodynamic electrochemical corrosion measurements in an aqueous 1-molar HCl solution with a PARSTAT 2273 electrochemical workstation using a three-electrode cell with the multicomponent alloy as the working electrode, a Pt foil counter-electrode and an SCE, scanning at 1 mV/s from −0.7 to 2 V relative to the open circuit potential. The multicomponent high-entropy $Al_{0.5}TiVCrNb_{0.5}$ alloy had a single-phase bcc structure and exhibited excellent corrosion resistance, with a much better combination of high breakdown pitting potential E_b and low corrosion current i_{corr} than conventional alloys, as shown clearly on the Ashby map in Figure 6.54. This was attributed to an effective passive surface film formed at the designed alloy composition, resistant to degradation in a low-diffusion multicomponent alloy and film. Figure 6.55 shows X-ray photoelectron spectra (XPS) from the passive surface film, obtained by using a Kratos AXIS Ultra DLD spectrometer on a scanning electron microscope, indicating that the film was a complex mixture of the oxides of all five alloy components, as shown in Table 6.17.

In summary, a number of multicomponent high-entropy metallic alloys have outstanding corrosion resistance in both acid and alkali environments, considerably better than well-known conventional corrosion-resistant alloys such as stainless steels and titanium alloys.

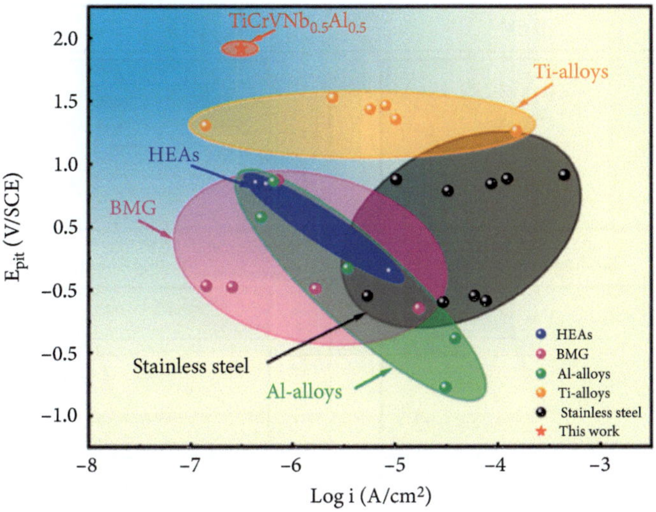

Figure 6.54 Ashby map of breakdown pitting potential E_b $(= E_{pit})$ and corrosion current i_{corr} $(= i)$ for multicomponent single-phase bcc $Al_{0.5}TiVCrNb_{0.5}$ and a variety of conventional alloys in aqueous 1 M HCl.

Reproduced from Li et al. (2021). Evaluation of corrosion resistance of the single-phase light refractory high-entropy alloy $TiCrVNb_{0.5}Al_{0.5}$ in chloride environment. *Journal of Alloys and Compounds 857*: Article 158278.

This is a consequence of two effects: firstly, a wide range of different components distributed evenly throughout a single phase high-entropy alloy supports the formation of a dense and highly corrosion-resistant multicomponent passivating surface oxide film; and secondly, continued growth and degradation or pitting of the passivating surface film are all effectively suppressed by the wide range of different local atomic environments and corresponding slow diffusion rates in both the multicomponent underlying alloy and the multicomponent surface oxide film. Achieving high levels of corrosion resistance depends, however, on careful tuning of the alloy composition to maximise these two effects.

4 Summary

Multicomponent high-entropy materials have been studied continuously since their discovery in the late twentieth century and the publication of the first two key papers announcing this discovery in detail in 2004. At first, scientific interest in multicomponent high-entropy materials was fairly limited, but it has grown strongly in the last decade or so, concentrating initially on the surprising simplicity of the structural features of materials containing large numbers of components, their surprising relative ease of manufacture, and the complexity of developing strategies for exploring and rationalising the vast and bewildering number of resulting new materials thus made available for investigation. More recently however, and not at all surprisingly, attention has turned increasingly towards

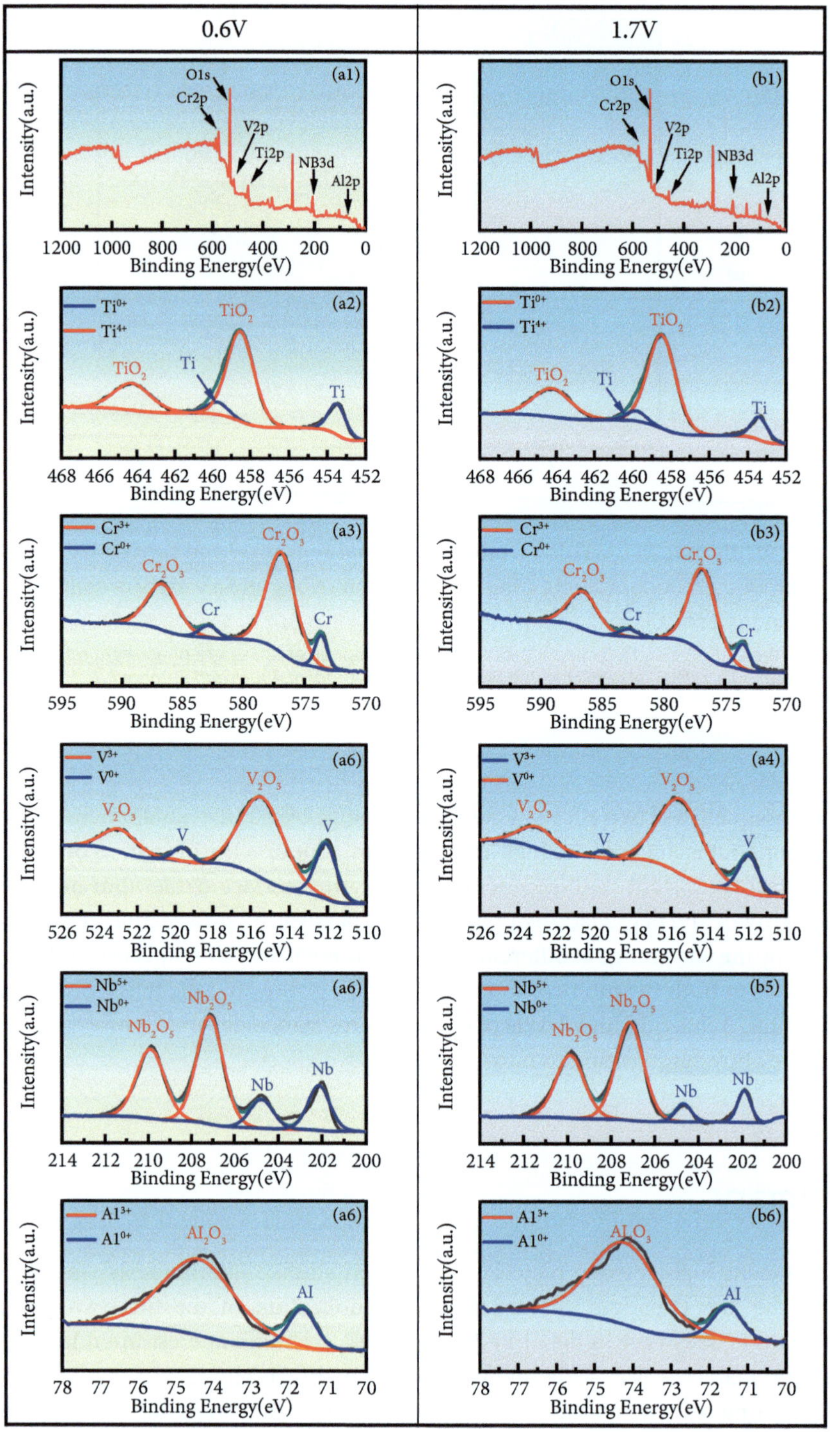

Figure 6.55 XPS spectra from the passive surface oxide film on multicomponent single-phase bcc $Al_{0.5}TiVCrNb_{0.5}$ after applying an electrolytic potential of 0.6 and 1.7 V.

Reproduced from Li et al. (2021). Evaluation of corrosion resistance of the single-phase light refractory high-entropy alloy $TiCrVNb_{0.5}Al_{0.5}$ in chloride environment. *Journal of Alloys and Compounds 857*: Article 158278.

Table 6.17 Composition of passive surface oxide film measured from XPS spectra on multicomponent single-phase bcc $Al_{0.5}TiVCrNb_{0.5}$ after applying an electrolytic potential of 0.6 and 1.7 V (after Li et al.[155]).

species	content at 0.6 V (%)	content at 1.7 V (%)
Al^{3+} (Al_2O_3)	12.6	15.5
Al	2.9	2.4
Ti^{4+} (TiO_2)	24.6	20.2
Ti	3.6	3.0
V^{3+} (V_2O_3)	10.6	11.8
V	3.5	2.5
Cr^{3+} (Cr_2O_3)	13.0	15.1
Cr	4.4	2.4
Nb^{5+} (Nb_2O_5)	20.9	23.0
Nb	3.8	4.0

studying the remarkable properties and corresponding potential applications of multicomponent high-entropy materials, at first mainly the mechanical properties of single-phase fcc Cantor alloys and then single-phase bcc Senkov alloys, subsequently moving on most recently to other, more functional properties and other materials, notably multicomponent ceramics.

Despite this mushrooming scientific interest, investigation of the properties and applications of multicomponent high-entropy materials is, of course, still very much in its early infancy. Despite a rapidly growing amount of research, there are still literally trillions of multicomponent materials that have never been made, let alone studied, and most of the properties of those that have been studied are yet to be explored fully. Nevertheless, some remarkable properties and potential applications have already been identified, although in all cases the new materials and their properties are still very far from being optimised. To be more specific, multicomponent high-entropy materials have already been shown, as discussed throughout this chapter, to have outstanding mechanical, catalytic, radiation-resistant and corrosion-resistant properties that are significantly better than current conventional materials; and they have also been shown to have excellent magnetic, thermal and wear-resistant properties that are closely competitive with current conventional materials. In all these cases, further improvements in material compositions, manufacturing methods and microstructural manipulation hold out great potential for further development. And there still remain enormous numbers of other multicomponent high-entropy materials, properties and applications yet to be explored.

References

1. Bunsell AR, Renard J. (2005). *Fundamentals of Fibre Reinforced Composite Materials.* Bristol: IOP Publishing, 191–275.
2. Clyne TW, Hull D. (2019). *Introduction to Composite Materials*, 3rd edn. Cambridge: Cambridge University Press, 31–42.

3. Otto F, Diouhý A, Somsen C, Bei H, Eggeler G, George EP. (2013). The influence of temperature and microstructure on the tensile properties of a CoCrFeNiMn high-entropy alloy. *Acta Materialia 61*: 5743–55.
4. Li Z, Zhao S, Ritchie RO, Meyers MA. (2019). Mechanical properties of high-entropy alloys with emphasis on face-centered cubic alloys. *Progress in Materials Science 102*: 296–345.
5. Gludovatz B, Hohenwarter A, Catoor D, Chang EH, George EP, Ritchie RO. (2014). A fracture-resistant high-entropy alloy for cryogenic applications. *Science 345*: 1153–8.
6. Tian YZ, Sun SJ, Lin HR, Zhang ZF. (2019). Fatigue behavior of CoCrFeMnNi high-entropy alloy under fully reversed cyclic deformation. *Journal of Materials Science and Technology 35*: 334–40.
7. George EP, Curtin WA, Tasan CC. (2020). High entropy alloys: A focused review of mechanical properties and deformation mechanisms. *Acta Materialia 188*: 435–74.
8. Li W, Xie D, Li D, Zhang Y, Gao Y, Liaw P. (2021). Mechanical behaviour of high-entropy alloys. *Progress in Materials Science 118*: Article 100777.
9. Gludovatz B, Hohenwarter A, Thurston KVS, Bei H, Wu Z, George EP, Ritchie RO. (2016). Exceptional damage-tolerance of a medium-entropy alloy NiCoCr at cryogenic temperatures. *Nature Communications 7*: Article 10602.
10. Li L, Chen Z, Kuroiwa S, Ito M, Kishida K, Inui H, George EP. (2022). Tensile and compressive plastic deformation behavior of medium entropy CrCoNi single crystals from cryogenic to elevated temperatures. *International Journal of Plasticity 148*: Article 103144.
11. Fu W, Gan K, Huang Y, Ning Z, Sun J, Cao F. (2021). Elucidating the transition of cryogenic deformation mechanism of CrMnFeCoNi high entropy alloy. *Journal of Alloys and Compounds 872*: Article 159606.
12. Zhang Z, Mao MM, Wang J, Gludovatz B, Zhang Z, Mao SX, George EP, Yu Q, Ritchie RO. (2015). Nanoscale origins of the damage tolerance of the high-entropy alloy CrMnFeCoNi. *Nature Communications* 6: Article 10143.
13. Bintu A, Vincze G, Picu CR, Lopes AB, Grácio JJ, Barlat F. (2015). Strain hardening rate sensitivity and strain rate sensitivity in TWIP steels. *Materials Science and Engineering A629*: 54–9.
14. Byun T, Hashimoto N, Farrell K. (2004). Temperature dependence of strain hardening and plastic instability behaviors in austenitic stainless steels. *Acta Materialia 52*: 3889–99.
15. Kireeva IV, Chumlyakov YI, Pobedennaya ZV, Kuksgaussen IV, Karaman I. (2017). Orientation dependence of twinning in single crystalline CoCrFeMnNi high-entropy alloy. *Materials Science and Engineering A705*: 176–81.
16. Xie J, Dong H, Hao Y, Fan Z, Wang CA. (2022). Exploring the formation mechanism of deformation twins in CrMnFeCoNi high entropy alloy. *Acta Metallurgica Sinica 35*: 1275–80.
17. Laplanche G, Kostka A, Reinhart C, Hunfield J, Eggeler G, George EP. (2017). Reasons for the superior mechanical properties of medium-entropy CrCoNi compared to high-entropy CrMnFeCoNi. *Acta Materialia 128*: 292–303.
18. Bajpai S, MacDonald BE, Rupert TJ, Hahn H, Lavernia EJ, Apelian D. (2022). Recent progress in the CoCrNi alloy system. *Materialia 24*: Article 101476.

19. Yang M, Zhou L, Wang C, Jiang P, Yuan F, Ma E, Wu X. (2019). High impact toughness of CrCoNi medium-entropy alloy at liquid helium temperature. *Scripta Materialia 172*: 66–71.
20. Liu J, Chen C, Xu Y, Wu S, Wang G, Wang H, Fang Y, Meng L. (2017). Deformation twinning behaviors of the low stacking fault energy high-entropy alloy: An *in-situ* study. *Scripta Materialia 137*: 9–12.
21. Zhang DD, Zhang JY, Kuang J, Liu G, Sun J. (2021). Superior strength-ductility synergy and strain hardenability of Al/Ta co-doped NiCoCr twinned medium entropy alloy for cryogenic applications. *Acta Materialia 220*: Article 117288.
22. Schneider M, Laplanche G. (2021). Effects of temperature on mechanical properties and deformation mechanisms of the equiatomic CrFeNi medium-entropy alloy. *Acta Materialia 204*: Article 116470.
23. Liu SF, Wu Y, Wang HT, He J, Liu JB, Chen CX, Liu XJ, Wang H, Lu ZP. (2018). Stacking fault energy of face-centred-cubic high entropy alloys. *Intermetallics 93*: 269–73.
24. Zaddach AJ, Niu C, Koch CC, Irving DL. (2013). Mechanical properties and stacking fault energies of NiFeCrCoMn high-entropy alloy. *Journal of Metals 65*: 1780–9.
25. Huang S, Li W, Lu S, Tian FY, Shen J, Holmström E, Vitos L. (2015). Temperature dependent stacking fault energy of FeCrCoNiMn high entropy alloy. *Scripta Materialia 108*: 44–7.
26. Laplanche G, Kostka A, Horst OM, Eggeler G, George EP. (2016). Microstructure evolution and critical stress for twinning in the CrMnFeCoNi high-entropy alloy. *Acta Materialia 118*: 152–63.
27. Li Z, Pradeep KG, Deng Y, Raabe D, Tasan CC. (2016). Metastable high-entropy dual-phase alloys overcome the strength–ductility trade-off. *Nature 534*: 227–30.
28. Li Z, Kormann F, Grabowski B, Neugebauer J, Raabe D. (2017). Ab initio assisted design of quinary dual-phase high-entropy alloys with transformation-induced plasticity. *Acta Materialia 136*: 262–70.
29. Li Z, Tasan CC, Pradeep KG, Raabe D. (2017). A TRIP-assisted dual-phase high-entropy alloy: grain size and phase fraction effects on deformation behavior. *Acta Materialia 131*: 323–35.
30. Su J, Raabe D, Li Z. (2019). Hierarchical microstructure design to tune the mechanical behavior of an interstitial TRIP-TWIP high-entropy alloy. *Acta Materialia 163*: 40–54.
31. He ZF, Jia N, Ma D, Yan HL, Li ZM, Raabe D. (2019). Joint contribution of transformation and twinning to the high strength-ductility combination of a FeMnCoCr high entropy alloy at cryogenic temperatures. *Materials Science and Engineering A759*: 437–47.
32. Bhadeshia H, Honeycombe R. (2017). *Steels: Microstructure and Properties*, 4th edn. Oxford: Butterworth-Heineman.
33. Seol JB, Bae JW, Li Z, Han JC, Kim JG, Raabe D, Kim HS. (2018). Boron doped ultrastrong and ductile high-entropy alloys. *Acta Materialia 151*: 366–76.
34. Wu R, Freeman AJ, Olson GB. (1994). First principles determination of the effects of phosphorus and boron on iron grain boundary cohesion. *Science 15*: 376–80.

35. Lejĉek P, Hofmann S, Paidar V. (2003). Segregation based classification of [100] tilt grain boundaries in α-iron and its consequences for grain boundary engineering. *Acta Materialia 51*: 3951–63.
36. Raabe D, Herbig M, Sandlobes Y, Li Y, Tytko D, Kuzmina M, Ponge D, Choi PP. (2014). Grain boundary segregation engineering in metallic alloys: a pathway to the design of interfaces. *Current Opinions in Solid State Materials Science 18*: 253–61.
37. Reed, RC. (2006). *The Superalloys: Fundamentals and Applications.* Cambridge: Cambridge University Press.
38. Wang ZG, Zhou W, Fu LM, Wang JF, Luo RC, Han XC, Chen B, Wang XD. (2017). Effect of coherent $L1_2$ nanoprecipitates on the tensile behavior of a fcc-based high-entropy alloy. *Materials Science and Engineering A696*: 503–10.
39. Zhang L, Zhou Y, Jin X, Du X, Li B. (2018). Precipitation-hardened high entropy alloys with excellent tensile properties. *Materials Science and Engineering A732*: 186–91.
40. Yang T, Zhao YL, Tong Y, Jiao ZB, Wei J, Cai JX, Han XD, Chen D, Hu A, Kai JJ, Lu K, Liu Y, Liu CT. (2018). Multicomponent intermetallic nanoparticles and superb mechanical behaviors of complex alloys. *Science 362*: 933–7.
41. Peng H, Hu L, Li L, Zhang W. (2020). Ripening of $L1_2$ nanoparticles and their effects on mechanical properties of $Ni_{28}Co_{28}Fe_{21}Cr_{15}Al_4Ti_4$ high-entropy alloys. *Materials Science and Engineering A772*: Article 138803.
42. Li Z, Fu L, Peng J, Zheng H, Ji X, Sun Y, Ma S, Shan A. (2020). Improving mechanical properties of an FCC high-entropy alloy by γ' and B2 precipitates strengthening. *Materials Science and Characterisation 159*: Article 109989.
43. Senkov ON, Wilks GB, Scott JM, Miracle DB. (2011). Mechanical properties of NbMoTaW and VNbMoTaW refractory high entropy alloys. *Intermetallics 19*: 698–706.
44. Senkov ON, Miracle DB, Chaput KJ, Couzinie JP. (2018). Development and exploration of refractory high entropy alloys—a review. *Journal of Materials Research 33*: 3092–3128.
45. Miracle DB, Tsai MH, Senkov ON, Soni V, Banerjee R. (2020). Refractory high entropy superalloys. *Scripta Materialia 187*: 445–52.
46. Zhou J, Cheng Y, Chen Y, Liang X. (2022). Composition design and preparation process of refractory high-entropy alloys: A review. *International Journal of Refractory Metals and Hard Materials 105*: Article 105836.
47. Ren X, Li Y, Qi Y, Wang B. (2022). Review on preparation technology and properties of refractory high entropy alloys. *Materials 15*: Article 2931.
48. Chen B, Zhuo L. (2023). Latest progress on refractory high entropy alloys: Composition, fabrication, post processing, performance, simulation and prospect. *International Journal of Refractory Metals and Hard Materials 110*: Article 105993.
49. Senkov ON, Gorsse S, Miracle DB. (2019). High temperature strength of refractory complex concentrated alloys. *Acta Materialia 175*: 394–405.
50. Qi L, Chrzan DC. (2014). Tuning ideal tensile strengths and intrinsic ductility of bcc refractory alloys. *Physical Review Letters 112*: Article 115503.
51. Senkov ON, Semiatin SL. (2015). Microstructure and properties of a refractory high-entropy alloy after cold working. *Journal of Alloys and Compounds 649*: 1110–23.

52. Couzinié JP, Lilensten L, Champion Y, Dirras G, Perrière L, Guillot I. (2015). On the room temperature deformation mechanisms of a TiZrHfNbTa refractory high-entropy alloy. *Materials Science and Engineering A645*: 255–63.
53. Lillensten L, Couzinié JP, Perrière L, Hocini A, Keller C, Dirras G, Guillot I. (2018). Study of a bcc multi-principal element alloy: Tensile and simple shear properties and underlying deformation mechanisms. *Acta Materialia 142*: 131–41.
54. Han ZD, Luan HW, Liu X, Chen N, Li XY, Shao Y, Yao KF. (2018). Microstructures and mechanical properties of Ti_xNbMoTaW refractory high-entropy alloys. *Materials Science and Engineering A712*: 380–5.
55. Zhang H, Zhao Y, Cai J, Ji S, Geng J, Sun X, Li D. (2021). High strength NbMoTaX refractory high-entropy alloy with low stacking fault energy eutectic phase via laser additive manufacturing. *Materials and Design 201*: Article 109462.
56. Wang M, Ma ZL, Xu ZQ, Cheng XW. (2021). Effects of vanadium concentration on mechanical properties of V_xNbMoTa refractory high-entropy alloys. *Materials Science and Engineering A808*: Article 140848.
57. Tseng KK, Juan CC, Tso S, Chen HC, Tsai CW, Yeh JW. (2019). Effects of Mo, Nb, Ta, Ti and Zr on mechanical properties of equiatomic HfMoNbTaTiZr alloys. *Entropy 21*: Article 15.
58. Yao H, Miao J, Liu Y, Guo E, Huang H, Lu Y, Wang T, Li T. (2021). Effect of Ti and Nb contents on microstructure and mechanical properties of $HfZrVTaMoWTi_xNb_y$ high-entropy alloys. *Advanced Engineering Materials 23*: Article 2100225.
59. Chen Y, Li Y, Cheng X, Wu C, Cheng B, Xu Z. (2018). Microstructure and mechanical properties of refractory high-entropy alloys with high plasticity. *Materials 11*: Article 208.
60. Wei W, Wang T, Wang C, Wu M, Nie Y, Peng J. (2021). Ductile $W_{0.4}MoNb_xTaTi$ refractory high-entropy alloys with excellent elevated temperature strength. *Materials Letters 2015*: Article 129753.
61. Gao X, Wang L, Guo N, Luo L, Zhu G, Shi C, Su Y, Guo J. (2021). In-situ development of MB_2 and their effect on microstructure and mechanical properties of refractory $Hf_{0.5}Mo_{0.5}NbTiZr$ high entropy alloy composites. *International Journal of Refractory Metals and Hard Materials 96*: Article 105473.
62. Kang B, Kong T, Dan NH, Phuong DD, Ryu HJ, Hong SH. (2021). Effect of boron on the microstructure and mechanical properties of refractory $Al_{0.1}CrNbVMo$ high-entropy alloy. *International Journal of Refractory Metals and Hard Materials 100*: Article 105636.
63. Senkov ON, Jensen JK, Pilchak AL, Miracle DB, Fraser HL. (2018). Compositional variation effects on the microstructure and properties of a refractory high-entropy superalloy $AlMo_{0.5}$ $NbTa_{0.5}$ TiZr. *Materials and Design 139*: 498–511
64. Guo NN, Wang L, Luo LS, Li XZ, Chen RR, Su YQ, Guo JJ, Fu HZ. (2016). Microstructure and mechanical properties of in-situ MC-carbide particulates-reinforced refractory high-entropy $Mo_{0.5}NbHf_{0.5}ZrTi$ matrix alloy composite. *Intermetallics 69*: 74–7.
65. Xu Q, Chen D, Tan C, Bi X, Wang Q, Cui H, Zhang S, Chen R. (2021). $NbMoTiVSi_x$ refractory high entropy alloys strengthened by forming bcc phase and silicide eutectic structure. *Journal of Materials Science and Technology 60*: 1–7.
66. Lilensten L, Couzinié JP, Bourgon J, Perrière L, Dirras G, Prima F, Guillot I. (2017). Design and tensile properties of a bcc-Ti-rich high-entropy alloy with transformation-induced plasticity. *Materials Research Letters 5*: 110–16.

67. Schuh B, Volker B, Todt J, Schell N, Perrière L, Li J, Couzinié JP, Hohenwater A. (2018). Thermodynamic instability of a nanocrystalline single-phase TiZrNbHfTa alloy and its impact on its mechanical properties. *Acta Materialia 142*: 201–12.
68. Xu QL, Liu KC, Wang KY, Lou LY, Zhang Y, Li CJ, Li CX. (2021). TGO and Al diffusion behavior of CuAl$_x$NiCrFe high-entropy alloys fabricated by high-speed laser cladding for TBC bond coats. *Corrosion Science 192*: Article 109781.
69. Ma X, Ruggiero P, Bhattacharya R, Senkov ON, Rai AK. (2022). Evaluation of new high entropy alloy as thermal sprayed bondcoat in thermal barrier coatings. *Journal of Thermal Spray Technology 31*: 1011–20.
70. Guo Y, Liu Q. (2018). MoFeCrTiWAlNb refractory high-entropy alloy coatings fabricated by rectangular-spot laser cladding. *Intermetallics 102*: 78–87.
71. Zhou L, Li F, Liu JX, Hu Q, Bao W, Wu Y, Cao X, Xu F, Zhang GJ. (2020). High-entropy thermal barrier coating of rare-earth zirconate: A case study on $(LaNdSmEuGd)_2Zr_2O_7$ prepared by atmospheric plasma spraying. *Journal of the European Ceramic Society 40*: 5731–9.
72. Guo Y, Shang X, Liu Q. (2018). Microstructure and properties of in-situ TiN reinforced laser cladding $CoCr_2FeNiTi_x$ high-entropy alloy composite coatings. *Surface and Coatings Technology 344*: 353–8.
73. Tian L, Feng Z, Xiong W. (2018). Microstructure, microhardness and wear resistance of AlCoCrFeNiTi/Ni60 coating by plasma spraying. *Coatings 8*: Article 112
74. Li K, Liang J, Zhou J. (2022). Microstructure and elevated temperature tribological performance of the CoCrFeNiMo high entropy alloy coatings. *Surface and Coatings Technology 449*: Article 128978.
75. Tan S, Liu X, Wang Z. (2022). Nanoindentation mechanical properties of CoCrFeNi high entropy alloy thin films. *Materials Technology 37*: 1097–1108.
76. Tüten N, Canadinc D, Motallebzadeh A, Bal B. (2019). Microstructure and tribological properties of TiTaHfNbZr high entropy alloy coatings deposited on Ti-6Al-4V substrates. *Intermetallics 105*, pp 99–106.
77. Wang J, Kuang S, Yu X, Wang L, Huang W. (2020). Tribo-mechanical properties of CrNbTiMoZr high-entropy alloy film synthesized by direct current magnetron sputtering. *Surface and Coatings Technology 403*: Article 126374.
78. Cui P, Li W, Liu P, Zhang K, Ma F, Chen X, Feng R, Liaw PK. (2020). Effects of nitrogen content on microstructures and mechanical properties of (AlCrTiZrHf)N high-entropy alloy nitride films. *Journal of Alloys and Compounds 834*: Article 155063.
79. Xia A, Dedoncker R, Glushko O, Cordill MJ, Depla D, Franz R. (2021). Influence of the nitrogen content on the structure and properties of MoNbTaVW high entropy alloy thin films. *Journal of Alloys and Compounds 850*: Article 156740.
80. Arshad M, Amer M, Hayat Q, Janik V, Zhang X, Moradi M, Bai M. (2022). High-entropy coatings for high-temperature applications: Materials, processing and properties. *Coatings 12*: Article 691.
81. Oses C, Toher C, Curtarolo S. (2020). High-entropy ceramics. *Nature Reviews Materials 5*: 295–309.

82. Zhang RZ, Reece MJ. (2019). Review of high entropy ceramics: Design, synthesis, structure and properties. *Journal of Materials Chemistry 7A*: 22148–62.
83. Xiang H, Xing Y, Dai F, Wang H, Su L, Miao L, Zhang G, Wang Y, Qi X, Yao L, Wang H, Zhao B, Li J, Zhou Y. (2021). High-entropy ceramics: Present status, challenges and a look forward. *Journal of Advanced Ceramics 10*: 385–441.
84. Cantor B, Chang ITH, Knight P, Vincent AJB. (2004). Microstructural development in equiatomic multicomponent alloys. *Materials Science and Engineering A375*: 213–18.
85. Yeh JW, Chen SK, Lin SJ, Gan JY, Chin TS, Shun TT, Tsau CH, Chang SY. (2004). Nanostructured high entropy alloys with multiple principal elements: Novel alloy design concepts and outcomes. *Advanced Engineering Materials 6*: 299–303.
86. Rost CM, Sachet E, Borman T, Moballegh A, Dickey EC, Hou D, Jones JL, Curtarolo S, Maria JP. (2015). Entropy-stabilized oxides. *Nature Communications 6*: Article 8485.
87. Miracle DB, Senkov ON. (2017). A critical review of high-entropy alloys and related concepts. *Acta Materialia 122*: 448–511.
88. Murty BS, Yeh JW, Ranganathan S, Bhattacharjee PP. (2019). *High-Entropy Alloys*. Amsterdam: Elsevier.
89. Tsai MH, Lai CH, Yeh JW, Gan JW. (2008). Effects of nitrogen flow rate on the structure and properties of reactively sputtered $(AlMoNbDiTaTiVZr)N_x$ coatings. *Journal of Physics D 41*: Article 235402.
90. Braic V, Vladescu A, Balaceanu M, Luculescu CR, Braic M. (2012). Nanostructured multi-element (TiZrNbHfTa)N and (TiZrNbHfTa)C hard coatings. *Surface and Coatings Technology 211*: 117–21.
91. Hsieh MH, Tsai MH, Shen WJ, Yeh JW. (2013). Structure and properties of two AlCrNbSiTi high-entropy nitride coatings. *Surface and Coatings Technology 221*: 118–23.
92. Braun JL, Rost CM, Lim M, Giri A, Olson DH, Kotsonis GN, Stan G, Brenner DW, Maria JP, Hopkins PE. (2018). Charge-induced disorder controls the thermal conductivity of entropy-stabilized oxides. *Advanced Materials 30*: Article 1805004.
93. Gild J, Samiee M, Braun JL, Harrington T, Vega H, Hopkins PE, Vecchio K, Luo J. (2018). High-entropy fluorite oxides. *Journal of the European Ceramic Society 38*: 3578–84.
94. Hu L, Zhang Y, Wu H, Li J, Li Y, McKenna M, He J, Liu F, Pennycook SJ, Zeng X. (2018). Entropy engineering of SnTe: Multi-principal-element alloying leading to ultralow lattice thermal conductivity and state-of-the-art thermoelectric performance. *Advanced Energy Materials 8*: Article 1802116.
95. Chen K, Pei X, Tang L, Cheng H, Li Z, Li C, Zhang X, An L. (2018). A five component entropy-stabilized fluorite oxide. *Journal of the European Ceramic Society 38*: 4161–4.
96. Roychowdhury S, Ghosh T, Arora R, Waghmare UV, Biswas K. (2018). Stabilizing n-type cubic GeSe by entropy-driven alloying of $AgBiSe_2$: Ultralow thermal conductivity and promising thermoelectric performance. *Angewandte Chemie 57*: 15167–71.
97. Chen H, Zhao Z, Xu W, Liu J, Zhou Y. (2020). High entropy $(YYbLuEuEr)Al_5O_{12}$: A novel high temperature stable thermal barrier material. *Journal of Materials Science and Technology 48*: 57–62.

98. Zhao Z, Xiang H, Chen H, Dai FZ, Wang X, Peng Z, Zhou Y. (2020). High-entropy $(NdSmEuYYb)_4Al_2O_9$ with good high temperature stability, low thermal conductivity, and anisotropic thermal expansivity. *Journal of Advanced Ceramics 9*: 595–605.
99. Lee KN, Fox DS, Bansal NP. (2005). Rare earth silicate environmental barrier coatings for SiC/SiC composites and Si_3N_4 ceramics. *Journal of the European Ceramic Society 25*: 1705–15.
100. Gai PL, Boyes ED. (2003). *Electron Microscopy in Heterogeneous Catalysis*. Bristol: Institute of Physics Publishing.
101. Prins R, Wang A, Li X, Sapountzi F. (2022). *Introduction to Heterogeneous Catalysis*, 2nd edn. London: World Scientific.
102. Zhang Y, Wang D, Wang S. (2022). High-entropy alloys for electrocatalysis: Design, characterization and applications. *Small 18*: Article 2104339.
103. Kusada K, Mukoyoshi M, Wu D, Kitagawa H. (2022). Chemical synthesis, characterization and properties of multi-element nanoparticles. *Angewandte Chemie 61*: Article 09616.
104. Li H, Lai J, Li Z, Wang L. (2021). Multi-sites electrocatalysis in high-entropy alloys. *Advanced Functional Materials 31*: Article 2106715.
105. Wang B, Yao Y, Yu X, Wang C, Wu C, Zou Z. (2021). High entropy alloys: From theory to experiment. *Journal of Materials Chemistry A 9*: 19410–38.
106. Yao Y, Huang Z, Xie P, Lacey SD, Jacob RJ, Xie H, Chen F, Nie A, Pu T, Rehwoldt M, Yu D, Zachariah MR, Wang C, Yassar RS, Li J, Hu L. (2018). Carbothermal shock synthesis of high-entropy alloy nanoparticles. *Science 359*: 1489–94.
107. Yusenko KV, Riva S, Carvalho PA, Yusenko MV, Arnaboldi S, Sukhikh AS, Hanfland M, Gromilov SA. (2017). First hexagonal close-packed high-entropy alloy with outstanding stability under extreme conditions and electrocatalytic activity for methanol oxidation. *Scripta Materialia 138*: 22–7.
108. Gao S, Hao S, Huang Z, Yuan Y, Han S, Lei L, Zhang X, Shahbazian-Yassar R, Lu J. (2020). Synthesis of high-entropy alloy nanoparticles on supports by the fast-moving bed pyrolysis. *Nature Communications 11*: Article 2016.
109. Qiu HJ, Fang G, Wen Y, Liu P, Xie G, Liu X, Sun S. (2019). Nanoporous high-entropy alloys for highly stable and efficient catalysts. *Journal of Materials Chemistry A 7*: 6499–6508.
110. Glasscott MW, Pendergast AD, Goines S, Bishop AR, Hoang AT, Renault C, Dick JE. (2019). Electrosynthesis of high-entropy metallic glass nanoparticles for designer multi-functional electrocatalysis. *Nature Communications 10*: Article 2650.
111. Rekha MY, Malik N, Srivastava C. (2018). First report on high entropy alloy nanoparticle decorated graphene. *Scientific Reports 8*: Article 8737.
112. Wang D, Liu Z, Du S, Zhang Y, Li H, Xiao Z, Chen W, Chen R, Wang Y, Zou Y, Wang S. (2019). Low-temperature synthesis of small-sized high-entropy oxides for water oxidation. *Journal of Materials Chemistry A 7*: 24211–16.
113. Okejiri F, Zhang Z, Liu J, Liu M, Yang S, Dai S. (2019). Room-temperature synthesis of high-entropy perovskite oxide nanoparticle catalysts through ultrasonication-based method. *ChemSusChem 13*: 111–15.
114. McCormick CR, Schaak RE. (2020). Simultaneous multication exchange pathway to high-entropy metal sulphide nanoparticles. *Journal of the American Ceramic Society 143*: 1017–23.

115. Was GS. (2017). *Fundamentals of Radiation Materials Science: Metals and Alloys*, 2nd edn. New York: Springer.
116. Jin K, Bei H. (2018). Single-phase concentrated solid-solution alloys: Bridging intrinsic transport properties and irradiation resistance. *Frontiers in Materials 5*: Article 26.
117. Yang T, Li C, Zinkle SJ, Zhao S, Bei H, Zhang Y. (2018). Irradiation responses and defect behavior of single-phase concentrated solid solution alloys. *Journal of Materials Research 33*: 3077–91.
118. Pickering EJ, Carruthers AW, Barron PJ, Middleburgh SC, Armstrong DEJ, Gandy AS. (2021). High-entropy alloys for advanced nuclear applications. *Entropy 23*: Article 010098.
119. Zhang Z, Armstrong DEJ, Grant PS. (2022). The effects of irradiation on CrMnFeCoNi high-entropy alloy and its derivatives. *Progress in Materials Science 123*: Article 100807.
120. Shi T, Lei PH, Yan X, Li J, Zhou YD, Wang YP, Su ZX, Dou YK, He XF, Yun D, Yang W, Lu CY. (2021). Current development of body-centred cubic high-entropy alloys for nuclear applications. *Tungsten 3*: 197–217.
121. Jin K, Lu C, Wang LM, Qu J, Weber WJ, Zhang Y, Bei H. (2016). Effects of compositional complexity on the ion-irradiation induced swelling and hardening in Ni-containing equiatomic alloys. *Scripta Materialia 119*: 65–70.
122. Barr CM, Nathaniel JE, Unocic KA, Liu J, Zhang Y, Wang Y, Taheri ML. (2018). Exploring radiation induced segregation mechanisms at grain boundaries in equiatomic CoCrFeMnNi high entropy alloy under heavy ion irradiation. *Scripta Materialia 156*: 80–4.
123. He MR, Wang S, Jin K, Bei H, Yasuda K, Matsumura S, Higashida K, Robertson IM. (2016). Enhanced damage resistance and novel defect structure of CrFeCoNi under *in situ* electron irradiation. *Scripta Materialia 125*: 5–9.
124. Granberg F, Nordlund K, Ullah MW, Jin K, Lu C, Bei H, Wang LM, Djurabekova F, Weber WJ, Zhang Y. (2016). Mechanism of radiation damage reduction in equiatomic multicomponent single-phase alloys. *Physical Review Letters 116*: Article 135504.
125. Nagase T, Rack PD, Noh JH, Egami T. (2015). *In-situ* TEM observation of structural changes in nano-crystalline CoCrCuFeNi multicomponent high-entropy alloy (HEA) under fast electron irradiation by high voltage electron microscopy (HVEM). *Intermetallics 59*: 32–42.
126. Xia SQ, Yang X, Yang TF, Liu S, Zhang Y. (2015). Irradiation resistance in $Al_xCoCrFeNi$ high entropy alloys. *Journal of Metals 67*: 2340–4.
127. Yang T, Xia S, Liu S, Wang C, Liu S, Fang Y, Zhang Y, Xue J, Yan S, Wang Y. Precipitation behavior of AlxCoCrFeNi high entropy alloys under ion irradiation. *Scientific Reports 6*: Article 32146.
128. El-Atwani O, Li N, Devaraj A, Baldwin JKS, Schneider MM, Soberiaj D, Wróbel JS, Nguyen-Manh D, Maloy SA, Martinez E. (2019). Outstanding radiation resistance of tungsten-based high-entropy alloys. *Science Advances 5*: Article 2002.
129. Parkin C, Moorehead M, Elbakshwan M, Hu J, Chen WY, Li M, He L, Sridharan K, Couet A. (2020). In situ microstructural evolution in face-centred and body-centred cubic complex concentrated solid-solution alloys under heavy ion irradiation. *Acta Materialia 198*: 85–99.
130. Kareer A, Waite JC, Li B, Couet A, Armstrong DEJ, Wilkinson AJ. (2019). Low activation, refractory, high entropy alloys for nuclear applications. *Journal of Nuclear Materials 526*: Article 151744.

131. Zhang W, Liaw PK, Zhang Y. (2018). A novel low-activation $VCrFeTa_xW_x$ high-entropy alloys with excellent heat-softening resistance. *Entropy 20*: Article 120951.
132. Sadeghilaridjani M, Ayyagari A, Muskeri S, Hasannaeimi V, Salloom R, Chen WY, Mukherjee S. (2020). Ion irradiation response and mechanical behavior of reduced activity high entropy alloy. *Journal of Nuclear Materials 529*: Article 151955.
133. Chang S, Tseng KK, Yang TZ, Chao DS, Yeh JW, Liang JH. (2020). Irradiation-induced swelling and hardening in HfNbTaTiZr refractory high-entropy alloy. *Materials Letters 272*: Article 127832.
134. Nagase T, Takeuchi A, Amiya K, Egami T. (2018). Solid state amorphization of metastable $Al_{0.5}TiZrPdCuNi$ high entropy alloy investigated by high voltage electron microscopy. *Materials Chemistry and Physics 210*: 291–300.
135. Cullity BD, Graham CD. (2009). *Introduction to magnetic materials*, 2nd edn. Hoboken: Wiley, 87–150.
136. Chaudhary V, Chaudhary R, Banerjee R, Ramanujan RV. (2021). Accelerated and conventional development of magnetic high entropy alloys. *Materials Today 49*: Article 13697021.
137. Kumari P, Gupta AK, Mishra RK, Ahmad MS, Shahi RR. (2022). A comprehensive review: Recent progress on magnetic high-entropy alloys and oxides. *Journal of Magnetism and Magnetic Materials 554*: Article 169142.
138. Huang EW, Hung GY, Lee SY, Jain J, Chang KP, Chou JJ, Yang WC, Liaw PK. (2020). Mechanical and magnetic properties of the high-entropy alloys for combinatorial approaches. *Crystals 10*: Article 10030200.
139. Zuo T, Gai MC, Ouyang L, Cheng Y, Feng R, Chen S, Liaw PK, Hawk JA, Zhang Y. (2020). Tailoring magnetic behavior of CoFeMnNiX (X = Al, Cr, Ga, Sn) high entropy alloys by metal doping. *Acta Materialia 130*: 10–18.
140. Li P, Wang A, Liu CT. (2017). Composition dependence of structure, physical and mechanical properties of $FeCoNi(MnAl)_x$ high entropy alloys. *Intermetallics 87*: 21–6.
141. Borkar T, Chaudhary V, Gwalani B, Chaudhary D, Mikler CV, Soni V, Alam T, Ramanujan RV, Banerjee R. (2017). A combinatorial approach for assessing the magnetic properties of high entropy alloys: Role of Cr in $AlCo_xCr_{1-x}FeNi$. *Advanced Engineering Materials 10*: Article 1700048.
142. Zuo TT, Li RB, Ren XJ, Zhang Y. (2014). Effects of Al and Si addition on the structure and properties of CoFeNi equal atomic ratio alloy. *Journal of Magnetism and Magnetic Materials 371*: 60–8.
143. Zhang Y, Zuo TT, Cheng YQ, Liaw PK. (2013). High-entropy alloys with high saturation magnetization, electrical resistivity and malleability. *Scientific Reports 3*: Article 01455.
144. Wang J, Zheng Z, Xu J, Wang Y. (2014). Microstructure and magnetic properties of mechanically alloyed FeSiBAlNi(Nb) high entropy alloys. *Journal of Magnetism and Magnetic Materials 355*: 58–64.
145. Wang W, Li H, Wei P, Zhang W, Chen J, Yuan S, Fan Y, Wei R, Zhang T, Wang T, Chen C, Li F. (2021). A corrosion-resistant soft-ferromagnetic high entropy alloy. *Materials Letters 304*: Article 130571.

146. Koželj P, Vrtnik S, Jelen A, Krnel M, Gacnik D, Dražic G, Meden A, Wencka M, Jezeršek D, Leskovec J, Maiti S, Steurer W, Dolinšek J. (2019). Discovery of a FeCoNiPdCu high-entropy alloy with excellent magnetic softness. *Advanced Engineering Materials 21*: Article 1801055.
147. Na SM, Lambert PK, Jones NJ. (2021). Hard magnetic properties of $FeCoNiAlCu_xTi_x$ based high entropy alloys. *AIP Advances 11*: Article 015210.
148. Feng X, Zheng R, Wu Z, Zhang Y, Li Z, Tan X, Xu H. (2021). Study on a new high-entropy alloy $Nd_{20}Pr_{20}La_{20}Fe_{20}Co_{10}Al_{10}$ with hard magnetic properties. *Journal of Alloys and Compounds 882*: Article 160640.
149. Shi Y, Yang B, Liaw PK. (2017). Corrosion-resistant high-entropy alloys: A review. *Metals 7*: Article 7020043.
150. Nascimento CB, Donatus U, Rios CT, Lopes de Oliveira MC, Antunes RA. (2022). A review on corrosion of high entropy alloys: Exploring the interplay between corrosion properties, alloy composition, passive film stability and materials selection. *Materials Research 25*: Article 20210442.
151. Scully JC. (1975). *The Fundamentals of Corrosion*, 2nd edn. Oxford: Pergamon Press.
152. Lee CP, Chang CC, Chen YY, Yeh JW, Shih HC. (2008). Effect of the aluminium content of $Al_xCrFe_{1.5}MnNi_{0.5}$ high-entropy alloys on the corrosion behaviour in aqueous environments. *Corrosion Science 50*: 2053–60.
153. Hsu YJ, Chiang WC, Wu JK. (2005). Corrosion behavior of $FeCoNiCrCu_x$ high-entropy alloys in 3.5% sodium chloride solution. *Materials Chemistry and Physics 92*: 112–17.
154. Kao YF, Lee TD, Chen SK, Chang YS. (2010). Electrochemical passive properties of $Al_xCoCrFeNi$ ($x = 0, 0.25, 0.50, 1.00$) alloys in sulfuric acids. *Corrosion Science 52*: 1026–34.
155. Li M, Chen Q, Cui X, Peng X, Huang G. (2021). Evaluation of corrosion resistance of the single-phase light refractory high entropy alloy $TiCrVNb_{0.5}Al_{0.5}$ in chloride environment. *Journal of Alloys and Compounds 857*: Article 158278.
156. Yan X, Guo H, Yang W, Pang S, Wang Q, Liu Y, Liaw PK, Zhang T. (2021). $Al_{0.3}Cr_xFeCoNi$ high-entropy alloys with high corrosion resistance and good mechanical properties. *Journal of Alloys and Compounds 860*: Article 158436.
157. Shang X, Wang Z, Wu Q, Wang J, Li J, Yu J. (2019). Effect of Mo addition on corrosion behavior of high-entropy alloys $CoCrFeNiMo_x$ in aqueous environments. *Acta Metallurgica Sinica English Letters 32*: 41–51.
158. Zhang JJ, Yin XL, Dong Y, Lu YP, Jiang L, Wang TM, Li TJ. (2014). Corrosion properties of $Al_xCoCrFeNiTi_{0.5}$ high entropy alloys on 0.5 M H_2SO_4 aqueous solution. *Materials Research Innovations 18*: 756–60.

Further Reading

Bhadeshia H, Honeycombe R. (2017). *Steels: Microstructure and Properties*, 4th edn. Oxford: Butterworth-Heineman.

Cullity BD, Graham CD. (2009). *Introduction to magnetic materials*, 2nd edn. Hoboken: Wiley.

Gai PL, Boyes ED. (2003). *Electron Microscopy in Heterogeneous Catalysis*. Bristol: Institute of Physics Publishing.

Jones DRH, Ashby MF. *Engineering materials: an introduction to properties, applications and design*, 4th edn. Oxford: Butterworth-Heinemann.

Miracle DB, Senkov ON. (2017). A critical review of high-entropy alloys and related concepts. *Acta Materialia 122*: 448–511.

Murty BS, Yeh JW, Ranganathan S, Bhattacharjee PP. (2019). *High-Entropy Alloys.* Amsterdam: Elsevier.

Prins R, Wang A, Li X, Sapountzi F. (2022). *Introduction to Heterogeneous Catalysis*, 2nd edn. London: World Scientific.

Reed, RC. (2006). *The Superalloys: Fundamentals and Applications.* Cambridge: Cambridge University Press.

Scully JC. (1975). *The Fundamentals of Corrosion*, 2nd edn. Oxford: Pergamon Press.

Trethewy K, Chamberlain J. (1995). *Corrosion for Science and Engineering*, 2nd edn. Harlow: Pearson.

Was GS. (2017). *Fundamentals of Radiation Materials Science: Metals and Alloys*, 2nd edn. New York: Springer.

7

Future Developments

1 Introduction

When Alain Vincent and I made the first multicomponent high-entropy materials in 1979 and 1980, and when, unaware of our previous work, Jin-Wei Yeh and his colleagues did the same thing in the mid-1990s,[i] we were all plunging into the dark. We were overturning the conventional way of making materials that had persisted for millennia,[ii] and we had no real idea of what we would find. We were about to uncover a cornucopia of possible new materials. We both published our first major papers on this new method of making materials in 2004, although I had already reported the basic idea in the previous two years. At first the rest of the world took little notice, but research into these new materials was beginning to grow by the end of the 2000s. It then took off and has not stopped since. After a decade and a half of intensive research it makes sense to pause and see what we have found, so as to provide a launch pad for the future. That is the purpose of the last chapter of this book.

In the next section we review our present understanding of multicomponent materials, essentially summarising the previous chapters in this book. We describe briefly the main areas where we have begun to achieve new understanding, indicate some of these areas that are particularly suitable for further research, and also identify some other areas that have been largely neglected. In the final section, we discuss very briefly future materials needs as the world continues to develop and grow, as well as the important role multicomponent materials will continue to play in fulfilling those needs and thus contributing to the health, wealth and wellbeing of future generations.

2 Multicomponent High-Entropy Materials

In this section, we first list the main discoveries and breakthroughs that have been made in our understanding of multicomponent and high-entropy materials:

1. Multicomponent phase space is truly massive and the total number of possible materials that could be made is mind-blowingly large. It is almost impossible to overestimate the importance of this fact. Even on the most conservative estimates (no more than sixty usable elements and for materials differing with composition differences of 1% or more), there is something like 10^{46} different materials that could be made. In fact, the true figure is many orders of magnitude greater still (since

[i] The discovery of multicomponent high-entropy materials is described in Chapter 2.
[ii] The millennia-long previous history of human development driven by the discovery of new materials made by conventional dilute alloying additions is described in Chapter 1.

Fundamentals of Multicomponent High-Entropy Materials. Brian Cantor, Oxford University Press. © Brian Cantor (2024).
DOI: 10.1093/9780191986710.003.0007

there are considerably more than sixty usable elements and since materials can differ dramatically with much smaller composition differences).[iii] And that is just the number of different materials themselves, before we begin to use our wonderful tools of microstructural control that can be tremendously effective in modifying and improving the structure and properties of each material.

2. We can describe multicomponent phase space as a sixty-cornered hyperpolyhedron in multidimensional space, with a different material (a total of 10^{46}) at every point in a hyperdimensional array of a hundred points in each dimension; or, more realistically perhaps, as an eighty-cornered polyhedron with a different material (a much larger total of 10^{123}) at every point in a hyperdimensional array of a thousand points in each dimension (for respectively sixty or eighty usable elements and for materials differing with composition differences of respectively 1 or 0.1%).[iv]
3. As a corollary, we have still hardly scratched the surface in exploring the field of materials and in finding the best new materials for our purposes. There are literally many billions of new materials yet to be made and studied. In the later years of his life, the great metallurgist Sir Alan Cottrell used to give a lecture entitled 'Shocks in Materials Science', in which he discussed the regularity with which materials scientists would find every few years some wonderful new group of materials—ductile irons, high-temperature superconductors, quasicrystals, metallic glasses, compound semiconductors, superplastic materials, lasers, piezoelectric materials, block co-polymers, solar energy convertors, fibre-reinforced composites, zeolites, etc. He noted this but didn't explain it, and wondered if it would continue. We now know that, by accident, materials scientists kept coming across new bits of multicomponent phase space. We also now know that there many more wonderful bits of multicomponent phase space yet to explore. Since we are now looking much harder, these discoveries of exciting new materials will certainly continue.
4. There is a surprisingly large number of single-phase materials in multicomponent phase space, and surprisingly few new and distinct multicomponent compounds. When more and more components are added to a material, it often turns out to be thermodynamically preferable for the different atoms and molecules to be added substitutionally into solid solution on the lattice of a pure single-component structure or on the sub-lattices of an existing binary or ternary compound structure, rather than to develop a completely new multicomponent compound structure with multiple sub-lattices for the different components. This is partly because compound formation is suppressed by the extra configurational entropy of mixing associated with a large number of components.[v] It is also partly because of the difficulty in our limited three-dimensional world of packing multiple different components onto multiple sub-lattices to form distinct new multicomponent compound structures.
5. The geography (or, perhaps, topology) of multicomponent phase space consists of vast single-phase multicomponent solid-solution regions (close-packed fcc, bcc and hcp regions; intermetallic aluminide, silicide, Laves phase and topologically

[iii] The number of possible materials is given by $\ln N = (c + n - 1)\ln(c + n - 1) - (c - 1)\ln(c - 1) - n \ln n$ for c possible components, x% difference between different materials, and $n = 100/x$, as discussed in detail in Chapter 3.

[iv] The structure of multicomponent phase space is discussed in detail in Chapter 3.

[v] So multicomponent materials are often called high-entropy materials.

close packed (TCP) phase regions; inorganic boride, oxide, sulphide and nitride regions; regions of alkali halides, alumino-silicates, spinels and perovskites; etc.). These single-phase regions are separated by equally vast multicomponent multiphase regions.[vi]

6. Exploring and charting multicomponent phase space is difficult, essentially because it is so vast and complex. Most of the tools we have developed for understanding materials are based on the behaviour of single components, with dilute alloying additions treated as relatively minor linearly additive perturbations. These tools are basically inadequate for concentrated multiple alloying additions in multicomponent high-entropy materials. Extended Hume-Rothery rules, thermodynamic modelling, *ab initio* atomistic modelling, machine learning and artificial intelligence techniques can all be helpful to some extent, but are generally unreliable because they depend on theories developed for single components or dilute alloys, they require large numbers of unknown parameters that cannot be estimated accurately, they do not take effective (or any) account of non-linear behaviour, and they only work well when interpolating within a known region.[vii] Ultimately, there is no real substitute for inspired exploration and discovery, i.e. for scientifically well-organised experimental investigation.
7. Multicomponent high-entropy materials are relatively straightforward to manufacture by most conventional processing methods, including solidification and casting, vapour deposition, additive manufacturing, annealing and heat treatment, and thermomechanical processing.[viii] The manufacturing difficulties traditionally associated with alloying additions arise from an initial increased tendency at relatively low concentrations to form embrittling compounds, but this tendency is often suppressed when more alloying components are added at higher concentrations.
8. There is a gigantic number of different local atomic and electronic structures in a single crystal of any multicomponent high-entropy solid solution. Considering just first and second near neighbours, there are literally trillions of different local atomic clusters in any 5- or 6-component material, with the variation in local structure extending over many tens of microns. The number of different clusters and the extent of the variation in local structure increases dramatically when more components are added or if third near neighbours are included. The different local clusters in multicomponent high-entropy solid-solution materials are associated with significant local mechanical distortions and strains in the underlying crystal lattice, as well as significant local electronic distortions and strains in the underlying band structure, with the magnitude of local distortions and strains typically up to 1–2% on average and several percent at maximum.[ix] The apparent regularity of an underlying crystal structure and electronic band structure is, therefore, somewhat misleading for multicomponent high-entropy solid-solution materials, with polycrystals of such materials in some ways being more like aggregates of large individual macromolecules.

[vi] The geography (or topology) of multicomponent phase space is discussed in detail in Chapter 3.
[vii] Strategies for exploring multicomponent phase space are discussed in detail in Chapter 5.
[viii] Methods for manufacturing multicomponent high-entropy materials are discussed in detail in Chapter 5.
[ix] The number of different local clusters, their spatial extent and their associated mechanical and electronic distortions and strains are discussed in detail in Chapter 4.

9. The wide variation in local atomic and electronic structure in multicomponent high-entropy solid-solution materials leads to similarly wide variation in the structure of defects in their underlying crystal lattices. There are again literally trillions of different vacancy structures, dislocation core structures, grain boundary structures and local electronic band structures in any multicomponent high-entropy material with five or more components. Variation in the structure of vacancies leads to relatively sluggish atomic diffusion at high temperatures; variation in the structure of dislocations leads to barriers to dislocation slip and enhanced yield strength and work hardening; variation in the structure of grain boundaries leads to relatively sluggish recrystallisation and reduced grain sizes during annealing at high temperature; and variation in electronic band structure leads to enhanced electron scattering and reduced electrical and thermal conductivity.[x]
10. Multicomponent high-entropy materials have exciting and valuable new properties of high strength, high ductility and fracture toughness, high radiation and corrosion resistance, high catalytic efficiency, low electrical and thermal conductivity, and many others yet to be studied.[xi] Many of these potentially valuable properties arise essentially from the wide variety of local atomic structures and associated local mechanical and electronic strains in the materials. The properties of multicomponent high-entropy materials can often be tuned very precisely by varying the composition of the material. Many multicomponent high-entropy materials are already being investigated for a range of engineering applications, but these are to date very far from being optimised and there are many more opportunities for engineering applications yet to be explored.

Much of the research into multicomponent high-entropy materials has concentrated on studying metallic alloy systems, and much of that research has been on the late transition-metal based single-phase fcc Cantor alloys and early transition-metal based single-phase bcc Senkov alloys. There is, of course, still an enormous amount of work to do on these materials, given their extensive composition range and wide potential for further microstructural manipulation. Important areas for further intensive scientific research include: atom probe studies of local structures and short-range order; photoelectron spectroscopic studies of electronic band structure; positron annihilation studies of vacancy formation and migration; high-resolution transmission electron microscope studies of dislocation dynamics; plasticity and fracture mechanics measurements in multicomponent metallic alloys; and electrical and optical measurements in multicomponent ceramic compounds. Important areas for further intensive engineering research include: optimising mechanical properties for high-temperature applications (improving the high-temperature strength of Cantor alloys and/or improving the low-temperature ductility of Senkov alloys); optimising the thermal, wear and corrosion resistance of spray-coated multicomponent high-entropy metallic and ceramic thermal barrier layers; optimising the manufacture and catalytic efficiency of multicomponent high-entropy metallic and ceramic nanoparticles; and optimising the strength and

[x] Variations in vacancy, dislocation, grain boundary and electronic band structures in multicomponent high-entropy materials are discussed in detail in Chapter 4.

[xi] Properties and potential applications of multicomponent high-entropy materials are discussed in detail in Chapter 6.

stability of radiation- and corrosion-resistant multicomponent high-entropy alloys. Experimental studies are prioritised in these suggestions: thermodynamic, *ab initio* and machine learning modelling methods are in many ways much easier to undertake but are unreliable and not an effective substitute for hard-won experimental data.

It is also worth pointing out that there are many aspects of multicomponent high-entropy materials that have hardly been investigated at all. Some of these can be listed as follows:

1. We have continued our millennia-long tradition of being conservative in alloying practice. Most research to date has concentrated on materials containing no more than three, four or five components. There has been very little investigation of materials containing six or seven components, and virtually nothing on materials with even more components. The effect of configurational entropy in suppressing compound formation is well established for materials with up to five components, but it is not clear if this continues when even more components are added, with an inevitable associated increase in chemical diversity and, therefore, increase in compound-forming tendency. Similarly, research has concentrated strongly on equiatomic multicomponent materials, and only recently begun to explore the very wide range of non-equiatomic variations.
2. There has been almost no investigation of dilute multicomponent materials, i.e. materials with large numbers of alloying elements in relatively dilute concentrations. There are a few conventional materials that have several alloying elements in dilute concentrations, but these are just a few special cases. There has been no general exploration of the effects of adding large numbers of alloying additions. As mentioned above, the configurational entropy of mixing suppresses compound formation in concentrated multicomponent materials, but we do not know if there is a similar, perhaps smaller, effect in dilute multicomponent materials. It is also not clear if there are noticeable non-linear effects caused by three-way, four-way, etc., interactions between dilute alloying additions, that build up with an increasing number of components.
3. There has recently begun to be some serious scientific investigation of multicomponent ceramic and inorganic compounds. There is a tremendous potential for investigating the compositional tuning of the range and mix of functional properties in multicomponent intermetallic and ceramic compounds.
4. There has recently begun to be some scientific investigation of multicomponent two-phase eutectic and matrix/precipitate materials, but almost no studies of multicomponent multiphase materials with three, four and more phases. Multicomponent phase space must contain many eutectic points with four, five, six or more eutectic phases, with the possibility of manufacturing and controlling complex multiphase structures. There is also a tremendous potential for optimising a multiplicity of material properties by microstructural control of multiple composite phases.

3 Future Materials

Humanity has moved forward over the last few millennia by developing new materials and thus continuing to liberate new opportunities for the creation of goods, wealth and comfort, health and longevity, civilisation, culture and beauty. The pace of development has accelerated in recent decades, as our scientific ability to discover and utilise new materials has grown. Towards the end of the twentieth century, however, it had seemed that our ability to discover new materials and continue to make progress might be at risk. We seemed to be investing ever-greater scientific effort into chasing ever-shrinking gains in material performance. But we were working on no more than a tiny corner (or, more accurately, several corners) of the full materials palette. The discovery of multicomponent high-entropy materials has completely removed the problem and opened up staggeringly large vistas of potential new materials on which we can base our future needs.

In summary, new materials with novel and improved functionality are essential if we are to find good solutions to our key twenty-first-century problems of poverty, ill-health and climate change. Better-quality power generation, transportation, personal security, drug delivery, telecommunications, diagnostic aids and much else all require new and improved materials with higher levels of performance and reduced impact on the planet. In the past, materials scientists and engineers have always been able to find suitable materials when they have been needed, as long as sufficient effort and resource has been put into their discovery and development. The discovery of multicomponent high-entropy materials now means that we can be confident that suitable and desirable new materials will always continue to be found somewhere in the vastness of multicomponent phase space. Multicomponent high-entropy materials will provide many new and improved properties to enhance our way of life; they will also reduce material consumption, with more generic compositions for multiple uses, and with greater resilience to impurity pickup during use and re-use. Overall, our discovery of multicomponent phase space and its plethora of multicomponent high-entropy materials gives us substantial cause for optimism about future developments for humankind through the twenty-first century and beyond.

Index